FROM ANIMALS TO ANIMATS

FROM ANIMALS TO ANIMATS

Proceedings of the First International Conference on Simulation of Adaptive Behavior

edited by Jean-Arcady Meyer and Stewart W. Wilson

A Bradford Book

The MIT Press
Cambridge, Massachusetts
London, England

Library of Congress Cataloging-in-Publication Data

International Conference on Simulation of Adaptive Behavior (1st : 1990 : Paris, France)

From animals to animats : proceedings of the First International Conference on Simulation of Adaptive Behavior / edited by Jean-Arcady Meyer and Stewart W. Wilson.

p. cm.

"A Bradford book."

Includes bibliographical references and index.

ISBN 978-0-262-63138-9 (pb : alk. paper)

1. Animal behavior—Simulation methods—Congresses. 2. Adaptation (Psychology)—Simulation methods—Congresses. I. Meyer, Jean-Arcady. II. Wilson, Stewart W. III. Title.

QL751.65.S55I58 1991

591.5'1—dc20 90-22881

CIP

CONTENTS

COGNITIVE MAPS AND INTERNAL WORLD MODELS

MOTIVATION AND EMOTION

ACTION SELECTION AND BEHAVIORAL SEQUENCES

ONTOGENY AND LEARNING

COLLECTIVE BEHAVIORS

EVOLUTION OF BEHAVIOR

PREFACE

An international conference entitled "Simulation of Adaptive Behavior: From Animals to Animats" took place in Paris on September 24-28, 1990. The object of the conference was to bring together researchers in ethology, ecology, cybernetics, artificial intelligence, robotics, and related fields so as to further our understanding of the behaviors and underlying mechanisms that allow animals and, potentially, robots to adapt and survive in uncertain environments.

"SAB90", as we called it, was the first major conference to test the hypothesis that people interested in understanding animal behavior, and people interested in simulating or constructing autonomous robots, would have important common interests and would welcome the chance to listen to and learn from each other. The conference further tested the somewhat more radical hypothesis that its focus constituted not only an intersection but a growing new *field* concerned, in both animals and "animats", with adaptive behavior.

By a variety of measures including size and international range of attendance, intellectual enthusiasm, quality and diversity of contributions, and degree of interaction among the participants, SAB90 offered strong support for both these hypotheses. Furthermore, the emergence of a field in its own right was signaled by the fact that while there was lively debate along many axes—e.g., top-down vs. bottom-up, learning vs. reflexes, hierarchical vs. flat, simulate vs. build, to mention a few—it was striking how people everywhere along the *animals to animats* axis were thinking about similar sets of problems.

These proceedings contain 62 papers, 59 that were actually presented at the conference, plus three whose authors could not attend. The book is divided into sections corresponding to the conference sessions. In each section, papers presented as talks are followed by related papers that were presented as posters.

The first section, The Animat Approach, contains papers on artificial animal research as a tool for understanding adaptive behavior and, indeed, as a new approach to artificial intelligence. The next sections—Perception and Motor Control, Cognitive Maps and Internal World Models, Motivation and Emotion, Action Selection and Behavioral Sequences, Ontology and Learning, Collective Behaviors, and Evolution of Behavior—contain papers on these themes from both the animal and animat perspectives. There follows a large section on Architectures, Organizational Principles, and Functional Approaches, containing several strong—and differing—theses on how to understand or achieve natural or artificial systems with adaptive behavior. The book concludes with a two-paper section, Animats in Education, that describes novel and uncomplicated robot and simulation technologies designed for teaching and research.

SAB90 could not have taken place without the assistance of many people and organizations. We are especially grateful to members of the Program Committee, whose conscientious reviewing selected the papers here from the more than 90 submitted, and who ably chaired the conference sessions. The Committee members were

Lashon Booker, MITRE Corporation, USA
Rodney Brooks, MIT Artificial Intelligence Lab, USA
Patrick Colgan, Queen's University at Kingston, Canada
Patrick Greussay, Université Paris VIII, France
David McFarland, Balliol College, Oxford, UK
Luc Steels, VUB AI Lab, Belgium
Richard Sutton, GTE Laboratories, USA
Frederick Toates, The Open University, UK
David Waltz, Thinking Machines Corp. and Brandeis University, USA

We thank each of the following sponsors of the conference, and mention particularly AFOSR which enabled us to provide substantial needed travel assistance, and the anonymous Corporate Donor whose early confidence in our project was a great boost.

Ecole Normale Supérieure
The Rowland Institute for Science
Ministère de la Recherche et de la Technologie
AFOSR (U. S. Air Force Office of Scientific Research)
C2V
Electricité de France
IBM France
OFFILIB
A Corporate Donor

The members of the Groupe de BioInformatique, URA 686, Ecole Normale Supérieure, worked long and hard to make the conference a success. We thank Anne Brelet, Eric Granjeon, Agnès Guillot, Jean-Louis Pennetier, Philippe Tarroux, and Pierre Vincens.

We wish to express our gratitude to the Ministère de la Recherche et de la Technologie for having generously placed at our disposal the Amphithéâtre Poincaré in which the conference sessions were held. We also express our particular thanks to Josiane Serre (Administrator of the ENS) and René Lafont (Director of URA 686) for their kind help in solving various administrative problems.

Finally, we are particularly indebted to Louis Bec and Wang Xueqing for the artistic conception of the SAB90 poster and the Proceedings cover.

We invite readers to enjoy and profit from the papers in this book, and look forward to the next SAB conference!

Jean-Arcady Meyer and Stewart W. Wilson
Conference Co-Chairs

THE ANIMAT APPROACH

Simulation of Adaptive Behavior in Animats: Review and Prospect

Jean-Arcady Meyer * and Agnès Guillot *,**

** Groupe de BioInformatique*
Ecole Normale Supérieure, 46 rue d'Ulm
75230 Paris Cedex 05 - France
Université Paris 6 - URA 686 CNRS
meyer@frulm63.bitnet

*** Université Paris-X*
200 avenue de la République
92000 Nanterre Cedex 01 - France
guillot@frulm63.bitnet

ABSTRACT

Following a general presentation of the numerous means whereby animats -- i.e. simulated animals or autonomous robots -- are enabled to display adaptive behaviors, various works making use of such means are discussed. This review cites 176 references and is organized into three parts dealing respectively with preprogrammed adaptive behaviors, with learned adaptive behaviors, and with the evolution of these behaviors. A closing section addresses directions in which it would be desirable to see future research oriented, so as to provide something other than proofs of principle or ad hoc solutions to specific problems, however interesting such proofs or solutions may be in their own right.

1. INTRODUCTION

In a changing, unpredictable, and more or less threatening environment, the behavior of an animal is adaptive as long as the behavior allows the animal to survive. Under the same conditions, the behavior of a robot is considered to be adaptive as long as the robot can continue to perform the functions for which it was built. Now, the survival of an animal is intimately involved with its physiological state and the successful operation of a robot depends upon its mechanical condition. Under these circumstances, it is obvious that one can associate with an *animat* -- whether the term indicates a simulated animal or an autonomous robot (Wilson, 1985, 1987a) -- a certain number of state variables upon which its survival or successsful operation depend, and that each of these state variables is characterized by a range of variation within which the animat's continued survival or operation are preserved. Such variables were referred to as *essential variables* by Ashby (1952) long ago. Their variation ranges describe a viability zone inside the given state space, and the animat can be referenced at any instant by a point within this zone. Under the influence of environmental or behavioral variations affecting the animat, the corresponding reference point moves and may at times come close to the limits of the viability zone. In this case, the animat's behavior can be called adaptive so long as it avoids transgressing the viability boundary (Figure 1) (Ashby, 1952; Sibly & McFarland, 1976).

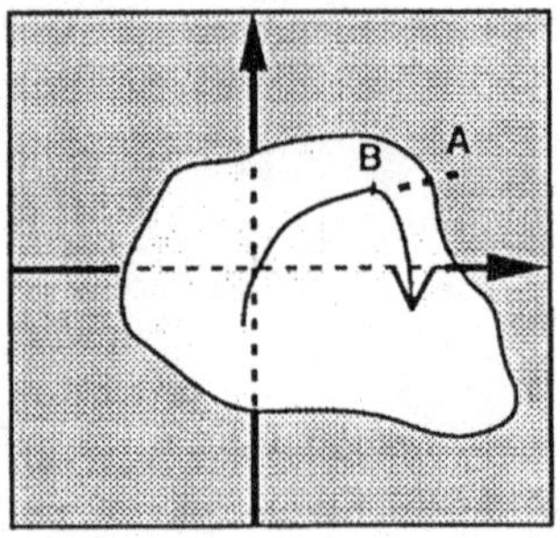

Figure 1. Viability zone associated with two essential variables, V1 and V2. The animat's behavior is adaptive because corrective action has been taken at point B, so as to avoid crossing out the corresponding viability zone at point A.

Such behavior can be generated by means of several different or complementary abilities and architectures. For example, the laws governing the animat's operation may rely upon various homeostatic mechanisms thanks to which, if the reference point alluded to earlier moves away from an adapted point of equilibrium -- adapted because it is suitably located within the viability zone --, this process tends to return it to its original position, thereby decreasing the risk that it will pass outside the limits of the zone. Other ways in which to lower this risk involve the use of high-quality sensory organs or motor apparatus that allow the animat to detect as early as possible that it is approaching these limits and/or to move away from them quickly and effectively. In this line of reasoning, it is obvious that the equivalent of a nervous system is mandatory in order to connect the animat's perceptions with its actions and that reflex circuits activated as quickly as possible increase the adaptive nature of its behavior. It is likewise clear that additional adaptive potential is afforded to any animat capable of responding with more than simple reflexes, particularly when it is able to choose from among several possible reactions the one that proves best adapted to the situation at hand. Such a behavioral control can in particular result from changes in the animat's motivation brought on by this situation. Lastly, it is understandable that the capacity of memorizing the perception/action sequences that have shown themselves to be useful or harmful in the past is of considerable adaptive value for any animat obliged to decide what to do in a given situation, and that this adaptive value is enhanced should the animat also be capable of more or less sophisticated forms of planning.

The intent of this article is to review significant approaches that make use, separately or in combination, of these various adaptive solutions, whether they were proposed by ethologists, computer scientists or roboticists. We hope that such a variety of concepts and techniques will encourage a dialogue between specialists with different scientific perspectives and that, from an interdisciplinary approach of adaptive behavior, both a better understanding of animal behavior and the conception of really autonomous robots will arise. This review will be concerned with computer simulations and real robots and will deal with innate -- because they are programmed or wired in -- and acquired -- because they result from learning or evolution -- adaptive solutions. However, a simple inventory of these approaches will be made here, as more detailled descriptions thereof are given in Meyer and Guillot (1990).

2. PREPROGRAMMED BEHAVIORS

Many simulation models exhibit adaptive behaviors because they were purposely programmed that way. The most classical models belonging to this category are cybernetic models that use suitable feedback loops to maintain an animat's essential variables inside a specific region of its viability zone. Thus Booth's feeding behavior model (Booth, 1978) is founded on the hypothesis that the initiation and termination of feeding in rats depends on the intantaneous value of the energy flow entering or exiting from lean tissues. An hysteresis loop explicitly governs the margin of variation of this flow, and therefore the animal's hunger. Likewise, in the drinking behavior model proposed by Toates & Oatley (1970), another hysteresis loop restricts the variations of the thirst signal. Numerous other instances of this type of cybernetic model are to be found in McFarland (1971, 1974, 1985) and Toates (1980, 1986).

Work by various authors in the field of biomimetic signal processing (Busnel & Fish, 1980; Nachtigall & Moore, 1988; Rivamonte & Dral, 1990) show promise as to the possibility of increasing the adaptive capacities of an animat by providing it with sensors that work more or less like the sensory organs of animals. For example, use has been made of knowledge gained about the compound eye of the fly (Franceschini *et al*, 1989) in the development of an on-board electro-optical system for the guidance of an autonomous mobile robot (Pichon *et al.*, 1990). Likewise, Wang *et al.* (1990) obtained insights into the vision system of the rattlesnake which they then used to carry out fusion of multisensor data in a robot.

A markedly different approach, but one that is also expected to improve the efficiency of an animat's sensory-motor system, is the one developed by Agre and Chapman (Agre, 1988; Agre & Chapman, 1987; Chapman, 1989) involving *deitic representations*. The basic idea is that it is unnecessary to equip the animat with a sensory apparatus capable, at all times, of detecting and distinguishing between all the objects in its environment in order to ensure its adaptive competence. All that is required is that it be able to register information only about objects that are relevant to the problem to be solved. That is, at any moment, the animat's internal representation should register only the features of a few key objects and ignore the rest. Also, those objects should be indexed according to the intrinsic features and properties that make them significant. Such an approach gives rise to the design of systems that actively control their sensory apparatus, since they must mark relevant objects and change their focus of attention as new objects make their appearance in the environment and/or that the objects previously marked cease to be of interest.

The efficiency of this approach is illustrated by the *Pengi* system, which is capable of attaining remarkably high scores on a popular video game. In this game, a penguin must survive as long as possible in a world containing ice blocks and bees, the ice blocks being moveable so as to serve as protection against a bee, or to smash it.

Various approaches achieve an adaptive sensory-motor integration by means of devices more or less directly suggested by natural nervous systems and notably from knowledge gained in the area of *neuroethology* (Camhi, 1984; Ewert, 1980, 1987; Ewert *et al*, 1983). These devices -- at least in their simulated version -- implement artificial neural networks similar, for the most part, to those made popular by the Parallel Distributed Processing Group (Rumelhart & McClelland, 1986).

As an example of the use of an artificial neural network for controling behavior, Beer (1989) has developed a model enabling an artificial insect to display a variety of behaviors --locomotion, wandering, edge following and feeding-- ensuring its survival in a simulated environment. Considering that the same neurons are involved in the initiation and control of these behaviors, it is evident that a proper organization must preclude the simultaneous occurrence of incompatible acts, that is those calling upon the same motor units or the same *behavioral final common path*, to use the expression of McFarland & Sibly (1975). The solution retained is a hierarchical organization where the consumatory part of feeding behavior takes precedence over the orientation-towards-food behavior, which in turn is dominant with respect to the obstacle-avoidance behavior. According to such an organization, exploration is the behavior engaged in by default, while locomotion is activated in the course of every behavior entering into this hierarchy.

This type of organization, in which a dual perception/action hierarchy is implemented, is quite similar to those proposed by Tinbergen (1951) in a purely ethological context, or by Simon (1982) in a more general perspective. Moreover, its capacity to generate adaptive behaviors has been studied at length by Albus (1979, 1981).

Other examples of hierarchical control are afforded by Jacobs (1972) and by Meyer (1986). This latter work describes how the rat *Psikharpax* makes use of a symbolic model of its world so as to survive in a familiar environment. Notably, *Psikharpax* is capable, in order to avoid dying of hunger, of fetching the keys that open the various doors giving access to its food. More generally, this work addresses the way in which artificial intelligence and ethology may contribute mutual insight, particulary in the realm of heuristic problem-solving.

Various studies of robot navigation also rely on explicit models of the outer world, these robots generally being run by appropriate planning programs to determine what path should be followed to go from one point to another (Giralt *et al.*, 1984; Koch *et al.*, 1985; Lin *et al.*, 1989; Nilsson, 1984). To solve the problems that can arise from such an approach when an unforeseen obstacle hinders the normal operation of the plan, Arkin proposes the *AuRA* (Autonomous Robot Architecture) architecture that is based both on an *a priori* world model and on mechanisms allowing rapid and effective reaction to changes occurring within the world (Arkin, 1989a; Arkin & Taylor, 1990). In particular, this architecture has recourse to *motor schemas* (Arbib, 1989a; Arkin, 1987) which are the basic units of motor behavior in *AuRA*. In Arbib and House (1987) and in Arbib (1989b) other applications of such schemas are to be found concerning models of worm acquisition by toads and frogs in an obstacle-cluttered environment that entails detour behavior.

For Brooks (1987), the use of a world model is not essential to the generation of intelligent behavior, in particular when carrying out navigational tasks. This viewpoint is illustrated by the construction of ten or so autonomous robots of highly varied sizes and missions (Flynn & Brooks, 1988), but all characterized by the same *subsumption architecture* (Brooks, 1986). This architecture represents a parallel and distributed computational formalism for connecting sensors to actuators in robots. Instead of fusing sensory data into an internal world model of the robot's environment and using this model to plan intelligent action, subsumption architecture provides an alternative way of writing intelligent control programs for robots in which sensors are more directly linked to action-suggestion modules. Fusion, here, happens at the actuator-plan level, rather than at the sensor-perception level.

One writes a subsumption program by specifying layers of networks of finite-state machines augmented with various timers and registers. For example, this architecture makes it possible for the robot *Genghis* to chase infrared sources over rough terrain (Brooks, 1989). Likewise, it permits *Squirt* -- "the world's largest one-cubic-inch robot" (Flynn *et al.*, 1989) -- to act as a "bug", hiding in dark corners and venturing out in the direction of noises only after the noises are long gone.

Another parallel and distributed structure is proposed by Maes (1989) to control the behavioral sequences of an animat. Instead of being governed by a hierarchy and according to a *top-down* design, the various acts making up the sequence follow each other in an order that is an emerging property displayed by a network of *competence modules*, according to a *bottom-up* design derived from Minsky's *Society of Mind* theory (Minsky,1986). For a comparison of such an approach with more traditional studies about behavioral sequences in animals, one should consult Guillot (1986, 1988).

The idea of using different agents that cooperate to achieve tasks without specific coordination or communication is carried to its limit in the proposition of sending to the moon a colony of small robots for building a permanent base (Brooks & Flynn, 1989). An example is to be found in Brooks *et al* (1990) of the simple behaviors that must be exhibited by each individual robot in order for the whole colony to be able, firstly, to select a promising area in which to construct the lunar base and, secondly, to level the ground in the construction site and pile up the soil in a small number of piles. Similar parallels between artificial self-organizing systems and insect societies are mentioned in Deneubourg & Goss (1989), Moyson & Manderick (1988) and Steels (1987, 1989).

3. LEARNED BEHAVIORS

Many simulation models implement individual learning processes by means of which an animal or a robot can improve the adaptivenesss of their behaviors.

The simplest sort of learning thus used is *conditioning* that modifies the behavior of the animat as a result of certain temporal relationships between events. Many models describe how such conditioning could arise at the single neuron level (Barto & Sutton, 1982; Sutton & Barto, 1987). In Sutton & Barto (1989), for example, such a model -- called the *temporal difference model* -- is compared with animal learning theories of reinforcement from Mowrer (1960) and Rescorla & Wagner (1972).

Other models make use of several neurons to simulate conditioning experiments. *LIMAX* (Gelperin *et al.*, 1985) is an example of a model enabling one to simulate Pavlovian conditioning in a slug which learns to avoid plants containing bitter or toxic chemical substances. Under these conditions, the plant's odor constitutes the *conditioned stimulus (CS)* while the bitter flavor is the *unconditioned stimulus (US)*.

Other varieties of learning can occur within artificial neural networks, as examplified by the work of Grossberg and his school (Grossberg, 1987, 1988; Grossberg & Kuperstein, 1989; Grossberg & Schmajuk, 1989) and by many other simulations.

In the work of Pomerleau (1989), for example, a *supervised learning* procedure has been used for training a three-layer neural network to govern the movements of *NAVLAB*, the Carnegie Mellon autonomous vehicle. Likewise, an *unsupervised learning* procedure has been used by Salu (1984) to simulate the manner in which a young cat learns to behave in its environment and how the corresponding pertinent information is memorized and recalled.

Many models call upon a different variety of learning, known as *reinforcement learning*, that prompts the animat to discover for itself which actions it must perform in order to maximize an external gain, or *reinforcement* signal.

The Barto & Sutton model (1981) pertains to this category. It allows an animat to learn to use odorous landmarks to orient itself in a two-dimensional universe. In this universe, a central landmark is surrounded by four others, each of which emits an odor the strength of which decreases with distance. The purpose of the animat is to make use of these odors to reach the central landmark. This task is carried out by means of an *Associative Search Network* that is endowed with special neurons drawn from Klopf's theories (Klopf, 1980). The point to be emphasized in connection with this is that the animat works out, in the course of its learning phase, a kind of rudimentary *cognitive map*, and that this capacity is to be compared with various experiments that may imply the use of such maps by animals (Ellen & Thinus-Blanc, 1987; Gould, 1986; Griffin, 1982; Menzel, 1978; O'Keefe & Nadel, 1978; Pearce, 1987; Roitblat, 1987; Tolman, 1948;). Two models, for example, are to be found in O'Keefe (1989) and in McNaughton (1989) that describe how spatial data might be memorized and used by the neuronal structures of the hippocampus.

In line with this notion of cognitive map, a number of papers describe adaptive networks of neuron-like elements that use and dynamically adjust an internal world model. Sutton & Barto (1981), for example, describe a simulation effecting a kind of *latent learning* in a network. Likewise, Barto *et al.* (1983) demonstrate how a single element can be used to model the reinforcement of the environment. Such a model makes it possible then to facilitate learning under conditions where the environment affords only an intermittent reinforcement signal. When this signal occurs after the animat has several successive actions, a fundamental *credit assignment* problem arises (Minsky, 1963): that of deciding what portion of the corresponding reward or punishment is to be assigned to each of these actions. Obviously, an elegant solution to this problem is afforded if a model can be used that will predict what reinforcement the environment could have furnished in response to each action.

The possibility for an animat to have access to a world model that allows it to carry out fictitious experiments and to plan its behavior through trial-and-error learning has, for example, been examined by Sutton (1990). Thus this author has studied a class of simple architectures called *DYNA* that apply various machine learning algorithms to approximate conventional dynamic programming control.

A *DYNA* architecture relies essentially upon four interacting components:

-The real world, that changes state in relation with the animat's behavior and that distributes rewards and punishments;

- The world model that the animat elaborates for itself and that is intended to mimic the one-step input/output behavior of the real world;

- The policy function relied on by the animat to determine what action to engage in response to each possible state of the world;

- The evaluation function to which the animat has recourse in order to assess the reward associated with each state, when its object is to maximize the long-term average reward per time step.

The world model, the policy function, and the evaluation function are progressively modified as the animat experiments with the operational laws of its world. These modifications depend upon two type of experiments that the animat may alternate between: actual experiments carried out on the real world -- that bring about updates by *temporal difference reinforcement learning* -- and fictitious experiments -- that make use of the model world and give rise to updates by *relaxation planning*. In either case, the algorithm applied is related to dynamic programming inasmuch as the credit is propagated from one step to the preceding one within action sequences.

Simulations effected on different variants of this architecture indicate that the corresponding animats are not only capable of learning how to move around in a labyrinth while minimizing their navigational errors by simple trial-and-error learning, but further that this learning is expedited when they also avail themselves of the planning possibilities afforded by their world models. What is more, certain of these architectures are easy to adapt for use in changing environments.

A similar approach has been put into practice by Thrun *et al.* (1990) on an adaptive connectionist planning method. In this work, an animat's world model is progressively elaborated in the light of interactions between this animat and its environment, and this world model is used to work out a look-ahead plan expected to maximize future reinforcements. A special *back-propagation* method has been conceived for the purpose, which effects a gradient descent in the space of all possible actions.

Another approach conceived in this line of thought is that of Schmidhuber (1990). In this work, two neural networks are used to govern the actions of an animat, with one of the networks being responsible for the sensory-motor integration, while the other models environmental changes and reinforcements resulting from the animat's actions.

Possibilities of learning through reinforcement as well as of planification are exhibited by the *AGAR* system (Travers, 1989), the architecture of which is not directly inspired by a neural network, but rather is based upon Minsky's *Society of Mind* theory cited earlier. In this system, behavior control is carried out by entities called *agents*, comparable to production rules in that they are characterized by conditions and actions. The conditions required to activate an agent may depend on the state of various sensors, as they may also depend on the activity of other agents. The actions arising from an activation can be any behaviors whatsoever, as long as they are capable of being expressed in a high level programming language such as LISP. They may, among other things, activate or supress other agents, activate a motor function, activate a *script* (Schank & Abelson, 1977), memorize the activation status of other agents, or create or modify agents. For example, a set of instructions is to be found in the paper of Travers that makes it possible to create the agents necessary to simulate the courting behavior of the stickleback.

The works that have just been mentioned are all based on the same implicit and simplifying hypothesis that the animat's sensory equipment allows it to elaborate a model of its world the states of which correspond unequivocally with those of the real world. Such an hypothesis is first of all more or less unrealistic and secondly results in models that are generally too complex to be usable in practice, save for certain simple academic cases. This is why Whitehead & Ballard (1990) have studied an adaptive control architecture that integrates active sensory-motor systems -- based on the *deitic representation paradigm* referred to earlier -- with decision systems based on reinforcement learning. Such an architecture learns not only how to solve a task like block manipulation, but also where to focus its attention in order to collect the necessary sensory information. In particular, it makes it possible to solve the problem of *perceptual aliasing* generated by the active perception mechanisms thus brought into play, problem that occurs because the animat's internal representation often confuses external world states with one another.

Reeke and Edelman's *Darwin III* model (Reeke and Edelman, 1988) likewise represents the mind as an unsupervised learner in which much of the clustering of stimuli into classes is not readily inherent in the stimuli and must be performed by the individual according to what is adaptive for its species and its own particular circumstances. Furthermore, the categories constructed by an organism aren't fixed, but constantly change in response to new experiences on its part of the environment. Thus, to carry out tasks involving categorization, recognition, generalization, and association with respect to moving objects, Darwin III uses a multi-joined arm and a moveable head with one or two eyes to refine its perceptions by means of simple motor actions, such as saccades and visual tracking, reaching, touching, and tracing. The corresponding architecture is a direct implementation of Edelman's *neuronal group selection* theory (Edelman, 1987), whereby the individual's behavioral repertoire arises out of a selection among diverse pre-existing groups of cells -- accomplished by differential modification of synaptic strengths -- according to what is of adaptive value to it in its econiche. This selection is elaborated without change in the connectivity pattern within and between cell groups, which has been fixed during the prenatal period of the brain's development.

Another organization, called a *classifier system*, was elaborated by Holland (1971) so that individual learning could be carried out. Many descriptions of this organization have been made, notably in Booker *et al.* (1989), Goldberg (1989), Grefenstette (1985,1987), Holland (1986), Holland *et al.* (1986) and Schaffer (1989).

However, because a classifier system is generally applied in conjunction with a genetic algorithm, specific simulations that implement such a system will be mentioned in the following section only. It should however be noted that a critical review of the way in which classifier systems function is given in Wilson & Goldberg (1989) and that changes in the basic structure of these systems are proposed by Wilson (1987b) in the interest of achieving a better control of behavioral sequences by a hierarchical structure. Such a structure, in addition to the similarities it presents with the hierarchical control models used by ethologists (Baerends *et al.*, 1970; Dawkins, 1976; Tinbergen, 1951), should materially accelerate the learning. Let us also mention that the work of Holland *et al.* (1986) interprets in terms of classifier systems the results of many classical experiments in animal conditioning and learning.

4. EVOLVED BEHAVIORS

The idea that an animal's behavior is partially determined by its genome and that this behavior is accordingly capable of evolving through natural selection implicitly underlies many optimization models proposed by ethologists and ecologists (Alexander, 1982; Clark, 1990; Krebs & McClery, 1984; Lendrem, 1986; Mangel & Clark, 1988; McFarland, 1982; Meyer, 1980), in the framework of what is known as a *functional analysis* of behavior. This type of analysis postulates that the behavior currently exhibited by an animal stems from successive improvements it has benefited in the past, from generation to generation. Provided the process has been going on a sufficiently long time, the behavior's selective value can be expected to have reached its maximum value or, in other words, to be *optimized.* In this context, the selective value of a behavior is bound up with its adaptive nature. Indeed, it estimates the extent to which this behavior influences the animal's capacity to produce a large number of fertile descendents, and this capacity obviously requires not only that the animal be able to keep itself within its viability zone, but further that it remain in a more restricted zone compatible with its reproductive needs.

In these perspectives, a static optimization model was, for example, proposed by Belovsky (1978) for predicting food selection in the moose. Likewise, a dynamic optimization model was proposed by McFarland (1977) to predict the optimum order of occurrence of three stereotyped acts in the courting behavior of the male newt: retreat displays, pursuit motions, and spermatophore transfers.

Other dynamic optimization models of behavior make use of dynamic programming techniques. This, for instance, is the case with McNamara & Houston's model (1986) which describes the relationships between short-term behavior and lifetime reproductive success for a bird, with Nollet's model (1988) which describes the queen's egg-laying cycle in bees and social wasps, with Guillot's model (1988) which describes ultradian activity cycles in mice, and with Mangel & Clark's model (1988) which describes the social hunting behavior of lions.

The optimization models just mentioned all call for just one decider. The extension of this approach to the case of two or more deciders entails, from a technical point of view, using of models that are formulated in terms of game theory and, from a theoretical standpoint, calling upon the important concept of *evolutionarily stable strategy* (Maynard-Smith, 1974).

This notion concerns a strategy which is optimized -- in the sense that it cannot be replaced by another more efficient -- and which has proven particularly fruitful in the area of behavioral ecology (Colgan, 1989; Maynard-Smith, 1982; McFarland, 1985; Parker, 1984, 1985). Houston & McNamara (1987) or Mangel & Clark (1988) are to be consulted for its extension to dynamic models.

If these optimization models make it possible to test the hypothesis under which a given animal's behavior is the result of a long evolutionary process -- in other words that, if it had been different, it would have been eliminated by natural selection -- it is clear that these models do not purport to explain by what means this behavior evolved. In particular, these models are unable to handle genotype-phenotype relationships that are instrumental in the evolutionary process thus hypothesized (Jamieson, 1986).

In this connection, the approach favored by Holland (1975) -- in search of an adaptive plan making it possible to improve the adaptation of any given system -- appears to be entirely complementary with respect to the preceding approaches, for it enables optimization problems to be solved by using explicit genotype-phenotype relationships. More specifically, this approach calls for a *genetic algorithm* which consists in the parallel management of a population of chromosomes such that each one of them codes -- generally in the form of a chain of binary symbols -- a possible solution to a particular optimization problem. Each of these chromosomes can therefore be assigned a fitness that assesses the corresponding solution. The application of the genetic algorithm accordingly consists in causing this population to evolve from generation to generation while rendering the probability of reproduction of each chromosome proportional to its fitness and using genetic operators such as mutations and crossing-over to give rise to new solutions in the population. Under these circumstances, this type of evolutionary process causes chromosomes of ever-increasing fitness to be generated until the optimum value is reached, or sufficiently nearly so for all practical purposes.

Many applications of algorithms of this kind are to be found in Goldberg (1989), Grefenstette (1985, 1987), or Schaffer (1989) and a critical review is available in Schaffer & Grefenstette (1990). A special application, designed for the study of the behavior of a population of interactive animats, has been elaborated by Sannier & Goodman (1987). In particular, this work shows that it is possible to evolve, through hybridization, a composite genome capable of producing and organizing a distributed system of independent animats which work towards a common goal. The model of Packard (1989) is also intended to simulate the evolution of animats in interaction in the framework of a spatial ecosystem. In this model, each animat's genome contains two genes, one coding the food threshold required for reproduction, and the other coding the number of offsprings. The results obtained indicate that the dynamic characteristics of the evolution of the simulated population strongly depend on the mode of driving the system with food, and notably on the size of the corresponding fluctuations.

Other applications of genetic algorithms, notably to the study of animat behavior, are presented in Koza (1990). In particular, this text describes how to generate a plan allowing an artificial ant to traverse a trail with gaps of various types. It also shows how to find the minimax strategy for the pursuer to catch an evader in the differential game of simple pursuit played on the plan.

Still other applications of genetic algorithms make it possible to simulate various interesting *co-evolution* phenomena (Axelrod, 1987; Hillis, 1990; Holland, 1990; Miller, 1989).

Likewise in the work of Viola (Viola, 1988) is to be found an original utilization of a genetic algorithm so as to cause a modification of the internal architecture of a robot to take place.

Lastly, let us mention the work carried out by Bertin (1990) who studies the evolution of behavior in aquatic animats called *paddlers*, that must seek out and feed on other animats called *glowballs*. Each paddler's behavior depends on a neural network the purpose of which is to translate the visual inputs affecting the animat into suitable navigational responses, in this case appropriate paddle frequencies. The architecture of these networks is invariable, but the behavior they give rise to is strongly dependent on a variety of numerical parameters characterizing each animat's visual system. The evolutionary experiments deal with these parameters, that undergo random mutations from generation to generation. However, in contrast with the procedures called on in a classical genetic algorithm, these experiments do not involve any crossing-over and therefore presuppose simple asexual reproduction.

The results obtained do not manifest any significant improvement of the animats' performances with time, due to the purely random exploration of the space of the parameters implied by these experiments. This point is obviously related to the secondary role these mutations would seem to play in the evolutionary process (Holland, 1975) and to the limited results achieved by other asexual mutation experiments that attempt to solve the problems of adaptive systems by merely "saving and mutating the best" (Dawkins, 1986; Fogel *et al.*, 1966; Hicklin, 1986).

As was mentioned previously, a classifier system is generally used in conjunction with a genetic algorithm. If an equivalent to this practice is to be sought in nature, it may be considered that -- within the so-called "Michigan approach" -- the genetic algorithm is used to recombine thoughts in a given mental model, whereas -- within the "Pittsburgh approach" -- its purpose is rather to simulate the evolution of a behavioral program coded in a genome.

Whatever the case, the first application of a classifier system to the simulation of an animat's behavior was that of Holland & Reitman (1978). In this work, an animat learns to solve a double survival problem: avoid dying of hunger or thirst. To do this, it must move around in a mono-dimensional universe containing food at one of its extremities and drink at the other. However, due to the fact that its hunger and thirst evolve at different rates, it must learn to adjust accordingly the frequency of its visits to the spots where food and drink are distributed.

The animat simulated by Wilson (1985) also makes use of a classifier sytem to find food and avoid obstacles on a two-dimensional grid.

Another classifier system has been developed by Booker (1982, 1988) to allow an animat called *GOFER* to learn to produce goal-seeking behavioral sequences through the use of rules that manipulate objects, goals, and object/goal associations, rather than rules that code simple stimulus/response associations. From this point of view, this work specifically calls upon a *cognitive* approach to behavior, as opposed to a *behaviorist* one (Bower & Hilgard, 1981).

In closing this review, let us mention the work of Ackley & Litman (1990) that explicitly combines the two natural processes by means of which behavior is optimized for survival: *learning* -- as the adaptation of the individual -- and *evolution* -- as the adaptation of a population. This work draws upon a strategy called *evolutionary reinforcement learning (ERL)* that makes use of genetic evolution techniques to allow effective learning based solely upon natural selection.

The corresponding simulations concern an animat population that struggles for existence in a simulated world where the only feedback mechanism is natural selection. The fate of these animats in their environment depends on their capacity to learn to find food, to avoid obstacles, and to escape from their predators. Their behavior is governed by a neural network and a learning process which relies on a special reinforcement algorithm. The network and the evaluation function used by the algorithm are coded in these animats' genome and accordingly may evolve as the simulation progresses.

Under these conditions, the *ERL* strategy displays better performance -- measured in terms of mean population survival times -- than control populations of randomly-moving animats, or than populations that employ just evolution or just individual learning. This result is obviously to be borne in mind when considering the various recent studies addressing the *Baldwin effect* (Belew, 1989; Duchateau *et al.*, 1990; Hinton & Nolan, 1987; Maynard-Smith, 1987; Nolfi *et al.*, 1990), or, more generally, the conjunction of a genetic algorithm and a neural network (Caudell & Dolan, 1989; Harp *et al*, 1989; Miller *et al*, 1989; Resnick, 1990; Todd, 1988; Whitley & Hanson, 1989).

5. OVERVIEW AND PROSPECT

The first remark that comes to mind at the end of this review is that the majority of the works alluded to were performed in the course of the past five years. It is therefore quite clear that the study of adaptive behaviors exhibited by simulated animals or autonomous robots is a theme of current concern. This sudden popularity can doubtless be explained by the fact that the subject promises to be fertile in theoretical innovations -- notably as concerns the relationships between individual learning and species evolution -- and practical accomplishments -- particularly as regards the construction of robots capable of carrying out various custodial services or exploratory tasks.

Furthermore, it so happens that natural animals afford the principle proof that adaptive behaviors -- certain of which may be extraordinarily efficient -- can be elicited by systems that have been properly organized, and that a great deal of knowledge has been accumulated by ethologists since they began to take an interest in the scientific study of animal behavior (Richard, 1984, 1985).

From this new-found interest in artificial animals or from this longstanding knowledge of natural animals, it can reasonably be hoped a growing interest in an interdisciplinary approach to adaptive behaviors.

It is probable, for instance, that ethologists will profit to a greater extent from the possibilities offered by simulation models. One of the advantages afforded by these models (Meyer & Guillot, 1986a) is indeed that their functioning laws are set forth in the corresponding programs in perfectly unambiguous form, whatever the scientific or philosophical biases of the one observing the behavior of these programs. Under these conditions, it can be expected that controversies as to the role of possible cognitive processes in the determinism of animal behavior (Colgan, 1989; Griffin, 1984; Hoage & Goldman, 1986) will soon no longer be hampered by mere statements of principle and ambiguous terminology, in which they have so often become mired down in the past. The availability of simulation programs makes it possible to identify which instructions devolve from a purely ethological interpretation and which from a cognitive one. An analysis of the program's sensitivity to these instructions and, more generally, a comparison of the program's behavior with that actually displayed by the animal should enable the validity of the corresponding interpretations to be assessed.

We are convinced that most of these controversies would automatically work themselves out as soon as interdisciplinary approaches will begin to prevail. For example, while Booker (1982, 1988) explicitly places his work in a cognitive perspective, our belief is that nothing in *GOFER's* program mentioned earlier is in contradiction with the classical concepts of ethology. Likewise, the expressions "world model" or "planning" used to describe a program's behavior often implement data structures and procedures that are probably less at variance with certain ethologists' concepts than these latter may at first be inclined to believe. After all, starting with Walter (1950, 1951) or Wiener (1961), it has long been accepted that very simple mechanisms can generate many "goal-directed" behaviors (Ackoff & Emery ,1972; Weir,1984).

Naturally this discussion is not unrelated with those devoted to symbolic and sub-symbolic approaches to artificial intelligence (Fodor & Pylyshyn, 1988; Lloyd, 1989; Smolensky, 1988; Waltz, 1988), nor is it unconnected with the controversy that divides the proponents of reactive control from those of knowledge-based planning in the area of robot navigation (Arkin, 1989b). In all these respects, the animat approach should eventually prove enlightening, if only through the experimentation it allows and the opportunity it affords to ascertain which model or which minimal architecture results in what type of adaptive behavior being exhibited.

It nevertheless appears to us that these advancements would be all the more explicit and useful if the aforesaid interdisciplinary approaches were to center around a minimal number of anchor points, the groundwork of which has yet to be laid.

In view of the extreme diversity of the approaches reviewed in this text, it indeed seems essential to seek a typology of the problems dealt with, on the one hand, and a typology of the proposed solutions, on the other hand, in order to find out -- among other things -- whether or not simple correspondances exist between the former and the latter. In this perspective -- which is closely allied with Wilson's proposal (1990) for a theory of environments -- it would certainly be interesting to discover that one specific type of problem is analogous to another specific type for such and such a reason and that, accordingly, the said problem stands a good chance of being solved by a specific type of solution or architecture rather than by another one.

However, for it to be likely that such knowledge will be available some day, current practice will definitely have to be superceded in which each published article is just embodying a proof of principle. Besides the fact that it is never established that the corresponding solution is minimal, nor that any given adaptive capacity is expressly due to a specific global architecture rather than to a particular operational detail, the limits of expression of these capacities are rarely explored.

Nothing short of a systematic comparison of several different implementations of a same type of solution with a range of problems as wide as possible is liable to reveal the generic properties of the solution considered. Conversely, only a systematic comparison of several different versions of the same problem with a range of solutions as varied as possible can allow an evaluation of the respective advantages and the degree of originality of these solutions.

This last type of knowledge is, in particular, the one that would prove the most useful to ethologists for interpreting a given animal's behavior as mobilizing the same solution as that which enables an animat to solve the same type of adaptive problem. Among the very large assortment of solutions that might on first thought be possible, obviously those that have already demonstrated their usefulness in similar fields of application should be tried out first. However, as suggested by Braitenberg (1984) when alluding to the "law of uphill analysis and downhill synthesis", it is bound to be more difficult for the ethologist than for the roboticist or the modeller to derive benefit from knowledge gained from animats. This will be true as long as the former is unable to be certain of the uniqueness of the solutions that are to be envisaged and as the latters' problem is simply to generate an *ad hoc* solution without attaching any particular importance to the manner in which it is brought about.

It is nevertheless clear that the times are rapidly changing and it is quite probable that the robotic solutions considered will not long remain equivalent. In particular, one may foresee that costs will be associated with robot behaviors as they are with animal ones (Meyer & Guillot, 1986b) and that the success of certain architectures will depend not only on such quantitative costs (McFarland, 1990) but also on more subtle and qualitative factors like the nature of the various emotions arising within such architectures (Albus, 1981; Dyer, 1987; Frijda & Swagerman, 1987; Gehm, 1989; Sloman & Croucher, 1981).

ACKNOWLEDGMENTS

The authors are greatly indebted to Stewart Wilson and Jody Mohammadioun for their valuable help in the conception and realization of this paper and wish to express to them their extreme gratitude.

They also want to thank all the people who helped provide this review by sending papers or technical reports. Corrections and additional references will be much appreciated.

REFERENCES

Ackley, D.H. & Littman, M.S. (1990). Learning from natural selection in an artificial environment. Proceedings of The Second Artificial Life Conference. Santa Fe.

Ackoff, R.L. & Emery, F.E. (1972). On purposeful systems. Aldine.

Agre, P.E. (1988). The dynamic structure of everyday life. Technical Report (1085). MIT Artificial Intelligence Lab.

Agre, P.E. & Chapman, D. (1987). Pengi: an implementation of a theory of activity. Proceedings of The Sixth National Conference on Artificial Intelligence. Seattle.

Albus, J.S. (1979). Mechanisms of planning and problem solving in the brain. Mathematical Biosciences. 45: 247-293.

Albus, J.S. (1981). Brains, behavior and robotics. Byte Books.

Arbib, M.A. (1989a). The metaphorical brain 2. Neural networks and beyond. Wiley.

Arbib, M.A. (1989b). Interactings subsystems for depth perception and detour behavior. In Arbib & Amari (Ed.), Dynamic interactions in neural networks: models and data. Springer Verlag.

Arbib, M.A. & House, D. (1987). Depth and detours: an essay on visually guided behavior. In Arbib & Hanson (Ed.), Vision, brain and cooperative communication . The MIT Press.

Arkin, R.C. (1987). Motor schema based navigation for a mobile robot: an approach to programming by behavior. Proceedings of The IEEE Conference on Robotics and Automation. Raleigh.

Arkin, R.C. (1989a). Integrating behavioral, perceptual, and world knowledge in reactive navigation. To appear in Robotics and Autonomous systems. Elsevier/North-Holland.

Arkin, R.C. (1989b). Towards the unification of navigational planning and reactive control. Proceedings of The AAAI Spring Symposium on Robot navigation. Stanford.

Arkin, R.C. & Taylor, D.T. (1990). Reactive behavioral support for qualitative visual navigation. Proceedings of The IEEE Conference on Intelligent Motion Control. Istambul.

Ashby, W.R. (1952). Design for a brain. Chapman & Hall.

Axelrod, R. (1987). The evolution of strategies in the iterated prisoner's dilemma. In (Davis, Ed.), genetic algorithms ans simulated annealing. Pitman.

Baerends, G.P., Drent, R.H., Glas, P. & Groenewold, H. (1970). An ethological analysis of incubation behaviour in the herring gull. Behav. Suppl. 17: 135-235.

Barto, A.G. & Sutton, R.S. (1981). Landmark learning: an illustration of associative search. Biological Cybernetics. 42: 1-8.

Barto, A.G. & Sutton, R.S. (1982). Simulation of anticipatory responses in classical conditioning. Behavioural Brain Research. 4: 221-235.

Barto, A.G., Sutton, R.S. & Anderson, C.W. (1983). Neuronlike elements that can solve difficult learning control problems. IEEE Trans. on Systems, Man and Cybernetics. SMC-13(5): 834-846.
Beer, R.D. (1989). Intelligence as adaptive behavior: an experiment in computational neuroethology. Technical Report (TR 89-118). Center for Automation and Intelligent Systems Research. Case Western Reserve University.
Belew, R.K. (1989). Evolution, learning and culture: computational metaphors for adaptive algorithms. Technical Report (CSE TR #CS89-156). Computer Science and Engineering. Univ. of California.
Belovsky, G.E. (1978). Diet optimization in a generalist herbivore: the moose. Theor. Pop. Biol. 14: 105-134.
Bertin, R.J. (1990). Experiments with simulated eyes, brains and hypothetical animals. Of Paddlers and Glowballs. Unpublished manuscript. Vakgroep Vergelijkende Fysiologie. Rijksuniversiteit te Utrecht.
Booker, L.B. (1982). Intelligent behavior as an adaptation to the task environment. PhD Thesis. Dept. of Computer and Communication Sciences. Univ. of Michigan.
Booker, L.B. (1988). Classifier systems that learn internal world models. Machine learning. 3: 161-192.
Booker, L.B., Goldberg, D.E. & Holland, J.H. (1989). Classifier systems and genetic algorithms. Artificial Intelligence. 40: 235-282.
Booth, D.A. (1978). Prediction of feeding behavior from energy flows in the rat. In Booth (Ed.), Hunger models: computable theory of feeding control. Academic Press.
Bower, G.H. & Hilgard, E.R. (1981). Theories of learning. Prentice-Hall.
Braitenberg, V. (1984). Vehicles. Experiments in synthetic psychology. The MIT Press.
Brooks, R.A. (1986). A robust layered control system for a mobile robot. IEEE Journal of robotics and Automation. RA-2(1): 14-23.
Brooks, R.A. (1987). Intelligence without representations. Proceedings of The workshop on Foundations of Intelligence. MIT. Endicott House.
Brooks, R.A. (1989). A robot that walks; emergent behaviors from a carefully evolved network. Neural Computation. 1: 253-262.
Brooks, R.A. & Flynn, A.M. (1989). Fast, cheap and out of control. Journal of the British Interplanetary Society. 42(10): 478-485.
Brooks, R.A., Maes, P., Mataric, M. & More, G. (1990). Lunar base construction robots. To appear in IROS-90, Japan.
Busnel, R.G. & Fish, J.F. (1980). Animal sonar systems. Plenum.
Camhi, J.M. (1984). Neuroethology: nerve cells and the natural behavior of animals. Sinauer Assoc.
Caudell, T.P. & Dolan, C.P. (1989). Parametric connectivity: training of constrained networks using genetic algorithms. In Schaffer (Ed.), Proceedings of The Third International Conference on Genetic Algorithms. Morgan Kaufmann.
Chapman, D. (1989). Penguins can make cake. AI magazine. 10(4): 45-50.
Clark, C.W. (1990). Modeling behavioral adaptations. To appear in Behav. Brain Sci.
Colgan, P. (1989). Animal motivation. Chapman & Hall.
Dawkins, R. (1976). Hierarchical organization: a candidate principle for ethology. In Bateson & Hinde (Ed.), Growing points in ethology. Cambridge Univ. Press.
Dawkins, R. (1986). The blind watchmaker. Longman.
Deneubourg, J.L. & Goss, S. (1989). Collective patterns and decision-making. Ethology, Ecology and Evolution. 1: 295-311.
Duchateau, G., Meyer, J.A. & Tarroux, P. (1990). Life histories and evolution of primitive organisms. Proceedings of The Second Artificial Life Conference. Santa Fe.
Dyer, M.G. (1987). Emotions and their computations: three computer models. Cognition and emotion. 1(3): 323-347.
Edelman, G.M. (1987). Neural Darwinism: the theory of neuronal group selection. Basic Books.
Ellen, P. & Thinus-Blanc, C. (1987). Cognitive Processes and spatial orientation in animal and man. Martinus Nijhoff.
Ewert, J.P. (1980). Neuroethology. Springer Verlag.
Ewert, J.P. (1987). Neuroethology of releasing mechanisms: prey-catching in toads. Behavioral and Brain Sciences. 10: 337-368.
Ewert, J.P., Capranica, R.R. & Ingle, D.J. (1983). Advances in vertebrate neuroethology. Plenum.
Flynn, A.M. & Brooks, R.A. (1988). MIT mobile robot: what's next? Proceedings of IEEE Robotics and Automation. Philadelphia.
Flynn, A.M., Brooks, R.A., Wells, W.M. & Barrett, D.S. (1989). Intelligence for miniature robots. Sensors and actuators. 20: 187-196.
Fodor, J.A. & Pylyshyn, Z. (1988). Connectionism and cognitive architecture: a critical analysis. Cognition. 28: 3-71.
Fogel, L.J., Owens, A.J. & Walsh, M.J. (1966). Artificial Intelligence through simulated evolution. Wiley.
Franceschini, N., Riehle, A. & Le Nestour, A. (1989). Directionally selective motion detection by insect neurons. In Stavenga & Hardie (Ed.), Facets of vision. Springer Verlag.
Frijda, N.H. & Swagerman, J. (1987). Can computers feel? Theory and design of an emotional system. Cognition and emotion. 1(3): 235-257.
Gehm, T.L. (1989). Neural networks and research on emotions: a mutual perspective. Unpublished manuscript. Department of Psychology. University of Giessen.
Gelperin, A., Hopfield, J.J. & Tank, D.W. (1985). The logic of Limax learning. In Selverston (Ed.), Model neural networks and behavior. Plenum.
Giralt, G., Chatila, R. & Vaisset, M. (1984). An integrated navigation and motion control system for autonomous multisensory mobile robots. In Brady & Paul (Ed.), Proceedings of Robotics Research, The First International Symposium. The MIT Press.
Goldberg, D.E. (1989). Genetic algorithms in search, optimization, and machine learning. Addison Wesley.
Gould, J.L. (1986). The locale map of honey bees: do insects have a cognitive map? Science. 232. 861-863.

Grefenstette, J.J. (1985). Proceedings of the First International Conference on Genetic Algorithms and their Applications. Lawrence Erlbaum Assoc.

Grefenstette, J.J. (1987). Genetic algorithms and their applications. Proceedings of The Second International Conference on Genetic Algorithms. Lawrence Erlbaum Assoc.

Griffin, D.R. (1982). Animal mind - Human mind. Springer-Verlag.

Griffin, D.R. (1984). Animal thinking. Harvard University Press.

Grossberg, S. (1987). The adaptive brain I & II. Elsevier/North-Holland.

Grossberg, S. (1988). Neural networks and natural intelligence. The MIT Press/ Bradford Books.

Grossberg, S. & Kuperstein, M. (1989). Neural dynamics of adaptive sensory-motor control. Pergamon Press.

Grossberg, S. & Schmajuk, N.A. (1989). Neural dynamics of adaptive timing and temporal discrimination during associative learning. Neural networks. 2: 79-102.

Guillot, A. (1986). Revue générale des méthodes d'étude des séquences comportementales. Etudes et analyses comportementales. 2(3): 86-106.

Guillot, A. (1988). Contribution à l'étude des séquences comportementales de la souris: approches descriptive, causale et fonctionnelle. Doctorat Thesis. Biomathématiques. Université Paris 7.

Harp, S.A., Samad, T. & Guha, A. (1989). Towards the genetic synthesis of neural networks. In Schaffer (Ed.), Proceedings of The Third International Conference on Genetic Algorithms. Morgan Kaufmann.

Hicklin, J.F. (1986). Applications of the genetic algorithms to automatic program generation. Master of Science Thesis. Dept. of Computer Science. Univ. of Idaho.

Hillis, W.D. (1990). Co-evolving parasites improve simulated evolution as an optimization procedure. Proceedings of The Second Artificial Life Conference. Santa Fe.

Hinton, G.E. & Nowlan, S.J. (1987). How learning can guide evolution. Complex systems. 1: 495-502.

Hoage, R.J. & Goldman, L. (1986). Animal Intelligence. Insights into the animal mind. Smithsonian Institution Press.

Holland, J.H. (1971). Processing and processors for schemata. In Jacks (Ed.), Associative information processing. Elsevier.

Holland, J.H. (1975). Adaptation in natural and artificial systems. Univ. Michigan Press.

Holland, J.H. (1986). Escaping brittleness: the possibilities of general purpose machine learning algorithms applied to parallel rule-based systems. In Michalski, Carbonell & Mitchell (Ed.), Machine learning: an artificial intelligence approach. Volume 2. Kaufmann.

Holland, J.H. (1990). ECHO: explorations of evolution in a miniature world. Proceedings of The Second Artificial Life Conference. Santa Fe.

Holland, J.H., Holyoak, K.J., Nisbett, R.E. & Thagard, P.R. (1986). Induction. Processes of inference, learning, and discovery. The MIT Press.

Holland, J.H. & Reitman, J.S. (1978). Cognitive systems based on adaptive algorithms. In Waterman & Hayes-Roth (Ed.), Pattern-directed inference systems. Academic Press.

Houston, A.I. & McNamara, J.M. (1987). Singing to attract a mate -a stochastic dynamic game. J. Theor. Biol. 129: 57-68.

Jacobs, W. (1972). How a bug's mind works. In Robinson & Knight (Ed.), Cybernetics, artificial intelligence, and ecology. Spartan Books.

Jamieson, I.G. (1986). The functional approach to behaviour: is it useful? Amer. Natur. 127(2): 195-208.

Klopf, A.H. (1980). The hedonistic neuron: a theory of memory, learning and intelligence. Hemisphere.

Koch, E., Yeh, C., Hillel, G., Meystel, A. & Isik, C. (1985). Simulation of path planning for a system with vision and map updating. Proceedings of IEEE International Conference Robotics and Automation. St. Louis.

Koza, J.R. (1990). Genetic programming: a paradigm for genetically breeding populations of computer programs to solve problems. Technical Report (STAN - CS - 90 - 1314). Dept. of Computer Science. Stanford Univ.

Krebs, J.R. & McCleery, R.H. (1984). Optimization in behavioural ecology. In Krebs & Davies (Ed.), Behavioural ecology: an evolutionary approach. Blackwell.

Lendrem, D. (1986). Modelling in behavioural ecology: an introductory text. Timber Press.

Lin, L.J., Mitchell, T.M., Philips, A.R. & Simmons, R. (1989). A case study in robot exploration. Technical Report (CMU-RI-TR-89-1). Carnegie Mellon University.

Lloyd, D. (1989). Simple minds. The MIT Press/Bradford Books.

Maes, P. (1989). How to do the right thing. Connection Science. 1(3): 291-323.

Mangel, M. & Clark, C.W. (1988). Dynamic modeling in behavioral ecology. Princeton Univ. Press.

Maynard Smith, J. (1974). The theory of games and the evolution of animal conflicts. J. Theor. Biol. 47: 209-221.

Maynard Smith, J. (1982). Evolution and the theory of games. Cambridge Univ. Press.

Maynard Smith, J. (1987). When learning guides evolution. Nature. 329: 761-762.

McFarland, D.J. (1971). Feedback mechanisms in animal behaviour. Academic Press.

McFarland, D.J. (1974). Motivational control systems analysis. Academic Press.

McFarland, D.J. (1977). Decision making in animals. Nature. 269: 15-21.

McFarland, D.J. (1982). Functional ontogeny. Pitman.

McFarland, D.J. (1985). Animal behaviour. Psychobiology, ethology and evolution. Longman Scientific & Technical.

McFarland, D.J. (1990). What it means for robot behaviour to be adaptive (this volume).

McFarland, D.J. & Sibly, R.M. (1975). The behavioral final common path. Phil. Trans. Roy. Soc. London. 270: 265-293.

McNamara, J.M. & Houston, A.I. (1986). The common currency for behavioural decisions. Amer. Natur. 127: 358-378.

McNaughton, B.L. (1989). Neuronal mechanisms for spatial computation and information storage. In Nadel, Cooper, Culicover & Harnish (Ed.), Neural connections, mental computations. The MIT Press/Bradford Books.

Menzel, E.W. (1978). Cognitive mapping in chimpanzees. In Hulse, Fowler & Honig (Ed.), Cognitive processes in animal behavior. Lawrence Erlbaum Assoc.

Meyer, J.A. (1980). Techniques d'étude des stratégies adaptatives: approches mathématiques et informatiques. In Barbault, Blandin & Meyer (Ed.), Recherches d'écologie théorique. Les stratégies adaptatives. Maloine.

Meyer, J.A. (1986). Intelligence artificielle et comportement naturel. Etudes et Analyses comportementales. 2(3): 154-169.

Meyer, J.A. & Guillot, A. (1986a). La simulation numérique du comportement. Etudes et analyses comportementales. 2(3): 120-140.

Meyer, J.A. & Guillot, A. (1986b). The energetic cost of various behaviors in the laboratory mouse. Comp. Biochem. Physiol. 83(3): 533-538.

Meyer, J.A. & Guillot, A. (1990). From animals to animats: everything you wanted to know about the simulation of adaptive behavior. Technical Report (BioInfo- 90-1). Ecole Normale Supérieure, Paris.

Miller, G.F., Todd, P.M. & Hegde, S.U. (1989). Designing neural networks using genetic algorithms. In Schaffer (Ed.), Proceedings of The Third International Conference on Genetic Algorithms. Morgan Kaufmann.

Miller, J.H. (1989). The co-evolution of automata in the repeated prisoner's dilemma. Technical Report (89-003). Santa Fe Institute.

Minsky, M.C. (1963). Steps towards artificial intelligence. In Feigenbaum & Feldman (Ed.), Computers and thought. McGraw-Hill.

Minsky, M.C. (1986). The society of mind. Simon and Schuster.

Mowrer, O.H. (1960). Learning theory and behavior. Wiley.

Moyson, F. & Manderick, B. (1988). The collective behavior of ants: an example of self-organization in massive parallelism. Technical Report (AI Memo 88-7). Artificial Intelligence Lab., Vrije Universiteit Brussels.

Nachtigall, P.E. & Moore, P.W. (1988). Animal sonar: processes and performance. Plenum.

Nilsson, N.J. (1984). Shakey the robot. Technical Report (323). SRI AI Center, Stanford University.

Nolfi, S., Elman, J.L. & Parisi, D. (1990). Learning and evolution in neural networks. Technical Report (CRL TR 9019). Center for Research in language. Univ. of California.

Nollet, P. (1988). Contribution à l'étude des stratégies adaptatives des insectes sociaux. Doctorat Thesis. Biomathématiques. Université Paris 7.

O'Keefe, J. (1989). Computations the Hippocampus might perform. In Nadel, Cooper, Culicover & Harnish (Ed.), Neural connections mental computations. The MIT Press/Bradford Books.

O'Keefe, J. & Nadel, L. (1978). The hippocampus as a cognitive map. Oxford Univ. Press.

Packard, N. (1989). Evolving bugs in a simulated ecosystem. In Langton (Ed.), Artificial Life. Addison-Wesley.

Parker, G.A. (1984). Evolutionarily stable strategies. In Krebs & Davies (Ed.), Behavioural ecology: an evolutionary approach. Blackwell.

Parker, G.A. (1985). Population consequences of evolutionarily stable strategies. In Sibly & Smith (Ed.), Behavioural ecology. Ecological consequences of adaptive behaviour. Blackwell.

Pearce, J.M. (1987). An introduction to animal cognition. Lawrence Erlbaum Assoc.

Pichon, J.M., Blanes, C. & Franceschini, N. (1990). Visual guidance of a mobile robot equipped with a network of self-motion sensors. To appear in Wolff and Chun (Ed.), Mobile robots.

Pomerleau, D.A. (1989). ALVINN: an autonomous land vehicle in a neural network. Technical Report (CMU-CS-89-107). Computer Science Dept. Carnegie Mellon University.

Reeke, G.N. & Edelman, G.M. (1988). Real brains and artificial intelligence. In Graubard (Ed.), The artificial intelligence debate. False starts, real foundations. The MIT Press.

Rescorla, R.A. & Wagner, A.R. (1972). A theory of Pavlovian conditioning: variations in the effectiveness of reinforcement and nonreinforcement. In Black & Prokasy (Ed.), Classical conditioning II. Appleton-Century-Crofts.

Richard, G. (1984). Histoire des idées en Ethologie (1ère partie). Etudes et analyses comportementales. 2(3): 131-190.

Richard, G. (1985). Histoire des idées en Ethologie (2ème partie). Etudes et analyses comportementales. 2(4): 195-254.

Rivamonte, A. & Dral, A.D. (1990). Computer simulation of dolphin vision. Unpublished manuscript. Radiation Standards and Development Laboratory. Redstone Arsenal.

Roitblat, H.L. (1987). Introduction to comparative cognition. Freeman and Co.

Rudnick, M. (1990). A bibliography of the intersection of genetic search and artificial neural networks. Technical Report (CS/E 90-001). Department of Computer Science and Engineering. Oregon Graduate Institute.

Rumelhart, D.E. & McClelland, J.L. (1986). Parallel distributed processing. Explorations in the microstructure of cognition. The MIT Press/Bradford Books.

Salu, Y. (1984). Theoretical models and computer simulations of neural learning systems. J. Theor. Biol. 111: 31-46.

Sannier, A.V. & Goodman, E.D. (1987). Genetic learning procedures in distributed environments. In Grefenstette (Ed.), Genetic algorithms and their applications: Proceedings of the Second International Conference on Genetic Algorithms. Lawrence Erlbaum Assoc.

Schaffer, J.D. (1989). Proceedings of the Third International Conference on Genetic Algorithms. Morgan Kaufmann.

Schaffer, J.D. & Grefenstette, J.J. (1990). A critical review of genetic algorithms. Unpublished manuscript. Philips laboratories. North American Philips Corporation.

Schank, R.C. & Abelson, R.P. (1977). Scripts, plans, goals and understanding. An inquiry into human knowledge structures. Lawrence Erlbaum Assoc.

Schmidhuber, J.H. (1990). Making the world differentiable: on using supervised learning fully recurrent networks for dynamic reinforcement learning and planning in non-stationary environments. Technical Report (FKI-126-90). Institut für Informatik. Technische Universität München.

Sibly, R.M. & McFarland, D.J. (1976). On the fitness of behavior sequences. Amer. Natur. 110: 610-617.

Simon, H.A. (1982). The sciences of the artificial. The MIT Press.

Sloman, A. & Croucher, M. (1981). Why robots will have emotions. Proceedings of The Seventh International Joint Conference on Artificial Intelligence. Vancouver.

Smolensky, P. (1988). On the proper treatment of connectionism. Behavioral and Brain Sciences. 11: 1-23.

Steels, L. (1987). Self-organisation through selection. Technical Report (AI Memo 87-5). Artificial Intelligence Lab., Vrije Universiteit Brussels.

Steels, L. (1989). Cooperation between distributed agents through self-organisation. Technical Report (AI Memo 89-5). Artificial Intelligence Lab., Vrije Universiteit Brussels.

Sutton, R.S. (1990). Integrated architectures for learning, planning, and reacting based on approximating dynamic programming. Proceedings of The Seventh International Conference on Machine Learning.

Sutton, R.S. & Barto, A.G. (1981). An adaptive network that constructs and uses an internal model of its world. Cognition and Brain theory. 4(3): 217-246.

Sutton, R.S. & Barto, A.G. (1987). A temporal-difference model of classical conditioning. Proceedings of Ninth Conference of the Cognitive Science Society.

Sutton, R.S. & Barto, A.G. (1989). Time-derivation model of Pavlovian reinforcement. In Gabriel & Moore (Ed.), Learning and Computational Neuroscience. The MIT Press.

Thrun, S., Linden, A. & Moller, K. (1990). Adaptive look-ahead planning. To be published in (Dorffner, Ed.), Proceedings OEGAI. Springer Verlag.

Tinbergen, N. (1951). The study of instinct. Clarendon Press.

Toates, F.M. (1980). Animal behaviour. A systems approach. Wiley.

Toates, F.M. (1986). Motivational systems. Cambridge Univ. Press.

Toates, F.M. & Oatley, K. (1970). Computer simulation of thirst and water balance. Medical and Biol. Engineer. 8: 71-87.

Todd, P.M. (1988). Evolutionary methods for connectionists architectures. Unpublished manuscript. Psychology Dept. Stanford University.

Tolman, E.C. (1948). Cognitive maps in rats and men. Psychological Review. 55: 189-208.

Travers, M. (1989). Animal construction kits. In Langton (Ed.), Artificial Life. Addison Wesley.

Viola, P. (1988). Mobile robot evolution. Bachelor Thesis. Dept. of Electrical Engineering and Computer Science. MIT.

Walter, W.G. (1950). An imitation of life. Scientific American. 182(5): 42-45.

Walter, W.G. (1951). A machine that learns. Scientific American. 185(2): 60-63.

Waltz, D.L. (1988). The prospects for building truly intelligent machines. In Graubard (Ed.), The artificial intelligence debate. False starts, real foundations. The MIT Press.

Wang, G., Nonon, H., Diep, D., Dubant, O. & Esquirol, P. (1990). A possible approach for the fusion of multisensors data. Proceedings of The Third International IPMU Conference. Paris.

Weir, M. (1984). Goal-directed behaviour. Gordon and Breach.

Whitehead, S.D. & Ballard, D.H. (1990). Learning to perceive and act. Technical Report (TR 331). Dept. of Computer Science. University of Rochester.

Whitley, D. & Hanson, T. (1989). The Genitor algorithm: using genetic algorithms to optimize neural networks. Technical Report (CS-89-107). Dept. of Computer Science. Colorado State Univ.

Wiener, N. (1961). Cybernetics. 2nd edition. The MIT Press.

Wilson, S.W. (1985). Knowledge growth in an artificial animal. In Grefenstette (Ed.), Proceedings of The First International Conference on Genetic Algorithms and their applications. Lawrence Erlbaum Assoc.

Wilson, S.W. (1987a). Classifier systems and the animat problem. Machine Learning. 2: 199-228.

Wilson, S.W. (1987b). Hierarchical credit allocation in a classifier system. In Davis (Ed.), Genetic algorithms and simulated annealing. Pitman.

Wilson, S.W. (1990). The animat path to AI (this volume).

Wilson, S.W. & Goldberg, D.E. (1989). A critical review of classifier systems. In Schaffer (Ed.), Proceedings of The Third International Conference on Genetic Algorithms. Morgan Kaufmann.

The Animat Path to AI

Stewart W. Wilson
The Rowland Institute for Science
100 Cambridge Parkway
Cambridge, MA 02142 USA
(wilson@think.com)

Abstract

A research methodology is proposed for understanding intelligence through simulation of artificial animals ("animats") in progressively more challenging environments while retaining characteristics of holism, pragmatism, perception, categorization, and adaptation that are often underrepresented in standard AI approaches to intelligence. It is suggested that basic elements of the methodology should include a theory/taxonomy of environments by which they can be ordered in difficulty—one is offered—and a theory of animat efficiency. It is also suggested that the methodology offers a new approach to the problem of perception.

1. Introduction

There are two broad approaches to the scientific understanding of intelligence, or how mind arises from brain. One is the natural science approach, analyzing and experimenting with phenomena of life, mind, and intelligence as they exist in nature. In this there are two main branches: physiology and especially neurophysiology, in which living systems are subject to detailed internal investigation; and experimental psychology, including studies of animals, in which living systems are studied through their external behavior. Related to the latter, but more observational, are fields such as linguistics and anthropology.

In contrast, the second broad approach to intelligence may be termed synthetic and computational, in which the objects studied are constructed imitations of living systems or their behavior. In "Computing machinery and intelligence", Turing (1950) suggested two possible directions for the computational approach:

> We may hope that machines will eventually compete with men in all purely intellectual fields. But which are the best ones to start with? Even this is a difficult decision. Many people think that a very abstract activity, like the playing of chess, would be best. It can also be maintained that it is best to provide the machine with the best sense organs that money can buy, and then teach it to understand and speak English. This process could follow the normal teaching of a child. Things would be pointed out and named, etc.

Turing's first proposed direction led to "standard AI" or computational cognitive science. Standard AI is basically competence-oriented, modelling specific human abilities, often quite advanced ones. However, while many AI programs exhibit impressive performance, their relevance for the understanding of natural intelligence is, in several respects, limited.

In addressing isolated competences, AI systems typically ignore the fact that real creatures are always situated in sensory environments and experience varying degrees of need satisfaction. Furthermore, the systems attach less importance to such basic natural abilities as perception, categorization, and adaptation than they do to algorithmic processes like search and exact reasoning. This leads eventually to problems connecting the arbitrary symbols used in internal reasoning with external physical stimuli ("symbol grounding" (Harnad, 1990)), and "brittleness" (Holland, 1986), the tendency for AI systems to fail utterly in domains that differ even slightly from the domain for which they were programmed.

AI systems also have an arbitrariness: it is often not clear why one program that exhibits a certain intellectual competence is to be preferred over some other one exhibiting the same competence, especially since the field has not agreed on—or too much sought—a clear definition of intelligence. In a sense, the programmer's facility for imitating a high-level fragment of human competence is a kind of trap, since from a natural science perspective there is usually no strong relation to nature.

Turing's second proposal, for a "child machine", received, over forty years, little attention or resources, perhaps because it seemed fantastic. Yet the child machine was to be situated from the start in a real sensory environment and was to learn through experience. It would have emphasized precisely the abilities that standard AI minimized. Turing's proposal is in fact close to what we have called the "animat" (or artificial animal) approach (Wilson, 1985a), and the child machine is an advanced form of animat. Rather than isolated competences, the animat approach is holistic, focusing on complete systems (simulated or, when possible, realized) that, like animals, exist in realistic environments and must cope with the varied problems that they present.

Obviously, we can't yet simulate human intelligence holistically. But the basic *hypothesis* of the animat approach is that by simulating and understanding complete animal-like systems at a simple level, we can build up gradually to the human. At each point we will be careful to include full connection with a sensory environment, together with maximum use of perception, categorization, and adaptation. Thus when we reach the human level these crucial abilities will not be missing. We hope to reach human intelligence "from below", instead of piecemeal through high-level competences as in Standard AI.

The animat approach also brings with it a needed element of pragmatism (Holland, Holyoak, Nisbett & Thagard, 1986). Survival needs and their derivatives are evidently the principal drivers of animal behavior, and so, at bottom, they must be for human beings. The effect is that needs have a powerful influence on the formation of percepts and concepts—in machine learning terms, they set the *inductive bias* (Mitchell, 1980)—yet this has been little acknowledged in AI work. The animat approach explicitly brings in needs by making them the drivers of system behavior.

Introduction of needs opens the way to operational definitions of intelligence since the efficiency of need satisfaction is in principle quantifiable. For example, some years ago van Heerden (1968) summarized his observations on human intelligence as follows:

> Intelligent behavior is to be repeatedly successful in satisfying one's psychological needs in diverse, observably different, situations on the basis of past experience.

With suitable changes, this definition can be applied from human to very simple animal levels. It brings in perception, categorization, and adaptation, and it bases degree of intelligence on rate of need satisfaction.

Our aim in this paper is to outline themes in the animat approach to AI—that is, to suggest "how to go about it", at least in first approximation. A number of efforts already exist [for a review, see Meyer & Guillot (1990); also see Smith & Wilson (1989)], and the interest is accelerating. We shall explain our view of what is needed, and suggest potential directions of formalization.

2. The Basic Animat Strategy

The basic strategy of the animat approach is to work toward higher levels of intelligence "from below"—using minimal ad hoc machinery. The essential process is incremental and holistic: given an environment and an animat with needs and a sensory/motor system that satisfies these needs to some criterion, increase the difficulty of the environment or the complexity of the needs—and find the minimum increase in animat complexity necessary to satisfy the needs to the same criterion. Alternatively, the environment could stay the same but the needs satisfaction criterion might be increased; again find the minimum animat complexity increase. In either case it is vital (1) to maintain the realism and whole-ness of the environment, however simple it is, so as to avoid special-purpose solutions; (2) to maximize physicality in the sensory signals, so as to avoid predefined symbolic inputs; and (3) to employ adaptive mechanisms maximally, to minimize the rate of introduction of new machinery and maximize understanding of adaptation.

Note that the strategy has a "problem side" (harder environments, increased efficiency) and a "solution side" (new architecture: sensory/motor, internal, adaptive). Changes in the problem side can be due to the experimenter, but also to (co-)evolutionary effects if the environment is evolving. Similarly, changes in the solution side can be deliberate or evolved, based on a genotype and selection.

Research on animat-like systems has tended to emphasize the solution side. A certain experimental environment is selected as being in some sense interesting, but most of the work goes into testing and refining a particular architecture in that environment. The result is often a successful system, but accompanied by insufficient insight from a formal point of view into the properties or difficulty of the environmental problem that has been solved, and with what efficiency. A major aim of this paper is to suggest the need for a more systematic understanding of environments.

We now discuss the problem and solution sides in more detail.

3. The Problem Side

3.1 Environments

Environments differ enormously in their complexity, uncertainty, and degree of reinforcement. Needed is a formal theory and taxonomy that will order environments and reveal their differences in difficulty. For example, some environments (e.g., some food concentration gradients) can be thought of as pure stimulus-response: the local environmental signal directly indicates the optimal action and provides reinforcement immediately. In others, the reinforcement is deferred, though the optimal action is still knowable from the stimulus. An example would be an odor gradient that reliably pointed toward the location of food.

In a somewhat more complicated environment, information-bearing stimuli are not as simply related to reinforcement as odor is related to food. Instead they may consist of more or less arbitrary cues like stimuli from a certain kind of bush that prey like to hide in or, to mention a human context, a certain kind of golden arch! In still more complicated environments, the optimal action is no longer knowable from the immediate sensory stimulus. Consider leaving your office and turning in the correct direction in accordance with a phone call received five minutes earlier. Or the environment of a stalking animal in which the current objective is temporarily out of sight. At such times the immediate sensory stimulus may contain no information at all relevant to attaining the objective. Further complexity is of course introduced by the presence of competing creatures with similar or different needs. All cases become more difficult if environment characteristics are statistical, or stimuli or reinforcements contain noise. These are just examples, but they suggest a bit of the range and subtlety of real environments.

Given this variety, a more formal characterization is desirable. A start can be made by noting that from an animat's (or animal's) point of view, the environment is a kind of machine that (in general) responds with a new sensory stimulus (which may include reinforcement) whenever the animat executes a motor action. One formal way to describe such an environment is as a *finite-state machine* (FSM) for which the motor actions are inputs and the sensory stimuli are outputs (Riolo, 1987; Rivest & Schapire, 1987). The behavior of a finite-state machine is defined by two equations (Minsky, 1967):

$$Q(t+1) = F(Q(t), A(t))$$
$$E(t+1) = G(Q(t), A(t)),$$

where A is the machine's input (in this case the animat's motor action), E is the machine's output (in this case the sensory stimulus), and Q represents the machine's (the environment's) current "state". Time t is assumed to be discrete. The variables A and E are in general vectors.

The first equation says that the environment's next state is a function F of its current state and the motor action. The second equation says that the next sensory stimulus to the animat is a function G of the current state of the environment and the motor action. The FSM formalism captures the idea that actions in a given environment result in new sensory stimuli; the state variable Q makes it possible for the machine to respond differently to the same action in different circumstances, a common property of real environments. That the FSM is "finite-state" means essentially that the number of different possible outputs for a given input is finite (though possibly very large), or equivalently, that the number of possible values of the state variable Q is finite. A strict finiteness property for real environments is perhaps debatable, but since large FSMs provide a good approximation in many problems of interest, the debate can be left for another occasion.

Besides reacting to animat actions with new sensory stimuli, real environments also sometimes present new stimuli in the absence of action (e.g., the clouds move while you gaze at them, other animats move in your field of view, etc.). This important property is not captured by the FSM formalism, and needs to be included in a fuller environment theory.

Before continuing, it is necessary to be quite careful about the meanings of E and A. Knowledge about the environment comes only through the use of the sensory and motor apparatuses, each acting as a kind of communication channel. Because these tend to be fixed in phylogeny ("hardwired") it is often useful to define the sensory and motor channels as part of the environment. Then, a particular environment of interest might consist, for example, of a (physical) maze as detectable by two eyes of a certain retinal description and manoeuverable by four legs of a certain musculoskeletal description. From this, one would proceed to establish the appropriate functions F and G, treating the retinal outputs as E and the motor command signals as A.

Alternatively, one could treat the sensory and motor channels as part of the animat and not as part of the environment. Such a division might be desirable in problems in which the sensory and motor equipment was subject to an evolutionary process. However, for our current purposes we shall use the former approach in which the two channels are regarded as fixed and part of the animat's environment.

The FSM formalism has advantages and disadvantages. An advantage is that the environmental description can be as precise as desired, and it is necessary to be precise in order to program a simulation. A disadvantage is that the FSM description has a certain opaqueness from the point of view of understanding levels of environmental difficulty. A further disadvantage is that animats deal in stimuli and actions while the FSM also contains the state variable Q, which the animat can't detect. Though in some sense the animat should learn "the reality behind appearances", there is merit in examining an environmental formalism from which Q is absent.

Suppose we try to express the next sensory stimulus directly in terms of the current stimulus and current action. Examination of some FSMs will show that the result is in general not determinate, as indicated by the following relation:

$$\{E(t+1)\} = f(E(t), A(t)).$$

Here the next stimulus is indeed a function of the current stimulus and action, but, as indicated by curly brackets, $E(t+1)$ will be one of a *finite set* of possible stimuli, not unique. Although the set is determined by the function f, the particular member of the set that occurs is not. Thus the above relation expresses a *non-determinacy* of the environment with respect to the variables E and A.

We might term this kind of environmental description a *sensory-state machine* (SSM). In fact, for every FSM there is an SSM that can be derived by straightforward (though perhaps tedious) examination of the FSM state diagram. The SSM, trades the determinacy of the FSM for a formalism that expresses the environment—or the animat's problem—solely in terms of variables that the animat knows about. Furthermore, the SSM's non-determinacy is familiar: the reaction of an environment to an action is very often not fully predictable from knowledge of that action and one's immediate sensory situation.

Let us note that the SSM as defined above is an *incomplete* description of the environment. It can be derived from the environment's FSM, but the FSM cannot be derived from it (in general). Nevertheless, the SSM appears to be a more useful construct for understanding levels of environmental difficulty, as we now attempt to show.

Consider an environment which an animat detects through extremely limited sensory apparatus. For example, the animat might have only a single small touch detector, pointed straight ahead. The SSM for this environment-cum-sensory-apparatus would be extremely non-determinate, since a large number of object shapes would be consistent with stimulation of the single small touch detector. Should, for example, the animat turn 30 degrees to the right, subsequent stimulation of the detector would be nearly unpredictable.

On the other hand, consider an animat in the same environment but having elaborate stereoscopic vision. In this case the SSM would contain little non-determinacy, since for example the visual stimulation subsequent to the same 30 degree turn or most other actions would be a unique function of the current image and therefore predictable.

Predictability of the results of actions in the context of sensory stimulation is the foundation of an animat's survival and, indeed, prosperity. Attainment of reinforcement depends on the ability to choose actions that lead to reinforcement, whatever the sensory circumstances. The examples above suggest that the degree of non-determinacy of an environment's SSM is an important measure of the environment's relative difficulty.

A simple and tentative taxonomy of environments can be constructed based on SSM non-determinacy.

Class 0. Environments with completely determinate SSMs and in which for every sensory stimulus there exists at least one action which if taken will result in positive reinforcement. This might be called a pure stimulus-response environment, meaning that the optimal action in each situation is a function only of the current stimulus. The "landmark" environment of Barto & Sutton (1981) is an example of a Class 0 environment.

Class 1. Environments with completely determinate SSMs in which for only some sensory stimuli does there exist at least one action which will result in positive reinforcement. This could be called a stimulus-response environment with sparse or deferred reinforcement. The 288-state environment of Grefenstette (1988) and the maze environment of Sutton (1990) are examples of Class 1 environments.

Class 2. Environments with partially non-determinate SSMs. In contrast to Classes 0 and 1, reliable prediction can no longer be based on the current sensory stimulus and action. The environment "WOODS7" of Wilson (1985a) is an example of a Class 2 environment, as is the "Little Prince" environment of Rivest & Schapire (1987).

In many cases it will be possible to reduce or eliminate the non-determinacy of a Class 2 environment by taking into account some degree of recent history. For example, suppose that for a particular Class 2 environment we construct the *second-order* SSM:

$$\{E(t+1)\} = f_2(E(t), A(t), E(t\text{-}1), A(t\text{-}1)).$$

It may well be the case that this SSM is less non-determinate that the first-order one, the additional context of the prior time-step's stimuli and actions serving to reduce the uncertainty. We can further imagine that for *some* order of SSM, the non-determinacy is eliminated. Let that order be k. Then we could describe the environment in question as being of Class 2.k, with higher values of k standing for greater difficulty.

This concludes our discussion of environments, in which we developed the idea that environments could be ordered in difficulty according to the non-determinacy of their SSMs.

One complication that we have not mentioned, but will take up in Section 4.2, is the fact of "stimulus profusion" in real environments. The environment may not have a tricky SSM, but at the sensory interface it always has a *very* large one. Thus realistic environments pose a problem of selection of relevant data. At higher levels this can be a problem of "pattern recognition" and quite complex.

3.2 Efficiency

The second part of the problem side has to do with animat needs and the efficiency with which they are satisfied. An animat can have (1) somatic needs (food, shelter), (2) reproductive needs, (3) additional needs like play, exploration, and prediction. A particular problem may address just one or a few of these, depending on how reinforcement is defined. In addition, hierarchies of secondary needs can in principle result from the primary ones, though this should be explicated experimentally. In the end, satisfaction of all needs can be viewed as in the service of reproduction/survival.

Efficiency of needs satisfaction is the grounds for choosing one solution over another. The best way to do this is probably using a competitive, evolutionary approach in which solutions have costs, niches exist or can form, etc. Though data exist from natural science, determining costs will be difficult. Simulations may tend to use computational in contrast to true somatic costs. This could ultimately turn the animat approach away from nature and toward artificial worlds, where the implications for natural intelligence may not be clear. In any case, one would like to have a reasonable theory of animat efficiency in terms of need satisfaction that will take into account costs and provide criteria for preferring solutions.

4. The Solution Side

4.1 Architecture

Here there appears to be a great deal of choice. However, the animat approach (going slowly "upward") should permit a strong criterion of "necessary and sufficient". The progression should at some level of abstraction parallel what actually exists in nature, but that is conjecture; the parallels are unlikely to be obvious, given the apparent role of accident in evolution (Gould, 1989). We can expect, however, that if the SSM description of environments is valid and useful then the best architectures at each stage should be those that most efficiently cope with increasing environmental non-determinacy.

For example, stimulus-response (Classes 0 and 1) environments should imply any of a set of associative memories, which could be implemented with networks, etc. However, as soon as the sensed environment does not uniquely characterize its state (Class 2), the animat can only reach optimal performance using some form of short-term memory, which suggests recurrent networks, classifier systems, etc. Further Class 2 complication will occur when reinforcements are highly delayed and the system must form and retain an intention, and its subordinate intentions, etc., until reinforcement is obtained. Efficiency may then require the introduction of higher-order internal states, modularity, etc. The animat strategy offers a way to bring these in naturally.

One interesting hypothesis is that the most efficient systems will be those that convert every frequently encountered important situation to one of "virtual stimulus-response" in which internal state (intention, memory) and sensory stimulus together form a compound stimulus that immediately implies the correct next intention or external action. This would be in contrast to a system that often tends to "figure out" or undertake a chain of step by step reasoning to decide the next action. The latter more contemplative system would presumably possess increased flexibility in the face of an uncertain environment. However, the present hypothesis is that greater overall efficiency will be found in systems that set up generalized S-R methods in the above sense. The motivation for the hypothesis is that in animals and people, even complex behavior, if frequent and important enough, tends to become reflexive. Standard AI has addressed the question of whether knowledge should be "interpreted" or "compiled" (Laird, Rosenbloom & Newell, 1986). The animat approach offers a new and perhaps more natural context in which to address it.

4.2 Perception

The preceding discussion of architecture bypassed the issue of sensory profusion, tacitly assuming inputs are few and well-defined, as in Standard AI. Perception—which might be defined as knowing what in the environment is relevantly the case—has proved very difficult to imitate computationally. It has a chicken and egg quality: How do you know what aspects of a complex profuse input to select or combine into patterns until you know how to view the input so as to find them, which in turn means knowing where or what they are in the first place.

One approach with some success is a combination of top-down and bottom-up processing in which, iteratively, fragmentary data from below suggest candidate remembered percepts above which in turn guide the lower search for confirming or disconfirming additional data [see, e.g., Grossberg (1987)]. Most uses of these and the related relaxation techniques [e.g., Geman & Geman (1984)] have occurred in the analysis of scenes or images from specialized domains, and so are subject to Standard AI's brittleness. In contrast, the animat approach, retaining stimulus-profuse environments but aimed at simpler percepts, should permit the development of more general and adaptive top-down/bottom-up strategies.

Woods-like environments containing simple objects ("tree", "food", etc.) offer an interesting test-bed for such strategies if the animat receives a stimulus that simulates natural light reflected from the objects. Stimuli could be as profuse as desired depending on the variety of the objects, the resolution and spectral sensitivity of the light detectors, and the range of vision. In such environments, the experimental procedure should be alert to the possibility that the system might form effective percepts quite different from, e.g., "tree 90 degrees to the right", (as stimuli are often directly given in Standard AI). Among other things, the percepts would reflect the bias of the animat's needs.

Besides top-down/bottom-up methods, there is another class of techniques for dealing with stimulus profusion. Since the animat has effectors that can change the sensory input (as when the animat moves), he can learn ways of doing so that select information contingently. This ability becomes useful in cooperation with detectors that condense stimuli to different degrees in different parts of the sensory field. Then the animat can move so as always to place the highest-resolution detectors at the point of greatest interest, leaving the rest of the field relatively diffuse.

The focal-peripheral vision of some animals is an example of stimulus condensation and coordinated action that should be investigated using animats. In primates, for instance, the environment is seen in detail in a central region but with falling definition toward the visual periphery [see, e.g., Wilson (1983)]. This permits vast reduction of stimulus detail over most of the visual field, a primitive form of generalization, but the reduction can always be reversed by moving the eyes. Such a move would be triggered by current needs and internal state in combination with a stimulus cue sufficient to get through the peripheral generalization (Wilson, 1985b).

Perception is one of the hardest human abilities to understand. Progress in machine perception has been slow. The animat approach offers a fresh perspective because well-defined experimental mechanisms can be investigated in contexts that retain essential characteristics of real organisms and environments.

5. Summary

This paper has outlined "the animat path to AI", a strategy for progressively understanding intelligence or the relation of mind to brain that differs significantly from Standard AI, and from the natural science approaches to the same problem. The approach is not new, in that examples of prior work exist and are somewhat known. This paper however attempts to bring out the value of the approach, calls for a more systematic effort, and offers some working themes.

Fundamentally, the animat approach advocates maintaining the holism of the situation of real animals in real environments, while progressively but efficiently increasing animat complexity only as necessary. The approach's hypothesis is that this program is feasible, and will ultimately lead to understanding of intelligence, adaptation, and perception at high levels. In support of the program, the paper proposes:

(1) Creation of a theory/taxonomy of environments, based on a "sensory-state machine" formalism;

(2) Establishment of criteria of animat efficiency in terms of need satisfaction and costs;

(3) The hypothesis that efficient animats will have architectures that deal with frequent, important situations by "virtual stimulus-response";

(4) The suggestion that the problem of "stimulus profusion" can be reduced through stimulus-condensing sensoria and contingent action.

References

Barto, A. G. & Sutton, R. S. (1981). Landmark learning: an illustration of associative search. *Biological Cybernetics*, **42**, 1-8.

Geman, S. & Geman, D. (1984). Stochastic relaxation, Gibbs distributions, and the Bayesian restoration of images. *IEEE Transactions on Pattern Analysis and Machine Intelligence PAMI-6*, 721-741.

Gould, S. J. (1989). *Wonderful Life.* New York: W. W. Norton.

Grefenstette, J. J. (1988). Credit assignment in rule discovery systems based on genetic algorithms. *Machine Learning* **3**, 225-245.

Grossberg, S. (1987). Competitive learning: from interactive activation to adaptive resonance. *Cognitive Science*, **11**, 23-63.

Harnad, S. (1990). The symbol grounding problem. *Physica D*, **42**(1-3), 335-346.

Holland, J. H. (1986). Escaping brittleness: the possibilities of general-purpose learning algorithms applied to parallel rule-based systems. In R. S. Michalski, J. G. Carbonell & T. M. Mitchell (Eds.), *Machine learning, an artificial intelligence approach. Volume II.* Los Altos, California: Morgan Kaufmann.

Holland, J. H., Holyoak, K. J., Nisbett, R. E., and Thagard, P. R. (1986). *Induction: Processes of Inference, Learning, and Discovery.* Cambridge, MA: MIT Press.

Meyer, J.-A. & Guillot, A. (1990). Simulation of adaptive behavior in animats: review and prospect. *These Proceedings.*

Minsky, M. (1967). *Computation: Finite and Infinite Machines.* Englewood Cliffs, NJ: Prentice-Hall.

Mitchell, T. M. (1980). *The Need for Biases in Learning Generalizations* (Technical Report CBM-TR-117). New Brunswick, NJ: Rutgers University, Dept. of Computer Science.

Laird, J. E., Rosenbloom, P. S., and Newell, A. (1986).

Chunking in Soar. *Machine Learning*, **1**, 11-46.

Riolo, R. L. (1987a). Bucket brigade performance: I. Long sequences of classifiers. *Genetic Algorithms and Their Applications: Proceedings of the Second International Conference on Genetic Algorithms* (pp. 184-195). Hillsdale, New Jersey: Lawrence Erlbaum Assoc.

Rivest, R. L. & Schapire, R. E. (1987). A new approach to unsupervised learning in deterministic environments. *Proceedings of the Fourth International Workshop on Machine Learning* (pp. 364-375). Los Altos, CA: Morgan Kaufmann.

Smith, S. J. & Wilson, S. W. (1989). Rosetta: toward a model of learning problems. *Proceedings of the Third International Conference on Genetic Algorithms* (pp. 347-350). San Mateo, CA: Morgan Kaufmann.

Sutton, R. S. (1990). Integrated architectures for learning, planning, and reacting based on approximating dynamic programming. *Machine Learning: Proceedings of the Seventh International Conference* (pp. 216-224). Palo Alto, CA: Morgan Kaufmann.

Turing, A. M. (1950). Computing machinery and intelligence. *Mind*, October, **59**, 433-460. Reprinted in *Computers and Thought*, E. A. Feigenbaum and J. Feldman, eds. (pp. 11-35), 1963.

van Heerden, P. J. (1968). *The Foundation of Empirical Knowledge*. Wassenaar, The Netherlands: Wistik.

Wilson, S. W. (1983). On the retino-cortical mapping. *Int. J. Man-Machine Studies*, **18**, 361-389.

Wilson, S. W. (1985a). Knowledge growth in an artificial animal. *Proceedings of the First International Conference on Genetic Algorithms and Their Applications* (pp. 16-23). Hillsdale, New Jersey: Lawrence Erlbaum Associates.

Wilson, S. W. (1985b). Adaptive "cortical" pattern recognition. *Proceedings of the First International Conference on Genetic Algorithms and Their Applications* (pp. 188-196). Hillsdale, New Jersey: Lawrence Erlbaum Associates.

WHAT IT MEANS FOR ROBOT BEHAVIOUR TO BE ADAPTIVE

BY DAVID MCFARLAND
Dept. Zoology. University of Oxford, U.K.

ABSTRACT

There is an analogy between animal and product design that can be formulated as an exact mathematical analogy. The success of a biological design is measured by the success of the genes that produce it, and this depends upon the ability of those genes to increase their representation in the population in the face of competition from rival genes. Similarly, when a variety of products is under consideration, they vary in the period required for product development, in the chance of failure in the market place, and in the expected returns from sales if the product is successful. The development period refers to the period before any return is achieved on investment. For animals this is the period between birth and reproduction, and for products it is the period prior to time that financial return accrues to the investor. The success of a design is evaluated by the net rate of increase of the genes coding for it (i.e. the return on investment) in the animal case, or, in the case of a product launched into the marketplace, of the money invested in it.
If we are to take the biological approach to robot design seriously, then we should first consider the ecological (or market) niche that a proposed robot is to occupy. Is the robot to be a toy, a brick-laying robot, or a bomb-disposal robot? Just as there are no general-purpose animals, so there should be no general-purpose robots. For robot behaviour to be adaptive, in terms of the analogy, it must optimise with respect to the selective pressures of the market place. Other forms of adaptation, such as acclimatisation and learning, are subject to the same criteria.

Acclimatization to altitude. Adaptive changes in a man breathing rarefied air for 4 days, followed by 6 days at sea level.
V = lung ventilation, E = serum erythropoietin, H = rate of hemoglobin synthesis, R = fraction of red blood cells (after Adolph, 1972)

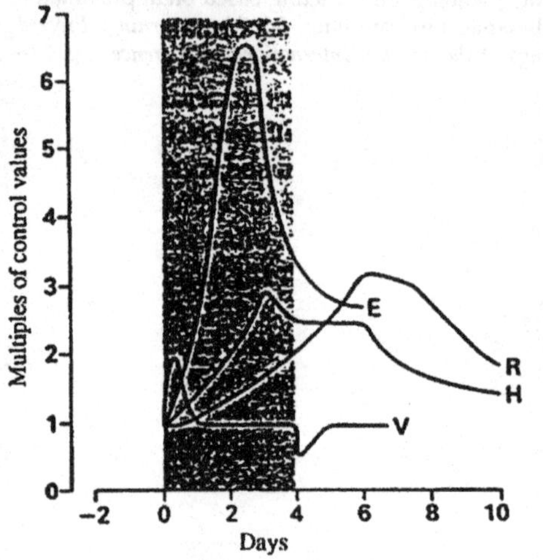

Fig. 1 Adaption by acclimatisation. The physiological changes that occur in acclimatisation to altitude run through a spectrum, ranging from fast but costly processes to slow-acting processes that are cheap in energetic terms.

The term adaptation, as used in biology has a number of meanings: Biologists usually distinguish between (1) evolutionary adaptation, which concerns the ways in which species adjust genetically to changed environmental conditions in the very long term; (2) physiological adaptation, which has to do with the physiological processes involved in the adjustment by the individual to climatic changes, changes in food quality, etc.; (3) sensory adaptation, by which the sense organs adjust to changes in the strength of the particular stimulation which they are designed to detect; and (4) adaptation by learning, which is a process by which animals are able to adjust to a wide variety of different types of environmental change.

Adaptation implies cost reduction, as can be seen from the example in Figure 1. In animal behaviour, real costs relate to Darwinian fitness. So, in considering the usage of the term adaptation in robotics, we should ask if there a concept equivalent to the fitness of a robot?

I will argue that there is such an equivalent concept, and that the analogy between animal and product design can be formulated as an exact mathematical analogy. Briefly, the success of a biological design is measured by the success of the genes that produce it, and this depends upon the ability of those genes to increase their representation in the population in the face of competition from rival genes. How does this relate to product design? Suppose that a variety of products is under consideration. They vary in the period required for product development, in the chance of failure in the market place, and in the expected returns from sales if the product is successful. The development period refers to the period before any return is achieved on investment. For animals this is the period between birth and reproduction, and for products it is the period prior to time that financial return accrues to the investor. The success of a design is evaluated by the net rate of increase of the genes coding for it (i.e. the return on investment) in the animal case, or, in the case of a product launched into the marketplace, of the money invested in it. This argument is outlined mathematically in Figure 2., and is to be spelled out fully by McFarland and Sibly (in prep).

ANIMALS

All gene-carriers reproduce at age t, and then produce n offspring. The chance of a juvenile surviving to breed is S. The instantaneous per capita rate of increase of genes coding for this life cycle is

$$r = \frac{1}{t} \log_e (S \frac{n}{2})$$

NB The representation of a single gene after a period t is e^{rt}.

The expected number of surviving copies of a gene is $\frac{1}{2nS}$.

Therefore

$$\frac{1}{2nS} = e^{rt}$$

ROBOTS

All sales occur at the end of the development period t, and the development costs are k. Total returns are n, but these are realised only if the product successfully reaches the marketplace, which occurs with change S. Hence expected returs, allowing for the change of product failure, are Sn, and expected net returns are Sn - kt.

The rate of return is given by

$$r = \frac{1}{t} \log_e \frac{Sn}{k}$$

NB The expected gross income = Sn (the expected sales in money terms).

The return ration (gross returs/initial investment) = e^{rt}, assuming instantaneous r to compare investments over different periods of time.

$$\text{Therefore, } \frac{Sn}{k} = e^{rt}.$$

Fig. 2. Outline of the mathematical analogies between the life cycles of animals and robots (From McFarland and Sibly, in prep).

Figure 3 summarises the analogies between the design cycles of animals and robots. Development period t refers to the period before any return is achieved on investment. For animals this is the period between birth and reproduction, and for products it is the period before any financial returns accrue to the investor. S measures the chance of a product (animate or inanimate) surviving to enter the gene pool or marketplace. n refers to the return on investment. For animals this is the number of offspring produced when the individual reproduces (i.e. when the gene is copied into half the offspring on average, in accordance with Mendel's laws). In the case of man-made products, n refers to the income from sales. The success of the design is evaluated by the net rate of increase of genes coding for it, in the animal case, or the money invested in it, in the case of man-made products.

The cases of longer adult life of sales period, can be analysed by an extension of the above method. In the simplest such case the adult achieves a constant breeding performance, producing n offspring every time it breeds, which occurs at intervals t apart. Similarly a product might achieve sales at a constant rate over a given period. If the animal is subject to mortality at a constant rate, the expected breeding performance is likely to follow a negative exponential curve. Similarly expected product sales might decline with time, taking into account the chances of failure to sell as a result of market competition or loss of market appeal.

ANALOGIES BETWEEN ANIMAL AND ROBOT LIFE CYCLES

return on investments	n	number of offspring	gross sales income assuming no failures
reproductive probability	S	chance of juvenile surviving to breed	chance of product reaching market
development period	t	age at breeding	development cost = k
design success (rate of return)		net rate of increase of genes coding for design ("fitness")	net rate of increase of money invested in design (instantaneous interest rate)

Fig. 3. Table of analogies between animal and robot life cycles

1. THE ECOLOGICAL NICHE

If we are to take the biological approach to AI design seriously, then we should first consider the ecological niche that a proposed robot is to occupy. In animal ecology, the niche is the role that the animal plays in the community in terms of its relationship both to other organisms and to the physical environment. Thus a herbivore eats plant material and is usually preyed upon by carnivores. The species occupying a given niche varies from one part of the world to another. For example, a small-herbivore niche is occuppied by rabbits and hares in northern temperate regions, by the agouti and viscacha in South America, by the hyrax and mouse deer in Africa, and by wallabies in Australia.

Niche occupancy usually implies ecological competition. When animals of different species use the same resources or have certain habitat preferences or tolerance ranges in common, niche overlap occurs. This leads to competition between species, especially when resources are in short supply. The competitive exclusion principle states that two species with identical niches cannot live together in the same place at the same time when resources are limited. The corollary is that, if two species coexist, there must be ecological differences between them.

In thinking about designing robots, we should first think about what market niche we envisage. Is the robot to be a toy, a brick-laying robot, a bomb-disposal robot, or what. Just as there are no general-purpose animals, so there should be no general-purpose robots. If our robot is to be employed as a brick-layer, then it must be able to compete in the market place with human brick-layers. Of course, the two species of brick-layers will not be alike in every (brick-laying) respect. Employers will value one for certain qualities, and the other for other qualities. In other words there will be only partial niche overlap. Nevertheless, employers will apply roughly the same cost-efficiency criteria to each species, and these criteria will supply the selective pressures characteristic of the brick-layer niche. These selective pressures provide the main ingredients of the cost function that is characteristic of the environment in which the robot is to operate.

2. COST FUNCTIONS

In biology, the cost function "specifies the instantaneous level of risk incurred by (and reproductive benefit available to) an animal in a particular internal state, engaged in a particular activity in a particular environment" (McFarland, 1977).

An animal in a particular state runs a (specifiable) risk of incurring real costs. It is obvious that it is more risky to be in some states than others. For example, it is obviously dangerous to allow hunger to approach lethal levels if the food supply is not guaranteed. Thus the risk of death increases steeply the nearer a variable is to it's lethal boundary. It has been shown (McFarland and Houston, 1981) that the risk function will generally be quadratic for state-variables that have a lethal boundary. In animal behaviour studies, the term 'cost' is generally used to signify the risk of death plus other factors that may lead to a decrement in Darwinian fitness, such as reduction of reproductive potential.

Such cost considerations apply to robots as well as animals. The bomb-disposing robot that runs out of fuel may as well be dead as far as its usefulness is concerned. Such a robot would have low fitness in the sense that it would not fare well in the market place. In considering the question of the fuel supply of our bomb-disposing robot, we assume that the robot has some sort of on-board energy store, and that robot activity is an energy-consuming business. The robot is equipped with a sensor for monitoring the fuel level, but this is not perfectly accurate.

In Figure 4 we assume that x_f is the state of the fuel reserves at a particular time and that the probability P of the state x_f moving to a new position along the state axis is given by a normal distribution. As the robot goes about it's business, the state x_f moves towards the boundary B at which the fuel reserves are zero. The area A represents the probability of crossing the boundary B. The area A increases as the square of the distance ox. In other words the risk of crossing the boundary B increases as the square of the fuel deficit, as shown in Figure 5.

The exact parameters of this function will depend upon the ecology of the refuelling situation. The frequency and reliability of refuelling opportunities will obviously be important factors influencing the risk of running out of fuel. The cost of the refuelling behaviour must also be taken into account. In the present case, we assume that the robot refuells by replacing its old rundown battery with a new recharged one. So it is the availability and location of the new batteries that matters.

In animal behaviour, real costs can be studied experimentally, and this is what some animal behaviourists spend their time doing. To obtain similar information about

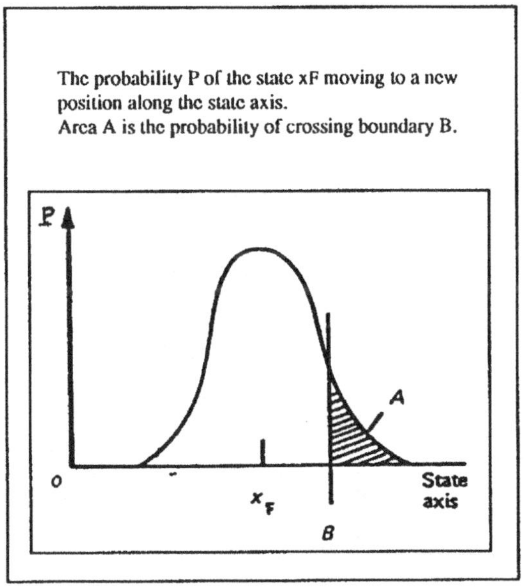

Fig. 4. The risk of fuel reserves falling below the lethal limit.

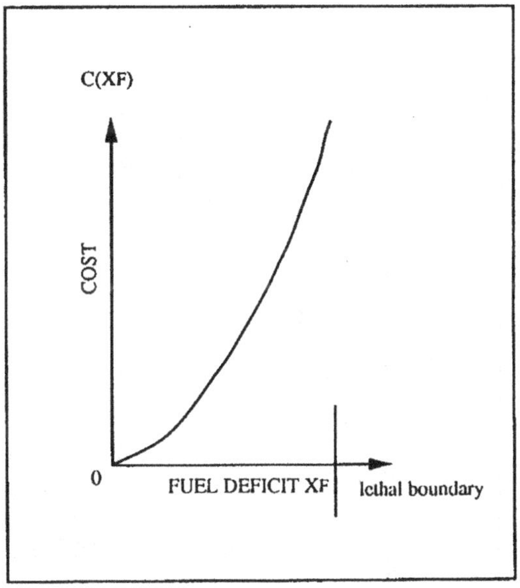

Fig 5. The cost of the fuel state as a function of the deficit.

the equivalent costs for a robot, it would be necessary to do field studies. Reliability studies of motor cars provide an example of this sort of approach.

Of course empirical study is unlikely to provide all the answers, and a good deal of computer modelling is required to fill in the gaps (For an example see Sibly and McCleery, 1985). Similarly, to evaluate the risks, costs and benefits of a robot launched into the market place, it would be necessary to do some ecological modelling that resulted in a cost function.

Let us now look at an example of a simple cost function that has received considerable attention from animal behaviourists.

Consider the two-dimensional quadratic cost function:

$$C(x) = Kx_1^2 + Lx_2^2 + Mu_1^2 + Nu_2^2 \qquad (1)$$

where C(x) is the instantaneous cost, x_1 and x_2 are state variables, u_1 and u_2 represent the rates of performing two activities, and K, L, M and N are scaling parameters, which we will assume to have value one, for our present purposes. The behaviour of an automaton would conform to this cost function if it obeyed the following simple rule:

if $x_1u_1 > x_2u_2$ perform activity u_1, else perform activity u_2.

We will call this RULE B.

The effect of applying this rule can be seen in Figure 6. In this figure RULE B is constrasted with an even more simple rule (RULE A), which is to perform that activity for which x is larger. Figure 6 shows that the two rules give different behaviour (shown here in terms of the reduction in x), and different cululative costs, as calculated on the basis of the cost function. RULE B is less costly and comes close to the optimal solution for this problem (Sibly and McFarland, 1976).

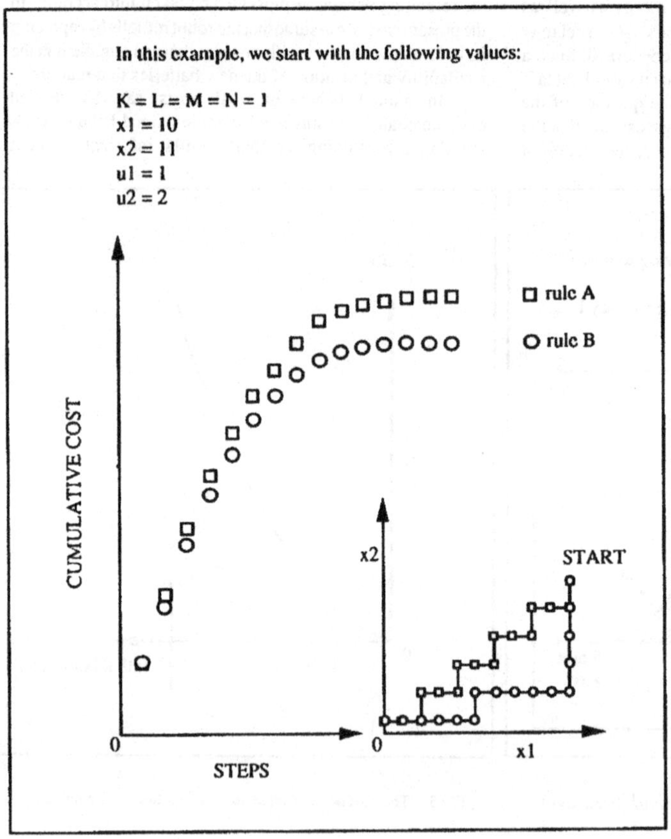

Fig. 6

The cumulative cost of following rules (large graph), and the changes in state that result (small graph)

Notice that it is possible for the rules governing behaviour to conform to some extremal or optimality principle, without there being any obvious sign of this in the formulation of the control system. Of course, the cost function can be explicitly represented (in which case McFarland and Houston (1981) would call it a goal fuction), as would be necessary in a planning system.

As a matter of fact, a simple planning approach to the same problem gives a very similar cost profile. (McFarland, unpublished). If we plan one step ahead (PLAN1), then we simply ask what next activity will result in the lower cost as calculated from the cost function. In planning two steps ahead (PLAN2) we ask which subsequent activity yields the lower cost.

If we compare PLAN1 and PLAN2 we find, not surprisingly that PLAN2 results in lower cost than PLAN1. If, however, we compare with our automaton (ABOVE), we find that PLAN2 < AUTOMATON < PLAN1. The performance differences are very small, but there are vast differences in the capital costs involved. The planner requires some reference to a representation of the cost function (called the goal function). The planner has to compute the notional cost of each of its various options, and then compare them. This all requires investment in hardware. It is difficult to know exactly how to make these calculations. I have tried various ways, and I have come to the conclusion (based on comparison of analogue simulations) that PLAN2 = PLAN1/9 = AUTOMATON/90. In other words it would take a brain 90 times the size of the AUTOMATON brain to improve on the automaton performance by planning.

The cost function provides a performance index against which the behaviour produced by a particular brain architecture can be evaluated objectively - that is in terms of the robot ecology, or in terms of the real world. In comparing different brains for a given robot designed to operate in a given ecological niche - what we should be looking at is value for money. Figure 7 shows a plot of value for money versus price for the four different ways of minimising the quadratic cost function discussed above. The price is calculated on the basis of the development costs (assuming no profit). So this is the absolute minimum price that could be charged without incurring a loss. The actual price charged would, of course, depend upon the expected sales.

In terms of value for money, RULE B is by far the best buy. Of course, this is a very simple problem, with only 2 dimensions. For more complex problems we would expect a planning approach to result in some savings in brain size, because a planner can re-use certain brain areas over and over again, whereas an automaton's brain is made up of circuits dedicated to particular procedures. Moreover, a brain that has some representation of the cost function (i.e. a goal function) should be capable of learning. So we might expect some improvement in performance.

Overall, we should expect that, with increasing complexity, the value for money provided by an automaton should decline, because the increasing cost of the hardware will mean that the selling price will have to increase. The value for money provided by a system capable of planning and

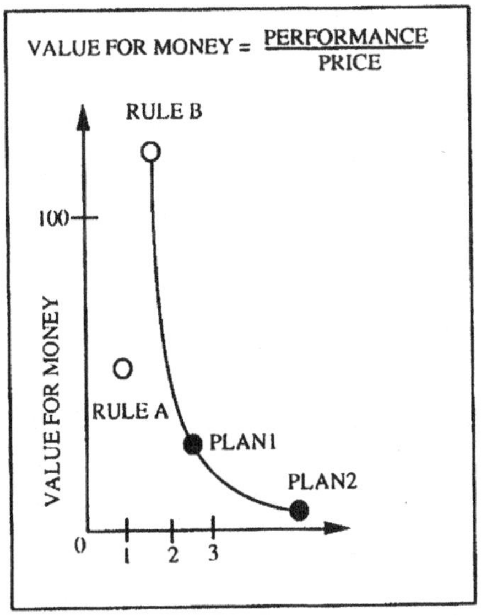

Fig. 7. Value for money of different rule-following and planning solutions to minimising the 2 dimensional quadratic cost function.

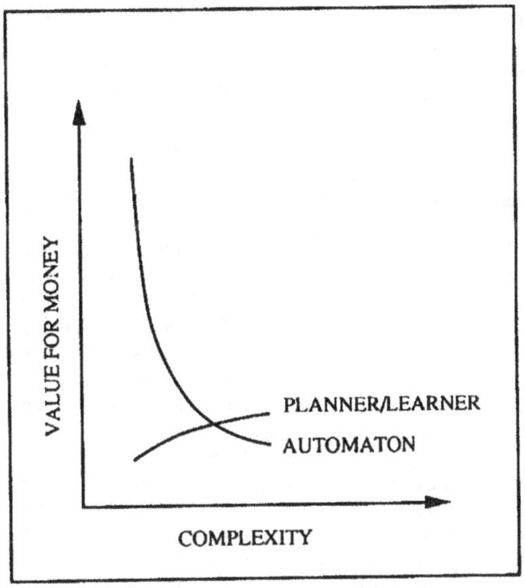

Fig. 8. Value for money of automation (rule following) and planning brain architectures as a function of the complexity (dimensionality of the niche) of the agent's life-style.

learning should increase, because of improvements in performance, as illustrated in Figure 8. We should remember, however, that in a multitask agent, the automaton solution may be better for some tasks, and the planning solution for others. Therefore, complex multitask agents should have a modular architecture, and I see the kind of exercise that I have sketched here, as leading to methods of costing different brain architectures.

3. LEARNING

So far we have concentrated upon evolutionary adaptation and its industrial equivalent. Let us now turn to adaptation of the individual. The equivalent of physiological adaptation is familiar to engineers as adaptive control.

An adaptive control system has the capacity to modify its strategy of seeking some optimum, or set of performance criteria, either by reference to some idealised form as represented by a model, or by learning based on the remembered results of past behaviour. Such adaptive control systems occur in animals. (See for example McFarland, 1971, pp 109-116).

Adaptation of individual behaviour also occurs in the form of learning. True learning requires some kind of planning architecture, which embodies a representation of the cost function. An automaton has no representation of the cost function, and can, therefore, engage only in preprogrammed forms of learning. The planning architecture embodies a goal function representation (which we call the goal function), and is therefore in a position to learn, in the true sense of the word. True learning requires some access to a representation of the cost function. Trial and error implies an experiment on the part of the agent, the results of which have to be evaluated. The results can be evaluated only by reference to an immutable set of values. For learning to be adaptive (i.e. of advantage) this set of values must reflect the cost function that it characteristic of the ecological niche. The more the set of values, or goal function, resembles the cost function then the better adapted will be the agent (see McFarland & Houston (1981) for a discussion). Note that it is not possible for the agent to modify the goal function to make it more like the cost function. Learning requires feedback about success and failure, and this feedback must be based upon an immutable set of values. This set of values form part of the goal function.

Learning has to operate in relation to a representation of the cost function, which we call the goal function. So the imprivement in performance that results from learning is an improvement in relation to the goal function. Now the goal function may or may not be a good representation of the cost function that is characteristic of the robots ecological niche. In animals it can never be a perfect representation, because of evolutionary lag, genetic deversity, and ecological competition. In robots the same logic holds. This means that learning is not always adaptive. Learning in robots is adaptive only in so far as the goal function resembles the cost function.

4. REFERENCES

McFarland, D. (1971) Feedback mechanisms in animal behaviour. Academic Press, London.

McFarland, D (1977) Decision-making in animals, Nature, 269, 15-21.

McFarland, D. and Houston, A. (1981) Quantitative Ethology: the state-space approach. Pitman Books, London.

Sibly, R.M. and McCleery, R.H. (1985) Optimal decision rules for gulls. Animal Behaviour, 33, 449-465.

Sibly, R.M. and McFarland, D.J. (1976) On the fitness of behaviour sequences. American Naturalist 110: 601-617.

Computational Neuroethology: A Provisional Manifesto

Dave Cliff
University of Sussex School of Cognitive and Computing Sciences
Brighton BN1 9QN, England, U.K.
E-mail: davec@uk.ac.sussex.cogs

Abstract

This paper questions approaches to computational modeling of neural mechanisms underlying behaviour. It examines the "simplifying" (connectionist) models used in computational neuroscience and concludes that, unless embedded within a sensorimotor system, they are meaningless. The implication is that future models should be situated within closed-environment simulation systems: output of the simulated nervous system is then expressed as observable behaviour. This approach is referred to as "computational neuroethology". Computational neuroethology offers a firmer grounding for the semantics of the model, eliminating subjectivity from the result-interpretation process. A number of more fundamental implications of the approach are also discussed, chief of which is that insect cognition should be studied in preference to mammalian cognition.

1 Introduction

This paper questions approaches to computational modeling of the neural mechanisms underlying behaviour. It examines the relationship between computational neuroscience [26] and that style of modeling popularly referred to as "connectionism", "parallel distributed processing", or "neural networks"[1] which has recently been subject to renewed attention in the fields of cognitive science and artificial intelligence (see e.g. [25, 23]).

Connectionist models are characterised by their simplified nature and concomitant inattention to biological data, and it is argued here that such "simplifying" computational neuroscience has serious inadequacies. A different approach is suggested which pays far more attention to the sensorimotor system and hence to behavioural interactions with the external environment. This approach involves computational modeling of the neural mechanisms underlying behaviour, in a manner akin to that used in connectionism. Such an analysis of behaviour as a product of neural activity is properly the domain of the field of neuroethology, and the new approach is therefore referred to as "computational neuroethology". Meaning is supplied to the models by embedding them in simulated environments which supply visual feedback without human intervention, that is they close the external feedback loop from motor output to sensory input.

The advantage of computational neuroethology is that the semantics of the network are well grounded, and thus results are generated by observation rather than by interpretation. That is, the fruits of computational neuroethology simulations are "hard" objective measurements rather than "soft" subjective ones. At a metatheoretical level, it is argued that the computational network simulation of cognitive processing should pay much more attention to the evolutionary history of those faculties it wishes to replicate. In particular, a conclusion of this paper is that the study of linguistic processes using network models is wildly premature. The study of insects is advocated as the most fruitful path for future research

As the reader will probably already have detected, this paper is intentionally polemic. It is aimed at an interdisciplinary audience, and the author is no polymath. For that reason, this paper is offered as a provisional manifesto in the hope that it provokes some interesting discussion. The argument is based on previous work by a number of authors. Because of its disputatious nature, there are more direct quotes in this paper than is common. There is no denying that this is a *selective* review of the literature. This paper is abridged from [9].

The paper opens with a discussion of computational neuroscience, distinguishing it from neural engineering, and identifying two classes of model: realistic and simplifying. Following this, the connectionist paradigm is briefly summarised. Next, criticisms of connectionism are discussed, with particular attention to the argument that connectionist models have no semantic grounding without behavioural linkage to a sensorimotor system. Then, a remedy to this objection is proposed: the adoption of the computational neuroethology approach. Computational neuroethology is defined, and a specific technique for providing a behavioural linkage is discussed. This approach has some important implications for future research, the most significant of which

[1] In this paper, these three terms will be treated as synonymous, and referred to collectively as "connectionist" models.

is that it encourages an evolutionary approach to understanding intelligence.

2 Neurosomething

The recent excitement in the cognitive science and artificial intelligence communities about "neural" issues has been motivated by a number of interests, and consequently the terminology has not been developed in a uniform manner. The first requirement is to distinguish between issues of science and issues of engineering. To this end, two important terms, computational neuroscience and neural engineering, are informally defined below.

- *Computational Neuroscience* is the study of real brains or nervous systems as computational systems, "...in the sense of representing, processing, and storing information." [20, p.3]. Sejnowski and Poggio define it as:

  > "...an approach to understanding the information content of neural signals by modeling the nervous system at many different structural scales, including the biophysical, the circuit, and the systems level." [20, series forward]

  Most often, this entails modeling the system under study on a digital computer. The term *computational neurobiology* [1, p.186] is here taken to have meaning identical with *computational neuroscience* [*ibid.* p.203].

- The construction of intelligent systems and massively parallel (fine-grain) digital computers inspired by principles found in naturally occurring neural systems has been referred to as *neural computing* [1, p.186] and *neural engineering* [1, p.203]. In this paper only the latter term will be used. The most important feature that distinguishes neural engineering from computational neuroscience is that in neural engineering the onus is on optimality. That is, neural engineering usually has the goal of creating an artefact that is (near-) optimal with respect to some measure of efficiency.

Computational neuroscience and neural engineering are not mutually exclusive endeavours: "...the two subjects have different goals. Nonetheless, they overlap and are mutually stimulating." [1, p.186]. But the science/engineering distinction is an important one:

> "Engineering is often based on science, but its aim is different. A successful piece of engineering is a machine which does something useful. Understanding the brain, on the other hand, is a scientific problem." [10, p.132].

This paper is concerned only with science; not with engineering. In particular, it is concerned with the relationship of computational neuroscience to cognitive science. It is not concerned with any attempt to create 'intelligent' artefacts that employ optimal processing strategies: that is a matter for artificial intelligence (AI); the processing strategies of naturally occurring cognitive activity may well be suboptimal, and (for the purposes of this paper) cognitive science will be considered as studying only naturally plausible models of cognitive processing. That is, I'll assume that the creation of what Dennett [11] calls "cognitive wheels"[2] is properly only the domain of AI; although, of course, some cognitive wheels may be lurking within the cognitive science canon, yet to be refuted.

3 Computational neuroscience

The advantages claimed [26] for modeling and simulation in computational neuroscience over conventional experimental techniques include:

1. Increased accessibility to the consequences of complicated nonlinear brain systems with many interacting components.
2. The possibility of discovering new phenomena by comparing experimental results to the predictions of simulation, and using these predictions as the basis for the design of new experiments.
3. Facilitating experiments (such as selective lesioning or ablation of particular channels, synapses, neurons or pathways) which would be difficult or even impossible to perform on living tissue.

Practitioners of computational neuroscience employ two classes of brain model: *realistic* models and *simplifying* models; these are discussed in more detail below. These two classes of model represent the extreme points of a continuum: in practice, models are likely to have features of both, and many different types of model will be required to span all levels of analysis.

3.1 Realistic Models

A realistic model "...consists of a very large scale simulation that tries to incorporate as much of the cellular detail as is available" [26, p.1300]. The realism of the model, while offering the advantages listed above, introduces two important weaknesses [26, p.1300]:

1. As more parameters and variables are added to the model to increase its realism, the complexity of the model grows, and so there is an increasing danger of

[2] "...an elegant but unnatural solution to a problem of natural design." [6, p.65]

the simulated nervous system being as poorly understood as the real thing.

2. The results may be invalidated by the inadvertent exclusion of important features, because all the cellular details are not yet known.

Probably the best known realistic models are the Hodgkin-Huxley neuron model [16] and the Hartline and Ratliff model of lateral inhibition in the eye of *Limulus* [24]. The realistic modeling approach is most appropriate when knowledge of the circuit to be modeled is almost complete down to the biophysical level, and the function of the circuit is already known [22, p.361].

3.2 Simplifying models

The use of "simplifying" models is an approach applied at the network level of nervous systems: the procedure involves starting with a function such as a perceptual ability and designing "simplified neural circuits that can perform the function within the constraints of the state of knowledge." [22, p.361].

Proponents of simplifying models claim that they offer greater conceptual clarity, and that they fulfill a perceived need for models capturing important principles. Their supporters propose [26, p.1300] that they are analogous to aspects of the physics literature such as textbook examples that admit exact solutions:

> "These models abstract from the complexity of individual neurons and the patterns of connectivity in exchange for analytical tractability." [26, p.1301].

Sejnowski *et al.* state [26, pp.1300–1301] that simplifying models include those currently being investigated under the general headings of connectionist models, parallel distributed processing models, and "neural networks"; they note that some of the models reported on under these headings are used not for brain research, but for neural engineering or for the study of purely psychological processes. It is those connectionist models used for brain research that are discussed here.

4 Connectionism

Again, this paper concentrates only on connectionist models which are also simplifying computational neuroscience models: engineering or purely psychological models are ignored. Examples of all three styles can be found in [25, 23].

It is is not within the scope of this paper to attempt a comprehensive definition of connectionism, nor is it possible to fully examine here the position of connectionism in the context of past practice in cognitive modeling. However, both of these tasks have been performed by Smolensky [27], whose definitions are quoted below. They are presented here as representative of the field, rather than as the view of a small group of Smolensky cohorts.

> "Connectionist models are large networks of simple parallel computing elements, each of which carries a numerical *activation value* which it computes from the values of neighboring elements in the network, using some simple numerical formula. The network elements, or *units*, influence each other's values through connections that carry a numerical strength, or *weight*. ...connections carrying positive weights are called *excitatory* and those carrying negative weights are *inhibitory*.
>
> "In a typical connectionist model, input to the system is provided by imposing activation values on the *input units* of the network; these numerical values represent some encoding, or *representation* of the input. The activation on the input units propagates along the connections until some set of activation values emerges on the *output units*; these activation values encode the output the system has computed from the input. In between the input and output units there may be other units, often called *hidden units*, that participate in representing neither the input nor the output.
>
> "The computation performed by the network in transforming the input pattern of activity to the output pattern depends on the set of connection strengths; these weights are usually regarded as encoding the system's knowledge." [27, p.1; original emphasis]

Smolensky describes connectionism's relation to the two main levels of analysis in previous work: the neural (as in realistic computational neuroscience) and the symbolic (as in nearly all work in cognitive modeling and artificial intelligence (AI) up until the early 1980's). Smolensky posits connectionism as constituting a new level of analysis, intermediate between the symbolic and neural levels. This he refers to as the *subsymbolic* level. The argument for the existence of this new level is involved, and will not be repeated here; see [27].

5 From frying pan to fire?

Connectionism has not been received with universal acclaim. Criticisms have been made both of specific models and of the methodology as a whole. Presented below are summaries of critiques by some other authors, along with some criticisms of my own. Again, this paper is only interested in connectionism *qua* simplifying computational neuroscience. I am aware that some criticisms have been replied to by proponents of connectionism, but a full review of the debate would be an unwelcome digression here. The four criticisms made below are those which I feel connectionism has most consistently failed to acknowledge or answer.

5.1 *The Hunting of the Snark*

A forceful critique of the connectionist approach was made by Crick [10]. He argues that connectionist "models" do not correspond sufficiently closely to real neural systems to be regarded as models in the usual sense: "In another context they might reasonably be referred to as existence proofs. As such they have a certain use." [10, p.131]. He attacks connectionist research on methodological and ontological grounds.

His methodological objection is that connectionists employ mathematics as an intellectual prop: he suggests "...that within most modellers a frustrated mathematician is trying to unfold his wings. It is not enough to make something that works. How much better if it can be shown to embody some powerful general principle for handling information, expressible in a deep mathematical form, if only to give an air of intellectual responsibility to an otherwise rather low-brow enterprise." [10, p.132]. Lehnert made a similar criticism, she argues that "the interdisciplinary appeal of connectionism is not so much a computational appeal, as it is an appeal based on theorem envy. ...connectionism has come to the rescue of a new generation of psychologists who are really closet mathematicians and physicists." [27, p.40].

Crick criticises past work in psychology and linguistics for its preference for simple "intelligible" models, an approach rooted in the belief that the brain is intractably complex. He notes that proponents of such an approach are generally not concerned by the absence of criteria of biological or psychological feasibility, and characterises their philosophy thus: "If it describes, in a succinct way, some of the psychological data, what can be wrong with that? Notice, however, that by using such arguments, one could easily make a good case for alchemy or the existence of phlogiston" [10, p.131].

Crick's ontological objection is rooted in his methodological criticism. He argues that the search for mathematical expression is highly questionable because there is no guarantee of deep general principles being embodied in the functioning of the brain. He offers the genetic code as a good example of a complex natural system which is not easily characterised by a small set of general principles: the brain may achieve its aims using a "series of slick tricks" [10, p.132]. He contends that further research should be aimed at resolving this issue:

> "Learning about neurons, their behaviour and their connections, will not by itself solve our problems, but will at least suggest the sort of answer to look for and can be used, often rather decisively, to disprove false theories." [10, p.132].

5.2 *Biological studies of cognition*

This preoccupation with elegant mathematics has lead, I believe, to connectionists ignoring a significant school of thought in neuroscience: that of Humberto Maturana and his colleagues. Maturana's work is concerned with understanding how phenomena such as cognition and language are rooted in biological processes. His work is challenging, using a large specialised vocabulary, and a full review of it is way beyond the scope of this paper: it will suffice here merely to nod in his direction. The notes that follow are based on an overview of his work by Winograd and Flores [28, pp.38–53].

The most germane of Maturana's arguments is that the nervous system should not be treated as an input-output device. But most connectionist models do exactly that. That is, they treat the neural function to be modeled as being implemented on a 'pipeline' processor. The most likely reason for this is the mathematical tractability of "feed-forward" network models, where each unit sends output only to units in subsequent layers in the network (i.e. later stages of processing). For instance, the "back-propagation" network learning algorithm [25, ch.8], the subject of much attention in the literature, operates only on feed-forward networks and is defined solely in terms of the input-output profile of the function at hand.

This maths-driven nature of connectionist architecture is acknowledged by Smolensky: "In the drive for more computational power, architectural decisions seem to be driven more and more by mathematical considerations and less and less by neural ones." [27, p.9].

So, the view (implicit in much connectionism) of nervous systems as input/output pipeline devices is mistaken and should be avoided.

Maturana further argued that there is no clear modular separation between an entity and its 'environment' [28, p.43]: an entity is defined only in terms of its surrounding environment. This is an issue ignored by virtually all current connectionist models. Most treat the portion of the nervous system under study as capable of being modeled *in vacuo*. At best, connectionist models are extremely poor approximations to *in vitro* studies.

5.3 *Microscope Envy*

Unfortunately, the connectionist disregard for past work in biology does not stop at the philosophical level of Maturana. Lehnert's diagnosis of connectionists as closet physicists suffering from theorem envy was noted above. I suggest that the complementary case also applies: some connectionists are physicists or mathematicians who are closet biologists suffering from microscope envy. Specifically, it appears that one concrete achievement of the connectionist paradigm has been to recast old problems of the symbolic paradigm at a new, biological, level.

This is a problem rooted in the cavalier attitude most proponents of connectionism adopt towards biological reality. Only one example of this is presented here: see [9] for another.

5.3.1 Micro-worlds, again

This example concerns the micro-world trap in symbolic cognitive modeling. This is about previous work in the symbolic AI paradigm which proposed to have created systems which 'understood' concepts or which were claimed to exhibit learning or 'discovery' of new concepts. There is a lot of such work, and a comprehensive literature review is out of the question here: this is an eclectic overview of the argument.

Essentially, the criticism of such models (which dogged AI research through the early 1970's; see [12] for a discussion and pointers to the micro-world literature) was that all the 'understanding' was actually being performed by the programmers, because in creating the system it was necessary to create a micro-world for the system to work on. Such toy worlds implicitly incorporated a vast amount of pre-processing.

Similarly, systems which were purported to learn or discover new concepts make their discoveries by "...working on data presented in notational formats that represent the fruits of centuries of human labor. Manipulating these representations could be the tip of the iceberg; creating them and understanding them may constitute the unseen bulk." [6, p.14].

One of the appeals of the connectionist approach listed by its proponents is the availability of autonomous learning procedures for tuning the weights in the network, thereby 'programming themselves' to perform the task at hand. At first glance, the supposed autonomy (no human intervention is required to specify the changes or to actually alter the weights) seems to indicate that connectionism does not suffer from micro-world problems.

Any learning network requires a source of input vectors — specifications of the activities of the input units of the network. And there's the rub. Unless the network is involved in the first stages of vision or hearing, the input vectors have to be prepared "off-line", invariably by humans rather than by other connectionist networks. That is, connectionist models rely on pre-processed information and are thus susceptible to the same problems as beset the micro-world studies of the symbolic paradigm. Smolensky acknowledges this problem. He notes [27, pp.7–8] how important the choice of representation for input (micro)features is, and that a model's performance depends crucially on the input and output representations chosen by the modeler.

Probably the most celebrated example of this is the Rumelhart and McClelland verb past-tense learning system [23, pp.216–271]: its 'wickelfeature' representation scheme (among other things) has been attacked, most notably in the critique by Prince and Pinker (e.g. [27, pp.46–47]) where the model is described as a fairy-tale account of the actual cognitive process.

5.4 Grounding the semantics

This final criticism of connectionism continues the theme of the *in vacuo* problem, further demonstrating the extent to which connectionism exists within a biological vacuum. This criticism is Lakoff's critique of the semantics of connectionist systems; Harnad [15] makes a similar point. Lakoff [21] criticises Smolensky's [27] account of the connectionist paradigm as making a huge omission in ignoring the body. He states:

> "The neural networks in the body do not exist in isolation; they are connected to the sensorimotor system. For example, the neurons in a topographic map of the retina are not just firing in isolation for the hell of it. They are firing in response to retinal input, which in turn is dependent on what is in front of one's eyes. An activation pattern in the topographic map of the retina is therefore not merely a meaningless mathematical object in some dynamical system; it is *meaningful*. A different activation pattern over those neurons would mean something different. One cannot just arbitrarily assign meaning to activation patterns over neural networks that are connected to the sensorimotor system. The nature of the hookup to the body will make such an activation pattern meaningful and play a role in fixing its meaning.
>
> "Compare this, for example, with a string of symbols in a ... computer program. The symbols are not meaningful in themselves. They have to be "given meaning" by being associated with things in the world. If the symbols are to stand for categories, those symbols must be given meanings by being associated with categories that are out there in the world." [21, p.39]

The essence of Lakoff's argument is thus that if neurons are appropriately located relative to the sensorimotor system then activation patterns over a network of neuron are meaningful *in themselves*. That is, the activation patterns do not have to be "given meaning" in the same way that symbol-strings do. But surely any connectionist model that is *not* connected to a sensorimotor system *has* to be "given meaning" in much the same way as symbolic systems do: whether you are giving meaning to one discrete symbol, or to a configuration of activations over a number of input or output units, you still have to associate these (sub)symbols with things in the world; you are still the source of meaning.

Lakoff continues:

> "In a full-blown connectionist theory of mind activation patterns over neurons are meaningful in

themselves by virtue of what they are connected to. The intractable problem of assigning meanings to symbols does not arise here.

"It is also important to remember that the isolated models connectionists build to study the properties of networks are not full-blown connectionist theories of mind. They vastly oversimplify, or totally ignore, sensorimotor input and output, assuming that, for the purpose of the study at hand, one can just as well use feature names, to which the model-builders must assign meanings. This is a crucial difference between isolated models and a full-blown theory.

"...it is vital to bear in mind that a full-blown connectionist theory of mind is a lot more than just an information-processing system." [21, p.40]

Lakoff's argument that the linking of neural models to a sensorimotor system provides an automatic grounding for the semantics of the model is persuasive. Pursuing this approach prevents theorists from inventing the semantics of patterns of activity (as they are free to while the models remain in their current isolated status). Smolensky's reply to this ("[it] seems an important philosophical point, but one that cannot really do any modeling work until the gap is bridged (...) between the subconceptual [i.e. connectionist] and neural levels..." [27, p.66]) is distinctly *mezza voce*. Again, the attack here isn't *ad hominem*, but Smolensky's views are, I believe, representative of the field. Acceptance of *in vacuo* modeling is widespread within connectionism.

5.5 Summary

Several criticisms of connectionism have been discussed. It has been argued that the connectionist paradigm is biologically *in vacuo* and in this sense is no advance on the symbolic paradigm: connectionism has acted merely as a palliative for several of the maladies of symbolism. The solution to this problem, as identified by Lakoff, is linking the model neural network to the external world via a sensorimotor system. Such an approach is more in line with the philosophy of the Maturana school. Such an approach is embodied in computational neuroethology.

6 Computational Neuroethology

6.1 Towards a definition

Computational Neuroethology (as far as I am aware, I originated this term, in [7]) is proposed here as a method of eliminating the *ad hoc* semantics of contemporary connectionist models. The definition of the term "computational neuroethology" builds on the definition of "neuroethology", given below.

Neuroethology (e.g. [4, 14]) is a young discipline within the biological sciences: it is where ethology (the study of behaviour) meets neuroscience. Its youth puts it in a situation not dissimilar to cognitive science; its precise aims and methods are still a subject of debate (see e.g. [18, 14]). Put most simply, neuroethology is the study of the neural mechanisms of animal behaviour [18, p.384]. As Clarac remarks, "By definition, neuroethology is a "vertical" science, whose main interest is to link the results obtained from many different levels of complexity in the nervous system." [18, p.384]. That is, neuroethology spans most levels of analysis in neuroscience.

The relevance of neuroethology to the *in vacuo* problem in connectionism is neatly captured by Grossberg:

> "Neuroethology teaches us that neural circuits are organised to generate adaptive goal-oriented behaviors. Without a behavioral linkage, no amount of superb neurophysiological experimentation can lead to an understanding of brain design, because this type of work, in isolation, does not probe the functional level on which an organism's behavioral success is defined." [13, p.389].

It is in this respect that connectionist computational neuroscience has most to learn from neuroethology, and it is this belief in the importance of behavioural linkage that most distinguishes computational neuroethology. Computational neuroethology replaces the *in vacuo* approach of connectionism with a (simulated) *in vivo* approach; and in doing so, the semantics of the model are automatically grounded.

So, computational neuroethology can be provisionally defined as the study of neuroethology using the techniques of computational neuroscience. This definition intentionally admits many classes of model, but the significant aspect is the increased attention to the environment that the neural entity is a component of. The vertical nature of neuroethology precludes restriction purely to "simplifying" (connectionist) models: modeling techniques from realistic computational neuroscience should also be applicable to computational neuroethology.

Furthermore, this broad definition is noncommittal on the interpretation of the word "computational": a strict interpretation would allow only neuroethological models that focus on behaviour as a result of computation; but the vertical nature of neuroethology surely indicates that models with no *direct* reference to computation should still be of interest, and the Maturanian anti-pipeline argument outlined above additionally indicates that sole focus on information processing may omit important factors. So "computational" can simply imply a reliance on computerised experimental techniques, as in e.g. computational physics.

And that, for the time being, is the working definition of computational neuroethology. It is a broad definition, but this is a provisional manifesto, and ensuing debate may refine the definition. What is required is guidance on *how* the connectionist model should be connected to a sensorimotor system, thereby granting the required behavioural linkage. This is a subissue of the field, and is discussed at length in the next section.

As I mentioned above, the term "computational neuroethology" was coined to describe my work in studying visual control of insect flight [7, 8], but plenty of other research in the literature could be classed as work in computational neuroethology.

6.2 The behavioural link

This section discusses my personal attitude towards the issue of providing connectionist models with a behavioural linkage. As such, it is only one of many possible approaches within computational neuroethology.

The need for a behavioural linkage is satisfied by recognising in the model the importance of the environment within which the neural entity is a component. The methods of connectionist computational neuroscience can be adapted to computational neuroethology in a fairly straightforward way: by embedding the network model within a ***closed-environment*** simulator. A closed environment simulator is one which provides a data-path that models the external feedback loop provided by the environment.

That is, closed-environment models eliminate any "humans in the loop": current connectionist models have a data-path that resembles the architecture of an open-loop control system; humans are responsible for feeding data (and meaning) into the input end and collecting data (and assigning meaning) at the output end. Closed-environment models are closer in nature to closed-loop control systems: the feedback loop in a closed-loop controller corresponds to the sensorimotor feedback provided by the simulated environment in a closed-environment model. The feedback properties of the environment have long been recognised by (neuro)ethologists.

Closed-environment simulator systems already exist, in the domain of aeroplane flight simulators. Here the neural network is real (and human) but that is not material: motor commands are issued (to the cockpit controls such as joystick and throttle) and the simulator performs the computations necessary on a model environment to generate the sensory input (e.g. dynamic 3-D graphics, etc.) which provides sensorimotor feedback.

Closed-environment simulators for neuroethology can be fashioned in much the same manner. The simulation of the medium thus involves *two* models: the model network and the model environment.

The use of simulated environments is open to the accusation of falling into the micro-world trap, discussed above. Closed-environment simulators are not subject to micro-world problems because they are not "toy" systems reliant on human interpretation. They are real-world models for a certain definition of real. Consider:

> "There is great methodological danger in tackling AI problems in toy worlds. The definition of "toy" here includes all worlds where it is not up to the AI system itself to do all the understanding of the world itself without relying on a human interpreter. Likewise a world is a "toy" world if the AI system is not responsible for carrying out its actions in the world without a human agent to interpret its responses. Such requirements on an AI system thus force it to be part of a robot system acting in a ***real world*** for some definition of real. This is a much stronger definition of real world than is normally used." Brooks [3, p.5, original emphasis].

Even under such a definition of real world, it is possible to work on real-world problems using closed-environment simulators rather than robotics (thus ignoring some of the more problematic engineering issues of sensorimotor transduction). An example is the SYCO simulator-model of visual processing in a hoverfly [8].

The advantage in using simulated "real" worlds rather than *the* real world (i.e. a robotics approach) is that the real real world has many superfluous degrees of freedom: many experiments in e.g. visual neuroethology take great care to minimise the dimensionality of the experimental animal's world by creating visually impoverished environments. Such precautions are also necessary, at least during development and testing, in perceptual robotics; but in simulated "real" worlds *identical* conditions can be recreated as many times as is required.

The closed-environment simulator approach thus provides a behavioural linkage for a simulated neural network. It grounds the semantics of the network in the semantics of the simulated environment. And this is the fundamental point: if the simulated environment is "real" in the above sense, then the model ceases to demand *interpretation*; results are *observed*. That is, discussion of the model network ceases to be *subjective* and becomes *objective*. Soft data is replaced by hard data. It is possible to talk of computational neuroethology as a falsifiable scientific endeavour.

6.3 Finale: Implications

The implications of computational neuroethology for cognitive science are, I believe, more fundamental than simply tightening up a slack discipline. I briefly outline some thoughts below.

The need for a behavioural linkage, and its satisfaction using a closed-environment simulation system, raises a supplementary issue: if there is so much emphasis placed on the preservation of the data-path from sensory input to motor output, how is it possible to model high-level cognitive functions where there is believed to be significant processing performed between input and output? Surely the sequence of functional units between input and output in complex tasks such as speech understanding or 3D reasoning from vision is so long as to make the modeling task prohibitively complex?

If this problem cannot be solved, at least in principle, then closed-environment simulation is only useful for a very small class of problems: problems which traditionally have been peripheral to cognitive modeling and artificial intelligence.

Fortunately, help is at hand. This problem is resolved by careful attention to two assumptions implicit in its formulation: first, that a sequential functional decomposition of cognitive processes is the best approach; second, that what have traditionally been 'core' topics in cognitive modeling really are core topics, i.e. that problems amenable to closed-environment simulation *really* are peripheral issues. Both of these assumptions have been questioned and found wanting by Brooks.

6.3.1 Subsumption Architecture

Brooks [2] has developed an approach to the design of control architectures for mobile robots operating in the real real world, which can readily be adopted for use in the simulated real worlds of closed-environment simulators. The key to Brooks's approach is the concept of a *layered* control system, which leads to control architectures and strategies radically different from convention.

The conventional approach to constructing a mobile robot control system is to decompose the problem into a set of *functional units*, creating an essentially sequential linear data-path which starts at the sensory transducers and ends at the motor actuators.

Brooks proposes that the primary decomposition of the control problem should be into *task achieving behaviours*. That is, he advocates dividing the desired intelligent behaviour of a system into a collection of simpler behaviours; behaviourally complex systems are thus decomposed into a number of computational systems achieving simpler behaviours. The intention is that the simpler systems should be independent, but some degree of overlap is likely in practice.

Brooks organises the behavioural decomposition on the basis of two concepts: task achievement and competence levels. Task achievement dictates that: "Each behavior should achieve some task. I.e., there should be some observable phenomenon in the total behaviour of the system which can be used by an outside observer to say whether the particular sub-behaviour is operating successfully." [3, p.6]. Competence levels specify overall system performance in a rather informal manner: "A set of task achieving behaviours together provide the robot with some level of competence. They should be designed so that as new task achieving behaviours are added to the system, the level of competence increases." [*ibid.*].

The different task-achieving behaviours form *layers* of a control system. The layers form what Brooks [2, p.7] calls a *subsumption architecture*: initially, the complete robot control system is constructed to achieve level 0 competence — this is referred to as the zeroth layer control system; once it is completely debugged it is never subsequently altered. Next, the first layer control system is constructed: it can receive input from the layer 0 system, and additionally it can suppress, or *subsume*, the output of the zeroth layer. The zeroth and first layers together implement level 1 competence. Similarly, higher layers are constructed to realise higher levels of competence.

A particular layer always runs unaware of the activities of all higher layers, any of which might interfere with its activity. Each level of competence includes as a subset all earlier levels of competence [2, p.7], so successive layers can be viewed as representing increasingly more sophisticated and constrained classes of allowable behaviour. In Brooks's original robot, each layer is built from a number of asynchronous modules, and each module is a finite state machine [2, p.9].

So the central ideas in the Brooksian philosophy are: that the control of a mobile robot can be viewed in terms of behaviours rather than functional modules; that the subsumption architecture allows for incremental construction and debugging of complex mobile robot control systems; and that there is no need for centralised control: "The control system can be viewed as a system of agents each busy with their own solipsist world." [2, p.19].

The intention then in closed-environment simulator computational neuroethology is to create a subsumption architecture where the modules forming each layer are not formal (symbolic) finite state automata, but small artificial neural networks. A network model constructed according to a subsumption architecture is an explicit recognition of the inhomogenous nature of natural nervous systems.

A guiding principle in Brooks's work that is particularly relevant to such an approach is:

> "Complex (and useful) behavior need not necessarily be a product of an extremely complex control system. Rather, complex behaviour may simply be the reflection of a complex environment.... It may be an observer who ascribes complexity to an organism — not necessarily its designer." [2, p.3]

6.3.2 Arthropod Imperialism

Moreover, Brooks argues that there are severe limitations on the generally accepted methodologies of artificial intelligence, and that the currently accepted conceptual decompositions and static representations are wrong. He argues for a shift to process-based modeling, contending that: "...mobility, acute vision, and the ability to carry out survival related tasks in a dynamic environment provide a necessary basis for the development of true intelligence." [3, p.2]. Brooks's argument is based in part on the evolution of intelligent beings: that intellectual capabilities such as writing and 'expert' knowledge are, on the evolutionary timescale, very recent developments indicates to him that the creation of artificial entities truly possessing such abilities will be relatively straightforward "...once the essence of being and reacting are available. ... This part of intelligence is where evolution has concentrated its time — it is much harder." [3, p.1]. Clark makes a related point:

> "...intelligence has evolved as a means of satisfying our basic survival requirements. It has *not* been selected for its capacity to achieve the high-level mental feats which so much work in AI is dedicated to modeling. If we *can* perform such feats... it is only *in virtue* of our being endowed with a set of low-level capacities which just happen to facilitate the higher-level activity.
>
> ...
>
> "The right microworlds to study are not fragments of the sophisticated human achievements, but the less sophisticated achievements of the various animal intelligences, ranged across the phylogenetic tree." [5, pp.4–5; original emphasis]

In particular, Brooks advocates the construction of robotic insects:

> "Insects are not usually thought of as intelligent. However, they are very robust devices. They operate in a dynamic world, carrying out a number of complex tasks... No human-built systems are remotely as reliable... Thus I see insect level behavior as a noble goal for artificial intelligence practitioners. I believe it is closer to the ultimate right track than are the higher level goals now being pursued." [3, p.7].

Similarly evolutionist arguments have been proposed in neuroethology. One of the most passionate advocates of restricting study to arthropods (the animal class to which insects belong) was Hoyle:

> "Many neurophysiologists... express no, or very little interest in invertebrates. Their goal is to understand the higher mammalian nervous system, period. ... The invertebrates have nervous systems which will certainly be understood first in cellular and connectivity terms... by definition, the general principles must be those features which vertebrates possess in common with invertebrates... The study of specifically human nervous systems is an applied, not a general, science." [17, p.17].

From the point of view of neural modeling, a fundamental advantage lies in insect neuroarchitecture. There is ample data in the biology literature demonstrating that insect (and other arthropod) nervous systems are constructed with an economical use of neurons.

The smaller number of nerve cells involved does not necessarily indicate that the neural principles of operation are simpler in insects than in higher animals. However, it does mean that a computational network simulation of insect nervous function is closer in architectural terms to biological reality than any corresponding model of mammalian or primate function, by several orders of magnitude.

So, perhaps the best approach to studying intelligence is, phylogenetically speaking, bottom-up rather than top-down. All previous work in cognitive modeling has focused on advanced animals, near the top of the phylogenetic hierarchy, i.e. humans (and, in vision research, other mammals). Yet the only case for such anthropocentricity is an *a priori* one. Studying intelligence by modeling insects might seem counterintuitive, but then again it might yield some useful results. Perhaps we have been missing something.

Studying animals lowlier than ourselves is not without precedent in other fields. The most notable example is in genetics, where countless person-years of research have been expended on the fruitfly *Drosophila melanogaster*; with the human genome project commencing only very recently. As John McCarthy (a founder of AI) puts it:

> It may be that AI is a problem that will fall to brilliance so that all we lack is one Einstein. [but] I think this is one of the difficult sciences like genetics, and it's conceivable that it could take just as long to get to the bottom of it." (McCarthy, quoted by Johnson [19, p.13]).

Perhaps cognitive science research should go the same way as genetics: we should cease to concentrate on high-level functions and get the basics right first; following a development path strongly influenced by the evolutionary history of natural intelligence.

Even if the study of insects provides us with no data specific to higher animals, it will help clarify the status of the general principles that connectionists search for — perhaps the answer is not to study simple models of large brains, but to study large models of simple brains.

This shift to an antianthropocentric focus for AI has manifest implications for the connectionist study of language. This is a core cognitive function [27, p.62]. But if the arguments listed above are accepted, then (*pace* the researchers involved) all current connectionist models of language are wildly premature. Language will be best understood as a very high layer in a subsumption architecture: how it interacts with lower layers could be of vital importance, and we should study these lower layers first.

If we are still yet to determine the subsymbolic or neural basis underlying the dance-language of bees, how then are we supposed to study such aspects of human language at anything but the most gross level of neuroanatomy (i.e. studies of lesioned patients)? We simply do not know enough.

7 Conclusion

The significance of computational neuroethology is its spanning of many levels of analysis in the brain and behavioural sciences. The prospects for computational neuroethology look good; it will never answer all of the questions asked of computational neuroscience, but it will provide "simplifying" network models with well-grounded semantics, and hopefully lead to a consequent rejection of interpretation in favour of observation. Subjectivity will give way to objectivity.

The use of closed-environment simulators, coupled with the rejection of the phylogenetically top-down study of intelligence, hails a new approach. This new approach is evolutionist and antianthropocentric. It focuses attention on the true issues underlying intelligence. Language must wait. Discuss.

Acknowledgements

Thanks to Harry Barrow, Inman Harvey, David Hogg, David Pickles, Aaron Sloman, and David Young for their comments on earlier versions of this paper, and to Hilary Tunley for proof-reading. This work was supported by UK SERC studentship grant #87306795.

References

[1] M. A. Arbib. Schemas and neural networks for sixth generation computing. *Journal of Parallel and Distributed Computing*, 6:185–216, 1989.

[2] R. A. Brooks. A robust layered control system for a mobile robot. A.I. Memo 864, M.I.T. A.I. Lab, September 1985.

[3] R. A. Brooks. Achieving artificial intelligence through building robots. A.I. Memo 899, M.I.T. A.I. Lab, May 1986.

[4] J. M. Camhi. *Neuroethology: Nerve Cells and the Natural Behaviour of Animals.* Sinauer Associates Inc., Sunderland, Mass., 1984.

[5] A. Clark. Artificial intelligence and the biological factor. CSRP 049, University of Sussex School of Cognitive and Computing Sciences, 1985.

[6] A. Clark. *Microcognition: philosophy, cognitive science, and parallel distributed processing.* M.I.T. Press — Bradford Books, Cambridge MA, 1989.

[7] D. T. Cliff. A closed-environment computational network model of visual processing performed by an airborne insect, 1988. Unpublished D.Phil. Research Proposal. University of Sussex School of Cognitive and Computing Sciences.

[8] D. T. Cliff. The computational hoverfly; a study in computational neuroethology. In *Proceedings SAB90.* M.I.T. Press — Bradford Books, 1990.

[9] D. T. Cliff. Computational neuroethology: A provisional manifesto. CSRP 162, University of Sussex School of Cognitive and Computing Sciences, May 1990.

[10] F. Crick. The recent excitement about neural networks. *Nature*, 337:129–132, January 1989.

[11] D. Dennett. Cognitive wheels: The frame problem of AI. In C. Hookway, editor, *Minds, Machines, and Evolution.* Cambridge University Press, Cambridge, 1984.

[12] H. L. Dreyfus. From micro-worlds to knowledge representation: AI at an impasse. In J. Haugeland, editor, *Mind Design: Philosophy, Psychology, Artificial Intelligence*, pages 161–204. M.I.T. Press — Bradford Books, Cambridge MA, 1981.

[13] S. Grossberg. Neuroethology and theoretical neurobiology. *The Behavioral and Brain Sciences*, 7(3):388–390, 1984.

[14] D. M. Guthrie, editor. *Aims and Methods in Neuroethology.* Manchester University Press, Manchester, 1987.

[15] S. Harnad. The symbol grounding problem. In *CNLS Conference on Emergent Computation*, Los Alamos, May 1989. Submitted to Physica D.

[16] A. L. Hodgkin and A. F. Huxley. A quantitative description of membrane current and its application to conduction and excitation in nerve. *J. Physiol.*, 117:500–544, 1952.

[17] G. Hoyle. Neural mechanisms underlying behaviour of invertebrates. In M. S. Gazzaniga and C. Blakemore, editors, *Handbook of Psychobiology*, pages 3–48. Academic Press, New York, 1975.

[18] G. Hoyle. The scope of neuroethology. *The Behavioral and Brain Sciences*, 7:367–412, 1984.

[19] G. Johnson. *Machinery of the Mind.* Microsoft Press — Tempus Books, Redmond, Washington, 1986.

[20] C. Koch and I. Segev, editors. *Methods in Neuronal Modeling: From Synapses to Networks.* M.I.T. Press — Bradford Books, Cambridge MA, 1989.

[21] G. Lakoff. Smolensky, semantics, and the sensorimotor system. *The Behavioral and Brain Sciences,* 11(1):39–40, 1988.

[22] S. R. Lehky and T. J. Sejnowski. Simplifying network models of binocular rivalry and shape-from-shading. In C. Koch and I. Segev, editors, *Methods in Neuronal Modeling: from Synapses to Networks,* pages 361–396. M.I.T. Press — Bradford Books, Cambridge MA, 1989.

[23] J. L. McClelland and D. E. Rumelhart, editors. *Parallel Distributed Processing. Volume 2: Psychological and biological models.* M.I.T. Press — Bradford Books, Cambridge MA, 1988.

[24] F. Ratliff. *Studies on Excitation and Inhibition in the Retina.* Rockerfeller University Press, New York, 1974.

[25] D. E. Rumelhart and J. L. McClelland, editors. *Parallel Distributed Processing, Volume 1: Foundations.* M.I.T. Press — Bradford Books, Cambridge MA, 1986.

[26] T. J. Sejnowski, C. Koch, and P. S. Churchland. Computational neuroscience. *Science,* 241:1299–1306, September 1988.

[27] P. Smolensky. On the proper treatment of connectionism. *The Behavioral and Brain Sciences,* 11:1–74, 1988.

[28] T. Winograd and F. Flores. *Understanding Computers and Cognition: A New Foundation for Design.* Addison-Wesley Publishing Co. Inc., 1986.

ON THE FEASIBILITY OF COMPUTATIONAL ARTIFICIAL LIFE

A REPLY TO CRITICS

Harold C. Morris
P.O. Box 657
Los Alamos, NM
87544 USA

Abstract

The thrust of this paper is to meet the objections of what may be termed a philosophical school, whose principals are Robert Rosen, Howard Pattee, and Peter Cariani. The objection that a computational universe is a flat "pseudo-world," because it is "all syntax and no semantics," is inquired into and refuted, as is the claim that nothing really new can evolve within such an artificial universe. It is concluded that no persuasive reasons have been advanced as to why computational artificial life is not feasible.

A convergence of several fields has resulted in the new discipline of Artificial Life ("AL") research. And just as AL science has several sources, it is moving in several distinct directions. Life-like entities are being developed as biochemical "wetware," robotic hardware, and as computer software. This paper shall, however, deal only with controversies surrounding the third variety, namely the computational AL form.

Generally stated, the AL program is to develop life-like organisms in the medium of choice. For myself and some other AL researchers, the computer is our medium of choice. Our objective is to implant or evolve individuals or colonies in automaton universes, to observe instances of propagation, adaptation, or communication, such as one usually associates with life forms.

Since the pioneering work of von Neumann (1966), cellular automata have been much used as computational media for AL research. In recent years, more sophisticated systems, for example the ComMet Computational Metabolism of Marek Lugowski (1989), have expanded on the basic cellular automaton theme.

A computational medium along these lines is my own Robotix World (Morris, 1990). In Robotix World, virtual robots rove over a grid, getting from fixed dispensers chunks of information, "modules," containing instructions written in the robots' programming language. The acquiring robot may then assimilate this module into its own internal program, to enact the instructions subsequently. Alternatively, it may stack such modules atop each other to build a brand new robot. These robots are thus self-programming, and propagating. Since robots can steal from or cannibalize modules off other robots, they can compete quite directly with one another. Robots can build other robots for later consumption or service, as we husband plants and animals. Since robots can receive instructional modules from other robots, and set down modules for future generations to discover, they are capable of communicating messages, directing the actions of others, and of leaving maps or other records for the edification of posterity. This system would seem to have most of the potentialities desirable in an AL universe.

Nevertheless, there are "naysayers" who would have us believe that nothing really deserving of the appellation "living" could arise in or inhabit any computational medium. One of these critics, Robert Rosen, has very recently stated that, in spite of all our contemporary efforts and advances, "we are in fact not an inch closer" to the fabrication of an organism than were our predecessors (Rosen, 1990). With colleagues Howard Pattee and Peter Cariani, Rosen is especially vehement in his denial that computational approaches to artificial life hold any promise of success. Let us proceed to their objections.

The first of these, expressed by Howard Pattee (1989), essentially says that a simulation necessarily refers to something outside itself and by definition cannot become something over and above a merely referential entity. A paper dragon can never become a real dragon. A simulation of a dairy cow can never yield a drinkable beverage, no matter how vivid or accurate a model it is.

Pattee has been moved to make this objection on noting the loose terminology of some AL researchers, who do speak of simulating life. However, let's be clear on one thing: There must be grounds independent of what a researcher says he is doing, in deciding what he has created. It makes no more sense to say that something cannot, by definition, be alive because it has been (perhaps carelessly) termed a simulation, than it makes sense to say that something must be alive, because it has been termed a realization of life.

Actually, there is no reason why an automaton cannot be simultaneously modelling some possibility for life, while also realizing a new type of organism. By close analogy, an artificially engineered crystal is a crystallographic model, i.e. does display a possible structure of atoms, even while it is also an artifact, something new under the sun. Of course, a geologist may not be interested in this new crystal as an artifact, but he can hardly afford to ignore it as a possible arrangement that atoms may take. And by the same token, a (natural) biologist may not consider an artifact to be within the purview of his field, but should not ignore those possibilities of propagation, adaptation, etc., brought to light by the computational AL researcher.

The idea that the same program may be treated on the one hand as a referential model, and on the other hand as a realization, should not be any more disturbing to one's sensibilities than is our tendency, for example, to treat a Voyager communicated photograph of Neptune as, on the one hand, an illustration of an aspect of nature, to be judged by its fidelity to the original, and on the other, as an aesthetic object, to be judged on its aesthetic merits.

1. The Measurement Problem.

A seemingly more substantive criticism of the Rosen/Pattee/Cariani school involves what Pattee terms the "incomprehensible" measurement problem. Members of this school hold that a living thing must be able to make measurements, but that measurement is a process that cannot be simulated by physical processes alone, and cannot be realized computationally at all. Cariani (1989) has made much of this.

It is easy to agree that we should like for our AL forms to have the capacity to make at least rudimentary observations, or to have something on the order of the venus fly-trap plant's sensitivity.

However, the "measurement problem" Cariani et al. have in mind has been borrowed from quantum physics, and as far as this author is concerned no computational AL researcher is seriously bent upon creating little quantum physicists.

Carpetlayers, and other scientists for that matter, do manage to make measurements without having to cope with mind-boggling philosophical obstacles. Cariani et al. must carry a heavy burden of proof in establishing the direct relevance of the measurement problem, as it is known in quantum physics, to the AL enterprise. Certainly nothing in Pattee's or Cariani's writings stands as sound objection to the creation of a computational entity with the sensitivity of a venus fly trap.

Related to this "measurement problem" is a directly relevant logical problem associated with a part observing the whole of a system it inhabits. To frame the problem in the form of a diagonal argument: We may infer that for an artificial life form or organ existing within a larger environment <u>to observe all the parts (of that universe) that cannot observe themselves</u>, is to do the impossible, as one realizes on considering the matter of <u>self-observation</u> (this being, of course, a variant on the Russell Paradox).

One can also formulate an infinite regress problem. To observe the entire universe is to entertain some model internally that corresponds to what is observed. Again, self-observation becomes problematic, in that it involves making a model of a model, ad infinitum, of the self. Rucker (1983) considers a closely analogous paradox, which he attributes to turn of the century philosopher Josiah Royce. Rucker suggests a possible "out" for the human mind infinitely introspecting, by hypothesizing that the brain may be itself infinitely divisible. Clearly this will not work, though, for a digital computer, which has a "lowest" level of discrete components.

At any rate, though these philosophically interesting conundrums may establish some limitations on the capacity of an AL form to arrive at an absolutely all-encompassing knowledge of its environment, all the life forms with which we are acquainted in nature are similarly limited. Such limitations can hardly be taken as decisive on the question of whether something is alive or not! And the modest, practical objective of obtaining "sensitive" organs or organisms, perhaps even entities that can recognize some patterns occurring in the world they occupy (or represent those patterns, however crudely), is not shown infeasible by these types of arguments.

Related to the above, Cariani (1989) has set as a requirement for an artificial life form, that it construct for itself a new "device for measurement."

Now, only human beings, among known natural life forms, create new measuring devices for themselves, in what I take to be customary usage of the terms "create for themselves" and "measuring devices." What Cariani must mean is that he would like to see a genealogical line evolve a new kind of sense organ. As far as that goes, I think many AL researchers would like to work toward realizing just such a prospect. The only question is whether this is a good criterion for admitting something to be a life form or not.

To be sure, this may come down to one's intuitions. Suppose a form evolves within a computational medium that can somehow sense the general excitation level of its immediate surroundings, as we sense heat and hear air vibrations as sounds. Ignore the fact that Cariani et al. believe that evolutionary emergence cannot occur in a computational environment, since that objection will be dealt with shortly. Then suppose that I, the AL researcher, having observed with delight the appearance of this new sensory capacity, figure out how it works, and imbue other creatures with it.

Why should we think that the latter creatures, those which had the sensory capacity built in by the experimenter, are less "entitled" to being called living (cr life-like), than those creatures that acquired it adventitiously. My intuitions insist that the main thing is whether the entity has the capacity or not.

Of course, in the somewhat parallel debate over artificial intelligence, some skeptics simply rule out mathematical calculating ability as evidence of machine intelligence, since, "after all, the machine was built to be a computer." Perhaps this same attitude will prevail in AL. One undesirable pragmatic consequence of this will be a temptation to "fudge," however. Again, referring to the "thermal" sensitivity example, the AL researcher who has evolved something with this sense may be reluctant to admit that the built the capacity into subsequent AL forms, if he knows that a certain ideology idscounts such pre-designing of sensory capacities.

2. A Semantic Void

Pattee and Cariani view a computational medium as intrinsically, irredeemably sterile, because it is, they claim, all syntax and no semantics. Cariani alludes to the semiotics of Charles Morris, in defining syntax as the rules governing a language or formal system, and semantics as the relationships holding between a language and the world. According to Cariani, a computational universe is, so to speak, a flat pseudo-world supporting no real semantics. Symbols are shuttled about lawfully but meaninglessly; lacking real world efficacy or reference, they can never really amount to a living thing in a real environment. Cariani evidently pictures the fruit of computational AL research as being something like a genetic blueprint for which no epigenetic rules exist, a blueprint for something that possesses sense organs which have nothing available to sense, in short, a mere string among strings.

Actually, there is just no reason to think of a computational universe as necessarily isolated from the world beyond the computer. Many if not most of us doing computational AL work at least entertain the possibility of allowing input into our computational universe from peripheral sources. However, I will address the case Cariani feels his objection most applies to, namely the self-contained computational medium that does not receive input from the external world.

Cariani's misconception lies in his understanding of semantics, and specifically his forgetting that a language may be an important part of the world it talks about. Thus a language may talk about itself. Quine invented his use/mention distinction to allow us to conceptualize the situation clearly. A syntactical configuration may consist in part of mention of syntactical entities, facts, or rules. And once such embedding is permitted, it can lead to indefinitely rich nestings of mention. The essential point is that a rich language can have semantics even without reference to anything but itself.

Granting that at one level of description, an artificial life form, say a virtual robot or bug or fish or what have you, is a syntactical string or configuration, that is not to say that it is semantically empty, for substrings within it may be instances of mention of other strings external to this entity but within the computational universe. In other words, reference is perfectly possible. The outside observer (e.g. the AL researcher) can then make interpretations of these instances of mention in terms of sensations, or memories, or even thoughts, entertained by the AL form.

3. Evolvability

Yet another contention put by Cariani and Pattee is the requirement that life forms be "evolvable." This contention is meant to go hand in hand with their contention that genuine evolution is impossible within a computational medium. Obviously, these two contentions add up to the proposition that artificial life cannot arise within a computational universe.

It actually seems a bit absurd to speak of "evolvability" as an affirmative property possessed by life forms. Both animate and inanimate things change under environmental pressure. It is precisely life's relative resistance to change, achieved through adaptation and self-replication, that distinguishes it from the nonliving. Affirmative capacities of life such as these are in fact under intensive study by AL science today.

Rather than speak of "evolvability" as a criterion for identifying life forms, then, let us speak instead of the evolvability of an artificial universe. I am confident that most of us in computational AL believe that we have realized evolutionary processes within computer media. Let us see why Cariani and Pattee object to our characterizations of these processes (and their fruits).

Cariani (1989) begins with the inarguable premiss that a computational medium can be described at one level as a formal system, with a finite alphabet of primitive symbols and a fixed set of rules of inference/substitution. But he then leaps to the conclusion that this means that nothing really new can arise in such a system, because no new primitive symbols can arise--only new combinations of the given primitives.

The fallacious jump here is in thinking that an artificial life form would be a new primitive symbol, or necessarily be composed at least in part of such new primitive symbols. There is no reason to think this. An organism or organ arising in a computational medium's evolution will be a new combination of the given primitive symbols. That's what "something new" is: a combination not present at the starting up of the evolutionary process. In most systems, the potential number of new combinations is going to be infinite.

The bad reasoning here is quickly exposed by some analogies. Would we tell an aspiring composer that he cannot possibly come up with any new, original musical compositions, because he's using a finite number of scales, each of which has a finite number of notes(i.e. primitive symbols)? A new musical composition is just a new combination, isn't it? And would we imagine that a child who has learned the English alphabet now "knows" all the works of literature that have been or ever will be written in that language, because, after all, those works of literature are only combinations of the same old primitives? As for the natural universe itself, are life forms unable to arise in evolutionary systems that have only finite types of elementary particles (e.g. kinds of quarks)? Is it absurd to refer to an artificial compound as "new" if it is, after all, only another combination of the same old elements?

I have endeavored in this paper to refute certain objections posed against the computational artificial life program, by Drs Rosen, Pattee, and Cariani. I wish to conclude with words of rapprochement.

First and most generally, I hope that all philosophers and scientists are aware of the value of criticism. We move forward blindly if we move forward with unchecked and unchallenged assumptions and an attitude of infallibility. The critics to whom I have made rebuttal here are to be thanked for their sincere efforts to make us in computational AL keep our feet on the ground and headed down the right path. Certainly, in trying to meet their objections, I have been forced to think more clearly and generally about what we are about in this field, and I am thus indebted to these critics.

Finally, I would remind us all that refutations of objections raised against the feasibility of computational artificial life do not have affirmative force in establishing the feasibility of such "software life forms." For that, we must continue to strive to obtain the kind of impressive results that will most effectively win over the skeptics and agnostics. It may be hoped that this conference will be looked back upon as a milestone in the demonstration of such results.

Bibliography

Cariani, Peter. On the Design of Devices with Emergent Semantic Functions. Unpublished dissertation, State University of New York at Binghamton, 1989.

Lugowski, Marek. "Computational Metabolism," in Artificial Life, C. Langton, Ed. Addison-Wesley, New York, 1989.

Morris, Harold. Formal Sciences Approaches to Artificial Life. Book manuscript on submission, 1990.

Pattee, Howard. "The Measurement Problem in Artificial World Models," paper delivered at first international workshop on artificial life, Los Alamos National Laboratory, 1987.

Pattee, Howard. "Simulations, Realizations, and Theories of Life," in Artificial Life, C. Langton, Ed. Addison-Wesley, New York, 1989.

Rosen, Robert. "What Does It Take to Make an Organism?" Paper delivered at second international workshop on artificial life, Santa Fe Institute, 1990.

Rucker, Rudy. Infinity and the Mind. Bantam, New York, 1983.

The Animat and the Physician

Alexandre Parodi
Indodyne SARL
10, rue de la Paix
75002 Paris
Fax: (33) (1) 40 20 98 98
Tel: (33) (1) 69 81 88 41
E-mail: (uucp) parodi@lri.lri.fr

Pierre Bonelli (LRI)
Equipe I&A
LRI, bâtiment 490
Domaine Universitaire de Paris-Sud
91405 Orsay cedex
Tel: (33) (1) 69 41 64 07
E-mail: (uucp) bonelli@lri.lri.fr

Abstract

Animat research has already produced interesting concepts and algorithms. In this paper, we analyze how this research can be applied to human intelligence understanding and to reproducing of some expert behaviors. To support these ideas, we experiment with an improvement of Boole, a Genetic Based Learning algorithm from animat research, in a medical domain of expertise. We experimentally demonstrate that our system obtains good results on a well known realistic medical diagnosis task, and we analyze its potential ability to solve more complicated problems.

Introduction

There has been much debate about how one can consider that a system is intelligent, most of the time according to how it processes information (rules, neurons etc.) in connection with the human brain. However, in [Wilson, 85] Wilson developed the idea that we could probably learn more from ethology, and he introduced the concept of animats, which are autonomous systems which learn how to survive and expand in a given environment.
We propose to discuss how this research can be profitable to the understanding of human intelligence, and how animat algorithms can be used to reproduce some intelligent human behavior.

1. Intelligence hierarchy

Let us consider the following intelligent systems hierarchy based on how explicit the input knowledge from the environment must be: systems that learn by being told, systems that learn by complete examples, and animats, i.e. systems that learn by reward.

1st level: Systems that learn by being told

Most computers get their knowledge by being given programs, i.e. a list of instructions to be executed in a specified order; the processing is completely explicit in the input knowledge .

Production systems without learning ability get their knowledge from rules which are used to reason about the input data and conclude about the output data to provide. They are somehow more intelligent because the order in which rules are executed depends on the data: control is data driven. This means that a few rules implicitly specify many different reasoning traces.

Specifically built neural networks get their knowledge from a set of predetermined weights which indicate how a formulated hypothesis (micro-feature represented by one neuron) influences another. However, without learning, they are not really much more intelligent than production systems, but some experiments tend to show that they are less brittle and noise sensitive and can exhibit even richer behavior than can be expected because they use parallel analog processing.

2nd level: Systems that learn by complete examples

Learning systems can manage with even less formalized knowledge: they only need examples which contain both the input and the corresponding desired output; they take care of extracting the appropriate knowledge that is needed to generalize the sampled behavior to new inputs. Such systems can be either rule based systems, connectionist networks, or classifier systems. A number of learning algorithms exist for each of these, and there is much debate about whether connectionist systems [Bounds, 89], rule based systems [Kodratoff, 89], homogeneous hybrid systems [Parodi & Khouas 89], or inhomogeneous hybrid systems [Hendler, 88] should be used.

3rd level: Animats. Systems that learn by reward

However, according to our hierarchy, animats are even more intelligent since their only source of knowledge is the reward (payoff or penalty) that they receive from the environment according to their behavior in each situation.
Therefore, they are not given explicit knowledge about the adapted behavior corresponding to each encountered situation: they have to infer it from past experience, and to generalize it for new situations.

In addition, they cannot be given all their knowledge at once: they need to learn it *incrementally*, and they cannot memorize it as it is: they have to extract and memorize useful knowledge which is diluted into massive, diffuse, inaccurate and inconsistent information.
It appears that this kind of intelligence is needed by autonomous systems in nature, such as animals, animal societies, or animal species which all learn to survive in a hostile environment by being given more or less fulfillment of their basic needs, expansion of their territories, or their total population.

It seems at the present time, that Genetics Based Machine Learning is best suited to fulfill these needs.

2. Animat and human intelligence

How can AI benefit from research in the animat domain ? This is the question we will try to answer in this section.

For some human activities, reasoning can be well conceptualized because people have access to complete consistent knowledge about the domain; additionally, the associated reasoning is conscious. In these cases explicit models of both knowledge (rules) and reasoning (meta-rules) can be build and permit expert systems (which "learn by being told") to be effective in behaving like humans for these tasks.

This tends to be true for instance in the case of computer configuration, machine trouble checking and diagnosis, etc., and in a number of man created domains.

However, in some other domains, man cannot really conceptualize the needed knowledge because it is not available or beyond his understanding.

A typical case is medical diagnosis where physicians interact with the human body which is extremely complex, not yet completely understood, and where input symptom data has a complicated relationship with the diagnostic output. Of course, academic (conceptualized) knowledge helps, but doctors claim that past experience (set of examples) is most important: it develops the intuition capability.

In this sense, acquiring the intuition that is needed for medical diagnosis is merely learning by examples.
However, a system which does medical work and behaves as a physician has to be more than a simple "learning by examples" system. Indeed, the information is very diffuse, and doctors induct knowledge from it incrementally during a very long period of time, with no specific memorization of each and every case, as is needed for animats.

In addition, medical researchers also design therapies which involve new treatments and drugs against diseases that are not yet very well understood: they are not given examples that contain what treatment to give for an infered diagnostic, or a given set of symptoms.

They rather guess and try, using their incomplete present knowledge to treat a patient, and then check whether the patient evolves toward a better or worse condition; from this point they update their knowledge iteratively. Therefore, improving therapies might be considered as learning by reward; this last quantity is directly related to the patient's condition improvement.

It clearly appears that this last activity is close to those performed by animats.

Hence, when attempting to build a machine which behaves as a physician, it seems quite natural to use genetic algorithms to learn the medical "intuition" for both diagnosis and therapy. However, one might say that classifier systems mechanisms do not have much to do with the human brain, neither at the conscious reasoning level nor at the neural one. However, "intuition" merely consists in being able to "feel" the correct solution unconsciously, sometimes with no possibility to explain or even justify it absolutely. Since modeling is not possible, our approach is "behaviorist"; i.e. we do not care if knowledge is coded inside our machine in the same (unknown) way as in a human being who has the desired behavior. Hence, the idea of identifying the animat problem and the doctor's problem and use a genetic algorithm to perform learning in this context is sound.

3. Experimental program

In the context of this paper, we restricted our experiments to a medical diagnosis task, which seems

a natural step toward therapy learning. In addition, the data base with which we experiment has been tested with a number of learning systems, which allows to validate our ideas quantitatively.
We therefore adapted Boole [Wilson 87] a classifier system that emerged from animat research to a medical diagnosis problem.
We modified it in a way so that it gathers its knowledges in the form of 0^+ order production rules: this allows some expert checking, but also human use for knowledge that cannot be easily acquired by humans such as therapy.

4. Quick overview of the NEWBOOLE system and its improvements.

In this section, we briefly describe the NEWBOOLE Classifier System used to perform the medical diagnosis task described in the experimental section of the paper.

NEWBOOLE is a modified version of Wilson's system BOOLE, which was initially designed to fulfill the animat problem constraints.

* BOOLE is a simplified version of the standard Classifier System (CS) which learns difficult boolean functions.
Like any CS, Boole maintains a population of classifiers (which can be thought of as bit-level zero order rules) according to Darwinian evolution principles. However, classifiers are not chained; they directly provide an output and the decision is made within a single step during recognition; consequently there is no message list nor Bucket Brigade Algorithm. Thus each classifier consists of a condition (taxon) and an action which are fixed length strings over the {0,1,#} alphabet.
Like other CS, BOOLE contains three components, namely a performance component, a reinforcement component, and a discovery component.

* NEWBOOLE integrates changes to both the performance and reinforcement components, which were made in order to get a faster convergence of the error rate in the context of learning from examples.

The changes resulted in the following differences with BOOLE:

<u>1/ General output:</u>

NEWBOOLE is no longer a specialized CS for Boolean functions: the number of classes can exceed 2.

<u>2/ Learning rate:</u>

NEWBOOLE obtains much faster learning rates then its predecessor, partly through the use of a new reinforcement component that we developed jointly with Wilson and Sen.

It is important to mention that BOOLE's generalization capabilities were maintained in the new system. For more details about the NEWBOOLE system, please refer to [Bonelli, Parodi, Sen & Wilson, 90].

<u>3/ Initialization:</u>

For each experiment, the classifier population was initialized by directly applying a modified version of Wilson's create operator.

<u>4/ Rule crossover:</u>

In addition, for the following experiments, NEWBOOLE has been improved by enhancing the crossover component of the genetic algorithm.
Indeed, the knowledge of the system is appropriately represented by a set of explicit 0+ order rules rather than a set of strings of bits as is the case with classifier systems. Each rule consists of a condition part, which is an *unordered set* of (attribute, value) couples, and a conclusion part, which provides the output value.
Therefore, the crossover that is used takes into account general properties of this representation. More generally, our approach consists in directly apply modified genetic algorithms onto the appropriate representation of the knowledge, rather than attempting to convert the knowledge into a classifier representation which can be processed through classical genetic algorithms.

Our first results tend to show that this feature further enhances the generalization quality, i.e. the ability to find appropriate answers for examples that the system has never seen before.
The error rate on such examples is reduced by about one third, hence providing results which are better than ID_3, and identical to neural networks using back-propagation, Michalski's logic reduction system or statistical Bayesian systems.

5. Description of the algorithms

In this section, we describe the algorithms that were used in the experiments.

<u>5.1. The BOOLE Classifier System</u>

We now present BOOLE with some detail concerning the parts that have been changed, in order to explain the improvements in the following paragraphs. Boole includes three components:

1/ *Performance component:* in the performance cycle, an input string is presented to the system, the match set M of all classifiers whose taxa match the input string is formed, and a single classifier from M is selected (using a probability that is proportional to its strength) whose action is output as the system's decision.

2/ *Reinforcement component*: this component modifies the strengths of classifiers according to performance level:

a/ Form the action set [A] consisting of classifiers from [M] whose action is the same as the chosen action; the remaining members of [M] form the set Not[A];

b/ Deduct a fraction e from the strengths of all classifiers in [A];

c/ * If the system's decision was correct, distribute a payoff quantity R to the strengths of [A]; but

* If the decision was wrong, distribute a payoff quantity R' (where $0 \leq R' < R$) to the strengths of [A] and deduct a fraction p from the strengths of [A] (at least one of R' and p is equal to 0);

d/ Deduct a fraction t from the strengths of Not [A].

The distribution of payoff is done so that rules which have many # 's (thus more general) are favored.

3/ *Discovery component,* which modifies the classifier population according to Holland's (1975) genetic algorithm and employs reproduction, genetic operations (crossover and mutation), and deletion.

BOOLE's version of the genetic algorithm is quite particular in the sense that only one offspring is added per invocation of the genetic algorithm. In this context, the parameter ρ will represent the average number of invocations of the genetic algorithm per cycle (i.e. the number of offspring added per cycle). For more details, please see [Wilson, 1987].
Here $\rho = 4$.

5.2 Reinforcement component enhancement

BOOLE's reinforcement component, under the "payoff-penalty" reinforcement regime ($p \neq 0$) adjusts classifier strengths in the following way:

- if the system's decision is correct, distribute a quantity R to the strengths of the Actionset [A].
- if the system's decision is false, penalize the strengths of [A] by deducting a fraction p from their values.
- finally, whether the system's decision is correct or not, deduct fractions e and t respectively from the strengths of [A] and Not[A].

Thus, following each performance cycle, *only the strengths of [A] are adjusted according to the system's performance.*

However, once we know that [A] contains accurate classifiers, we also know that Not[A] only contains inaccurate classifiers; in this case it would make sense to penalize the rules in Not[A]. This acknowledgement led us to a *symmetrical payoff-penalty* algorithm, in which we respectively reward and penalize the accurate and inaccurate classifiers present in the Matchset.

The new reinforcement component is the following:

1/ Form the subset of [M] consisting of those classifiers whose action is accurate; this is the correct set [C]. The remaining members of [M] form the set NOT[C].

2/ Deduct a fraction e from the strengths of [C].

3/ Since [C] contains the accurate classifiers, distribute a payoff quantity R to the strengths of [C].

4/ Since Not[C] contains the inaccurate classifiers, deduct a fraction p from the strengths of Not[C].

Thus, the effect of the reinforcement component can be written as:

$$S_{[C]}(t+1) = (1-e) \times S_{[C]}(t) + R \qquad (1)$$

$$S_{Not[C]}(t+1) = (1-p) \times S_{Not[C]}(t) \qquad (2)$$

where $S_{[C]}$ and $S_{Not[C]}$ are respectively [C]'s and Not[C]'s total strengths.

This new algorithm constitutes a clear departure from Boole: indeed, if we have several possible output values then the knowledge of the correctness of the output of each classifier from the match set is used. This information can be provided by the knowledge of the correct output for each example, as is done in most learning systems. However, this does not make any difference with boolean functions such as the Multiplexer since only two values are possible: if one is known as wrong, then the other one is right.

As in Boole, the payoff R to [C] is distributed by a biased distribution function D, which

favors more general rules (i.e. with many "don't cares" #) as follows.
First, the generality of each classifier n° i of length L is computed as:

$$g_i = \frac{\text{number of \#'s in i}}{L} \qquad (3)$$

Let us define:

$$d_i = 1 + G \times g_i \qquad (4)$$

where G is a "generality emphasis" parameter.

Then, the portion of reward R_i that is given to classifier i becomes:

$$R_i = D(i) \times R = \frac{d_i}{\sum_i d_i} R \qquad (5)$$

In addition, some the following improvements were made for the experiments which are considered in this article.

5.3 Initialization algorithm

The reason for choosing a different initialization procedure is that the search space to be explored is far too vast to permit a random initialization process since the population classifiers are supposed to recognize examples from the Learning Base which represents a microscopic fraction of the search space.

Therefore, the initial classifiers are generated in the following way:

a) An example from the learning base is chosen at random.

b) The condition part of the new classifier is a copy of the symptomatic part of the example, except that "don't care" symbols are inserted with a probability of 0.5.

c) The conclusion part of the new classifier is chosen randomly among the 4 possible diagnoses.

d) Steps a), b), c) are repeated P times, where P is the size of the population.

5.4 Symbolic discovery

In this section we describe how we address symbolic processing in our system, in order to use symbolic data, as is needed for medical work.

. *Usual representation*

We assume that the examples are described with a collection of attributes. Each attribute can take a small set of discrete mutually exclusive values. For instance, in the weather domain, we consider four attributes {SKY, TEMP, WIND, OUTLOOK} which can respectively take the following values:

SKY: {blue, grey, rainy};
TEMP: {cold, mild, warm, hot};
WINDY {true, false};
OUTLOOK: {mistral, east wind}.

A weather description example might be:

(SKY = blue) & (TEMP = cold) &
(WINDY = true). (6)

In this case, the following rule would fire:

(SKY= blue) & (WINDY = true) →
(OUTLOOK = mistral) (7)

The straight forward GA coding of such a description would consist in a mapped concatenation of fixed length binary strings, each of them representing an attribute's value. In the same weather domain, the attributes values could be coded in binary as:

SKY: (blue = 00, cloudy = 01, grey = 10, rainy= 11);
TEMP: (cold = 00, mild = 01, warm = 10, hot = 11);
WINDY: (false = 0, true = 1);
OUTLOOK: (mistral = 0, east wind = 1).

The same weather description (6) would then be represented by the data :

00001 (8)

and the classifier corresponding to rule (6) is :

00##1 / 0 (9)

The problem here is to check whether we could apply GBML directly on the attribute-value representation without preliminary coding.

Firstly, the number of attributes to be treated for each rule would be smaller than the number of bits in the corresponding CS classifier, resulting in faster processing.
Moreover, classifiers have many explicit don't care positions, which must be processed, resulting in a waste. Using the attribute-value representation, attributes whose values do not have any incidence in the conclusion of a rule will not be present in its premise, and will not be explicitly processed.

Set representation

The way rules are coded does change the matching algorithm which is then close to what is used in O^+ expert systems, but the reinforcement algorithm does not have to be changed at all.

However, it is obvious that the discovery component will have to be modified in order to make it work on this symbolic representation.

Since a condition is actually a conjunction and because the logical "AND" is commutative, we can note that it actually is a variable size unordered set of (attribute, value) couples; for instance the logic condition:

(SKY=blue) & (TEMP=cold) & (WINDY=true) (10)

can be represented as the unordered set of three couples:

{(SKY,blue), (TEMP,cold), (WINDY,true)} (11)

We developed an algorithm which exploits this representation by performing crossover and mutation directly on these sets without any conceptual reference to strings. This algorithm will be presented in an upcoming article.

6. Experiments and discussion of results.

We tested NEWBOOLE's ability to perform medical diagnosis using data on the lymphography domain, provided by the Institute of Oncology in Ljubljana, Yugoslavia. We chose this particular set of data because numerous other inductive learning systems have been applied to it in the past, hence enabling quantitative comparisons.

This example base consists of 148 examples separated into 4 distinct diagnoses. Each example is completely described by 18 discretely valued attributes representing the symptoms and one of 4 classes representing the corresponding diagnosis.

6.1 Conditions of the experiments

Our main goal was to test the quality of the generalization, i.e. the induction capability.

For each experiment, 70% of the examples (i.e. 103 examples) were randomly chosen for learning, and the remaining 30 % (i.e. 45 examples) for testing. Hence the system is tested with examples it has never seen, thus enabling quantitative measurement of its performance for induction.

During each learning cycle, an example from the learning base is chosen at random and presented as input to the system. The system then makes its decision according to its past experience and applies the reinforcement and discovery components according to the experimental parameters.

Learning was performed over 10,000 cycles to ensure that proper generalization has been achieved. Indeed, even after the error rate on the learning data base has converged to almost zero, the system still improves its generalization capability.

6.2 Experimental Results

Results below are the average of 4 independent experiments. Figure 1 shows Symbolic Newboole's performance over the Test Base, i.e. its induction ability. The graph plots the smoothed system's ***average score*** which is the percentage of correct decisions over the past 100 cycles versus the number of cycles since the experiment began.

This plot shows that the system more or less stabilizes its performance after a few thousand cycles at approximately 82 %.
The average result for the test experiment was 37 right answers out of 45, which represents approximately 82 % of correct classifications. This score is to be compared with an estimate of 85 % of correct diagnoses by specialists, 76 % with NewBoole without enhanced crossover, and 76% for assistant-86 which is an ID_3 based learning system.

We provide here after a table which compares these results in induction with other learning systems which are used in the same conditions.
Of course, these results have to be considered with the idea that they are the average of four independent estimations of probabilities of around 80 % with 47 samples each. Hence, we get a mean square root error of about 2.5 %. Therefore, the differences between Bayes, BP and Newboole with rule crossover are insignificant here.

6.3 Discussion of the results

These results clearly validate our intuition that a Classifier System can provide high classification rates in a domain such as medical diagnosis.
This is amplified by the fact that no other inductive, incremental learning system obtains such results besides Connectionist Systems. However, Classifier Systems have the advantage of being equivalent to 0^+ rule based systems which provide comprehensive solutions, and thus contain an explanation capacity, whereas neural networks merely provide sets of numerical coefficients without any clear semantic meaning.

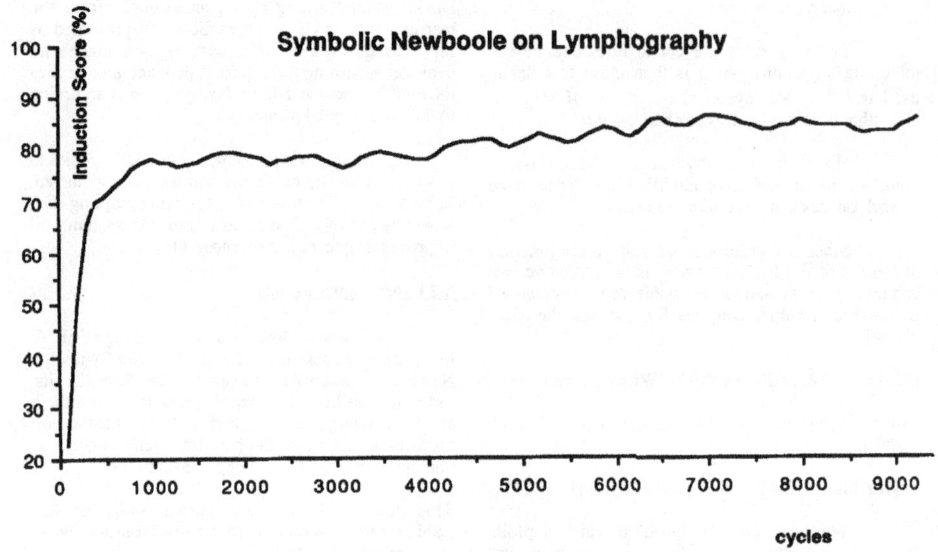

Figure 1: Symbolic Newboole convergence over test base

(smoothed over 3 points: one point every 100 cycles)

Name	Source	Inc.	Principle	Represent.	Induction
Bayes	[Clar., 87]	NO	Statistical	Probabilities	83 %
BP	([1])	YES	Connectionist	Weights	82 %
Assistant-86	[Cestnik et al., 87]	NO	Entropy min.	Decision tree	76 %
AQ15	[Michalski et al., 86]	NO[2]	Logic reduc.	Rules	81 %
Newboole	[Bonelli, Parodi Sen & Wilson]	YES	Genetic	Classifiers	76 %
Sym. Newboole		YES	Genetic Rules	Rules	82 %
Experts	[Michalski et al., 86]	YES	Unknown	Unknown	85 %

Table 1: Comparison of various learning systems using the lymphography data base:

This table presents the source of this data, learning incrementality, principle of the algorithm, knowledge representation and induction rate.

[1] We conducted this Back-Propagation test with a 18:60-40-20-20-4 (18 inputs, 4 hidden layers and 4 outputs) network architecture and only 3 trials rather than 4.

[2] This system acquires rules from examples which are considered one after the other; however, it only adds generalized logic formula and cannot revise its past knowledge accordingly to compress it. In this sense, it is not incremental as Back-Propagation, ID6 and GBL systems are.

More importantly, these systems have the potential to do therapy design as well, since they possess all the animat qualities.
Indeed, even Newboole learns by reward internally, once the example is known.

However, it is strickening that the best learning systems all do the same score on this medical diagnosis problem. This result also happens to be true in other domains such as phoneme recognition. This suggests that there is a limit to the quality of induction that a system can get by learning from a given data base; this might simply be the implicit knowledge that is contained inside it. This also supports the idea that if humans do slightly better on this problem, it is mainly because they had access to other knowledge before, in the form of other examples from a neighboring domain, or academic courses. This last remark illustrates how background knowledge could be useful in such a system. The ease to describe it in terms of rules partly motivates the need to use such a representation in lieu of classifiers.

Conclusion

In this paper, we have developed the idea that animat research helps to understand and reproduce complex human capabilities such as medical work.
We have experimentally shown that an improvement of Boole can learn to do medical diagnosis as well as other learning algorithms, and we discussed its unique potential to do therapy design. Moreover, our results suggest that genetic algorithms can be successfully applied directly on other knowledge representations than strings of bits.
Some interesting extensions of this work would be to check whether this remains true in other domains, apply our algorithm for medical therapy design, and to include background knowledge in the system.

Bibliography

[Bonelli, Parodi, Sen & Wilson 90], "NewBoole: A Fast GBML System", Bonelli P., Parodi A., Sen S., Wilson S., Proceedings of the International Conference on Machine Learning, June 1990, Austin, Texas.

[Bounds, 89], "Expert Systems and Connectionist Networks", Connectionism in Perspective, Pfeifer R. et al., North-Holland 1989

[Cestnik et al., 87] Cestnik G., Konenenko I., & Bratko I., "Assistant-86: A Knowledge-Elicitation Tool for Sophisticated Users". In Bratko & Lavrac (editors) Progress in Machine Learning, 31-45, Sigma Press.

[Clark & Niblett, 87] Clark, P. & Niblett, T. (1987), "Induction in noisy domains". In I. Bratko & N. Lavrac (Editors) Progress in Machine Learning, 11-30, Sigma Press.

[Hendler, 88] "Activation Spreading Inconsistencies: On the need for hybrid symbolic/connectionist models", Proceedings of the AAAI 1988 Symposium on Parallel Models of Intelligence

[Kodratoff, 89], "Enlarging Symbols to more than Numbers or Artificial Intelligence is the science of explanations", p. 157-172

[Michalski et al., 86] Michalski, R., Mozetic, I., Hong J., & Lavrac N., (1986). "The multi-purpose Incremental Learning System AQ15 and its Testing Applications to three medical domains". In proceedings of the fifth National Conference on Artificial Intelligence, 1041-1045. Philadelphia, PA: Morgan Kaufmann.

[Parodi & Khouas, 89], "Parodi, A., Khouas, S., "Adaptive Approach for Learning Abstractions in Semantic Networks", Neuro-Nîmes 89, International Workshop on Nueral Networks and their Applications, Nîmes (France), 13-16 Nov, 1989.

[Wilson, 85] Knowledge growth in an artificial animal, Wilson S. W., Proceedings of the first International Conference on Genetic Algorithms and their Applications, July 24-26, 1985, CMU, Pittsburg, PA, p. 16-23, Lawrence Erlbaum associates, publishers, Hillsdale, New Jersey

[Wilson, 87] "Classifier Systems and the animat problem", Machine Learning 2, 199-228, 1987

Extended classifiers for simulation of adaptive behavior

Renaud Dumeur
Département Informatique,
Université de Paris-8,
2, rue de la Liberté,
93256 St Denis cedex 02
red@litp.ibp.fr,
C[2]V,
82, Boulevard Haussmann,
75008 Paris,
renaud@ccv.fr

Abstract

In this paper, we propose the use of the animat approach to automatically generate animation scripts for computer synthesized movies. We want to design a system where animats are actors able to improvise from a few contextual informations. We plan to implement such animats with extended classifiers allowing a compact encoding of behavior rules while preserving their ability to be modified by inductive genetic operations.

1 Introduction

When computer animated characters become more and more realistic in their rendering, the problem of specifying more realistic individual and collective behaviors also appears. We think that animats can be used to solve this new problem and that behavior simulation based systems can change the way computer animations are designed. We propose a "programming by environment" approach using animats and evolution simulation which can drastically reduce the work of animation script writing.

Animats are computer simulated entities, exhibiting animal-like autonomous individual or collective behavior. We want to use them as low-cost credible crowd artists in order to ease the writing of animation scripts. In the world of film making, crowd artists are employed to give the audience background informations about the time and location where the action of the movie takes place. The movie director gives to these actors fuzzy indications on how to behave, then they are left nearly without control during the filming. On the other hand, as the central character behaviors may not easily infered from the movie context, the corresponding actors are more precisely directed and have less freedom than crowd artists.

2 Animation automation

Consider traditional animation movies: drawn by hand, they require a frame-by-frame precision level script for each character of the animation. A first level of automation consists in drawing a few frame images by hand and apply an interpolation procedure to generate the missing frames required for a smooth animation. Improvements from the traditional paper based technique only reside in greater flexibility of the drawing tools and reduction of involved manpower.

By using computers, we can automate the generation of the script by writing procedures in a computer language. Object oriented [6] languages seem to match the requirements of animation programming. In such languages, an object is specified by a local state (a set of state variables) and a set of procedures (the object's methods). An object's behavior is implemented by its methods. As its methods process the object's local state, it is easy to obtain a wide range of behaviors from a few procedural specifications, by varying the content of the entities local states.

If object oriented programming systems facilitate the behavioral specification of large groups of simulated actors, they do not give any assurance that the resulting acting will meet the desired goals. This work is still under the responsibility of the designer who must keep in mind a model of the potential interactions between actors.

A few fixed procedures can simulate complex behaviors. Even a cellular automaton, Conway's *game of life* [2], can generate primitive animats – *gliders* and *glider guns*, for instance – with interesting behaviors. In the frame of computer animated graphics, we are moreover helped by spectators who will tend to interpret the events occurring in the movie. This fact is highlighted in V. Braitenberg's book [3] where simple animats (called *vehicles*) are involved. These animats are carts, endowed with captors (photosensible cells) and effectors (propellers such as wheels), which are wandering in an environment containing light sources. By changing the disposition and the properties of the different vehicle brain compo-

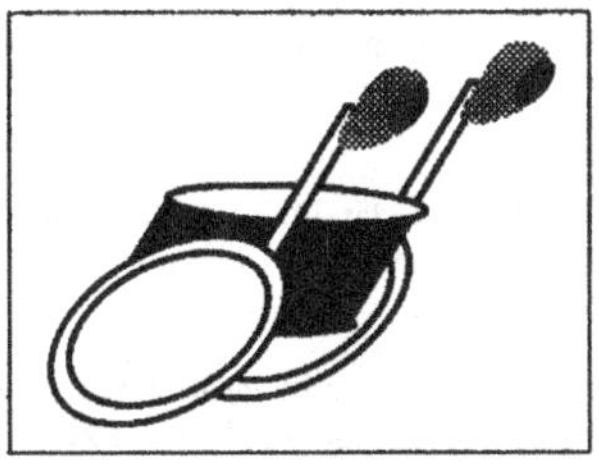

Figure 1: An animat speeding up.

Figure 2: An animat hits the brakes.

nents, one can produce different behaviors, such as goal seeking strategies or avoiding reactions. Sometimes, these behaviors can be interpreted as the results of the animat "feelings", because the candid spectator, ignorant of the involved internal machinery, applies human rules for analyzing behavior. This effect can be amplified if we choose a cartoon-like animat representation. For instance, the figure 1 shows how we could represent an accelerating animat.

The same way, when a roving animat notices an attack or an obstacle, we can amplify this fact with a representation as shown by the figure 2.

If we make our animat interact with other object such as light sources, we can use classical rendering techniques to amplify the animat action. The figure 3 shows an animat locating a light source, its shade being projected on the ground.

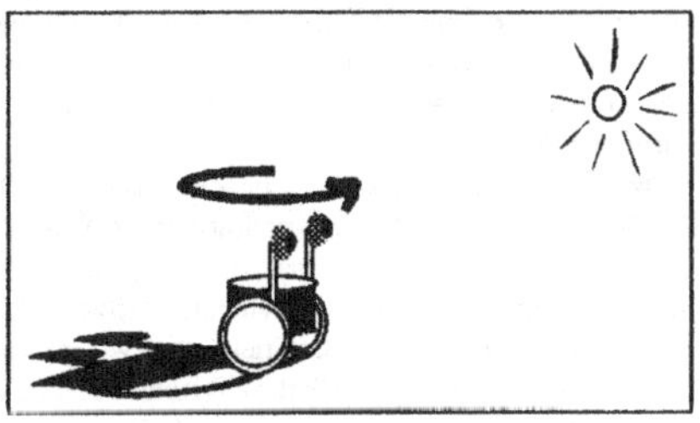

Figure 3: An animat locating a light source.

3 Adaptive behavior

For the spectator, the interest of an animated character also consists in the way it interacts with its environment. The more different environmental responses the character will have, the more the spectator will be speculating on the progress of the animation and will keep his attention focused on the movie. His motivation will last until he has a full model of simulated characters' way of reacting to environmental stimuli.

The most straightforward way of achieving variety in animat environmental response would be to encode large set of rules in the animats, each rule associated with a specific environmental input. Such a design will only extend to a fixed amount the duration of the spectator interest. If the environmental stimuli are fluctuating enough, they will exhaust the set of possible behavior of the animat. It is still possible to choose a stochastic rule selection scheme in order to introduce some noise in the animat response. If such a solution will confuse the spectator, maybe enough to keep his attention, it will still allow him to find the underlying determinism of the rules. Moreover, if this noise is too high, the spectator will perceive the character's actions as groundless and may be bored.

When large groups of animats are put into play, the problem of an initial rule set does is not as important as it is for a few animats. The collective behavior not only depends from the rule set contained in each animat, but also from animats interactions. Compared to a single animat rule set, this complexity is a better match to both the environmental variety and the analytic abilities of the spectator. Real world entities such as ants give goods example of adaptive behavior emerging from interaction between primitive elements [10] which can be transposed into a computer simulation. Work on bird and fish flocks [11, 1] have also produced attractive results. Nevertheless, in both examples, the the relationships – repulsion or attraction for instance – between the different animation actors must still be defined by the animation designer. As we wish to overcome this kind of constraints when generating animation scripts, we must to use inductive mechanisms that allows the generation of new behavioral rules.

4 Adaptive vehicles

For our experiment, we want to use animats similar to V. Braitenberg vehicles. In these animats, environmental events are transformed by the captors, then processed by a network of fibers linking threshold devices [9] reaching effectors, and eventually transformed by the effectors in action on the environment. In the case of our very simple animat, this action is a motion caused by the rotation of powered wheels.

Such a design is based on analogic signal processing which does not match formalisms human beings are used to deal with. As one of our goal is the evolution of a predefined set of rules, we prefer a different model: production rules. Such approach has been introduced by J. Holland [7] with its Genetic Algorithm (GA) processed classifier systems (CFS) and used for adaptive behavior simulation by [12].

The use of production rules for computer animation allows the animation designer to specify some basic behavior and let them be processed by an evolutionary mechanism.

5 Genetic Algorithms

Genetic Algorithms are search and optimization techniques based on the mechanics of natural evolution as described by the neodarwinian theory [5]. GA's main goal is to solve combinationally exploding search problems with processes drawn from biology. The relation between the two domains becomes clear if we look at biological evolution as a search process for the possible life forms in the space of the possible chemical constructs.

The RNA-DNA couple is the basic material containing informations which specify the way life forms are built and are functioning. This large and complex chemical construct is a product of natural evolution initiated from simpler molecules[1]. Alterations (random *mutations*) of this molecule have consequences on the physical characteristics (phenotype) of the life form it builds. Another operation, the *crossover* which, occurs during sexed reproduction, allows genetic characteristics to be quickly spread in a population.

Simplified genotypes and genetic operations can be simulated on a computer. In simulated genetics, genomes are reduced to a vector of features which are modified and diffused in the population by random mutation and crossover. A selection pressure is also simulated which corresponds to a reproduction probability for the different genotypes. As fitting individuals will reproduce more often that the others, their genetic characteristics will be spread faster in the population. Consequently, the average fitness of the population population will tend to an optimum, the stochastic character of the genetic operators allowing the system to escape from local maxima of the search space.

In the case of animats, we can use GAs to explore the space of possible behaviors. To do this, it is necessary to code behavioral descriptions under a form which supports the application of genetic operators.

6 Classifiers

A classifier is a linear string of symbols from $\{0, 1, \#\}$. This string is divided into two parts : a *condition* part and an *production part* of equal sizes. A classifier system (CFS) is a group of classifiers exploiting messages comming from a mailbox. A message is a linear string of symbols from $\{0, 1\}$ which size is half the classifier's one.

Briefly, a CFS run is a repetition of a cycle decomposed as follows:

1. construct an environmental message and put it in the mailbox. Mark it specifically as an *input* message to give to classifiers the opportunity to distinguish it from other messages.
2. collect matching classifiers. A matching classifier has a *condition* part identical to the message, except that both 0 and 1 match #.
3. evaluate the bid made by each matching classifier. The bid usually depends on the classifier's strength (past utility) and expresses the ability of a classifier to solve the current situation.
4. choose matching classifiers for message production (the higher the classifier bid is, the more probably the classifier will be chosen). Post the output message which will be inserted in the CFS mailbox for the next step. This output is a copy of the *action* part of the classifier, except that each # symbol has been substituted by the identically placed value in the input message. The result of the rewriting is marked as an *output* message so the environment will be able to interpret it.
5. remove *output* messages and interpret them as actions on the environment.
6. revaluate each classifier strength coefficient. This coefficient can be modified by:
    - an external payoff which is the evaluation of the CFS performance by the environment,
    - by incomes from other classifiers which have received previously posted messages,
    - and by penalties associated to the posting of these messages during the current cycle. Other penalties, such as life costs, may help to put tune the selection pressure on the classifiers.

The whole system can be viewed as a microeconomy, where external events are communicated with messages to elementary agents. Messages can be passed between agents via the CFS mailbox and can form activation chains. Those chains are usually maintained by strength revision schemes such as J. Holland's *Bucket Brigade algorithms.*

Without any induction mechanism, a CFS will select, depending on environmental constraints, the subset rules which is relevant for the environment and have been proved useful to the system (their actions have received a payoff from the environment).

If we apply a GA to a CFS, it becomes able develop new rules. Mutations in the classifier conditions create new pattern matching abilities and also create new responses to stimuli when they touch the action part. When Crossover recombines features from two existing classifiers, the two resulting *offsprings* have good chances to fit the environmental constraints because their conditions and actions are derived from existing features owned by selected individuals, thus reflecting some

[1] it is also suggested that DNA has been generated by clay crystals [4]

aspects of the environment. This aspect has been thoroughly studied in [8].

6.1 A CFS based animat

Consider a minimal CFS based animat brain. Captors put binary representations of stimuli in specific places of the environmental message stored in the mailbox. Suppose we want to specify a basic set of rules for this animat, to make him attracted by light. We choose to set three one-bit quantities representing the animat perception in the environmental message. The two first values of this message are function of the light intensities received at each captor :

- If one eye receives more light than the other, the corresponding bit in the message is set to one.
- When no light is received by any of the eyes, the two bits are set to zero.
- When both eyes receive the same amount of light, both bits are set to one.

The third bit indicates if the animat has reached its goal.

On the effector side, the first bit and the second bit of the output message respectively command the left and the right wheel motors. When an effector bit is set to true, then the corresponding motor is on.

Using the classifier semantics, we can implement a light-seeking behavior with the following rules:

```
if not(LEFT_EYE) and RIGHT_EYE and not(TOUCH)
then
  LEFT_MOTOR = true
  RIGHT_MOTOR = false
endif

if LEFT_EYE and not(RIGHT_EYE) and not(TOUCH)
then
  LEFT_MOTOR = false
  RIGHT_MOTOR = true
endif

if LEFT_EYE and RIGHT_EYE and not(TOUCH)
then
  LEFT_MOTOR = true
  RIGHT_MOTOR = true
endif

if TOUCH
then
  LEFT_MOTOR = false
  RIGHT_MOTOR = false
endif
```

The above rules can be coded in the following four classifiers:

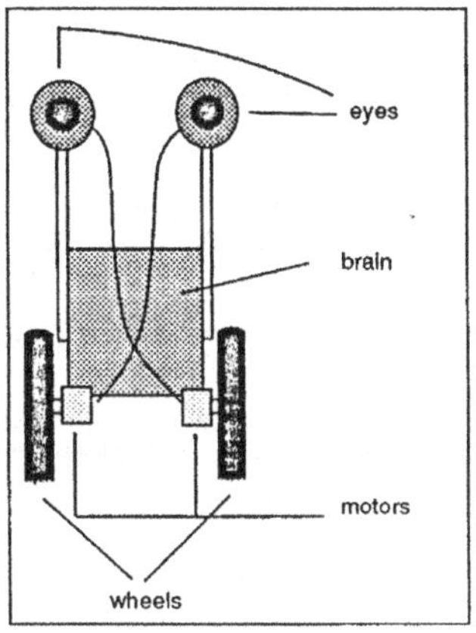

Figure 4: A light seeking animat built with analogic fiber technology (front view).

Conditions			Actions	
left eye	right eye	touch	left motor	right motor
0	1	0	1	0
1	0	0	0	1
1	1	0	1	1
#	#	1	0	0

Such a simple behavior requires four rules because of the implicit information used by the message rewriting process: the positioning of # symbols.

Using the other way, based on fibers/threshold devices technology, a similar behavior is easily implemented if we connect each animat's captor to the opposite motor (cf. figure 4).

Of course, we can arrange the animat interface so that each motor effector in the classifier action part matches the opposite captor in the classifier conditions. But in this case, things become more complicated if we decide to change the behavior of our animat.

In order to reduce the number of rules needed by so simple behaviors we can endow the classifiers with a connection map. This map specifies paths between input features detected by the rule condition filter and the output features which will be stored in the output message.

7 Rules with connection maps

Paths between input and output classifier filters are implemented by adding a *projection area* to each classifier. This projection area is a temporary work memory where some input message features are stored when the classifier is fired. Thus, the size of this memory is the same as a CFS message: half the size of a classifier. The projected features are then stored in the output message. The *connection map* specifies the paths followed by input features, filtered by the classifier conditions, stored into the projection area, then transferred in the output message through the rule actions filter.

The message rewriting process can be decomposed in two phases:

1. fill each projection area slot with input features pointed from the projection map. For each projection slot i:

$$projection[i] = input[map_{conditions}[i]]$$

2. fill each output message slot with projected features pointed by the projection map. For each output message slot i:

$$output[i] = projection[map_{action}[i]]$$

When an extended classifier input filter slot contains #, it means that both 0 or 1 can be stored in the associated projection area slot. The same way, if a classifier output filter slot also contains # then it will store the content of the corresponding projection area slot in the output message.

Note that the part of the map covering the conditions is of type

message position → projection area position

although the part used for output projection is of the form

projection area position → message position

Such a scheme ensures that no output message will be underspecified and that crossover and mutation will never produce invalid maps. Moreover, the proposed rewriting process can be performed in parallel because it avoids conflicts during memory transfers.

It is interesting to note that GA driven modifications on rule connection maps and condition/action patterns do not alter their meaning in a disrupting way. For instance, if we consider two # symbols placed respectively in the condition and in the action part of a rule. If they are linked by the connection map, it means that we have a *variable*, local to the rule, which is bound at rule fire time. If one of these symbols is turned by mutation into 1 or 0, the variable disappear and is replaced by a constant. In the later case we can consider the transformation – as it is the case of classifier systems – as a rule specialization. Differently, if the connection map is altered but the variable maintained, that is one of the variable binding site have been changed, then we can say that the variable has been substituted by an other one.

7.1 *Animat's rules with connection maps*

The rewriting, using connection maps, of our last example of animat behavior will lead to the following rule set :

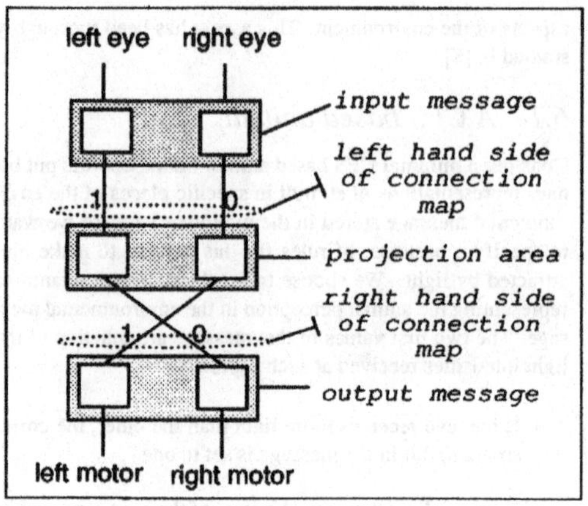

Figure 5: Connections for a light seeking animat.

```
if not(TOUCH)
then
   LEFT_MOTOR = RIGHT_EYE
   RIGHT_MOTOR = LEFT_EYE
endif

if not(TOUCH)
then
   LEFT_MOTOR = false
   RIGHT_MOTOR = false
endif
```

The figure 5 displays how the connection map creates crossing links between the input part and the output part of the light seeking rule. Note that the first three rules given in section 6.1 are replaced by only one equivalent rule.

Although we have no experimental results on the performances of such rule system, one can see that the size of the rule space is dramatically increased by the adjunction of connection maps to rules. The set of every possible rules of length N becomes $(\frac{3N}{2})^N$ instead of 3^N for the traditional classifiers.

8 Conclusion

About GA processing, the exploration of such huge amount of rules can be shorten if we initialize the system with a set of predefined rules. We also hope that the greater message rewriting abilities of the rules will allow us to reduce the size of the rule population. This reduction is of great interest if we wish to disassemble the classifiers to a human readable form.

On the GA side, many issues must be fixed. For instance, which kind of crossover is preferable? Is it a good thing to separate crossovers on the connection map from crossovers on rule condition/action filters?

The automated production of systems able to challenge the human analytic abilities of modeling behavior, to surprise people and to keep them interested may become a growing field. The long term consequences of work on this domain could have a strong impact on fields, such as the entertainment business, which seems to be reserved to human beings. In fact, it would be a natural evolution for this kind of activities, which nowadays are industries, to take advantage of automated creation.

References

[1] I. Aoki. Internal dynamics of fish schools in relation to inter-fish distance. *Bulletin of the Japanese Society of Scientific Fisheries*, 50(5):751–758, 1984.

[2] E. Berlekamp, J.H. Conway, and R. Guy. *Winning Ways for Your Mathematical Plays*. Academic Press, New York, 1982.

[3] V. Braitenberg. *Vehicles: Experiments in Synthetic Psychology*. MIT Press, Cambridge, 1984.

[4] A.G. Cairns-Smith. The first organisms. *Scientific American*, 252(6):90–100, June 1985.

[5] R. Dawkins. *The Blind Watchmaker*. W.W. Norton, New York, 1986.

[6] Adele Goldberg and A. Robson. *Smalltalk-80: The language and its implementation*. Addison-Wesley, Reading, Massachussets, 1983.

[7] J. H. Holland. *Adaptation in Natural and Artificial Systems*. University of Michigan Press, Ann Arbor, 1975.

[8] John H. Holland, Keith J. Holyoak, Richard E. Nisbett, and Paul R. Thagard. *Induction*. The MIT Press, 1986.

[9] W.S. McCulloch and W. Pitts. A logical calculus of the idea immanent in nervous activity. *Bulletin of Mathematical Biophysics*, 5:115–133, 1943.

[10] F. Moyson and B. Manderick. *The Collective Behavior of Ants: An Example of Self-Organization in Massive Parallelism*. Technical Report, Vrije Universiteit Brussels, 1988.

[11] Symbolics and Withney Demos Production. Stanley and stella in breaking the ice. Film, 1987.

[12] S. Wilson. Classifier systems and the animat problem. *Machine learning*, (2):199–228, 1987.

PERCEPTION AND MOTOR CONTROL

Biomimetic Sonar Processing: From Dolphin Echolocation to Artificial Neural Networks

H. L. Roitblat[1], P. W. B Moore[2], P. E. Nachtigall[2], & R. H. Penner[2]
[1]Department of Psychology, University of Hawaii, Honolulu, HI 96822 USA
herbert@uhccux.uhcc.hawaii.edu
[2]Naval Ocean Systems Center, Hawaii Laboratory, Kailua, Hawaii, 96734 USA
pmoore@nosc.mil

Abstract

Analysis of animal performance can provide important cues about the design of automated artificial biomimetic systems. On the basis of behavioral observations, we have been developing models of dolphin echolocation ability that have applicability to the design of biomimetic sonar systems. A dolphin was trained to perform an echolocation delayed matching-to-sample task. The clicks the animal generated during task performance were recorded and digitized along with the echoes returned by the stimulus objects. The dolphin's performance was then modeled using artifical neural networks.

1. Biomimetics

The study of animals can provide a very important adjunct to formal analyses in the design of automated systems such as robots and autonomous vehicles. Animals have evolved in a real world, solving real problems, such as gathering and interpreting essential information. Evolution supports the emergence of solutions that are well adapted to the animal's ecological niche, but provides no guarantee that the evolutionary solutions an animal derives are the best possible solution to a given problem (see Gould & Lewontin, 1979). Evolution merely asserts that, in light of the competing demands presented by the animal's evolutionary history, its ecology, and its other needs, a solution (vis a vis the whole organism and all its adaptations and constraints) is better, or at least no worse, than any other that its ancestors had achieved (Roitblat, 1982).

Although formal analyses have undoubtedly been successful in developing solutions to many automatic process problems such as those encountered in designing robots, many other problems have resisted solution. Solutions to scientific and engineering problems are inspired by many sources, but are ultimately derived from the intuition of the engineer, as formalizations of folk physics, folk psychophysics, folk psychology, etc. Folk science is the set of generally held beliefs that people employ in their ordinary activities. For example, many college students recognize that a ball rolled out of an inclined tube will fall some distance in front of the tube. Most of these same students, however, often mistakenly believe that if they drop a ball while walking, the ball will fall directly under the point at which it was released (McCloskey, 1983; see also Holland, Holyoak, Nisbett, & Thagard, 1986).

Systems involving falling balls are well analyzed so anyone with training in mechanics can see clearly the difference between the folk beliefs concerning falling balls and formal scientific beliefs (we may call these "facts"). Our intuitions have been trained to correspond with the analyzed facts, rather than with unanalyzed apparent perceptions. In situations in which formal scientific analyses have not yet been fully applied, we have no assurance that our scientific intuitions similarly avoid the pitfalls of our naivete. A creative scientist or engineer will apply his or her folk-science intuition to a difficult unanalyzed problem and if successful, will then test those hunches and discard a majority of them as unsuitable. Investigation of animals provides an alternative source for engineering hunches.

Animals that solve certain problems, such as visual perception or echolocation, provide an existence proof that such solutions are available. They also embody the specific mechanisms by which such solutions are achieved. To the extent that those solutions and mechanisms are understood, they can provide an alternative source of potential engineering solutions that are of proved utility. The study of animal behavior can suggest solutions the way other metaphors and analogies do, but it provides the added incentive of being demonstrably valid, at least for one system.

Neural network theories provide a prime example of the application of a biological investigation to the solution of serious computational problems. Artificial neural networks perform what has been called "brain-style" computation. The parallel computation of the brain appears to provide a better metaphor for certain kinds of computation than does the standard von Neumann computer architecture. Model systems based on this metaphor have proved to be highly successful at a number of pattern recognition problems, for example, that have resisted solution by more traditional computational approaches. The analogy of a generalized brain has served the development of powerful computational devices.

Our research is in part an attempt to apply the same method of analogy more narrowly focused on the mechanisms of echolocation object recognition. Our goal is to investigate the means by which dolphins use their echolocation abilities and to develop a model of those abilities that can be used by engineers to develop automatic systems capable of similar performance.

2. Biological Sonar

Several species are known to use biological sonar to obtain information about the world around them. Among these, bats and dolphins have received the most intense study. Investigations of bat sonar extend at least to the end of the 18th Century (Spallanzani, 1793). In modern times, studies of bat sonar were reinvigorated by Griffin (for a review of the early work, see Griffin, 1958). The hypothesis that dolphins also use ultrasonic vocalizations to echolocate was proposed by McBride (1956). Basic field and observational studies, followed by controlled experimental studies confirmed this hypothesis. The first experimental investigations of dolphin echolocation were mainly obstacle avoidance studies (reviewed by Moore, 1980). Later experiments, including many from this laboratory (e.g., Nachtigall, 1980) and Soviet laboratories (e.g., Sokolov, 1978) have investigated the psychophysics of dolphin echolocation.

Bottlenose dolphins (*tursiops truncatus*) can readily identify many characteristics of submerged objects through echolocation (see Nachtigall, 1980, for a review). Dolphin echolocation clicks emerge from the rounded forehead or melon as a highly directional sound beam with 3 dB (half power) beamwidths of approximately 10° in both the vertical and horizontal planes (Au, Moore, & Pawloski, 1986). Echolocation clicks are very broad band, have peak energy at frequencies ranging from 40 to 130 kHz with source levels of 220 dB re: 1 uPa at 1 m, (Au, 1980; Moore & Pawloski, 1990). The time between successive clicks depends on the distance between the animal and the target it is scanning. The average time between emitted clicks in a train is typically 15 - 22 msec longer than the time required for the click to travel through the water to the target and return as an echo (Morozov, Akopian, Zaytseva, & Sokovykh, 1972; Penner, 1989). Bottlenose dolphins have excellent directionally selective hearing (Au & Moore, 1984), spanning over 7 octaves, and can detect frequencies as high as 150 kHz (Johnson, 1966). The specific processes by which the dolphin extracts acoustic information about the targets is unknown and particularly interesting questions concern how the animal performs feature extraction from a set of returning echoes.

Although both bats and dolphins use sonar, the characteristics of the signals they emit and of the neurological apparatus they use to interpret echoes differ substantially (Zook, Myron, Ilya, & Morgane, 1988; Zvorykin, 1959, 1963). The dolphin echolocation signal is a very brief (about 50 useconds) broadband click (Au, 1980). In contrast, the bat echolocation signal is comparatively long-duration, containing both constant frequency and FM modulated components in many species (Bellwood, 1988; Fenton, 1988 Suthers, 1988). These differences suggest that dolphins and bats use substantially different mechanisms and strategies that are adapted to their specific environments. They illustrate the variety of solutions animals can derive to similar problems.

3. Behavioral methods

We have been studying and modeling dolphin echolocation performance (e.g., Nachtigall, 1980; Nachtigall, Patterson & Bauer, 1985; Roitblat et al., 1990a; Moore, et al., 1990) on a delayed matching-to-sample (DMTS) task. In this task a sample stimulus is presented underwater to a blindfolded dolphin. The dolphin is allowed to echolocate on this object ad lib. The object is then removed from the water, and after a

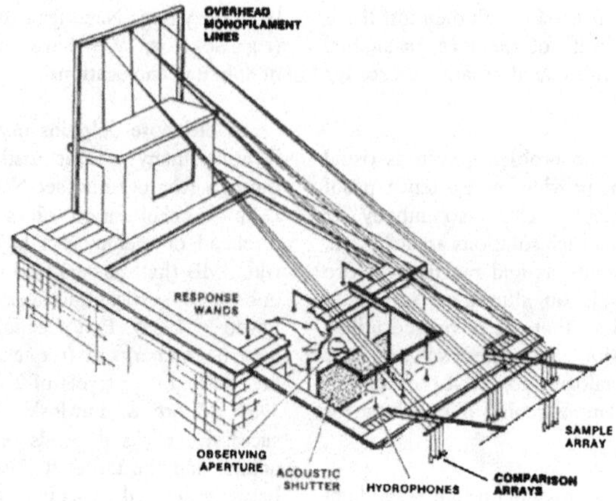

Figure 1. A diagram of the apparatus used in the echolocation matching-to-sample task. The enclosure is located in Kaneohe Bay, Oahu, Hawaii.

short delay, three alternative objects are presented (the comparison stimuli). One of these objects is identical to (matches) the sample object, and the dolphin is required to indicate the matching stimulus by touching a response wand directly in front of it. Figure 1 shows a diagram of the apparatus. The object that serves as sample and the location of the correct match vary randomly from trial to trial, so the dolphin must select the correct alternative by identifying the sample and the locating the correct match.

The subject of our continuing investigation of echolocation performance is a highly experienced (e.g., Nachtigall, Patterson, & Bauer, 1985) male bottlenose dolphin, housed in a floating enclosure in Kaneohe Bay at the Hawaii Laboratory of the Naval Ocean Systems Center. During the daily echolocation delayed matching-to-sample tests the animal's eyes were covered with soft removeable eyecups that completely occlude his vision. The dolphin stationed 100 cm under water with his head in the center of an observing aperture. The sample array (i.e., the set of three stimuli that could serve as the sample) was located 5.3 m directly in front of the aperture. Three sets of comparison targets were suspended to the left, right and in front of the animal. These arrays were hung from a bar located 4.3 m from the aperture. The left and right arrays were located approximately 1.6 m (22°) to the left and right respectively of the center comparison array. A neoprene-covered aluminum panel located 25 cm in front of the observing aperture served as an acoustic shutter preventing the animal from echolocating on either the sample or the comparison stimuli when the screen was raised into position. Echolocation clicks were detected by B&K 8103 hydrophones located 2 m from the observing aperture in the direct path between the aperture and the targets. Echolocation clicks were recorded from the center hydrophone. Echoes from the targets presented in the center positions were recorded using a custom-built hydrophone (flat to 200 kHz) placed next to the animal, at the observing aperture. Recordings were made using a RACAL store-4 tape recorder, with a 300 kHz dynamic range, from which clicks and echoes were digitized at 1 MHz. For additional behavioral details see Roitblat, et al., (1990).

The present studies employed three targets: (a) a PVC plastic tube open at both ends (15 cm long, 7.5 cm diameter, 30 mm wall thickness), (b) a water-filled stainless steel sphere (5 cm diameter), and (c) a solid aluminum cone (10 cm diameter base, 10 cm height), each presented approximately 100 cm below the water's surface. Four examples of each stimulus were used, one as sample, and the other three as alternative comparison stimuli.

At the start of each trial the dolphin was signaled to enter the observing aperture. The sample stimulus was lowered into the water and the acoustic shutter was opened. The dolphin was allowed to scan the sample. The acoustic screen was then raised back into the closed position, the sample stimulus was removed from the water and three alternative stimuli were presented, each

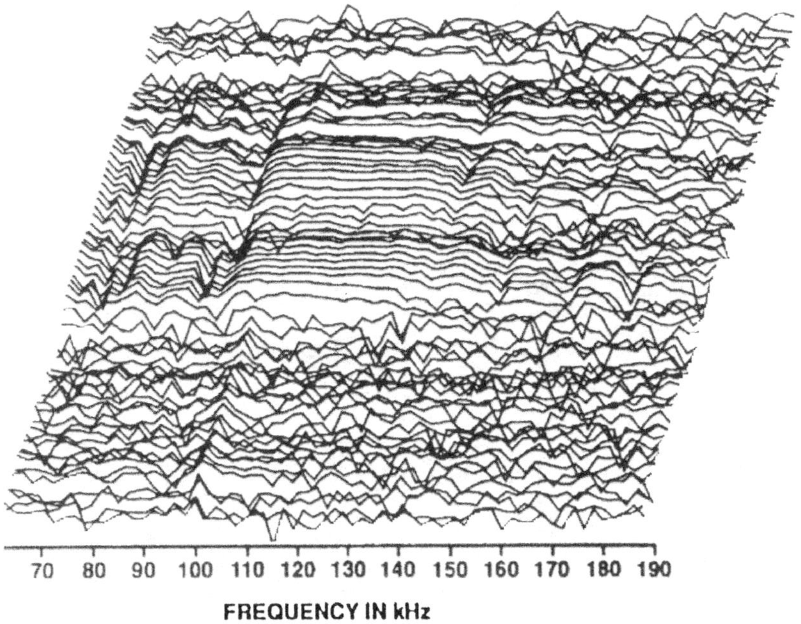

Figure 2. The spectra resulting from a train of successive echoes returning from the cone stimulus. The first echo is at the bottom of the figure, the final echo is at the top.

about 100 cm below the surface of the water. The same three alternatives were present on each trial, but their positions varied randomly between trials. The screen was then again lowered and the dolphin was allowed to echolocate on the comparison stimuli. The animal indicated his choice by contacting a small ball on the end of a response wand approximately in contact with the water surface, and directly in front of each comparison stimulus array. Contacting the ball in front of the comparison stimulus that matched the sample was designated the correct response and was reinforced with 3 Columbia River smelt.

The dolphin averaged about 94% correct DMTS performance on these targets, emitting an average of about 37 echolocation clicks in 4.2 scans to the comparison stimuli per trial (Roitblat, et al., 1990a). A scan is a train of echolocation clicks directed at one object, terminated either by a cessation of clicking, or by the start of a scan to another object. The number of clicks depended reliably on the stimulus being scanned. More clicks were required to recognize the cone and sphere than to recognize the tube. Furthermore, the dolphin tended to scan multiple stimuli in selecting the matching comparison, and at least some of the targets were scanned more than once in a trial. Therefore, his search strategy was not self-terminating.

4. Biomimetic Sonar Models

4.1 Echo analysis using a counterpropagation network

We developed a series of artificial neural network systems to model the dolphin's echolocation performance. Our initial work was directed at demonstrating the feasibility of such systems for this purpose. Our later work is directed at developing systems that incorporate more dolphin-like processing features.

INTEGRATOR GATEWAY NETWORK

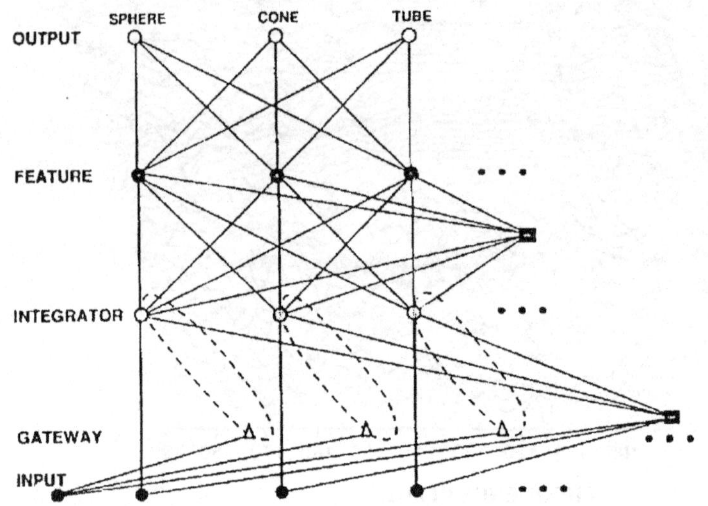

Figure 3. A schematic of the Integrator Gateway Network.

In our first study (Roitblat, Moore, Nachtigall, Penner, & Au, 1989), we selected a relatively small sample of echoes and a "standard" network architecture in order to demonstrate that a network could be trained to discriminate them. We used a counterpropagation network (Hecht-Nielsen, 1987, 1988), because these networks "learn" representations of stimulus classes that are relatively easy to interpret.

We trained the counterpropagation network with a sample of ten echoes from each of three objects. The inputs to the network were echo spectra divided into 20 bins. Each bin represented the amplitude of a specific, 4.95 kHz wide, frequency band in the echo.

A counterpropagation network contains four layers of processing elements. The bottom layer consists of 20 input units, to which the 20 bins of spectral information were presented. The next layer (the normalization layer) contains one unit more than the input layer. This layer normalizes the inputs so that all vectors are of constant length (Hecht-Nielsen, 1987). The third layer consists of a so-called Kohonen layer. The units in this layer compete for activation. Each element in this layer has variable strength connections to every element in the normalization layer. The activation of a unit in this layer depends on the match between the levels of activation of the units in the preceding layer and the corresponding weights of the connections transmitting this activation to the units in the Kohonen layer. The unit that most closely matches the normalized input pattern wins the competition and produces an output of 1.0. The other units in this layer are then inhibited and produce an output of 0.0. The winner of the competition also "learns" the input pattern that caused it to become active by adjusting the weights on its inputs to more closely approximate the pattern of activity over the units in the normalization layer. This algorithm for weight adjustment has a fortunate property for purposes of understanding the features of the echo that are important to discriminating objects in that each Kohonen unit automatically adjusts its input connection weights to approximate the centroid or prototype of one category. Each unit partitions the set of input patterns and functions as a basin of attraction for similar patterns. The winning Kohonen unit is selected on the basis of the similarity between the input pattern and its learned prototypes. If the number of units in the Kohonen layer is equal to or greater than the number of input patterns to learn, then the Kohonen layer learns to perform a nearest-neighbor classification. If there are fewer Kohonen units than the number of input patterns, the network learns a prototype classification. The pattern of weights on the input to each Kohonen unit is a measure of the representation that unit has learned.

The output of the Kohonen layer leads to the classification layer, which contains one unit for each of the nominal target categories (tube, sphere, and cone). The elements in this layer each connect to all of the elements in the Kohonen layer. The connection weights to these elements are adjusted during training to map the classifications determined by the Kohonen layer onto the desired nominal classes. One of these units was supposed to be activated when the cone was presented, a second when the sphere was presented and the third when the tube was presented.

After training, the counterpropagation network classified the spectral information from the training echoes and from an additional set of 30 novel echoes with better than 95% accuracy. The ability of this network to accurately classify the echoes suggests that spectral information was sufficient and only a single echo was necessary to identify the targets on which the dolphin was echolocating. This finding argues strongly against echolocation models that claim that time-domain information is the critical feature of echo identification or that claim that correlations between highlight patterns of successive echoes (e.g., time separation pitch) is required for recognition.

Although the network could identify the target with only a single echo, the dolphin concurrently performing the same task emitted many more clicks in identifying the same targets. We also noticed that the dolphin was more variable in terms of the number of clicks and the number of scans used to identify the correct match than was predicted by a sequential sampling model of his performance (Roitblat, et al., 1990). This sequential sampling model assumes that each echo is a sample from a random distribution of possible echoes, that each echo is obtained at some cost or effort, but supplies some stochastic information about the identity of the stimulus being scanned, and that the dolphin continues to obtain samples of information from additional echoes until the stimulus can be identified with sufficient confidence (see Roitblat, et al., 1990, for more details).

The increased number of clicks and the increased variability in their number, relative to the models used to describe the dolphin's performance, suggested that the models were based on inappropriate assumptions about the quality of the echoes that the dolphin received. The sequential sampling model assumed that the echoes were drawn from a stationary distribution, which may have been an inappropriate assumption in light of the variability in the dolphin's click production. Typically, the echolocating dolphin emits clicks that vary in source level and peak frequency (Moore & Pawloski, 1990). Emitted clicks in a train have been observed to have both uni- and bi-modal energy distributions, with peak energy in both the low frequency range (40 to 60 kHz) and the high frequency range (90 to 130 kHz). This plasticity in echolocation signal production results in echoes differing both in signal-to-noise ratio and spectral composition.

Because of the sampling procedure we used in selecting echoes for analysis by the counterpropagation model, it is possible that the echoes submitted to the neural network were not typical of the population of echoes the dolphin used. This possibility could have led to an overestimate of the ability of the models to recognize targets on the basis of dolphin echolocation returns. Other possible causes for overestimating the success of the model are considered later in the paper.

In response to these considerations we extended our analysis to include every echo available to the dolphin (up to the limits of our equipment to capture them). In contrast to our previous studies concerning the classification of echoes, in which echoes were selected for inclusion if they were sufficiently intense, in the present study we captured the echo resulting from every click the animal emitted in the sampled trials. The clicks producing these echoes were analyzed separately, and were found to vary in intensity and frequency within a click train. These clicks did not, however, vary systematically as a function of the stimulus that was being examined (Roitblat, Penner, & Nachtigall, in press). Figure 2 shows the spectra resulting from a train of successive echoes returning from the cone stimulus. As can be seen in this figure, the echoes also varied substantially as a function of their position in the click train.

Sequential sampling models typically assume that the observer combines information from successive samples when identifying the target. This model thus explains the multiple clicks emitted by the dolphin as

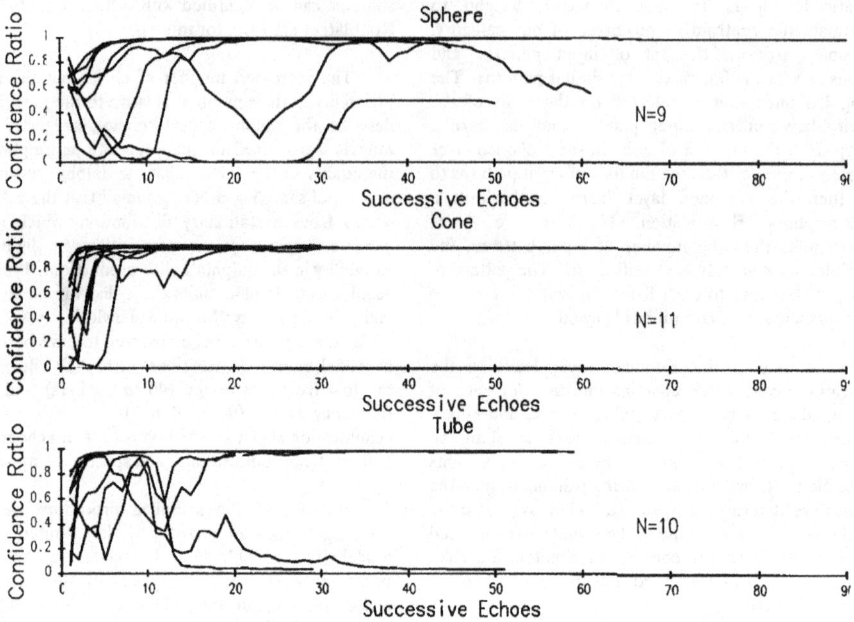

Figure 4. Results of generalization testing of the network in the form of the confidence of the network in assigning the echo train to the proper category. "Confidence" was defined as the ratio of the activation level of the correct classification unit versus the total output of the three classification units.

attempts to accumulate sufficient information to reach a confidence criterion. An alternative model holds that the dolphin does not combine information from successive echoes, but emits multiple clicks until a sufficiently "good" echo is returned to allow adequate target identification. We developed two neural networks in order to investigate this hypothesis. One network, the integrator gateway network incorporated the assumption that the dolphin combines information from successive echoes. As a control for this network we also trained a standard backpropagation network with the same stimuli. The backpropagation network does not combine successive echoes, and treats each one as an independent event. The inputs to both networks were echo spectra divided into 30 bins. Each bin represented the amplitude of a specific, 1.95 kHz wide, frequency band in the echo from 31.25 kHz to 146.5 kHz. The outputs of the networks were stimulus classes. One output corresponded to each stimulus class (sphere, cone, and tube). The resulting activations of each of these output classes were taken to be an estimate of the likelihood that the echo resulted from the particular stimulus type (Qian & Sejnowski, 1988).

4.2 The integrator gateway network

Our new integrator gateway network incorporates the assumption that the dolphin averages or sums spectral information from successive echoes and continues to emit clicks and collect returning echoes until it can classify the target producing those echoes with sufficient confidence. Figure 3 shows schematic of the Integrator Gateway Network.

In addition to the 30 bins of spectral information, each echo was also marked as to whether the echo was (1.00) or was not (0.00) at the start of an echo train. Recall that the dolphin directs a series of clicks to one target at a time, so it seemed plausible to include information marking the start of a click train. The first (leftmost) input to the network contained the start-of-train marker, the remaining bins contained the amplitude of a specified frequency range. The frequency inputs were then passed to a scalar unit and to the integrator layer. The integrator layer also contained 30 units, connected to the frequency units in the input layer in a corresponding one-to-one pattern. The connections to the scalar unit were fixed at 1/n, where n is the number of frequency inputs. The weights to the integrator layer were fixed at 1.00. The output of the scalar unit, which was simply the sum of all of its inputs, was passed to each of the units in the integrator layer via a fixed weight of -1.00. The effect of this scalar unit was to subtract the average activity of the input layer (neglecting the start-of-train marker) from the inputs to the integrator layer. This subtraction preserved all of the relative activity information present in the inputs, but kept the inputs within a manageable range.

The elements in the integrator layer computed a cumulative (running) sum of the inputs they received. The role of this layer was to accumulate and integrate information from successive echo spectra. The outputs of the integrator layer were passed via fixed connections with 1.00 weights to corresponding units in the gateway layer. The integrator layer and the gateway Layer each contained the same number of units. Each unit in the gateway layer acted as a reset for the corresponding unit in the integrator layer. The units in the gateway layer contained multiplicative transfer functions that received one input from its corresponding unit in the integrator layer and a second input from the start-of-train marker. Because this marker had 1.00 activity at the start of a click train and 0.00 activity otherwise, it functioned as a reset signal, causing the units in the integrator layer to be reset to 0.0 at the start of every scan.

The output of the integrator layer also led via variable-weight connections to each of the elements in the feature layer. The outputs of feature layer led via variable-weight connections to the classification layer. The elements in these two layers contained sigmoid transfer functions and were trained using a standard cumulative backpropagation algorithm with the epoch duration set to the number of training samples (60).

The network was trained with six sets of ten successive echoes each, selected from the ends of haphazardly chosen echo trains. An equal number of cone, tube, and sphere echoes were used.

4.3 Integrator gateway results and discussion

Figure 4 shows the results of generalization testing of the network in the form of a derived confidence measure. The network was given all 30 scans (10 scans of each target for a total of 1,335 sequential echoes), and was required to classify each echo train. "Confidence" was defined as the ratio of the activation level of the correct classification versus the total output of the three classification units. A confidence ratio of 1.00 indicates that only the correct unit is active. Confidence of 0.00 indicates that the correct unit is entirely inactive.

Table 1
Number of Clicks to Network Confidence Criterion

Target Scanned					
Sphere	Cone	Tube	Sphere	Cone	Tube
Integrator Gateway			Backpropagation		
16S	20C	40C	1S	1C	3S
9S	4C	18C	6S	30C	1S
7S	2C	20T	1S	1C	11S[1]
6S	6C	23T	5S	2C	1T
19S	14C	5T	14S	2C	14T
19S	6C	4T	14S	30S[1]	1T
34S	6C	4T	3S	32S[1]	1T
7S	4C	4T	1S	57S	1T
23C	6C	5T	40T	22S	2T
	3C	4T		22S	1T
	11C			27T	

Note: Entries are the number of clicks needed by the network to achieve the 0.96 confidence criterion. C indicates a Cone decision, S indicates a Sphere decision, T indicates a Tube decision. [1]Indicates that the dolphin stopped echolocating before the network reached its confidence criterion. On these scans, the decision is the one with the highest confidence at the end of the scan.

Intermediate confidences correspond to intermediate likelihood ratios (Qian & Sejnowski, 1988).

Recall that echo trains varied in length under control of the dolphin. Therefore, it is not entirely clear how to measure the performance of this network. According to sequential sampling theory (e.g., Laming, 1968; Smith & Vickers, 1988; Roitblat, et al., 1990) a rational decision maker would continue to collect echo evidence only until a sufficiently confident classification is available and then stop. Table 1 shows the number of clicks in each train that were required to reach a confidence ratio of 0.96 and the classification that the network derived. Some of the scans ended before the network could achieve this confidence level. An average of 11.67 clicks was required to reach this level of confidence. Three erroneous classifications were made (90% correct).

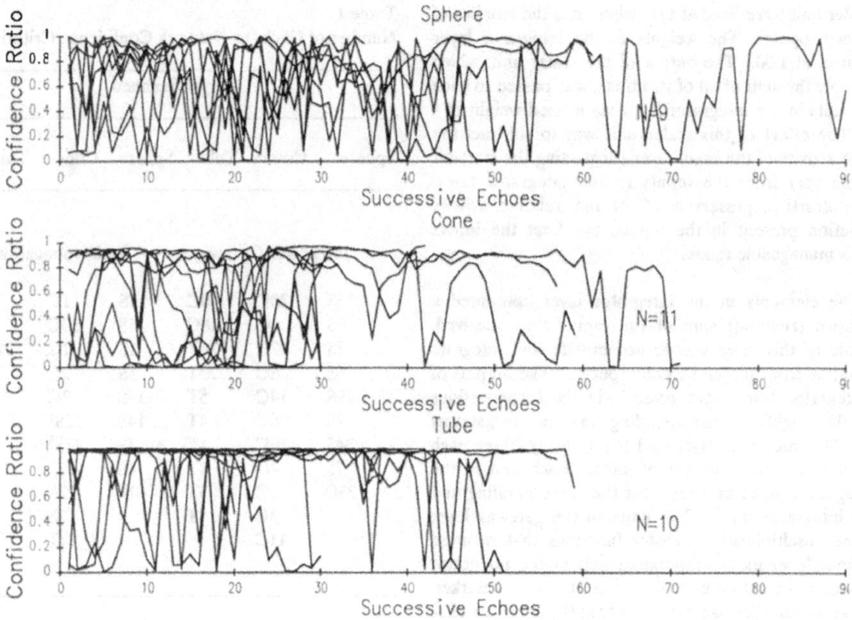

Figure 5. Confidence of the backpropagation network in assigning the echo train to the proper category as a function of the number of echoes received.

Overall, the animal's performance is better than that of our network. Roitblat, et al. (1990) reported that the dolphin was 94.5% correct at selecting the correct match. This level of performance required the animal to identify both the sample and the correct comparison stimuli. The probability of both occurring was observed to be 0.945. Because a correct choice depends on the probability of correctly identifying the sample and on the same trial correctly identifying the matching comparison stimulus, the probability of a correct choice is:

$$p(\text{correct choice}) = p(\text{sample id}) * p(\text{comparison id}) \quad (1)$$

The simplest assumption is that both stimuli were identified with the same probability. On this assumption, the probability of each identification can be estimated at *p* (id) = W0.945 = 0.972 (assuming that each occurred with equal probability). By no measure was our network 97.2% accurate at identifying the stimuli, but when it did identify the stimuli it tended to do so with fewer echoes than were used by the dolphin. Apparently, the dolphin is using information that is not being used by our network or (more likely) the dolphin has more precise representations of the echo prototypes than were developed with our small training set (only 60 of the 1,335 available echoes). On the other hand, because of the dolphin's extensive experience in echolocation in general and echolocation matching-to-sample in particular, the dolphin may be attempting to identify the stimuli in this experiment not just relative to the three stimuli we presented, but relative to a much broader class of potential targets, which may require more echoes. Our current work in progress includes a broader range of targets to investigate this possibility.

4.4 A simple backpropagation network

The integrator gateway network reflects the assumption of sequential sampling theory that the dolphin combines information from successive echoes in deriving its identification. In contrast, a standard backpropagation network does not integrate over successive echoes and instead attempts to identify each echo independently. A backpropagation network can be used as a model of a system that emits multiple clicks because the echoes vary in quality. Rather than integrating the echoes it simply waits for a single adequate echo that allows it to meet its confidence criterion.

We trained a backpropagation network on the same data that were submitted to the integrator network in order to determine whether the additional structure of the integrator network contributed to its performance accuracy. The network contained exactly the same number of inputs, hidden units, outputs, and adjustable connections as the integrator network. The only difference between the networks was the presence of the integration apparatus in the integrator network and its absence in the backpropagation network. The backpropagation network was trained using the fast-backpropagation algorithm to adjust the weights (Samad, 1988) to the same 0.05 RMS error criterion using the same training parameters and then tested on the full set of echoes.

4.5 Backpropagation results

Figure 5 shows the confidence of the backpropagation network in assigning the echo train to the proper category as a function of the number of echoes received. Compared to the categorization performance of the integrator network, the backpropagation network was much more variable. As Figure 5 illustrates, the individual echoes were highly variable, and frequently assigned to an erroneous category.

The performance of the backpropagation network when judged by the standards of sequential sampling theory are shown in Table 1. This table shows the number of clicks necessary to first reach a classification with greater then 0.96 confidence. On average the backpropagation network (11.57 echoes) reached its confidence criterion in the about the same number of clicks (t (df = 58) = 0.03, p> .05) as the integrator network (11.67 echoes), but it produced more errors (X^2(df=1) = 5.96).

These data suggest that the integrator network added significantly to the ability to classify sequentially produced echoes. In other words, by implementing a signal "averaging" mechanism in the neural network we allowed the system to take advantage of the redundancy inherent in the use of multiple echoes from the same source and in the stochastic properties of the noise in which those echoes are embedded. In contrast, the backpropagation network is required to process not only the characteristics of the echoes themselves, but also the characteristics of the noise. This results in many spurious classifications. Presumably, if a larger training set had been employed, the backpropagation network would have learned to "abstract" the salient properties of the echoes, but within the constraints of a relatively small training set (60 of 1,335 or just 4% of the total number of echoes), the integrator network does a much better job of separating the signal from the noise.

5. Conclusion

The gateway integrator network adds a level of complexity to the standard backpropagation network architecture that contributes substantially to its performance. Its design is inspired by properties of the dolphin's performance and it represents one step along a development path that seeks to include more and more of the mechanisms that we can identify from the neurobiology of echolocation (e.g.,Suga, 1990) and from the performance of dolphins in their aquatic environment.

References

Au, W. W. L. (1980). Echolocation signals of the Atlantic bottlenose dolphin (*tursiops truncatus*) in open waters. In R. G. Busnel & J. F. Fish (Eds.), *Animal sonar systems* (pp. 251-282). New York: Plenum Press.

Au, W. W. L. (1988). Detection and recognition models of dolphin sonar systems. In P. E. Nachtigall & P. W. B. Moore (Eds.), *Animal Sonar: Processes and performance* (pp. 753-768). New York: Plenum Press.

Au, W. W. L., Moore, P. W. B. & Pawloski, D. (1986). Echolocating transmitting beam of the Atlantic bottlenose dolphin. *Journal of the Acoustical Society of America*, 80, 688-691.

Bellwood, J. (1988). Foraging behavior, prey selection, and echolocation in *phyllostomine* bats (*phyllostomidae*). In P. E. Nachtigall & P. W. B. Moore (Eds.) *Animal sonar: Processes and performance*. (pp. 601-605). New York: Plenum.

Fenton, M. B. (1988). Variation in foraging strategies in five species of insectivorous bats--implications for echolocation call design. In P. E. Nachtigall & P. W. B. Moore (Eds.) *Animal sonar: Processes and performance*. (pp. 607-611). New York: Plenum.

Griffin, D. R. (1958). *Listening in the dark: The acoustic orientation of bats and men*. New York: Dover.

Hecht-Nielsen, R. (1987). Counterpropagation networks. *Applied Optics*, 26, 4979-4984.

Hecht-Nielsen, R. (1988). Applications of counterpropagation networks. *Neural Networks*, 1, 131-139.

Holland, J. H., Holyoak, K. J., Nisbett, R. E., & Thagard, P. R. (1986). *Induction: Processes of inference, learning, and discovery*. Cambridge, MA: MIT Press.

Johnson, C. S. (1966). *Auditory thresholds of the bottlenosed porpoise, Tursiops truncatus (Montague)*

(Naval Ordnance Test Station Technical Publication No 4178). Naval Ordnance Test Station.

Laming, D., (1968). *Information theory of choice reaction times*. New York: Academic Press.

McBride, A. F. (1956). Evidence for echolocation by cetaceans. Deep-Sea Research, 3, 153.

McCloskey, M. (1983). Intuitive physics. *Scientific American*, 24, 122-130.

Moore, P. W. B. (1980). Cetacean obstacle avoidance. In R. G. Busnel & J. A. Fish (Eds.), Animal sonar systems: Biology and bionics. New York: Plenum.

Moore P. W. B. & Pawloski, D. (1990). Investigations on the control of echolocation pulses in the dolphin (*Tursiops truncatus*). In J. Thomas & R. Kastelein (Eds.) *Sensory abilities of cetaceans*. New York: Plenum Press (in press).

Morozov, V. P., Akopian, A. I., Zaytseva, K. A., & Sokovykh, Y. A. (1972). Tracking frequency of the locations signals of dolphins as a function of the distance to the target, *Biophyzika*, 17, 139-143.

Nachtigall, P. E. (1980). Odontocete echolocation performance on object size, shape, and material, In R. G. Busnel & J. F. Fish (Eds.), *Animal Sonar Systems*, pp. 71-95, New York: Plenum Press.

Nachtigall, P. E., Patterson, S. A., & Bauer, G. B. (1985). Echolocation delayed matching-to-sample in a bottlenose dolphin. Paper presented at the Sixth Biennial Conference on the Biology of Marine Mammals, Vancouver, B.C., Canada. November.

Penner, R. H. (1989). Attention and detection in dolphin echolocation, In P. E. Nachtigall and P. W. B. Moore (Eds.), *Animal Sonar: Processes and Performance*, New York, Plenum Press. 707-713.

Qian, N. & Sejnowski, T. J. (1988). Predicting the secondary structure of globular proteins using neural network models. *Journal of Molecular Biology*, 202, 865-884.

Roitblat, H. L., Moore, P. W. B., Nachtigall, P. E., Penner, R. H., & Au, W. W. L. (1989). Dolphin echolocation: Identification of returning echoes using a counterpropagation network. *Proceedings of the First International Joint Conference on Neural Networks*. Washington, DC: IEEE Press.

Roitblat, H. L., Penner, R. H., & Nachtigall, P. E. (1990). Matching-to-sample by an echolocating dolphin (*Tursiops truncatus*). *Journal of Experimental Psychology: Animal Behavior Processes*, 16, 85-95.

Roitblat, H. L., Penner, R. H., & Nachtigall, P. E. (1990). Attention and decision making in echolocation matching-to-sample by a bottlenose dolphin (*Tursiops truncatus*): the microstructure of decision making. In J. Thomas & R. Kastelein (Eds.), *Sensory abilities of cetaceans*. New York: Plenum Press (in press).

Smith, P. L. & Vickers, D. (1988). The accumulator model of two-choice discrimination, *Journal of Mathematical Psychology*, 32: 135-168.

Samad, T. (1988). Back propagation is significantly faster if the expected value of the source unit is used for update. International Neural Network Society Conference Abstracts.

Sokolov, V. Ye. (1978). Morskiye mlekopitayushchiye: Rezul'taty i metody issledovaniy. Moscow: Nauka Publishing.

Spallanzani, L. (1793). Letter to Vassalli. Cited in D. R. Griffin, *Listening in the dark: The acoustic orientation of bats and men*. New York: Dover.

Suga, N. (1990). Cortical computation maps for auditory imaging. *Neural Networks*, 3, 3-21.

Suthers, R. (1988). The production of echolocation signals by bats and birds. In P. E. Nachtigall & P. W. B. Moore (Eds.) *Animal sonar: Processes and performance*. (pp. 23-45). New York: Plenum.

Zook, J. M., Myron, S. J., Glezer, I., & Morgane, P. (1988). Some comparative aspects of auditory brainstem cytoarchitecture in echolocating mammals: Speculations on the morphological bases of time-domain signal processing. In P. E. Nachtigall & P. W. B. Moore (Eds.) *Animal sonar: Processes and performance*. (pp. 311-316). New York: Plenum.

Zvorykin, V. P. (1959). Morphological basis of locative and supersonic abilities in bat. Arkh. Ant. Gistol. Embriol. 36, 19.

Hierarchical Dishabituation of Visual Discrimination in Toads[1]

DeLiang Wang
Michael A. Arbib
Center for Neural Engineering
University of Southern California
Los Angeles, CA 90089-2520, U.S.A.
wang@rana.usc.edu, arbib@pollux.usc.edu

Abstract

Habituation is a basic form of learning in which animals come to respond less and less to repeated presentation of a given stimulus. Studies of habituation in *Aplysia* have yielded important insights into basic mechanisms of synaptic plasticity, while studies in humans reveal stimulus-specific habituation with mutual dishabituation by pairs of different stimuli. Studies in toads suggest a new phenomenon which leads us to new models of vertebrate learning which are subject to experimental test. Instead of mutual dishabituation for different worm-like stimuli, toads exhibit a dishabituation *hierarchy*, in which stimulus *A* may dishabituate *B*, but not vice versa. We offer a model of this hierarchy in which the toad's visual discrimination is reflected in different firing rates in some higher visual center, hypothetically anterior thalamus. This theory, developed through neural simulation based on an extensive model of toad retina, predicts that retinal R2 cells play a primary role in the discrimination while R3 cells refine the feature analysis by inhibition.

The theory predicts new dishabituation hierarchies based on reversing stimulus-background contrast and shrinking stimulus size. After the predictions were made, several were tested by behavioral experiments. In particular, we selected a pair of stimuli whose ordering in the dishabituation hierarchy we predict to be changed by contrast reversal, and the experimental result is as predicted. A size shrinking prediction failed to be validated, and further experiments suggest that visual pattern discrimination in toads is relatively unaffected by stimulus size. Finally, we discuss new insights into a network, offered as a preliminary model of the medial pallium, that can express the dishabituation hierarchy of toads.

1. Background

Habituation is an elementary form of learning in which response to a stimulus will diminish with repeated presentation of the stimulus if there is no punishment or reward associated with the presentations. In the marine mollusc *Aplysia*, habituation has been much studied for insight into molecular mechanisms of synaptic plasticity (see Bailey and Kandel, 1985, for a review). Here, habituation seems to be independent of the specific patterning of stimuli, whereas habituation in mammals is stimulus specific. Given two different patterns *A* and *B* (e.g., two tones of different pitch, volume, or duration), the animal can exhibit this specificity by the phenomenon of *dishabituation* (i.e., stimulus B can release behavior despite habituation to A). Moreover, this dishabituation is mutual *in mammals*: If stimulus *A* can dishabituate stimulus *B*, then stimulus *B* can dishabituate stimulus *A* (Thompson and Spencer 1966; Sokolov 1975).

---

[1] Th e research described in this paper was supported in part by grant no. 1RO1 NS 24926 from the National Institutes of Health (M.A.Arbib, Principal Investigator). We wish to express our sincere thanks to Prof. J.-P. Ewert with whom one of us (W.-D.L.) conducted the experiments that greatly furthered the modeling reported here.

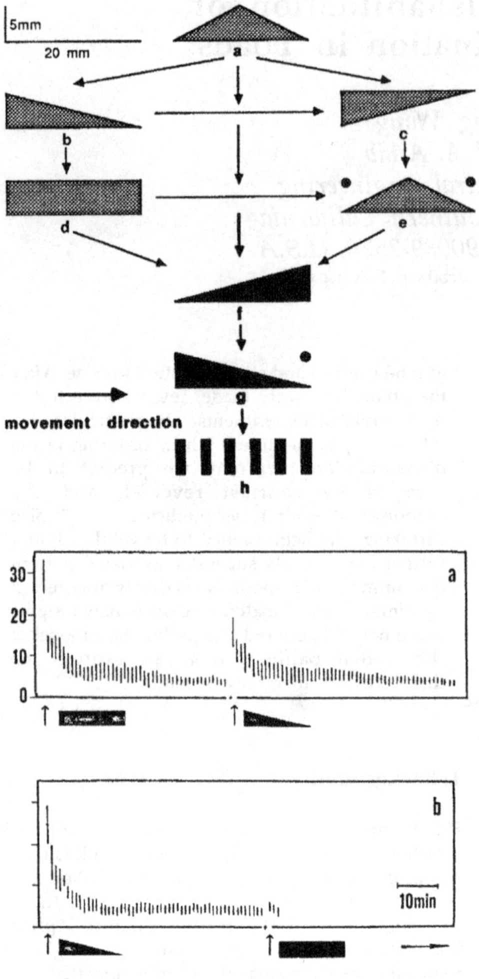

When repeatedly presented visually with the same prey dummy, toads also orient less and less frequently to the stimulus but, in striking contrast to mammals, this habituation exhibits *hierarchical stimulus specificity* (Ewert and Kehl 1978). After an extensive study with 700 toads, they found the dishabituation hierarchy shown in Fig.1, which we summarize symbolically as

$$a > b \approx c > d \approx e > f > g > h.$$

A worm-like pattern higher in the hierarchy can dishabituate another worm pattern lower in the hierarchy, whereas a stimulus lower in the hierarchy cannot dishabituate habituation to a stimulus higher in the hierarchy. At the same level in the hierarchy, the left one can slightly dishabituate the right one, but not vice versa.

Lara and Arbib (1985) were the first to offer a model of the neural mechanisms that might underlie the dishabituation hierarchy. They started from the idea of Sokolov (1975) that stimulus-specific habituation was based on the formation of a neural model of the habituated stimulus, with behavior being released to the extent that the current stimulus differs from the stored model. However, such a model yields symmetric dishabituation, and so Lara and Arbib suggested that the habituated stimulus was stored as a model of *intensity* of some brain variable, with a stimulus being habituated to the extent that its encoding is *weaker* than the stored model. This clearly yields hierarchical dishabituation.

The challenge, then, was to offer a plausible neural network whose output strength reflects the ordering shown in Fig.1. This Lara and Arbib did, but their model was flawed because it was based on an *ad hoc* model of the toad retina. Subsequent work in our group has led to a neural network retinal model (Teeters 1989) which is detailed enough to yield distinctive spatiotemporal firing patterns for the various stimuli. This provided the basis for Wang and Arbib (1990) to model how the outputs of the retina could be combined to drive cells whose firing strength will accurately reflect the ordering shown in Fig. 1.

Figure 1. Dishabituation hierarchy for the worm patterns used in stimulus-specific habituation in toads. Between any pair of stimuli in the hierarchy, one stimulus (a black dummy moving against a white background) is presented repeatedly until habituation to the stimulus, then another stimulus is presented until it elicits no response. In the typical example at the bottom of the figure, each vertical bar indicates the number of orienting responses per minute made by the toad on a trial in which a given stimulus is repeatedly moved from left to right on a screen in front of the animal. In the hierarchy shown above, a stimulus can dishabituate all the stimuli below or to the right of it, but not vice versa. Before any habituation, all the stimuli are about equally strong in releasing orienting responses. (Redrawn from Ewert and Kehl 1978.)

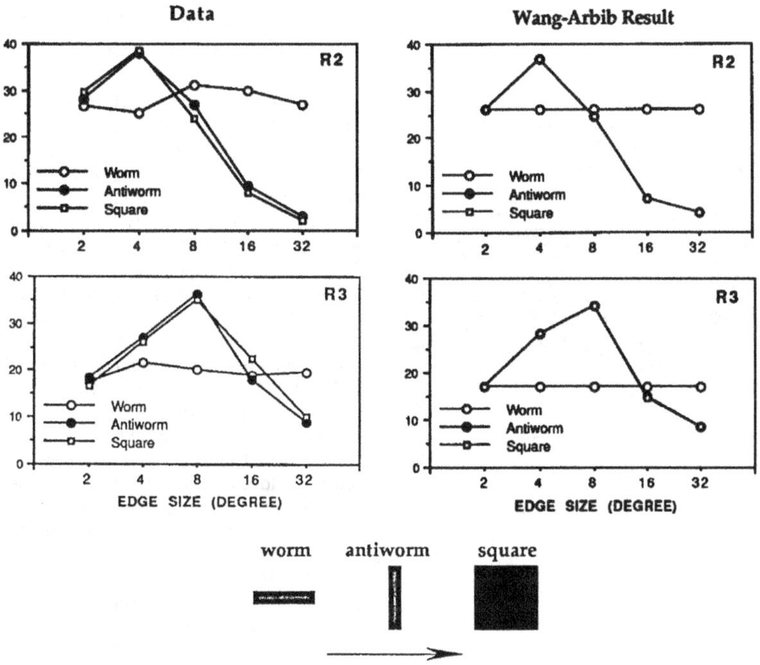

Figure 3. Left: The experimental data on R2 and R3 responses to worm, antiworm and square (redrawn from Ewert 1976); **Right:** The corresponding model responses. A worm stimulus here is a rectangle with its elongated edge parallel to the direction of movement; an antiworm stimulus is a rectangle with its elongated edge perpendicular to the direction of movement. Each point in the figure represents a temporal average firing rate.

In quantitative terms, a moving stimulus is directly mapped onto the receptor layer, and the response of a neuron in a later layer is formed by a matrix of cells from the previous layer according to the *leaky integrator* model (following the style of modeling in Cervantes-Perez et al. 1985). The simulated responses of the retina model to various worm, antiworm and square stimuli are presented in Fig. 3 for the R2 and R3 ganglion cells, together with the electrophysiological data (Ewert and Hock 1972; Ewert 1976). The average firing rate of a cell is computed by the temporal integration of its instantaneous firing rate divided by the time period during which a non-zero firing rate occurs consecutively. The figure shows a close match between the model responses and experimental data.

Having fixed the retinal model to achieve the results of Fig. 3, we then experimented with various connectivity matrices for the input to ATH until we obtained ones yielding the results shown in Fig. 4. This figure shows the average firing rates of a single ATH neuron to the 8 worm-like stimuli of Fig. 1, where all moving objects are 20 mm long and 5 mm high, which correspond to 16° and 4° visual angles, respectively, from the viewing distance of 70 mm. The dots which are added to the triangular objects (see Fig.1) are 1 mm in diameter which is about 1°. Because all the moving objects have the same length and height, the critical cues are (1) leading edge (or the angle subtended by a leading edge); (2) trailing edge; (3) isolated dot; and (4) striped pattern. The open symbols represent the response of the full model, while for a comparison the filled symbols provide the response without R3 inhibition. The result clearly matches the ordered dishabituation hierarchy in Fig.1. Not only do stimuli higher in the hierarchy generate larger ATH responses, but the stimulus pairs b–c and d–e which are on the same level in the hierarchy generate nearly equal responses. Note that the number of arbitrary

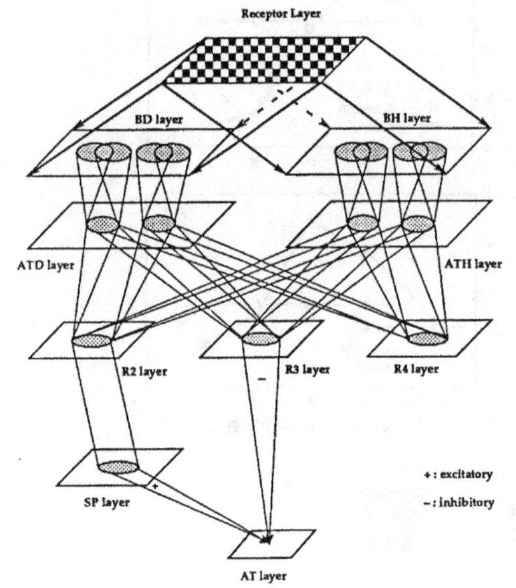

Figure 2. Anatomy of the entire model for ordered discrimination of wormlike stimuli. It incorporates retina, tectum, and anterior thalamus, with discrimination represented by the relative level of firing in anterior thalamus.

The structure of the rest of this paper is then as follows: We recall the salient details of the Wang - Arbib model and note that it makes detailed predictions about *changes in the dishabituation hierarchy* with changes in contrast or size of the stimuli. We next review experiments conducted to test these predictions. A prediction on contrast reversal was confirmed. A prediction on changing size was not confirmed. We argue that this does not invalidate our model, but does challenge us to extend the model by adding a mechanism for size invariance in prey-discrimination. Finally, we offer a tentative model for habituation on the basis of the ordered encoding, and show how it may be refined on the basis of new behavioral data.

2. A Model of Ordered Visual Pattern Discrimination in Toads

The optic tectum discriminates prey from predator, but in our model we assume that it does not have an important role in "subworm" discrimination (i.e., in determining the discriminations in Fig. 1) since the naive toad is equally responsive to all these stimuli. Instead, it will serve as a relay of information from retina to anterior thalamus (AT). The small pear cells (SP) of the optic tectum receive inputs from the R2 cells and project ascendingly to AT (Lázár et al. 1983), which also receives R3 and R4 retinal projections (Scalia and Gregory 1970). Among other ascending projections to telencephalon, AT has a direct projection to the medial pallium (Scalia and Colman 1975; Neary and Northcutt 1983) which seems critical for habituation (Finkenstädt and Ewert 1988). Based on the particular position of AT in the learning loops of the toad brain (Ewert 1989), we hypothesize that it is in the anterior thalamus that the subworm pattern discrimination is finally exhibited.

Due to the lack of more detailed data, we presently use an array called ATH for modeling the computational function of the anterior thalamus. ATH neurons receive excitatory-center inhibitory-surround inputs from tectal SP cells, and direct inhibitory inputs from R3 cells. The neural model (Wang and Arbib 1990) is summarized in Fig. 2, which incorporates retina, tectum and anterior thalamus. Our detailed retina model is based on the Teeters (1989) model. In the figure, conical projections represent on-center off-surround convergence, while the cylindrical projection from the R2 layer to the SP layer represents a 1-to-1 mapping. The connections from the receptor layer to both the depolarizing bipolar cell (BD) layer and the hyperpolarizing bipolar cell (BH) layer constitute a small many-to-one convergence.

The receptor layer contains 140x140 cells which correspond to a 70°x70° visual field. Bipolar and amacrine cell layers (ATD: on-channel, ATH: off-channel) consist of 140x140 cells respectively, in correspondence with the receptor layer. Three types of ganglion cells, R2, R3 and R4, have been modeled, each consisting of 25x25 cells which correspond to a 70°x70° visual field since the ganglion cells have 20° receptive fields (RF) and lie 2° apart. The R2 layer projects to the SP layer in the tectum, and the SP layer and the R3 layer together converge on the AT layer in the anterior thalamus, where the discrimination of worm-like patterns is finally achieved. The entire simulation contains about 100,000 cells.

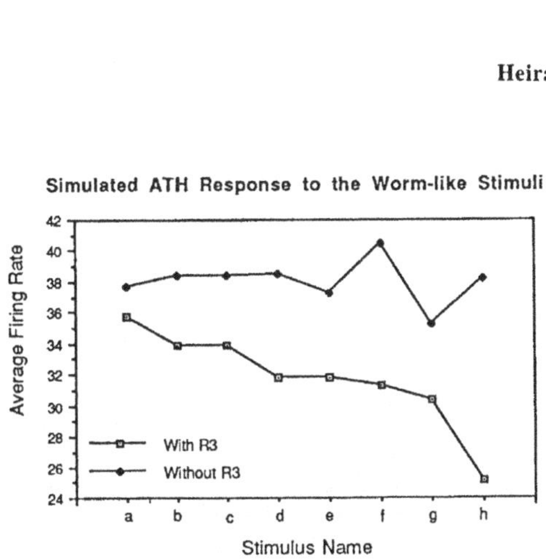

Figure 4. ATH response to the 8 worm-like stimuli shown in Fig. 1. The 8 average firing rates are ordered, which corresponds to the ordered hierarchy in Fig. 1. For comparison, the figure also shows ATH response to the stimuli without inhibitory projection from R3 cells.

hierarchies out of the 8 stimuli is 8! (40,320), and only one is supported by the data.

This model represents an integration of behavioral, physiological, anatomical, and theoretical studies on the brain in order to understand the neural mechanisms for pattern discrimination in toads. Based on the modeling, our explanation of the dishabituation hierarchy is:

1) Both the leading edge and the trailing edge of a worm stimulus have to be taken into consideration.

2) The receptive field of ATH neurons (25° ERF and 50° IRF in our model) is big enough to "see" both the leading and the trailing edge. Stimulus **a** elicits the biggest response because both diagonal edges elicit strong responses in R2 cells and these responses can be best integrated in ATH cells due to the relatively small distance between the midpoints of its leading and trailing edges.

3) Stimuli **b** and **c** are preferred to stimulus **f** because the inhibition of R3 cells, which have off-channel preference, is bigger for **f** than for **b** and **c**.

4) Stimuli with dots appear lower in the hierarchy because they elicit smaller R2 responses due to IRF interaction.

5) A striped pattern elicits the smallest response in ATH neurons because of R3 inhibition. This is particularly clear when we compare the two curves in Fig.4.

This level of modeling enables us to offer specific predictions which form strict testbeds for the theory. Since, in the model, both the on-channel and off-channel responses are considered important, it is interesting to know whether the dishabituation hierarchy will be changed if stimulus-background contrast direction is reversed. We thus presented the same set of stimuli to the simulation as in Fig.1, but *reversed the contrast direction*. On the basis of the simulated ordering of ATH responsiveness, a different dishabituation hierarchy from Fig.1 is predicted. In fact (Wang and Arbib 1990), the a to h ordering of Fig. 1 becomes

$$\mathbf{f} > \mathbf{a} > \mathbf{d} \approx \mathbf{e};\ \mathbf{d} > \mathbf{b} > \mathbf{h};\ \mathbf{e} > \mathbf{c} > \mathbf{h};\ \text{and}\ \mathbf{h} > \mathbf{g} \qquad (1)$$

The response of R2 cells is about the same with respect to contrast reversal, but now R3 cells show a trailing edge preference.

The current model also predicts that different hierarchies are produced based on different sizes of worm-like stimuli. As an example, we shrank the size of all the stimuli to 10 mm long and 2.5 mm high (this corresponds to 8° by 2°, given the same distance to the stimulus) and tested these stimuli. The dishabituation hierarchy predicted by the model is

$$\mathbf{h} > \mathbf{d} > \mathbf{e} > \mathbf{a} > \mathbf{f} > \mathbf{g} \approx \mathbf{b} \approx \mathbf{c} \qquad (2)$$

Our explanation of the remarkable difference is that since the stimulus size is halved compared to Fig. 1, the previous IRF interaction in the R2 receptive field is converted into the ERF interaction which strengthens overall response. This ERF interaction is particularly manifested by the striped pattern **h**. Note that the R3 inhibition in ATH neurons is relatively smaller than the excitation from SP cells, and thus cannot prevent stimulus **h** from inducing a strong ATH response which reflects the final integration.

3. Testing the Model

After the modeling was completed, Wang and Ewert conducted several behavioral experiments to test predictions of the model. Wang, Arbib and Ewert (1990) describe the experimental methods, and provide explicit figures to show the results which we here summarize. To determine whether the animal was able to discriminate two visual dummies *A* and *B*, the prey-catching orienting response of the toad was first habituated to stimulus *A*; then the response to *B* was tested and also habituated. Experiments were also repeated in the reverse order: first *B* and then *A*. All experiments were repeated with 10 different toads to reach stable statistical results.

Contrast Reversal

In order to test the prediction of the stimulus-background contrast effect on the dishabituation hierarchy, we selected a pair of stimuli **b**/**f**, which showed a strong change of order between the black experimental hierarchy in Fig. 1 and the predicted white hierarchy in (1). For this pair, we used stimuli of the same size as used by Ewert and Kehl, using black stimuli on a white background to confirm their finding, and white stimuli on a black background to test our prediction. With this selected pair, the experimental result is as predicted by the theory.

Size Effects

To test the prediction of size reduction shown in (2), we selected a pair of stimuli **b**/**d** (called small **b** and small **d** hereafter) which yields a strong change of preference of dishabituation in (2) from that in Fig. 1. In the experiment, both small **b** and **d** were 2.5 mm high and 10 mm long. However, we found that even though the stimulus size was halved, the *original* preference is established by toads

small **b** > small **d** (3)

at least for this particular pair of stimulus configurations. The model failed to conform to this situation.

What went wrong with the model? Basically, the model predicts that the dishabituation hierarchy changes with stimulus size. One reasonable explanation for the failure of the model would be that toads may exhibit the same dishabituation hierarchy within a certain range of stimulus size. To test this conjecture of *size invariance*, Wang et al. (1990) tested the dishabituation ordering for toads confronted with two different size stimuli of the same configuration, namely **b**. It was found that

big **b** > small **b** (4)

in the dishabituation ordering (though no obvious difference in strength was found between big **b** and small **b** in releasing prey-catching behavior). No mutual dishabituation was found.

Further experiments tested dishabituation preference between small **b** (10x2.5 mm^2) and big **f** (20x5 mm^2). Two effects take place in this particular pair. Based on the *size* effect shown (4), one might expect that big **f** has a preference to small **b**, but from the perspective of *configurational* cues one would expect shape **b** to have preference to shape **f** as demonstrated in Fig.1. What actually happened is

small **b** > big **f** (5)

The only explanation we can offer based on the results in (3), (4), and (5) is that visual object discrimination in toads is somewhat unaffected by object size. This poses an important challenge for future extensions of our model: to explicitly represent the interaction of size cues with the preference ordering to explain how the toad can exhibit a measure of *size invariance* in the dishabituation hierarchy.

4. A Model for Habituation and Dishabituation

We have demonstrated that the firing rate of a class of thalamic cells in response to different wormlike stimuli exhibits the same order as shown in the dishabituation hierarchy. We now suggest a learning mechanism for habituation suggested by synaptic mechanisms of habituation in *Aplysia*, and then report how experimental data have led to a refinement of the model.

The Role of Medial Pallium

Preliminary physiological studies show that visual telencephalic neurons exhibit long-term adaptation, while in tectal and thalamic-pretectal neurons neither neuronal adaptation nor facilitation effects lasted longer than 90 to 120 seconds (Ewert 1984). Bilateral transection of medial pallium (MP) in telencephalon destroys the aftereffects in previously habituated toads (Finkenstädt et al. 1986). In addition, both the effects of conditioning and the associative learning ability of naive animals are abolished by MP-lesion. Recently, three types of MP cells were identified by single-cell recording. Spontaneously active visually sensitive MP1 neurons strongly increase, MP2 neurons do not alter, and MP3 neurons decrease, their discharge rates in response to 30 minutes of repetitive stimulation with a visual moving object traversing their excitatory receptive fields (Finkenstädt 1989). These results plus the result from 2DG studies (Finkenstädt and Ewert 1988) support the idea that the medial pallium, the structure homologous to the mammalian hippocampus, is the site of synaptic changes underlying habituation.

Functional Modeling

We thus propose a neural model for this role of the medial pallium in habituation, structuring the model so that a stimulus higher in the hierarchy of Fig.1 can dishabituate habituation to a stimulus in the lower part. A basic component of the model is shown in Fig. 5a, motivated by the finding in *Aplysia* that synapses that were functionally inactivated due to profound habituation could be restored by a sensitizing stimulus (Carew et al. 1971). We thus show a *habituation synapse* that will decrement its efficacy as a result of repeated firing, and a presynaptic *sensitization* input whose activity will restore the habituation synapse to its full efficacy. Motivated by our earlier modeling of habituation and sensitization in *Aplysia* (Wang and Hsu 1990), we thus made such a synapse the basic ingredient in our

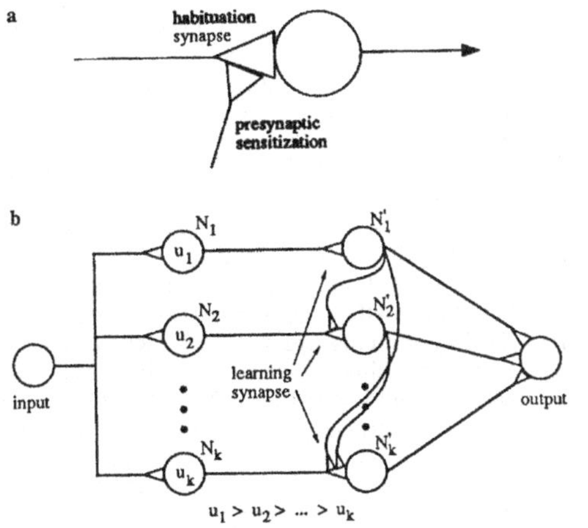

Figure 5. (a) A *habituation synapse* that will decrement its efficacy as a result of repeated firing; activity of the presynaptic *sensitization* input will restore the habituation synapse to its full efficacy (Wang and Hsu 1990). (b) A preliminary model of medial pallium. The dishabituation hierarchy is explained by having stronger input signals from anterior thalamus excite, and thus habituate, more rows of cells than do weaker signals. Recent experiments suggest that the presynaptic sensitization (which was motivated by data from *Aplysia*) is not appropriate for a model of habituation in toads.

initial model of the medial pallium shown in Fig. 5b.

The input from AT is distributed to k neurons N_1, N_2, ..., N_k with the property that they have decreasing "thresholds" $u_1 > u_2 > ... > u_k$ such that if a stimulus can trigger N_i, it can also trigger N_j for $i < j \leq k$ Thus a stimulus with a certain intensity triggers a population of cells which contains the population triggered by a less intensive stimulus. (A similar form of population coding has been proposed by Grossberg and Kuperstein 1986, pp.160-167, and Grossberg and Schmajuk 1989.) We extend the model of MP with another layer of learning neurons, with each N_i connected to N_i' $(1 \leq i \leq k)$ through a synapse which generalizes the synapse of Fig. 5a as follows: the synapse can be habituated by repeated activation, but it is also acted on by presynaptic sensitization from neurons N_1' to N_{i-1}'. We assume that dishabituation is due to a sensitization process by presynaptic facilitation, so that after habituation to a given stimulus, not only can a sufficiently stronger stimulus find a path via some N_j connected to N_j', but the activity in N_j' will sensitize any habituated synapses of N_i for $i < j$.

Back to the toad

In the model of Fig. 5b, dishabituation will erase previously learned effects of habituation, as suggested by the finding in *Aplysia* that synapses that were habituated could be restored by a sensitizing stimulus. However, *Aplysia* is not a toad. Groves and Thompson (1970), analyzing data from *mammals*, argued that sensitization, while releasing a new response, does not affect the trace of habituation. Could it be that, in *toads*, the memory trace of the previously habituated stimulus *A* survives the release of responses by the novel stimulus *B* ?

Wang and Ewert (unpublished observations) investigated the matter by testing the toad's response to a habituated stimulus soon (30 seconds) after the presentation of a dishabituating stimulus. Two black stimuli **b** and **f** against a white homogeneous background were used for this purpose. As shown in Fig. 1, black **b** has preference over black **f** in dishabituation. In the new experiment, **f** was first presented to the toad, and immediately following full habituation to **f**, stimulus **b** was presented. The experimental results are shown in Fig. 6. When the dishabituating stimulus, black **b** in this case, was presented, the toad showed a remarkable increase in the orienting response — the circle in Fig. 6 represents the average number of orienting turns for 0.5 min while all other black dots represent the number for 1 min. Yet when stimulus **f** was presented again after the toad had responded to **b** for 30 seconds, all 10 tested toads showed a very sharp decrease in the orienting response. To further test this effect, Wang and Ewert switched back again to stimulus **b** after the toad failed to release a response to the second presentation of stimulus **f**. All the toads quickly recovered the response to the second presentation of stimulus **b**. Once again, when stimulus **f** was switched back, the toads immediately stopped the response. These results clearly demonstrate that dishabituation does not erase the effects of previous

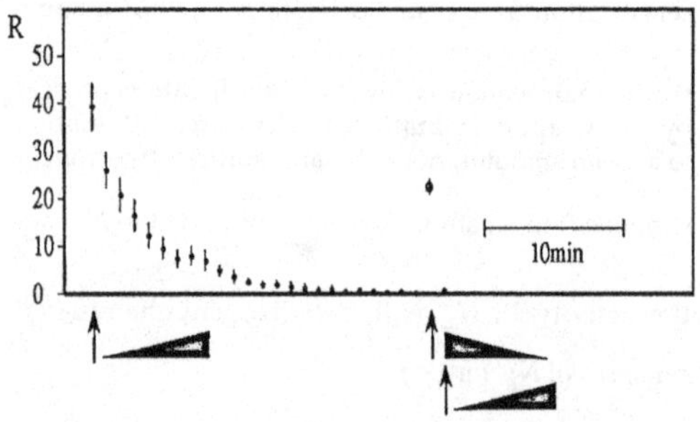

Figure 6: Data opposed to the inclusion of sensitization synapses in the model of Fig. 5b. After full habituation to **f**, the toad immediately responded to **b** (dishabituation), yet when stimulus **f** was presented again after 30 seconds, the toad showed that it was still habituated to **f**. (Wang and Ewert, unpublished observations.)

habituation, but rather has a separate neuronal process.

The striking result in Fig. 6 conforms with the dual process theory of habituation (Groves and Thompson 1970), which (not surprisingly!) makes amphibians closer to mammals in terms of learning. Therefore, the neural mechanisms previously proposed for habituation in *Aplysia* (Wang and Hsu 1990) have to be modified when modeling the toad. The ability to separate different learning traces resulting from different stimulus patterns makes it possible to study directly the toad's capacity for visual pattern discrimination. To do this, we need to study quantitatively the retention of each acquired pattern by the dishabituation method.

5. Discussion

In modeling anterior thalamus, we have suggested a pattern recognition paradigm for toads which uses intensity coding for representing different worm-like stimuli. Our analysis of the present model suggests that pattern recognition in toads is based on visual cues like leading edge, trailing edge, dots, or striped patterns. This may be related to the fact that the retinal ganglion cells of these animals respond to quite complex features of the stimulus.

It has been found behaviorally and physiologically that the edge preference (the leading edge or the trailing edge) of toads in prey catching behavior switches with the reversal of contrast direction (Burghagen and Ewert 1982; Tsai and Ewert 1987). Our model predicts that the dishabituation hierarchy also changes with respect to stimulus-background contrast, and this has been experimentally validated with respect to the particular pair of objects, b and f. The full test of the entire predicted hierarchy needs more extensive behavioral investigations.

The size variance prediction of our model (2) was not validated experimentally. The toad's discrimination of stimulus shapes as revealed by the dishabituation method is, at least to some extent, unaffected by stimulus size. Of course, toads will not exhibit prey-catching behavior if a stimulus is too big, and toads will ignore a stimulus if it is too small. Thus new results from experiments (Wang, Arbib, and Ewert 1990) challenge further modeling, and our model of toad visual pattern discrimination has to be adjusted to explain size invariance. Do toads form concepts in recognizing visual objects? To what extent have they developed size invariance? Why do toads form the specific shape preference of Fig. 1 in discriminating visual objects by dishabituation?

Finally, our preliminary modeling of habituation and dishabituation mechanisms suggests the need for neurophysiological studies of the medial pallium to test the model of Fig. 5b *as modified by the removal of presynaptic sensitization*: When a habituating stimulus (stimulus f in Fig. 6) is presented repeatedly, a population of cells is first triggered and later habituated due to the habituation synapses in Fig. 5b. Later when a dishabituating stimulus (stimulus b) is presented, a new response is released because it activates some neurons in the cell population which have not been triggered, and thus not been habituated, by the habituating stimulus. Shortly after the presentation of the dishabituating stimulus, the animal will stop its prey-catching behavior if the habituating stimulus is switched back. This is because habituation caused by the repeated presentation of the habituating stimulus is not weakened by a brief presentation of the dishabituating stimulus.

Much further work remains. Only preliminary modeling has been done (Lara 1989) to explain the fact that toads can recognize different visual objects by dishabituation or conditioning (Cott 1936; Eibel-Eibesfeldt 1952; Brower and Brower 1962). Such observations suggest the continuing fruitfulness of the toad as a subject for integrated modeling and experimentation to probe the neural mechanisms of visual pattern discrimination and learning.

References

Bailey, C.H., and Kandel, E.R., 1985, Molecular approaches to the study of short-term and long-term memory, in *Functions of the Brain* (C.W.Coen, Ed.) Oxford University Press, pp.98-129.

Brower, J.V.Z., Brower, L.P., 1962, Experimental studies of mimicry. 6. The reaction of toads (*Bufo terrestris*) to honeybees (*Apis mellifera*) and their dronefly mimics (*Eristalis vinetorum*), *Amer Natural.* 96: 297-307.

Burghagen, H., Ewert, J.-P., 1982, Question of 'head preference' in response to worm-like dummies during prey-capture of toads *Bufo bufo*. *Behav Proc.* 7: 295-306.

Carew, T.J., Castellucci, V.F., and Kandel, E.R., 1971, An analysis of dishabituation and sensitization of the gill-withdrawal reflex in *Aplysia*. *Int. J. Neurosci.* 2: 79-98.

Cervantes-Perez, F., Lara, R., and Arbib, M.A., 1985, A neural model of interactions subserving prey-predator discrimination and size preference in anuran amphibia", *J. Theor. Biol.*, 113:117-152.

Cott, H.B., 1936, The effectiveness of protective adaptations in the hive-bee illustrated by experiments on the feeding reactions, habit formation and memory of the common toad (*Bufo bufo bufo*). *Proc Zool Soc*, pp.111-133.

Eibel-Eibesfeldt, I., 1952, Nahrungserwerb und Beuteschema der Erdkrote (*Bufo bufo L.*). *Behav* 4: 1-35.

Ewert, J.-P., 1976, The visual system of the toad: behavioral and physiological studies on a pattern recognition system, in *The amphibian visual system: a multidisciplinary approach* (K. Fite, Ed.), Academic Press, pp.141-202.

Ewert J.-P., 1984, Tectal mechanisms that underlie prey-catching and avoidance behaviors in toads, in *Comparative neurology of the optic tectum*, (H. Vanegas, Ed.), Plenum, pp. 246-416.

Ewert, J.-P., 1989, The release of visual behavior in toads: Stages of parallel/hierarchical information processing, in *Visuomotor Coordination: Amphibians, Comparisons, Models, and Robots* (J.-P. Ewert and M.A. Arbib, Eds.), Plenum Press, pp.39-120.

Ewert, J.-P., Hock, F.J., 1972, Movement sensitive neurons in the toad's retina. *Exp Brain Res* 16: 41-59.

Ewert, J.-P., Kehl, W., 1978, Configurational prey-selection by individual experience in the toad *Bufo bufo*. *J Comp Physiol* 126: 105-114.

Finkenstädt, T., 1989, Stimulus-specific habituation in toads: 2DG studies and lesion experiments, in *Visuomotor Coordination: Amphibians, Comparisons, Models, and Robots*, (J.-P. Ewert and M.A. Arbib, Eds.), Plenum, pp.767-797.

Finkenstädt, T., Adler, N.T., Allen, T.O., and Ewert J.-P., 1986, Regional distribution of glucose utilization in the telencephalon of toads in response to configurational visual stimuli: A 14C-2DG study. *J.Comp.Physiol.* 153:99-110.

Finkenstädt, T., and Ewert J.-P., 1988, Stimulus-specific long-term habituation of visually guided orienting behavior toward prey in toads: a ^{14}C-2DG study. *J Comp Physiol* A 163: 1-11.

Grossberg, S., and Kuperstein, M., 1986, *Neural dynamics of adaptive sensory-motor control*. North-Holland.

Grossberg, S., and Schmajuk, N.A., 1989, Neural dynamics of adaptive timing and temporal discrimination during associative learning, *Neural Networks* 2: 79-102.

Groves, P.M., and Thompson., R.F., 1970, Habituation: A dual-process theory. *Psychol. Rev.* 77: 419-450.

Lara, R., 1989, Learning and memory in the toad's prey/predator recognition system: A neural model, in *Visuomotor Coordination: Amphibians, Comparisons, Models, and Robots* (J.-P. Ewert and M.A. Arbib, Eds.), Plenum Press, pp.833-855.

Lara, R., and Arbib, M.A., 1985, A Model of the Neural Mechanisms Responsible for Pattern Recognition and Stimulus Specific Habituation in Toads, *Biol. Cybern.*, 51:223-237.

Lázár, Gy., Toth, P., Csink, Gy., and Kicliter, E., 1983, Morphology and location of tectal projection neuron in frogs: a study with HRP and Cobalt-filling, *J Comp Neurol* 215: 108-120.

Neary T.J., and Northcutt, R., 1983, Nuclear organization of the bullfrog diencephalon. *J Comp Neurol* 213: 262-278.

Scalia, F., and Colman, D.R., 1975, Identification of telencephalic efferent thalamic nuclei associated with the visual system of the frog. *Neurosci. Abstr* 1: 65.

Scalia, F., and Gregory, K., 1970, Retinofugal projections in the frog: location of the postsynaptic neurons. *Brain Behav Evol* 3: 16-29.

Sokolov, E., 1975, Neuronal mechanisms of the orienting reflex, in *Neuronal mechanisms of the orienting reflex*, (E. Sokolov and O.Vinogradova, Eds.) Erlbaum, pp.217-235.

Teeters, J.L., 1989, A simulation system for neural networks and model for the anuran retina. Technical Report 89-01, Center for Neural Engineering, University of Southern California.

Tsai, H.J., and Ewert, J.-P., 1987, Edge preference of retinal and tectal neurons in common toads (*Bufo bufo*) in response to worm-like moving stripes: the question of behaviorally relevant "position indicators", *J Comp Physiol* A 161: 295-304.

Thompson, R.F., and Spencer, W.A., 1966, Habituation: A model phenomenon for the study of neuronal substrates of behavior, *Psychol Rev* 73: 16-43.

Wang, D.L., and Arbib, M.A., 1990, How does the toad's visual system discriminate different worm-like stimuli? *Biol.Cybern.* (in press).

Wang, D.L., Arbib, M.A., and Ewert, J.-P., 1990, Visual Pattern Discrimination in Toads: A Dialog between Modeling and Experimentation, in *Visual Structures and Integrated Functions*, (M.A. Arbib and J.-P. Ewert, Eds.), *Research Notes in Neural Computing*, Springer-Verlag.

Wang, D.L., and Hsu, C.C., 1990, SLONN: A simulation language for modeling of neural networks. *Simulation* 55: 69-83.

The Computational Hoverfly; a Study in Computational Neuroethology

Dave Cliff
University of Sussex School of Cognitive and Computing Sciences
Brighton BN1 9QN, England, U.K.
E-mail: `davec@uk.ac.sussex.cogs`

Abstract

Studies in computer vision have only recently realised the advantage of adding a behavioural component to vision systems, enabling them to make programmed 'eye movements'. Such an *animate vision* capability allows the system to employ a nonuniform or *foveal* sampling strategy, with gaze-control mechanisms repositioning the limited high-resolution area of the visual field. The hoverfly *Syritta pipiens* is an insect that exhibits foveal animate vision behaviour highly similar to the corresponding activity in humans. This paper discusses a simulation model of *Syritta* created for studying the neural processes underlying such visually guided behaviour. The approach differs from standard "neural network" modeling techniques in that the simulated *Syritta* exists within a closed simulated environment, i.e. there is no need for human intervention: such an approach is an example of computational neuroethology.

1 Introduction

For an animal (or an autonomous robot) to adapt and survive in uncertain environments, a sense of sight is undoubtedly a useful thing to have. The creation of seeing machines has been the topic of much research in artificial intelligence and computer vision. Unfortunately, most such research has ignored the behavioural contexts in which natural vision occurs. Recently, a research paradigm known as *animate vision* has emerged, where the seeing machine is given the ability to make programmed 'eye movements', allowing it to look around: animate vision acknowledges the behavioural contexts of natural vision.

One such context is the need to control gaze when the image sampling strategy uses nonuniform resolution, i.e. where only a restricted area of the field of view is high-acuity, as in foveal vision commonly found in predatory animals. Foveal vision offers a number of advantages for any real-time visual system, whether artificial or natural (robot or animal). Animate foveal vision is not an exclusively mammalian trait. The hoverfly *Syritta pipiens* exhibits animate foveal vision behaviour which is remarkably similar to corresponding behaviour in humans.

The research project described here is a 'neural network' simulation-model study of the mechanisms underlying animate vision in *Syritta*. The study of the neural basis of behaviour is properly the domain of the field of neuroethology, and this project (relying as it does on computer simulation) is thus a form of "computational neuroethology".

The simulation models the hoverfly living in a closed dynamic environment: activity in the model nervous system is expressed as flight behaviour of the model fly, which in turn generates new visual input for the simulated eyes, which feed the model nervous system, thus completing a visual feedback loop from sensorimotor output to photoreceptor input. The model nervous system is not 'hard-wired' but is created using techniques from current 'connectionist' network learning theory and from adaptive filter theory. The method by which the network is created is incidental: it is the capability of the mature network that is of interest. A number of different strategies for creating the network are being explored; the optimal solution (in the engineering sense) is not necessarily the most biologically interesting.

This paper gives a brief review of the past research on which this project builds, and then presents an overview of the simulator system. The project draws on literature in a number of fields: a more complete account of the background literature and the simulator is given in [13], which this paper is abridged from. Computational neuroethology [12] is discussed here only in passing.

The work is at an early stage. So far the most significant results are in the design of the simulator, especially the generation of the view through *Syritta*'s eyes.

2 Rationale

Work of this nature does not have a long academic pedigree. Therefore, the notes below concentrate on a few papers in some depth, rather than superficially skimming many. The animate vision paradigm is described, followed by a discussion of nonuniform sampling. The argument underlying the interest in insects is then reviewed.

2.1 Animate Vision

These notes on animate vision are based on the publications of Ballard (e.g. [3, 4]) and Brown (e.g. [7]), concerning the Rochester Robot [8]. The definition of animate vision given below is, however, my own interpretation and therefore possibly not wholly representative of the Rochester view.

The term "animate vision" denotes visual processing performed in circumstances where the image-formation apparatus is a subsystem, whose parameters are under dynamic control of the overall visual processing system. In nature, the image-formation apparatus consists usually of two eyes, linked to musculature allowing for dynamic resampling of the optic array, i.e. seeing creatures are capable of looking around. It is the capability of looking around that distinguishes animate vision from previous work in computer vision, where degrees of freedom associated with the camera (e.g. focus, position in space) are rarely under dynamic control. Ballard [4, p.1635] states: "We term the collection of different mechanisms for keeping the fovea over a given spatial target *gaze control*. The single most distinguishing feature of an animate vision system is high-speed gaze control mechanisms."

The rationale underlying animate vision is that one should first question *why* we want to use vision: "...*animate vision* argues that vision is more readily understood in the context of the visual *behaviors* that the system is engaged in, and that these behaviors may not require elaborate categorical representations of the 3-D world." [4, p.1635, original emphasis].

Moreover, the aim in animate vision is not simply to mimic human visual behaviour: there are cogent computational reasons for adopting the animate approach in preference to the conventional static methods. These advantages are discussed in detail in [3] and summarised in [13]. The advantages are dependent on the ability to control gaze, and they are negated if gaze control itself is found to be prohibitively difficult. Results so far indicate that this is not a devastating problem.

Brown [7] has attempted to solve the complications introduced by time delays and interactions of gaze control subsystems by using signal synthesis adaptive control. Brown's main interest is in artificial (robotic) gaze-control, to which he finds such methods well suited; yet his results are often consistent with physiological data.

2.2 Nonuniform Sampling

Natural animate vision is, to a certain extent, a solution to the problems caused by employing a nonuniform sampling strategy. That is, if the imaging resolution is higher in one area of the field of view than in the remainder, then the ability to control the view-direction of the high-resolution area is highly advantageous.

The overwhelming majority of work in computer vision employs uniform-resolution image acquisition (URIA), where the imaging resolution of all pixels in the image is assumed to be identical; commonly the pixels are arranged on a rectangular grid.

Nevertheless, interest is growing in non-URIA approaches, where the imaging resolution is in inverse proportion to the distance from the central point of sampling. Such an arrangement is found in the human visual system, where the central high-resolution area of the retina is named the *Fovea centralis retinae*. Thus, non-URIA vision is commonly referred to as "foveal" vision.

Young's work [28] indicates that there are significant computational advantages to be gained when using a non-URIA scheme. He works with a technique referred to as the *logarithmically sampled image* (LSI). Young demonstrates a number of useful properties of the LSI which arise because it allows for simple representation of image transformation about the centre of sampling. One notable property is the ease of straight line detection. Young demonstrates that the technique can be extended to discover the transformation that best matches one arbitrary image to another.

Nevertheless, these advantages of the LSI are largely counteracted in practice by the restricted area of high resolution. Young [28, pp.149–150] remarks that such problems would be obviated by linking LSI processing to an animate vision component, thus providing a resampling mechanism; strategies resembling saccadic eye-movements would be required for analysis of static images, while analysing image sequences will require movements akin to tracking.

2.3 The Antianthropocentric Context

The animate and non-URIA paradigms share a common feature: they draw inspiration from psychophysical and biological studies of human (or more generally, primate) visual processing. This marked preference towards human visual processing is, however, an unfounded bias: visual processes at the levels addressed by these paradigms are not species-specific. The anthropocentricity of these theories cannot be justified. Specifically, there is a functional isomorphism between the visual behaviours of humans and certain invertebrates. This isomorphism was described by Land [18].

2.3.1 Land's Analogy

Land considers four types of visually mediated behaviour, selected because they "...are sufficiently elemental and common for comparisons between species to be easily interpreted." [18, p.51]. These four are:

Keeping still. When the image of the background slips across the retina because of (unintentional) movement of the animal, counter-rotations are evoked in order to stabilise the animal.

Visual Search. Despite the mechanisms for keeping still, animals are capable of "looking around", i.e. making voluntary turns without an obvious stimulus.

Target Location. A specific target (i.e. important stimulus), appearing in the visual field at a nonzero eccentricity (i.e. away from the centre of the field of view), causes a turn of the correct direction and amplitude to make the stimulus appear directly ahead (i.e. at zero eccentricity).

Target Tracking. Once a target is located at near-zero eccentricity, it often needs to be tracked against a dynamic background. Smooth tracking movements are made in order to minimise the target eccentricity over time: a kind of "target lock" mechanism.

These four visual behaviours relate to human eye movements (see e.g. [27]) in the following manner:

- Optokinetic nystagmus[1] (OKN) and the vestibulo-ocular reflex[2] (VOR) are aspects of the human mechanism for keeping still.
- Saccadic movements[3] correspond to target location.
- Smooth tracking[4] corresponds to target tracking.
- Visual search is effected in humans using combinations of saccadic movements and smooth tracking, which will most likely have to override the counter-movements evoked by OKN and VOR.

Land's argument centres on studies [15] of the male *Syritta pipiens* hoverfly. In terms of the four movement types listed above, male *Syritta*'s animate vision is a good analogue of human animate vision, but with fewer degrees of freedom: the anatomy of *Syritta* is such that movement of the body corresponds to eye movement. This is because the eyes are fixed to the head and there are no significant neck movements: body orientation is thus equivalent to view direction.

Syritta's eyes produce non-URIA images (resolution at the centre of the field of view is roughly three times better than that at the periphery), with an area of binocular overlap; *Syritta*'s retinal angle of acceptance is around 280–300°. Also, the number of distinct photoreceptors in *Syritta*'s eyes is less than half the 65536 pixels in a 256 × 256 image (a standard size in computer vision), whereas the human eye has about 127 million receptors.

[1]The involuntary following by the eyes of any large moving pattern.

[2]Eye movement in the opposite direction of head movement.

[3]High speed relocations of direction of gaze.

[4]Slow and accurate following of a target by the fovea.

Syritta flight behaviour has been described by Collett and Land [15]. The hoverfly can maintain stationary hovering for extended periods of time; it flies at any angle to the long axis of the body; and it can fly in directions with zero or negative forward components (i.e. sideways or backwards). While some of *Syritta*'s behaviour is concerned with activity in the vertical, the interesting behaviour occurs only horizontally. Collett and Land recorded only three degrees of freedom of flight: yawing angular velocity; forward velocity, i.e. horizontal velocity in the direction of the fly's long axis (head-to-tail); and sideways velocity normal to the long axis.

The reduced number of degrees of freedom, and especially the small number of receptors make *Syritta* particularly suitable for the study of low-level non-URIA animate vision via simulation. Ballard contends that simulation models of animate vision processes should be rejected in favour of robotic systems because vision simulators cannot operate in real-time; yet surely this is a problem of engineering, not science.

So, study of *Syritta* allows for examination of a subset of issues in primate eye movement (and hence also animate vision), along with a convenient reduction in the number of degrees of freedom.

2.4 Summary

The particular attraction of insects in the context of this project is their amenability to study by simulation. The comparatively small number of units in *Syritta*'s eye makes synthesis of *Syritta*'s retinal image on a pixel-per-receptor basis a realistic possibility. This makes *Syritta* particularly appealing for the study of computational 'neural' network models of animate vision with nonuniform sampling. Additionally, the Collett and Land studies of *Syritta* provide a significant body of data against which the success of the simulation can be judged.

The construction of the network simulation nevertheless raises a number of issues. The dynamic nature of animate vision makes motion processing a core topic, and at higher layers there is the issue of distinguishing between self-generated image motion and that which results from unexpected movements. Also, the problem of creating a network to dynamically control movement, particularly if the controller is to be adaptive, is a current research issue. The background literature to these issues is briefly reviewed in the next section.

3 Theoretical Foundations

This section examines the literature on which this project will build. A selection of work on insect vision (the optomotor response and motion detection) is discussed, followed by an overview of relevant recent work in 'neural network' theory. The notes in this section are necessarily brief, see [13] for further details.

3.1 The Optomotor Response

The optomotor response is the classic visually induced behaviour studied in flying insects. It is a closed-loop visual control system, acting to maintain a stable path by correcting unexpected movements of the body. The main cause of such movements is turbulence in the air, but the system is also operative in walking insects.

On travels of reasonable distances (and assuming the fly is not flying sideways), minor *translational* deviations will have a much less significant effect than minor *rotational* deviations, which can lead to the insect flying round in circles. A precise method for detecting and correcting against unplanned rotations about the three axes (yaw, roll, pitch) is a necessity. This function is performed by the optomotor response.

The vast majority of all experimental work on the optomotor response has studied only one rotational dimension at a time: the yaw response is most commonly studied. It is generally assumed that results for yaw can be applied to pitch and roll, so the discussion below is phrased purely in terms of yaw.

Figure 1 illustrates apparatus for a typical optomotor-yaw experiment. The fly is secured on the axis of a large cylinder which has a pattern on its internal wall; often this consists of vertical stripes. The cylinder is motorised and can be rotated, independent of the fly. The placing of the fly on the axis of the cylinder ensures that the optic flow field (i.e. image slip across the retina) is uniform. The fly is monitored by apparatus which enables its yaw torque to be measured.

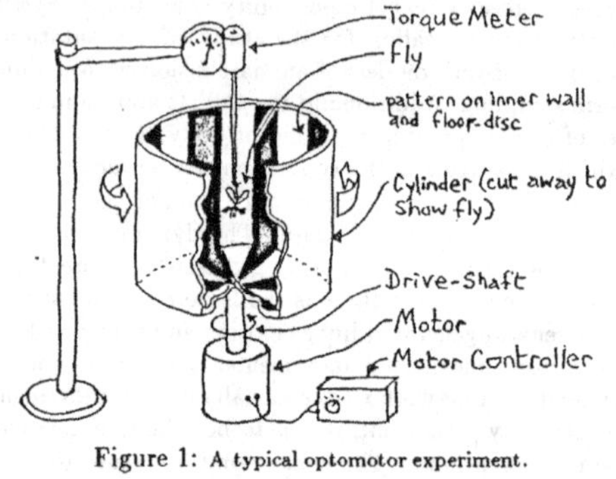

Figure 1: A typical optomotor experiment.

The significant result of experiments with such apparatus is that, as cylinder rotates, the fly turns in the same direction and at approximately the same speed. This turn attempts to reduce the movement of the pattern across the fly's retina to a minimum.

The uniform optic flow is, to the fly, an unambiguous indicator of self-motion. The fly interprets this as indicating that it is unintentionally rotating in a stationary environment, and it turns in order to restore its previous orientation.

For the fly to do this, the optic flow has to be interpreted as movement, i.e. the fly's visual system must detect motion. We will assume here that there are neurons, or groups of neurons, which act as motion detectors. The motion detectors report on speed of motion in their preferred direction.

So, the optomotor response poses two main questions: how are the outputs of motion detectors combined in order to provide information on the sign and magnitude of different motion components (i.e. rotation and translation); and how is the optomotor response overridden to allow the fly to make *intentional* rotations? Both of these are open questions, despite many plausible hypotheses. Motion detection is discussed in the next section, the notes below discuss the control-override issue.

The optomotor response does not block intentional turns. There is no conclusive explanation of this mechanism, but there is a widely accepted hypothesis known as the efference copy hypothesis (ECH hereafter), commonly attributed to von Holst and Mittelstaedt (see [11, p.342]). No studies have provided experimental verification of the ECH [14], and several alternatives to the ECH have been proposed.

A radical alternative has been tentatively proposed by Bruce and Green [9, pp.248–251]. While the ECH proposes a hierarchical control strategy, the Bruce and Green model is more heterarchical, lacking a "higher centre" issuing control commands. They propose that the optomotor response is overridden by competition from a number of other regulatory subsystems, which switch into and out of control under the influence of environmental, behavioural, and random factors: they argue that perception should not be treated in isolation from action. As such, their model is more in tune with the subsumption architecture style of computational neuroethology discussed in [12]. Subsumption architecture controllers were originally proposed by Brooks [5].

Brooks proposes that the primary decomposition of the control problem should be into *task achieving behaviours*. That is, he advocates decomposing behaviourally complex systems into a number of computational systems achieving simpler behaviours.

The intention in this project will be to attempt to effect each task-achieving behaviour using small artificial neural networks. A network model constructed according to a subsumption architecture is an explicit recognition of the inhomogenous nature of natural nervous systems.

3.2 Visual processing: noise and motion

Visual processing in insects has been studied from a number of aspects. This project concentrates on two aspects, motion detection and filtering against noise, which have previously been treated largely as separate issues. A unified treatment of the two is a major aim of this project.

3.2.1 Noise filtering

Any nervous system is subject to intrinsic noise (spontaneous discharges, fluctuations in spike frequency or neurotransmitter concentrations at synapses, etc.). Visual perception is further affected by noise introduced at the transduction stage by the photoreceptor. Recent work by Srinivisan, Laughlin, and Dubs [25] provides a persuasive argument for the case that the centre-surround antagonism in the receptive fields of visual interneurons (i.e. lateral inhibition) is best explained as a safeguard against noise.

This view challenges the previously suggested explanations for lateral inhibition, e.g. that it cancels DC bias in the signal, or that it is a filter for enhancing edges.

The advantage of this approach is that it is based on statistical estimation theory, using linear mean-squared estimation to derive the receptive fields for receptors in both the spatial and temporal domains. The receptive fields so arrived at explain both lateral inhibition and the phasic responses of fly interneurons. The importance of the statistical basis is that it quantitively explains the spatiotemporal receptive profile of fly interneurons without recourse to *a priori* notions of which components of the signal are most important for subsequent processing: the suggestion is that the early stages of visual processing are concerned primarily with protection against noise, rather than extracting specific image features.

3.2.2 Motion Detection

Motion detection and processing has received intense study: there is a lot of literature on this topic, and the notes that follow provide only a cursory listing of the past research relevant to this project.

Probably the most widely known model of retinal motion detection is the long-established Reichardt model (see e.g. [20, p.163]): this performs a local correlation (i.e. multiplication) across space and time.

The Reichardt detector has been extended by van Santen and Sperling [26], whose model is referred to as the elaborated Reichardt detector (ERD). It computes the temporal covariance of the input, and its creators demonstrate that, given the right filters, it is equivalent to several different types of detectors proposed by others as alternatives to the Reichardt model.

Adelson and Bergen [1] discuss an alternative approach which views motion detection as orientation-detection in 3D ($x-y-t$) space. Their detector extracts motion information by responding to oriented spatiotemporal energy. This 3D-orientation nature of the energy model bypasses explicit considerations of correlation or correspondence, but is equivalent to the ERD. Advantages of the energy model over the ERD include: that its components are consistent with known (human) physiology and psychophysics; and that it is composed of linear filters, thus offering the tractability of linear systems theory.

The Reichardt model was originally proposed as a motion-detector in insect vision [16], but the ERD and the spatiotemporal energy model are offered primarily as models of human motion detection. Further recent studies of human motion processing by Mather *et al.* [21] have indicated that the human system employs gradient-based motion detectors (which compare spatial and temporal luminance gradients): their results indicate that the Reichardtian model is, at least for human motion processing, implausible.

As far as I know, no investigation has yet been made of the validity of the Reichardtian model for insect motion detection in light of Mather's results for humans. Yet the phasic temporal response of fly visual interneurons explained by predictive coding theory provides a signal on which a spatial gradient detector could operate. The phasic response effectively provides a temporal derivative of the image, which decays over time: areas of the image which are static or low-frequency in the spatial domain would largely be filtered out; any moving high-frequency (edge) components would leave 'smears' on the image, with the spatial extent of the smear being proportional to the magnitude of the velocity component normal to the edge.

3.3 Neural networks: control and learning

The past decade has seen an explosion of interest in so-called 'connectionist' models of processing, which will for the purposes of this paper be assumed to be synonymous with 'neural-network' and 'parallel distributed processing' models.

There are two aspects of connectionist literature that are particularly germane to this project: control and learning. The construction of a model neural control system for the optomotor response is hoped to elucidate details of the neural basis of optomotor control in *real* nervous systems. The application of connectionist techniques to control problems typically focuses on *adaptive* controllers which adjust internal parameters in response to error signals, thus dynamically improving performance. Most work in such connectionist control research pays little attention to biological reality, either using the crudest references to neural structure or biologically indefensible learning algorithms such as back-propagation [24, ch. 8] with no attempt to relate the resultant mature networks to biological reality.

A disregard for biology is acceptable if the purpose of the neural model is to produce an engineering solution to the problem at hand, but the aim here is for a scientific study of real nervous systems via simulation. Therefore, the architecture of the model should be a close reflection of biological reality, and if a biologically implausible learning algorithm is used to create an adaptive network, then the mature network should at least be compared to relevant neurological results from studies of real nervous systems.

But this begs the question as to why poorly understood neural-network learning algorithms, such as back-propagation, should be used when there are well-established techniques that are more powerful or more general and are better understood. Increasingly, the results of learning algorithms or network architectures originally proposed for 'neural-network' modeling are being explained and extended by reference to previous work in long-established fields of mathematics (e.g. [23]). Of particular interest to this project are learning techniques based on adaptive filter theory [17].

Such techniques have been demonstrated to provide near-optimal results for simple learning tasks [19]. Paulin, Nelson, and Bower [22] applied related techniques to analysis of human VOR eye movement control, and discovered that the *a priori* assumptions of what constituted 'ideal' control (i.e. unity gain and 180° phase shift) in previous models were non-optimal in the sense of stochastic optimal estimation theory (Kalman filters). More significantly, they discovered that the optimal-estimation model qualitatively reproduced experimental results that classical control models failed to account for. That is, what were previously thought to be shortcomings actually constitute a sophisticated control strategy.

This use of statistical estimation theory to analyse a system and produce a model which better accounts for biological data is very similar in nature to the work, discussed above, of Srinivisan *et al.* on retinal processing and also of Brown on controlling animate vision subsystems. It appears that the adaptive filter approach offers more promise of biologically relevant results than any other current approach to network learning.

4 Approach

The preceding notes have demonstrated that the optomotor response, despite its apparent simplicity, provides a context for the study of a number of issues at different levels. At the lowest level, there is the problem of how motion detection is effected and acted upon. At a higher level, there is the control issue of how the response interacts with intended motion. It is these two issues that this project is currently examining.

This section gives a description of the simulator, and elaborates on the research being conducted. It is important to note that the simulator is not the sole product of this research: the project includes a theoretical analysis of the issues addressed. Nevertheless, the simulator is emphatically more than a pretty graphics system. As Buchner notes, the two central questions of insect visual physiology are: "*What* does an insect see when it walks or flies around? And *how* does its visual system abstract the relevant information from the light signals in the photoreceptors?" [10, p.562]. The simulator goes quite some way towards answering the first question, and in this respect the construction of the simulator may be viewed as a result in itself. Further implications of this simulation approach are discussed at length in [12].

The simulator is primarily a model of male *Syritta pipiens* in that the optics of *Syritta* are used to create the simulated fly's eyes. For this reason, the simulated fly is referred to as *Syritta computatrix*[5], the computational hoverfly, or SYCO for short[6]. However, results from studies on other insects have also influenced parts of the design of SYCO.

The essence of the simulation is a model of *Syritta*'s optics, situated in a dynamic visual environment. The model of the optics is used to sample the visual environment, creating simulated retinal input. This retinal input feeds a flight control system (FCS) which is responsible for detecting and interpreting motion in order to effect the optomotor response. The output of the FCS alters the position of the simulated fly's eyes in the environment, thus closing the visual feedback loop. The model environment includes turbulence and wind, which must also be overcome by the FCS. The simulator is much like commercial flight-simulator systems, with the exception that there is no human control in the SYCO simulator: all control signals are generated by the FCS.

The FCS is where the 'neural network' resides. The intention is to use the neural-network learning techniques discussed above to allow the FCS to adapt to a variety of types of stimulus. The mature networks should detect motion and be capable of generating an optomotor response with override.

Only the yaw response (panning) is studied here, both alone and in interaction with translational movement.

Implementation details are not given here. Conceptually, the simulator is composed of two parts: the FCS, which can be considered for the moment as a black box; and everything else, which is referred to as EBB (from everything-but-the-box). Figure 2 shows a schematic diagram of the simulator, introducing the terminology; all boxes other than FCS are components of EBB.

The notes that follow discuss the simulator. First the components of EBB are described, then work to date on the FCS is discussed.

[5] The nomenclature was inspired by Arbib's computational frog *Rana computatrix* [2].

[6] "*Syritta*" here refers solely to *S. pipiens*: "SYCO" abbreviates *S. computatrix*.

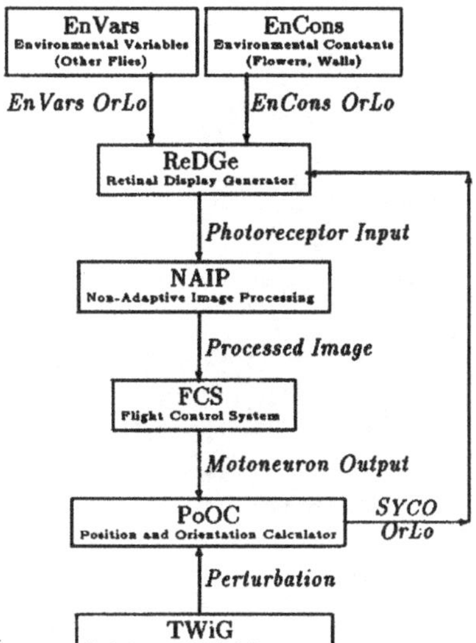

Figure 2: Schematic diagram of the SYCO simulator.

Figure 3: View from inside the cylinder, about one third of a radius from the centre and at an altitude about one quarter of the vertical extent of the cylinder. The angle of view corresponds approximately to that given by a 17mm lens on a 35mm camera.

4.1 EBB

4.1.1 The environment

The environment within which SYCO resides is completely under experimental control: the simulator has been constructed to allow for arbitrarily complex visual environments to be defined. A primary distinction is made between animate and inanimate objects in the environment. Animate objects are referred to as EnVars (environmental variables), and inanimate (stationary) objects as EnCons (environmental constants).

Despite the potential for complexity, the simulator has so far only been used for studies of flight activity in visually impoverished environments, similar to those used in the conventional optomotor experiments discussed above. All experiments to date have used cylindrical walls, which can be rotated (linearly or periodically) relative to the fly. Visual patterns are mapped onto the internal surfaces of the cylinder. The patterns used are rectangular gratings and random-dot mosaics. Figure 3 shows an image traced using conventional optics from inside the cylindrical environment with a rectangular grating on the cylindrical surface extended to a radial pattern on the floor-disc; the ceiling-disc is blank white.

4.1.2 ReDGe

The retinal display generator, ReDGe, simulates *Syritta*'s 300° non-URIA sampling of the optic array as closely as possible. Full details of the eye-model are given in [13]. Figure 4 illustrates the view through the SYCO ReDGe eye-model.

4.1.3 NAIP

The NAIP module implements non-adaptive image processing. It can be considered as a model of the processing performed by the first few layers of nervous tissue in the fly. In reality these initial layers apply massively parallel processing to the retinal image. This is simulated by using conventional image processing techniques, on images produced by ReDGe. NAIP models the temporal smearing introduced as a result of the time-course of the photoreceptors, and convolves the smeared image with a mask which represents the effects of lateral inhibition.

This processing is nonadaptive in the sense that in all experiments with the simulator, the parameters of the receptor time-course and lateral inhibition cannot be altered by the learning process (adaptation). However, they may be altered between experiments in order to gauge the sensitivity of the FCS to these parameters.

Figures 5 and 6 illustrate the operation of NAIP on a view from inside a rotating cylinder. Figure 6 clearly shows stripes 'smeared' to the right. This is indicative of edges moving to the left, and is the sort of image feature which the spatial-gradient based motion detector (based

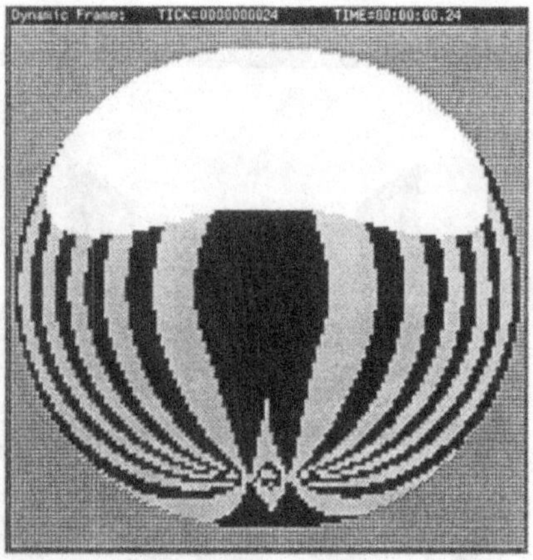

Figure 4: View through the SYCO ReDGe eye-model; location in cylinder is similar to that of the previous figure. Resolution of this image is approximately that of *Syritta*'s eyes. The centre of the floor disc appears twice because it is within the binocular area of the field of view. The increased resolution towards the centre of the image leads to the 'bulging' appearance of the walls.

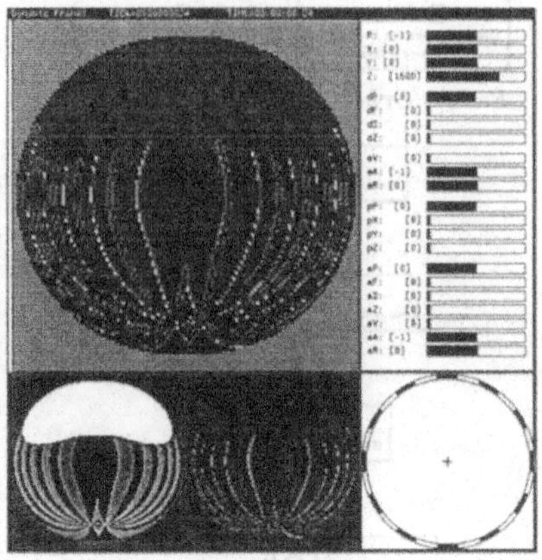

Figure 5: One frame from a SYCO movie. Upper-right window is system variables. Lower-right window is plan view (fly on cylinder axis). Bottom-left window is retinal image, i.e. input to photoreceptors. Bottom-centre window is photoreceptor output, i.e. decaying time-smeared image: grey represents zero; white is positive excitation; black is negative. Main (top-left) window is photoreceptor output after lateral inhibition.

on Mather's work discussed above) operates on. Such smears are not an ***unambiguous*** indicator of motion: results from lateral inhibition (spatial) processes have to be combined with those from the smearing (temporal) processes in order to unambiguously extract motion information.

4.1.4 PoOC and TWiG

The output from the FCS is a vector specifying the angular and translational velocities of the fly. Such a representation was chosen in the first instance for its amenability to comparison with Collett and Land's original data for *Syritta*.

The Position and Orientation Calculator (PoOC) uses the FCS output vector to update the fly's OrLo (orientation and location in the world). In the current implementation, the vector is simply added to the previous OrLo, yielding the new OrLo. It is when the OrLo is updated by PoOC that the clock ticks, i.e. the new OrLo signifies the transition from time t to time $t + \Delta t$.

The Turbulence and Wind Generator (TWiG) supplies a perturbation vector which is also added in by PoOC when creating the new OrLo. The vectors supplied by TWiG in a simulation session may be random (i.e. turbulent air), or may be more structured (e.g. a constant cross wind), as required.

4.1.5 Discussion

The simulator is designed according to the principles of computational neuroethology discussed in [12]; three of the most salient issues are summarised here.

- The simulator recognises the views of Maturana and his colleagues, that a nervous system is not best characterised as separate from the environment in which it resides. They argue that an entity exists not as a separable object inside some environment but as a *part* of the environment. Such a view implies that nervous systems modeled *in vacuo* (as is traditional in computational neuroscience) are missing the point, treating closed systems as input-output devices.

- A possible criticism of SYCO is that the simulator is nothing more than a 'toy' world. This argument is seductive because the use of a visually impoverished environment, such as that found in optomotor experiments, bears a superficial similarity to the conceptually impoverished environments created for (e.g.) blocks-world reasoning. This argument is refuted at length in [12]: basically, the SYCO simulator is not a toy world because the simulated fly is not reliant on human intervention to either help it understand the world or to interpret its responses, thus satisfying Brooks's definition of a 'real' world [6, p.5].

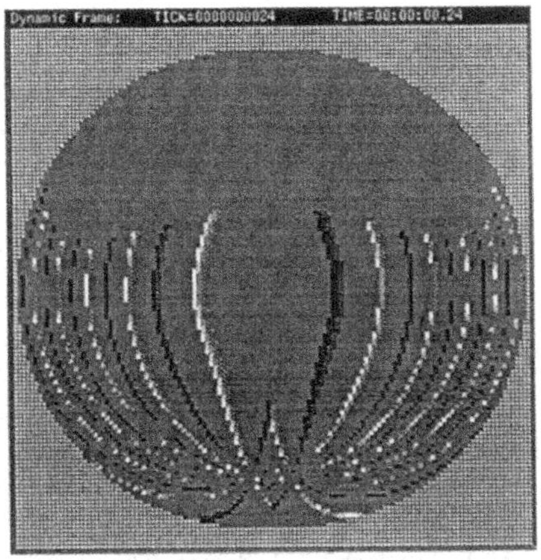

Figure 6: Enlarged version of the photoreceptor output (bottom-centre window in previous figure).

- The crucial aspect of the simulator is that activity on FCS outputs at time t will have a direct, predictable effect on the FCS inputs at time $t+\Delta t$. Thus the environment is "closed" in the sense that the FCS will have the ability to control aspects of optic flow over time (other aspects, such as image flow due to other flies or turbulence, are not under FCS control). The use of a closed-environment simulator offers well-grounded semantics for the symbols in the model. That is, study of activity in the FCS is a matter of (objective) *observation* and not (subjective) *interpretation*: the concept of simulated retinal input is not new in connectionist modeling, but the closing of the loop from output to input would appear to be novel.

4.2 FCS

The intention is to create trained networks which can replicate, to some degree of accuracy, the classic optomotor reflex and target-tracking behaviour of real flies. Trained networks can operate as closed-loop servo controllers in the hoverfly simulator; their outputs will represent the degrees of freedom of flight used in gaze-control by hoverflies. This allows for direct objective comparison with natural hoverfly activity: an unusual feature in research of this type. To date, no completely successful FCS network has been formulated.

The approach in the first instance was to employ the "backprop" network learning algorithm [24, ch. 8] to create a network which took as input single images from NAIP and output signals representing the direction and speed of the corresponding optomotor-yaw response. The intention was that the network derived this information from the extent to which high-frequency components in the ReDGe image were time-smeared by NAIP, i.e. a form of spatial gradient-based motion detection, similar to Mather's work discussed above. Quantisation noise introduced by the relatively low resolution of the images limited the operational velocity range of such a technique. Also, the high dimensionality of the input space meant that even the most simple of networks required extensive training: the results on the few occasions that the network converged were poor. In most cases direction was reported correctly, but speed was never adequately interpolated.

Unsupervised learning is yet to be explored. The current approach is to employ adaptive filter techniques on small subnetworks which will be integrated within a layered control system [6]. This work is at an embryonic stage and there are no conclusive results to report.

5 Conclusions

Animate foveal vision offers significant benefits for the construction of robotic systems. The intuitive inclination is to study naturally occurring animate foveal vision in humans and other primates, but the antianthropocentric approach offers a much simpler model which captures the essential features.

The SYCO project currently focuses particularly on the neural mechanisms underlying motion detection and optomotor control. The use of a closed-environment simulator offers all the advantages of the 'real world', but with less complexity and hence increased tractability.

No claims of biological validity are attempted in justifying the use of backprop or adaptive filter theory to create the FCS. Rather, these techniques are employed as a useful method of searching for high-order statistical correlations in the input-space of the network (cf. [23]). The units found in the mature trained networks will then be compared to known neurological data concerning visual guidance of flight in flies. That is, it is the end result of learning that is the main interest. However, the techniques by which the mature networks are arrived at is of partial interest, as the work of Paulin *et al.* demonstrated [22]: that an optimal filter theory model best accounts for observed data on primate VOR eye movements indicates that the primate system is, in this mathematical sense, optimal.

Buchner's [10, p.562] question of *what* an insect sees as it flies around is, to a first approximation, answered by SYCO. The more challenging question of *how* the visual system extracts information remains to be answered.

The research possibilities offered by the SYCO simulator are only just being explored. Experiments impossible in living tissue or impracticable in robotics can routinely be performed on SYCO.

The work continues. There is much to do.

Acknowledgements

Thanks to my supervisors David Hogg and David Young, and to Glenn Kramer and Hilary Tunley for help with LaTeX and proofreading. This work was supported by U.K. SERC studentship grant #87306795.

References

[1] E. H. Adelson and J. R. Bergen. Spatiotemporal energy models for the perception of motion. *J. Opt. Soc. Am. A*, 2(2):284–299, Feb. 1985.

[2] M. A. Arbib. Levels of modelling of mechanisms of visually guided behaviour. *Behav. Brain Sciences*, 10:407–465, 1987.

[3] D. H. Ballard. Eye movements and spatial cognition. TR 218, University of Rochester Computer Science Department, Nov. 1987.

[4] D. H. Ballard. Reference frames for animate vision. In *Proc. IJCAI*, pages 1635–1641, 1989.

[5] R. A. Brooks. A robust layered control system for a mobile robot. A.I. Memo 864, M.I.T. A.I. Lab, Sept. 1985.

[6] R. A. Brooks. Achieving artificial intelligence through building robots. A.I. Memo 899, M.I.T. A.I. Lab, May 1986.

[7] C. M. Brown. Predictive gaze control. In *Proc. Fifth Alvey Vision Conf.*, pages 103–108, 1989.

[8] C. M. Brown, editor. The Rochester Robot. TR 257, University of Rochester Computer Science Department, Aug. 1988.

[9] V. Bruce and P. R. Green. *Visual Perception: physiology, psychology and ecology.* Lawrence Erlbaum Associates, London, 1985.

[10] E. Buchner. Behavioural analysis of spatial vision in insects. In M. A. Ali, editor, *Photoreception and Vision in Invertebrates*, pages 561–621. Plenum Press, New York, 1983.

[11] J. M. Camhi. *Neuroethology: Nerve Cells and the Natural Behaviour of Animals.* Sinauer Associates Inc., Sunderland, Mass., 1984.

[12] D. T. Cliff. Computational neuroethology: A provisional manifesto. In *Proceedings SAB90.* M.I.T. Press — Bradford Books, 1990.

[13] D. T. Cliff. Network control for animate vision with nonuniform sampling. CSRP 163, University of Sussex School of Cognitive and Computing Sciences, 1990. Forthcoming.

[14] T. S. Collett. Angular tracking and the optomotor response: An analysis of visual reflex interaction in a hoverfly. *J. Comp. Physiol. A*, 140:145–158, 1980.

[15] T. S. Collett and M. F. Land. Visual control of flight behaviour in the hoverfly, *Syritta pipiens* L. *J. Comp. Physiol.*, 99:1–66, 1975.

[16] M. Egelhaaf, K. Hausen, W. E. Reichardt, and C. Wehrhahn. Visual course control in flies relies on neuronal computation of object and background motion. *Trends Neurosci.*, 11(8):351–358, 1988.

[17] S. Haykin. *Adaptive Filter Theory.* Prentice Hall International, Englewood Cliffs, NJ, 1986.

[18] M. F. Land. Similarities in the visual behaviour of arthropods and men. In M. S. Gazzaniga and C. Blakemore, editors, *Handbook of Psychobiology*, pages 49–72. Academic Press, New York, 1975.

[19] M. R. Lynch and P. J. Rayner. The properties and implementation of the non-linear vector space connectionist model. In *First IEE Int. Conf. Art. Neural Networks*, pages 186–190, London, Oct. 1989. IEE.

[20] D. Marr. *Vision.* W. H. Freeman, New York, 1982.

[21] G. Mather, B. Moulden, and A. O'Halloran. Polarity specific adaptation to motion: evidence for gradient detectors, 1990. preprint.

[22] M. G. Paulin, M. E. Nelson, and J. M. Bower. Dynamics of compensatory eye movement control: An optimal estimation analysis of the vestibulo-ocular reflex. *Int. J. Neural Sys.*, 1(1):23–29, 1989.

[23] T. Poggio and F. Girosi. A theory of networks for approximation and learning. A.I. Memo 1140, M.I.T. A.I. Lab, July 1989.

[24] D. E. Rumelhart and J. L. McClelland, editors. *Parallel Distributed Processing, Volume 1: Foundations.* M.I.T. Press — Bradford Books, Cambridge MA, 1986.

[25] M. V. Srinivasan, S. B. Laughlin, and A. Dubs. Predictive coding: a fresh view of inhibition in the retina. *Proc. R. Soc. Lond. B*, 216:427–459, 1982.

[26] J. P. H. van Santen and G. Sperling. Elaborated Reichardt detectors. *J. Opt. Soc. Am. A*, 2(2):300–320, Feb. 1985.

[27] A. L. Yarbus. *Eye Movements and Vision.* Plenum Press, 1967.

[28] D. S. Young. Logarithmic sampling of images for computer vision. In A. G. Cohn, editor, *Proc. Seventh SSAISB Conf.* , pages 145–150, London, Apr. 1989. Pitman/Morgan Kaufmann.

Can Allied Reflexes Promote the Integration of a Robot's Behavior?

by

Philip Teitelbaum, Vivien C. Pellis, and Sergio M. Pellis

Psychology Department, University of Florida, Gainesville, FL 32611

If we wish a robot to behave adaptively, it may be useful to design it according to principles enabling animals and people to do so. We use brain damage and drugs in animals to break behavior down into its subcomponents, and stages of recovery to reveal levels of reintegration. By partial transection in the brain, or by drugs, we appear temporarily to inactivate central motor programs involved in spontaneous behavior (20). As in earlier work using complete transection (19), behavior breaks down into reflexes; spontaneous environmentally-directed behavior is absent. But by our procedures, allied reflexes operate as intermediate-level submodules. By studying how they interact in recovery, and by analyzing controls over individual reflexes, some principles emerge, perhaps useful for robot design. In addition, approaching the subject via phenomena produced by pathology may yield insight into imperfections in a robot that may arise from imbalance in adaptive systems.

(1) Allied Reflexes can Act as Isolated, Adaptive Submodules

Extensive lateral hypothalamic (LH) damage produces a simplification of motivated behavior, that we have called, after Magnus (10), a "zero-condition". All spontaneous environmentally oriented behavior is temporarily abolished. The animal lies motionless, virtually comatose. However, its autonomic system remains relatively intact -- if such an animal is tube-fed, it lives and recovers (20). Within a couple of days, somnolence usually disappears, but the animal remains for several days in a state of catalepsy and akinesia, symptoms often seen in Parkinson's disease (16).

For instance, an LH-damaged cat remains for many minutes with its forelimbs spread widely apart, or one foreleg placed up on its back at quite an extreme angle. Similarly, a cataleptic rat remains with its hindlegs on a raised platform, forelegs on the floor, in an awkward downward-tilted posture. These symptoms support the generally held view that dopamine-deficiency produces an inability to initiate movement. It therefore seems paradoxical when, as shown in fig. 1, a rat made cataleptic by the dopamine receptor blocker haloperidol will, if pushed from behind, leap vigorously into the air (2, 12). This paradox is resolved by detailed analysis of the rat's responses leading to the jump. As the cataleptic animal is pushed forward (Fig. 1A), it braces against such displacement by shifting its weight backwards. When its hind legs begin to slip (B), a leap is triggered (C and D), away from the surface where it is unstable. When it lands on the horizontal table top, it immediately resumes immobility. Thus, cataleptic leaping is merely an allied postural support defensive reflex, triggered by postural instability. The animal does not suffer from a general inability to initiate movement -- its isolated support submodule simply does not do so when the animal is in a stable posture, even if awkward.

The static postural support submodule consists of an aggregate of allied reflexes (including standing, crouching, bracing, clinging, stability-related stepping, righting, and jumping), all of which homeostatically maintain support or regain upright unmoving stability. It is isolated from other submodules, individually involved in locomotion, turning, head-scanning, orienting, and ingestion, which are inactivated (20). When a submodule is isolated, inhibitory controls are lost. We suggest this leads to exaggerated spread, buildup, and afterdischarge (19) of its allied reflexes, compared to when it is integrated with other systems. In incomplete development, systems are also more isolated than in the adult. For instance, in human infants, reflexes spread, seemingly inappropriately. If a very bright light is shone into an infant's eyes, they close, shutting out the light. As the same time, the mouth closes tightly (13). Similarly, when emerging from a dark movie theater into bright sunlight, some children sneeze. Mouth closure and sneezing seem irrelevant, as far as light is concerned. If exposed to dust or noxious fumes, however, it is adaptive for all three avoidance reflexes to work together. Thus, allied reflexes spread, adaptively increasing intensity of response to intense stimuli, even if the context is sometimes inappropriate.

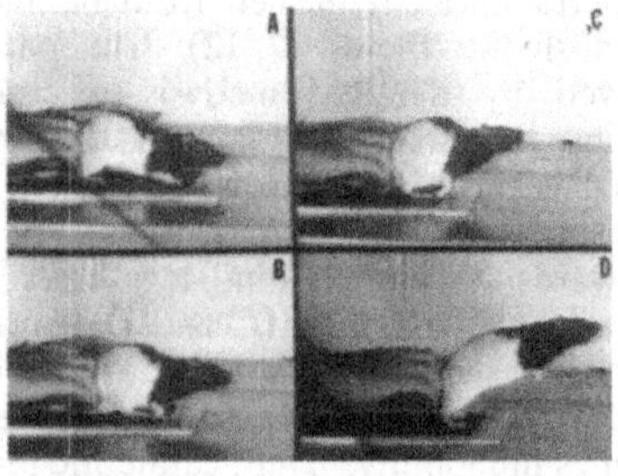

Fig. 1. A jump by a haloperidol (2.5 mg/kg i.p.) -treated rat being pushed forward. See text. (Figs. 1-3 are photos taken from the video monitor).

(2) Additional controls over a submodule: Evolution by side-effect

If a normal undrugged rat is immersed in cold (10°C) water for 4-6 minutes, it becomes hypothermic (body temperature as low as 27°C). It crouches unmoving, clings, braces, and rights, very much like a rat made cataleptic by dopamine depletion (18). Static postural support remains intact under such hypothermia and helps counteract it. The hypothermic animal only crouches and shivers, but that helps it to rewarm. In the rigidity of dopamine-deficiency, limb extensors and flexors contract simultaneously when standing (3). This appears to be an exaggeration of the positive supporting reflex, necessary to transform the limb into a rigid pillar while standing (11). We suggest that this reflex, useful for support on dry land for creatures evolved from the sea, was co-opted in warm-blooded animals for shivering thermogenesis. This also explains why, when isolated by dopamine-deficiency, the entire postural support submodule is exaggeratedly sensitive to inhibition by skin-warming (18). If an LH-damaged cataleptic rat has its skin warmed slightly, then, unlike a normal adult, it lies flat, without support, and does not even right itself on the ground. In normal infant rats, similar loss of all the allied reflexes involved in static postural support can be produced by slight skin-warming. We suggest that this exaggerated spread of allied reflexes indicates that the integration of their postural support submodule with other systems is still incomplete.

Evolution thus appears to seize upon a side-effect of the action of a subsystem to adapt it for other purposes. The trembling of simultaneously excited flexors and extensors involved in standing generates heat, so linking them as a unit to a thermodetector produces a control system that can increase body temperature by shivering. Conversely, skin-warming can now inhibit standing, leading to a flattened posture that dissipates heat effectively. As an additional side-effect, the crouched immobility involved in hypothermia is adaptive for avoiding the attention of predators. So it is not surprising that static postural support is the submodule released most commonly in rodents that "freeze" in fear when exposed in an open field. And so on, the same submodule presumably being co-opted as a whole for one use after another. Thus, complex functions evolve, co-opted from simpler ones, often seemingly unrelated (17). During traumatic breakdown, the process is reversed: simpler submodules sustain life until complex functions recover.

(3) Allied Reflexes Shape and Direct a Behavioral Chain

During a jump, ground-contact is absent. The stimuli maintaining the continuity of the act can then be more easily analyzed. If a rat without labyrinths, its vision blocked by opaque contact lenses, is made cataleptic by haloperidol, and is pushed from behind, then, like an intact cataleptic rat, when its hindlegs slip, a jump is triggered (Fig. 2). Therefore, the jump is a proprioceptive-tactile reaction to stimuli originating in the body and legs.

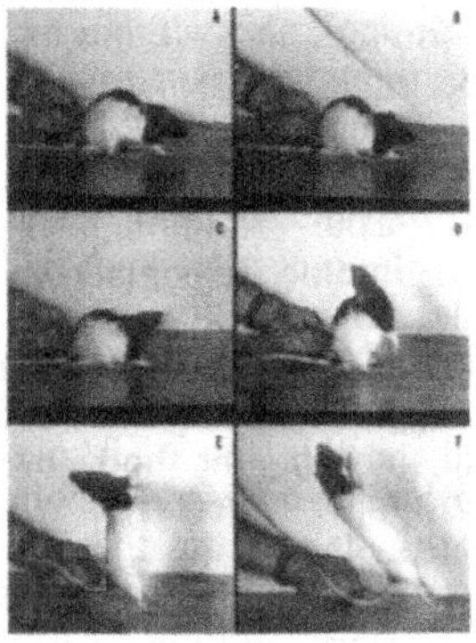

Fig. 2. A backward jump by a labyrinthectomized haloperidol-treated rat being pushed forward. (To facilitate jumping, 50 mg/kg methysergide i.p. was also injected). See text.

However, without vestibular control, its form and direction are different. In an animal with intact labyrinths, the jump follows a smooth forward trajectory, guided by the head (Fig. 1). Without the labyrinths to dampen its amplitude, to guide it, and to coordinate it with the body, the head is thrown backward, seemingly as far as it can go (Fig. 2D, E). Head-neck dorsiflexion triggers the upper torso to lift upward, shifting the animal's body weight on to its hindlegs, in turn triggering the legs to kick, propelling the body into the air. With only proprioceptive stimuli from the neck operating, while the body is whirling in the air, the exaggeratedly dorsiflexed head and neck come back into a partially ventroflexed position (Fig. 2F). This shows that a proprioceptive "zero" reference level exists for the alignment of head with body, which is overridden by tactile stimuli from the ground. But even with vision, without the labyrinths to guide the head, the animal is thrown upward and backward, often performing a backward somersault. Therefore, after an initial reflex segment of a behavioral chain is triggered by stimuli in one modality, allied reflexes based on other sense modalities shape its form, chain its components together, and guide it through the environment.

(4) The Hierarchical Encephalization of the Jump Reaction

A fragment of the jump can be elicited in a decerebrate cat (7). An upward jump can be triggered merely by allowing the weight of the animal's upright body to induce flexion of the hind legs. Head dorsiflexion facilitates this. Unlike the presumably more encephalized haloperidol-treated animal, foot slippage is not necessary to trigger the jump. We conclude that as hierarchical encephalization proceeds, a simple reflex becomes a more elaborate chain, each successive segment recruited as an allied reflex by the preceding one, released and controlled by a more specialized stimulus. This is illustrated by comparing the jumps by haloperidol-treated versus undrugged intact rats when placed on a board that is slowly raised from the rear (12). Unlike haloperidol-treated rats, most undrugged rats do not jump from the slope as their head-downward tilt increases. Before their hindlegs slip, they either walk down, or they turn and climb upwards. However, a few normal rats do jump from the slope, so the question can be answered. If the slope is placed in front of a wall, or facing over the edge of a table, a haloperidol-treated rat, even with its eyes occluded, leaps straight ahead, regardless of outcome. Not so for the normal. Without vision, it does not jump; instead, it flattens its body and slides down. With vision, faced by a wall or a visual cliff, it either inhibits the jump, or turns and leaps over the side to the nearby horizontal surface of the table (Fig. 3). Note also that head-dorsiflexion is not a necessary trigger for a jump to the side. Thus, the normal jump is less reflexive, because at this level of control, leg-kick and head-dorsiflexion become inhibitable.

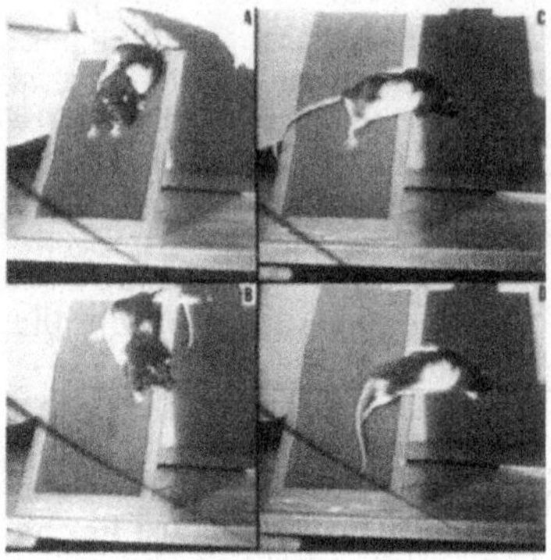

Fig. 3. Reorientation of a jump from an increasing slope by an undrugged intact rat. See text.

(5) <u>Allied Reflexes Promote Recovery from Brain Damage</u>

(a) Redundant Reflexes can Directly Activate the Same Function

Some parkinsonian patients have difficulty turning over in bed. They cannot rotate axially, but instead use their limbs to push their body over. Under appropriate slight sensory stimulation, or in times of great stress, they <u>can</u> roll over, so the disability is an "axial apraxia", reflecting not a paralysis, but an inability to call up the appropriate motor programs (9). Using LH lesions, we mimicked this phenomenon in rats for a day or so. By also denervating the labyrinths, however, we slowed the recovery for as long as 21 days. Note that a labyrinthectomized rat with brain intact is able to use axial rotation, starting with the shoulders and followed by the pelvis, to right itself on the ground. Therefore, proprioceptive and tactile stimuli from ground contact are sufficient to activate axial rotation. Clearly, LH damage produces a massive deficit in proprioceptive-tactile activation of axial rotation. This is compensated for by redundant allied vestibular reflexes. Such redundancy thus decreases the likelihood of loss of function after damage and promotes a more rapid recovery (15).

(b) Indirect Triggering by an Allied Reflex can Compensate for Inadequate Activation

Slow recovery isolates subcomponents of righting, not previously described. When contact-righting reappears, it is accomplished only by the hindlegs, which push the pelvis, followed by the rest of the body, over to prone. Later, the forelegs, also without axial rotation, initiate the righting. Rostrocaudal contact-righting, involving axial rotation, appears still later. Thus, three subcomponents (pushing with the limbs, axial rotation,, and cephalic dominance) interact in contact-righting (15).

During the slow recovery, a stage in righting appears, in which, even though axial rotation cannot be triggered directly by body contact with the ground, pushing the body over with the limbs is no longer necessary. When the paw merely places lightly on the ground, axial rotation is triggered indirectly as an allied reflex (Fig. 4). Thus, indirect triggering of a subcomponent in a behavioral chain via an allied reflex is also a way to compensate for inadequate direct activation.

For instance, people with Parkinson's disease often stand frozen, unable to initiate locomotion. Some of them learn to keep little wads of paper in their pockets, and when unable to will themselves to walk, they drop a little ball of paper on the floor in front of them. The sight of the ball initiates a step over it, which in turn indirectly triggers the initiation of locomotion. Similarly, they can walk forward readily over a series of lines spaced a foot or so apart on the floor, but they freeze when they reach the end of the series. Analogously, climbing up stairs is easy for them, whereas "simple" walking is not (17). Following a parade is easy, facilitated by the sight of the people walking directly ahead of them and by the rhythmic beat of the band. In sum, many seemingly paradoxical kinesias of parkinsonism are based on indirect triggering of locomotion by allied reflexes.

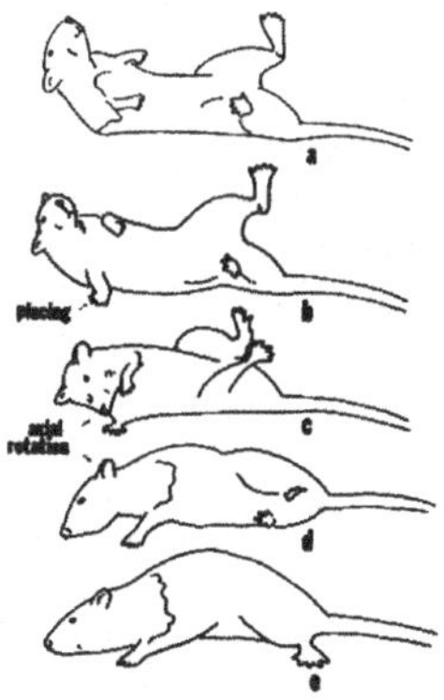

Fig. 4. Tracings from a 16 mm movie (24 fps) showing a stage of recovery following LH-damage in a labyrinthectomized rat, in which forepaw contact with the ground triggers shoulder rotation. See text.

Indirect triggering also provides temporary allied reflex bridges that help escalate to the terminal segments of a sequence of behavior when adequate activation is lacking. For instance, when an adult, experienced, highly aroused male rat mates with a female, he typically mounts, clasps, thrusts, and achieves intromission. Snout-contact with the female's neck rarely or never occurs, so such contact has not been highlighted as essential. In contrast, nape-contact occurs often in "play" in juvenile rats, which rarely leads to mating, so the two behaviors are thought to be unrelated. However, in the transition period preceding sexual maturity, nape-contact is an important link in the escalation of "play" to mating (14). Similarly, failure to maintain snout-to-nape contact by adult male decorticate rats prevents successful mounting and intromission. After the first successful intromission, aided by snout-to-nape contact, such contact is no longer necessary in subsequent mating (22). (The "warm-up" involved in the repetition of sequences of "play" also promotes such escalation [23].)

(6) <u>"Warm-up" -- The Escalation of Activation by Repetition of the Initial Segments in a Behavioral Chain</u>

Late in recovery from LH damage, the rat walks around and appears to explore its environment relatively normally. It stops and eats palatable food it encounters, giving the impression that it is hunger-motivated. However, if it walks into a corner, it is trapped, often for long periods, performing stereotyped head-scanning movements and stepping patterns (6). Such a "behavioral trap" can indicate that the animal's behavior is so disintegrated that it is disconnected from its usual adaptive outcome. We therefore used the Eshkol-Wachman movement-notation system, which allows one to record and interpret the movement of the body parts that compose the whole, without having to label each act by its adaptive outcome (5). Films were made of the movements of each animal throughout recovery from akinesia to relatively normal locomotion. These films were analyzed frame-by-frame, and written "musical scores" of movement sequences were prepared. In this way, it was possible to isolate the dimensions of movement along which recovery occurred day-by-day, and its amount (6).

When an LH-damaged animal is cataleptic, its head, body, and legs are used in a coordinated whole-body reaction for defending and maintaining upright stable support. As exploratory lateral and longitudinal head-scanning movements appear and are repeated with increasing amplitude ("warm-up"), they have to recruit the front legs and more caudal segments of the body away from support into scanning. A tug-of-war goes on. While the front legs remain rooted to the ground, dominated by the static postural support submodule, head-scanning reaches a degree of exaggerated bending and stretching. This "strait-jacket" phenomenon (6) reflects the action of two independent submodules, acting at cross-purposes. Later in recovery, the front legs and torso are recruited into head-scanning, and the exaggerated bending and stretching shifts to the pelvis, as only the hindlegs remain rooted, still resisting the scanning actions of the forequarters. Thus, in recovery from dopamine deficiency, due to the conflict between the recovering control system (tactile scanning) and the previously dominant mode of control (static postural support), a form of behavioral disintegration exists.

Long after the LH-damaged animal has recovered the ability to walk forward freely, it freezes into immobility for a few moments each time it is set down in an open field. Its static postural support submodule is dominant, and the animal cannot walk forward. It must recapitulate the recovery sequence of scanning and pivoting before it can walk. With repetition, the amplitude and cephalocaudal recruitment of the scans grow in amplitude, suggesting that "warm-up" promotes escalation along the sequence (6). An analogous warm-up sequence can be seen when infant rats start to walk in an open field (4).

(7) Allied Reflexes Account for Stages of Recovery of Setpoint Regulation

LH lesions cause rats and cats to stop eating and to die of starvation (1). If such rats are offered palatable food, they eventually eat enough to keep themselves alive, and recover regulatory intake. At the same time, such rats do not drink water. This recovers also, but slower than feeding. The interaction of these two differentially recovering regulations yields stages of recovery, which have helped to assess severity of damage, and how recovery is progressing (21). However, a mechanistic understanding of such stages has been lacking. We suggest that the action of allied reflexes provides one. For instance, in the first stage of "aphagia and adipsia", the animal refuses to eat food or drink water. However, pinching its tail can induce biting of solid food, which leads to the allied reflex of swallowing. Alternatively, reflexive suckling of milk via a nipple can lead to sufficient ingestion. In the second stage, the animal eats palatable foods, but not enough to maintain its weight, and does not drink water; this is "anorexia and adipsia". Here, reflexes related to smell and taste of food promote ingestion, but without a link to the internal environment they cannot maintain adequate intake. In the third stage, if the animal is offered a liquid diet, calorie and weight regulation reappear; we suggest that this depends on allied reflexes related to gastric emptiness and fullness. Here, the animal fails to eat only because it does not drink; this is "adipsia and dehydration-aphagia". Finally, the "recovery" stage appears, in which the animal drinks water, eats dry food and survives. However, its eating and drinking are totally dependent on allied reflexes: its eating is reflexively tied to the absence of light, and the animal drinks only when it eats dry food (21). Furthermore, its regulatory adjustment is slow, remarkably like the time course of respiratory response to exercise when the allied reflexes involved in muscle proprioception have been denervated (8). These facts suggest that the precision and rapidity of adjustment of the set points involved in caloric regulation, and in body weight, are determined by allied reflexes.

Conclusion

Historically, using the single reflex as the unit of behavior, it has not been very fruitful to try to build an understanding of how complex behaviors are integrated from it. In the design of a robot, a submodule of allied reflexes may be a more suitable unit. Recovery in animals and people can be valuable to discover successively more complex levels of integration. Then one can figure out how each level reduces to the one below. In this process, an understanding and investigation of allied reflexes, based on the principles outlined above, should prove helpful.

Acknowledgement: We thank Kibbutz Merhavia and the members of the library of Mossad Misra in Israel for generously allowing P.T. to use their facilities while writing part of this paper. The work was supported by a grant from the Whitehall Foundation.

References

1. Anand, B.K. and Brobeck, J.R. (1951). Hypothalamic control of food intake. Yale Journal of Biology and Medicine, 24: 123-140.

2. DeRyck, M., Hruska, R.E. and Silbergeld, E.K. (1982). Estrogen and haloperidol-induced versus handling-related catalepsy in male rats. Pharmacology, Biochemistry, and Behavior, 17: 1027-1035.

3. DeRyck, M. and Teitelbaum, P. (1983). Morphine versus haloperidol catalepsy in the rat: an electromyographic analysis of postural support mechanisms Experimental Neurology, 79: 54-76.

4. Eilam, D. and Golani, I. (1988). The ontogeny of exploratory behavior in the house rat (Rattus rattus): the mobility gradient. Developmental Psychobiology, 21: 679-710.

5. Eshkol, N. and Wachman, A. (1958). Movement Notation. London: Weidenfeld and Nicolson.

6. Golani, I., Wolgin, D., and Teitelbaum, P., (1979). A proposed natural geometry for recovery from akinesia in the lateral hypothalamic rat. Brain Res., 164: 237-267.

7. Guzman, F.C. and Del Pozo, E.C. (1953). "Jump reflex" in hypothalamic cat. Neurophysiology, 16: 376-380.

8. Horrobin, D.F. (1970). Principles of Biological Control. Letchworth: Garden City Press.

9. Lakke, J.P.W.F., (1985). Axial apraxia in Parkinson's disease. Journal of Neurological Sciences, 69: 37-46.

10. Magnus, R. (1926a). On the co-operation and interference of reflexes from other sense organs with those of the labyrinths. Laryngoscope. 36: 701-712.

11. Magnus, R. (1926b). Some results of studies in the physiology of posture. Lancet. 211: 531-536, 585-588.

12. Morrissey, T.K., Pellis, S.M., Pellis, V.C. and Teitelbaum, P. (1989). Seemingly paradoxical jumping in cataleptic haloperidol-treated rats is triggered by postural instability. Behavioural Brain Research, 35: 195-207.

13. Peiper, A. (1963). Cerebral Function in Infancy and Childhood. New York: Consultants Bureau.

14. Pellis, S.M., Pellis, V.C. and Dewsbury, D.A. (1989). Different levels of complexity in the play-fighting by muroid rodents appear to result from different levels of intensity of attack and defense. Aggressive Behavior, 15: 297-310.

15. Pellis, S.M., Pellis, V.C., Chen, Y-c., Barzci, S. and Teitelbaum, P. (1989). Recovery from axial apraxia in the lateral hypothalamic labyrinthectomized rat reveals three elements of contact-righting: cephalocaudal dominance, axial rotation, and distal limb action. Behavioural Brain Research, 35: 241-251.

16. Sacks, O.W. (1974). Awakenings. New York: Doubleday.

17. Satinoff, E. (1978). Neural organization and evolution of thermal regulation in mammals. Science, 201: 16-22.

18. Schallert, T., Whishaw, I.Q., DeRyck, M. and Teitelbaum, P. (1978). The postures of catecholamine-depletion catalepsy: Their possible adaptive value in thermoregulation. Physiology and Behavior, 21: 817-820.

19. Sherrington, C.S. (1906). The Integrative Action of the Nervous System. Forge Village, MA: Murray Printing Co. (Issued as a Yale Paperbound, 1961).

20. Teitelbaum, P. (1986) The lateral hypothalamic double disconnection syndrome: A reappraisal and a new theory for recovery of function. In: One Hundred Years of Psychological Research in America: G. Stanley Hall and the Johns Hopkins Tradition. Edited by S.H. Hulse and B.F. Green, Jr.. The John Hopkins University Press.

21. Teitelbaum, P. and Epstein, A.N. (1962). The lateral hypothalamic syndrome: recovery of feeding and drinking after lateral hypothalamic lesions. Psychological Review, 69: 74-90.

22. Whishaw, I.Q. and Kolb, B. (1985). The mating movements of male decorticate rats: Evidence for subcortically generated movements by the male but regulation of approaches by the female. Behavioural Brain Research, 17: 171-191.

23. Wolgin, D.L. (1982). Motivation, action, and behavioral integration. In R.L. Isaacson and N.E. Spear (Eds.), The Expression of Knowledge. New York: Plenum Press, pp. 243-290.

COORDINATION OF LEG MOVEMENT IN WALKING ANIMALS

Holk Cruse
Dept. of Biol. Cybernetics, Fac. of Biology, and Center of Interdisciplinary Research, University of Bielefeld, Fed. Rep. of Germany
E-mail: UBIOF140 at DBIUNI11

ABSTRACT

When a robot has to move in a locally uncertain environment, propulsion by means of walking legs is advantageous compared to a wheel driven system. However, the control of walking legs is more complicated. The question of how the movement of the different legs is coordinated will be investigated here. Three different solutions which have been developed during natural evolution will be compared. These are the walking systems of an insect, of a crustacea, and of a mammal. The results show that coordinating mechanisms differ considerably in these animals.

1. INTRODUCTION

An autonomous robot which has to move in an uncertain environment has to deal with the problem of how to perform a goal-oriented behaviour. This problem is a global one, meaning that the robot has to deal with the detection of possible paths and to decide which of these it should take. This includes the problem of obstacle avoidance. When the environment is "locally certain", as for example a semi-artificial environment with flat surfaces, the technical problem of how to move the body forward is comparatively simple and can be solved by driving the robot with wheels. If, however, the local structure of the environment is uncertain, i.e., consists of irregular terrain, a robot with walking legs is advantageous. Several attempts have been made to construct such a walking machine. Nevertheless, comparing the walk of a six-legged robot with that of an animal such as an insect, immediately reveals differences. The walking of an animal is much more versatile, and it appears to be more efficient and elegant. Thus it is useful to consider biological control mechanisms in order to apply these or similar mechanisms to the control of walking legs in machines. In the past, engineers have pointed out that little information is available on the biological control mechanisms, but this situation has changed recently. This paper provides a summary of the recent biological results focussing on data obtained from insects, crustaceans, and cat by means of behavioural methods.

2. CONTROL OF THE INDIVIDUAL LEG

The results to be described here mainly concern the coupling mechanisms between legs, i.e. those mechanisms that produce a proper coordination of the walking legs even when walking is disturbed. However, for this purpose it is necessary also to consider briefly how the movement of an individual leg is controlled. The mechano-neuronal system that produces this movement might be called the step pattern generator. To avoid a possible misunderstanding it should be stressed that it is completely open whether this step pattern generator contains an endogenous central pattern generator.

For simplicity, only forward walking will be discussed. The cyclic movement

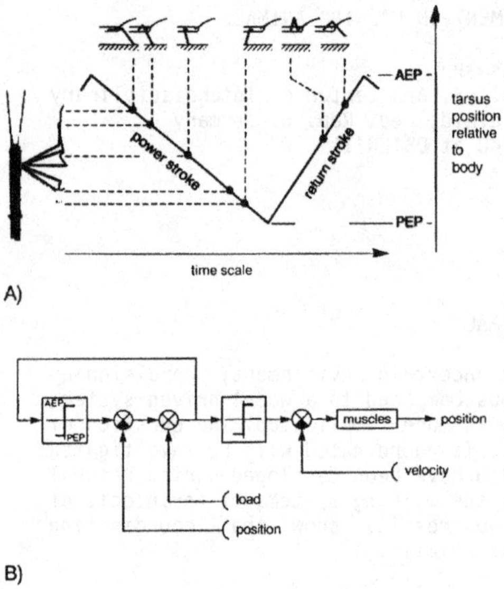

Fig. 1. (A) The cyclic movement of a leg consists of the power stroke during which the leg supports and propels the body and the return stroke during which the leg is swung back to the initial position. In forward walking the latter is called the anterior extreme position (AEP). The power stroke which then corresponds to a retraction movement of the leg ends at the posterior extreme position (PEP). Position of the leg always means the position of the leg tip relative to the body.
(B) Circuit model of the movement of a leg. The left hand part represents a hierarchically upper-level circuit which decides which of the two states, power stroke or return stroke, should be adopted by the system. The left relay characteristic produces the two alternative target positions, AEP or PEP, when its input value is positive or negative, respectively. The value of the target position is compared with the actual leg position. The result is also affected by signals from load-sensitive organs. The output of the decision circuit is considered as reference input for the second, lower-level circuit (right), a velocity controlling feedback system.

of a walking leg consists of two parts (Fig 1A), the power stroke (also stance phase or support phase) and the return stroke (also swing phase or recovery phase). During the power stroke the leg is on the ground where it can support and propel the body. In a forward walking animal this corresponds to a retraction movement of the leg. During the return stroke the leg is lifted off the ground and swung to the starting position for the next power stroke. In forward walking this corresponds to a protraction movement of the leg. The question of how to coordinate the movement of the individual joints of a leg will not be considered here. In the stick insect, which has been investigated in detail, the movement of the leg during both the power stroke and the return stroke is controlled by at least one servo-system. By means of negative feedback this servo-system regulates a parameter corresponding to the velocity of the leg (Cruse 1981, Cruse 1985b, Cruse and Pflüger 1981, Cruse and Schmitz 1983, Dean 1984, Schmitz 1985). This view was later shown to be true even for a single joint (Weiland and Koch 1987, Bässler 1988). With respect to the temporal pattern of stepping, the critical question is how the step pattern generator decides whether or not the transition from one mode (power or return stroke) to the other should be made. Experimental results have shown that for insects the most critical point is the transition from power stroke to return stroke. This seems to be intuitively clear as this transition terminates the support function of the leg; if performed at the wrong moment the animal may be left unsupported which may be hazardous. Nevertheless, the crayfish was found to use another solution (see below). In the stick insect, three parameters influence this transition: position, load, and phase in the step cycle of the other leg. The first two depend on proprioceptive information describing the state of the leg itself; these peripheral influences ensure that the leg only lifts off the ground when its position relative to the body is far enough to the rear and that it only lifts off when the load under which it stands is small enough (Bässler 1977, Cruse 1985c).

These results can be summarized by the circuit model shown in Fig 1B (Cruse 1985d). It consists of two parts. The first is a hierarchically superior circuit that decides which of the two states, power stroke or return stroke, the system should adopt. The relay characteristic produces the two alternative target positions AEP and PEP (anterior and posterior extreme positions, Bässler 1972) when its input value is positive or negative, respectively. The value of the target position is compared with the actual leg position. The result is also affected by signals from load-sensitive organs. The output of the decision circuit is considered as reference input for the second, inferior circuit, a velocity-controlling feedback system. A positive output value of the decision circuit corresponds to a movement in the anterior direction (positive velocity, return stroke) and a negative value to movement in the posterior direction (negative velocity, power stroke). Results reported by Bässler (1988) support the assumption that load-sensitive organs also affect the system on the lower level, which is however not shown in Fig 1B. The model could easily be reformulated in neuronal terms. The decison part can, for example, be represented by a bistable circuit as described by Land (1972) or Bässler (1986). Investigations of an intersegmental reflex - the TOT-reflex first described by Graham (1979) - show that the part of the system which decides on power or return stroke is hierarchically superior to that part which decides whether the individual leg walks forward or backward (Schmitz and Haßfeld 1989). This was also assumed to be the case for crustacea (Ayers and Davis 1977).

3. COORDINATION BETWEEN LEGS

The third parameter is of major importance here because it depends on coordinating influences: the transition from power stroke to return stroke of one leg also depends on the relative timing or the phase of the leg's movement to that of the other legs. In addition, direct peripheral influences from the sense organs of neighbouring legs might exist which are not involved in the step pattern generators of these legs. Not only are coordinating mechanisms known that influence the power-return stroke transition, but also mechanisms that influence the return-power stroke transition at the anterior extreme position (AEP) and mechanisms that influence the strength of the motor output during the power or the return stroke. The experimental data show no essential differences in the intraleg, peripheral influences among insects, crustacea, and even mammals. In contrast, comparative investigations have shown that the coordinating mechanisms are quite different for crustaceans, insects, and mammals (Cruse and Müller 1986, Cruse and Schwarze 1988, Cruse and Knauth 1989, Müller and Cruse 1991a, Cruse and Warnecke 1991). Thus the results obtained for the three most thoroughly investigated species, the crayfish, the stick insect, and the cat are discussed separately.

4. THE CRAYFISH

As a common feature it can be stated that in arthropods coupling between ipsilateral legs, i.e. legs on the same side of the body, is generally stronger than coupling between contralateral legs, i.e. legs on opposite sides of the body (rock lobster: Clarac and Chasserat (1983); crayfish: Müller and Cruse (1991a); stick insect: Dean and Wendler (1982), Foth and Bässler (1985b), Graham (1977a); cockroach: Delcomyn and Cocatre-Zilgien (1988); tarantula spider: Seyfarth and Bohnenberger (1980)). Therefore ipsilateral mechanisms are easier to investigate and will be treated first.

To begin with the crayfish, two coordinating mechanisms are found acting between neighbouring ipsilateral legs (Cruse and Müller 1986). These two influences are schematically presented in Fig 2. The two traces represent the movement of the two legs. Several phase situations are plotted in each sketch. They might be produced by a disturbance of the normal walk for example due to an irregular or slippery ground. One coordinating mechanism is only rostrally-directed. This is illustrated in Fig 2A: as long as the posterior leg performs a power stroke, the anterior leg has to perform or continue a return stroke. In addition the velocity of the movement during the return stroke is also decreased to some extent. Thus the return stroke can be prolonged so that normal coordination is regained in the next step. The horizontal striped bar indicates the time during which this mechanism is active. The intensity of the influence is roughly indicated by the thickness of the bar. The second influence is caudally-directed. This is illustrated in Fig 2B: when the anterior leg is near the end of its power stroke or at the beginning of its return stroke, an influence with increasing intensity has the effect of ending the return stroke and beginning the power stroke of the posterior leg, thus shortening the return stroke of this leg. As indicated by the stippled wedge, this influence ends abruptly about 200 ms after the beginning of the return stroke of the anterior leg. Both coordinating mechanisms influence the transition from return stroke to power stroke.

A crayfish has four ipsilateral legs. Thus two of these legs have both an anterior and a posterior neighbour. These "middle" legs might therefore obtain both influences which, depending on their temporal relation, could superimpose. When the middle leg obtains the signal from the anterior leg at the beginning of the power stroke of the posterior leg then in the middle leg a long return stroke can be produced which is interrupted by a short power stroke. This is shown for one example in Fig 2C. This indicates that in this situation the caudally directed influence is somewhat stronger. In this case the transition from power to return stroke is affected, which means that the PEP is shifted to a more rostral position. In any case both mechanisms act together in the same sense so that an incorrect phase relation between two legs is immediately corrected within the following step. Although for convenience the mechanisms are described as influencing the transition between the two states, this example also shows that the influence could also be described as facilitating one of the two states. As could be shown by means of a model calculation, each mechanism alone would be sufficient to produce a coordinated walk. In this respect the connections can be considered redundant.

An additional influence acting between adjacent ipsilateral legs of the lobster was described by Müller and Clarac (1990a,b) and was also found in the crayfish (Müller 1990). These authors electrically stimulated a leg nerve which containes fibres from a mechanical sense organ at the tip of the leg. The stimulus caused the adjacent leg to prolong its power stroke, or, if it was in return stroke, interrupt the return stroke and begin a power stroke. This coordinating influence was found to act in both anterior and posterior direction but it is open whether it plays an important role during normal walking.

To couple contralaterally neighbouring legs, i.e. legs of the same segment, the crayfish uses a mechanism which very much resembles the ipsilateral, caudally directed influence (Müller 1989, Müller and Cruse 1991 a,b). In contrast to the ipsilateral mechanism where the influence is only active during a small fraction of the whole step cycle (see

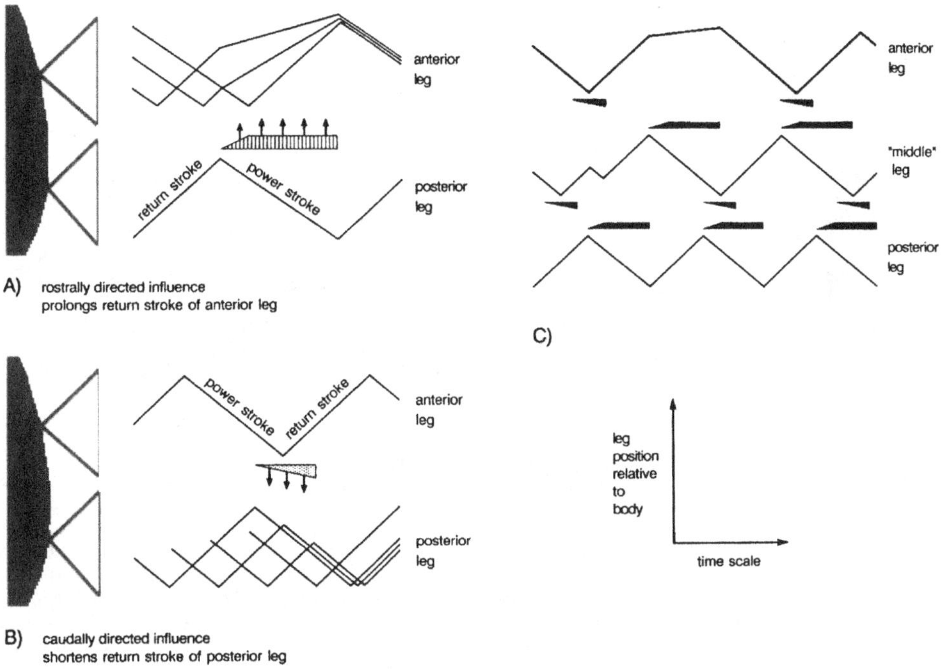

Fig. 2. Coordination between the ipsilateral legs of a crayfish (A, B). The upper trace of each shows the anterior leg. Each schema is drawn as if only one of the two coordinating mechanisms existed. In each case the influencing leg is drawn only once. For the influenced leg several traces are presented to show the effect of the coordinating mechanism. The duration and the intensity of the influences are roughly indicated by the length and the thickness of the wedges, respectively.
A) The rostrally directed influence is active during the power stroke of the posterior leg. It prolongs the return stroke of the anterior leg and can also decrease the speed of the limb movement.
B) The caudally directed excitatory influence is active at the end of the power stroke and the beginning of the return stroke of the anterior leg. It "excites" the start of a power stroke in the controlled, posterior leg.
C) Coordination between three ipsilateral legs when both forward and caudally directed influences are active. The top trace shows the most anterior leg. The "middle " leg obtains a signal from the anterior leg which excites a power stroke. At the same time the middle leg obtains an influence from the posterior neighbour to perform a return stroke. As the latter signal at the very beginning is weak, the signal from the anterior leg "wins" until this influence ceases and the rostrally directed influence dominates (after Cruse and Müller 1986).

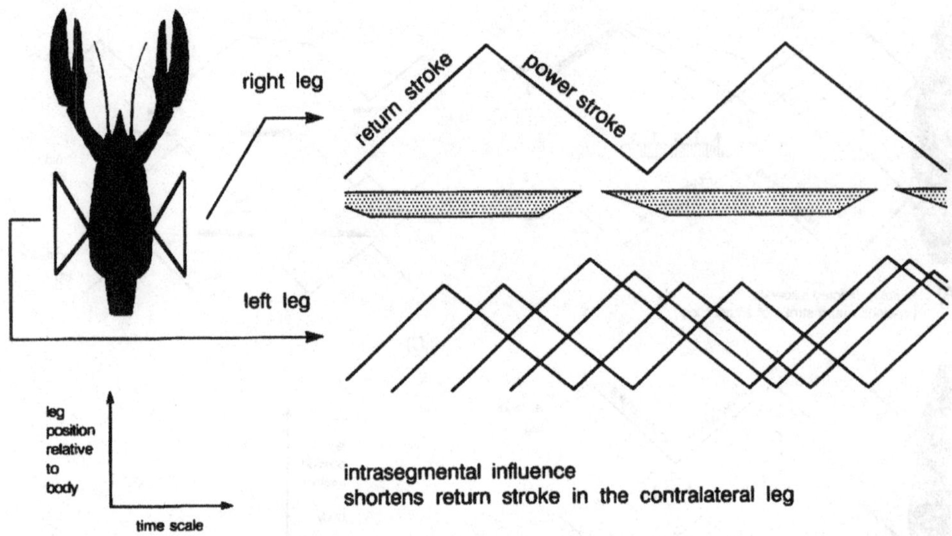

Fig. 3. Influences between two contralateral legs of a crayfish. The influences are assumed here to act only in one direction, from right to left. As in the ipsilateral caudally directed coupling mechanism (Fig 2B) the controlled leg is influenced to start a power stroke. Compared to the ipsilateral case the influence continues for a much longer part of the step period but the influence is weaker. Thus several steps could be necessary to retain the normal coordination and relative coordination might eventually be observed (after Müller and Cruse 1991a).

Fig 2B), the contralateral influence continues during most of the cycle (Fig 3). With the exception of a small phase range the return stroke of the other leg is always shortened. The effect is much weaker than ipsilateral influences. While in ipsilateral legs the effect of a disturbance is compensated within the following step, because of the weaker coupling in contralateral legs a disturbance results in a so called gliding coordination (the relative coordination of von Holst (1939)). This means that several steps are necessary to regain normal coordination. This contralateral influence acts in the same way in both directions but one leg might be somewhat more dominant. This dominance seems to be inherent to some extent but can additionally be influenced experimentally by inducing different turning tendencies by means of an optical stimulus. A gradient along the longitudinal axis was found in the way that anterior leg pairs show less strong coordination than more posterior pairs. In some cases the coupling between diagonally neighbouring legs was found to be stronger than between contralateral legs of the same segment. This result, however, does not require direct coupling connections between diagonal neighbours but can be explained on the basis of the above mentioned intersegmental contralateral influences (Cruse and Müller 1991b).

5. THE STICK INSECT

As mentioned above, quite different results were found concerning the coordinating mechanisms in the stick insect. Three different mechanisms influence the PEP, i.e. the transition from power stroke to return stroke. They are shown schematically in Fig 4. Two influences are rostrally directed.

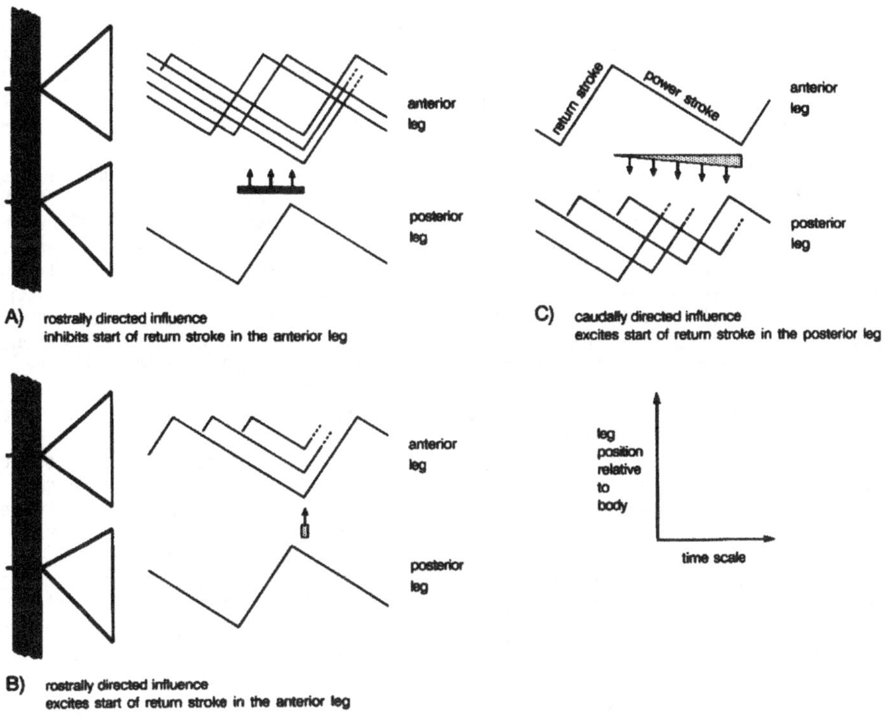

Fig. 4. Coordination between two ipsilateral legs of a stick insect (A,B,C). The upper trace of each shows the anterior leg. Each schema is drawn as if only one of the three coordinating mechanisms existed. In each case the influencing leg is drawn only once. For the influenced leg several traces are presented to show the effect of the coordinating mechanism. The duration and the intensity of the influences are roughly indicated by the length and the thickness of the bars and wedges, respectively.
A) The rostrally directed inhibitory influence is active during the return stroke of the posterior leg.
B) The rostrally directed excitatory influence is active at the beginning of the power stroke of the posterior leg.
C) The caudally directed excitatory influence is active during the power stroke of the anterior leg.
The terms "excitatory" and "inhibitory" are applied according to the effect on the start of the return stroke in the influenced leg.
The duration of the return stroke could not be shown to be influenced by these coordinating mechanisms. As it is not known in detail how the varying amplitudes are compensated return strokes are only roughly indicated (after Cruse and Schwarze 1988).

One (A) hinders a leg from starting a return stroke as long as the posterior leg performs its return stroke and may continue for about another 100 ms. This influence can prolong the power stroke (Cruse and Epstein 1982, Cruse and Schwarze 1988, Dean and Wendler 1982, Foth and Bässler 1985a). Experimental results indicate that this is done by shifting the threshold for beginning a return stroke to a more posterior position (Cruse 1985a). In contrast to this "inhibitory" influence the following two have an "excitatory" effect. The second rostrally directed influence (B) can elicit a return stroke in a leg when the posterior leg starts a power stroke (Cruse and Schwarze 1988). This influence shortens the duration of the power stroke. A third influence is caudally directed (C): the start of a return stroke of a leg is more strongly excited, i.e. occurs earlier, the farther the anterior leg is moved rearward during its power stroke (Cruse and Schwarze 1988, Dean and Wendler 1982, Foth and Bässler 1985b, Bässler et al. 1987). This has the effect that the posterior leg performs a return stroke before the anterior leg begins its return stroke. In the model shown in Fig 1B these influences can be imagined to change the AEP or PEP value of the relay characteristic, or, logically equivalent, to sum with positive or negative sign to the input of this characteristic. These three mechanisms are redundant in the sense that they produce the same effect: the nearly immediate re-establishment of coordination in the case of any disturbance.

In addition three other, less important coordinating mechanisms are found in the stick insect (Fig 5). One is probably a special development for an animal climbing on branches as it is not found in other animals such as locusts (Pearson and Franklin 1984) or lobsters (Cruse, Clarac and Chasserat 1983). The position where the hind leg finishes its return stroke is given by the actual position of the tarsus of the ipsilateral middle leg: the hind leg tarsus is placed immediately behind the middle leg tarsus (Cruse 1979a, Dean and Wendler 1983). Thus information on the spatial position of the middle leg tarsus relative to the body is given to the neuronal system controlling the movement of the hind leg. The same "targeting" occurs between middle and front leg and also between front leg and antennae (unpubl. observation). The main function of these mechanisms is presumably to help a leg find ground contact, an essential function when climbing on branches but trivial when walking on a continuous surface. However this mechanism also contributes to coordination. If any irregularities occur in the step of a leg, the targeting mechanism can shorten or prolong the return stroke of the next posterior leg in order to retain a normal coordination. It was shown in a neural network simulation that this mechanism can be realized by a simple system consisting of nine neurons,i.e., three sensory, three moto- and three interneurons (Dean 1990).

Another coupling mechanism is the so-called "coactivating" influence which affects the motor output during the power stroke (Bässler 1979, Cruse 1985a). An increase in the motor output of a leg induced, for example, by experimentally decreasing the speed of the retraction movement immediately leads to an increase in the motor output of neighbouring legs during their power stroke. In contrast to the influences discussed earlier which act only in one, either forward or rearward direction, the coactivating influence acts in both directions. It is therefore a mutual influence between all immediately adjacent legs with the single exception of the two hind legs. Thus, these coactivating mechanisms provide an increase in the propulsive force of the whole animal.

If the hind leg steps onto the tarsus of the middle leg, the hind leg often lifts off again and is replaced

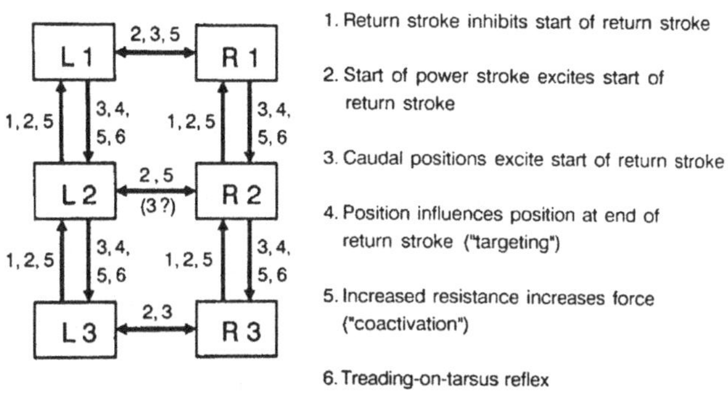

Fig. 5. Summary of the coordinating mechanisms operating between the legs of a stick insect (modified after Dean 1989)

slightly posterior without an interruption of walking. This treading-on-tarsus or TOT reflex can also be elicited experimentally by touching the middle leg tarsus with a fine brush. If the tarsal stimulus on the middle leg is applied during the early return stroke of the hind leg, this return stroke is prematurely terminated (Graham 1979, Schmitz and Haßfeld 1989). The TOT reflex was also found between middle and front legs.

The coactivating influences are not the only contralateral effects. As was found for the crayfish, coupling between contralateral legs is weaker than that between ipsilateral legs. However, two of the three mechanisms found for ipsilateral legs also act between contralateral legs of the same segment. These are the two "excitatory" mechanisms shown in Fig 4B,C for ipsilateral legs; both elicit the start of a return stroke in the other leg (Cruse and Knauth 1989). The first influence increases the probability that the contralateral neighbour will start a return stroke when one leg has started a power stroke (corresponding to Fig 4B). This influence exists in all three pairs of legs: front, middle and rear. The second influence corresponds to that shown in Fig 4C. The farther back one leg has moved during its power stroke, the higher is the probability that the contralateral neighbour starts a return stroke. Indications of this mechanism have only been found for front and for hind legs. For contralateral legs the first mechanism seems to be of greater importance and evidence for the latter is only indirect.

As in the crayfish these contralateral influences, unlike the ipsilateral coupling, seem to act in both directions between the two legs of one segment. Asymmetrical stepping patterns, i.e. patterns where contralateral legs do not walk in an exact alternating mode (phase of 0.5) but show phase values of about 0.4 or 0.6, commonly occur in the stick insect (Graham 1972). This asymmetry can be influenced by altering the load of the legs (Foth and Graham 1983b). This indicates that in insects, too, one side can dominate the other. As the motor output is increased when the leg walks under load, this result also indicates that a step pattern generator which produces a stronger motor output also produces coordinating signals to other legs which are of higher intensity. This is supported by experiments considering the amplitude of force oscillations of a leg standing on a fixed platform while the other legs are walking (Cruse and Saxler 1980). The correlation between strength

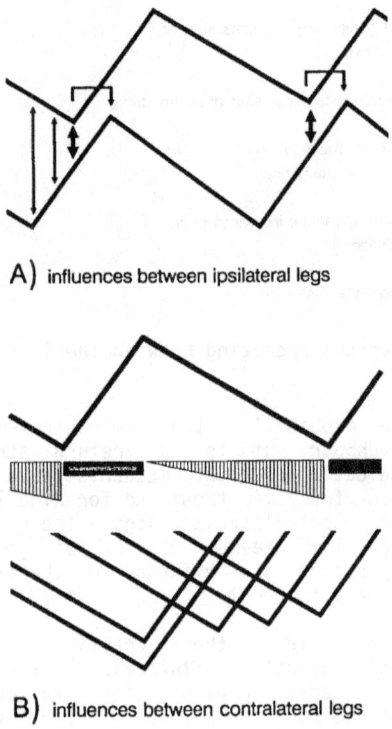

Fig. 6. Schematic representation of the coupling influences between (A) ipsilateral legs and (B) contralateral legs of the cat.
(A) The coupling mechanisms between front (upper trace) and ipsilateral hind leg (lower trace). The distance between both legs (two-way arrows) has to be smaller than a given threshold to start the return stroke of the front leg. After some delay the hind leg finishes return stroke and starts power stroke (bent arrow).
(B) The coupling mechanisms assumed to exist between contralateral legs are shown as if they acted only in one direction, from the leg shown in the upper trace to that shown in the lower trace. Several traces are shown for the influenced, lower leg to illustrate the effects resulting from the coupling mechanisms. During the return stroke of the influencing leg (black bar) the power stroke of the influenced leg may be prolonged because the start of the return stroke is inhibited. During the power stroke of the influencing leg the influenced leg is excited to start the return stroke, which will occur the earlier, the farther the position of the influencing leg has moved backwards (wedge).

of motor output and of coordinating influences can also explain the following result obtained in the crayfish (Cruse and Müller 1986). When the movement of one leg was briefly stopped during the power stroke, the normal coordination was regained in a way that shows that the coordinating influences from this single leg dominate the neighbouring legs. This was astonishing at first sight because one would expect that the combined influences of all the other, well-coordinated legs should dominate the deviant leg.

6. THE CAT

Investigation of slow-walking cat (Halbertsma 1983, Cruse and Warnecke 1991) showed the following results. Influences between contralateral legs, i.e. both front legs or both hind legs, are stronger than those occurring between ipsilateral legs, i.e. front and hind leg of the same side. As mentioned earlier, this finding is markedly different to the findings in arthropods. Influences which coordinate front legs seem to be of the same kind as those for the hind legs. These influences are in general symmetrical, which means that the same type of influence acts from right to left leg and in the reverse direction. Two types of influences are described for contralateral legs: first, when the influencing leg performs a return stroke, the influenced leg is prevented from starting or continuing a return stroke. Second, when the influencing leg performs a power stroke, the probability that the influenced leg starts a return stroke is the higher, the farther the influencing leg has moved backwards during its power stroke. This is shown in Fig 6 and qualitatively agrees with the finding in the stick insect. The "inhibitory" influence which in the stick insect is assumed to inhibit only the start of the return stroke, in the cat also seems to inhibit the continuation of an ongoing return stroke, thus "exciting" the start of a power stroke. As in arthropods, there seems to be some asymmetry in the sense that the leg of one side may be somewhat more dominant, which would mean that there is no qualitative, but possibly a quantitative difference between the legs of a contralateral pair.

In contrast to contralateral coupling, the ipsilateral mechanisms are asymmetrical, i.e., a qualitatively different influence acts from front to hind leg than does in the reverse direction. The front leg is influenced to start a return stroke when the hind leg has approached the front leg to a given distance. It is not the position of the legs in the body-fixed coordinate system, but the position of both legs relative to each other that constitutes the decisive criterion for the front leg to finish its power stroke. The hind leg is influenced to start a power stroke after the front leg has begun its return stroke. The signal for the hind leg to finish return stroke may either be calculated from the position relative to the front leg, or, another possible interpretation may be that the start of the front leg return stroke as such serves as a signal for the hind leg to finish return stroke. In any case, the switch in the hind leg follows after some delay, thereby producing the "understepping" of the hind legs and leading to an obvious overlapping between return strokes of adjacent legs.

7. MODELS

A model of the coordinated movement of walking legs can be used in three ways. First, as a short description of the experimental results. Second, it can serve as a tool to predict experimental results which occur, for example, after a disturbance of the normal leg movement. Third, it can be used to test whether a hypothetical assumption concerning the nature of the underlying mechanism does not contradict natural behaviour.

The models proposed up to now for different arthropods (for review see Cruse and Graham 1985) do not describe the detailed movements in each joint of the legs but are concerned only with the projection of the line between the leg basis and the leg tip on the horizontal plane. This means that only the forward-backward component of the movement is taken into account. Each leg is considered as a relaxation oscillator which in all models corresponds qualitatively to the one shown in Fig 1B. The earlier models were hierarchically organized in the sense that the flow of information between two neighbouring oscillators was only unidirectional (Wendler 1968, Graham 1972,1977, Cruse 1979b). A model with mutual, though symmetrical influence was proposed by Wilson (1966) for the cockroach (see also Pearson and Iles 1973). For the rock lobster Chasserat and Clarac (1986) proposed a model for the coupling between two legs where both legs influence each other. This was used for coupling between ipsilateral as well as contralateral legs. A model which is based on detailed behavioural investigations was presented by Cruse and Müller (1986) for the ipsilateral legs of the crayfish which has recently been completed by Müller and Cruse (1991b) also for the contralateral legs. This model contains the asymmetrical influences between ipsilateral legs (Fig 2) and the symmetrical influences between contralateral legs (Fig 3), and is probably the first comprehensive model based on behavioural data.

On the basis of measurements of the above-mentioned force oscillations (Cruse and Saxler 1980) a hypothetical model was proposed for the stick insect by Cruse (1980a,b) which contains several types of influences running in different directions. At least on a qualitative level most of these assumptions could be supported by later experimental investigations as described above. Using these experimental results (see Fig 5 for a summary) an extensive computer simulation has recently been successfully completed by Dean (in press). This simulation contains all mechanisms shown in Fig 5 except for mechanism no. 5 and 6. As mentioned, all these models describe only the horizontal projection of the leg movement. Very recently a complete kinematic and kinetic 16-body model of the stick insect was constructed by Pfeiffer et al. (1989) which can now be used as a basis for modelling the real movement of the legs. However, this task requires further information not only on the vertical component of the movement (see following section), but also on the question of how to control the movement of the individual joints of a leg.

8. STARTING, TURNING AND HEIGHT CONTROL

An animal does not always walk. Therefore it has to be possible to activate and inactivate that neuronal circuitry that produces the walking movements. This circuitry can be imagined to consist of a system schematically represented by Fig 1B which has to be completed by the coordinating influences. An animal can start walking spontaneously or by being excited by different external, e.g. tactile stimuli. Some crustacea were found to be easily stimulated to walk when the legs of the animal are moved rearwards by moving the underground relative to the animal using a motor driven belt (Ayers and Davis 1977). Other species respond by a resistive force to this situation (U. Müller, unpublished observation) and the same was found for the stick insect.

When the standing stick insect is stimulated to walk, for example, by a tactile stimulus, all legs that are on the ground immediately start with caudally directed forces (Cruse and Saxler 1980, Bässler et al. 1985, for locust: MacMillan and Kien 1983). This is supported by observations of the starting behaviour (Dean and Wendler 1984). This can be interpreted as showing that the neural circuitry of each leg always starts in the state of power stroke when activated after a pause. This is, however, not generally the case. Land (1972) reported for the jumping spider that a leg can also stop the movement within a return stroke and later continue walking by completing the interrupted return stroke.

An animal does not always walk in a straight line. How is turning possible? When a stick insect walks in a moderate curve the coordination between right and left legs can be maintained and only the step amplitude, i.e. the distance between AEP and PEP, increases for the legs on the outside and decreases for those on the inside of the curve (Graham 1972). When the curve is narrow, right and left legs can become uncoupled and the outside legs walk with higher frequency (for stick insect: Jander (1985); for cockroach: Camhi (1988)). In extreme cases legs on one side can walk forward and those of the other side walk backward so the insect turns in place (for honey bee: Zolotov et al. (1975), for stick insect: own observation). However, for the cockroach turning in place only changes in step amplitudes have been recorded (Franklin, Bell and Jander 1981). They also found higher step frequencies in both front legs compared to middle and hind legs but no apparent decoupling between right and left legs.

The coordination of legs in a walking animal is discussed here only with respect to the movement in the forward-rearward direction, i.e. the horizontal component of the movement. When walking over an uneven surface, however, the legs have to act together in order to control the height of the body and coordinate the movement in the vertical direction. The behaviour of a stick insect walking over different obstacles (Cruse 1976) or standing on different forms of substrate (Cruse, Riemenschneider and Stammer, 1989) can be quantitatively described according to a hypothesis that each leg acts as a vertical "spring" which is independent of the other legs; thus the mechanical connections via the body are sufficient and no neuronal coupling between the legs is necessary to control body height.

9. CONCLUDING REMARKS

The results, which up to now have been mainly based on behavioural data, show that each walking leg can be considered as an oscillating unit with the properties of a relaxation oscillator. While the internal organization of this oscillator seems to reflect a hierarchical structure, the coupling between the different oscillators is organized in a non-hierarchical manner. The coupling between two such oscillators is not continuous as is the case in many mathematical descriptions of coupled oscillators. In contrast, the coupling between walking legs seems to be active only within selected phase ranges and "switched off" at other times. Ipsilateral influences are asymmetrical in the sense that rostrally directed influences are in general qualitatively different to caudally directed ones. Contralateral mechanisms seem to be symmetrical although quantitative differences can be found such that one side is able to dominate the other to some extent. There is a redundancy in the sense that different mechanisms produce basically the same result, namely a proper coordination between legs. The existence of several different mechanisms provides the possibility of a faster recovery of normal coordination after a disturbance. It also achieves a higher stability of the whole system. To obtain stable relationships earlier theoretical considerations on coupling mechanisms which acted only in one direction had to assume the existence of a hierarchy of eigenfrequencies in the different oscillators. This means that in insects for example front legs, if uninfluenced by coordinating signals, walked faster than middle legs and these in turn walked faster than hind legs (see Cruse and Graham (1985) for review). Because of the combination of different caudally and rostrally directed coupling mechanisms this assumption is no longer necessary.

The mechanisms seem to be quite different in different animals. Comparative studies showed that the mechanism in the crayfish and in the rock lobster are very similar (Clarac 1985). Forward-directed inhibitory influences were found for the stick insect, for the grasshopper (Graham 1978) and probably exist in the cockroach. One might speculate that the differences found between crayfish and stick insect reflect those between decapods and insects. What might be the reason for the different strategies used for the coordination of legs in crayfish and in stick insects? A basic difference is that maintaining the stability of the body position is much less a problem for the crayfish walking under water and using eight legs than for the stick insect walking on land with only six legs and climbing on branches. Therefore the stick insect is under much more constraint to keep as many legs on the ground as possible at each moment. Thus one might speculate that in stick insects coordination has to be obtained by a shorter or longer prolongation of the power stroke, whereas the crayfish can afford to solve the coordination problem by maintaining legs for a longer time in the return stroke. The latter might be the simpler task from the viewpoint of computation because (a) stability problems arising when changing the end of the return stroke might generally be less critical than those when changing the end of the power stroke and because (b) the control of the leg movement is easier during the return stroke when the movement of the leg is not mechanically coupled to the other legs.

A qualitative difference found between arthropods and the cat is that in the latter contralateral coupling is stronger than ipsilateral coupling. This might result from the fact that in the cat the coordination between ipsilateral legs allows both legs to perform overlapping return strokes, something that has been rarely observed in arthropods. What might be the reason for this difference? Simultaneous lifting of two ipsilateral neighbouring legs, in principle causes an instability of the body, particularly in a four-legged animal. Therefore one should assume that this situation is avoided by the system controlling leg movement. Nevertheless, the instability is much less of a problem for mammals, such as cats, than for the crayfish or the stick insect. In mammals generally right and left legs are arranged below the centre of body mass, in complete contrast to arthropods. Thus, in the cat, the horizontal projection of the lever arm, by which gravity rotates the body, is much shorter, and the body is, to a larger extent, also supported by the legs of the other side of the body. Thus, the cat can afford coordinating mechanisms that have the effect of ipsilaterally overlapping return strokes.

ACKNOWLEDGEMENTS

I wish to thank A. Baker for proof reading the English manuscript.

REFERENCES

Ayers JL Davis WJ 1977. J comp Physiol 115, 1-27
Bässler U 1972. Kybernetik 10, 110-114
Bässler U 1977. Biol Cybern 25, 61-72
Bässler U 1979. Physiol Entomol 4, 193-199
Bässler U 1986. Biol Cybern 54, 65-69
Bässler U 1988. J exp Biol 136, 125-147
Bässler U Dübner C Fahrig T 1987. Zool Jb Physiol 91, 393-401
Bässler U Foth E Breutel G 1985. J exp Biol 116, 301-311
Camhi JM 1988. Experientia 44, 401-408
Clarac F 1985. In: Barnes, W.J.P. & Gladden, M.H. (Eds.) Feedback and motor control in invertebrates and vertebrates, Croom Helm, London, pp 379-400
Clarac F Chasserat C 1983. J exp Biol 107, 189-217
Clarac F Chasserat C 1986. Biol Cybern 55, 171-185
Cruse H 1976. Biol Cybern 24, 25-33
Cruse H 1979a. Physiol Entomol 4, 121-124
Cruse H 1979b. Biol Cybern 32, 107-113
Cruse H 1980a. Biol Cybern 37, 131-136
Cruse H 1980b. Biol Cybern 37, 137-144
Cruse H 1981. J exp Biol 92, 87-95
Cruse H 1985a. J exp Biol 114, 513-519
Cruse H 1985b. J exp Biol 116, 343-355
Cruse H 1985c. J exp Biol 116, 357-362
Cruse H 1985d. In: Gewecke,M. & Wendler,G. (Eds.) Insect Locomotion, Parey, Hamburg, Berlin pp 19-26
Cruse H Clarac F Chasserat C 1983. Biol Cybern 47, 87-94
Cruse H Epstein S 1982. J exp Biol 101, 161-170
Cruse H Graham D 1985. In: Bush, B.M.H. & Clarac,F. (Eds.) Co-ordination of Motor Behaviour, Soc. for Experimental Biology Seminar 24, Cambridge University Press, Cambridge, pp 283-301
Cruse H Knauth A 1989. J exp Biol 144, 199-213
Cruse H Müller U 1986. J exp Biol 121, 349-369
Cruse H Pflüger HJ 1981. J exp Biol 92, 97-107
Cruse H Riemenschneider D Stammer W 1989. Biol Cybern 61, 71-77

Cruse H Saxler G 1980. Biol Cybern 36, 159-163
Cruse H Schmitz J 1983. J exp Biol 102, 175-185
Cruse H Schwarze W 1988. J exp Biol 138, 455-469
Cruse H Warnecke H 1991. Exp Brain Res (submitted)
Dean J 1984. J comp Physiol A 155, 771-781
Dean J 1989. J exp Biol 145, 103-131
Dean J 1990. Biol Cybern 63, 115-120
Dean J 1991. Biol Cybern (submitted)
Dean J Wendler G 1982. J comp Physiol 148, 195-207
Dean J Wendler G 1983. J exp Biol 103, 75-94
Dean J Wendler G 1984. J exp Biol 110, 203-216
Delcomyn F Cocatre-Zilgien JH 1988. Biol Cybern 59, 379-384
Foth E Bässler U 1985a. Biol Cybern 51, 313-318
Foth E Bässler U 1985b. Biol Cybern 51, 319-324
Foth E Graham D 1983. Biol Cybern 48, 149-157
Franklin R Bell WJ Jander R 1981. J Insect Physiol 27, 249-255
Graham D 1972. J comp Physiol 81, 23-52
Graham D 1977. Biol Cybern 26, 187-198
Graham D 1978. J exp Biol 73, 159-172
Graham D 1979. Biol Cybern 32, 139-145
Halbertsma JM 1983. Acta Physiol Scand Suppl 521, 1-75
v.Holst E 1939. Ergebn Physiol 42, 228-306
Jander JP 1985. In: Gewecke,M. & Wendler,G. (Eds.) Insect locomotion, Parey, Hamburg, Berlin, pp 33-42
Land MF 1972. J exp Biol 57, 15-40
McMillan D Kien J 1983 Proc R Soc Lond B 218, 287-308
Müller U 1990 Thesis. Fac of Biology, Univ of Bielefeld
Müller U Clarac F 1990a. J exp Biol 148, 89-112
Müller U Clarac F 1990b. J exp Biol 148, 113-128
Müller U Cruse H 1991a. Biol Cybern (submitted)
Müller U Cruse H 1991b. Biol Cybern (submitted)
Pearson KG Franklin R 1984. Int'l J Robotics Res 3, 101-112
Pearson KG Iles FJ 1973. J exp Biol 58, 725-744
Pfeiffer F Weidemann HF Danowski P 1990. Proc of the 1990 IEEE Int Conf 3, 1458-1460
Schmitz J 1985. In: Gewecke,M. & Wendler,G. (Eds.) Insect locomotion, Parey, Hamburg, Berlin, pp 27-32
Schmitz J Haßfeld G 1989. J exp Biol 143, 373-388
Seyfarth EA Bohnenberger J 1980. Proc Intern Congr Arachnol 8, 249-255
Weiland G Koch U T 1987. J exp Biol 133, 137-156
Wendler G 1968. In: Marko,H. & Färber,G. (Eds.) Kybernetik 1968, Oldenbourg, München, Wien, pp 68-74
Wilson DM 1966. Ann Rev Entomol 11, 102-123
Zolotov V Frantsevich L Falk EM 1975. J comp Physiol 97, 339-353

A biological visuo-motor system: how dissimilar maps interact to produce behavior

Gerhard Manteuffel
Brain Research Institute (FB 2)
University of Bremen
POB 33 04 40
D-2800 Bremen, FRG

Abstract

Results from electrophysiology, anatomy, and simulations show that the tectal sensori-motor system in salamanders can be understood as a system that preserves retinal coordinates on both tectal hemispheres and establishes a head-centered three-dimensional coordinate system by a combination of the bilateral visual maps. Thus, the outputs of the sensory maps are results of retinal inputs and intertectal signal transfer. The sensory maps are connected to respective motor maps organized in the coordinates of the neck muscles. The bilateral distribution of excitation in the whole system enables the animal to perform head saccades directed toward a stimulus at an arbitrary location within its egocentric frame of reference.

1. The biological system

Research on the system controlling saccades in amphibians has largely concentrated on the visual input side. It is well established that moving stimuli, like squares of moderate size or bars elongated in the direction of movement, elicit saccades of the head with a high probability. The probability is reduced if stimuli of other shapes are used.

As in other vertebrates, saccades are triggered by the optic tectum and can be released by electrical stimulation of this area. Tectal output units are well activated by visual stimuli like squares or horizontal rectangles which are moved across their receptive fields (c.f. Himstedt et al., 1987). The tectum receives an orderly arranged input from the eyes, establishing a retinotopic map on which the nasal visual field is represented rostrally and the temporal visual field caudally; the superior visual field is represented close to the dorsal midline, whereas the ventro-lateral margin is stimulated by visual objects in the lower part of the visual field.

Recently we were able to show that an ipsilateral visual map exists as well. It is established by an information transfer from the other tectal hemisphere (that receives input from the contralateral eye). The ipsilateral map represents only the binocular part of the visual field and, therefore, covers only the rostral half of the optic tectum. The ipsilateral map is pointsymmetrical to the contralateral one and, thus, constitutes the ability to calculate binocular disparity (Manteuffel et al., 1989).

In comparison to the amount of knowledge on visual properties of tectal neurons, there are only few data on sensori-motor coupling in the saccade control system. It is known that electrical stimulation of the toad's tectum evokes saccades toward a location that roughly corresponds to the retinotopic map established by afferents from the contralateral eye (Ewert, 1967). However, recent stimulation experiments have shown that there exists no simple correlation between stimulation site, stimulation strength and evoked saccade (Jordan et al., 1990). It is likely that the motor map of the tectum in salamanders is not arranged in the coordinates of external space. The bimodal distribution of the populations of efferent neurons in a dorsal and a ventrolateral group with increasing density toward caudal tectal levels (Fig. 1) rather indicates a recruitment system arranged in the coordinates of the neck muscle system (Manteuffel, 1990; Naujoks-Manteuffel and Manteuffel, 1990). According to this principle, larger saccades would be evoked when more premotor neurons become activated. In fact large saccades are necessary toward temporal goals which are stimulating more posterior sites of the optic tectum in the visual domain.

The motor system executing the saccades in salamanders is comparatively simple. The basic structure of the arrangement of the muscles responsible for head movements has most likely been inherited from fish ancestors with a basic equipment of paired epaxial and hypaxial muscles (i.e. above and below the axis of the spinal column). With a joint intercalated between the posterior pole of the head and the first vertebra, vertical and horizontal articulations can occur. Therefore, the resulting head movements can be described best in the coordinates given by the directions of the forces of the two pairs of muscles. The bilateral epaxial muscles (m. intertransversarius capitis superior) insert at the ear capsule and the hypaxial m. rectus cervicis at the os triangulare. Both pairs of muscles insert caudally at transverse processes of the second and third vertebrae.

Saccades are largely ballistic movements in salamanders (Werner and Himstedt, 1985), in general falling short with larger horizontal angles. Therefore two ore more saccades are often needed to bring a target into the center of the visual field. The animals approach a prey by executing a saccade followed by a few steps of straight walk. If necessary, this sequence can occur repetitively

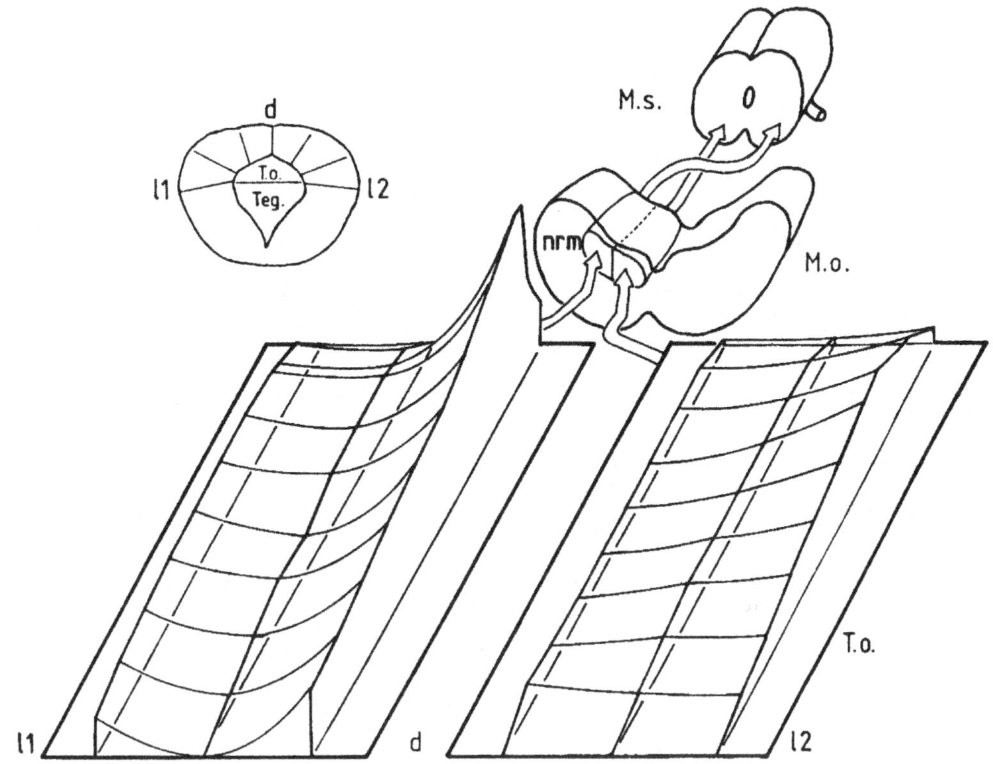

Fig. 1 Tectal motor map and the respective efferent descending projections to the bulbar reticular formation. The upper left cross section through the midbrain shows the optic tectum (T.o.) with the dorsal junction of both hemispheres (d) and the lateral tectal borders (l1, l2). The distribution of efferent neurons across and along both tectal hemispheres is shown schematically in the main diagram. The neuronal population giving rise to the crossed tecto-bulbar pathway is shown on the right, the population of the uncrossed pathway on the left side. The tectofugal pathways end at different portions of the nucleus reticularis medius (nrm). These portions have also different projections to the spinal cord (M.s.). According to these wiring, the tectal output reaches always the contralateral spinal motoneurons. Functionally, this means that each tectal hemisphere is able to control headmovements to the contralateral side.

if neccessary.

Monocular salamanders, however, regularly choose an indirect course toward a prey. They first deviate toward the side of the seeing eye and then curve backwards when further approaching the prey. This results in a "hooked" path that is particularly pronounced in some species of salamanders with a comparatively large binocular field of vision (Roth, 1987).

2. The model of the system

The knowledge on tectal structure and function summarized above was used to develop a model of this system. It was intended to use only physiological properties and parameters. In the cases where they were not known for salamanders, the assumptions were derived from data of similar systems.

The model was designed as shown in figure 2. It was planned as a computer simulation model, so no attempt was made to develop a complete mathematical description. The function of the system is described in the following.

The direction of the head is adjusted by the actions of the bilateral epaxial (E) and hypaxial (H) neck muscles driven by their motoneuron pools (M). These muscles allow essentially two-dimensional head movements within a coordinate system relative to the trunk (x", y"). Within this coordinate system the eigenvectors of the muscles have directions of 45°, 135°, 225°, and 315°.

The eyes are fixed in the head such that they are positioned obliquely to each other and to an orthogonal coordinate system given by the horizontal plane of the head and the respective vertical (x,y,z). The retinae of both eyes therefore represent two independent coordinate systems (x'r, y'r and x'l, y'l). Thus, an object located within the binocular field that is determined horizontally by the convergence angle alpha is represented differently in each coordinate system. The eyes of salamanders are well approximated by spheres where the node of the dioptric apparatus is in the center. Therefore, the retina is a hemispherical surface which can be interpreted as a coordinate system where the angles of light beams falling into the eye are directly transformed into two-dimensional coordinates. This ocular two-dimensional coordinate system is simplfied as a plane in figure 2.

Efferent retinal projections to the midbrain terminate in the optic tectum and transfer each of the ocular coordinate systems onto a single hemisphere. As a result of retino-tectal topography, the contralateral retinal coordinates are transferred to the visual layers of the optic tectum. Each tectal hemisphere also receives an input from the ipsilateral eye but with a reversed retinotopy. This is achieved by an intertectal transfer. The ipsilateral and contralateral topographies are arranged point-symmetrically as found in an experimental study (Manteuffel et al, 1989). Therefore, a more posterior excitation in one tectal hemisphere results in a more anterior excitation of the other hemisphere and vice versa. However, studies in frogs (Gaillard et al., 1988) showed that direct excitation from the contralateral eye is stronger than the indirect input from the other tectal hemishere. As a result, under binocular conditions each tectal hemishere receives a direct retinal input from the contralateral eye and a weaker input from the ipsilateral eye via the other tectal hemisphere.

In one hemisphere the indirect and the direct excitation are at the same site if the object evoking the input signal is at a certain distance (the system's horopter) and their disparity in the visual layer may be a measurement of object distances farther than the horopter. In the output layers, however, it is assumed that the contralateral excitation suppresses the ipsilateral one by means of lateral inhibition so that it is ensured that each visual hemisphere will have only one output. Under monocular conditions, where there is no direct retinal input to one tectal hemisphere, the indirect signal survives such that both tectal hemispheres then have an output.

The efferent system of the tectum has been analysed by neuroanatomical tracing (Naujoks-Manteuffel and Manteuffel, 1988, 1990). On both tectal sides the density of output cells increases from the rostral pole up to a very caudal level and then decreases rapidly (see also Fig. 1). The distribution is also non-homogeneous in a mediolateral direction. Efferent cells of the uncrossed tectobulbar tract are clustered medially and laterally and are sparse in the intermediate sector. Those giving rise to the crossed tract are sparse medially and increase in number toward the lateral margin.

It is assumed that the lateral and medial tectal efferent neurons are premotoric to the hypaxial and epaxial muscles, respectively. The rationale for this is that these neurons are located at a site of the visual map where objects that are obliquely above or below the animal evoke a neural response. This is also the direction of the eigenvectors of the epaxial and hypaxial muscles.

Since dendrites of tectal output cells arborize considerably over some 100 micrometers (Roth et al, 1990), they can receive input from more than one visual column. As a result, the intermediate visual zone of the tectum which is responsive to stimuli close to the visual horizon, will have an efference to both the pre-hypaxial and the pre-epaxial output neurons.Thus, the motor map of the tectum is apparently a map in the coordinates of the neck muscle system. Likewise, it is a parametric map because the number of premotor neurons at a particular site of the map can be considered as a representation of muscular contraction of a specific neck muscle. In the model, the numbers of efferent cells along and across the tectal output map was taken as changing continuously according to functions matching the empirical data from neuroanatomy.

The bilateral visual layers of the tectum may act locally (i.e. at the site of activation) and independently upon the respective motor maps. Any activity at a certain site on the visual map may recruit the respective

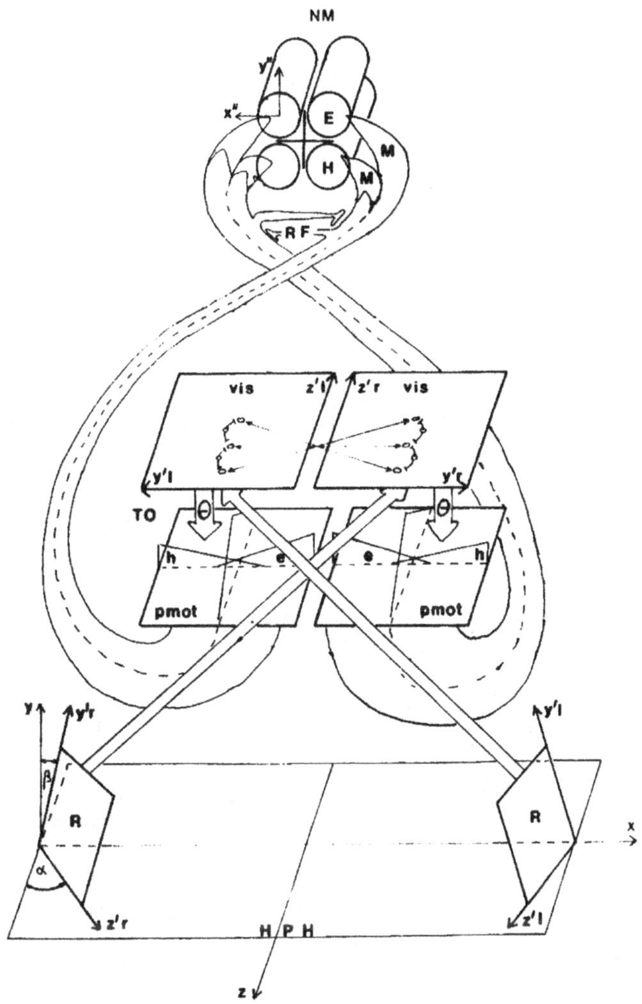

Fig. 2 Diagram of the model of the visual saccade generation in salamanders. Abbreviations (from sensory side to effectors): HPH: horizontal plane of head constituting the head centered coordinate system x,z, and the respective vertical y; R: retinal coordinate system y',z'of right (r) and left (l) eye (these coordinate systems are shifted against the head coordinate system horizontally <alpha> and vertically <beta>); vis: tectal visual system establishing the sensory map in retinal coordinates (arrows indicate order of tecto-tectal information transfer and the small circles with adjacent lines symbolize intercolumnar inhibition); pmot: premotor map of the optic tectum (TO), receiving input from the visual layer above via a threshold element (theta), (e,h: epaxial and hypaxial recruitment function across tectum, perpendicular to the recruitment function along the motor map); RF: reticular formation with mutual inhibition between both efferent pathways serving the "winner takes all" activation of only a single side; M: motoneuron population driving the epaxial (E) or hypaxial (H) neck muscles (NM) which constitute the motor coordinate system x", y" of head movements.

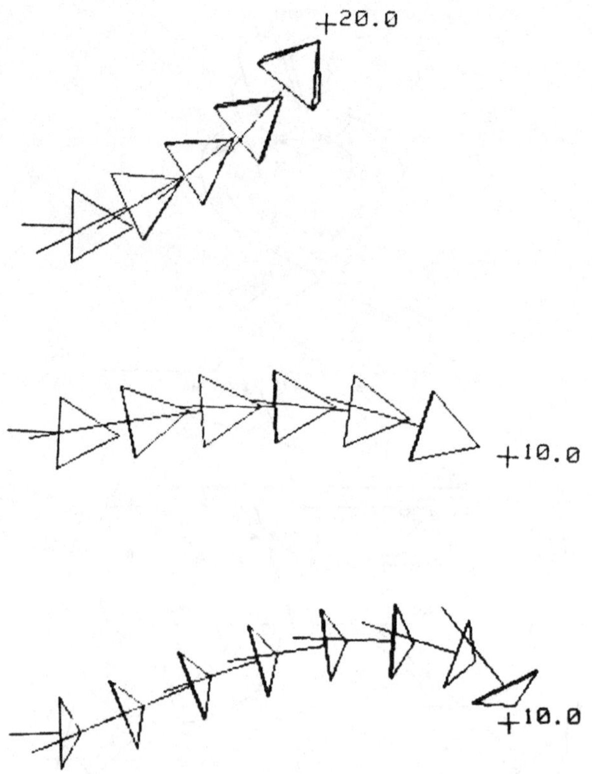

Fig. 3 Top: Binocular model salamander approaching and reaching a target (+) that is above body level (=10 in arbitrary units). The head is raised during approach. Middle: Monocular salamander (right eye removed) with moderate eye convergence approaching a target on a devious course. Bottom: Same situation as previous but the salamander has a more pronounced eye convergence (the convergence angle alpha is indicated at the base of the triangle symbolizing the head).

premotor neurons, whereby the degree of activity in the visual layer determines whether or not a threshold intercalated between both maps can be surpassed. However, this excitation determines only the probability of a saccadic movement to occur and not its precision (i.e. the turning angle). It is likely that the threshold is variable and controlled by motivational afferents arising from the amygdala which projects to the tectum (Finkenstädt et al., 1983; Rettig, 1988).

According to this hypothesis, both tectal sides convey independent outputs that may, however, differ in strength: a more posterior excitation within the visual map of one tectal hemisphere will result in a stronger premotor activity (since more neurons are recruited) than in the other premotor map triggered by a visual excitation located more rostrally. The two output channels are then considered to inhibiteach other , so that activation finally occurs only in the more active one.

The simulation of the model was performed in a three dimensional space. The program consisted of 8 basic steps:

1. Calculate object position within the coordinate systems of the eyes (object on retinal maps). Stop program if object is not represented in any of the two maps.
2. Transfer to tectal coordinates (bilateral tectal maps).
3. Calculate tectal coordinate on other hemisphere if monocular (tecto-tectal transfer).
4. Calculate recruitment strength according to the empirically grounded functions on the motor map at the location beneath visual map coordinate.
5. Find winner of antagonistic pairs of muscle recruitments.
6. Calculate compound vector of head movement.
7. New graphics.
8. Stop program if object is reached or calculate next foreward step of model salamander in the direction of the previous head angle rel. ground and continue.

Step 1 is actually consisting of several substeps since the positions of the object on the retinal coordinate systems depend on several parameters as horizontal and vertical angle of eye position relative to the head and head elevation (nose up or down). The program is ended if the object is out of sight of both eyes. (An angular coordinate system was chosen to mimic the physical projection pattern of the optics of the eye on the halfspherical retinal surface. Thus, only coordinates between 0 and 180 degrees were allowed, coordinates calculated outside of this range were interpreted as out of sight of the particular eye.)

The simulation reveals the typical head movements and lines of approach to a prey found in normal or monocularly blinded salamanders (Fig. 3). Most impressive, the strange approach of monocular salamanders toward prey is precisely reproduced. Such animals do not walk straight ahead to an object presented at a short distance exactly in front of them, but first deviate toward the side of the seeing eye. Somewhat later, when approaching closer, they curve sideways and eventually reach the object (Roth, 1987). This can be explained by the type of intertectal transfer between the visual maps of both sides. At the starting point, the tectum contralateral to the open eye receives a relatively caudal excitation and the transferred signal to the other tectum occurs rostrally. Hence, the output of the contralateral tectum is stronger and evokes turning away from the object (toward the seeing eye). During the approach, the object will stimulate more and more temporal positions of the retina and this results in an increasingly rostral position of excitation on the contralateral tectum. As a result of intertectal transfer, the tectal side ipsilateral to the eye will then be stimulated more and more caudally and will therefore have an increasing output. When the ipsilateral tectal output becomes stronger than the contralateral one, the animal turns to the side of the blinded eye and, thus, toward the object.

Horizontal and vertical as well as oblique saccades are properly produced by the model. Thus, objects at any position in the three-dimensional simulation space can be reached (as long as they are not elevated too high above ground level). Turning to objects which are positioned very temporally occur in several steps as it is found in living salamanders.

The model is extremely robust to changes of "somatic" parameters. Variations of the form parameters of the head are not able to disturb the performance of the model significantly. Not only is it possible to vary the interocular distance to a large degree but also changes in eye convergence angles do not have visible consequences on object reaching. An exception to this rule is found only under monocular conditions. The more the eyes converge (i.e. take a frontal position) the larger is the deviation of the course during monocular approach. Exactly the same is observed in salamanders, where species with more frontal eyes show the most extreme deviations when approaching a prey object compared to those with more lateral eye positions.

The robust properties of the system have mainly three causes. First, the system contains a feedback loop that is closed via the environment. Second, the intertectal transfer enables the system to keep on working even if it is occasionally monocular. Finally, the efferent recruitment system is not very critical at exact locations of excitation on the sensory map since the recruitment parameters do not change dramatically with various loci. Thus, fairly correct saccades are produced under a large variety of input vectors.

The basic philosophy of the presented model was to avoid any observer centered viewpoints and parameters. For example, there are no parameters like "retinal error" or "motor error" in the model. Such parameters are observer centered since only the observer can perceive "error" in relation to what she or he thinks to be the goal of the system. The model shows that it is unnecessary to assume the explicit internal representation of these types of errors. The power of the entire system to correct gaze errors rather is an emergent property of the physiological circuits which form specific maps.

In general, it may well be that even "higher" cognitive abilities can in principle be explained by interactions of spatial and parametric sensory, motor, and associative maps. This would imply that the so-called cognitive maps are distributed over many of the above mentioned specific and multimodal maps, thus representing emergent properties of interactions rather than real physical maps. These interactions can be controlled by comparatively crude motivational ("limbic", "instinctive", etc.) signals that depend on internal states and innate assessments. This will result in a rich set of possible behaviors with a finite set of internal maps, as it is the case in all developed invertebrate and vertebrate animals.

References

EWERT, J.-P. (1967) Aktivierung der Verhaltensfolge beim Beutefang der Erdkröte (*Bufo bufo* L.) durch elektrische Mittelhirnreizung. Z. vergl. Physiol., 54: 455-481.

FINKENSTÄDT, T., EBBESSON, S.O.E., EWERT, J.-P. (1983) Projections to the midbrain tectum in *Salamandra salamandra* L. Cell Tiss. Res., 234: 39-55.

GAILLARD, F., GARCIA, R., ROUSSEL, H. (1988) First neurophysiological approach to the retino-tectal projections in *Discoglossus pictus* (Anura). J. comp. Physiol., 162: 435-441.

HIMSTEDT, W., HELLER, K., MANTEUFFEL, G. (1987) Neuronal responses to moving visual stimuli in different thalamic and midbrain centers of *Salamandra salamandra* (L.). Zool. Jb. Physiol., 91: 243-256.

JORDAN, M., MATSUSHIMA, T., ROTH, G. (1990) Does the toad's tectum have a motor map? In: Brain, Perception, Cognition (N. Elsner & G. Roth, eds), p 86, Thieme Verlag, Stuttgart.

MANTEUFFEL, G. (1990) Interaction of tectal sensorimotor maps in salamander. In: Brain, Perception, Cognition (N. Elsner & G. Roth, eds), p 24, Thieme Verlag, Stuttgart.

MANTEUFFEL, G., FOX, B., ROTH, G. (1989) Topographic relationships of ipsi- and contralateral visual inputs to the rostral tectum opticum in the salamander *Plethodon jordani* indicate the presence of a horopter. Neuroscience Lett., 107: 105-109.

NAUJOKS-MANTEUFFEL, C., MANTEUFFEL. G. (1988) The origins of descending projections to the medulla oblongata and rostral medulla spinalis in the urodele *Salamandra salamandra* (Amphibia). J. comp. Neurol., 273: 187-206.

NAUJOKS-MANTEUFFEL, C., MANTEUFFEL, G. (1990) Quantitative distribution of descending tectal efferent cells in salamanders. Neuroscience Lett., in press.

RETTIG, G. (1988) Connections of the tectum opticum in two urodeles *Salamandra salamandra* and *Bolitoglossa subpalmata* with a special reference to the nucleus isthmi. J. Hirnforsch., 29: 5-16.

ROTH, G. (1987) Visual behavior in salamanders. Studies of Brain Function, Vol. 14, Springer, Berlin.

ROTH, G., NAUJOKS-MANTEUFFEL, C., GRUNWALD, W. (1990) Cytoarchitecture of the tectum mesencephali in salamanders: a Golgi and HRP study. J. comp. Neurol., 291: 27-42.

WERNER, C., HIMSTEDT, W. (1985) Mechanism of head orientation during prey capture in salamander (*Salamandra salamandra* L.). Zool. Jb. Physiol., 89: 359-368.

Biological and Computational Stereo Vision

Stephen T. Barnard

Artificial Intelligence Center
SRI International
333 Ravenswood Ave., Menlo Park, CA 94025

Abstract

A computational model of stereo fusion is examined in the light of biological and psychophysical knowledge of stereo vision in humans and other mammals. Several analogies are suggested, including the use of independent spatial-frequency channels with one-octave separation, the role of vergence and the limits of fusion, sensitivity to vertical disparity, and the use of 3 pools of disparity detectors. It is argued the similarity between the morphology of the visual cortex and the fine-grained, SIMD architecture exploited by the computational model leads to similar constraints on the computation of stereo disparity in both milieus, and therefore naturally leads to processes with similar properties.

1 Introduction

Scientific investigation of stereo vision in humans and other animals has an extensive history in neurobiology and psychology, dating from Wheatstone's discovery of the phenomenon in 1838 [19]. Recently computational modelers have made substantial progress in simulating the process of stereo fusion on the computer. This paper examines one such model in detail, and in particular points out some striking similarities between the model and current knowledge of stereo vision in higher mammals.

1.1 Stereo geometry

Stereo vision is a way of interpreting and exploiting visual information that is relatively well understood, in animals as well as machines. The reason is clear: compared to other perceptual cues for depth, the problem is well defined. Once the images are brought into point-to-point correspondence, recovering the third dimension is a straightforward application of trigonometry.

*The work described in this article was supported under DARPA contracts MDA903-86-C-0084, DACA76-85-C-0004, and 89F737300. Use of the Connection Machine was provided by DARPA.

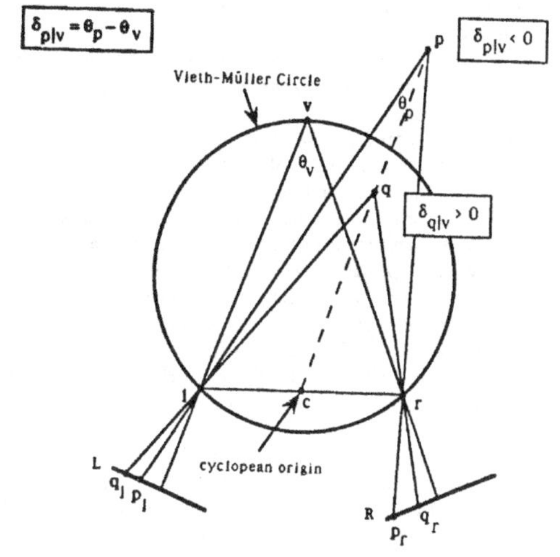

Figure 1: Basic Stereo Geometry

The geometrical principle behind stereo vision, illustrated in Figure 1, is quite simple. Assume that two cameras form images through left and right centers of perspective **l** and **r**, onto planes L and R. (In practice these would be imperfect optical lens systems, but for this discussion we assume ideal "pinhole" projections.) Furthermore, assume that the cameras are fixed upon point **v**, which is to say that the two rays perpendicular to the image planes passing through the centers of perspective (the *principle rays*) intersect at **v**. Let θ_v be the angle between these principle rays. We say that the *absolute* disparity of **v** is θ_v. Now consider another point **p** projected onto image planes L and R as shown, and let the angle between these rays be θ_p. We say that the *relative* disparity of p with respect to v is $\delta_{p|v} = \theta_p - \theta_v$. Relative disparity is the more commonly used definition.

The circle through **l**, **r**, and **v** (actually a sphere) is called the *Vieth-Müller circle* (closely related to the horopter discussed below in Section 1.3). Note that relative disparity $\delta_{p|v}$ is positive for points inside, negative for points outside, and zero for points on the circle. Positive and negative relative disparities are sometimes referred to as *crossed* and *uncrossed*, respectively.

Clearly, disparity is related to distance. Let $\mathbf{c} = (\mathbf{l} + \mathbf{r})/2$ be the midpoint between the focal points (the *cyclopean origin*), and $\mathbf{p}$ and $\mathbf{q}$ be two points on a line with $\mathbf{c}$; if $|\mathbf{p} - \mathbf{c}| > |\mathbf{q} - \mathbf{c}|$ (i.e., $\mathbf{p}$ is farther from the observer than $\mathbf{q}$) then $\delta_p < \delta_q$. The two projections of a single point in the scene are called *conjugate points*, and they determine the location of the point in the scene uniquely.

This property of conjugate points establishes the importance of the *correspondence problem*: how can one determine which pairs of points in the two images correspond to actual points on surfaces in the scene? The scene is presumably made of light-reflecting surfaces, and individual markings on these surfaces will, in some sense, look about the same in the two images. In realistic imagery the correspondence problem is much harder than it may first appear, however, due to many complicating factors such as occlusion (in which some parts of the scene are seen only by one camera), periodic surface markings, surface areas with no markings at all, distortion of the surface markings in the images dues to perspective projection, sensor noise, optical distortion, etc.

Knowledge of the relative orientation of the two cameras makes the correspondence problem much more tractable by reducing the dimensionality of the search space. Any point in three-dimensional space, together with the centers of projection of the two camera systems, defines a plane called an *epipolar plane* (Figure 2). The intersection of an epipolar plane with an image plane is called an *epipolar line*. Every point on an epipolar line in one image must correspond to a point on the corresponding epipolar line in the other image. This constraint, often called the *epipolar constraint*, therefore limits the search for the match of any point in the first image to a one-dimensional neighborhood in the second image, as opposed to a two-dimensional neighborhood, with an enormous reduction in computational complexity.

1.2 Stereo vision in higher mammals

Much is known about the biological and psychological basis for stereo vision, but there is currently no complete explanation for how the brain exploits binocular information. (See Kandel [12] and Kuffler [13] for detailed information.) The two retinas of the human visual system signal their response to visual stimuli through the optic nerves. The optic nerves partially cross in the optic chiasma, carrying information from the left half of the visual field to the right half of the brain, and information from the right half of the visual field to the left half of the brain. The visual field projects to an area on the back surface of the brain called the primary visual cortex (also called striate cortex or area 17). The primary visual cortex is almost certainly the place where the first stereoptic computation is performed.

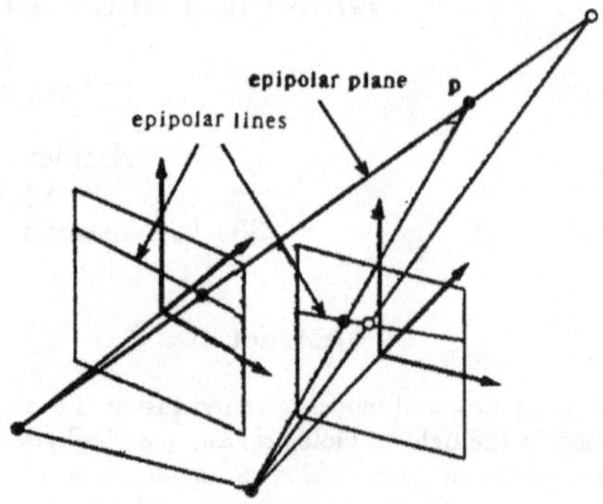

Figure 2: The epipolar constraint.
Two camera systems with different orientations are shown. A point p in space and the two centers of projection define an epipolar plane, which intersects the images in two epipolar lines. Any other point in the epipolar plane (for example, the point shown as an open circle) must project to image locations on the epipolar lines. Therefore, searches for corresponding points in the images can be restricted to one dimension.

The visual cortex is a sheet of half a dozen distinct layers of cells. It is morphologically uniform across its entire expanse: groups of cells on one part of the sheet appear essentially the same as groups of cells anywhere else on the sheet. The visual stimulus is *topographically mapped* to the cortex: each part of the cortex responds to stimulation at specific location on the retina, and there are no discontinuities except for the separation of the visual field into left and right halves.

Visual stimulation enters the cortex at one layer (layer IV), and then projects to other layers *locally*, spreading only about 1mm to neighboring cortical cells. The cells in layer IV, (so-called "simple" cells) are strictly monocular: they respond exclusively to one input from one eye. The "complex" cells in the subsequent layers, however, can be stimulated by both eyes to greater or lesser degrees. Globally, the visual cortex is organized into *ocular-dominance columns*. The preference for one eye (a complete preference in layer IV) entends through the width of the cortex (i.e., through a column of cells normal to the surface of the cortex), and eye-preference is also highly correlated between neighboring columns. Ocular dominance columns are thought to play an important role in stereopsis, but the mechanism is not understood. To a large degree this organization is innate,

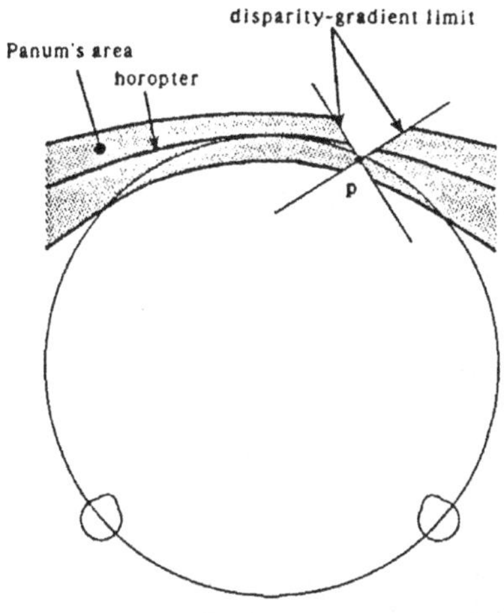

Figure 3. Limits to fusion.

but it does undergo further elaboration and sharpening after birth. Depriving one eye of stimulation after birth causes the formation of deviant ocular dominance columns [10] dominated by the active eye. In addition, there is a psychological effect that is probably related: the development of stereo vision passes through a critical period of postnatal development. If an animal is not exposed to stereoscopic stimulus during this relatively short period, it will never be able fully to develop stereo perception.

The visual cortex is organized in a similar way with respect to *orientation*. Simple cells in layer IV are not sensitive to orientation, but the complex cells they stimulate are. Furthermore, orientation specificity varies in an orderly, continuous pattern across neighboring cortical columns. The orientation system and the ocular dominance system seem to be independent of one another [10].

1.3 Limits to stereo fusion and depth perception

The locality of interaction on the cortex apparently limits the fusion of the two visual fields in human stereopsis. The range over which fusion occurs is called *Panum's area* (Figure 3) [15]. Points inside Panum's area are seen as single points (fusion), while points outside are seen double (diplopia). It is known that while fusion is an important feature of stereo vision, it is not absolutely necessary — even diploptic stimuli can be a cue for depth within limits. The largest disparity that still gives rise to a single, fused image varies from 6 minutes of arc at the center of the visual field to 20 minutes of arc at a peripheral angle of 6 degrees [7]. The thresholds for diploptic stereo perception are about four times those for fusion. The center of Panum's area, shown in the figure as a curve passing through the fixation point, is called the *horopter*. Points on the horopter project to identical positions on the two retinas (i.e., they have zero disparity). Note that the horopter does not follow the Vieth-Müller circle precisely because the geometry and optics of the ocular system don't quite satisfy the ideal model of Figure 1.

Experiments with stabilized images, in which the subjects were prevented from using eye movements to change vergence, have revealed that the limits to fusion are actually much more complex than Panum's area [7]. For example, fusion exhibits a time dependence, called *hysteresis*: as the disparity of fused, stabilized images is increased, they remain fused far beyond Panum's area, until they suddenly "break away" into diplopia. Also, the limits to fusion for stimuli rich in features, such as random dot stereograms (see Section 1.4), is much greater than for simple stimuli such as individual lines.

Another limit on fusion is the *disparity-gradient limit* [2], which states that for fusion to occur the rate of change of disparity with respect to visual angle (a dimensionless number) must be bounded. Most observers have a limit of about one. While Panum's area is an absolute limit of fusion, the disparity-gradient limit is more subtle. It implies that fusion of one point excludes the possibility of fusion of some nearby points, even if they fall within Panum's area. As shown in Figure 3, if a point p is fused then no points in a cone centered on p can be fused simultaneously. It has been shown [18] that if an imaged surface has disparity-gradient limit of less than two, then the left and right images are *topologically equivalent*; that is, all points in one image are visible in the other (i.e., the surface is not self-occluding), and furthermore the ordering of points is the same in both images. The disparity-gradient limit has been used explicitly to limit search for correspondences in at least one computational model [16].

1.4 The Modularity of Stereo Perception

The question of whether stereopsis is inextricably linked to other modes of visual perception, or whether it can stand alone as an independent mode, was clarified by the invention of the random-dot stereogram [11]. A random-dot stereogram consists of two synthetic images of random points, which are constructed to depict perspective views of the same virtual surface (Figure 4). A random dot stereogram can be constructed as follows. The points in, say, the left image are randomly placed. The right image consists of these same points, properly displaced to provide disparity values consistent with another view of some scene model. Each image by itself usually contains no depth information because it consists only of

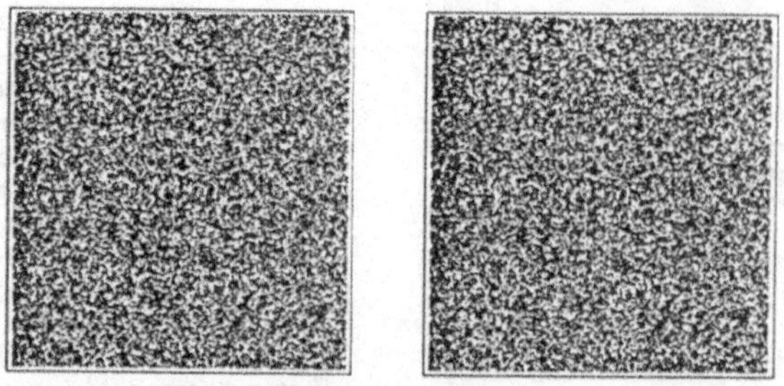

Figure 4: Random-dot stereogram.

random dots. When the pair are viewed stereoscopically, however, the surface is readily perceived. Stereo vision is therefore capable of producing depth perception independently of any monocular cues such as size constancy, texture gradients, shading, and so on. Random-dot stereograms enable systematic comparison of human and machine performance because their parameters (such as noise, dot density, and the shape of the virtual surface) can be controlled precisely.

1.5 Computational approaches

Historically, the computational modeling of stereo vision has been driven by two motivations. The practical applications of automated stereo are so important, especially in cartography and robotics, that many engineering-oriented approaches have been tried. These often use "correlation" techniques: patches of intensities in one image are searched for in the other image by maximizing a measure of correlation or minimizing a measure of error. The other motivation is the desire to model biological stereo, and these approaches are typically feature-based: discrete local features (usually edges) are matched across images. Another approach, which could be called the optimization approach, represents a dense disparity map *directly* as the state variable of a system (usually defined on a two-dimensional grid, or lattice). Stereo fusion is then formulated as an optimization problem: find the best disparity map by maximizing an objective function that measures the "quality" of the map, or, equivalently, minimizing an "energy" function that measures the lack of quality.

2 Stereo as Optimization (CYCLOPS)

Optimization approaches to stereo usually involves minimizing some variant of the following function, the *stereo constraint equation*:

$$\mathcal{E} = \int\int [(\mathcal{L}(x-\frac{\mathcal{D}}{2},y)-\mathcal{R}(x+\frac{\mathcal{D}}{2},y))^2+\lambda(\nabla\mathcal{D})^2]dxdy\,,$$

where $\mathcal{L}$ and $\mathcal{R}$ are functions of image intensities (usually bandpassed images), $\mathcal{D} = \mathcal{D}(x,y)$ is a disparity map (a cyclopean map in this case), and λ is a regularization constant. The first term in the integrand is a measure of the photometric error associated with $\mathcal{D}$, and the second term (proportional to the squared magnitude of the gradient of disparity) is a measure of the first-order complexity of $\mathcal{D}$. By minimizing $\mathcal{E}$ one finds the flattest map with least photometric error, with the tradeoff between complexity and error specified by the regularization constant λ. Note that the epipolar constraint is used because only horizontal disparities are allowed.

CYCLOPS is a multigrid, stochastic-optimization system implemented on the Connection Machine. It is intended as a practical, fast, and reliable system for automated stereo, for application in cartography and, eventually, real-time robot vision. Although the motivation behind CYCLOPS is pragmatic, there are several remarkable similarities to biological stereo vision.

2.1 Bandpass pyramids

The multigrid strategy of CYCLOPS starts by converting each image into a Laplacian pyramid [3], in which the image is separated into band-limited components separated by one octave. Furthermore, each band is sampled at a rate half that of the next higher-frequency band, producing a pyramid structure. There is a particularly efficient method of computing a Laplacian pyramid using a small 1D recursive kernal that is approximately Gaussian and exploiting the separability of the Gaussian [3]. Bandpass components are then produced as difference-of-gaussians (DOG), which is a close approximation to the $\nabla^2 G$ kernal which is commonly used in other models. The common feature of these kernals is that they have the "Mexican hat" shape characteristic of the center-surround receptive fields in the early visual system.

There are several benefits to using the Laplacian pyramid:

1. Disparity has a linear scaling property: if the sampling rate of the images is reduced by a factor of 2, then the range of disparity is also reduced by a factor of 2. Therefore, at some very coarse sampling rate, disparity is negligible. A matching system can start at a coarse level of resolution, find a coarse disparity relatively quickly, and then use it to initialize the search at the next higher octave, etc. This is exactly the way the multigrid search of CYCLOPS works.

2. Raw image intensities are not usually invariant across stereo images for a variety of reasons, including sensor bias and illumination effects. However, these systematic sources of error primarily affect the low-frequency components of the image. Using bandpassed images provides a kind of automatic gain control, similar to the effect of lateral inhibition in neural systems.

3. It may be impossible to eliminate vertical disparity completely. That is, the epipolar constraint may be satisfied only approximately. Nevertheless, vertical disparity will always be much smaller than horizontal disparity, and at the coarser resolutions will be negligible. The resolution pyramid therefore allows a degree of tolerance to vertical disparity.

CYCLOPS constructs truncated Laplacian pyramids from each epipolar-corrected image:

$$\{(L_l, R_l) | l = 0, \ldots, l_{max}\} \ .$$

The lowest level of resolution, l_{max}, is determined by the size of the original images and by a limit on the range of disparity between them. For typical aerial images with which we work the lowest resolution in the pyramids would be 32 × 32. If the original images were 1024 × 1024, the pyramids would be six levels deep and the system would accommodate a disparity range of 127 pixels.

2.2 The three-pools mechanism

CYCLOPS uses a version of simulated annealing to minimize a discrete version of the stereo constraint equation at each level of resolution, l:

$$\begin{aligned} E_l &= \sum_i \sum_j (L_l(i - \left\lfloor \frac{D_l}{2} \right\rfloor, j) - R_l(i + \left\lceil \frac{D_l}{2} \right\rceil, j)^2 \\ &+ \sum_i \sum_j \lambda \, |\nabla D_l(i,j)|^2 \qquad (1) \end{aligned}$$

where L_l and R_l are components of the lth components of the Laplacian pyramids, D_l is a discrete disparity map, and the operator $|\nabla \cdot|^2$ is now a discrete approximation of the squared gradient magnitude computed over four nearest neighbors.

The discrete disparity map, can be represented within the system in different ways. In earlier versions of CYCLOPS, D was represented as a grid of 8-bit signed integers and could attain any values (within this range) at any level. In the current system, however, a representation called the *three-pools mechanism* is used to obtain more efficiency without sacrificing dynamic range. It works as follows: At each level of the resolution pyramid the disparity map D is the sum of a constant *base* disparity (inherited from the previous level) and a variable *incremental* disparity:

$$D_l(i,j) = D_l^0(i,j) + \delta_l(i,j)$$

with $\delta_l(i,j) \in \{-1, 0, 1\}$. Now δ_l is the state variable of the system and is represented as a signed 2-bit integer. The local photometric errors (the terms in the first sum in equation (1)) are tabulated into the three possible values they can attain at each level. Note that incremental disparity is interpreted as a correction to the coarse answer obtained from the previous level. In this way the system can handle a substantial range of disparity, but the search space at each level of the pyramid is kept as small as possible.

The three-pools mechanism is illustrated in Figure 5. The optimization algorithm assigns disparities at the coarsest level as shown by the filled-in circles in 5(a). It then sets the base disparity (the shaded circles) at the next higher level 5(b) of the resolution hierarchy as specified by the following equation:

$$D_l^0(i,j) = 2D_{l+1}(\lfloor i/2 \rfloor, \lfloor j/2 \rfloor) \ .$$

At this level the incremental disparity is allowed to vary between -1 and 1 during the next optimization phase, as indicated by the short arrows in the figure. After the optimization determines the best incremental disparity, it uses the composite disparity in 5(c), $D_l^0(i,j) + \delta_l(i,j)$, to set the base disparity of the next level.

Note that if the difference in composite disparity between neighboring grid points becomes too large the points become decoupled. This is caused by a threshold in the squared-gradient-magnitude operator:

$$|\nabla D(i,j)|^2 = \sum_{\{i',j'\} \in \mathcal{N}(i,j)} \min\left(1, (D(i,j) - D(i',j'))^2\right)$$

where $\mathcal{N}(i,j)$ specifies the four nearest neighbors of $\{i,j\}$.[1] Once a component of the squared disparity gradient becomes greater than 1, further separation of the disparity values between neighboring points has no effect on the second term in equation (1). The effect of this decoupling is that the system can establish sharper estimates for disparity along occlusion boundaries.

[1] A value of $\lambda = 64$ seems to work well for this operator, assuming that the images are digitized properly into 256 grey levels.

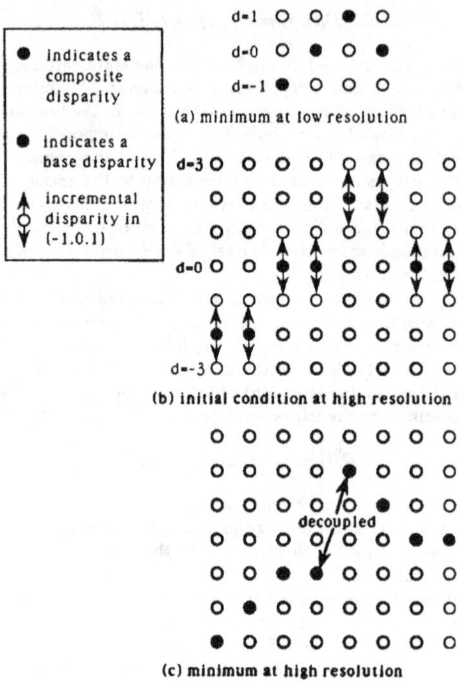

Figure 5. The three-pools mechanism.

2.3 Stochastic optimization

The method for minimizing E at each level, called microcanonical annealing, is somewhat different from the standard forms of simulated annealing. It has been adapted from a result in statistical physics [4] to be particularly efficient on data-parallel computers with no floating-point hardware. Perhaps the most interesting property of microcanonical annealing is that, unlike the standard forms, it doesn't use temperature as a control parameter. Instead, it represents kinetic energy explicitly with a grid of demons. Temperature is *measured* as a statistical feature of the system (the average demon energy). Details can be found in [1].

Like conventional simulated annealing methods, the microcanonical algorithm generates a sequence of states, $\{S_k\}$, that in the limit converges to a Boltzman distribution:

$$Pr(S) \propto e^{-E_S/\beta},$$

where β is temperature (in units of energy), which characterizes the rate of decrease of the probability of a state with increasing energy. Simulated annealing works by gradually reducing β,[2] causing the distribution of states to "relax" from a high-temperature, highly uncertain ensemble into a minimum-energy state at $\beta = 0$. In practice, the algorithm does not find the absolute minimum, but rather a good approximation to the minimum with high probability. The major advantage of simulated annealing over deterministic gradient-descent methods is that, by permitting "uphill" moves to states of higher energy, it can avoid getting trapped in locally stable states.

Subpixel estimates of disparity can be obtained by averaging over many states. When the finest level of resolution has been obtained the system is not cooled to $\beta = 0$; instead, it is heated to a a fairly high temperature and a sequence of disparity maps is generated. The average disparity (over this sequence) quickly converges to a good subpixel estimate. In terms of the discussion to follow, this technique averages over samples from the posterior distribution.

2.4 A Bayesian interpretation

One view of the computation is that of maximum *a posteriori* (MAP) estimation [8]. We observe (L, R) which contains information about D. The Bayesian approach is to find D that maximizes $Pr\{D|(L, R)\}$. Applying Bayes' Rule we have

$$Pr\{D|(L,R)\} = \frac{Pr\{(L,R)|D\}Pr\{D\}}{Pr\{(L,R)\}}$$

The probability in the the denominator, $Pr\{(L, R)\}$, is constant, so the posterior probability $Pr\{D|(L, R)\}$ is

[2] Or, in the case of microcanonical annealing, reducing kinetic energy.

proportional to the numerator:

$$Pr\{D|(L,R)\} \propto Pr\{(L,R)|D\}Pr\{D\} .$$

$Pr\{(L,R)|D\}$ is determined by the photometric error in the stereo constraint equation. (Given a disparity map, image pairs with less error are more likely.) The prior probability $Pr\{D\}$ is determined by the variational term. (Disparity maps that are flatter are more likely.)

The state variable of CYCLOPS, that is to say the discrete cyclopean disparity map D, is an MRF whose distribution of energies is guaranteed to be Boltzman in the limit. Writing the probabilities in terms of the distributions we have

$$\begin{aligned} Pr\{D|(L,R)\} &\propto \left(e^{-E_1(L,R,D)/\beta}\right)\left(e^{-E_2(D)/\beta}\right) \\ &\propto e^{-(E_1+E_2)/\beta} \\ &\propto e^{-E/\beta} \end{aligned}$$

where E_1 and E_2 are the error and the variational terms in E, respectively. Therefore, Bayes' Rule and tells us that by minimizing E we are maximizing the posterior probability $Pr\{D|(L,R)\}$.

3 Analogies to Biological Stereo

The design of CYCLOPS was intended to perform dense stereo fusion as efficiently as possible on massively-parallel, locally-connected hardware; there was no conscious intent to model biological stereo vision. Nevertheless, there are several nontrivial analogies to be drawn between this computational model and current knowledge of the human visual system. CYCLOPS is by no means unique among computational models in this respect. The MPG (Marr, Poggio, Grimson) model, in particular, which was expressly based on neurobiological and psychophysical evidence [14, 9], shares many of the features of CYCLOPS.

3.1 Independent spatial-frequency channels

There is considerable evidence that the human visual system has at least four, and possibly more, independent channels tuned to different spatial frequencies separated by about one octave [20]. The Laplacian pyramid in CYCLOPS and similar structures in other models play the same role. The reasons why this encoding is important in CYCLOPS, discussed in Section 2 (efficiency of coarse-to-fine search, automatic gain control, and less sensitivity to vertical disparity), suggest functions these channels may perform in human vision.

3.2 Vergence eye movements and limits to fusion

The existence of Panum's area, which is a quite severe limit, means that the human visual system must resort to vergence eye movements to build a stereo interpretation in scenes with large ranges of depth. CYCLOPS has a kind of Panum's area of its own: if we use, say, five levels of hierarchy, then the system can accomodate at most 63 pixels of disparity. In typical cartographic images of rough terrain, we could expect to be within range of disparity using typical sampling rates, so CYCLOPS can work in the application for which it is primarily intended without vergence control. If it is to be applied to the oblique ground-level scenes found in robot vision applications, however, it will have to use vergence control. This would seem to require a *selective fusion* ability: the system should be able to distinguish those parts of the visual field that are fused from those that aren't.

3.3 Vertical disparity

Human stereo vision can deal with a small amount of vertical disparity, but it resorts to eye movements to align the visual fields as closely as possible [5, 6]. CYCLOPS uses information about the camera orientation to remove vertical disparity. Error in the camera model may cause some residual vertical disparity, but the simulated-annealing search is robust enough to handle this under reasonable conditions. Apparently, the search-space reduction of the epipolar constraint is so useful that vertical disparity is to be avoided as much as possible, but must be tolerated to degree.

3.4 Three pools of disparity

Studies of anamolous stereo vision suggests that there are three "pools" of disparity detectors, corresponding roughly to crossed, uncrosssed, and near-zero disparity [17]. CYCLOPS uses the same representation, coupled with several octaves of hierarchy, to achieve efficiency without sacrificing dynamic range. At every level in the hierarchy, components of the incremental disparity can assume only three local values: -1 (crossed), 1 (uncrossed), and zero. The base disparity, however, can grow by a factor of two across every level. The search space in any one level is therefore kept to a minimum, while the final composite disparity is allowed to have a substantial range.

3.5 Visual cortex

The visual cortex has a remarkably uniform structure, as descibed by Hubel and Weisel:

> **Given what has been learned about the primary visual cortex, it is clear that one can consider an elementary piece of cortex to be a block about a millimeter square and two millimeters deep. To know the organization of this chunk of**

> **tissue is to know the organization for all of area 17; the whole must be mainly an interated version of this elementary unit.** [10]

As a computational substrate, the visual cortex is quite similar to a large grid of locally-connected processors. Since the cortical columns appear to be anotomically and functionally equivalent, they can be thought of as "running the same program" on different parts of the visual field, and therefore can be simulated efficiently by a SIMD (single-instruction, multiple-data) architecture.

4 Conclusions

It is no accident that we find many parallels between computational stereo models and current knowledge of stereo vision in higher mammals. Historically, the formulation of models was often guided by biological knowledge (in particular, the MPG model). From a pragmatic standpoint, machine models and biological neural systems operate under similar constraints, and this is increasingly so with the introduction of massively parallel architectures so similar to cortical structure.

I don't suggest that the cortex is literally performing a simulated-annealing computation, but rather that the cortex is a system of locally connected, essential identical components that can be *described* and *simulated* with statistical methods, and that minimizing the stereo constraint equation is a good hypothetical candidate for one computation that it performs.

References

[1] Stephen T. Barnard, "Stochastic stereo matching over scale," *Int. J. Comp. Vision*, vol. 3, 17-32, 1989.

[2] P. Burt and B. Julesz, "A disparity gradient limit for binocular fusion," *Science*, vol. 208, 615-617, 1980.

[3] P. Burt, "The Laplacian pyramid as a compact image code," *IEEE Trans. on Communications*, vol. 31, 532-540, 1983.

[4] M. Creutz, "Microcanonical Monte Carlo simulation," *Phys. Rev. Lett.* 50, 1411-1414, 1983.

[5] A. L. Duwaer and G. van den Brink, "What is the diplopia threshold?" *Vision Res.*, vol. 20, 295-309, 1981.

[6] A. L. Duwaer and G. van den Brink, "Diplopia thresholds and the initiation of vergence eyemovements," *Vision Res.*, vol. 21, 1727-1737, 1981.

[7] Fender, D. and B Julesz, "Extension of Panum's fusional area in binocularly stabilized vision, J. of the Opt. Soc. of Amer., Vol. 57, No. 6, 819-830, June 1967.

[8] Geman, S. and D. Geman, "Stochastic relaxation, Gibbs distributions, and Bayesian restoration of images," *IEEE Trans. PAMI*, vol. 6, 721-741, 1984.

[9] W. Eric L. Grimson, "Computational experiments with a feature based stereo algorithm," *IEEE Trans. Patt. Anal. and Mach. Intell.*, vol. PAMI-7, no. 1, January 1985.

[10] David H. Hubel and Torsten S. Weisel, "Brain mechanisms of vision," in *the Brain*, W. H. Freeman and Co., New York, 1979.

[11] B. Julesz, *Foundations of Cyclopean Perception*, Univ. of Chicago Press, Chicago, 1971.

[12] Eric R. Kandel and James H. Schwartz, *Principles of Neural Science*, 2nd ed., Elsevier Science Publishing Co., New York, 1985.

[13] Stephen W. Kuffler, John G. Nicholls, and A. Robert Martin, *From Neuron to Brain*, 2nd. ed., Sinauer Associates Inc., Sunderland, Massachusetts, 1984.

[14] D. Marr and T. Poggio, "A theory of human stereo vision," *Proc. Roy. Soc. London*, vol. B 204, 301-328, 1979.

[15] Ogle, K. N., *Researches in Binocular Vision*, Saunders, Philadelphia, 1950.

[16] Pollard, S. B., J. E. W. Mayhew, and J. P. Frisby, PMF: A stereo correspondence algorithm using a disparity gradient limit, *Perception*, Vol. 14, 449-470, 1985.

[17] Whitman Richards, "Anomalous stereoscopic depth perception," *J. Opt. Soc. Amer.*, vol. 61, no. 3, 410-414, March 1971.

[18] Trivedi, H. P. and S. A. Lloyd, The role of disparity gradient in stereo vision, *Perception*, Vol. 14, 685-690, 1985.

[19] C. Wheatstone, *Phil. Trans. Roy. Soc. B* (London) 371, 1838.

[20] Hugh R. Wilson and James R. Bergen, "A four mechanism model for threshold spatial vision," *Vision Res.*, vol. 19, 19-32, 1979.

MODELING AND SIMULATION OF ANIMALS' MOVEMENTS (*)

Simon BENHAMOU & Pierre BOVET
Centre National de la Recherche Scientifique
Laboratoire de Neurosciences Fonctionnelles
F-13402 Marseille Cedex 9 (BOVET@FRMOP11)

Abstract

Probabilistic models were developed to represent animals' movements. The simplest one makes it possible to compute the sinuosity of an animal's search path and to determine some basic properties such as its diffusion. Applying this model in the framework of optimal foraging theory led us to determine the sinuosity value which minimizes the path length of a central place forager. More complex models, integrating cybernetic controls of the sinuosity and the velocity as a function of environmental stimulations, show how animals can orient themselves in relation to a stimulation gradient or exploit patchy environments using simple klino- and ortho-kinetic mechanisms. Another type of movement model was developed to study orientation mechanisms based on an egocentric spatial memory.

1. Introduction

Animals often exhibit random search paths: take for example the paths of foraging ants, which anybody can observe. This intrinsic randomness does not however prevent the animals from orienting efficiently towards specific goals and/or aggregating in the the most suitable areas of their environment. To understand space-use mechanisms (those whereby an animal regulates the time it spends in the various areas of the environment) and orientation mechanisms (those whereby an animal moves towards a specific goal), it is necessary to first model the search paths. It is afterwards important to determine which environmental cues are relevant to animals and which kinetic parameters they have to regulate to be efficient. Using modeling and computer simulation of animals' movements, we have attempted to formalize some of the mechanisms involved in movement control. In this context, animals have been taken to be probabilistic self-directed mobile agents.

Here we present a general overview of the theoretical studies we have published over recent years in the field of modeling animals' movements. These models deal with a large range of natural spatial behaviours, from random foraging to oriented movements based on spatial memory. Some of these models link up with the optimal foraging theory: they are an attempt to determine which movement strategies maximize the efficiency of food searching in a stochastic environment. Other models were devised with a view to explaining orientational performances on the basis of elementary sensorimotor probabilistic mechanisms. Since a mathematical approach would be too complex to be practicable here, the properties of our models were established using computer simulations.

(*) This text is a revised version of a communication previously presented at a workshop on Modeling, Analysis and Simulation of Biological Motion, Bonn 1989.

2. Definition of the sinuosity of search paths

The simple random walk model used in physics to represent the movements of brownian particles is not suitable for representing the search paths of most animals because their cephalo-caudal polarization and their bilateral symmetry cause a tendency to go forward that this model does not account for. A first order correlated random walk model is able to account for these basic properties. Any search path is then represented as a sequence of steps with length P, and changes of direction randomly drawn from a normal distribution (wrapped around the circle) with a null mean and a standard deviation σ: the correlation between the direction of successive steps is expressed by $r = \exp(-\sigma^2/2)$.

Dividing a path into discrete steps is a useful but often arbitrary way of representing it. Consequently, we look for a measure of the "amount of turning" associated with a given path length which did not depend on the discretization step. We have shown that the expected standard deviation of the distribution of changes of direction σ_R obtained after rediscretization with a step R of a path simulated with parameters σ and P is given by $\sigma_R = 0.85\,\sigma/\surd(R/P)$. Separating what belongs to the model (σ and P) from what relates to the rediscretization (σ_R and R), we define the sinuosity of a path as (Bovet & Benhamou 1988):

$$S = \sigma/\surd P = 1.18\,\sigma_R/\surd R$$

The first part of this formula is a theoretical expression for the sinuosity, whereas the second part is a practical expression for quantifying the sinuosity of actual paths. Indeed, the rediscretization with a constant step length (R) of an actual path yields a discretized path which can be analysed using our model. Inasmuch as the observed distribution of changes of direction fits the basic assumptions of the model, the second part of the above formula can be directly used to measure the sinuosity of the path analysed.

It is possible to demonstrate that the expected diffusion of a search path, i.e. the beeline distance between the beginning and the end of this path, is given by:

$$E(D) = 1.77\,\surd L/S$$

where S is the sinuosity and L is the length of the path. A dynamic expression for the expected diffusion is obtained by replacing the path length by the product of the travelling time and the mean velocity.

3. Optimal sinuosity in central place foraging movements

The case of a forager searching for immobile, pinpoint, randomly distributed prey items from a central place and homing straight back was examined to determine the optimal sinuosity of the search path (no foraging occurs during the homing path). The optimal policy which was considered consists of minimizing the expected total (searching plus homing) path length required to survey a given area A (depending on the number of prey items the animal is able to load and the prey density) with a search path width W (corresponding to the prey detection range on both sides of the animal).

For this purpose, the animal during its search must on the one hand reduce any overlapping between search path loops, in order to avoid searching several times at the same place, and on the other hand, reduce its diffusion distance from the central place, which corresponds to the distance it will have to home when it has finished searching. To fulfill the first requirement, the search path must have a low sinuosity, because the overlapping obviously increases with the sinuosity, whereas the second requirement means that the search path must have a high sinuosity, because the expected diffusion distance is inversely proportional to the sinuosity. Using computer simulations, it can be shown that the optimal sinuosity S^* is given by (Bovet & Benhamou 1990):

$$S^* = (1.20 - 0.11 \, Ln(A/W^2))/ W$$

Obviously, this formula can apply only if the search path can be suitably represented as a first order correlated random walk and if the animal is able to accurately memorize the location of the central place to which it must home straight back.

4. Kinetic adaptations to the heterogeneity of the environment

Klinokinesis and orthokinesis have been classically defined as mechanisms regulating the angular velocity (rate of change of direction) and the linear velocity, respectively, of a random movement as a function of either the local value of the stimulus intensity (in the absolute mode, i.e. "without sensory adaptation"), or its variation (in the differential mode, i.e. "with sensory adaptation"). In order to make a formal distinction between the spatial and the temporal components of movements, klinokinesis and orthokinesis have been recently redefined as mechanisms regulating the sinuosity and the velocity intensity, respectively (Benhamou & Bovet 1989).

Absolute klinokinesis and orthokinesis are elementary space-use mechanisms. They allow animals to adapt their movements to the spatial heterogeneity of the environment. An animal regulating its sinuosity (klinokinesis) and/or its velocity (orthokinesis) as a function of the suitability of the area where it is moving can thus increase the time spent in the most suitable areas of the environment, and reduce the time spent travelling between them. For this purpose, the animal needs only to move with a sinuosity as high and/or a velocity as low as the suitability of the area crossed is high in comparison with the mean suitability of the environment. This is simply a consequence of the diffusion law: the time spent in a given area is proportional to the square of the sinuosity and inversely proportional to the velocity intensity.

One important application of absolute klinokinesis and orthokinesis relates to the adaptation of foraging behaviour to the contagious distribution of the local density of prey items occurring in patchy environments: increasing the sinuosity after finding a prey item leads an animal to concentrate its search effort in higher prey density areas, and decreasing the velocity leads it to increase both the probability of detecting encountered prey items and the time allotted to feeding. Absolute klinokinesis and orthokinesis can thus be used to formalize the "area-restricted searching" behaviour mentioned in the framework of optimal foraging theory.

5. Klinokinetic orientation in a stimulation gradient field

Differential klinokinesis constitutes a sinuosity regulating mechanism which can be used by animals to efficiently orient themselves in a stimulation gradient field (i.e. an environment containing a stimulus the intensity of which varies monotonously in space) without perceiving the gradient direction: they need only to measure the variations in the stimulus intensity occurring during their movements. In an axial gradient field (where the stimulus intensity varies monotonously along an axis), this mechanism can be formulated as (Benhamou & Bovet 1989):

$$S = S_b\,(1\text{-}k.\cos\Theta)$$

where S is the local sinuosity of the path, S_b the basic sinuosity, k (ranging between -1 and 1) the klinokinetic factor and Θ the orientation of the ongoing step in relation to the gradient direction (cos Θ expresses the normalization between -1 and 1 of the intensity variation perceived during a step: no perception of the gradient direction is therefore required). Using computer simulations, we have shown that this elementary mechanism efficiently leads to a drift by the animals in the gradient direction. The expected drift $E(D_k)$ is practically equal to the product of the path length L and the klinokinetic factor k:

$$E(D_k) = k.L$$

whether the sinuosity regulation is carried out by varying the standard deviation of the distribution of changes of direction or the step length.

Differential klinokinesis was applied by Bovet (1984) to account theoretically for the light-orientation behaviour of organisms endowed with primitive eyes which are able to determine local illumination, but not the light direction. Another application of differential klinokinesis, involving a radial stimulation gradient field (where the stimulus intensity varies monotonously with the distance from the centre of the field), has been carried out by Benhamou (1989) to account for the olfatory orientation of a mammal in its home range based on its own scent marks. Computer simulations showed that this model is able to account for both the stabilization of a mammal's movements within its home range and the maintenance of the radial olfactory gradient field which was postulated by the model.

6. Orientation relying on an egocentric spatial memory

Another type of orientation is exhibited by an animal which is able to return to a particular place in the absence of any gradient or landmark indicating the location of this place. This animal has to memorize the location of this place by means of an egocentric coding process, i.e by relating the distance and the direction of the place to the animal's own location and orientation by processing the information collected "en route" about its ongoing path (path-integration). This type of orientation mechanism requires the animal to continuously update the egocentric coding of the location of its goal (distance and direction) during its movements. After the ith change of direction (rotation) α_i, estimated by $\hat{\alpha}_i$, and the $i+1^{th}$ step with length (translation) P_{i+1}, estimated by $\hat{P}_{i+1}$ (after i+1 steps, the animal has performed only i changes of direction), the new egocentric coding of the place to which the animal intends to return is recurrently expressed by the estimations of the distance to this place, D_{i+1}, and its direction in relation to the animal's antero-posterior axis, Ω_{i+1} (Benhamou et al 1990):

$$\hat{D}_{i+1} = \surd(\hat{D}_i^{\,2} + \hat{P}^2_{i+1} - 2\hat{D}_i.\hat{P}_{i+1}.\cos(\hat{\Omega}_i - \hat{\alpha}_i))$$
$$\hat{\Omega}_{i+1} = \arctan(\sin(\hat{\Omega}_i - \hat{\alpha}_i)/(\cos(\hat{\Omega}_i - \hat{\alpha}_i) - \hat{P}_{i+1}/\hat{D}_i)) + k\pi$$

with k=0 when the denominator is positive and k=1 otherwise.

Because of the inaccuracy of animals' measuring systems, the memorized location of the place is subject to random errors which tend to accumulate during the process owing to its iterative nature. We performed computer simulations in order to quantify the accuracy of the egocentric coding of the place where the animal intends to return as a function of the sinuosity and length of the outward path and the type and magnitude of route-based information estimation errors: the coding accuracy was found to be very sensitive to the random errors occurring in idiothetic (without external reference) estimations of rotatory movements but not in estimations of translatory movements or in allothetic (compass-based) estimations of rotatory movements. It is therefore not surprising that numerous arthropods which have been found to use a sun-compass are able to achieve better orientational performances than mammals which have to idiothetically estimate their changes of direction during their movements.

7. Conclusion

Two different types of mechanism controlling movements have been described here: space-use mechanisms (absolute klinokinesis and orthokinesis) and orientation mechanisms (differential klinokinesis and route-based navigation). These two types of mechanism which are both of considerable adaptive value in animals' exploitation of space have not always been clearly distinguished in the past. Modeling and simulating animals' movements thus make it possible to clearly formalize these mechanisms and to compute their efficiency in animals' spatial behaviours.

REFERENCES

Benhamou S. 1989. An olfactory orientation model for mammals' movements in their home ranges. *J. Theor. Biol.* 139, 379-388.

Benhamou S. & Bovet P. 1989. How animals use their environment: a new look at kinesis. *Anim. Behav.* 38, 375-383.

Benhamou S., Sauvé J.-P. & Bovet P. 1990. Spatial memory in large scale movements: efficiency and limitation of the egocentric coding process. *J. Theor. Biol.* 145, 1-12.

Bovet P. 1984. Modèles clinocinétiques. In: P. Clément & R. Ramousse (eds.), *La vision chez les invertébrés*. Paris: C.N.R.S., 232-236.

Bovet P. & Benhamou S. 1988 Spatial analysis of animals' movements using a correlated random walk model. *J. Theor. Biol.* 131, 419-433.

Bovet P. & Benhamou S. 1990. Optimal sinuosity in central place foraging movements. *Anim. Behav.*, in press.

COGNITIVE MAPS AND INTERNAL WORLD MODELS

Schemas for Prey-Catching in Frog and Toad[1]

Michael A. Arbib and Alberto Cobas [2]
Center for Neural Engineering
University of Southern California
Los Angeles, CA 90089-2520, U.S.A.
arbib@pollux.usc.edu

Abstract

This paper first outlines the methodology of schema theory (Arbib 1981), which integrates perception and action by decomposing an overall behavior into the interaction of functional, neurally explicable, units called *schemas*. It offers comparisons with other methodologies from Artificial Intelligence (AI) and Brain Theory (BT), and reviews the RS (Robot Schema) language and the Arbib and House (1987) model of detour behavior in *Rana computatrix* which associates potential fields with objects — an attractant for the prey; a repulsor for the fencepost; and a forward field for the toad itself which are then combined to create more complex fields which determine the trajectory of the animal.
However, rather than analyze detour behavior here, the remainder of the paper presents a schema-theoretic model for the decision-making mechanisms which control prey-catching behavior in frog and toad; the extension of the work to model predator-avoidance is discussed elsewhere. It thus contributes to *Rana computatrix*, an evolving set of models of anuran visuomotor coordination (e.g., Arbib 1987). Our new model of prey-catching is rooted in recent experimental data on the behavior of animals with and without brain lesions. These data motivate the model's use of independent processing of the different parameters that define the stimulus position (horizontal eccentricity, elevation and distance). The model, which emphasizes action generated by the concurrent activity of multiple motor schemas rather than the serial activity of such schemas, predict new behaviors for experimental test.

1. An Introduction to Schema Theory

Schema theory (Arbib 1975, 1981) provides a way to tame the complexity of large systems that are to function in the real world, offering an approach explicitly designed to bridge between cognitive science and brain theory (BT), as well as to contribute to distributed artificial intelligence (DAI). Schemas are *active* modular entities, each involving data structures and control:

a) Schemas serve to represent, at least, perceptual structures and distributed motor control. Schemas are ultimately defined by interaction with a physical environment rather than (as in most AI systems) by cross-references in some logical formalism.

b) Schema theory provides a distributed model of computation. The brain can support many concurrent activities for recognition of different objects, and the planning and control of different activities. Thus schema theory views the use, representation, and recall of knowledge as mediated through the activity of a network of interacting computing agents, schema instances. This activity may involve passing of messages, changes of state (including activity level), instantiation to add new schema instances to the network, and deinstantiation to remove instances.

c) The activity level of an instance of a perceptual schema represents a "confidence level" that the object represented by the schema is indeed present; while that of a motor schema may signal its "degree of readiness" to control some course of action. A schema network does not, in general, need a top-level executor since schema instances can combine their effects by distributed processes of competition and cooperation

[1] The research described in this paper was supported in part by grant no. 1RO1 NS 24926 from the National Institutes of Health (M.A.Arbib, Principal Investigator) and Fulbright/MEC fellowship FU88-35011116 (Spain) to A.C.

[2] Present address: Center for the Neurobiology of Learning and Memory, University of California at Irvine, Irvine, CA 92717, USA.

(i.e. interactions which, respectively, decrease and increase the activity levels of these instances), rather than the operation of a serial inference engine on a passive store of knowledge.

The RS (Robot Schema) language offers a message passing formalism (Lyons and Arbib 1989), but, more generally, processes of cooperative computation enable schemas to modulate one another to compensate for the incompleteness of information available to each one (Arbib 1989, e.g., Chapters 5 and 7). The discussion of "Path-Planning and Detour Behavior" below shows how motor schemas may cooperate by summing their effects on the motor apparatus, while the use of schemas in computer vision (e.g., Draper et al. 1989) demonstrates both the competition between schema instances offering discordant interpretations of a region of an image, as well as the cooperation between schema instances which provide contextual cues for one another.

d) In BT, a given schema, defined functionally, may be distributed across more than one brain region; conversely, a given brain region may be involved in many schemas. A top-down analysis may advance specific hypotheses about the localization of (sub)schemas in the brain, and these may be tested by lesion experiments. In DAI, schema instances must be allocated to a limited set of processors, thus raising issues of (distributed) scheduling.

e) A given schema may have many different implementations, either biologically or technologically. In BT, once a schema-theoretic model of some animal behavior has been refined to the point of hypotheses about the localization of schemas, we may then model a brain region by seeing if its known neural circuitry can indeed be shown to implement the posited schema. In DAI, individual schemas may be implemented by artificial neural networks, or in some programming language on a "standard" (possibly distributed) computer. In the former case, the development of schema theory for DAI is much influenced by BT; in the latter case, schema theory converges with work on concurrent object-oriented programming.

f) For BT, the analysis of interacting computing agents called *schema instances* is intermediate between the overall specification of some behavior and the neural networks that subserve it. For DAI, schemas provide a form of knowledge representation which differs from frames and scripts by being of a finer granularity. A schema is more like a molecule than an atom in that schemas may well be linked to others to provide yet more comprehensive schemas.

The style of cooperative computation is far removed from serial computation and the symbol-based ideas that have dominated conventional AI. The representation of the world is *the pattern of relationships between all its partial representations*. There is no executive center. Rather, the brain (or a DAI system designed according to schema theory) has many different subsystems whose states constitute a distributed representation of the world, and this representation mediates the competition and cooperation whereby the subsystems find the appropriate pattern of interaction with the world.

1.1. Comparisons

Schema theory provides a knowledge representation protocol which is related to the frames and scripts of AI, but more related to studies in distributed artificial intelligence rooted in work on actors and the HEARSAY speech understanding system. As the name "frame" suggests, frames tend to "build in" from the overall framework, while schema theory is more generative. Where some AI workers would represent a birthday party by a single frame or script with specific slots to be filled in, schema theory would build up to such a representation through an assemblage of schemas for salient objects and actions.

Much work in BT and AI contributes to schema theory, even though the scientists involved do not use this term. For example, Minsky (1985) espouses a "Society of Mind" analogy in which "members of society" play a role analogous to schemas. Brooks (1986) builds robot controllers using layers made up of asynchronous modules that can be considered as a version of schemas. This work shares with schema theory, with its mediation of action through a network of schemas, the point that no single, central, logical representation of the world need link perception and action; while sharing with Grey Walter (1953) and Braitenberg (1984) the study of the "evolution" of simple "creatures" with increasingly sophisticated sensorimotor capacities.

1.2. The RS Approach

RS (the Robot Schema language: Lyons and Arbib 1989) formalizes each schema as a port automaton with a set of input and output ports through which it can communicate with other instances. Instantiation and deinstantiation operations capture the notion that, as action and perception progress, certain schema instances need no longer be active (they are deinstantiated), while new ones are added as new objects are perceived and new plans of action are elaborated (schemas are instantiated as new schema instances which are linked into the network). A basic schema definition includes a behavior specification, a sequence of statements forming a C-like program which makes explicit how an instance of the schema will read inputs, write outputs, change internal state, and execute (de)instantiations. An assemblage also has ports, but its behavior is defined through the interactions of a network of schema instances.

Schemas may be formed hierarchically. A schema assemblage may itself be considered a schema for further processes of assemblage formation; and the network itself will be dynamic, growing and shrinking as various instantiations and deinstantiations occur. The RS syntax for an assemblage tells us how to put schemas together in a way which does not depend on how the behavior specification is given, whether directly as in a basic schema, or indirectly, as when the schemas constituting an assemblage are themselves schema assemblages. (It is thus simple to extend RS to include neural net specifications by making it possible, in the basic schemas, to define the behavior directly in terms of a neural network as well as by a C-like program.)

1.3. Path-Planning and Detour Behavior

Work on *Rana computatrix* (e.g., Arbib 1987), "the frog which computes", models data (e.g., that reviewed by Ewert 1989) on visuomotor coordination in frog and toad to study the integration of action and perception in distributed systems. To see how work in BT contributes concepts to AI as well, we briefly consider the motor schemas involved in detour behavior. A toad, viewing a vertical paling fence through which it can see a worm, may either snap directly at the worm, or may detour around the barrier. However, it will not move when it sees a barrier if there is no worm behind it. Thus, it is the worm that triggers the animal's response, but there is a complex trajectory dependent upon the relative spatial position of worm and barrier. Arbib and House (1987) offer a model of this. Study of the toad's brain led us to posit separate schemas to provide the depth maps for prey and barriers, and to ask how their output can be used to guide the animal's behavior. We associate with each point of each depth map a vector that indicates the preferred direction of the animal were it to follow a path through that corresponding position. The set of all these arrows forms a "potential field" showing the potential movement of the animal at each location represented in the depth map. A single prey will set up an attractant field in which, from every point in the animal's representation of space, there is an arrow suggesting a choice of movement toward the prey. Fenceposts have a local repulsive field. A fence post has no effect upon the toad's behavior unless it is within a body width of the toad. We also associate a field with the toad itself which just "urges" the toad forward. By summing up all the vectors corresponding to objects that the toad senses, the model forms a potential field from which it may compute the appropriate course of action.

Application of the "neural technology" of this model to a mobile robot (Arkin 1989) provides a case study of how thinking about schemas for brain modeling may help us in developing perceptual robotics. The robot, which can plan its way around a building or across a given space, has hierarchical control provided by AURA, an AUtonomous Robot Architecture. A subsystem called the Navigator conducts off-line path planning on the basis of a terrain map using relatively conventional AI techniques. Another subsystem, called the Pilot, extracts more specific features, e.g., landmarks, to elaborate the plan into motor schemas, with perceptual schemas embedded within the motor schemas. Such schemas include stay-on-path, move-to-goal, move-ahead, find-landmark, and avoid-obstacles, and are each defined by a potential field. The path along which the robot is going is represented by a flow field. The whole vector field gives a sense of what the dynamics will be if the world remains constant as the animal moves. However, the whole field need not be computed in advance, but only the field for the locale of the robot at the time of sensing. If an obstacle appears as the robot moves along its path, then, by integrating the repulsive potential field from the obstacle, the robot will detour around the obstacle, but the potential field for the planned path will then recapture it to move it along its way.

To conclude this introduction, we note that Lara et al. (1984) have provided schemas for a wide range of toad behavior, but these do not meet the lesion data reviewed below. Moreover, where ballistic movement is obtained in the Lara et al. (1984) model by a sequence of several schema activations, the new model achieves this by activating several motor schemas concurrently, and with a variable intensity. This can yield more flexible and varied behavioral patterns than the purely sequential approach.

2. The Overall Organization of the Simulation System

Our schema-based simulation of prey-catching has been written in C to run on a Sun workstation. The system consists of an ideal environment in which a variable number of predators, prey and a toad move independently and interact with each other. Each object (toad, prey, or predator) is represented as a data structure encoding the object's position in the environment, its dimensions and other properties. Toads are given schema programs explicating the neural interactions underlying behavior; predators and prey move following simple rules giving, for each

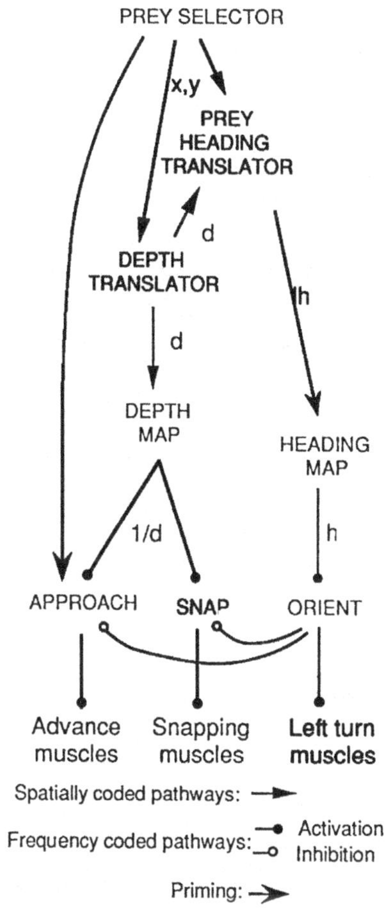

Figure 1: A unilateral view of the schema model for prey capture. (x,y = retinotopic coordinates, d = distance, 1/d = closeness, h = heading, lh = left-side heading.)

step, a movement distance and heading at random within a preprogrammed range. In the simulation, toads only react to moving stimuli, following their natural behavior. In the simulation, we thus find two classes of functions: (i) environment functions that are in charge of the maintenance of the simulated environment and its interface with some of the schema functions and (ii) functions that implement the schemas, and that communicate with each other through specific ports.

2.1. Main Module

The Main Module is an environment function. This is the function that is executed first, starting up the simulation by calling all the rest of the functions. It then starts the actual simulation by executing a loop in which each cycle consists in: (i) activation of all prey objects via the schema ActivatePrey (ii) activation of all predator objects via the schema ActivatePredator, and (iii) the complex schema-based activation of the toad object, via the schema ActivateToad which is the heart of the model. Each object performs one 'ballistic' behavior episode and returns the distance and heading moved, relative to the object's own position and orientation. Additionally, the Activate Toad reports whether it has attempted a snap at a given prey. The Main Module updates the environment according to the values returned by each object and executes a new cycle unless all prey have been "eaten".

2.2. ActivatePrey and ActivatePredator

Activate Prey is the only function for prey objects. Each prey has a specific preset probability to move, a range of distances to move to, and a probability to re-orient. If a first random value is greater than a certain threshold, a moving_distance within its specific range is selected at random, otherwise moving_distance is set to 0 (i.e., it remains unchanged from its default value). If a second random value exceeds the next threshold, a change_heading value is selected at random (otherwise its absolute orientation is not changed from the previous one). For completeness (although we avoid predator avoidance in the present paper - see Cobas and Arbib, to appear, for details), we note that ActivatePredator is the only schema function for predator objects. It is identical to ActivatePrey save for the fact that it is applied to predator objects, and so may be set with different parameter values.

2.3. ActivateToad and Imager

ActivateToad is the main toad schema function. It first activates the Imager schema which picks up from the environment all objects that moved since the last time frame, without any regard of their identity. In our simulation the Imager acts as a mere interface between the environment and the toad, collecting data on the other objects (prey, predators) that recently changed position or orientation. The positions of all these objects are transformed so they are relative to the toad's position and orientation, and the list is returned in the Imager's output port so that it may serve as input to various schemas in the toad's prey-capture and object-avoidance systems. If no moving objects are detected by the Imager, the

ActivateToad schema finishes and no movement is elaborated for the toad on this cycle. Otherwise, a move is elaborated using the schemas to be presented in the next sections.

The use of the Imager as a mere processor of lists of moving objects is part of the encapsulation strategy of schema theory. Our task in this paper is to probe the neural interactions revealed by behavioral data on prey-capture in normal and lesioned animals, and so we do not need to probe the interior details of the Retina, PreyRecognizer, Depth-mapper or Predator Recognizer which provide input to the schemas whose interactions are our present concern. Detailed models of these four systems do exist, however (see Teeters 1989, Cervantes-Perez et al. 1985, House 1989, and Liaw and Arbib 1990, respectively), and the refinement and reconciliation of these models on the basis of the insights gained from the present schema analysis is a challenge for current research.

3. A schema model for prey capture

In this section, we integrate an exposition of the schemas which constitute our model of the pre-capture in frog and toad with an exposition of the data on the behavior of normal and lesioned animals that undergird the model.

Activating the Prey-Capture System: The first assumption of our model is that a PredatorRecognizer schema is activated should there be any moving object detected by the Imager. If it detects a predator from the objects passed by the Imager, the predator-avoidance system (Cobas and Arbib, to appear) is enabled. If not, the prey-capture system (which we describe in detail) is activated instead.

The notion of a PredatorRecognizer (in pretectum) whose activity can inhibit the prey-selection system is based on the observation that escape behavior fails to occur when the pretectal region is destroyed, whereas prey-catching is "disinhibited" and the animal snaps at objects much larger than normal prey.

An overview of the prey-capture model can be seen in Fig. 1. For clarity in the representation, we only show the schemas for one half of the visual field (Cobas and Arbib give the full bilateral model). If no moving predator is recognized, the first schema to be activated is the PreySelector schema. Since all moving objects passed to it by the Imager schema are prey (the Predator Recognizer was not activated), the PreySelector schema extracts a unique prey object.

The "midbrain hemifield deficit". An important feature of the model is that it segregates different coordinates of prey location to different pathways in the brain. The data for this are as follows. Unilateral removal of the entire tectum causes a scotoma, both for prey capture and predator-avoidance, corresponding to the regions of the visual field that are solely represented in the missing tectal hemisphere, i.e., it spares the binocular field (see Ewert 1989 and Ingle 1983 for related reviews). However, unilateral lesions of a defined tract caudal to the tectum in the medulla (Kostyk & Grobstein 1987) do not produce any scotoma, although the animal responds abnormally to stimuli located in the *entire half of the visual field* (including the binocular field) ipsilateral to the lesion. The effects evoked by this kind of lesion have been termed a "hemifield medullary deficit". For prey in the affected hemifield, frogs respond always with a forward directed movement. The animal is incapable of executing visually elicited left turning movements when the prey is in the left hemifield and, in its place, it would either advance or snap always at the front. These movements are independent of the stimulus horizontal position (eccentricity), but change depending on the stimulus elevation and distance. One conclusion is that the representation in the medulla is different from that in the tectum, in the sense that this representation is relative to the toad's body and no longer to the toad's eye. It can also be concluded that *different components of the stimulus position are handled though different pathways that can be disturbed independently.*

Back to the Model: We thus postulate a DepthTranslator schema which obtains the distance, and a HeadingTranslator schema which obtains the angle from the toad to the selected prey. This angle is used by the Orient schema which computes and returns the toad's new heading. The Snap schema uses the closeness to the prey and the toad's heading to decide whether to aim a snap at the prey, and the Approach schema, using also the toad's heading angle and the closeness to the prey returns the distance to advance. We now discuss each of these schemas in somewhat more detail.

3.1. Depth and Heading

Population coding of closeness: With small lesions in the ventral part of the medulla, selectively interrupting the pathway that carries depth information (Grobstein & Staradub 1989), the frog turns more or less appropriately but fails to snap and instead advances an incorrect distance, *overestimated* in relation to the extent of the lesion. This suggests a population coding for depth — but since fewer fibers codes greater depth, we will postulate that the total firing of the "DepthMap" correlates with "closeness", 1/d, rather than directly with depth d.

The *"turn/snapping" deficit* is observed when the pathways that carry the depth and eccentricity information are both interrupted by a large lesion in the ventral part of the medulla. The result is that the frog will always respond with a small advancing movement, no matter the actual prey distance, without ever snapping or turning.

Back to the Model: Once a single target is selected, the PreySelector schema triggers a cascade of events. It directly primes the Approach motor schema to make it responsive to further stimulation from other schemas (see 3.2). Further, the PreySelector transmits the selected coordinates to the HeadingTranslator and the DepthTranslator schemas. A full model of the DepthTranslator (e.g., House 1989) is bilateral, estimating a depth value by matching the disparity obtained from the coordinates of both tecta with the corresponding lens accommodation. In the present simplified version, the DepthTranslator simply gets the data structure of the prey-object and puts its distance from the toad in its output port. The distance estimate is output to the HeadingTranslator schema, and also to the Depth Map schema. The DepthMap schema provides the transition from a spatially-coded representation to a frequency-coded information suggested by the findings of Grobstein & Staradub, and has "closeness" for output.

(In a more general model in which touch and hearing supplement vision, the DepthMap schema would be a convergence point for parameter estimates obtained by other sensory systems, holding a depth value referenced to a body or head coordinate system, and no longer to any particular sensory modality. In this way, for each parameter there would be only one DepthMap (for prey — predators and barriers would have other depth maps, cf. Arbib and House 1987) but several Translator schemas, each one specialized in linking a particular sensory system with the general reference coded in the map.)

The HeadingTranslator computes the *body-centered* angle to the prey-object from the toad, using both the retinotopic coordinates x,y and the distance d as input, and puts this value in its output port. If this heading value corresponds to, say, a left ipsilateral heading, this is sent to the Left Heading Map schema in the medulla, while if the heading corresponds to the right side (as when coming from the contralateral part of the binocular visual field) its output is sent to the Right Heading Map schema. These contralateral connections are omitted from Fig. 1, but not from our simulation.

3.2. Orient, Snap, and Approach

The Snap-Hop Transition. In normal conditions, the frog snaps when prey is located nearby, but reorients and hops toward a stimulus that is further (but not too far) away. The transition between the two patterns occurs at a distance of about one body length for a lateral stimulus and two body lengths for a frontal stimulus (Ingle 1970). We now wish to model in more detail the dependence of this "choice" on prey heading. The way we do so is constrained by data on frogs with the medullary deficit. For such animals, not only does the animal not orient to prey in the affected hemifield (though it will still snap or hop), but the transition boundary is invariably at twice the body length when the stimulus is located anywhere in that hemifield (Kostyk & Grobstein 1987), the same as if the stimulus had been located right in front, which is also the same direction to which the animal always advances.

Back to the Model: The Orient schema acquires from its input h (which we assume to be equivalent to 0° if the medullary pathway is cut) an *activation level*. To address these data, we posit that this value will not only establish the turning angle of the toad toward the selected prey, but will also enter into the interactions of the Orient schema with the Snap and Approach schemas, determining the overall movement of the toad.

The more strongly the Orient schema is activated, the more strongly will it stimulate the pool of motoneurons controlling the turn muscles in the spinal cord. In our simulation, the Orient schema receives a heading estimate directly from the HeadingTranslator schema (0 heading is straight ahead for the toad). The activation level of the Orient schema increases by the product of the absolute value of the angle and a predetermined weight factor. If the final activation is greater than or equal to the schema threshold, then the orient command is enabled, the toad's new heading is made equal to the prey angle (positive to the left, negative to the right) and that heading is returned in the output port.

Similarly to the HeadingMap, the Depth Map is differentially connected to both the Snap and Approach schemas in such a way that the activity level of these two motor schemas will increase more vigorously in response to closer objects. If the level of activation of the Snap and Approach schemas is enough, they activate the corresponding pool of motoneurons in the spinal cord that actually carry out the action.

The Snap schema determines if the prey is within the snapping zone. This evaluation is an increasing function of the closeness of the prey (1/d) and a decreasing function of the toad's heading toward the prey as calculated by the Orient schema.

(Eq. 1) activation_level
= 2*closeness - 0.01*|new_heading|

The snapping command is made effective if and only if the final activation is greater than or equal to the Snap schema's threshold.

In a similar way, the Approach schema determines the distance for the toad to advance as a function of the closeness to the selected prey and the activation level of the Orient schema. In order for it to be effective, the threshold has to be reached or surpassed. The presence of the prey by itself raises the activation level to a default value. The closeness of a selected prey contributes positively to the activity level, while the heading angle contributes negatively.

(Eq. 2) activation_level =
7*closeness - 0.1*|new_heading|

If the approach command is finally effective, the actual moving distance is made no greater than the maximum the toad is estimated to make in a single ballistic episode.

The schemas of the motor pattern generator circuitry in the spinal cord are mutually interconnected, in such a way that both Snap schemas would activate each other, as do the Approach schemas, while both Orient schemas would inhibit each other. Further details, including pseudo-code for all the schemas, may be found in Cobas and Arbib, to appear). Several of these motor schemas may be active at the same time. In this way, we could obtain a snapping movement directed to a prey located at a short distance at the right side of the toad by the activation of the Approach and the Snap schemas due to the short distance (or high closeness), and the Orient schema using the required right heading value. In this way, the final motor behavior will result from the combined activity of these motor schemas. Their coordination would be regulated by the appropriate transmission delays from one schema to another and from one schema to the controlled musculature, in such a way that that the whole group of muscles would be activated with the required synchronization.

4. Performance of the Model

4.1. Normal function

We now examine the inputs to the three motor schemas and their interactions under normal conditions. The weight of the interaction among these schemas will determine the shape of a series of "motor regions" (Fig. 2), which would delimit those areas of space where a prey would elicit an orienting movement, an approach movement, a snapping movement, or some combination of these. For example, the weight of the Depth map input to the Snap schema determines the extension of the snapping zone to the front, and the weight on the input from the Orient schema determines the width of the zone laterally. The snapping zone boundary shown in Fig. 2 follows from Eq. 1 for the Snap schema. The units used in the model are body lengths, and the weight factors are chosen to result in an activation level that, after comparing with the threshold, yields a realistic response. The border of the approach zone follows from Eq. 2 for the Approach schema.

In Fig. 2 we see that the snapping zone resulting from our model extends 2 body lengths to the front and 1 body length at 90° to one side, for the normal situation. These values more or less correspond with those reported in the literature (cf. our earlier discussion of "The Snap-Hop Transition"). No similar experimental values for the Approaching zone existed in the literature, and thus we estimated 7 body lengths in the frontal direction and 1 in the lateral direction as a foil for later experiments. Note that the precise shape of the boundary line results just from the simple linear interaction of the heading and closeness to the prey, in the above equations.

Since the actual shapes of the fields are vague or undefined, especially for the approaching zone and the posterior field of the toad, we consider the actual performance to be satisfactory. Other shapes could be obtained using more sophisticated interactions between these parameters, but this must await further experiments on the dimensions and shape of the snapping and approaching zones.

In the following paragraphs we analyze how the model would resolve several prey configurations. Certain of the results depend on the linkages between the two sides of the brain that have only been briefly mentioned above.

Prey at front within the snapping zone. The small distance to the prey would be enough to activate the Snap schema over its threshold and increase the activity of the Approach schema in such a way that a short frontal advancing movement combined with a snapping movement would be obtained. The Orient schema would not reach the threshold and would not have any effect. The activity of both sides would be equivalent.

Prey at right far outside the snapping zone. The conditions would be the same as in the previous case with the addition that a strong right heading value would intensely activate the Orient schema, which in turn would decrease the activity of the Snap and Approach schemas below their respective threshold. The result wouldbe a simple right turning movement.

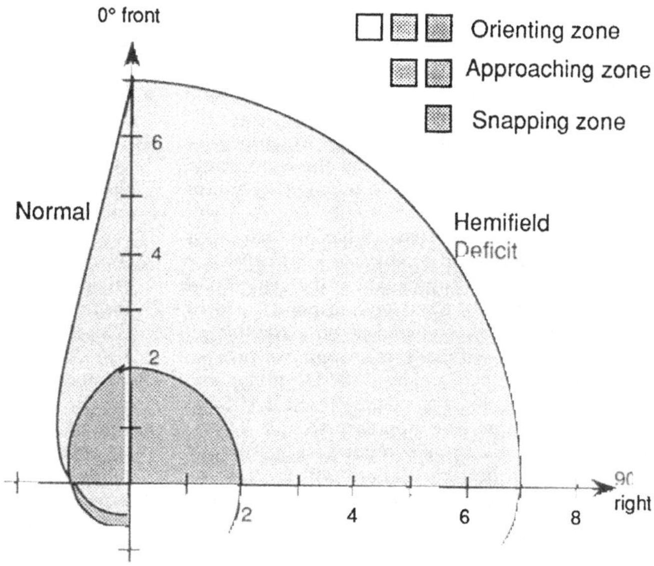

Figure 2: The model yields motor zones in prey acquisition which match those for the normal animal and for a toad with a medullary hemifield deficit. →

Prey at right inside the snapping zone. The increase of the closeness value (decrease of distance to the prey) over the preceding situation would suffice to activate the Approach and Snap schemas over their threshold to yield a snapping movement aimed to the right. In this way, the action of a pronounced right heading requires a higher closeness value to obtain a snapping movement as compared to a frontal situation, yielding a snapping zone wider at the front than at the sides, as has been observed in behavioral experiments.

Prey at front-left outside the snapping zone. The closeness parameter will be too weak in this case to activate the Snap schema over its threshold, and a mild heading to the left would be enough to activate the left Orient schema over its threshold. The Approach schema would be sufficiently activated to remain over the threshold after the small inhibition by the Orient schema, and so an advance movement directed a little bit to the left would be obtained.

4.2. "Lesion" effects

The "midbrain hemifield deficit". This deficit results from the lesion of one side of the descending output pathways of the tectum, at the level of the medulla. In our model, the corresponding "simulated" lesion in the left side of the model severs the ipsilateral output of the left HeadingTranslator and the contralateral output of the right HeadingTranslator. The effect would be that the Left HeadingMap schema would be without the heading input and would therefore produce no stimulation of the left Orient schema for all the prey stimuli falling onto the left visual field (to the left of the animal's midline). The immediate consequence is that the animal would be incapable of executing visually elicited left turning movements when the prey is in the left hemifield and, in its place, it would always either advance or snap as if the prey was at the same actual distance but at the front.

In our simulation, this effect is replicated by resetting the heading value returned by the HeadingTranslator to 0. As a result, the activity and therefore the reported new_heading from the Orient schema are invariably also 0, no matter the actual heading to the prey. This results in no turning movement and cancels the influence of the Orient schema on the Snap and Approach schemas (their respective activation_levels will not be decremented), which now always perform as if the prey were located straight ahead.

In the hemifield deficit, the snapping zone in the affected side appears equally wide laterally as at the front; whereas, in the normal, although the exact shape of the curve is not known, the frontal and lateral extent are approximately equal to two and one body length, respectively. This is replicated in the model since what narrows the width of the snapping zone at the sides is the increasing inhibitory action effected by the Orient schema output.

The "snapping deficit". When the pathway that carries the depth information is lesioned at the level of the medulla, the frog turns more or less appropriately but fails to snap and instead advances an incorrect distance, overestimated in relation to the extent of the lesion. In our model the effect is produced by the inability to excite the Snap schema above its threshold due to the diminished input from the DepthMap. The Approach schema is also affected although not completely, due to the stronger influence the Depth Map has on it. The predicted reaction would be as if the prey were always outside the snapping zone (which after the lesion would be very much reduced or nonexistent), and therefore would always yield an advance movement perhaps combined with a turn.

The "turn/snapping deficit": When the pathways that carry the depth and eccentricity information are both interrupted by a large lesion in the ventral part of the medulla, the frog always responds to prey with a small advancing movement, no matter the actual prey distance, without ever snapping or turning. In our model, the elimination of the DepthMap and HeadingMap schemas leave the Orient and Snap schemas inactivated. The Approach schema, having been aroused by the priming of the PreySelector and without any further stimulation, will increase its own activity and constitute in this way the default response of the system. The predicted reaction would be as if the prey were always straight outside the snapping zone, and therefore would always be an advance movement.

5. Conclusions

The present model postulates the construction of behavior through the interaction of different motor schemas, in which there is no need for a unique schema to win the competition (although that might well be the result), but two or more schemas can simultaneously be active and cooperate to yield a more complicated motor pattern. Each one of these motor schemas is driven by specific internal maps *validated by experimental data* in which topographically-coded sensory data are converted to population-coded information suitable to command the motor schemas that in turn drive muscle activity. After the required motor command has been executed, the animal starts to process the next wave of sensory information. This model has allowed us to reproduce some of the data available on the behavior of both normal and lesioned animals, and to offer new predictions. This work offers one small example of how the paradigm of schema theory is evolving in a fashion well suited to contribute to both brain theory and perceptual robotics.

References

Arbib, M.A., 1981, Perceptual structures and distributed motor control, in *Handbook of Physiology - The Nervous System II. Motor Control* (V.B. Brooks, Ed.), Amer. Physiol. Soc. pp. 1449-1480.

Arbib, M.A., 1987, Levels of Modelling of Visually Guided Behavior, *Beh. Brain Sci.*, 10:407-465.

Arbib, M.A., 1989, *The Metaphorical Brain 2: Neural Networks and Beyond*, Wiley-Interscience.

Arbib, M.A., and House, D.H., 1987, Depth and Detours: An Essay on Visually Guided Behavior, in *Vision, Brain and Cooperative Computation*, (M.A. Arbib and A.R. Hanson, Eds.), A Bradford Book/The MIT Press, pp. 129-163.

Arkin, R.C., 1989, Neuroscience in motion: the application of schema theory to mobile robotics, in *Visuomotor Coordination: Amphibians, Comparisons, Models, and Robots* (J.-P. Ewert and M.A. Arbib, Eds.), Plenum Press.

Braitenberg, V., 1984, *Vehicles: Experiments in Synthetic Psychology*, The MIT Press.

Brooks, R.A., 1986, A robust layered control system for a mobile robot, *IEEE Journal of Robotics and Automation*, RA-2:14-23.

Cervantes-Perez, F., Lara, R., and Arbib, M.A., 1985, A Neural Model of Interactions Subserving Prey-Predator Discrimination and Size Preference in Anuran Amphibia", *J. Theor. Biol.*, 113:117-152.

Cobas, A., and Arbib, M.A., Prey-Catching and Predator Avoidance in Frog and Toad: Defining the Schemas, in preparation.

Draper, B.A., Collins, R.T., Brolio, J., Hanson, A.R., and Riseman, E.M., 1989, The Schema System, *International Journal of Computer Vision*, 2:209-250.

Ewert J.-P., 1971, Single unit response of the toad (Bufo americanus) caudal thalamus to visual objects, *Z. Vergl. Physiol.* 74: 81-102.

Ewert, J.-P., 1989, The release of visual behavior in toads: Stages of parallel/hierarchical information processing, in *Visuomotor Coordination: Amphibians, Comparisons, Models, and Robots* (J.-P. Ewert and M.A. Arbib, Eds.), Plenum Press, pp.39-120.

Grobstein, P. and Staradub, V., 1989, Frog orienting behavior: the descending distance signal, *Soc. Neurosci. Abstr.* 15:54

House, D., 1989, *Depth Perception in Frogs and Toads: A Study in Neural Computing, Lecture Notes in Biomathematics* 80, Springer-Verlag, Berlin.

Ingle, D., 1970, Visuomotor functions of the frog optic tectum, *Brain Behav. Evol*. 3: 57-71.

Ingle, D. J., 1982, Organization of visuomotor behaviors in vertebrates, in *Analysis of Visual Behavior* (D. J. Ingle, M.A. Goodale, and R. J. W. Mansfield, Eds.), pp. 67-109, MIT Press.

Ingle, D., 1983, Brain mechanisms of visual localization by frogs and toads, in *Advances in Vertebrate*

Neuroethology (J.-P.Ewert, R.R. Capranica and D.J. Ingle, Eds.) Plenum Press, pp. 177-226.

Kostyk, S.K., and Grobstein, P., 1987, Neuronal organization underlying visually elicited prey orienting in the frog. I. Effects of various unilateral lesions, *Neuroscience* 21:41-55.

Lara, R., Carmona, M., Daza, F., and Cruz, A. (1984) A global model of the neural mechanisms responsible for visuomotor coordination in toads, *J. Th. Biol.* 110:587-618.

Liaw, J.-S., and Arbib, M.A., 1990, A Neural Network Model for Response to Looming Objects by Frog and Toad, in *Visual Structures and Integrated Functions,* (M.A. Arbib and J.-P. Ewert, Eds.), *Research Notes in Neural Computing,* Springer-Verlag.

Lyons D.M., and Arbib, M.A., 1989, A Formal Model of Computation for Sensory-Based Robotics, *IEEE Trans. on Robotics and Automation,* 5:280-293.

Minsky, M.L., 1985, *The Society of Mind,* Simon and Schuster.

Teeters, J.L., 1989, A simulation system for neural networks and model for the anuran retina. Technical Report 89-01, Center for Neural Engineering, University of Southern California, Los Angeles.

Walter, W.G., 1953, *The Living Brain,* Penguin Books.

Mapbuilding using Self-Organising Networks in "Really Useful Robots"*

Ulrich Nehmzow and Tim Smithers†
Department of Artificial Intelligence
University of Edinburgh
5 Forrest Hill
Edinburgh EH1 2QL
Scotland
Telephone 031 - 667 1011 ext 2529/2517
Telex 727442 UNIVED G
E-mail:ulrich@uk.ac.ed.edai tim@uk.ac.ed.edai

September 1990

Abstract

The Really Useful Robots (RUR) project is seeking to understand how robots can be built that develop and maintain the task achieving competences they require for flexible and robust behaviour in variable and unforeseen situations, as opposed to these being installed by their designers. In this paper we present an experimental autonomous robot with a map building competence which uses a self-organising network. Map building forms a necessary step on the way to development of a navigational competence. Some encouraging initial test results are also presented.

1 Introduction

The traditional approach to control in (mobile) robots is to decompose the task into separate components, and implement these using standard control techniques, see [Levi 1987], for example. This we call an *analytical approach*. Alternatively, a control structure can be built 'bottom up', first building foundational competences (such as 'move around and avoid obstacles'), and later on top of these more complicated competences (such as 'explore', 'map building', and 'map using'). This we call a *synthetic approach*, see [Brooks 1986], for example.

At Edinburgh we have adopted a synthetic approach in what we call the 'Really Useful Robots' project (RUR) [Nehmzow et al 1989]. This project is attempting to develop a control architecture which supports the development of task achieving competences by the robot. In other words, we are trying to understand how a robot can sequentially acquire and maintain the behavioural competences it requires, rather than have them 'installed' by us as its designers. We believe that this *autonomous acquisition* of task achieving competences will lead to greater flexibility and robustness in the behaviour of robots with respect to variable and unforseen situations. In investigating this idea we are motivated and informed by the adaptive control mechanisms we see in simple animals which result in them having flexible, reliable, and robust competences well matched to the tasks they are responsible for achieving and to the environment in which they are exercised.

Trying to get a robot to acquire the skills it needs means that as many decisions as possible are left to the robot, rather than being predefined by the designer. *Alder*, the first of the 'Really Useful Robots' (see figure 6)[1], is able to adapt to a changing environment, and to acquire useful competences. It uses what we call *fixed* and *plastic* components in its control architecture

†Names appear in alphabetical order, with both being principal authors on this occasion.

[1]Alder is a mobile robot whose base is built with a Fischertechnik kit. It is about 25cm long, has an 8052 based microcomputer on board (16k RAM) and is equipped with up to seven tactile sensors plus odometer. In addition to this a sonar sensor is available, but has not been used to obtain the results presented in this paper. More information about Alder and the 'Really Useful Robots' approach can be found in [Nehmzow et al 1989].

to achieve this, fixed components being the mechanics of the robot, its shape (morphology), the control program, and its so-called 'instincts', the plastic component being an artificial neural net (in the current system). So far we have successfully demonstrated robot acquisition of simple obstacle-avoidance, dead-end-escape, and wall-following competences. The next stage is to get the robot to develop a navigational competence, using a similar approach. This first requires a mapbuilding competence.

As in *controlling* mobile robots, so in *mapbuilding* a traditional approach can be taken. The traditional approach either uses some kind of fixed structure that is filled with information as it is obtained by sensory activities,see, for example, [De Almeida 1989], [Moravec 1988], [Goto and Stentz 1987] or [Tsai et al 1990]. We call these 'self detailing schemes'. Alternatively, a complete 'ready-made' map may be provided from the start, see, for example, [Tsubouchi and Yuta 1987], which we call 'pre-installed map schemes'. In these traditional approaches the structure of the map is defined by the *designer*, and so is not necessarily the best to store and retrieve the knowledge the *robot* needs to have and can acquire about the world. Here, as well as in controlling robots, *predefining means limiting.* Limiting what is represented by a map is not of itself a problem, but knowing what needs to be represented and what doesn't is. To get around this problem we think that the process of constructing a map should be left, as far as possible, to the robot, not to the designer, the idea being, that a map built by the robot is likely to be more useful to it, than one given to it by its designer.

Alder uses a self-organising network (see [Kohonen 1988]) to construct internal representations of the world it experiences as it moves around. In this paper we describe how this mapbuilding competence is implemented and some early test results which we believe are encouraging.

1.1 The place of this paper within the Navigational System of Alder

The plan is to design a navigational system as shown in figure 1.

Fixed components (as mentioned above) within this structure are the sensors, including the higher level interpretation of incoming data (the top block in figure 1), and the behaviour[2] of the robot. *Plastic components* are the self-organising network, as described in section 2, and an artificial neural network which associates particular robot movements in the real world with changes

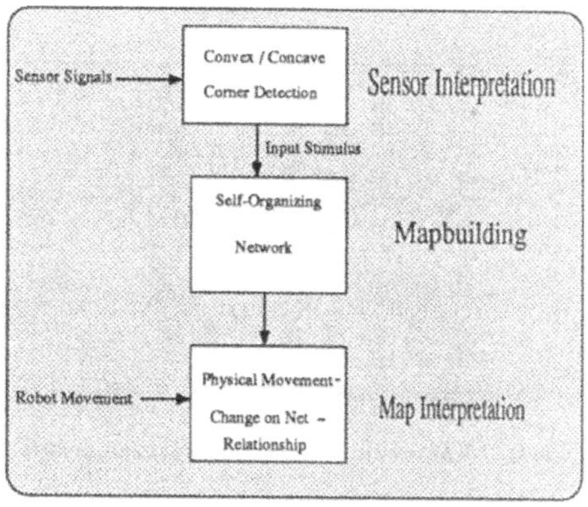

Figure 1: The proposed Navigational System of Alder

in excitation of the self-organising network (the bottom block in figure 1). The combination of all these components forms the navigation system.

This paper discusses only one aspect of this system: mapbuilding, using self-organising networks (the middle block of figure 1).

1.2 What is a map?

We take a map to be *any one-to-one mapping (bijection) of a state-space on to map-space.* Examples of such maps are the plan of the London Underground, the Edinburgh telephone directory, and a family tree. The connections between stations, phone owners and numbers, and family members are represented in all these cases. A map is therefore not just a representation of a 'birds-eye-view' of the world the robot inhabits.

In our RUR project the state-space that is represented using a self-organising network has nothing to do with actual physical locations in the real world, at least not directly. What *is* represented by these maps is the result of previous and present motor-sensory experience of the robot. The motor actions and sensory actions of the robot are tightly coupled, so tightly coupled in fact that they cannot be viewed independently of each other! Therefore the state space represented *does* in the end have something to do with actual physical locations of the robot in the real world, but only indirectly through motor-sensory behaviour of the robot.

Whenever the term 'map' is used in this paper it is used in the sense of the definition given above, i.e. a bijection of state-space on to map-space.

[2] 'Behaviour' here means any fixed pattern of motor-sensory activity, it is therefore a 'fixed' component of the robot.

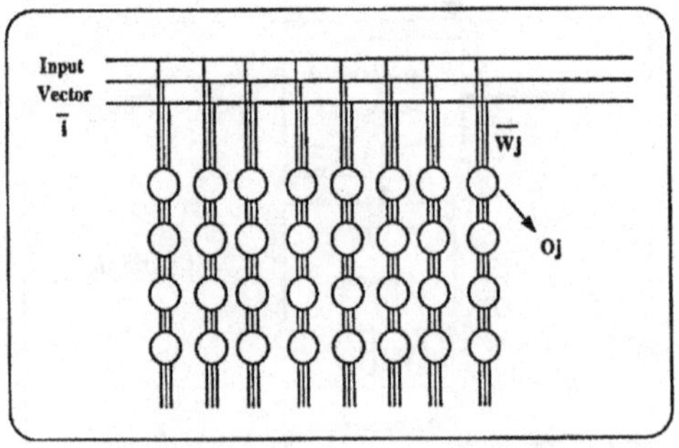

Figure 2: A two-dimensional self-organising network

2 Self-Organising Networks

Unsupervised learning can be achieved, using self-organising networks [Kohonen 1988]. These networks develop internal representations of the input space by mapping distinct input vectors onto distinct areas of the network (injection). One advantage of these networks is that they can accept redundancy in the input vectors. Redundant information is ignored by the net as long as the input vector contains *sufficient* information for a meaningful mapping to be generated. Surplus information does not impair performance. This fact makes it easier to define a suitable input space. Figure 2 shows an example of a two-dimensional self-organising network.

The input vector $\vec{i}$ is the same for all the cells. The output o_j of cell j is determined by the scalar product of input vector $\vec{i}$ and the weight vector $\vec{w}_j$ of cell j, and is given by:

$$o_j = \vec{w}_j \vec{i} \tag{1}$$

where $\vec{w}_j$ is the individual weight vector of cell j. The weight vectors are unit vectors: $||\vec{w}_j|| = 1$.

Initially, the elements of the weight vectors are set to randomly selected values such that they are all unique. Therefore one cell of the network is bound to respond most strongly to a particular input presented to the net. The weight vector of this maximally responding cell, as well as the weight vectors of all the cells within a defined neighbourhood around this cell, are changed according to the following equation:

$$\vec{w}_j(t+1) = \vec{w}_j(t) + \eta(\vec{i} - \vec{w}_j(t)) \tag{2}$$

where η is the so-called 'gain', a value that determines the amount of change (typically around 0.5). Weight vectors outside the specified neighbourhood [3] remain unchanged.

[3] The neighbourhood size is often chosen to be decreasing over time, in order to achieve quick changes in the beginning, and only little changes later.

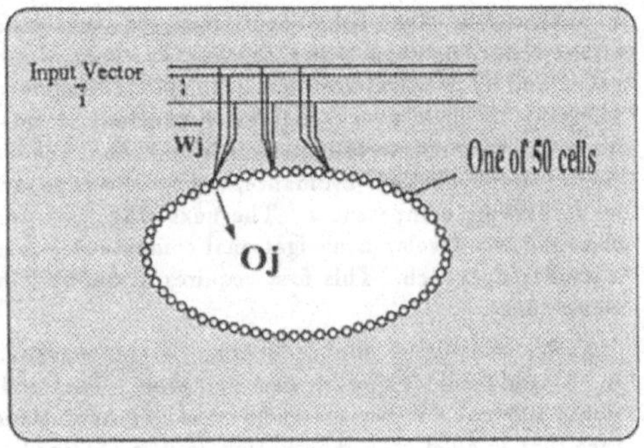

Figure 3: The self-organising network used on Alder

After several 'epochs', i.e. presentations of input vectors to the network, typical dissimilar responses appear for dissimilar input vectors. Thus a mapping arises, whereby different input vectors are mapped onto different regions of the network (bijection). This therefore is a *map* according to the definition given in section 1.2.

2.1 The Self-Organising Network used for Mapbuilding in Alder

Obviously, border effects can occur if the network has edges. This can be avoided by joining the opposite edges of the network. In the case of a two-dimensional network this yields a torus-shaped network.

As long as a network with a sufficient number of cells is used [4] it is not necessary to use a network of the same dimensionality as the space the agent acts in (i.e. although the robot moves in a two-dimensional world, a one-dimensional network can be used to obtain the self-organising state-space-representation). Since the computation for a one-dimensional network is much less [5], and the computational power of Alder is limited, we used a ring of 50 cells as self-organising network (see figure 3).

The behaviour of this network is as previously described in section 2, the neighbourhood within which weight vectors are updated is $\pm$ 2 cells (constant over time). A typical response of the network is shown in

[4] A rough calculation to determine the minimum number of cells required: If the network is to represent n distinct states, and the neighbourhood size within which weight vectors of cells are updated is m, then the minimum number of cells obviously is n · m. However, this is the *minimum* number, in practice a bigger number is needed.

[5] The number of necessary weight-vector changes is proportional to n^m, where n is the neighbourhood size and m the dimensionality of the map.

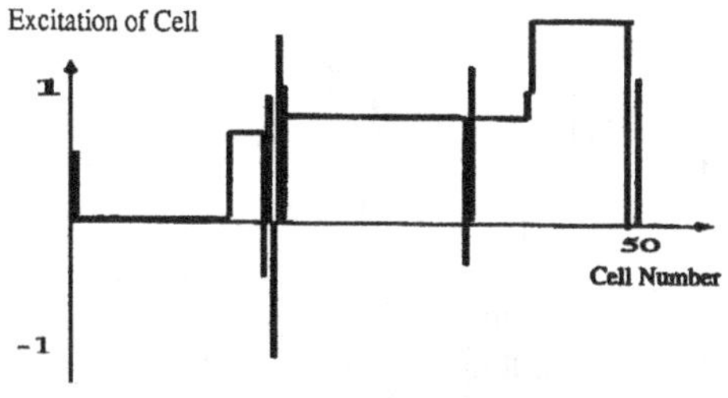

Figure 4: A typical response of the ring after having settled in

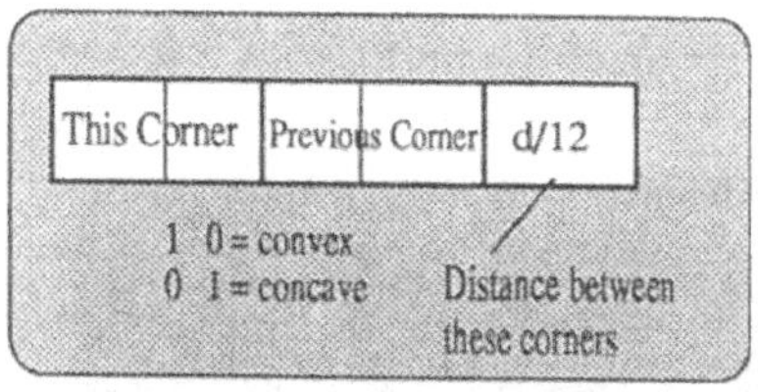

Figure 5: A typical Input Vector

figure 4 (in this figure the ring is cut and shown as a line).

The particular response of the network to a particular input stimulus (i.e. the activity pattern of all the cells of the net) can itself be viewed as a vector. This is convenient for later analysis, the particular response vector to a particular vector is called $\vec{\kappa}$ in this paper.

2.2 The Role of Robot Behaviour

Environment, agent (that is robot), and task are very tightly coupled and cannot be treated individually. This became obvious through the following observation:

The first behavioural pattern the robot showed was to wander around randomly and simply avoid obstacles. Distances travelled between obstacles and sensor readings obtained were used as input vectors to the net (see also section 2.3). The result was that the $\vec{\kappa}$s of the net in the settled state showed no strong correlation to particular physical locations. This is not surprising since if *we* were to wander randomly around a darkened room avoiding obstacles encountered by feeling them, we too would not build up a good 'picture' of the room and what is in it and where.

When we changed the robots behaviour to a wall-following behaviour the results became much more interesting: together with an improved input vector (see 2.3) the $\vec{\kappa}$s did indeed show a correlation with particular physical locations in the real world. By adopting wall-following behaviour the robot negotiated its way around the perimeter of its enclosure and the $\vec{\kappa}$s corresponded to corners.

2.3 The Input Vector

A self-organising network clusters the input information presented to it in a (statistically) meaningful way. If the input to the net contains no meaningful information (meaningful in relation to the task of constructing a state-space representation of the robot in its world, that is) the network will not develop any meaningful structure. The very first input vector we used simply contained information about whether the robot had "seen" something on its left or on its right hand side, plus information about two previous sensor readings (again only whether the obstacle was seen on the left or on the right), plus odometry information. Such information is too weak to construct a meaningful state-space representation (at least without requiring a great many input vectors), consequently the $\vec{\kappa}$s in the settled state of the net had little correlation with particular locations in the real world.

This was obviously due to a lack of sufficient structure in the input vector presented to the net. We therefore enriched the input vector by preprocessing the sensory information obtained. Instead of taking data straight from the sensors and feeding it into the net, sensor readings were used to detect convex and concave corners[6]. This information (whether a corner was encountered, and whether this corner was convex or concave) was then used as input to the net. The input vector eventually used to obtain the results shown in section 3 contained information about the present corner and previous corners encountered as well as the distance travelled between the present and the previous corner (An example of such an input vector is shown in figure 5)[7].

2.4 A brief summary of the whole mechanism

1. Initialise the self-organising network (i.e. the ring) by filling the weight vectors of all cells with randomly

[6] This is very easily achieved: If the time the robot needs to turn towards the wall exceeds a certain threshold time, then it is assumed that a convex corner is detected. Similarly, if the time it takes the robot to get away from a detected obstacle exceeds a certain threshold time, it is assumed that a concave corner is detected.

[7] Note that it is necessary to encode the information about convex and concave corners using 2 bits, because artificial neural networks can only make use of non-zero input lines (see also equation 2).

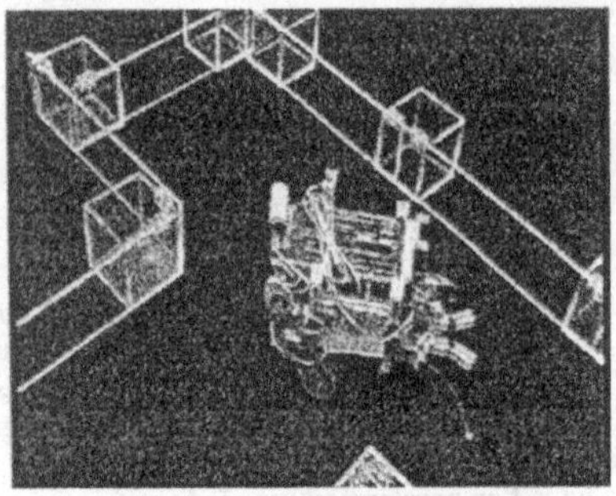

Figure 6: A typical environment for Alder

chosen values.

2. Normalise all the weight vectors.
3. Present an input stimulus to the ring (the input stimuli chosen for our experiments are desribed in section 2.3).
4. Determine the response to this stimulus for each cell of the ring according to equation (1).
5. Determine the unit that is responding most strongly.
6. Update the weight vectors of the five units within a neighbourhood of $\pm$ 2 cells of the maximally responding cell, according to equation (2).
7. Normalise those five weight vectors again.
8. Continue with step 3.

3 Experimental Results

The experiment was conducted as follows: The robot was placed within a fence (see figure 6) and allowed to explore it, following the wall for several rounds.

Whenever a convex or concave corner was detected an input vector as described in section 2.3 was created and presented to the ring. The gain η for updating weight vectors (see equation 2) was initially very high (5.0) and decreased by 5 per cent after every presentation of an input vector. The longer the robot went around in the enclosure, the more settled the ring became and the more precise the response to a particular input stimulus. After about three rounds a particular corner was marked by pressing the 'attention' button on the robot. The $\vec{\kappa}$ of the ring for that particular ('marked') corner was

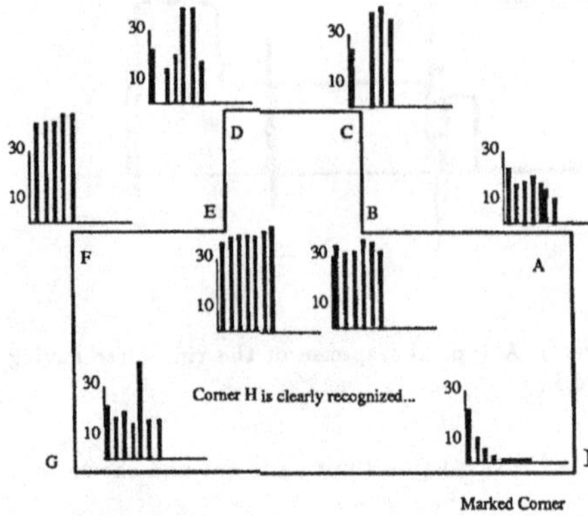

Figure 7: Recognizing corner H, looking at one previous corner

then stored ($\vec{\kappa_m}$), all subsequent $\vec{\kappa}s$ were compared with $\vec{\kappa_m}$ by calculating $||\vec{\kappa} - \vec{\kappa_m}||$; the smaller this value, the closer the $\vec{\kappa}$ at that particular corner to $\vec{\kappa_m}$. Obviously, if the robot is able to construct a meaningful internal state space representation this difference should be small when the robot is at the marked corner, and it should be noticeably larger at any other corner.

The input vector chosen will determine the quality of the internal representation obtained. Only information that is actually contained in the input vector sequence can eventually be found on the internal map! The results we obtained highlight this very clearly:

Originally we used an input vector that contained information about the current corner encountered and the previous corner to that one, as well as the distance travelled between these two corners (this vector is shown in figure 5). In figure 7 we can see that corner H is uniquely identified, using this input[8].

It is obvious that corner H is the easiest of all corners to recognize, because of the long distance travelled between corners G and H.

Looking at corners C and F, on the other hand, one might expect that these corners get confused on the internal map, because both C and F are concave corners, the previous corner in both cases is convex, and the distances between them is similar. Exactly this is observed (see figure 8).

The same applies to corners B and E: they are

[8] The bars in these diagrams show $||\vec{\kappa_m} - \vec{\kappa}||$. The smaller this value, the closer the response of the net to the response the net showed at the marked corner. The bigger this difference, the more distinct are the marked corner and the corner currently encountered. In other words: the smaller the value, the more alike are marked corner and present corner.

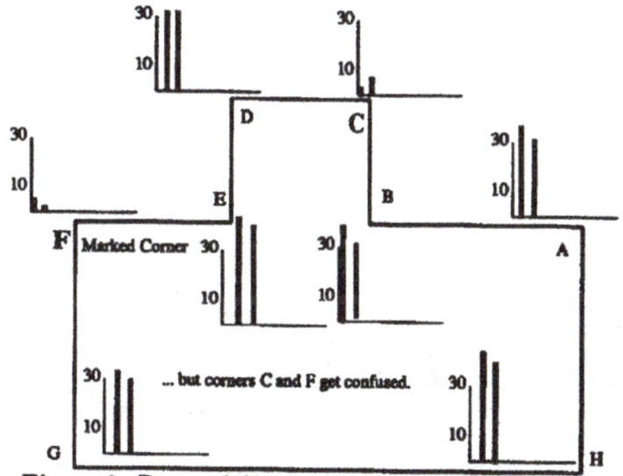

Figure 8: Recognising corner F, looking at one previous corner

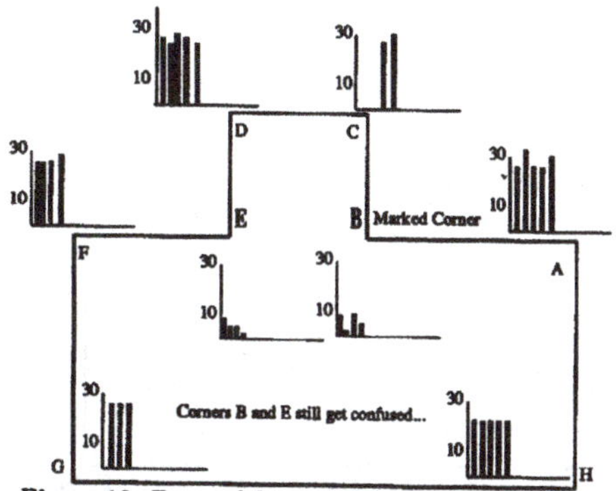

Figure 10: Recognising corner B, looking at two previous corners

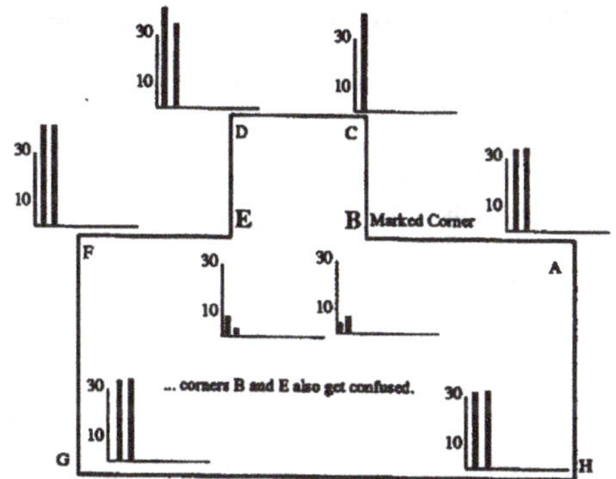

Figure 9: Recognising corner B, looking at one previous corner

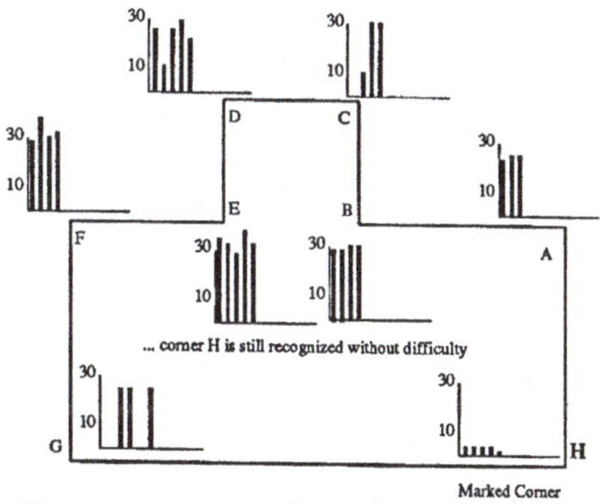

Figure 11: Recognizing corner H, looking at two previous corners

both convex, both have previous corners that are concave, and again the distances between them are similar. Again, our expectations that these two corners might have an identical representation on the resulting map are experimentally verified (see figure 9). This confusion of corners C and F and B and E respectively arises from the fact that the motor-sensory-space of the robot (input-space of the network) is not well coupled to the physical space the robot is acting in. According to the definition given in section 1.2 the resulting network is indeed a map, but a map of the motor-sensory space rather than the physical space[9].

The obvious answer to this problem is to increase the information contained in the input vector. Taking the example of corners B and E, even looking at *two* previous corners is not going to avoid confusion of these two corners, because for both corner B and E the respective previous two corners are of the same type, and the distances travelled are similar too.

To verify our theory we nevertheless used an input vector that contained information about the present corner as well as *two* previous corners: as expected corners B and E still cannot be distinguished (figure 10), and corners H gives no problems, as before (figure 11).

Only if the input vector contains information about the present corner as well as the *three* previous corners do corners B and E have different input vectors, therefore one would expect the map to be able to distinguish them. This is exactly what happens (see figure 12)!

[9] The mapping of the physical space onto map-space here is an injection, that of the motor-sensory space a bijection.

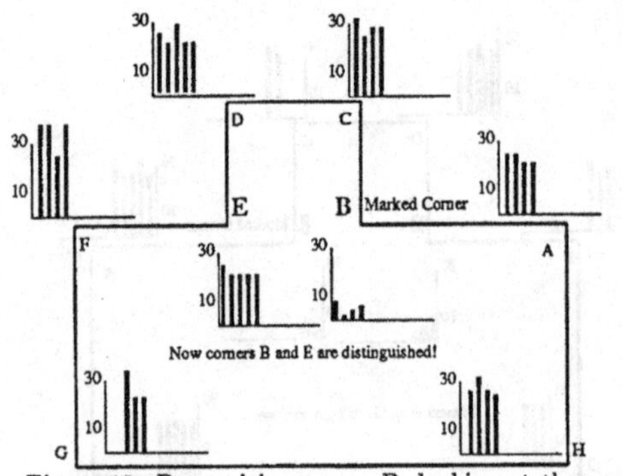

Figure 12: Recognizing corner B, looking at three previous corners

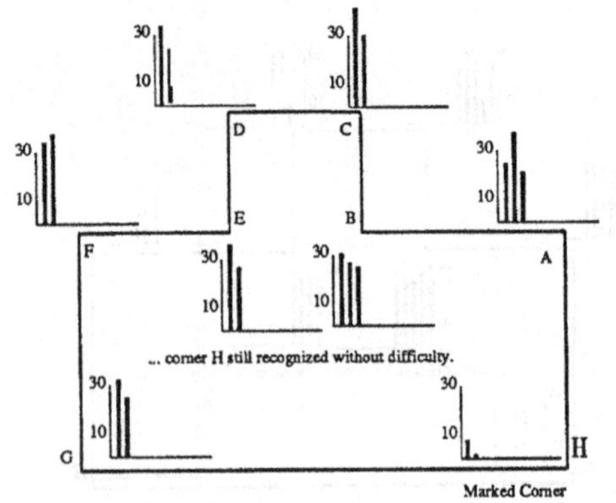

Figure 13: Recognizing corner H, looking at three previous corners

Again, as in all the previous cases, corner H is recognized without any problems (see figure 13).

These observations show the relationship between input vector and resulting internal representation very clearly, they also show that self-organising networks *can* be used to generate meaningful internal representations of the world.

Another observation we made is that the 'certainty' increases with increasing experience: $||\vec{\kappa_m}(t)-\vec{\kappa_m}(t+1)||$ decreases continuously as the robot's experience increases (see figure 14).

In summary, this is what happened: Having had sufficient time to explore the environment, the robot was able to recognize particular corners that had been marked by the experimenter, provided this was theoretically possible[10]. In cases where the input vector presented to the self-organising net contained insufficient information, Alder confused corners that looked alike, simply because their input vectors were (almost) identical. This is not surprising. The robot's ability to recognize the marked corner reliably increased with experience, $||\vec{\kappa_m}(t) - \vec{\kappa_m}(t+1)||$, the difference between excitation of the net at the marked corner at time t and excitation at the marked corner at time t+1, decreased continously. It eventually became zero.

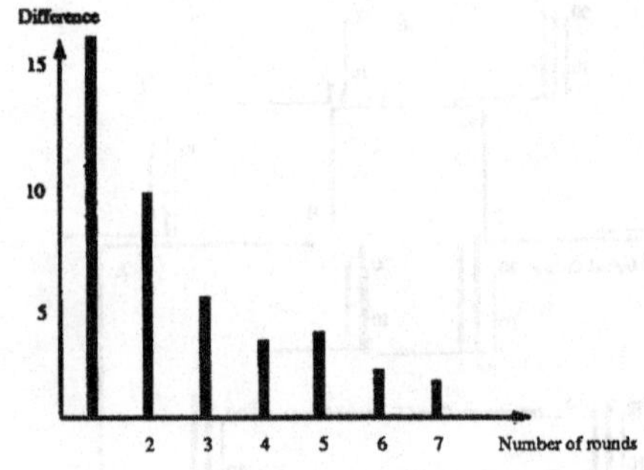

Figure 14: $||\vec{\kappa_m}(t) - \vec{\kappa_m}(t+1)||$

4 Comparison with Bee Navigation

As [Cartwright and Collett 1983] have found in their experiments, *Apis mellifera* use nearby landmarks to guide their way to a food source. "[Bees] do not find their way using anything analogous to a floor plan or map of the spatial layout of landmarks and food source. The knowledge at their disposal is much more limited, consisting of no more than a remembered image of what was present on their retina when they were at their destination. ... Bees find their way, the experiments seem to say, by continuosly comparing their retinal image with their snapshot and adjusting their flight path so as to lessen the discrepancy between the two." We do not claim that the mechanism used for landmark learning in Alder is identical to the one used in bees; however, there are similarities. Like bees, Alder does not generate a floor plan (a conventional map), instead it uses 'snapshots' (distinctive excitation patterns of the self-organising network as a response to the sensory input) to recognize locations.

[10] Alder signals the recognition of a marked corner by a swift turn, first to the left and then to the right.

Another similarity to bee navigation is the fact that the mechanism described in this paper is very robust and fairly immune to noise. 'Undetected' corners, varying distance measurements and even moving the robot to a different location (without the control algoithm realising this) do not impair Alder's ability to recognize locations. [Cartwright and Collett 1983]: "The bee's guidance system is immune to a considerable amount of noise".

5 Conclusion

Building an internal state space representation (a *map*) is an essential step towards developing a navigational competence in a mobile robot. We have shown that it can be achieved, using self-organising networks. Particular input stimuli result in distinct reponses of the (ring-shaped) self-organising network, distinct physical locations in the real world can thus be distinguished, provided that robot behaviour and sensing are well coupled to the environment and the task of mapbuilding. The mechanism presented is similar to the 'feature recognition' mechanism used by bees to navigate—bees compare stored retinal patterns to the current retinal pattern to calculate a flight direction vector ([Gould 1988] [Waterman 1989] and [Cartwright and Collett 1983]).

The next step in the RUR project is to develop a network that can learn the relationships between physical movements in the world and observed activity changes in the self-organising network (the bottom block shown in figure 1), which will make true navigation possible.

Acknowledgements

The work reported here is supported by a grant from the UK Science and Engineering Research Council (grant number GR/F/5852.3). We would also like to thank John Hallam and Peter Forster, our colleagues on RUR, and Mitch Harris for helpful comments on earlier drafts of this paper.

References

[De Almeida 1989] R, De Almeida, C. Melin: Exploration of Unknown Environments by a Mobile Robot, in: Proceedings of IAS 2, Amsterdam 1989

[Brooks 1986] Rodney Brooks: A robust layered Control System for a Mobile Robot, MIT AI memo 864, 1986

[Cartwright and Collett 1983] B.A. Cartwright and T.S.Collett: Landmark Learning in Bees, Journal of Comparative Physiology (1983) 151:521-543

[Goto and Stentz 1987] Y. Goto, A. Stentz: The CMU System for Mobile Robot Navigation, in:Proc. IEEE intern. Conference on Robotics and Automation 1987 (p. 99)

[Gould 1988] James L. Gould and Carol Grant Gould: The Honey Bee, Scientific American Library, New York 1988, p. 106

[Kohonen 1988] Teuvo Kohonen: Self-Organization and Associative Memory, Springer 1988

[Levi 1987] Paul Levi: Principles of Planning and Control Concepts for Autonomous Mobile Robots, in: Proc. IEEE intern. Conference on Robotics and Automation 1987 (p. 874)

[Moravec 1988] Hans P. Moravec: Sensor Fusion in Certainty Grids for Mobile Robots, AI magazine Summer 1988 (pp. 61-74)

[Nehmzow et al 1989] U. Nehmzow, J. Hallam and T. Smithers: Really Useful Robots, in Proceedings of IAS 2, Amsterdam 1989

[Tsubouchi and Yuta 1987] T. Tsubouchi, S. Yuta: Map assisted Vision System of Mobile Robots for Reckoning in a Building Environment, in: Proc. IEEE intern. Conference on Robotics and Automation 1987 (p. 1978)

[Tsai et al 1990] Jeffrey J.-P. Tsai, Mark Metea and John Cesarone: A knowledge-based navigation scheme for autonomous land vehicles; in: Applied Artificial Intelligence 4:1-14, 1990

[Waterman 1989] Talbot H. Waterman: Animal Navigation, Scientific American Library, New York 1989, p.183

THE CONTRIBUTION OF QUANTITATIVE MODELS TO THE LONG DISTANCE ORIENTATION PROBLEMS

Marc Jamon

Centre National de la Recherche Scientifique
Laboratoire de Neurosciences Fonctionnelles
France-13402 Marseille Cedex 9
e-mail: BOVET@FRMOP11

Abstract

Examples of simulation models applied to the long distance orientation of rodents, bees, salmon and pigeons are reviewed. In each case the models provide new insights which are contradictory to the current map hypothesis. Salmon and rodent performances are assumed to be consistent with a random process. Bees supposedly use non cognitive processes to perform map-like orientation procedures, and pigeons' homing is found to involve mainly stochastic processes. The efficiency of the simulated models as well as the fact that they point to hitherto underestimated aspects of orientation make them useful new tools for solving spatial problems.

1. Introduction

Authors using the theoretical approach to the long distance orientation in animals have consistently neglected the development of quantitative models. Consequently, the properties of random processes have been rarely formulated and their possible contribution to observed performances have been underestimated (Jamon, 1987). Nor have performances been attributed to any distinguishable orientation mechanisms. Indeed, long distance orientation has been largely interpreted in terms of analogies with human navigation techniques, leading to the explicitly formulated idea that "routine animals' movements are governed by a navigational process closely analogous to every day marine practice" (Gallistel, 1989). The reference to man made concepts of orientation is based on the "map" metaphor, which implies that animals can build up a mental representation of their environment. Two sorts of maps have been envisaged. When an animal moves in a familiar area, it is supposed to develop a mental representation of the landmarks (Wiltschko and Wiltschko, 1987, talk about animals' mental picture of the spatial distribution of the factors used), which has been variously called the mosaic map, familiar area map, topographical map or cognitive map. When the animal has to orient in an unfamiliar territory, it is supposed to use some sort of bico-ordinate map, formed by the intersection of at least two gradients along association with a compass, exactly as human navigators do. In this case, one talks about a grid map, gradient map or navigational map. This map and compass hypothesis was first formulated by Kramer (1953), but considerable difficulties have since been encountered in determining the map.

Recent developments in the theoretical approach to orientation have led to more attention being paid to quantitative models, owing to the development of computers and simulation techniques. These have thrown new light on orientation by predicting the performances of orientation mechanisms which were considered *a priori* to be implausible.

To show how these new techniques can change our view of animal orientation, I propose to review four examples where models have provided new insights contradicting the current map hypothesis. Each of these involves species suspected of being capable of long distance orientation. the first two are alternative models proposed in situations where homing by means of familiar area map was first envisaged: they concern rodents and bees. The last two deal with situations where the use of a navigational map has been hypothesized: these concern salmon and pigeons.

2. The case of rodents' homing ability.

Numerous experiments have shown that rodents transported a relatively long distance from their home can return to their previous territory with a probability of success which is higher than that predicted by randomness (Joslin, 1977) (fig. 1).

Various orientation mechanisms have been suggested to explain this good successful homing performance level. The animals have been said to rely on some specific orientation mechanism such as route-based navigation involving the use of magnetic cues (Mather and Baker, 1980, 1981), or to pilot by means of a large familiar area map (Furrer, 1973). In the latter case, they would have to build up a topographical representation of a large territory, extending beyond the limit of their actual home range during the course of exploratory trips.

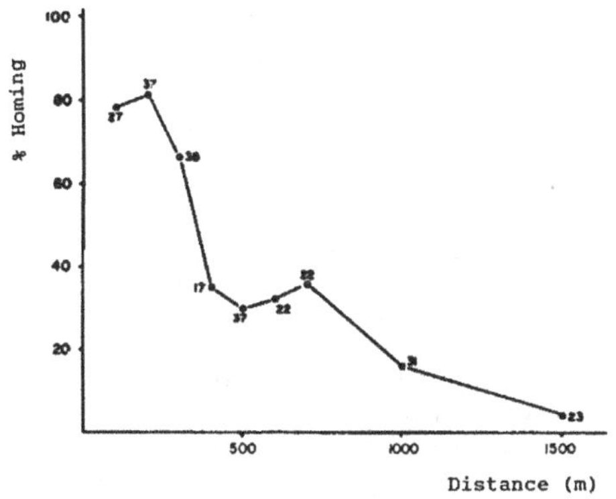

Figure 1. Relationship between distance at release and homing success in a rodent (Sigmodon hispidus). Figures along the curve correspond to the number of animals displaced at each distance. (from Bovet, 1987; copyright by Martinus Nijhoff Publishers)

One difficulty with these interpretations is that they are founded on partial information provided by successive recapture of the animals during a few days after their release or on initial orientation in the case of caged animals. When homing paths where recorded, the interpretation of orientation performance was quite different, as in the case of an experiment carried out in the Camargue (Jamon and Bovet, 1987).

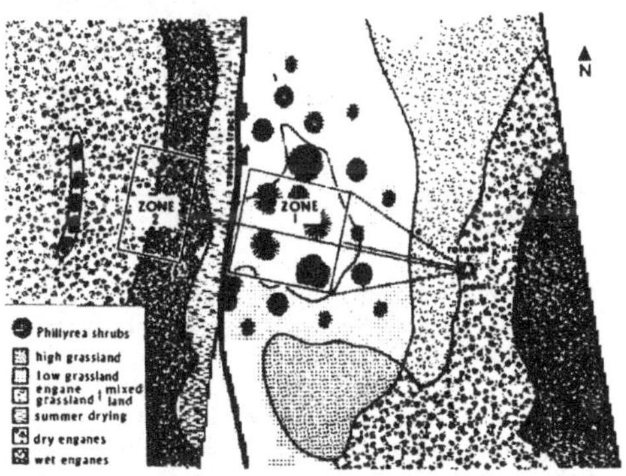

Figure 2. Overhead view of the study area with main landscapes. Squares delimit the trapping plots. Broken lines delimit the angular directions where orientation was assumed to be correct (Reproduced from Jamon and Bovet, 1987 with permission).

In this experiment, wood mice (*Apodemus sylvaticus*) trapped in two neighboring plots were equipped with chemoluminescent tags and released by night at a remote place, 120 m from plot 1 and 250 m from plot 2 (fig. 2).

Their movements were then observed from an observation post. Due to a particular vegetation configuration, the two plots had very different vegetation cover, in spite of their proximity, and the release place had intermediate vegetation cover so that landscape gradually evolved towards that of plot 1 in direction of the plots and towards that of plot 2 in the opposite direction.

The percentage of returns was 75 % in the case of plot 1 and 47% in that of plot 2 mice, which was consistent with the performances commonly obtained in other experiments, suggesting that these animals were endowed with an orientation ability. Examination of their paths revealed the existence of some quite different mechanism, however. When released, the mice from both plots wandered around in a random manner but progressively moved away in opposite directions (fig. 3). Plot 1 mice therefore moved homeward, whereas plot 2 mice moved away from home. Moreover, Those animals which orientated homeward homed during the first night, whereas the others needed several days to home.

It was assumed that the mice's movements resulted from a random walk process which was locally biased by local variations in the vegetation. Most mice from plot 1 were caused to move toward their home by the progressive landscape similarity so that they quickly homed, whereas the mice from plot 2, lured on by the environmental configuration encountered, moved in the wrong direction and homed only after several days, probably by chance.

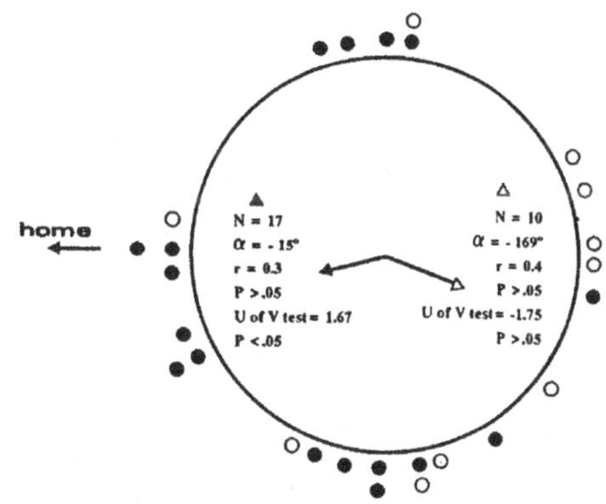

Figure 3. Wood mice's bearings 50 m from Release point.
Black dots: plot 1 mice; white dots: Plot 2 mice. ; arrows: r-vectors.
Circular Statistics are given on each side of the figure.
The distribution of bearings homewards and in the opposite direction differs significantly in the two species ($\chi^2 = 4.57$, $p < .05$).

In order to check the hypothesis that a small local bias may have progressively caused a general trend in the random movement in a particular spatial direction, a theoretical model was developed in which the random wandering was simulated using the Bovet's algorithm (Bovet and Benhamou, 1989). Each path was modeled with a sequence of linear segments with constant length and random changes of direction. The paths were entirely defined in terms of random distribution of changes of direction drawn at each successive step with a Gaussian distribution having a mean = 0 and standard deviation σ. The environmental bias was simulated by an axial gradient and was related to the random walk by associating an absolute klinokinetic effect to the basic movement pattern. In practice, the standard deviation (σ_i) of the changes of direction at every step i was related to the position along the gradient axis following the equation:

$$\sigma_i = ((\mathrm{arctg}(x/25)/\pi)+0.5).(V_\sigma)+\sigma_{MIN}$$

where x is the projection of the animal's position on the gradient axis, V_σ is the fixed range of variation of σ, and σ_{MIN} is the minimal value of σ.

In spite of the limited range of variation of σ and the low diffusion properties of the absolute klinokinetic effect (Bovet and Benhamou, 1989), the simulated paths progressively diffused as a function of the gradient and their distribution was consistent with that of observed paths. This confirmed that the good homing return may have resulted from a biased random process and therefore did not necessarily involve any orientation ability.

These results suggest a new interpretation of wood mice's homing performances. The high homing rate was found to include both the rapid returns of correctly oriented mice and the later returns of uncorrectly oriented mice, which more probably arrived by chance. Moreover, the proportion of correctly oriented mice was consistent with the hypothesis that they performed a random search which was locally biased by the spatial heterogeneity. The influence of random processes on the homing success therefore turns out to be more complicated than has generally been assumed (Furrer, 1973). In this study, the particular composition of the landscape was certainly responsible for the experimental results. Since events are rarely uniformly distributed, the locally-biased random search hypothesis can be extended to many other situations. In cases where all the animals are removed from the same plot, one can therefore expect the presence of a perceptible local heterogeneity to result in a high level of homing success.

3. The case of bees' orientation

Unlike rodents, honey bees possess orientation capacities which have been quite well elucidated. They have been shown to rely on two different orientation mechanisms for memorizing spatial orientation. First, they are able to accurately calculate their home vector by means of path integration (Mittelstaedt, 1985) using the sunlight as a reference cue. This type of vector navigation has been studied in detail in arthropods (Wehner and Wehner, 1986). Secondly, bees have been assumed to memorize 2-dimensional photographs of the environment (Wehner and Räber, 1979; Wehner, 1981; Collett and Cartwright, 1983) which they then consult serially to perform some sort of route-specific orientation.

Recently Gould considerably revitalized the whole issue by publishing experimental data suggesting that bees are able to build up a cognitive map of their environment (Gould, 1986). Most of Gould's arguments have been severely criticized (Dyer and Seeley, 1989); we shall focus here however on the crucial detour experiment to look at the contribution of modeling to this problem.

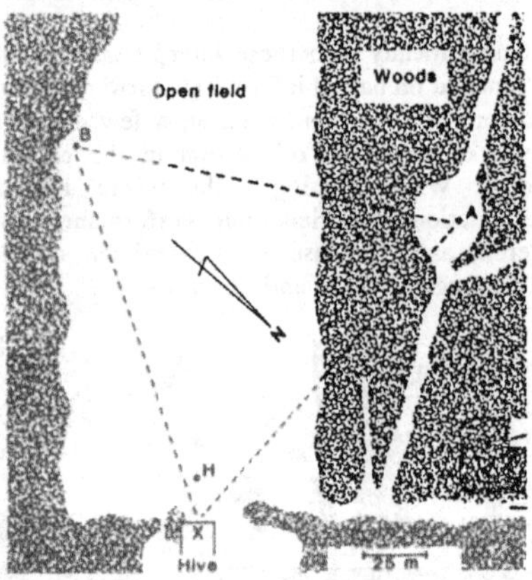

Figure 4. Plan view of the test area in Gould's experiment. : A: foraging site ; B and H release sites (from Gould 1986; copyright 1986 by the AAAS)

In Gould's experiment, individually marked honey bees (*Apis mellifera linguistica*) were trained to find food at site A, 160 m west of the hive (fig. 4). These foragers were then captured at the hive entrance and transported in the dark to site B, also 160 m away, but SSW of the hive. When released, they consistently flew directly toward A, as shown by their initial orientation (fig. 5, vector 1) and their flight times. As a control, bees transported halfway to site B and back to near the hive flew off toward site A as expected (fig. 5, vector 2), whereas unmarked bees released at site B dispersed in various directions (fig. 5, vector 4).

Gould interpreted bees' successful correction of their orientation after the imposed detour as reflecting a cognitive process. He stated that the bees, in the course of previous foraging flights, had acquired by latent learning a knowledge of their territory involving cognitive mapping, exactly as higher vertebrates are thought to do, thus invalidating the presumed vertebrate-invertebrate dichotomy.

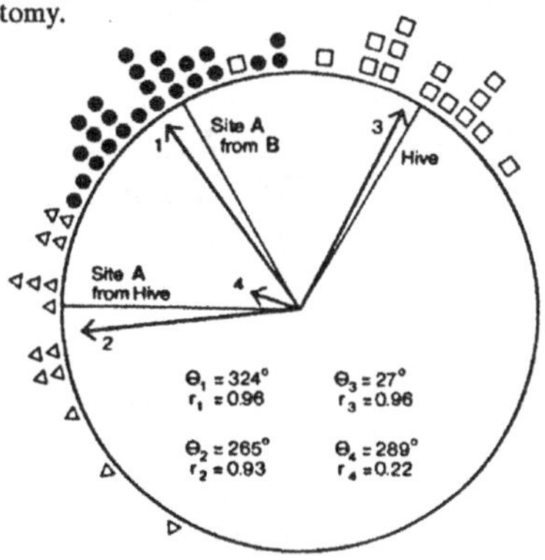

Figure 5. Vanishing bearings of bees.
(●) outgoing foragers released at B ; (△) outgoing forager released at H ; (□) incoming forager released at B; The short vector 4 represents the mean vector of bearings of outgoing unmarked bees released at B. (individual vanishing bearings not represented). (from Gould, 1986; Copyright 1986 by the AAAS)

Carthwright and Collett (1987) have adopted a completely different viewpoint. They assumed that "the small size of a bee's brain impresses upon us the need to keep Occam razor sharp", and they tried to develop a computational model capable of producing "map-like" behavior without making mental maps.

They supported their model on the two orientation mechanisms for which there exists good evidence in arthropods, i.e. the use of snapshots and vector navigation and examined their efficiency at producing map-like behavior.

Admittedly, no one of these orientation mechanisms can by itself account for the detour experiment: path integration does not correct for passive transport (Wehner and Srinivasan, 1981) and snapshots can only be used to reproduce previously travelled routes. The authors therefore added two further assumptions about the bee's orientation system: first, they assumed that the bees memorized two sets of snapshots: one of these included distance filtered snapshots, i.e. close landmarks were excluded from the memorization process. These filtered snapshots might allow the bee to compare a distant landmark arrangement with a memorized panorama. Secondly, they assumed that bees linked the current hive-vector, gained by path integration, to each filtered snapshot they memorized to make it suitable for navigation purposes.

In this way, a bee progressively memorizes a stack of oriented snapshots forming a hive-centered map, during the course of its movements. When transported to an arbitrary place within its home range, it searches for the snapshot which best fits with its current retinal image and follows the associated vector until a more closely matching snapshot provides another vector.

Cartwright and Collett (1987) checked the feasibility of this model using a two step simulation procedure (fig. 6).

In the first stage (the map building) a model bee was placed at the edge of its terrain and supplied with the home vector. It took a snapshot and linked it with the home vector. The bee then followed the home vector and updated it during its flight. When the fit between the snapshot and the current image fell below some criterion, the bee searched for a better fitting snapshot in its stack or, if not available, it took a new one (fig. 6b). This sequence was repeated until the whole terrain was mapped.

In the second stage, the efficiency of map use was tested by counting the successful returns of the model bee when it was released from the periphery without any home vector. In this case it took the available best matching snapshots at each cell and followed the associated vectors (fig. 6c). The frequency of arrival depended on the fit criterion used in the map. 63% and 77 % of successful returns occurred with a criterion of respectively 0.5 and 0.6

This 'hive-centered map' accounted for the ability of passively transported bees to find their hive, but not a remote place. In order to explain the detour experiment, Collett and Cartwright assumed, from several studies on arthropods, that bees could add vectors. Thus when bees were transported to site B, in Gould's experiment, they presupposedly linked the hive-vector associated with the memorized snapshot of the current retinal image and added the vector to site A from the hive, which they had learned during training.

Cartwright and Collett have shown that it is useless to look for hypothetical vertebrate capacities, as Gould did, to account for bee's detour performance in bees. They showed that invertebrates' orientation system, as far as we know, virtually possesses the properties required for performing map-like operations. Unlike cognitive maps, this hypothesis involves only limited assumptions, such as the ability to filter out close landmarks from snapshots, e. g. by means of the motion parallax (Collett and Cartwright, 1983), and to memorize linked pairs of snapshot and vectors. The simulation showed that other related spatial operations are also considerably simplified: the bee's location needs not be specified with great accuracy, since it suffices to have a coarse map consisting of a few snapshots each covering a large area. Moreover, complex manipulations of the spatial arrangement of landmarks are replaced by vector additions which are thought to occur in arthropods.

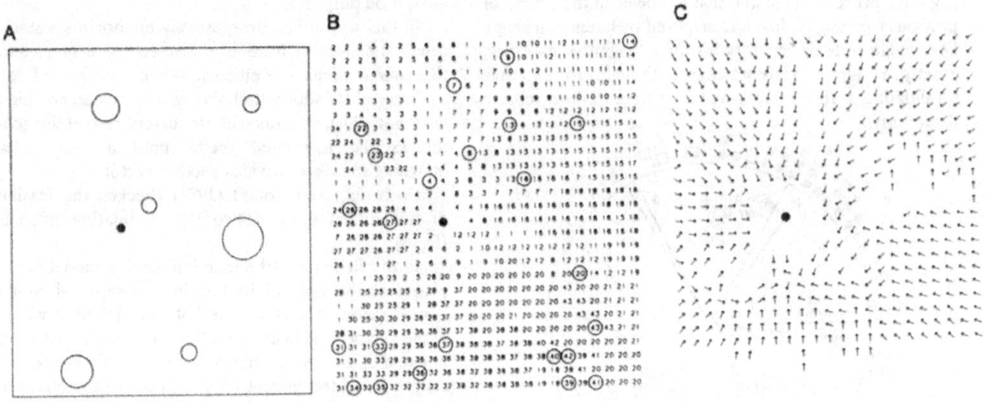

Figure 6. Simulation of the snapshot map.
a) part of the terrain. Open circles: landmarks ; Black dot: goal ; b) snapshot map. A new snapshot is taken when the goodness of fit between retinal image and current snapshot falls below 0.5. Circled numbers show cells where snapshots were taken. The other numbers show which snapshots are current. The number of snapshots to memorize depends on the criterion precision; with a criterion of 0.5 : 43 snapshots C) vector field associated with the selected snapshots during the homing simulation.The flight achieved when the bee arrived within two cells of the goal. (from Cartwright & collett, 1987; copyright by Springer-Verlag)

4. The return migration of salmon

In the course of their life, salmon leave their natal river as fry and migrate far out into the open sea before returning to their natal stream to spawn. The process that brings salmon thousands of kilometers back to their natal stream estuary has been diversely interpreted in terms of random wandering or bico-ordinate navigation ability.

Saila and Shappy (1963) made a major contribution to the random search hypothesis by showing that the return performance was consistent with a biased random walk model. Quinn and Groot (1984) recently showed however that further progress in our knowledge of salmon migration are inconsistent with this model. These authors pointed out that the following characteristics of the homing course cannot be accounted for by the random hypothesis:

- salmon commonly travelled at speeds of 40 to 50 Km/d.
- all the returns were concentrated in a very short period of time: according to Brugner (1980), 80% of the returns occurred within 13 days.
- the recapture rate (about 5 to 10%, and sometimes up to 20%), underestimated the actual return rate, due to many failures in tags reports.

In Quinn and Groot's opinion, covering 1000-2000 Km in 4-10 weeks requires a very high degree of orientation and is therefore more likely to have involved a navigation process, probably based on a magnetic map (Quinn, 1982).

Although these observations were actually no longer consistent with Saila and Shappy's model, they do not contradict the random hypothesis. Indeed, it was further shown that a very simple random search model was able to account for these characteristics (Jamon, 1990).

Basically, this model assumed that salmon swim fast and straight during the last stage of the open sea migration, but have no means of locating their home direction. Therefore they disperse with a uniform radial scattering pattern. Only a few fish could actually swim straight home in this way, but a lot of them (theoretically up to almost 50 %) are likely to reach the coast.

Upon reaching coastal waters, the salmon were assumed to wander up and down until they encountered their home stream by chance. This wandering behavior was simulated by a 1-dimensional random walk model defined as a sequence of constant linear segments and up and down orientations.

In the simulation scenario, the open ocean was taken to be a completely uniform area. The coast was defined as a straight line with an infinite length. The home stream was represented as a point on the coast and was surrounded on each side by a familiar area 74 Km wide. The salmon were assumed to home from places randomly selected within an area of 1000,000 Km^2 between 1000 and 2000 Km from the coast (fig. 7).

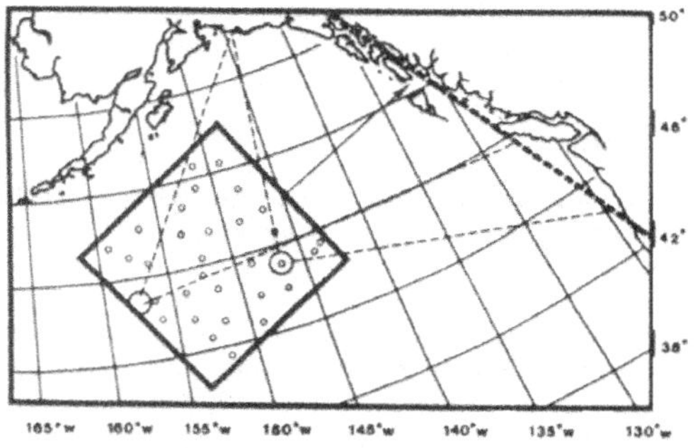

Figure 7. Schematic representation of the salmon homing model. Square defines the assumed distribution of salmon at see at the time of return. small circles: random distribution of salmons. Arrow: assumed homeward direction. Fine dashed lines: examples of maximum angular dispersion compatible with successful return. Thick dashed line: theoretical coastline.

Each journey lasted at most 1000 h (around 42 days). During this time the salmon were assumed to swim continuously at a constant speed of 3Km/h. Those which did not arrive within the endurance time allowed for homing were assumed to die.

The homing success depended greatly on the step length at the coast, thus varying from 3 to 16%, whereas the temporal pattern of return remained constant. Assuming a steplength at the coast of 37 Km, as did Saila and Shappy (1963), 6.33% of the salmon homed, having travelled at a rate of 59 Km/d and among these homing salmon, 80% homed within an interval of 18 days. A lower swimming speed of 2Km/h lessened the homing success to 4% and the rate of travel to 45 Km/d.

Resuming the Quinn and Groot's arguments, then:

1) The random search model accounted for rates of travel higher than 45Km/d, in the salmon speed range of 2 to 3Km/h.

2) 80% of the returns occurred within a time interval of 18 days.

3) The homing success was about 6% and could increase by up to 16% depending on the step length. This range of values is consistent with published tag reports.

The basic formulation of this random model unrealistically assumed the environment to be uniform. Accounting for the environmental heterogenity considerably improved the performance. For instance, assuming that the movements of salmon were restricted to 500 Km north and south of their home stream (which seemed a plausible assumption in view of salmon's distribution at sea) the homing success was doubled with a steplength of 37 Km, and reached up to 50 % assuming a linear path at the coast. Assuming now that salmon in the open sea are helped by some vague compass cues, we can expect their initial orientation to be homeward ± 90°. This would increase the homing success twofold, whereas the rate of travel would not be affected. Finally, salmon might both swim between reflecting barriers and use compass cues. Under this assumption, the simulated homing success would range from 11 to 99% depending on the assumed step length (28 % with L=37Km).

The observed homing performances were consistent with the uniform random orientation hypothesis. Moreover, moving straight ahead both at sea and at the coast turned out to be the most efficient random strategy. Therefore, straight paths are not necessarily directed (oriented). None of the proposed arguments thus provides evidence supporting navigation in salmon.

Moreover, accounting for the environmental irregularity considerably improved the performance level. This could be correlated with the currents which are known to be of paramount importance for fishes' migration. Indeed, they can influence fishes' movements in many ways, by increasing the swimming speed by drift, providing compass cues, or channelling them as narrow reflecting barriers would do, for instances, and they probably combine all these possibilities. This could provide a possible link between random movement processing and the enviro-regulation hypothesis (Neill, 1984) which assumes that fish swim in a preferendum area in order to maintain an optimal physiological state.

5. The long distance homing of pigeons

At variance with salmon, there is good evidence for the existence of a specific orientation process in pigeons.

Much information has been collected on the various cues pigeons rely on and their ability to use them as a compass. The supposedly associated map still remains elusive, however. Actually two different maps have been envisaged. The one includes the familiar area, supposedly ranging less than 70 km around the loft, and is assumed to be a mental representation of the terrain, forming a mosaic map. But, more attention has been paid to the other map, which is used when pigeons return from distances of several hundreds of kilometers across unfamiliar areas. At these distances, pigeons are assumed to rely on a gradient map (Wallraff 1974, Wiltschko and Wiltschko, 1978). According to this assumption, pigeons might be capable of perceiving small intensity variations of at least two gradients in olfactive, magnetic or unknown cues in their home range and to extrapolate some rules about their further course and direction several hundred of kilometers away from their home loft. When they are far from their home loft, they estimate their own coordinates in gradient intensities, and by comparing them with those memorized at the loft in relation to the compass direction of the gradient, they can determine the home direction. The arguments in favour of the map are indeed extremely flimsy and authors have presented contradictory arguments about the magnetic vs. olfactive nature of the map that make the whole idea seems questionable (Gould, 1982a,b; Wallraff, 1982, 1983, Schmidt-Koenig, 1987, Wiltschko and Wiltschko, 1989). One difficulty with the use of a gradient map is that large scale gradients need to vary over long distances in some predictable order and to be temporally stable to be compatible with efficient spatial localization.

Last year, Wallraff (1989) threw new light on this problem by devising a computer simulation model to check the plausibility of orientation mechanisms based on unreliable sources of information. The aim was to show that atmospheric compounds could generate usable long distance gradients, but the conclusions seem to be of more general interest in the context of gradient maps.

Wallraff assumed that pigeons at the home site memorized local intensities and directions of several gradients (for simplicity, axial gradients). When the model bird moved away, it was able to find the homeward direction of each gradient with a certain level of probability which depended on the distance *d* from the home, with the following relation:

$$Pd = 0.5 \pm d^b c$$

where b and c are fixed constants.

The pigeon then tossed for the homeward direction at each gradient and computed the mean direction of these vectors to determine its flying vector. This kind of direction making was repeated at given time intervals resulting in a sequence of linear segments and rotations. In fact, to reduce the chaotic appearance of the pigeon's course, the flying vector was weighted with the previous one.

Homing performances of model pigeons released at 100 Km from home under these preconditions depended on the number of presumed gradients and the importance of the conservative weighting value. A remarkably low value of P sufficed, however, to produce homing performances comparable to those observed in nature.

These results show, as in the case of bees, that no precise location for those animals is necessary to rapidly orient homeward. They also showed that the good homing performances of pigeons are consistent with a stochastic process. Although the gradient direction may have been so successfully obtained from the successive variations in gradient intensity encountered during flight, Wallraff opted for the hypothesis that pigeons learned a compass direction at the home (obviously for the sake of compatibility with the results of clock-shift experiments), and therefore he did not abandon the map concept, however He suggested considerable changes in the contents of the map: no overall arrangement of the gradients was assumed. Indeed, the flying vector was a compromise between independent pieces of gradient information. This should considerably change our view of pigeon's orientation.

6. Conclusion

The performances reviewed here have been generally assumed to reflect mapping ability without the cognitive processes involved in mapping being clearly defined. In each example, simulating orientation mechanisms successfully reproduced the observed performances, showing that maps were either not necessary, or that they are much more elementary than it might appear at first sight.

Admittedly, the efficiency of the models proposed to account for the performances does not necessarily make them a valid means of explaining orientation mechanisms. To provide a really convincing description of orientation, they would have to deal at least with biological properties and better still with animals' rules of thinking, but too many gaps exist about biological parameters and too many arbitrary assumptions have been made.

Nevertheless, each of these simulation models provides some relevant information about possible orientation mechanisms. The first of these was the idea that homing largely consists of stochastic processes. This has important consequences such as the idea, formulated by Wallraff, that specific errors, instead of reflecting different individual evaluations of the "map factors" could merely result from statistical noise.

Another point worth putting forward is the fact that environmental heterogeneity can have a strong influence on animals' search behavior. This was found to be so in the case of the random searching of rodents and salmon. Generally speaking, animals are rarely confronted with uniformly distributed events. This should prompt us to focus more on animal-environment interactions rather than on internal representations of space.

The models also shatter previous ideas about precision. Indeed, the animals do not need more than a coarse sense of location. By making successive corrections, they will rapidly keep to their home bearings, as was found to occur in bees and pigeons.

Simulation models can therefore considerably change our views about animal orientation. They allow performances to be attributed to alternative hypotheses.

In view of the strong anthropomorphic tendency pervading the study of animal orientation, they can serve as a heuristic tool for reconsidering long distance orientation theory.

REFERENCES

Benhamou S. ; Bovet P. 1989. How animals use their environment: a new look at kinesis. Anim. Behav. 38: 375-383.

Bovet J. 1987. Cognitive map size and homing behavior. in "Cognitive processes and spatial orientation in animal and man.". (Ellen P.; Thinus-Blanc C. eds.) Vol 1: Experimental animal psychology and ethology, Martinus Nijhoff Publishers, Dordrecht, 252-265.

Bovet P. ; Benhamou S. 1988. Spatial analysis of animals' movements using a correlated random walk model. J. Theor. Biol. 131: 419-433.

Brugner R. L. 1980. Some features of ocean migrations and timing of pacific salmon. in "Salmon ecosystems of the north Pacific.". (Mc Neil; Himsworth D.C. eds.). Oregon State University Press, Corvallis: 153-164.

Cartwright B. A. ; Collett T. S. 1987. Landmark maps for honeybees. Biol. Cybern. 57: 85-93.

Collett T. S. 1987. Insect maps. Trends Neurosci. 10: 139-141.

Collett T. S. ; Cartwright B. A. 1983. Eidetic images in insects: their role in navigation. Trends Neurosci. 6: 101-105.

Dyer F. C. ; Seeley T. D. 1989. On the evolution of the dance language. Am. Nat. 133 (4): 580-590.

Furrer R. K. 1973. Homing of Peromyscus maniculatus in the channelled scablands of east-central Washington. J. Mammal. 54 (2): 466-482.

Gallistel C. R. 1989. Animal cognition: the representation of space, time and number. Annu. Rev. Psychol. 40: 155-189.

Gould J. L. 1982a. The map sense of pigeons. Nature 296: 205-211.

Gould J. L. 1982b. The homing mechanism of pigeons: replies. Nature 300: 294.

Gould J. L. 1986. The locale map of honey bees: do insects have cognitive maps? Science 232: 861-863.

Jamon M. 1987. Effectiveness and limitation of random search in homing behavior. in "Cognitive processes and spatial orientation in animal and man.". (Ellen P.; Thinus-Blanc C. eds.) Vol 1: Experimental psychology and ethology, Martinus Nijhoff Publishers, Dordrecht, 284-294.

Jamon M. 1990. A reassessment of the random hypothesis in the ocean migration of Pacific Salmon. J. Theor. Biol. 143: 197-213.

Jamon M. ; Benhamou S. and Sauve J. P. 1986. Initial orientation and navigation in homing rodents. in "Orientation in space.". (Beugnon G. ed.) Privat, Toulouse, 45-55.

Jamon M. ; Bovet P. 1987. Possible use of environmental gradients in orientation by homing wood mice, Apodemus sylvaticus. Behav. Processes 15 (1): 93-107.

Joslin J. K. 1977. Rodent long distance orientation ("Homing"). Adv. Ecol. Res. 10: 63-89.

Kramer G. 1953. Wird die Sonnenhöhe bei der Heimfinderorientierung verwertet ? J. Orn. 94: 201-219.

Mather J. G. ; Baker R. R. 1980. A demonstration of navigation by small rodents using an orientation cage. Nature 284: 259-262.

Mather J. G. ; Baker R. R. 1981. Magnetic sense of direction in Wood mice for route-based navigation. Nature 291: 152-155.

Mittelstaedt H. 1985. Analytical cybernetics of spider navigation. in "Neurobiology of Arachnids.". (Barth F.G. ed). Springer-Verlag, Berlin, Heidelberg, Tokyo, 298-316.

Neill W. H. 1984. Behavioral enviroregulation's role in fish migration. in "Mechanisms of migration in fishes.". (Mc Cleave J.D.; Arnold G.P.; Dodson J.J.; Neill W.H. eds). Plenum Press, New York, London: 61-66.

Quinn T. P. 1982. A model for salmon navigation on the high seas migration. in "Proceedings of the salmon and trout migratory behaviour symposium.". (Brannon E.L.; Salo E.O. eds). School Fisheries, University Washington, Seattle, Washington, USA: 229-237.

Quinn T. P. ; Groot C. 1984. Pacific salmon (Oncorhynchus) migrations: orientation versus random movement. Can. J. Fish. Aquat. Sci. 41: 1319-1324.

Saila S. B. ; Shappy R. A. 1963. Random movement and orientation in salmon migration. J. Cons. Perm. Int. Explor. Mer 28: 153-166.

Schmidt-Koenig K. 1987. Bird navigation: has olfactory orientation solved the problem ? Q. Rev. Biol. 62 (1): 31-47.

Wallraff H. G. 1974. Das Navigationssystem der Vögel. Schriftenreihe "Kybernetic", Oldenbourg-Verlag, München, Wien.

Wallraff H. G. 1982. The homing mechanism of pigeons. Nature 300: 293.

Wallraff H. G. 1983. Relevance of atmospheric odours and geomagnetic field to pigeon navigations. What is the "map" basis ? Comp. Biochem. Physiol. 76 (A): 643-663.

Wallraff H. G. 1989. Simulated navigation based on unreliable sources of information (models on pigeon homing. Part 1). J. Theor. Biol. 137: 1-19.

Wehner R. 1981. Spatial vision in arthropods. in "Handbook of sensory physiology.". (Autrum H. ed). Vol VII/6C. Springer, Berlin, Heidelberg, New-York: 287-617.

Wehner R. ; Räber F. 1979. Visual spatial memory in desert ants, Cataglyphis bicolor (Hymenoptera: Formicidae). Experientia 35 (12): 1569-1571.

Wehner R. ; Srinivasan M. V. 1981. Searching behaviour of desert ants, Genus Cataglyphis (Formicidae, Hymenoptera). J. Comp. Physiol. 142: 315-338.

Wehner R. ; Wehner S. 1986. Path integration in desert ants: approaching a long-standing puzzle in insect navigation. Monit. Zool. Ital. 20 (3): 309-331.

Wiltschko R. ; Wiltschko W. 1989. Pigeon homing: olfactory orientation - a paradox. Behav. Ecol. Sociobiol. 24 (3): 163-174.

Wiltschko W. ; Wiltschko R. 1978. A theoretical model of migratory orientation and homing in birds. Oïkos 30: 177-187.

Wiltschko W. ; Wiltschko R. 1982. The role of outward journey information in the orientation of homing pigeons. in "Avian navigation.". (Papi F.; Wallraff H.G. eds.) Springer-Verlag, Berlin, Heidelberg, 239-252.

Wiltschko W. ; Wiltschko R. 1987. Cognitive maps and navigation in homing pigeons. in "Cognitive processes and spatial orientation in animal and man.". (Ellen P.; Thinus-Blanc C. eds.) Vol 1: Experimental psychology and ethology, Martinus Nijhoff Publishers, Dordrecht, 201-216.

Navigating With a Rat Brain: A Neurobiologically-Inspired Model for Robot Spatial Representation

Maja J Mataric
MIT Artificial Intelligence Laboratory
545 Technology Square, #721
Cambridge, MA 02139, USA
maja@ai.mit.edu

Abstract

This paper presents a neurobiologically-feasible spatial representation model. The model was implemented and tested on a physical autonomous mobile robot. It was shown to be both computationally simple and physically robust.

The described model is a possible interpretation of the organization and function of the rat hippocampus. The paper presents relevant biological, psychological, and neurobiological data, and gives a detailed set of comparisons between the physical hippocampus and our "synthetic" rat implementation. The implications of the many similarities are described. Finally, areas for future study in both biology and robotics are suggested.

1 Introduction

Most animals, including humans, spend much of their waking time in transit from one place to another [**Waterman 89**]. Purposefully moving about requires a system for spatial modeling integrated with the mechanisms for handling navigation, locomotion, and motivation. These systems have evolved to perform with impressive robustness. Understanding their function has long been a goal of cognitive scientists, biologists, and neuroscientists. More recently, this goal has been adopted by members of the Artificial Intelligence and robotics communities interested both in simulating biological systems and designing better artificial ones.

The question asked by both communities is: "What kind of spatial information is stored?" In order to answer it, experiments are designed to test where on the qualitative-to-quantitative scale the representation lies, and whether it is centralized or distributed. This paper describes a qualitative, distributed spatial representation tested empirically on a mobile robot.

2 Cognitive Maps

A *cognitive map* is a generic term for an internal representation of spatial information. The term has come to connote a very analytical, centralized representation. In this paper, we will use the term in its generic meaning, and analyze its variants.

A cognitive map is usually assumed to represent space with a set of landmarks, each of which is an element (object or feature) serving as a point of reference [**Presson and Montello 88**]. According to Piaget, a landmark is a spatial primitive, and thus a basic building block of spatial representations [**Piaget and Inhelder 67**]. Although most landmark studies concentrate on visual cues, the concept generalizes to any perceptable feature. Animals construct landmarks from auditory, olfactory, and tactile cues as well [**Gould 82**], taking advantage of their different characteristics [**O'Keefe 89**].

In this paper, we will analyze cognitive maps along two dimensions: 1) what information they encode and 2) how they encode it. The "what" dimension can vary from completely qualitative or topological to very quantitative or metric. The "how" dimension varies from totally global or centralized to entirely distributed or decentralized.

2.1 How Qualitative?

The nature of the representation determines the type and number of landmarks required for localizing. In a qualitative representation, an object can be remembered as being proximate to a landmark, defined within a radius around it. On the other end of the spectrum, the position of an object can be computed precisely from the known locations of three landmarks [**Pick, Montello and Somerville 88**]. The question is how much metric information is recorded.

The psychological literature is divided on this issue. Studies testing response times in object position recall indicate that adults are certainly capable of reconstructing Euclidian distance and constructing metric maps. While sufficient information is gathered and recorded to make metric inference possible, the computation appears to be relatively slow and performed only when needed. Experiments have shown that the produced metric inferences tend to be distorted based on local feature density and importance.

[Piaget and Inhelder 67] proposed, and later research supported, an almost exclusively topological theory of early spatial knowledge. The knowledge of children in the preschool period is believed to be fundamentally non-metric, and rely on the presence of landmarks. More specifically, the landmarks must be close, so that children can form relative distance relations among them. This introduces a paradox: while children do not seem to record metric information, they must be using it to form relations between landmarks. These studies are especially interesting in testing the limits of qualitative representations.

2.1.1 Experiments With Rats

Acquiring spatial information is one of the most highly developed cognitive abilities in rats **[McNaughton and Nadel 90]**. The following experiments were aimed at testing the types of information encoded in the rats' cognitive maps.

Maze-running experiments were used to test the animals' ability to construct internal maps **[Gallistel 89]**. After allowing the rats to familiarize themselves with a maze, the lengths of some corridors were altered. On subsequent trials, the rats ignored their sensory input and ran into the walls of the shortened corridors, and stopped before the ends of the lengthened ones, at locations corresponding to their previous length.

A more challenging experiment allowed the rats to learn a maze and then blocked the known path to the goal. The rats were able to find alternate routes and even generate short-cuts to the goal. They succeeded in doing so in the dark as well, presumably using an entirely different set of non-visual cues.

A well known series of experiments employed a radial maze. After allowing the rats to explore some of the arms, the maze was rotated. Subsequently, the rats entered already sampled arms without realizing the redundancy of their actions. This evidence points toward a metric representation of space, utilizing absolute angles and distances, as opposed to a topological representation utilizing adjacency relations between landmarks, in this case the arms of the maze. Instead of selecting an arm to explore based on whether its neighbor was already explored, the rats apparently made the selection based on the arm's absolute position. In order to find that position and establish global references, the rats used landmarks external to the maze, such as the position of overhead lights.

Another experiment showed that even when given perceptible sensory features, rats seem to prefer to find the goal using the cognitive maps they have developed. The experiment involved placing a rat into a circular vat of opaque liquid and allowing it to locate a submerged platform. In subsequent trials, even if the platform was moved to a new but visible location, the rat preferred to locate it based on previously established landmarks **[Morris 81]**. The rat was able to find the platform even if local cues were eliminated by moving the vat to a different room. This indicates the use of both local and global information about the environment.

The rat's spatial representation has proven to be very well adapted to its navigational tasks. Elaborate sensory systems are designed for functional redundancy. Special-purpose computational hardware performs various routine tasks, such as position and angular displacement from a reference point. Both topological and metric information appears to be represented and accessible.

The task of the navigation system, biological or artificial, determines the level of qualitativeness of the representation. In this paper we present a topological model containing only the metric information necessary for the task.

2.2 How Distributed?

The "hardware" of biological systems is fully distributed, but it is not yet known how the "software" maps onto this structure. In the case of cognitive maps, we are interested in the level on which the spatial representation is distributed over the neurons computing the relevant information.

Neurophysiological data gathered from rats indicate a distributed spatial representation on the level of landmarks in the hippocampus. Hippocampal neurons are believed to encode non-egocentric cognitive maps. The firing of these neurons, or *place cells*, is associated with the animal's absolute location in the environment. A *place field* is the part of the environment represented by maximal firing of a neuron. Rather than having specific neurons or network patches correspond to particular locations, the entire hippocampal network participates in representing various aspects of the mapped region **[Eichenbaum and Cohen 88]**. Place fields appear to be distributed evenly around the mapped environment. Consequently, a single neuron often has multiple place fields and codes for multiple locations in the environment **[O'Keefe 89]**. **[Eichenbaum and Cohen 88]** suggest that such a fully distributed, non-topological representation lends the system additional generality, allowing it to possibly store semantic as well as spatial relationships.

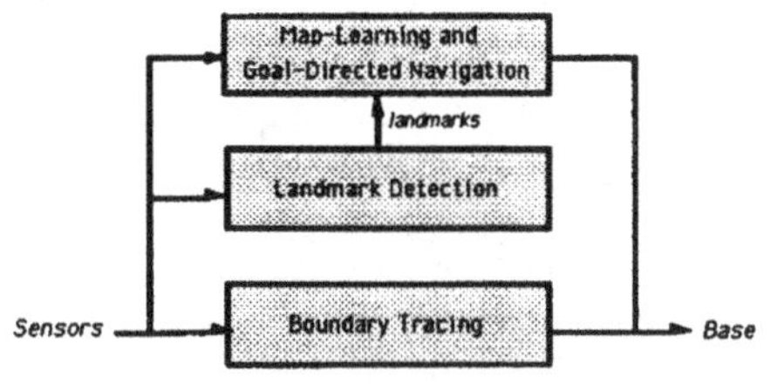

Figure 1: The three layers of competence on the robot: reflexive navigation, landmark detection, and map-building and path finding.

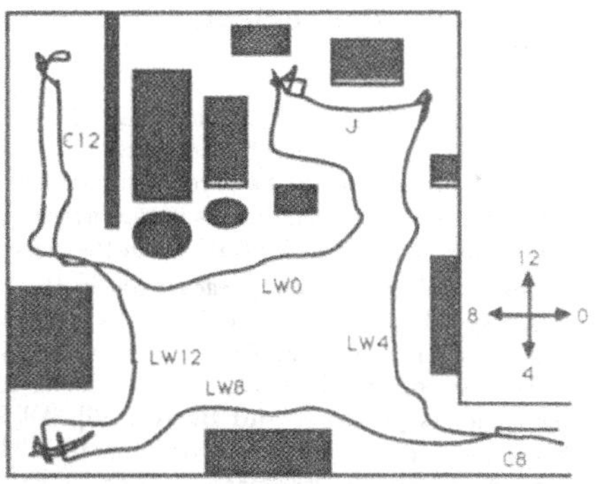

Figure 2: An example of the robot's reflexive navigation behavior in a cluttered office environment. The landmark labels include the landmark type and compass bearing (LW8 = left wall heading South, C0 = corridor heading North, J = long irregular boundary).

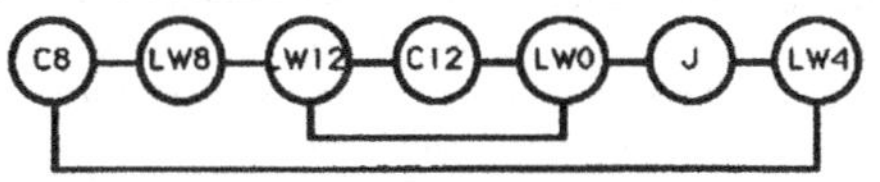

Figure 3: The map the robot constructed in the shown environment. Topological links between landmarks indicate their physical spatial adjacency.

The biological data is hotly debated, but the benefits of such a decentralized representation are worth exploring. In contrast, in robotics, the prevalence of single-processor von Neumann computers has so far overwhelmingly mandated the adoption of centralized models for spatial representation. We introduce a model for a very different, distributed representation, which was implemented and tested on a fully autonomous mobile robot.

3 A Robot With an Artificial Rat Hippocampus

The robot is equipped with sonar sensors and a compass. The sonars provide range information in twelve directions around the robot, and the compass provides roughly sixteen heading directions. The robot is programmed with a parallel, distributed control system based on the subsumption architecture. The subsumption architecture is an approach to robot control which proposes simple, tight reflex-like rules coupling the robot's sensory inputs and effector outputs [**Brooks 86**]. This organization allows the robot to act in real-time without having to perform time consuming processing. An important feature of this approach is that the reflexes are active in parallel, providing distributed, decentralized control of the robot. The robot's behavior is made more complex by incrementally adding more rules, combinations of which form *layers of competence*. Each layer can take advantage of those below.

Our robot consists of three layers of competence (figure 1). The bottom layer combines a set of simple reflexes which, when acting in parallel, result in following boundaries of objects while avoiding static and dynamic obstacles. This is the basic, reflexive and exploratory behavior of the robot.

The middle layer implements a proprioceptive system which detects landmarks in the environment by monitoring the way the robot is moving. The landmarks are defined as combinations of the robot's motion and its sensory input. For example, a corridor is a combination of moving straight and receiving short lateral distance readings. Figure 2 illustrates the robot's typical behavior in a cluttered office environment. Landmark labels consist of the type (LW = left wall, C = corridor, J = long irregular boundary, etc.) and the associated compass direction.

As they are discovered, the landmarks are sent to the third layer of control, which constructs a map of the environment while the robot is moving. Figure 3 shows the map the robot constructed for the shown environment.

The nature of the map representation and the path finding algorithm implemented on the robot will be compared to some interpretations of biological data gathered from analogous processes performed in the rat hippocampus.

3.1 The Distributed Spatial Representation

In our implementation, the distributedness of the representation lies at the level of landmarks. We use a network of landmarks as the basic representation structure. Physically adjacent landmarks in the environment are represented as immediate neighbors in the network.

Making a biological analogy, each node in the network corresponds to a place field of neurons in the hippocampus. Each node is implemented as a behavior, or a collection of real-time rules, corresponding to a unique landmark **[Mataric and Brooks 90]**. All network nodes are processing information in parallel. They receive inputs from the landmark detector, as well as from the sensors on the robot. Additionally, each landmark can communicate with its neighbors by *spreading activation*.

Our approach models a place field as a single landmark in the map. A more neurobiological model would distribute each landmark over a field of neurons. It would be interesting to explore the potential advantages of such an approach for both biological and artificial systems. In the hippocampus, multiple neurons may be necessary to represent a single landmark, but how a landmark is mapped over a place field is not known. The distributedness of the representation may utilize redundancy for increased fault tolerance.

The robot's location in the environment is indicated by an *activated* network node, analogous to a hippocampal place cell with high potential. In order to assure uniqueness of activation, the active node performs lateral inhibition. Additionally, the activated node spreads *expectation* to its neighbor in the direction of motion. Consequently, the next node to be activated is primed, or pre-activated. This is based on experiments showing that proximal landmarks are experienced together in the hippocampus in that they may share activation and are recalled simultaneously. In our implementation, expectation is a method of preserving minimal context which is useful for disambiguating similar landmarks. It also bears similarity to one of the functions of the head-direction neurons which respond to particular orientations of the rat's head **[McNaughton and Nadel 90]**.

Neurobiological data indicates that, in rats, the environment is represented by the changes in the activation levels of the neurons associated with motion, rather than their absolute, instantaneous firing rates **[O'Keefe 89]**. Our implementation simulates the activation levels by propagating different types of messages in the landmark network. The complexity of the analog signals is only roughly approximated by the binary signals for lateral inhibition and expectation. The messages spreading motivation to pursue a particular goal transmit eight bits of distance information.

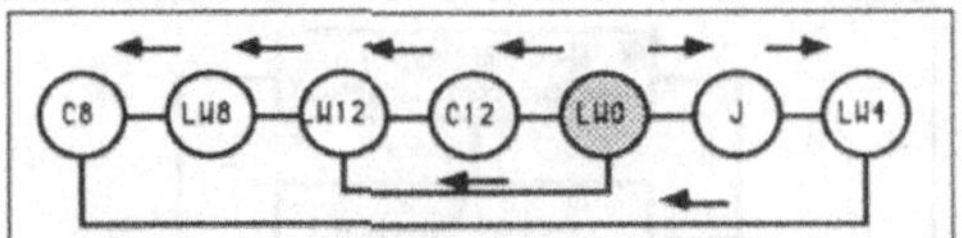

Figure 4: The map actively carries out path finding. The shaded node is the goal node. The arrows indicate spreading of activation from the goal.

3.2 Building a Map and Finding Paths

The process of building a cognitive map consists of the robot wandering through the environment and storing the encountered landmarks in a distributed network. This is analogous to spontaneous growth of the hippocampal potential resulting from exposure of animals to new environments **[Barnes 88]**.

The landmark map is used for localizing, i.e. determining the robot's position. Once localized, the robot can find and execute paths to desired goal locations. The robot's goals are currently selected by the user, but could be supplied by other reasoning modules associated with fulfilling tasks or satisfying the robot's needs. An analogous process is found in biology. According to **[O'Keefe and Nadel 78]**, goal locations in the hippocampus are selected by an external mechanism which acts in response to the animal's needs (hunger, thirst, etc.) and desires (shelter, mate, etc.).

The activity of the hippocampal place cells need not be stimulated by explicit sensory inputs. Cognitive activity, such as visualizing or "thinking about" the goal location, is sufficient **[Eichenbaum and Cohen 88]**. This process is modeled by path finding. Given a goal and a current location, path finding is implemented by *spreading of activation* from the goal landmark throughout the map as illustrate in figure 4. The activation is eventually propagated to all locations in the map. The direction from which activation arrives into a network node specifies locally the direction in which the robot should travel next from that landmark. This eliminates the need for replanning if the robot strays from the desired path since regardless of where the robot is located, it knows the optimal direction to pursue toward the goal **[Mataric 90b]**.

The process of activation spreading is equivalent to performing a network search. Each landmark in the map stores an estimate of its physical length. As the activation wave is propagated through the network, it accumulates the length values of the landmarks it passes. Consequently, at each landmark, the shortest path towards the goal corresponds to the direction with the smallest accumulated length. The robot takes the optimal direction locally, at each landmark, which produces the globally shortest path to the goal.

Very similar, purely local approaches to navigation are suggested by **[McNaughton 89]**. He identifies three possibilities. The first, most similar to our implementation, utilizes a transition matrix linking local views with movements connecting them. Another uses spatially invariant landmarks and forms permanent associations between them and the goal. This is equivalent to the result of activation spreading in our system, but has a fixed goal location. The third approach proposes a built-in method for computing motion sequences explicitly.

4 Using Motion as a Representation Tool

In **[McNaughton and Nadel 90]**, McNaughton describes experiments in which a rat's legs were bound to prevent motion. Such restraint prevented all activation change in the hippocampus as the rat was carried through both a known and a novel environment. When motion of the legs was allowed, neuronal activity resumed, even if the rat was not free to walk about. These experiments stress the importance of integration between the locomotion mechanism and the cognitive map, suggesting that motion is crucial in generating a map.

Similarly, the connections in our network representation do not carry any information. All of the data is contained in the landmarks, and in the robot's reflexive navigation behavior. Consequently, the landmarks are connected by motion directives. At each location, the robot knows where to move in order to reach the next landmark. This may model exactly the mechanism used in the hippocampus. **[McNaughton 89]** presents data suggesting that relationships between landmarks in the hippocampus are most likely encoded in terms of movement. **[Foster, Castro and McNaughton 89]** suggests that hippocampal activity reflects the association of movements with their spatial consequences. The relationship between two places is represented by the movements executed in getting from one to the other. As in our model, the rat knows what movement will take it from one landmark to another.

5 Evaluation

Computationally, our method is attractive for a number of reasons. The parallel implementation allows for very fast (constant time, $O(1)$) localization withing the map, regardless of its size. Shortest path finding to a goal location consists of a parallel network search from the goal to the activated node. The robot makes a greedy, locally optimal choice at each landmark which results in a globally optimal path to the goal. The algorithms runs in worst case linear time ($O(n)$) in the size of the map **[Mataric 90a]**.

Aside from using landmark length estimates to augment path finding, the method we present relies almost entirely on topological information. It could be made more general and biologically feasible by taking advantage of the available metric information for making geometric inferences. For example, discovering physical proximity between two landmarks can generate new neighborhood connections in the network. This would allow the robot to create shortcuts through reasoning only, without having to physically traverse the paths. Rats are known to possess this ability **[McNaughton 89]**.

Most robotics planning systems use some kind of a central control module which consults the map to make decisions about the robot's motion. In contrast, our approach does not use such a separate, centralized reasoning engine outside the network. All spatial reasoning necessary for building and using the map is in the map itself. This is also biologically rooted: the hippocampus performs all of the necessary computation independently, using its specialized hardware. In general, the notion of many specialized modules acting in parallel seems to be more biologically justified than the classical, sequential planning approach.

The controversy between global versus distributed spatial coding in the hippocampus remains to be resolved through further neurobiological study. The approach presented here, while distributed according to the standards of the robotics community, is nonetheless global in the biological sense. The structure of the distributed landmark network is isomorphic to the topology of the environment, and is thus a centralized structure, which can be accessed sequentially. Whether such an isomorphism exists in the hippocampal coding is not yet known.

6 Conclusions and Future Directions

The interpretation of experimental data usually suffers from the bias of the interpreter. Rat hippocampus studies remain controversial as the field is divided on the fundamental issues concerning the nature of the spatial representation involved.

In this paper we have taken a particular set of interpretations of the function of the rat hippocampus and compared it to our physical implementation of a parallel distributed algorithm for robot spatial mapping and path finding. While they do not prove the correctness of the interpretation, our results serve as an empirical proof of its feasibility and efficiency. A number of important questions remain:

How is motion information used to encode the relationships between different landmarks or place fields? How are they linked together into sequences generating shortest paths? The answers to those questions might resolve the distributed versus global map dilemma.

What are the benefits of distributing landmarks over multiple, overlapping place fields? How are multiple maps overlaid onto the same area? In the hippocampus, this mechanism allows for encoding many environments in finite space. Any hints toward the solution of this problem would certainly be useful in representation research in Artificial Intelligence.

Which tasks and architectures are well suited for the use of distributed spatial representations? When are they better suited than global maps? These questions can be answered by building more distributed systems, and evaluating their performance.

Rather than propose a theory of either rat of robot navigation, we presented a working alternative to the standard planning approach usually employed in Artificial Intelligence. Additionally, we suggested the usefulness of biological models in designing artificial systems, as well as the utility of computational models in analyzing biological ones.

Our approach is an example of a specialized subsystem performing self-contained computation in parallel. Biological system demonstrate that collections or societies of such computational agents are worth exploring in order to address global issues in Artificial Intelligence **[Minsky 86]** and robot control.

7 Acknowledgements

Geoffrey Hinton originally pointed out the work on the rat hippocampus and its relevance to the research presented here. Rodney Brooks suggested the idea of a distributed representation and provided comments on an earlier draft of this paper. Bruce McNaughton provided helpful information about current hippocampus studies.

The research reported here was done at the MIT Artificial Intelligence Laboratory. Support for this research was provided in part by the Hughes Research Laboratories Artificial Intelligence Center in Malibu, in part by the University Research Initiative under Office of Naval Research contract N00014-86-K-0685, and in part by the Advanced Research Projects Agency under Office of Naval Research contract N00014-85-K-0124.

8 References

[Barnes 88] "Spatial Learning and Memory Processes: The Search for Their Neurobiological Mechanisms In the Rat", C. A. Barnes, *TINS, Vol. 11, No. 4*, 1988.

[Brooks 86] "A Robust Layered Control System for a Mobile Robot", Rodney A. Brooks, *IEEE Journal of Robotics and Automation*, RA-2, April 1986, 14-23.

[Eichenbaum and Cohen 88] "Representation in the Hippocampus: What Do Hippocampal Neurons Code?", Howard Eichenbaum and Neal J. Cohen, *TINS, Vol. 11, No. 6*, 1988.

[Foster, Castro and McNaughton 89] "Spatial Selectivity of Rat Hippocampal Neurons: Dependence on Preparedness for Movement", Tom C. Foster, Carl A. Castro, Bruce L. McNaughton, *Science*, June 30, 1989, 1589-1582.

[Gallistel 89] "Animal Cognition: The Representation of Space, Time and Number", C. R. Gallistel, *Annual Review of Psychology 1989*, Vol 40, Palo Alto, California, 155-189.

[Gould 82] "Ethology: The Mechanisms and Evolution of Behavior", James L. Gould, *W. W. Norton and Company, New York*, 1982.

[Mataric 90a] "A Distributed Model for Mobile Robot Environment-Learning and Navigation", Maja J Mataric, *MIT AI Lab Tech Report #1228*, May 1990.

[Mataric 90b] "Environment Learning Using a Distributed Representation", Maja J Mataric, *Proceedings of 1990 IEEE International Conference on Robotics and Automation*, May 1990, 402-406.

[Mataric and Brooks 90] "Learning a Distributed Map Representation Based on navigation Behaviors", Maja J Mataric and Rodney A. Brooks, *Proceedings of 1990 USA-Japan Symposium on Flexible Automation*, Kyoto, July 1990.

[McNaughton 89] "Neuronal Mechanisms for Spatial Computation and Information Storage", Bruce L. McNaughton, *Neural Connections, Mental Computation*, Lynn Nadel, Lynn A. Cooper, Peter Culicover, R. Michael Harnish, eds., MIT Press, Cambridge, Massachusetts, 1989.

[McNaughton and Nadel 90] "Hebb-Marr Networks and the Neurobiological Representation of Action in Space", Neuroscience and Connectionist Theory, Mark A. Gluck and David E. Rumelhart, eds., *Lawrence Erlbaum Associates, Publishers*, Hillsdale, New Jersey, 1990.

[Minsky 86] "The Society of Mind", Marvin L. Minsky, *Simon and Schuster*, New York, 1986.

[Morris 81] "Spatial Localization Does Not Require the Presence of Local Cues", R. G. M. Morris, *Learning* *and Motivation*, Vol. 12, 239-261.

[O'Keefe 89] "Computations the Hippocampus Might Perform", John O'Keefe, *Neural Connections, Mental Computation*, Lynn Nadel, Lynn A. Cooper, Peter Culicover, R. Michael Harnish, eds., MIT Press, Cambridge, Massachusetts, 1989.

[O'Keefe and Nadel 78] "The Hippocampus as a Cognitive Map", John O'Keefe and Lynn Nadel, *Clarendon Press*, Oxford, 1978.

[Piaget and Inhelder 67] "The Child's Conception of Space", J. Piaget and B. Inhelder, *Norton*, New York.

[Pick, Montello and Somerville 88] "Landmarks and the Coordination and Integration of Spatial Information", Herbert L. Pick, Jr., Daniel R. Montello, and Susan C. Somerville, *British Journal of Developmental Psychology*, 1988:6, The British Psychological Society, 372-375.

[Presson and Montello 88] "Points of Reference in Spatial Cognition: Stalking the Elusive Landmark", Clark C. Presson and Daniel R. Montello, *British Journal of Developmental Psychology*, 1988:6, The British Psychological Society, 378-381.

[Waterman 89] "Animal Navigation", Talbot H. Waterman, *Scientific American Library*, New York, 1989.

Four Important Issues in Cognitive Mapping

W.K. Yeap and C.C. Handley
AI Laboratory,
Department of Computer Science,
University of Otago, New Zealand

coscwky@otago.ac.nz
chandley@otago.ac.nz

Abstract

We introduce the notions of Raw and Full Cognitive Maps and Absolute Space Representations (ASRs) and discuss (i) the value of computing a representation of the local environment (ASR), (ii) the need for a global representation of one's immediate surroundings, (iii) categorisation and the formation of concepts and (iv) the problem of planning a route.

0. Introduction

The problem of computing a cognitive map is fundamental to any autonomous mobile system, be it a rat, a human or a robot. When Tolman (1948) first suggested the idea of a cognitive map, he was probably referring to a "map" of the spatial layout of the environment (mazes in his case) but later it became clear, especially after Lynch's (1960) work, that the notion of a cognitive map is a complex one. In the early 70's, there was an outcry from geographers, urban planners and designers that a cognitive map is not a map (see [Downs and Stea, 1973]). A cognitive map is tied to our spatial behaviour and it is therefore affected by a wide variety of factors ranging from our mode of travel and past experiences to our preferences and attitude. One problem with this view is that it leads to a confusing use of the term and one is often left with the impression that a cognitive map holds one's entire knowledge. However, if the system is to adapt and survive in a (hostile) world with other agents in it, such factors must be considered.

Cognitive mapping is therefore a complex process which involves both one's perception and conception of the outside world. Studies which emphasised only one level, either the perceptual (e.g. work on autonomous mobile robots) or the conceptual (e.g. early models), were at best incomplete and very often asked many questions inappropriate at that level. For example, robotics researchers were concerned with how to partition the environment in terms of spaces large enough for the robots to plan a collision free path, but planning a collision free path is a local problem and an attempt to solve it at the path planning level is inappropriate (for more detail, see [Yeap et. al., 1990]).

Our past work has been the development of a computational theory of cognitive maps to explain what needs to be computed and why [Yeap, 1988, 1990; Yeap and Robertson 1990; Yeap et. al., 1990]. We stress the importance of studying the process as a whole, from perception to cognition and generally in that order. Our investigation of the process begins with Marr's (1982) computational theory of vision and, as in Marr's work, the notion of a *representation* is central to our study. The next section presents a brief overview of the theory, the main idea being that a cognitive mapping process should first compute a *raw* map and then a *full* map. Using this theory we discuss four important issues in cognitive mapping, two at each level: (i) the significance of computing a representation of the local environment (an Absolute Space Representation or ASR) and (ii) the need for a global representation of one's immediate surroundings; (iii) categorisation and concept formation and (iv) the problem of planning a route. These issues arise from the insights gained from the implementation of the theory and from further consideration of the nature of the cognitive mapping process.

1. A Computational Theory of Cognitive Maps

Although there are many factors which influence our conceptual view of the world, our representation of the world must begin from what we perceive. This observation suggests that the first step in a cognitive mapping process is to compute a representation of the physical environment. Since the different conceptual views of the world are but different ways of looking at what is already computed, this representation of the physical environment should be fairly independent of the conceptual representations that are developed later. Our theory therefore suggests that a cognitive mapping process should be studied as a process consisting of two loosely-coupled modules. An early cognitive mapping process computes a representation of the physical world as perceived by our senses. We call this representation a Raw Cognitive Map; indicating that the computed representation is not interpreted. A later cognitive mapping process computes a representation of the conceptual world. We call this representation a Full Cognitive Map; the word "full" indicates the full richness of the map as a cognitive representation.

The Raw Map is a network of Absolute Space Representations (ASRs), i.e. representations of individual local environments which have been visited (or entered) by the viewer. A link in the network indicates a direct connection between adjacent ASRs. When the viewer enters a new part of the environment, a boundary of the new ASR is computed immediately. This provides the viewer with a notion that she is (bounded) in a (local) environment. In addition, the boundary provides an estimate of the size of that environment and, more importantly, an independent frame of reference for describing other spatial elements in view [Huttenlocher and Presson, 1979]. The boundary of an ASR is formed from surfaces which are perceived as obstacles to the viewer's movements. The gaps between these surfaces are the exits from an ASR. Note that connections between different ASRs are based upon the viewer's experience of the environment and unless the viewer can remember all the ASRs visited and recognise them on later visits, the Raw Map created would not be isomorphic to the physical environment. (See [Beck and Wood, 1976] for an interesting experiment illustrating the tourists' perspective of two structures which are actually adjacent but are represented as being far apart). An algorithm for computing the boundary of an ASR from surfaces perceived at a given position (either using a limited or a full 360° view) has been implemented [Yeap, 1988]. More recently, a new algorithm has been implemented which also considers surfaces that have been perceived from earlier vantage points [Yeap et. al., 1990]. Figure 1 shows an example of the boundary computed for an ASR (using either algorithm).

When computing an ASR boundary, the following steps are taken:

From the input (i.e surfaces perceived),

(i) remove the insignificant surfaces. In general, these are small surfaces and those below the viewer's height. A boundary surface is often perceived as one which will obstruct the viewer's horizontal movement.

(ii) choose a starting surface for the boundary. In general this would be the exit (an imaginary surface) through which the viewer has just entered.

(iii) apply some heuristics to compute the boundary following an anti-clockwise scan of the remaining surfaces.

The main problem lies in determining which of the remaining surfaces should be connected to the current surface. This process is repeated until a closure is formed with the viewer in it. The two algorithms differ in the heuristics used. The first employs a set of heuristics based on the observation that a surface with an occluding vertex (i.e one which occludes another surface, example point a in Figure 1) would in general be useful to connect to another with a vertex

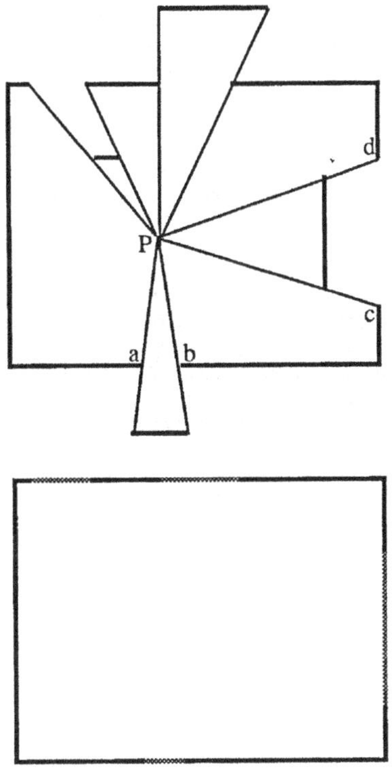

Figure 1
An example of the boundary of an ASR (*bottom*) computed from the environment (*top*). The viewer is located at point P and some vertices of the surfaces are labelled.

of the same type. This algorithm thus works by:

(i) extracting a list of vertices from the significant surfaces (from above)

(ii) arranging the vertices in the order in which they are perceived when scanning the view in an anti-clockwise manner. The first and last vertex in the list come from the starting surface.

(iii) labelling each vertex as occluding, occluded or connected.

The task is to find which of the in between vertices should be connected together to form the boundary following the (main) heuristics described (and others). However, when computing a raw map, each ASR is not computed from a single view of the environment nor from a 360° view. It was argued that surfaces perceived from each single view are used to build a representation of one's immediate surroundings and from this representation, one computes the ASRs. Thus, at any moment, a viewer could have more information than what is perceived from the current position. In particular, the viewer could know what is behind some of the surfaces. The earlier algorithm failed to make use of this "hidden" information since their vertices would not be labeled using the current scheme. In the earlier implementation, such parts of a surface were simply ignored.

The second algorithm does not require that surface vertices be labelled and instead computes all possible connections to the current surface by finding those surfaces it intersects somehow. Figure 2 shows some of the possible connections. Each of these is then given a weighting depending on various factors like how far the surfaces are separated, and the final boundary is one with the minimum

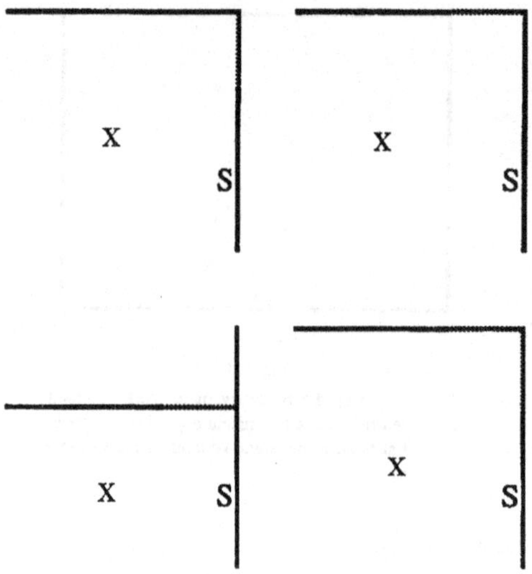

Figure 2

Examples of some possible connections between two surfaces when forming a boundary. X represents the region within which the viewer is located. S is the current surface and grey lines are the connections to be made. It is not necessary that the connection be made at a right angle.

total weight. This algorithm can be viewed as a search algorithm for finding the shortest path (or more appropriately, the least cost boundary) and could be implemented very efficiently (see [Yeap et. al., 1990] for details).

Except for its network structure, the Raw Map contains much information that has not yet been organised. The Full Map is proposed as a collection of representations which grows out of the Raw Map, each of which imposes a particular way of looking at (some of) the information in the Raw Map. One important notion is the concept of a "place" and in particular, the concept of "activity-places" [Steinitz, 1968]. Many experiments (e.g., [Chase, 1983; Hirtle & Jonides, 1985]) have demonstrated the hierarchical nature of a place representation. Using the ASRs, one could form a hierarchy of places by grouping one or more ASRs as a place, one or more places as a higher-order place, and so on. Figure 3 shows an example of a hierarchical place representation as computed in one of our computer implementations. To build such a representation, the names of the different places which our simulated robot enters or leaves are introduced to the robot via two commands as shown in Figures 4 and 5. The hierarchical relations themselves are computed automatically.

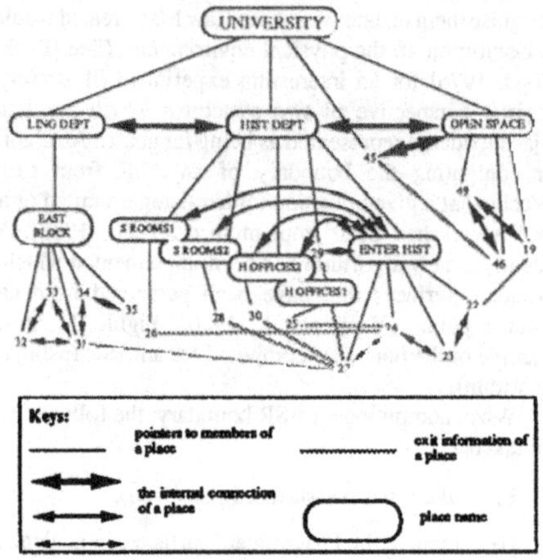

Figure 3

A hierarchy of place representations. The numbers at the tip of the hierarchy represent the ASRs in the Raw Map.

A significant use of the cognitive map is to plan a path from one part of the environment to another. This problem was investigated using a representation of the cognitive map computed thus far, that is the raw map plus a hierarchy of places. There are many ways in which a path could be computed and the following algorithm was implemented:

Plan_a_path (x, y)

(i) use the hierarchy of places to find the common parent of both ASRs, x and y

(ii) search down the hierarchy to produce a path between x and y. Note that this is a typical and a very efficient approach when computing a path from a hierarchical representation but the path produced might not be optimal.

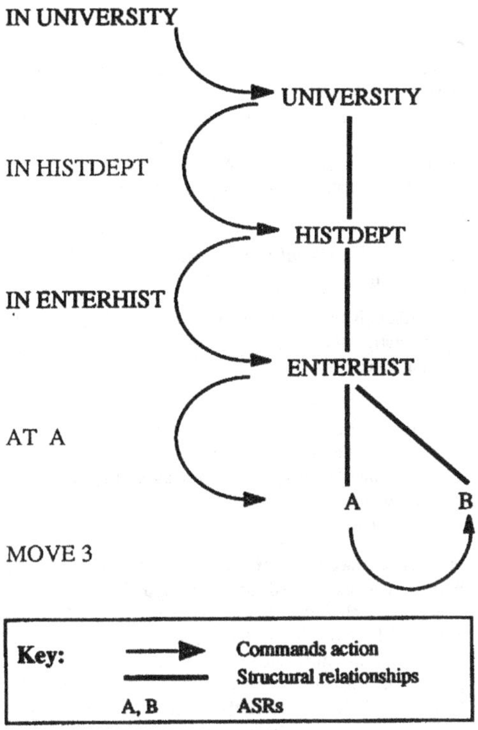

Figure 4

The IN command is used to introduce the names of the places a simulated robot enters. The names are given before specifying a move to the actual physical space. (Note that the AT command specifies the initial ASR and from then on, the MOVE command specifies the exit to which the robot will move).

(iii) using this path, simulate a walk and compute the relative position of y from x

(iv) use the direction obtained in (iii) to constrain a search in the raw map for the most direct path from x to y. This path may or may not be the same path as the one produced in (ii).

In the last step, we start from x and consider all its known exits. Each exit is weighted based on a calculation of how far the exit will deviate from the straight line joining x and y. This calculation is computationally expensive and is inaccurate given the information in the raw map. To reduce the need to calculate a weighting for each exit, each ASR is typed as a terminal space if it has only one exit, a corridor if it has two exits and an intersection if it has three or more exits. With this

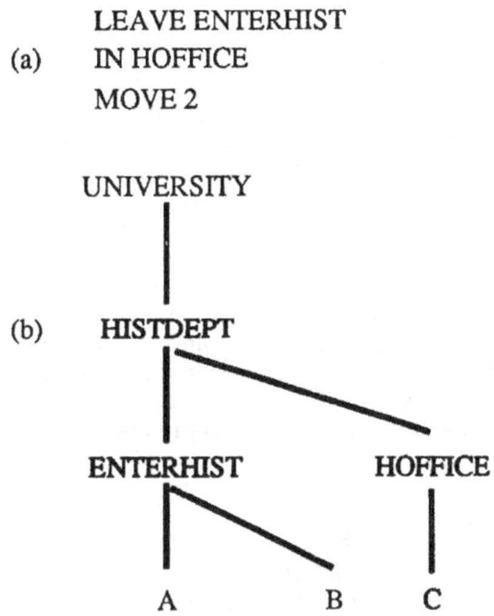

Figure 5

The LEAVE command specifies that the robot is no longer at a particular place: (a) Commands given after Figure 4, (b) the resulting representation computed. Note that one can specify Leave University without having to specify Leave Histdept.

typing information, any exit which leads to a non-goal terminal space can be ignored immediately. More importantly, if an exit leads to a corridor, a weight is only calculated at the end of the corridor which significantly reduces the number of calculations needed. Making the types of an ASR explicit enables one to make more efficient use of the information in the raw map. We are currently investigating what other kinds of information could be made explicit in our cognitive map representations, and how useful the information is for planning and re-planning a route (for further details, see [Yeap, 1990] and [Yeap and Robertson, 1990]). We discuss below the idea of object concept formation from surface information in the Raw Map.

2. On Computing the Raw Map

The Value of Computing an ASR

The basic idea of computing a representation of each local environment and connecting these together to form a representation of the larger environment is very simple and elegant. The idea has also been put forward, although with little detail, by previous workers on cognitive maps (e.g. [Byrnes, 1979]). More recently, robotics researchers have suggested various algorithms for partitioning the environment in terms of a network of empty spaces, for example, in terms of convex polygons [Chatila, 1982], primary convex polygons [Crowley, 1984] and primary convex regions [Rueb and Wong, 1987]. Once computed, these partitions remain unchanged. This is a problem in general since the way in which the environment is partitioned depends on the vantage point from which it is perceived. (An exception is an algorithm by Rueb and Wong (1987) but their resulting representation of the environment is so (unnecessarily) partitioned that it would not be useful as a cognitive map [Yeap et. al., 1990]). Most current approaches in robotics overcome this problem by simultaneously building a "map" of the whole environment with a global co-ordinate system while partitioning the environment. This map then allows the robot to keep track of its movements. It therefore avoids re-partitioning any part of the environment which has been partially or completely partitioned earlier but is now perceived from a different vantage point. Our algorithms for partitioning the environment (or in our terms, computing the ASRs) are based on the use of heuristics but we also note that it is difficult to partition the environment in the same way when viewed from different vantage points. How important is it to always partition a part of the environment in the same manner? Note that the environment itself does not remain constant and therefore the computed representation could, for one reason or another, be different from that in memory. More significantly, the importance of having a local space is to provide the viewer with a bounded region in which she can relate her current actions. We therefore argue that,

> *the problem in computing a useful representation of the whole environment lies not in how the environment should be partitioned into its local parts but on how successfully one can recognise where one is.*

In other words, the key to successfully computing a representation of the environment does not depend on how uniquely the environment can be partitioned but on how successfully its parts can be recognised. If we can recognise that (part of) the current space has been visited before and if its shape does not match that computed previously, then the problem is simply to decide on a final shape for the ASR. This shifts the central issue of computing a Raw Map from the problem of partitioning the environment to that of recognising the environment. The importance of our ability to recognise where we are has always been strongly emphasised by both environmental and developmental psychologists working on cognitive mapping [Acredolo et. al., 1975; Heft, 1979]. The observations that humans are often lost in environments with few distinctive features (such as a desert) and that rats could not learn a maze that had no constant landmarks [Snygg, 1935] not only show that recognition is important but also that humans and rats do not compute the global representation used in current robotic systems. Note that such a global representation is both unnecessary (e.g. we never need to remember precisely, in terms of the co-ordinate values, the position of a table in our office) and unable to be computed (since we do not compute absolute distance information with our senses). Since few motile animals have evolved with senses that compute absolute distance information, it is interesting to observe how successful these robotic researchers would be in their efforts to build autonomous mobile robots. We believe that their systems would be relegated to just (bigger) toy systems in the near future.

The Need for a Global Representation of one's Immediate Surroundings

Although we have argued above that a global representation of the whole environment is not computed, we are convinced that,

> *a global representation of one's local environment should be computed, so that successive views of one's surroundings can be integrated to provide a memory not only of the things in the surroundings but also one's position in them.*

From the characteristics of our own visual system, such as a limited view and limited range vision, and from our visual behaviour, such as the need for active scanning [Norton and Stark, 1971], psychologists have long suggested that we need to integrate the different views to form a more complete picture of what we perceive of our surroundings (e.g. [Gibson, 1979]). However, although it is possible to construct an ASR from successive views, we argue that an ASR is not a suitable representation of one's immediate surroundings. An ASR is computed to represent a local environment which is different from the viewer's immediate surroundings. For example, things perceived outside of an ASR are not represented as part of the ASR. A representation of the immediate surroundings, on the other hand, would provide a much richer description for computing an ASR than the successive views themselves and for example, this has prompted the design of the second algorithm for computing an ASR.

From an implementation standpoint, it is inefficient to compute a representation of the immediate surroundings using an egocentric frame of reference. This suggests that the representation should be described using an independent global frame of reference in which viewer's own position is made explicit. This would allow one to keep track of one's own movements in the surrounding space, and more importantly, to realise that one is re-visiting a part of the immediate surroundings rather than moving into a new space. In our simulation studies, the latter has played a useful part in providing an extra cue for learning the environment.

The need to integrate successive views, the usefulness of the knowledge of one's immediate position, and the consequence of having a rich description for computing the

ASRs all argue well for the computation of a representation of the immediate surroundings. The problem, however, remains as to how best to compute such a representation since our senses do not provide absolute distance information. Related issues include the size of the immediate surroundings and the frame of reference used. For example, should the immediate surroundings be centered upon the current ASR or fixed until the viewer moves out of its range? Again, a proper solution must await confirmation of the accuracy of range information provided by our visual system.

3. On Computing the Full Map

Categorisation and Concept Formation

Many early studies on cognitive maps in humans assumed that the inputs to the underlying process are objects and places which are heavily laden with meanings (e.g. [Kaplan, 1976], [Kuipers, 1978]). These inputs are a product of the pervasive thinking in terms of categorisation and concept formation. With our new model of cognitive maps, we argue that,

> *the converse is true. Cognitive maps actually provide the basis from which even the most basic concepts and categories (including objects and places) are developed.*

The emphasis on object (form) recognition in vision research, both in theoretical studies and in industrial applications, often give the misleading impression that objects are perceived as if they exist in their own independent forms. Our own intuitive feeling of what we perceive around us does not help either, since what we see tends to be described as which objects rather than where. It was not until Marr's (1982) seminal work on computer vision that any attention was paid to the simple fact that our eyes deliver descriptions of surfaces seen at various distances in front of us. That objects (and events) appear to us with a spatiality that we should not ignore has been raised sporadically in several areas of research. For example, Schulman (1983) discussed the importance of whereness of things in remembering them in one's memory and Nelson's (1986) work, which emphasised event knowledge as the basic mental representation from which one's concepts and categories are derived, also provides support in the sense that (object) concepts and categories are not learned without a context. In our work, these events would be encoded together with the spatial environment in which they occurred. In their summary of the work on how infants encode the events they observe, Huttenlocher and Smyth-Burke (1987) provide some evidence that infants beginning at 8 or 9 months old are aware of the larger spatial framework within which other objects are located.

We therefore propose that cognitive mapping, which captures the spatial layout of the local environment by noting the whereabouts of the surfaces, is fundamental to our thought processes. Interest in the individual groups of surfaces leads one to form object categories and later object concepts (see also [Kagan, 1979]). Such categories and concepts are formed not only by detecting similarities in the structural properties of elements but also by individual "theories" concerning their formation [Medin, 1989].

The Problem of Planning a Route

The problem of planning a route through an environment has been studied by both AI and robotics researchers but unfortunately they aim only at producing a (shortest) route from the representation (e.g. [Mitchell, 1988]). Their approach in general reduces the problem to a search over some form of a network of nodes and research in computer science and operation research has provided a variety of such algorithms (such as A*). We argue that,

> *the aim of using a cognitive map to plan a route to traverse an environment is not only merely to compute a path from one point to another but also to determine as much relevant information as possible from what is known about the environment.*

The availability of efficient search algorithms demonstrates that finding a (known) route is not a problem. Computer scientists are concerned with the efficiency of their algorithms when applied to a very large network, but for route planning in a real environment, the search is often constrained to only a small network (e.g. [Davis, 1986, p. 51]). However, unlike the simulation studies, many things can go wrong in the real world. To adapt and survive in such a hostile environment, the agent must be provided with vital information about the route rather than just a path to travel on. Our current work on planning and re-planning a route from our representation of a cognitive map has demonstrated that the map, even in its current form, contains lots of useful information [Yeap, 1990; Yeap and Robertson, 1990]. For example, the typing of ASRs into intersections, corridors, and dead-ends reveals their significance when planning a route and knowing the conceptual places containing the previous route is very useful when re-planning.

4. Conclusion

We are interested in developing a computational theory of cognitive maps in an environment which is as rich and varied as the one in which we dwell. We strongly believe that the current best approach is to draw our ideas from observations made about human (and animal) behaviour in their natural environment. Nature has much to tell us. It is part of the folklore of Computer Vision that a (computer) solution to the matching problem in binocular stereopsis below the level of object recognition would not have been discovered if nature had not demonstrated that it could be done. Similarly, we believe that we could not have understood the full complexity of the cognitive mapping problem without learning from nature.

However, in our approach it is equally important to stress the need to know what representations are computed from bottom up (i.e from the senses). This provides a clear idea of what is available and hence constrains what can be done. If the simulated behaviour does not match the observed behaviour, then the model needs to be improved. Furthermore, our approach enables all the internal representations computed by the system to be grounded in information from the outside world and this is important if one wants to develop a truly intelligent system [Harnad, 1990].

Acknowledgements

Research in the AI Laboratory is supported by grants from the Otago University and the University Grant Committee. We would also like to thank the referees for their comments and the organisers for a grant (from the US Air Force Office of Scientific Research) to attend the conference.

References

Acredolo L.P., Pick H.L. Jr. & Olsen M.G. (1975) Environmental Differentiation and Familiarity as Determinants of Children's Memory for Spatial Location. Developmental Psychology, 11, 4, 495-501.

Beck R.J. & Wood D. (1976) Cognitive Transformation of Information from Urban Geographic Fields to Mental Maps. Environment and Behavior, 8, 199-238.

Byrne R.W. (1979) Memory for Urban Geography. Quarterly Journal of Experimental Psychology, 31, 147-154.

Chase W.G. (1983) Spatial Representations of Taxi Drivers. In D. Rogers & J.A. Sloboda (Eds.), The Acquisition of Symbolic Skills (pp. 391-405). New York: Plenum Press.

Chatila R. (1982) Path Planning and Environment Learning in a Mobile Robot System. Proceedings of the European Conference on AI, France, 211-215.

Crowley J.L. (1984) Navigation for an Intelligent Mobile Robot. IEEE Journal of Robotics Research, 1(2), 31-41.

Davis E.S. (1984) Representing and Acquiring Geographic Knowledge. London: Pitman.

Downs R.M. & Stea D. (1973) Image and Environment: Cognitive Mapping and Spatial Behaviour. Chigaco: Aldine Publishing Coy.

Gibson J.J. (1979) The Ecological Approach to Visual Perception. Boston: Houghton Mifflin Company.

Harnad S. (1990) The Symbol Grounding Problem. Presented at CNLS Conference on Emergent Computation, in press.

Heft H. (1979) The Role of Environmental Features in Route-learning: Two Exploratory Studies of Way-finding. Environmental Psychology & Nonverbal Behaviour, 3, 3, 172-185.

Hirtle S.C. & Jonides J. (1985) Evidence of Hierarchies in Cognitive Maps. Memory &nd Cognition, 13(3), 208-217.

Huttenlocher J. & Presson C.C. (1979) The Coding and Transformation of Spatial Information. Cognitive Psychology, 11, 375-394.

Huttenlocher J. & Smyth-Burke T. (1987) Event Encoding in Infancy. In P. Salapatek & L. Cohen (Eds.), Handbook of Infant Perception vol 2, From Perception to Cognition (pp. 209-231). New York: Academic Press.

Kagan J. (1979) Structure and Process in the Human Infant: The Ontogeny of Mental Representation. In M.H. Bornstein & W. Kessen (Eds.) Psychological Development from Infancy: Image to Intention, (pp. 159-182). New Jersey: Lawrence Erlbaum Associates.

Kaplan S. (1976) Adaptation, Structure, and Knowledge. In G.T. Moore & R.G. Golledge (Eds.) Environmental Knowing: Theories, Research, and Methods (pp. 46-57). Stroudsbury, Pa: Dowden, Hutchinson and Ross.

Kuipers B. (1978). Modelling Spatial Knowledge. Cognitive Science, 2, 129-153.

Lynch K. (1960) The Image of the City. MIT Press, Cambridge, MA.

Marr D. (1982) Vision: A Computational Investigation into the Human Representation and Processing of Visual Information. New York: W.H. Freeman Coy.

Medin D.L. (1989) Concepts and Conceptual Structure. American Psychologist, 44(12), 1469-1481.

Mitchell J.S.B. (1988) An Algorithmic Approach to Some Problems in Terrain Navigation. Artificial Intelligence, 37, 171-201.

Nelson K. (1986) Event Knowledge: Structure and Function in Development. New Jersey: Lawrence Erlbaum Associates.

Norton D. & Stark L. (1971) Eye Movements and Visual Perception. Scientific American, 224, 34-43.

Rueb K.D. & Wong K.C. (1987) Structuring Free Space as a Hypergraph for Roving Robot Path Planning and Navigation. IEEE Transactions on Pattern Analysis and Machine Intelligence, 9(2), 263-273.

Schulman A.I. (1983) Maps and Memorability. In D. Rogers & J.A. Sloboda (Eds.), The Acquisition of Symbolic Skills, (pp. 359-367). New York: Plenum Press.

Snygg D. (1935) The Relative Difficulty of Mechanically Equivalent Tasks II. Animal Learning. Journal of Genetic Psychology, 47, 321-336.

Steinitz C. (1968) Meaning and the Congruence of Urban Form and Activity. Journal of the American Institute of Planners, 34, 233-248.

Tolman E.C. (1948) Cognitive Maps in Rats and Men. Psychological Review, 55, 189-208.

Yeap W.K. (1988) Towards a Computational Theory of Cognitive Maps. Artificial Intelligence, 34, 297-360.

Yeap W.K. (1990) Re-planning a Route – A Pragmatic Approach. Proceedings of the 5th Rocky Mountain Conference on AI, New Mexico, 99-103.

Yeap W.K., Naylor P.S. & Jefferies M. (1990) Computing a Representation of the Physical Environment – A Memory-based Approach. To appear in the 1st `Pacific Rim International Conference on AI, Nagoya.

Yeap W.K. & Robertson C.J. (1990) Planning a Route from a Cognitive Map. In R. Trappl (Ed.), Cybernetics and Systems '90, (pp. 895-902). Singapore: World Scientific Publishing.

Attracting Similar Shapes Towards Each Other

Mitch R. Harris

Department of Artificial Intelligence
University of Edinburgh
Forrest Hill, Edinburgh EH1 2QL, Scotland
email m.harris@uk.ac.edinburgh

Abstract

This paper draws an analogy between the way insects use vision to move *themselves* with respect to local landmarks, and the problem of moving *objects* relative to each other in vision-guided robotic assembly. In particular, an algorithm is presented for attracting similar shapes together which was directly inspired by a model of navigation in honeybees, and which shares the same characteristics of robustness and immunity to noise. In particular, the algorithm can rotate, translate and scale one 2D shape to align it with another despite the presence of significant distortion and missing or extraneous features.

1 Introduction

There are several reports[1] that insects are able to store retinal images and to later compare these stored images with a current one in order to facilitate homing with respect to a set of (proximal) landmarks. For example digger wasps can accurately fly to their tiny nest entrance by using local landmarks such as pine cones scattered up to a metre or more from the entrance itself (Tinbergen, 1932). Similarly, a honeybee can reliably relocate a food source such as a flower even though it is too small to see from a distance or can only be seen from above due to surrounding vegetation. Again, once they are in the approximate vicinity of the food it has been shown that bees can home into the exact location by using local landmarks (Anderson, 1977).

One of the most striking features of this type of homing mechanism is its robustness in the face of missing, additional or moved landmarks between visits. Cartwright and Collet proposed an elegant computational model of landmark learning in honeybees in which the bee stores an image (retinal snapshot) of its surroundings when it finds a food source (Cartwright & Collett, 1983). To re-find the this location the bee must first find the approximate vicinity (*ie* by some other form of navigation) so that the most of the relevant landmarks are visible. Given these starting conditions, the authors show how comparison of the snapshot and a current image can lead to the computation of a flight vector which, if followed, tends to bring the bee into a position where snapshot and new image are more similar. Thus as the (simulated) bee flies around it continually compares its snapshot with the image of its current surroundings and computes new flight vector and so on, thus bringing it ever closer to the 'memorised' location.

The main reason for the robustness of Cartwright and Collet's algorithm is that it does not need to solve the correspondence problem of which features in the image match which features in the snapshot. Computation of the flight vector is independent of the goodness of fit between image and snapshot and so no search is involved. The effect of the computation is that the bee moves towards things (or gaps between things) which appear too small and away from things which appear too big.

For simplicity, it is assumed that image and snapshot each consist of a 360° wide ring of dark and light segments. Each light or dark region in the image is paired with the nearest region of the same sign in the snapshot and the angular difference is found. These differences are simply averaged to find an overall rotation for the bee. Translational vectors are computed by comparing the angular sizes of dark and light areas. For example a dark area in the image is compared with the nearest dark area in the snapshot; if the image area is *smaller* then this suggest moving radially *towards* that part of the image. As with rotation, radial vectors generated in this way are simply averaged, giving an overall translation vector. Even if individual pairings of edges are wrong in terms of the correspondence, the overall flight vector tends to be roughly right and small errors are compensated for by the iterative nature of the process.

1.1 A robot guidance problem

It was decided to take a similar approach to the bee model to help solve an analogous problem which arises when guiding a hand-held object to bring it in line with a stationary object under camera guid-

[1] See (Cartwright & Collett, 1983) for review

ance. This problem is being researched as part of a project to build an automated robotic assembly system (Malcolm & Smithers, 1990). A prototype part mating system has been developed which uses a self-calibrating closed loop vision servoing to stack objects (wooden cubes) on top of each other (Conkie & Chongstitvatana, 1990). The servoing problem is to guide a cube held in a robot gripper into alignment with a stationary cube. There are two cameras each of which views both cubes to be aligned. The two images are processed separately to find the 2D relative displacements of the cubes. These displacements from the two images are finally combined and converted to 3D robot motions to iteratively reduce the displacements.

The image processing is thus effectively reduced to a two-dimensional problem. Currently, the silhouette of each cube is found using thresholding and a polygon approximator (Malcolm, 1983). This reduces the data (for one image) to two polygons. If these can be aligned in 2D then the system as a whole should be able to align the objects in 3D.

Currently the system performs the 2D alignment by finding the horizontal displacement between the centroids of the two polygons. When vertical alignment has been achieved the gripper is rotated until a particular edge in the lower polygon is parallel with a particular edge in the upper polygon. Although this method allows the stacking to be performed the alignment produced is somewhat inaccurate and has limited generality. The inaccuracies stem from the difficulty of extracting a complete and accurate outline from an image. As well as the usual problems with noise, shadows, object texture *etc*, there is the added problem that the gripper fingers may obscure part of the outline. Instead of using the rather crude centroid measure, it might be possible to align particular features of the polygons (*eg* leftmost vertical edge) but, in general, it is not possible to anticipate which features will come out clearly and which will be degraded.

In order to overcome these shortcomings, it was decided to develop an alignment algorithm which used *all* available features but weighted those which matched in the two polygons more strongly. The results of this are described below. Performance exceeded expectations and gave rise to the possibility that the algorithm may have more uses than the one described above. These are discussed afterwards.

2 A shape attracting algorithm

It soon became apparent that the algorithm which was developed for this purpose could easily be extended to do more than horizontal alignment of polygons. First, it is not restricted to polygons but operates on any two sets

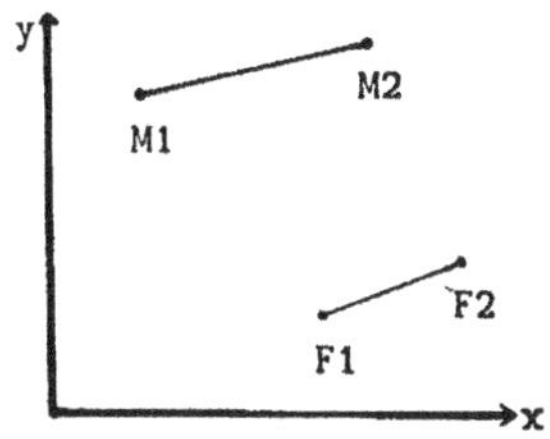

Figure 1: Mapping moving line onto fixed

of straight lines. It is thus suited to output from some kind of edge detector further processed by a straight line approximation algorithm. For the purposes of this paper we assume that each line is represented by an (unordered) set of two 2D coordinates. One set of (f) lines corresponds to the fixed object, the other set of (m) lines corresponds to the moving object. What the algorithm does is to compare every line in the moving object with every line in the fixed object (an $m \times f$ matrix) and compute an overall translation, rotation and scaling for the moving object. The intuition is that each line in the moving object is 'attracted' to each line in the fixed object to an extent which depends on the similarity of the two lines. The global movement emerges from summing the effects of the local attractions.

Every comparison between a line from the moving object M and a line from the fixed object F is done in the same way. We compute the transformation T to map M onto F and a closeness measure A between M and F. The global transformation T_g for n such comparisons (line pairs) is defined as the weighted sum of the individual transformations:

$$T_g = \frac{\sum_{i=1}^{n} T_i A_i}{\sum_{i=1}^{n} A_i} \tag{1}$$

To initially simplify the description, consider T_g as being simply a translation along the x axis bringing the moving object in vertical alignment with the fixed object. The technique will be extended to rotation and scaling subsequently. For each line pair we compute the x vector which brings the centres of the lines together. From the end coordinates this is:

$$x_v = \frac{1}{2}(F1_x + F2_x) - (M1_x + M2_x) \tag{2}$$

2.1 Computing the weights (A)

For each line pair the weighting factor, attraction (A), is also computed. This is defined as the 'similarity' of the two lines (S) divided by the distance between the lines (D) squared:

$$A = \frac{S}{D^2 + \epsilon} \qquad (3)$$

The constant ϵ is the smallest movement the system is interested in and just prevents division by zero when lines coincide[2]. S is a compound measure of how similar the two lines are in length and orientation. For parallel lines of the same length it takes the value 1, for perpendicular lines (of any length) it is zero. In order to compute S, however, it is first necessary to know how the moving line maps onto the fixed line, as explained below.

Distance (D)

The intuition here is that the closer two lines are together the more strongly they 'attract' each other. However, deciding how to measure the distance between two lines is not as simple as it first seems. In the first implementation, the distance between the centres of the two lines was used. However, this can lead to large A weights when the centres of two lines happen to be close even when they have differing lengths and orientations. A better distance measure combines the distances between the ends of the lines, for example the average separation between end-points. The problem is, we do not know which end of the moving line maps on to which end of the fixed line (the coordinates are unordered). The mapping we are interested in is the one which involves least rotation of the moving line ($M1 \rightarrow F1$ in figure 1). There are thus four relevant distances, call them a, b, c, d. The squares of these distances can easily be computed from the coordinates:

$$\begin{array}{ll} M1 \rightarrow F1: & a^2 = (F1_x - M1_x)^2 + (F1_y - M1_y)^2 \\ M2 \rightarrow F2: & b^2 = (F2_x - M2_x)^2 + (F2_y - M2_y)^2 \\ M1 \rightarrow F2: & c^2 = (F2_x - M1_x)^2 + (F2_y - M1_y)^2 \\ M2 \rightarrow F1: & d^2 = (F1_x - M2_x)^2 + (F1_y - M2_y)^2 \end{array}$$

In the example case the distances a and b correspond to the desired mapping and the distances c and d correspond to a 'crossed' mapping. Some simple geometry shows that $a^2+b^2 < c^2+d^2$ where a and b correspond to the mapping involving the least rotation. Furthermore, the quantity $a^2 + b^2$ provides a useful measure of the separation of the two lines:

$$D^2 = a^2 + b^2$$

[2] ϵ can be the smallest non-zero positive real that can be represented in the implementation language.

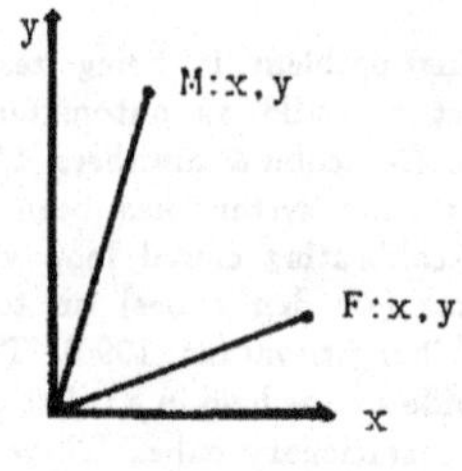

Figure 2: Normalised lines

The computation of D^2 thus provides us with a mapping between the end-points of the lines (except when they are perpendicular in which case the mapping is arbitrary). This mapping is needed for the next stage.

Similarity (S)

S is calculated from normalised end coordinates, ie each line is translated so that one end lies at the origin. The lines are normalised so that corresponding ends (in the mapping found above) are translated to the origin. This means that the angle between between the normalised lines (and the origin) is always acute. Given two normalised vector coordinates, x_m, y_m and x_f, y_f, the similarity measure is then defined as follows:

$$S = 1 - \sqrt{\frac{(x_f - x_m)^2 + (y_f - y_m)^2}{x_f^2 + y_f^2 + x_m^2 + y_m^2}}$$

The top half of the fraction, $\sqrt{(x_f - x_m)^2 + (y_f - y_m)^2}$, is simply the distance between the points M and F. For two parallel lines of the same length M and F coincide so this will be zero. If the vectors M and F were the same magnitude but perpendicular then the distance between the end points would be given by the denominator of the above fraction: $\sqrt{x_f^2 + y_f^2 + x_m^2 + y_m^2}$. As the angle between the line changes, the value of the fraction thus varies between zero (parallel) and 1 (perpendicular) in the desired manner. In addition, the greater the difference in line length (magnitude of M and F) the more the fraction tends towards 1. Finally, the fraction is subtracted from 1 so that S varies between zero and 1 in the right direction[3].

This completes the description of the computation of a global x translation as defined in equations 1, 2 and 3. The global y translation can be computed similarly and

[3] If the 'wrong' end of one of the lines is normalised so that the angle between them was between 180° and 90° then S turns out negative.

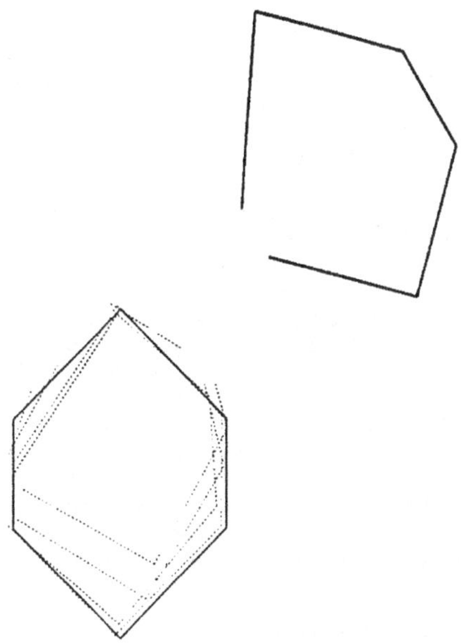

Figure 3: Attraction of simple figures. The top shape is a distorted version of the lower, with one line missing and another displaced. This attracted to the shape below resulting in perfect alignment of corresponding features. Broken lines show intermediate positions. All that is visible of the fifth position is the displaced line segment.

independently (but in the same program loop). The system described thus far may have applications in aligning correctly oriented parts, such as a 2D peg-in-hole problem.

3 Extending the range of transformations

Equation 1 contains the variable T referring to a transformation of some kind. This has already been applied to translation in x and y and can relatively easily be adapted to translation in three axes. T must be something for which it makes sense to take the weighted average. For example, $\frac{1}{2}T$ applied twice gives the same effect as T for translation, rotation or scaling. This would not necessarily be true for a compound transformation, for example if T was a matrix or a pair of complex numbers. This is why T is decomposed with rotation and translation components dealt with separately.

3.1 Rotation

Using the same equations as before we can find the weighted sum of the difference in orientation between line pairs. For each pair, therefore we compute θ, the minimum degree of turn which when applied to the moving line makes it parallel to the fixed line. Thus θ has a sign, say negative for clockwise and positive for anticlockwise and the weighted sum of these angles gives the overall rotation to be applied to the moving set of lines. This overall rotation can then be applied about .he centre of the moving object so that it doesn't interfere with any translation which might be going on simultaneously.

The desired angle θ is the one between the (normalised) vectors in figure 2. Unfortunately, usual methods of computing the angle between vectors do not give the sign of the smallest angle in the desired manner. The following method was therefore adopted. First the angles between each vector and the positive x axis are computed using arctangents. The difference between these angles is then brought into the desired range of $\pm\frac{\pi}{2}$ radians by applying the rewrite rules shown below. For example, a turn of three right angles clockwise is rewritten as a turn of one right angle anticlockwise.

1. If $\theta > \pi$ then $\theta \Leftarrow \theta - 2\pi$
2. If $\theta < -\pi$ then $\theta \Leftarrow \theta + 2\pi$
3. If $\theta > \frac{\pi}{2}$ then $\theta \Leftarrow \theta - \pi$
4. If $\theta < -\frac{\pi}{2}$ then $\theta \Leftarrow \theta + \pi$

For 3D (wire frame) shape attraction there would be three angles to compute for each pair of lines, one for each axis of rotation.

3.2 Scaling

Scaling is the multiplication factor applied to the moving lines to make them the same length as the lines of the fixed object. Like rotation, scaling is applied about the centre of the moving object. In order to make a moving line the same length as a fixed counterpart it can be multiplied by the ratio of the lengths of the two lines. This ratio is not suitable for weighted averaging, however, because the ratios are skewed between scaling up and scaling down and do not combine additively. The logarithm of the ratio does behave in the right manner:

$$T = \log\left(\frac{\text{length}(\mathit{fixed})}{\text{length}(\mathit{moving})}\right)$$

T is negative for shrinking, positive for magnification and zero for no change. The overall scaling factor is thus obtained by taking the antilog of the weighted sum of T for individual line pairs.

4 Performance

In order to test the algorithm a simple graphics interface was written to display two sets of edges and to apply rotations, scaling and translations to the moving set. The accompanying figures are the direct output of this interface and serve to illustrate how the moving shape appears to converge onto the fixed shape when the algorithm is applied iteratively. First the results of the algorithm without the scaling component are described.

4.1 Translation and Rotation

Figure 3 shows a simple example of the kind that the shape attractor was originally designed for. With two similar figures the algorithm will rotate and translate one to line up with the other in a few iterations with a 'capture angle' of just under $\pm 45°$. The capture angle may be less if the shape has a high order of (almost) rotational symmetry. Not surprisingly, for example, with shapes like slightly irregular stars a good fit may be found instead of the best one if the good fit involves less movement. In general however, the shape attractor is far less prone to being trapped in local minima than was anticipated, as the examples illustrate. The second example (figure 4) shows how the shape attractor can solve hidden figure (figure-ground) problems. The fixed set of lines consists of three intersecting and distorted shapes and the moving shape is a perfect version of one of the shapes. In each case the program can 'find' the hidden shape despite the presence of distracting information. The program can also often find partially occluded shapes by the same token, though performance obviously depends on how much of the target figure is missing.

Figure 5 shows a complex example in which a shape rotates and dilates and eventually finds a good fit with part of the landscape below. In general, however, the algorithm does not work so well when there are competing attractions from widely separated parts of the image.

Whilst the behaviour of the shape attractor is difficult to quantify, being dependent on the similarity of the two sets of lines, the examples do clearly demonstrate its intrinsic resistance to distortion, clutter and missing edges, which strongly suggests it may have applications in processing edge data. Some potential applications are discussed below.

4.2 Implementation and Extension

As already indicated the time complexity of the algorithm is $O(n \times m)$. No effort has yet been made to implement the algorithm efficiently but it is estimated that there should be no problem in getting the time per iteration down low enough for real-time applications, certainly where $n \times m < 100$. Some time saving can be made by computing S for each line pair first and doing no further processing for that pair of lines if the similarity is less than some threshold, *eg* < 0.5. In the iterative version, the algorithm is made to terminate when the change in orientation, position and scaling from the previous iteration are all below their respective (small) thresholds. It has not been proved that the algorithm will terminate under these conditions, but no counter-example has yet been found.

As the computation for each pair of lines is totally independent, the algorithm is obviously suited to parallel implementation if the images contained a sufficient number of lines to warrant it. This might arise if the image contained curves approximated by a large number of line segments. As such line segments would all have the same 'direction' this information could also be used to constrain the way in which lines can be mapped on to each other; neighbouring segments of similar orientation should not be mapped in opposite directions.

5 Potential Applications

5.1 Vision Guided Part Mating

The basic idea was explained earlier and so is not re-described here. However, it would be interesting to try and map the apparent 2D rotation difference between two objects onto 3D rotation of the moving object. For example, edges moving in depth perspective will produce pseudo-rotations in the 2D image as the outline changes. Without knowing the correct line-line mapping between the two sets of edges it will not be possible, even with two cameras, to unambiguously map the apparent 2D rotation into rotational components in three axes. However, the problem could be constrained by partial knowledge of the relative orientations of parts (*eg* we may know they only vary in the z-axis because they were both originally lying on a flat surface), by using robot motion itself to discover which rotations reduce the 2D difference, and by knowledge about the shapes of the parts themselves. The scaling component of the algorithm would probably not be needed in this application.

5.2 Guiding Mobile Camera

It may be possible to use the algorithm to instruct a mobile camera how to move in order to make the (changing) image approximate to a stored template. This is a similar idea to a model of navigation using landmarks in bees described earlier.

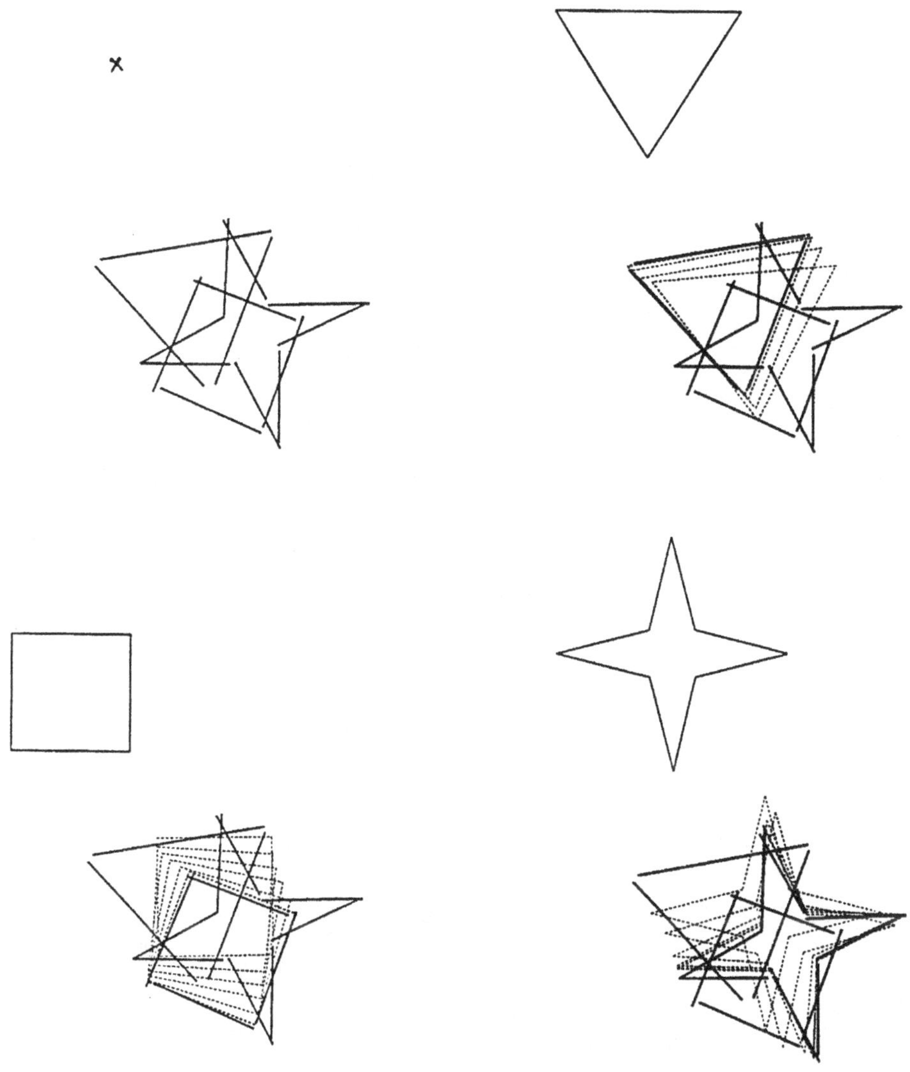

Jumbled figures (a), Attracting triangle, square and star (b,c,d). Note how the final alignment on the left spoke of the star figure is not so good due to the concentration of distracting lines.

Figure 4: Finding hidden figures

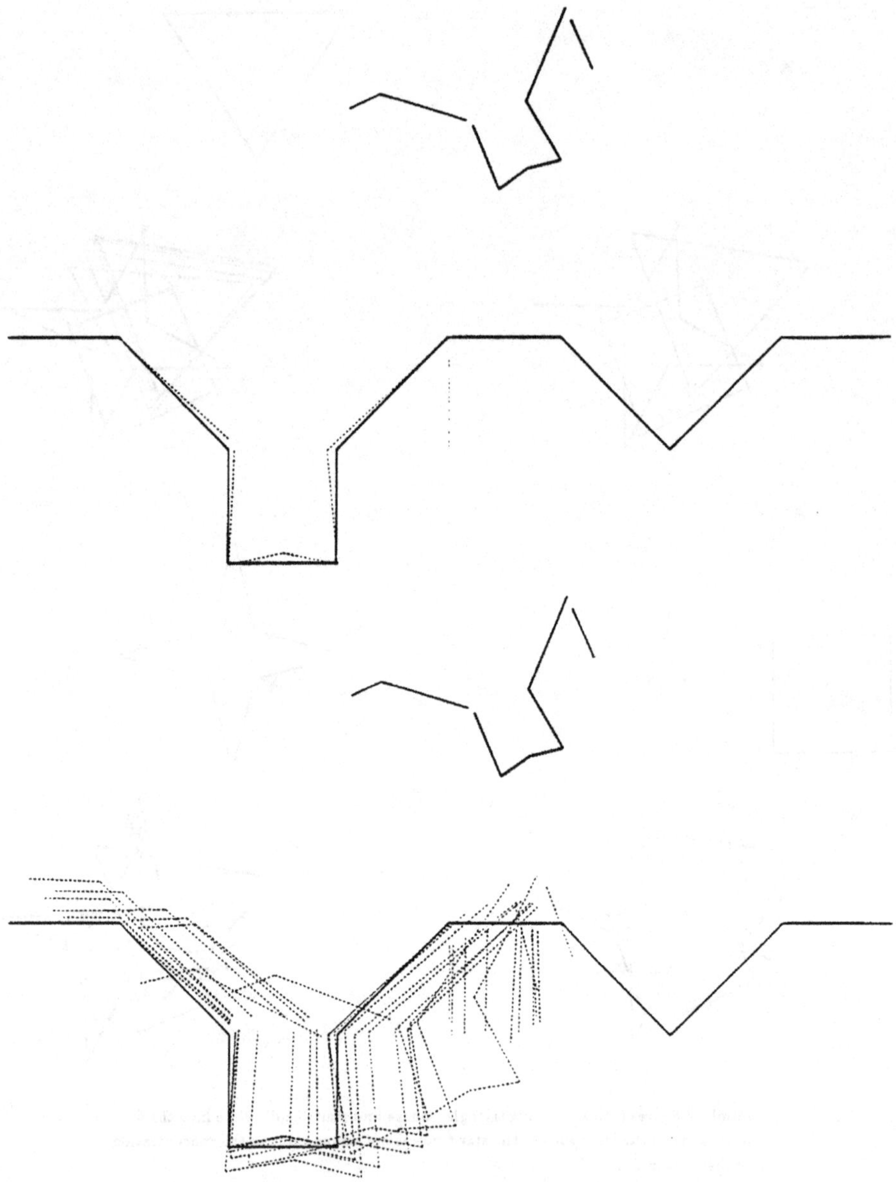

Figure 5: Rotation and scaling in complex figure. (a) First and final positions only, (b) intermediate iterations.

Similarly, instead of moving an object in front of a stationary camera as suggested above, it may be possible to guide the camera to find a particular view-point. For example, if we have a partially completed assembly in camera view we may want the camera to move to particular vantage points so that it is better able to guide the next stage of the assembly. If the camera is affixed to the robot wrist then it could possibly be used to guide the hand into a good pre-grasp attitude. Note that the scaling component could be used to give depth information here.

It may even be possible to do part recognition in this manner. A part could be 'memorised' by taking a number of snapshots (perhaps just the outline) of the object as the camera moves around it. If the object comes into view again then movements can be generated from each of the snapshots to see if a good match can be obtained. In this manner actual robot motion replaces computation of 2D viewpoints from 3D models so that only simple 2D matching is needed. In this manner part recognition and acquisition could be combined in one operation.

5.3 Static Template Matching

Using iterative transformation of a set of coordinates (as in the example figures) representing a template, the template can be matched against an image. In this case the scaling component of the algorithm may well prove useful. For example, the technique might be a useful for looking for specific shapes in cluttered 2D scenes such as X-rays.

As with the other applications, the approximate orientation of the target object needs to be known so that the template can be oriented within the capture angle of the algorithm. Without additional knowledge being built in it would be necessary to 'release' the template in several different orientations (six, say) and discover empirically which gives the best match. A cheaper method might be to use gross features of the two sets of edges to roughly align the major axes before attraction. For matching polygons, Mero (see discussion) uses the technique of aligning the points furthest from the centroid (Mero, 1979). The fact that the shape attractor tends to align closely matching features whilst ignoring mismatches suggests the possibility that it could be used for locating differences between a template and a target image.

5.4 Discussion

The shape attractor differs from previous approaches to shape comparison. Perhaps the main difference is that the output of the algorithm is in terms of transformations instead of similarity measures. This may make it particularly suited to applications involving the control of motion.

Of the other available techniques, the shape attractor is perhaps most similar in spirit to active contours (elastic boundary deformation), *eg* (Kass *et al*, 1987), in which local attractions cause a flexible curve to converge on a target boundary. In our case however, we are dealing with rigid transformations, and the effects of the local attractions are summed instead of acting independently. Also, we process discrete edges as opposed to a connected curve. A mathematical comparison of the shape attractor with active contour techniques is in preparation[4], and it may be possible to use it as a cheaper alternative in some applications, for example for tracking a moving object between image frames.

Another different technique is geometric matching. For example, Mero offers an effective method of matching a (model) polygon to an object outline (silhouette) derived from an image (Mero, 1979). This method is invariant for rotation and scaling of the boundary and was devised for recognising 2D industrial parts from geometric models. The method is tolerant to a certain amount of noise in the outline being analysed but the match fails totally if there is any significant difference in shape between the model and the outline. This is a desirable property for the identification application, but Mero's algorithm could not be readily used for locating missing or additional features, for example. In contrast, the current algorithm will try to superimpose any two sets of edges, however dissimilar they are, and so can solve some problems that pure geometric matchers cannot tackle.

In summary, an algorithm has been developed which is in many ways analogous to the one proposed by Cartwright and Collet for navigation in honeybees. In particular the non-selective use of features, vector summing and resultant noise immunity are common to both. The weighting system is unique to the present algorithm, however; Cartwright and Collet use only simple unit vectors and the proximity heuristic (adjacent areas 'attract' each other). Finally, it may well be the case that performance of the bee model could be improved by using all possible pairings and adopting a weighting system similar to the one presented here.

Acknowledgments

This work is supported by the ACME Directorate of the UK Science and Engineering Research Council, grant GR/E 68075. Thanks to principle investigator, Chris

[4]In collaboration with Manuel Trucco, Dept. AI Edinburgh University

Malcolm for critical comments and suggestions and to Piak Chongstitvatana for encouraging me to follow up the initial ideas. Thanks also to members of the Department of Artificial Intelligence Behaviour-based Systems Research Group for comments and discussion including Manuel Trucco, Bob Fisher, Tim Smithers, Alistair Conkie and Lykourgos Petropoulakis.

References

Anderson, A.M. (1977). A model of landmark learning in the honeybee. *Journal of Comparative Physiology*, 114:335–355.

Cartwright, B.A. and Collett, T.S. (1983). Landmark learning in bees. *Journal of Comparative Physiology*, 151:531–543.

Conkie, A.D. and Chongstitvatana, P. (1990). An uncalibrated stereo visual servo system. In Conkie and Chongstitvatana, *British Machine Vision Conference*, Also available as U. of Edinburgh DAI research paper No. 475.

Kass, M., Witkin, A. and Terzopoulos, D. (1987). Snakes: active contour models. In et al, Kass, *Proc. IEEE-ICCV*, pages 259–267, London.

Malcolm, C.A. and Smithers, T. (1990). Symbol grounding via a hybrid architecture in an autonomous assembly system. *Robotics and Autonomous Systems*, 6:123–144.

Malcolm, C.A. (1983). The outline corner filter. In Rooks, B., (ed.), *Proceedings of the 3rd International Conference on Robot Vision and Sensory Controls (ROVISEC)*, pages 61–68, North Holland Publishing Company, Amsterdam, Also available as University of Edinburgh DAI research paper No. 212.

Mero, L. (1979). *An Algorithm for Scale and Rotation Invariant Recognition of 2 Dimensional Objects*. Technical Report, Computation and Automation Institute, Hungarian Academy of Sciences, Budapest.

Tinbergen, N. (1932). Uber die orientierung des bienenwolfes (*Philanthus triangulum* fabre). *Z. Vergl Physiol*, 16:305–355.

MOTIVATION AND EMOTION

ETHOLOGICAL AND PSYCHOLOGICAL MODELS OF MOTIVATION- TOWARDS A SYNTHESIS

Frederick Toates,
Biology Department,
The Open University,
Milton Keynes
MK7 6AA,
United Kingdom.

Per Jensen,
Section of Ethology,
Swedish University of Agricultural Sciences,
POB 345,
Skara, S-53227,
Sweden.

Abstract

By looking at a variety of individual motivational systems, the first steps towards reconciliation of models in the psychological and Lorenz traditions is made. A model that contains features of both emerges. The relevance of this model to animal welfare issues is discussed.

1.0 Introdution

It seemed once that study of motivation theory was in serious decline; in psychology and ethology, the days of the grand theories (e.g. Hull, Lorenz) appeared to be over (Toates, 1986). Psychologists lost interest in the topic and ethologists had moved in their droves to the greener pastures of optimal foraging and sociobiology. Then applied ethology came to the rescue; a renewal of interest in motivation theory was prompted by considerations of animal welfare, suffering and the associated recommendations for legislation (Dawkins, 1980). To assess when an animal is suffering or why animals in captivity might spend much time in various bizarre activities, not observed in wild-living conspecifics, one needs motivation theory. However, applied ethologists encountered a major difficulty: fragmentation and contradictions in the literature. To some extent, psychological and ethological approaches differ in their assumptions and areas of interest. However, differences do not divide neatly along party lines. Thus, ironically, some writers in the psychological tradition (e.g. Gallistel, 1980; Glickman and Schiff, 1967; Herrnstein, 1977; Hogan, 1967; Toates, 1986) argued for the applicability of the best-known ethological model, that of Lorenz (1950), whereas, to some ethologists, it has been used for little more than to illustrate how motivation doesn't work (Dawkins, 1986; Archer, 1988). The issue is more than of academic interest; recommendations for animal husbandry depend upon which model one believes (Baxter, 1983; Hughes and Duncan, 1988a,b).

Some of the conflict seems to arises from semantic confusion, loose use of language and the assumptions of different perspectives, and we attempt to resolve some of this. We suggest where common ground can be found. First, it is necessary to try to sanitize the vocabulary.

1.1 The meaning of motivation

One reason for confusion seems to be the meaning different authors assign to the concept of "motivation". Even within psychology, the concept is used in very different contexts and meanings by different researchers (Toates, 1986). In psychology, "drive" and "motivation" are overlapping and sometimes identical concepts, while in ethology, since the influential work of Hinde (1959; 1960), "drive" has assumed a role much like a swear-word in church, while "motivation" has survived as a well-respected term. The term "motivation" is found in most ethology books, but rarely defined in a way that enables one to know what is meant by claiming that an animal is more or less motivated for certain behaviour. Although mainly used to discuss how responses may vary in different contexts, "motivation" is usually only a heading for the chapter in which this is found. For example, Slater (1985) explicitly takes this broad view, in saying, while discussing the variability of responses in animals: "...discovering just what it is that leads them to behave differently from one time to another presents some very interesting problems...These are the problems of motivation or, in other words, the mechanisms leading animals to do what they do when they do it." McFarland and Sibly (1975) provided a formal representation of motivation, the state-space approach. Although primarily introduced as a powerful alternative to unitary drive concepts, they also tried to tidy up the terminology. Thus, "motivational state" is the state value of all causal factors influencing a setup of functionally related behaviour patterns. The motivational state maps onto a "tendency" which is the strength of the behaviour in the competition for the motor apparatus (the "final common path").

This may be a convenient and fairly unambiguous way of talking about motivation, but for many psychologists, it doesn't relate to their conception of motivational problems. In the terminology of McFarland and Sibly, all behaviour of any animal will by definition be guided by a motivational state, determining its tendency. This leads to the danger of a circular argument, since by the definition, an animal will always be under the control of the behaviour for which the tendency is strongest, i.e. for which it is most motivated. So, if all behaviour reflects a hypothetical motivational state, which can only be deduced from observations of the behaviour performed, we run into difficulties of circularity. In appreciation of this, some psychologists have tried to find a definition which would distinguish behaviours that are guided by motivation and those that are not, which would turn the concept into an explanatory, rather than only a descriptive, one. For example, Toates (1986) argues that a motivational system is one that selects a goal (incentive) to be pursued and organizes the animal's commerce with it. Wise (1987) has suggested that we only talk about motivation for those behaviour patterns modulated by neurohormones and neuromodulators, leading to the conclusion that a behavioural repertoire comprises both motivated and non-motivated behaviour. However, Blackburn and Pfaus (1988) in a reply, settle for the same approach as most ethologists in saying: "We prefer to think of motivation as the study of why animals do what they do when they do it." They also reject emphatically the possibility of an animal showing non-motivated behaviour: "An animal without motivation is not a nonmotivated animal; it is a corpse."

So, in spite of different traditions, it is obvious that researchers feel a need for the concept of motivation, at least as a descriptive term to be used when discussing the variability of behavioural output. There is also a tendency to try to make the concept more explanatory and in this mixed tradition, various models have been developed. Probably different insights may be gained from several of these models. Therefore, we will discuss some behavioural phenomena starting from two very different models, emanating from the different traditions of psychology and ethology. We will not at this stage try to solve the disharmony in terminology.

In psychology, the most influential model of motivation has been that of physiological homeostasis. This has either formed the basis of the theory or has served as a starting point from which other theories have built (Grossman, 1967). In ethology, the model of Lorenz (1950), involving the hydraulic analogy and using energy concepts, also incorporates negative feedback and explicitly represents behaviour as maintaining an internal parameter within bounds. Both traditions have had to confront similar issues brought into focus by their models, e.g. the respective roles of internal and external causal factors, do the same factors that switch motivation on also serve to switch it off etc. In spite of certain similarities, none-the-less the degree of overlap and integration between these two approaches has been minimal. It is our contention that a more balanced perspective is appropriate in which features of both models can be integrated and that is the purpose of the present paper. First, we shall briefly outline a general description of a behavioural system, which we will use to analyse the validity of the two models.

1.2 The behavioral system

Different factors apparently affect different behaviour patterns in a very complicated, but still (at least in some well studied cases) fairly predictable, manner. Internal and external causal factors both control the performance of specific behaviour through some motivational mechanism. The actual motor program, which obviously has a varying degree of flexibility, will be left aside for the present (although in animal welfare, the variability of specific motor programs may be of utmost importance). The behaviour elicits some feedback on the causal factors. The external causal factors are mainly affected through the stimulus consequences of the behaviour (e g, food disappears when eaten, a nest arises when the animal performs nest building). The internal causal factors are affected in various ways including metabolic changes and proprioceptive feed-back. Also, there may be an effect of external causal factors on the internal, without any behavioural output being involved. This was shown for aggressive behaviour in cichlid fish by Heiligenberg (1974). Of course, this is the effect involved in the well-known heterogenous summation of stimuli. It may also be the process involved in priming (in the ethological sense), where one early exposure to some stimulus may change the probabilities for various behaviour later in the life of the animal.

2.0 Two Traditions

2.1 Lorenz

In the famous Lorenz model, action-specific energy accumulates as a function of time, represented by the flow into a tank. Height of fluid in the tank is analogous to the strength of drive. As fluid accumulates, the tendency to open the valve increases. Fluid flowing out is analogous to behaviour. A releasing stimulus, represented by a weight, also exerts a tendency to open the valve. The model explicitly incorporates the facility for behaviour to exhibit positive feedback, thereby representing "the avalanche-like way in which the endogenous activities tend to grow after being released" (Lorenz, 1950, p.260). In the sense that a combination of external stimulus and internal drive determines the strength of behaviour, the model captures rather well an aspect of how behavioural systems function (Hinde, 1960). In other aspects, the model is not generally applicable.

Faced with a number of new facts in motivational research, Lorenz reformulated his model. Among other things, he was concerned with the dichotomy between the role of external and internal stimuli. He also realised that there was enough evidence around to state that external factors might affect internal without any actual behavioral output. For example, Heiligenberg (1974) had shown that short-time exposure to certain social stimuli had a long-time effect on aggression tendency in cichlid fish. Therefore, Lorenz (1978) modified his model in the following way: apart from the continuously flowing endogenous "action-specific energy", there are other water taps, through which energy caused by activating stimuli (i.e. stimuli not having any immediate behavioural effect) is flowing. The releasing stimulus is represented, in the new model, by a large amount of liquid being poured into the container. Thus the clear dichotomy between the effect of external and internal causal factors has disappeared.

We shall argue that: (1) Although the specific details of the model might well be wrong in a number of respects, there are *features* of behaviour that require a formulation not unlike that of Lorenz, (2) Some motivational systems fit the model rather well and others less well and (3) The modifications to the original model have not been widely recognized.

2.2 Homeostasis in psychology

The principle that, when a physiological variable departs from a set-value, action is taken to correct the deviation is a corner-stone of physiology. In psychology, the principle has been used to explain the regulatory basis of feeding and drinking (McFarland, 1974). Some authors dichotomized motivations into those that fitted homeostatis, eg. feeding, and those that did not, e.g. sex (Grossman, 1967). It was argued that the former class were internally driven whereas the latter were externally driven. There were various modifications of the basic homeostatic model to fit the data: even homeostatic systems sometimes exhibit non-homeostatic behaviour. Others argued that incentive motivation theory could explain a wider number of results, and would enable sex and aggression to be discussed under the same rubric as feeding and drinking (Bindra, 1976; Toates, 1986).

In the following sections we review a number of behavioural phenomena, noting the applicability of the Lorenzian and psychological models.

3.0 The Behavioural Phenomena

3.1 Turning motivation/behaviour on - the internal factor

The accumulation of fluid is a controversial aspect of the Lorenz model. To help resolve this, it is necessary to divide the argument into logical parts. First, we ask whether, taking a black-box perspective, motivational systems behave as the model suggests. If so, which of them? All? Next, we ask - if a system does behave in this way, is it most usefully modelled in the way that Lorenz suggests and with his theoretical assumptions? Confusion has been caused by subsequent writers not being sure how literally to interpret the term 'energy'. Is it just a convenient metaphor or does Lorenz mean to imply that the living system incorporates some of the physical properties of energy (Hinde, 1960)? The unwary are often confused on comparing physiological and Lorenzian models, only to find that a dehydrated rat corresponds to a Lorenzian full tank!

Does the tendency to perform behaviour increase as a function of the time since it was last performed? First, consider systems that maintain physiological constancy. Tendencies to feed and drink increase as a function of the time since the behaviour was last performed. This has much to do with the metabolism of energy substrates in body tissues and the loss of water through various routes, respectively. Thirst is quickly induced in non-deprived animals by sodium chloride injections. Classical regulatory physiology can, at least in principle, better account for the relevant internal factor, and postulation of 'motivational energy' risks clouding the issue. The failure of theorists in the Lorenzian tradition to articulate this has been a source of unnecessary confusion. We shall argue that such extraneural physiological changes can explain much of the increase in tendency as a function of time, but not all. In other systems concerned with physiological regulation there is little evidence to suggest that behavioural tendency increases in an analogous way. Sodium ingestion might never be motivated normally but can quickly be induced by depletion of NaCl. Provided an appropriate ambient temperature is maintained, behavioural temperature control is not normally displayed at intervals. For example, pigs respond to high ambient temperatures by wallowing, but the behaviour is never seen when it is cold (Stolba and Wood-Gush, 1989).

Consider now systems not concerned with specific physiological homeostasis and with which a corresponding extraneural variable is not normally closely associated. For these, there is considerable variation in the tendency for behaviour to appear at regular intervals. There is little evidence that a tendency to perform aggression, the motivation most discussed in the context of Lorenz, increases as a function of the time since last performed (Archer, 1988). However, other systems do seem to programme behaviour at intervals in an unchanging environment. Rodents make regular patrols of their environment (Ely and Henry, 1978). The activity patterns of small nocturnal rodents with access to a running wheel seem to be intrinsically programmed (Kavanau and Rischer, 1968). Predation is motivationally distinct from aggression, in some cases occurs 'out of context' and appears to increase in tendency as a function of time since last performed (Polsky, 1975). A good example of where behavioural tendency increases as a function of time is dustbathing in domestic fowl (Vestergaard, 1980). Attempts to tease apart an endogenous neural factor from such factors as accumulation of a substance on the feathers has proven difficult. In laying hens, Nicol (1987) found a rebound in the frequency of performing such behaviours as wing stretching and flapping following a period of environment-imposed 'deprivation', and wrote: "....it is most parsimonious to argue that some continuous change occurs in the animal's state during the restrictive period".

A possible source of confusion and controversy concerns the locus at which any spontaneous build-up might occur. The examples given, dust-bathing and wing-stretching etc, that best lend support to Lorenzian notions consist in, if not 'fixed action patterns', then 'relatively inflexible' motor acts typical of a species and context. According to some criteria, these might be classed as non-motivated behaviours (Cf. Wise, 1987).

There is also evidence that the tendency to perform some specific behaviour patterns that normally serve physiological regulation increases as a function of time since they were last performed. This can give behaviour a certain autonomy from the physiological state that it normally serves. The examples that Lorenz

(1981) used refered to rather *specific acts*, these often being in the service of a particular motivation. Thus he speculated that "In a species that, like the European starling, acquires most of its food by performing one single motor pattern, the autonomous motivation of the latter is obviously sufficient to ensure adequate feeding and needs no appreciable additional drive from the side of tissue". To account for such phenomena, Gallistel (1980) spoke of *potentiation* of neural circuits underlying rather specific behaviours that occurs with deprivation.

In psychology theorists are concerned with the flexibility of goal-directed behaviour in which a number of different courses of action can be recruited to obtain a given goal-object (Bindra, 1976; Epstein, 1982). In such terms, the question is - can any spontaneous build-up of motivation, as modelled by an increase in pressure of Lorenz's tank, be shown in a variety of different ways, e.g. maze running, lever-pressing? If it is true that the locus of the change is within the circuitry underlying specific behaviours, one would not necessarily expect this to be revealed in general motivational indices but only in specific behaviours. We suggest that this is a potentially fruitful area for more investigation.

Gallistel (1980) presented a hierarchical model of the organization of behaviour, with motivation at the top and relatively 'simple acts' at the bottom. In his analysis, motivational states can 'selectively potentiate' the simple acts, but the bottom levels can develop a certain autonomy from the top. Gallistel notes: "Lorenz's important principle - the autonomous build up of a potentiated condition in a unit that coordinates some long uncalled upon act - has been lost sight of in the swirl of controversy over energy theories of motivation". He argues that "..an act may be in a high state of potentiation for no other reason than that it has been unperformable for some while. The only thing that reduces this state of potentiation is the occurrence of the potentiated act".

Gallistel (1980) and Polsky (1975) note the common claim that predation is always tied to hunger and, to refute it, cite several references showing an autonomy of predation from hunger. Gallistel also notes that, in cats, components of predation such as stalking and pouncing can gain some autonomy from predatory behaviour, often being performed in isolation. He writes: "The cat may pursue for some while and with considerable resourcefulness a pattern of action that finally yields something for it to pounce upon. Not, mind you, something of obvious biological significance such as food; just something to pounce upon, pouncing itself being the culminating act in the entire behaviour pattern". Polsky notes that the performance of killing in rodent killers has a self-satiating effect.

Whether one considers a change in the neural circuitry organizing a behaviour to be motivational, there are other results that point to its motivational significance. In support of Lorenzian ideas, Gallistel discusses the notion of *relative hierarchy*, noting that for, functional units at, or towards, the top of the hierarchy, there can be a reversal of what is normally in the service of what. For example, a rat denied access to water will run in a wheel for the reward of water. However, the contingency can be reversed: a satiated rat will drink if rewarded with the opportunity to run in a wheel (Premack, 1965), which he interprets as support for the notion of action-specific energy. Gallistel argues that it is not easy to see what is the biological need for such running and continues: "....it may be gratuitous to postulate such a need; but whether or not the need exists, the neural circuitry that organizes this activity is a focus for the autonomous accumulation of potentiation". Gallistel cites several other examples, all of which lead him to the same conclusion: (1) that behavioural acts (e.g. paper shredding) can show autonomy from the motivational system with which it is normally functionally related (in this case, nest building) and (2) other acts will be recruited if they are rewarded with the opportunity to engage in the behaviour in question.

3.2 Turning motivation/behaviour off

'Non-Lorenzian' theorists assume that behaviour is terminated by certain extrinsic *consequences*, i.e. by effects beyond the neural circuitry organizing the behaviour. For example, as Hinde (1960) notes "....cessation of activity may be due to the animal encountering a 'goal' stimulus situation, and not to the performance of an activity. If this goal stimulus situation is encountered abnormally early, the behaviour which normally leads to it may not appear at all". Such terminating consequences can lie more or less near to the boundary of the animal's own body. For example, aggression is said to be terminated by the removal of an intrusion (Archer, 1988), feeding by the passage of nutrients through the alimentary tract and the anticipation of nutrient gain (McFarland, 1974), nest building by the sight or sensation of a completed nest (Baxter, 1983). On the other hand, a factor serving to terminate behaviour might be intrinsic to the nervous system. For example, Gallistel (1980) argues that activating a neural circuit can reduce the level of the autonomous potentiation. We shall take a number of systems in turn to see what the evidence is.

Hughes, Duncan and Brown (1989) investigated whether nest building in the domestic fowl *Gallus gallus domesticus* was terminated by the appearance of the completed nest. They tried short-circuiting the natural sequence leading to nest-building by presenting the birds with a completed nest. The birds still performed a certain amount of nest-building behaviour. Although this points towards a strictly internal causal factor involved in nest-building in hens, studies by Hinde and co-workers of nest-building in canaries (Hinde and Steel, 1966) show a high degree of flexibility and continuous regulation from the consequences of behaviour.

In rats, nutrients ingested orally are much more effective than the same quantity given by intragastric or intravenous infusion (see Toates, 1986). Even large infusions are often relatively unsatiating. This leads to two possible and not mutually incompatible explanations: (1) oral and gastric stimulation are necessary for normal satiety and (2) performance of ingestive behaviour *per se* contributes to satiety. That sucking in rat pups is not regulated by ordinary satiety mechanisms (i.e. milk in the stomach), but by some independent neural pathway (Kenny et al, 1979), is evidence in favour of the second suggestion. The same has been observed in piglets (Stephens, 1975), and Castren, Algers and Jensen (1989) found that, in pigs kept in a semi-natural environment, the time between two consecutive sucklings was not affected by whether there was any milk ejection in the first suckling.

Another way of short-circuiting behaviour is to present the goal object 'free' in an operant situation. Birds and rats will still work in a Skinner box for food in the presence of free food (Gardner and Gardner, 1988).

Glickman and Schiff (1967) observed paper-shredding in the Mongolian gerbil *(Meriones unguiculatus)*. The behaviour continued even though adequate warmth, food, burrowing material and a place of retreat were provided. It was shown at a high rate "even after their cages were virtually filled to the brim with the shredded remnants of previous days". Deprivation of the opportunity to chew potentiated the behaviour.

We conclude that motivation is turned off by some combination of the proprioceptive feed-back from performing behaviour and the extrinsic consequences of the activity. It appears rarely possible to inhibit behaviour expression completely by presentation of the goal-state. This points towards a model of behavioural control based partly upon an obligatory motor program, which will be more or less flexible in some of its parts and altered in response to environmental feedback.

Traditionally, the issue of what switches motivation off has been connected with the psychological concept of reinforcement. For instance, Hogan (1967) and Glickman and Schiff (1967) suggest that, if one were able to show that behaviour *per se* could serve as a positive reinforcer, this would seem to be compatible with the Lorenz model. The next section addresses such issues.

4.0 Use of the terms Affective, Hedonic and Reinforcement

To some, terms such as 'reinforcement', 'affective value' and 'hedonic' seem intuitively appropriate in describing the behaviour of advanced vertebrates (Epstein, 1982). They are sometimes used to mean little more than, given the opportunity, an animal tends to behave in such a way as to achieve something or that it enjoys achieving or doing something. For example, Hughes, Duncan and Brown (1989) suggest that "....the performance of nest-building behaviour, regardless of its consequences, is reinforcing and is thus important to the fowl". Under impoverished laboratory conditions, Mongolian gerbils run the equivalent of 12-15 miles a night, and Hogan and Roper (1978) suggest that: "..running can become divorced from any regulatory function, so that it simply expands to fit the available time".

Motivation can be seen as a conjoint function of internal states and external stimuli. It then plays a part in guiding the animal's commerce with an incentive. Food, water and sexual contact are variously described as positive reinforcers or as having hedonic, incentive or affective value (Cabanac, 1971). Motor performance has usually been seen as (a) flexible in that a variety of different responses can be recruited to reach a given goal (Bindra, 1976) and (b) hedonically neutral, meaning the animal is moved into action by the pull of incentives or associations with them (see Herrnstein, 1977). However, in spite of the undoubted flexibility of behaviour and the motivational significance of incentive contact, there is also evidence that (a) certain response patterns are particularly associated with certain motivational states (Gallistel, 1980) and (b) behaviour *per se* plays a role in the reinforcement process (Glickman and Schiff, 1967; Herrnstein, 1977).

Animals can readily perform certain tasks if they are able to use species-typical behaviour, i.e. 'constraints on learning' (Bolles, 1970). Tasks that are at odds with such behaviour are more difficult to learn and animals sometimes 'degenerate' into species-typical behaviour even after apparently solving the problem. These phenomena are relevant to motivation theory: certain species-typical behaviours are heavily primed in an appropriate motivational state. Hughes and Duncan (1988) suggest that: "...these animals had a marked tendency to perform different appetitive elements of feeding behaviour, and that the more strongly the primary behaviour is motivated, the more obvious and fixated these secondary and fragmented behaviour patterns become".

As Glickman and Schiff note, a theory of reinforcement should address, for example, running and burrowing as much as feeding and copulation. They propose: "An obvious and circular solution is that the performance of these responses is reinforcing. The circular solution, if scientifically meaningless, is not necessarily wrong".

Some flexibility of motor output is self-evident; otherwise, rats and pigeons could never master arbitrary instrumental tasks as well as they do. However, this coexists with a certain prescription of species-specific and motivation-specific behaviour. Such specified behaviour has motivational value, or, if one prefers the expressions, is affectively or 'hedonically' loaded (Epstein, 1982; Herrnstein, 1977). Behaviour characteristic for a species and motivational state 'intrude' upon the experiment even when their effect on gaining food is a negative one (Gardner and Gardner, 1988). In pigeons, placed in a context associated with food delivery, the behaviour of pecking a lighted key will emerge even if to do so *delays* the arrival of food. So long as they get a free pellet once in a while to maintain the motivational significance of the apparatus, pigeons peck even though there is a negative contingency between behaviour and reward. Such results cannot be explained in conventional terms. They can only be subsumed under existing psychological principles if one assumes that pecking is not "hedonically neutral" (Herrnstein, 1977). By the criterion that food is reinforcing, one would have to conclude that behaviour also forms a part of the reinforcement complex. Normally, of course, pecking and the gain of food would coexist to form a complex. In the laboratory, one can sometimes tease apart the components. Whether this is analogous to what can happen in the abnormal behaviour of domestic animals is an issue of profound importance (see later).

Traditionally in psychology, reinforcement has been used to describe a process of strengthening that underlies learning. Exactly what reinforcer can reinforce what response (Bolles, 1970) and how the mechanism of reinforcement actually works (Bindra, 1976) have been topics of lively debate, but the concept of strengthening a tendency is implicit throughout. Most experimentation and theorizing has concerned reinforcers that are external to the animal, e.g. food, sex or a change in the environment. To establish that behaviour *per se* can be positively reinforcing one has to devise an experiment in which the opportunity to behave is made contingent upon a response but for which no extrinsic consequence is derived. Such an experiment is not easy to design. In behaviour that involves a sequence, one can prevent the goal being reached and see what happens to earlier parts of the sequence but this does not necessarily prevent behaviour from effecting some change.

Hinde and Steel (1972) investigated nest-building in canaries, noting the sequence that normally occurs (1) pecking at nest material and gathering, (2) later, carrying material to the nest site and (3) later still, nest construction. They were interested in seeing what role reinforcement played in the sense of later events in the sequence reinforcing earlier ones, using the term 'reinforcement' simply to mean "events in the sequence which are necessary for the maintenance of earlier events". They found that events late in the sequence were not necessary for the earlier components to be maintained. Subjects that could only gather maintained this behaviour for weeks.

For rats, by the criterion of whether an animal will perform an operant to gain something, running in a wheel is positively reinforcing (Hogan and Roper, 1978). As Herrnstein (1977) notes, in female rats, the reinforcement value varies with the phase of the estrous cycle.

Aggression brings this particular issue sharply into focus. Can the opportunity to perform aggression serve as a positive reinforcer even when such behaviour does not change the external world? (a question related to this is - can an appetitive aspect of aggression be shown?) The best-known and most reliable demonstration of an animal learning an operant task for the reinforcement of aggressive opportunity is the Siamese fighting fish *(Betta splendens)* swimming through a hoop for access to a mirror in which it can perform its aggressive display (Hogan, 1967). Hogan suggests that such behaviour *per se* is self-reinforcing. However, even here the effect is not a robust one nor is the interpretation completely unambiguous (Hogan and Roper, 1978). Some writers argue that repeated exposure becomes aversive, presumably because the 'rival' fails to respond appropriately to the aggressive display (Rhoad, Kalat and Klopfer, 1975).

When placed in an aversive motivational state, monkeys can learn an operant response for a target to attack, even though this does not change the external situation (Azrin, Hutchinson and McLaughlin, 1965). Caution is needed here. Under certain conditions, i.e. when an animal is in what might be termed an 'aversive state' as a result of pain or frustration (Cohen, Looney, Campagnoni and Lawler, 1985), performance of aggression might be said to constitute positive reinforcement. However, that is not to say that animals have some intrinsic need, a spontaneous internal drive, to behave aggressively (Cherek, Thompson and Heistad, 1973).

5.0 Effect of Behaviour on Physiological State

Whether behaviour *per se* has a positively reinforcing effect, does it affect physiological state? In the context of the so-called stress hormones, this issue has potential relevance to welfare.

Placed in a frustrating situation, pigs having the facility to pull a chain appear less stressed by the index of corticosteroid levels (Dantzer and Mormede, 1981). Rats housed in pairs and subjected to electric shocks normally fight immediately after the onset of shock, termed 'shock-elicited aggression'. Williams and Eichelman (1971) looked at blood pressure in the tail. Rats subjected to shock showed an elevation in blood pressure. However, rats subjected to the same shock but paired, hence showing shock-induced aggression, showed a *fall* in tail blood pressure. Conner, Vernikos-Danellis and Levine (1971) and Levine, Weinberg and Brett (1979) measured plasma ACTH and adrenocortical steroid levels in rats following electric shock, comparing individual housing with paired housing (facilitating shock-induced aggression). For both groups, plasma ACTH was higher following shock than control animals. However, animals able to fight were significantly lower in ACTH than those shocked alone. They argued that "This mechanism can serve an adaptive function by dampening the internal response which may be excessive and have damaging consequences". Henry (1976) discusses what common-sense, rather than an understanding of coping, might have predicted: "Since paired animals fight at the onset of shock, it might have been expected that the combination of shock with fighting would have elicited a greater pituitary adrenal response than either stimulus alone". However, in terms of coping: "It would appear that pairing changed the animals' perceptions. Instead of experiencing helplessness and depression, the fighting was associated with arousal of the classic Cannon sympathetic adrenal-medullary response". Levine, Weinberg and Brett argue that "...the reason shock-induced fighting reduces arousal is that, like eating or drinking, fighting may also be a consummatory response...." In a similar vein, Weinberg, Erskine and Levine (1979) suggest that there is an "element of self-reinforcement" to fighting, analogous to the act of feeding.

For gastric ulceration, Weiss, Pohorecky, Salman and Gruenthal (1976) compared three conditions of receiving shock: (1) alone, (2) with a companion allowing aggression and (3) with a companion but with a glass partition dividing the two rats. The latter condition allowed the rat to perform aggressive-type behaviour but with no consequence to the opponent. Rats in condition 1 had much more severe gastric ulceration than in conditions 2 or 3. That 2 and 3 had a similar degree of ulceration indicates that it was the performance of the aggressive response that was instrumental. Weiss *et al* consider that aggressive behaviour in this situation should be termed a *coping response.* They add that for a non-circular definition of a response as 'coping', it must provide "relevant feedback". They list 3 possible ways in which fighting might give such feedback: 1) the animal would have been fighting at the time of shock termination and therefore might have formed an association between its behaviour and shock termination. 2) fighting might in some way distract the animal and 3) "Certain highly prepotent responses in danger situations, such as fighting, might inherently produce their own relevant feedback".

That the *performance* of certain defensive actions ameliorates the effect of stressors, even though the response does not yield useful information, emerged from experiments of Guile and McCutcheon (1980). They suggested that the animal is *prepared* to perform such responses in an aversive situation. One such is struggling in response to shock, and preventing such struggling by restraint increased ulceration. Conversely, allowing a block of wood for the animal to chew reduced ulceration. It was suggested that chewing is a natural defense reaction for a restrained rat in an aversive situation and would, under natural conditions, serve the function of forming an escape route. Anisman (1978) refered to fighting as a *species-specific defence reaction* (Bolles, 1970), performance of which might be expected to have comparable physiological effects to fleeing.

A model of the relationship between noxious stimulation, behaviour and brain biochemistry was proposed by Stolk, Conner, Levine and Barchas (1974). It represents what happens when rats are subjected to footshock, alone or together. In the together condition, rats engage in fighting, which can be viewed as a coping response. In the 'shock with partner' condition, aggressive behaviour is associated with an increased turnover of norepinephrine which, in turn, facilitates further shock-elicited aggression. By contrast, placing rats in the 'shock without partner' condition is followed by a decrease in turnover of norepinephrine, which biases future behaviour away from displaying shock-elicted aggression.

6.0 Some effects of External Causal Factors on Motivation

External causal factors may change the readiness of an animal to behave in a specific manner, without releasing any immediate behavioral response. For example, exposing female mice to the smell of male urine does not immediately change their behaviour, but increases the sexual behaviour when a male is later introduced. Social stimuli also leads to a synchronisation of oestrus and an earlier sexual maturity in female pigs (Hemsworth, Salden and Hoogerbrugge, 1982). Sexual behaviour of adult boars is significantly affected by whether they are reared alone or in a group, and whether or not they are exposed to stimuli from adult females before puberty (Hemsworth, Beilharz and Galloway 1977).

In such cases, we are probably dealing with a stimulus effect on endocrinological state, i.e. the external stimuli cause a hormonal change, which in turn alters the responsivity of the animal. However, there are also cases where distinct exposures may cause a more direct effect, perhaps affecting the internal state of the animal directly via the CNS. The most obvious situation is the well-known imprinting effect, where a short exposure during a critical phase may profoundly affect later behaviour, e.g. towards what object or individual it will direct its behaviour, or its later sexual preferences. But there are more short-term changes in behavioural tendency, caused by external stimuli, that might be working through a direct nervous effect sooner than via hormonal changes. The "old" ethological concept of summation of stimuli (Seitz, 1940; Tinbergen 1953) describes precisely that effect: when an animal is exposed to several "releasing stimuli", specific to a certain behaviour, they act together in an additive manner. When they are presented after each-other, they gradually increase the behavioural tendency of the animal, and finally the behaviour is released (so-called temporal summation).

Heiligenberg (1974) postulated, that the difference between the ethological concepts of "motivation" and "release" is partly a question of the temporal relationship between stimulus and performance of behaviour. According to Heiligenberg, we tend to say that the motivation for a certain behaviour has changed when we have a long-term change in the probability of it being observed. On the other hand, when there is a short and distinct change in probability, we tend to say that the behaviour has been released. But in fact, there is a continuum of time-probability effects, which makes the dichotomy less likely to be a reflection of true processes in the animal.

7.0 Vacuum Activities

Lorenz argued that behaviour can be exhibited in a vacuum, e.g. a starling catching non-existent insects. Critics have pointed out that one cannot be certain that a vacuum was present. A ray of light might have simulated an insect or the bird might have hallucinated one. Without splitting hairs over whether a subliminal stimulus was present, the observation suggests that, even in the absence of tissue need, deprivation of behavioural opportunity can strongly accentuate food-getting behaviour (Gallistel, 1980).

Within applied ethology, phenomena that would fit the concept of vacuum-activities have commonly been described. For instance, hens in battery cages frequently perform dust-bathing movement on the wire floor (Black and Hughes, 1974), tethered sows without straw perform elements of nest building before farrowing on the bare concrete floor (Baxter, 1982) and nest building movements are observed before egg laying in caged layers

(Duncan, 1970). Sham chewing in tethered pregnant sows may also be interpreted as vacuum movements.

8.0 Welfare Implications

We dispute Baxter's (1983) claim that behaviour is merely something that arbitrarily serves end-states. Baxter suggested that we use the term "psychological needs" (or "cognitive needs", Baxter, (1989)), to emphasize that behaviour serves only the fulfillment of the internal representation, which constitutes the actual need. According to Baxter, a term like "behavioural need" (introduced by Thorpe in the Brambell report, (Command Paper 2836 (1965)), would be misleading.

We instead suggest, that in many cases, control of the motor program of an animal may to some extent be emancipated from the tissue needs. The cause of the behaviour, and the factors that stop it, are usually dependent on a combination of internal and external factors, acting together. Thus, an animal may well be said to possess a behavioural need, meaning that the performance of a species- and motivation-specific behaviour may be necessary for reduction of the motivation controlling this particular behaviour. This closely resembles the conclusion reached at a workshop on behavioural needs: "Animals have behavioural needs, that arise when an animal is motivated. Welfare problems arise when an animal is motivated to perform behaviour which, because of an unsuitable environment, it cannot." (Hughes, 1988). To this, one may add, that welfare problems may also arise when the animal can perform the normal behaviour, but the functional consequence of the behaviour is impossible to achieve. For instance, Baxter (1982) observed one tethered sow performing so much rooting movement on the concrete floor during the nest-building period before farrowing, that the nose disc was bleeding. The frequency of this behaviour was strongly increased compared to loose housed sows with straw, indicating that turning off nest building was prevented. As another example, hens deprived of dust have an impaired plumage (Van Liere and Bokma, 1987). In spite of the fact that the hens are able to perform dust-bathing movements, this does not provide the functional consequences of cleaning of the feathers.

Hughes, Duncan and Brown (1989) note that it might be insufficient to provide the consequences of behaviour and the animal might have a *need* to perform the behaviour that would normally lead to the consequences. Duncan and Kite (1989) report "severe frustration" shown by hens in battery cages that are unable to perform nesting in the pre-laying period, arguing that it is important to the hen to have a hollow so as to be able to express such behaviours as crouching and scraping the feet out.

For an abnormal behaviour such as cross-suckling in calves, we suggest the relevance of species-typical and motivation-typical behaviour. The feeding situation is abnormal, involving high nutrient yield for little behaviour. The positive reinforcement inherent in behaviour might then gain some autonomy from its normal nutrient association and the animal might get locked into a positive feedback loop. Since both stereotyped bar-biting in sows and cross-suckling in calves are more frequent after feeding (de Passilleé *et al* 1990; Jensen, 1988) and at least bar-biting has been shown to be related to the low feeding level usually applied with pregnant sows (Appleby and Lawrence, 1987), this may perhaps be a fruitful approach.

9.0 General Discussion

Lorenz' model serves to remind us that in some motivational systems, but not necessarily all, factors endogenous to the nervous system can account for some of the variability of behaviour. To what extend one feels that a hydraulic model is a useful representation of such factors is perhaps a matter of taste. We might prefer individual explanations rather than a generic analogy. For example, with our understanding of biological clocks (Kavanau and Rischer, 1968) and cognitive mechanisms, we might prefer to explain running-wheel activity in terms of clocks and patrolling in terms of updating a cognitive representation.

Some confusion arises from different levels of analysis of motivational systems. The behavioural 'black-box' type of model of Lorenz does not necessarily map neatly onto underlying neural structure (Cf. Hinde, 1960). Allied to this are differences as to exactly how much of the system should be described by the term 'motivation', all of the tendency to express a particular behaviour or part? To a contemporary psychologist, motivation would most likely refer to a c.n.s. property that can (a) be revealed in a variety of flexible goal-directed behaviours, recruiting various motor programmes, and (b) compete for expression with other motivations by inhibiting them (Bindra, 1976; Toates, 1986). Does motivation increase in strength with deprivation, as a result of intrinsic nervous system processes and is it lowered by behaviour *per se?* (1) This appears to happen for some motivations. (2) There is neural circuitry that organizes some specific actions that are normally recruited in the service of a motivational system. This circuitry can spontaneously recover and hence contribute to the expression of the behaviour (Cf. Sinclair, 1978). Whether one considers the latter contribution to be one of increasing motivation is somewhat a matter of linguistic taste. However, if, in McFarland's (1974) terms, *causal factor space* defines all of the causal factors that influence a decision to act, then the state of recovery of certain neural circuits or time since last acting would need to be included.

From an adaptive perspective, why should behaviour *per se,* rather than simply incentive contact, form part of reinforcement? (1) It could be adaptive for a mechanism to exist that strengthens an association between, on the one hand, a species-typical and motivation-typical behaviour and, on the other, incentive commerce that results from its expression. (2) Some species-typical behaviours need to be expressed early and before any benefit of trial-and-error experience with incentives can be derived. A young bird needs to peck and only later might form an association between pecking and the gain of nutrients (Hogan, 1977). Pecking in young chicks develops in skill and precision without any positive associations between pecking and the gain of food being necessary (Hess, 1956). A frightened rat could be well served by an innate tendency to freeze (Bolles, 1970). If such behaviours fail to realize appropriate effects or if other factors do not serve to maintain the motivational significance of the context, then their reinforcement value might be expected to decrease. Thus Rhoad, Kalat and Klopfer (1975) found that repeated elicitation of aggression in Siamese fighting fish becomes aversive if the display fails to effect a change in the eliciting situation. Placed on extinction conditions and never given free food in the Skinner box, a pigeon will ultimately stop pecking the key. If behaviour *per se* were too strongly reinforcing the animal could get locked inextricably into a positive feedback loop (Cf. Sinclair, 1978). This might indeed happen in abnormal behaviour divorced from normal incentive contact.

Why do some behaviours have a strong endogenous component and post-deprivation rebound (e.g. dustbathing), whereas others (e.g. aggression) lack this feature? Why is dustbathing intrinsically programmed to appear at intervals? Possibly the ability to detect changes in the feathers that might cue such behaviour is not so reliable. By contrast, one might expect the tendency to aggression not to build up as a function of deprivation and to be lowered by performing aggression. Aggression is only adaptive if there is some threat. Any threat can presumably be detected through sensory channels, is unlikely to oblige by appearing at regular intervals and might reappear very quickly. However, one would also expect that, placed in an aversive state, aggression would have positive reinforcement value (Glickman and Schiff, 1967). Therefore we reject the idea of an intrinsic increase in tendency towards aggression while giving an important qualification to the idea with which it is so often associated that aggression is positively reinforcing.

The argument that something is reinforcing usually leads to the argument that the behaviour shows an appetitive phase. It might be argued that there is no true appetitive phase of aggression even in the case of the Siamese fighting fish swimming through a hoop for the reward of its image. One might argue that the training procedure classically conditions aggressive behaviour and that the fish is simply swimming up a gradient of cues each more powerfully eliciting aggressive motivation. However, a similar argument could be made for any other operant situation and its reinforcer (Bindra, 1976; Hogan and Roper, 1978). As Hogan and Roper so aptly express it:

> "At this point the argument becomes semantic, and one's opinion seems to be a matter of taste".

Is motivation more or less internally or externally caused (Hughes, 1980)? This discussion can be as misleading as arguing whether a behaviour depends more upon genetics or the environment (Hogan and Roper, 1978). It is misleading to attach weightings of importance to a factor. However, one can argue as to the source of *variation* in behaviour. For example, both feeding and aggression depend upon external and internal factors; the variance in each might be explicable in terms of variations in either factor. Given a standard and undisturbed laboratory cage with food *ad lib*, variations in feeding might be explained almost entirely in terms of variations in internal state. At other times, feeding might be aroused in a 'satiated' rat by presenting a cue paired in the past with food presentation (Weingarten, 1984), and variability in behaviour could be attributable to change in the external world. One imagines that in most cases the variance in fear and aggression is to be explained by variations in the environment. In some cases though variations in an internal factor, e.g. seasonal fluctuations in testosterone levels, would play a role in variations in aggression. The scale developed by Hughes (1980), running from external to internal weighting of responsibility might therefore be conceived in terms of the normal source of variability. Focussing upon variations in the internal contribution to motivation, these might arise from tissue needs (e.g. temperature control) or from factors intrinsic to the nervous system (e.g. dustbathing) or a combination (e.g. feeding).

In the Hughes/Duncan model, behaviour is shown as exerting positive feedback on motivation (Hughes and Duncan, 1988a). Two of the examples that they give (feeding and copulation) do not enable us to conclude that behaviour *per se* exerts positive feedback since their consequences act through conventional reward pathways (oral and genital stimulation). On the other hand, neither are they incompatible with the idea that a component of positive feedback is generated intrinsically. However, their argument that abnormal behaviours not having functional consequences, such as bar biting in pigs and tongue playing in calves, represent the animal getting locked into an inextricable positive feedback loop does imply that behaviour *per se* exerts positive feedback upon motivation. They suggest that: "Welfare considerations will arise in situations when motivation reaches high levels, yet the consummatory behaviour cannot be performed, nor does the environment provide the appropriate functional consequences" and note that this is diametrically opposed to Baxter's (1983) conclusion that there is no behavioural need to behave.

References

Anisman,H. (1978) Neurochemical changes elicited by stress. In *Psychopharmacology of Aversively Motivated Behaviour*, (eds. H.Anisman and G.Bignami), New York, Plenum Press, pp.119-172.

Appleby,M.C. and Lawrence,A.B. (1987) Hunger as a cause of stereotyped behaviour in tethered sows. *Applied Animal Behaviour Science*, 17, 377.

Archer,J. (1988) *The Behavioural Biology of Aggression.* Cambridge University Press, Cambridge.

Azrin,N.H., Hutchinson,R.R. and McLaughlin,R. (1965) The opportunity for aggression as an operant reinforcer during aversive stimulation. *Journal of the Experimental Analysis of Behaviour*, 8, 171-180.

Baxter,M.R. (1982) The nesting behaviour of sows and its disturbance by confinement at farrowing. *Hohenheimer Arbeiten*, 121, 101-114.

Baxter,M.R. (1983) Ethology in environmental design. *Applied Animal Ethology*, 9, 207-220.

Baxter,M.R. (1989) Philosophical problems underlying the concept of welfare. *Proceedings of the 3rd International Symposium on Poultry Welfare*. Tours, France, 23-33.

Bindra,D. (1976). *A Theory of Intelligent Behaviour*. Wiley, New York.

Black,A.J. and Hughes,B.O. (1974) Patterns of comfort behaviour and activity in domestic fowls: A comparison between cages and pens. *British Veterinary Journal*, 130, 23-33.

Blackburn,J.R. and Pfaus,J.G. (1988) Is motivation really modulation? A comment on Wise. *Psychobiology*, 16, 303-304.

Bolles,R.C. (1970) Species-specific defense reactions and avoidance learning. *Psychological Review*, 77, 32-48.

Cabanac,M. (1971) Physiological role of pleasure. *Science*, 173, 1103-1107.

Castren,H., Algers,B. and Jensen,P. (1989) Occurrence of unsuccessful sucklings in newborn piglets in a semi-natural environment. *Applied Animal Behavioural Science*, 23, 61-73.

Cherek,D.R., Thompson,T. and Heistad,G.T. (1973) Responding maintained by the opportunity to attack during an interval food reinforcement schedule. *Journal of the Experimental Analysis of Behaviour*, 19, 113-123.

Cohen,P.S., Looney,T.A., Campagnoni,F.R. and Lawler,C.P. (1985) A two-state model of reinforcer-induced motivation. In *Affect, Conditioning and Cognition: Essays on the Determinants of Behaviour*, (eds. F.R.Brush and J.B. Overmier), Lawrence Erlbaum, Hillsdale, pp.281-297.

Command Paper 2836 (1965) *Report of the Technical Committee to Enquire into the Welfare of Animals kept under Intensive Livestock Husbandry Systems*. London, HMSO.

Conner, R.L., Vernikos-Danellis,J. and Levine,S. (1971) Stress, fighting and neuroendocrine function. *Nature*, 234, 564-566.

Dantzer,R. and Mormede,P. (1981) Pituitary-adrenal consequences of ajunctive activities in pigs. *Hormones and Behaviour*, 15, 386-395.

Dawkins,M.S. (1980) *Animal Suffering*. Chapman and Hall, London.

Dawkins,M.S. (1986) *Unravelling Animal Behaviour*. Longman, Harlow.

Duncan,I.J.H. (1970) Frustration in the fowl. In B. M. Freeman and R. F. Gordon (eds.) *Aspects of Poultry Behaviour*, British Poultry Science, Edinburgh.

Duncan,I.J.H. and Kite,V.G. (1989) Nest site selection and nest-building behaviour in domestic fowl. *Animal Behaviour*, 37, 215-231.

Ely,D.L. and Henry,J.P. (1978) Neuroendocrine response patterns in dominant and subordinate mice. *Hormones and Behaviour*, 10, 156-169.

Epstein,A.N. (1982) Instinct and motivation as explanations for complex behaviour. In: D.W.Pfaff (ed.) *The Physiological Mechanisms of Motivation*, Springer, New York, pp.25-58.

Gallistel,C.R. (1980) *The Organization of Action: A New Synthesis*. Lawrence Erlbaum, Hillsdale.

Gardner,R.A. and Gardner,B.T. (1988) Feedforward versus feedbackward: An ethological alternative to the law of effect. *The Behavioural and Brain Sciences*, 11, 429-493.

Glickman,S.E. and Schiff,B.B. (1967) A biological theory of reinforcement. *Psychological Review*, 74, 81-109.

Grossman,S.P. (1967) *A Textbook of Physiological Psychology*. John Wiley, New York.

Guile,M.N. and McCutcheon,N.B. (1980) Prepared responses and gastric lesions in rats. *Physiological Psychology*, 8, 480-482.

Heiligenberg,W. (1974) A stochastic analysis of fish behaviour. In D.J. McFarland (ed.) *Motivational Control Systems Analysis*, Academic Press, London, 87-118.

Hemsworth,P.H., Beilharz,R.Z. and Galloway,D.B. (1977) Influence of social conditions during rearing on the sexual behaviour of the domestic boar. *Animal Production*, 24, 245-251

Hemsworth,P.H., Salden,N.T.C.J. and Hoogerbrugge,A. (1982) The influence of the post-weaning social environment on the weaning to mating interval of the sow. *Animal Production*, 35, 41-48.

Henry, J.P. (1976) Mechanisms of psychosomatic disease in animals. *Advances in veterinary Science and Comparative Medicine*, 20, 115-145.

Herrnstein,R.J. (1977) The evolution of behaviourism. *American Psychologist*, 32, 593-603.

Hess,E.H. (1956) Space perception in the chick. *Scientific American*, 195, 71-80.

Hinde,R.A. (1959) Unitary drives. *Animal Behaviour*, 7, 130-141.

Hinde,R.A. (1960) Energy models of motivation. *Symposium of the Society of Experimental Biology*, 14, 199-213.

Hinde,R.A. and Steel,E. (1972) Reinforcing events in the integration of canary nest-building. *Animal Behaviour*, 20, 514-525.

Hogan,J.A. (1967) Fighting and reinforcement in the Siamese fighting fish *(Betta splendens)*. *Journal of Comparative and Physiological Psychology*, 64, 356-359.

Hogan,J.A. (1977) Development of food recognition in young chicks. IV. Associative and nonassociative effects of experience. *Journal of Comparative and Physiological Psychology*, 91, 839-850.

Hogan,J.A. and Roper,T.J. (1978) A comparison of the properties of different reinforcers. *Advances in the Study of Behaviour, Vol. 8*, Academic Press, New York, pp.155-255.

Hughes,B.O. (1980) The assessment of behavioural needs. In: *The Laying Hen and its Environment.* (ed. R.Moss), The Hague, Martinus Nijhoff, pp.149-166.

Hughes,B.O. (1988) Discussion 3. Proceedings of Workshop on behavioural needs in farm animals. *Applied Animal Behavioural Science*, 19, 339-386.

Hughes,B.O. and Duncan,I.J.H. (1988a) Behavioural needs: can they be explained in terms of motivational models? *Applied Animal Behaviour Science*, 19, 352-355.

Hughes,B.O. and Duncan,I.J.H. (1988b) The notion of ethological 'need', models of motivation and animal welfare. *Animal Behaviour*, 36, 1696-1707.

Hughes,B.O., Duncan,I.J.H. and Brown,M.F. (1989) The performance of nest building by domestic hens: is it more important than the construction of a nest? *Animal Behaviour*, 210-214.

Jensen,P. (1988) Diurnal rhythm of bar-biting in relation to other behaviour in pregnant sows. *Applied Animal Behaviour Science* 21, 337-346.

Kavanau,J.L. and Rischer,C.E. (1968) Program clocks in small mammals. *Science*, 161, 1256-1259.

Kenny,J.T., Stoloff,M.L., Bruno,J.P. and Blass,E.M. (1979) Ontogeny of preference for nutritive over non-nutritive suckling in albino rats. *Journal of Comparative and Physiological Psychology*, 93, 752-759.

Levine,S., Weinberg,J. and Brett,L.P. (1979) Inhibition of pituitary-adrenal activity as a consequence of consummatory behaviour. *Psychoneuroendocrinology*, 4, 275-286.

Lorenz,K.Z. (1950) The comparative method in studying innate behaviour patterns. *Symposium of the Society for Experimental Biology*, 4, 221-268.

Lorenz,K.Z. (1978) *Vergleichende Verhaltensforschung: Grundlagen der Ethologie*. Springer, Wein.

Lorenz,K.Z. (1981) *The Foundations of Ethology*. Springer, New York.

McFarland,D.J. (1974). *Motivational Control Systems Analysis*. Academic Press, London.

McFarland,D.J. and Sibly,R.M. (1975) The behavioural final common path. *Proceedings of the Royal Society of London*, 270, 265-293.

Nicol,C.J. (1987) Behavioural responses of laying hens following a period of spatial restriction. *Animal Behaviour*, 35, 1709-1719.

de Pasilleé,A.M.B., Metz,J.H.M., Mekking,P. and Wiepkema,P.R. (1990) Does milk stimulate the calf to suck? *Proceedings of the Society for Veterinary Ethology Summer Meeting*, Montecatini, Pistoia, p.10.

Polsky,R.H. (1975) Hunger, prey feeding and predatory aggression. *Behavioural Biology*, 13, 81-93.

Premack,D. (1965) Reinforcement theory. In *Nebraska Symposium on Motivation* (ed. D.Levine) University of Nebraska Press, Lincoln, pp.123-180.

Rhoad,K.D., Kalat,J.W. and Klopfer,P.H. (1975) Aggression and avoidance by *Betta splendens* toward natural and artificial stimuli. *Animal Learning and Behaviour*, 3, 271-276.

Seitz,A. (1940) Die Paarbildung bei einigen Cichliden, *Zeitschrift fur Tierpsychologie*, 4, 40-84.

Sinclair,J.D. (1978) A theory of behaviour, based on "rest principle", control of the strength of neural connections. *Neuroscience and Biobehavioural Reviews*, 2, 357-366.

Slater,P.J.B. (1985) *An Introduction to Ethology*. Cambridge University Press, Cambridge.

Stephens,D.B. (1975) Effects of gastric loading on the sucking response and voluntary milk intake in neonatal piglets. *Journal of Comparative and Physiological Psychology*, 88, 796-805.

Stolba,K. and Wood-Gush,D.G.M. (1989) The behaviour of pigs in a semi-natural environment. *Animal Production*, 48, 419-425.

Stolk,J.M., Conner,R.L., Levine,S. and Barchas,J.D. (1974) Brain norepinephrine metabolism and shock-induced fighting behaviour in rats: Differential effects of shock and fighting on the neurochemical response to a common footshock stimulus. *The Journal of Pharmacology and Experimental Therapeutics*, 190, 193-209.

Tingergen,N. (1953) *The Study of Instinct*. Oxford University Press, London.

Toates, F. (1986) *Motivational Systems*. Cambridge University Press, Cambridge.

Van Liere,D.W. and Bokma,S. (1987) Short-term feather maintenance as a function of dust-bathing in laying hens. *Applied Animal Behavioural Science*, 18, 197-204.

Vestergaard,K. (1980) The regulation of dustbathing and other behaviour patterns in the laying hen: A Lorenzian approach. In: *The Laying Hen and its Environment.* (ed. R.Moss), The Hague, Martinus Nijhoff, pp. 101-113.

Weinberg,J., Erskine,M. and Levine,S. (1979) Shock-induced fighting attenuates the effects of preshock experience in rats. *Physiology and Behaviour*, 25, 9-16.

Weingarten,H.P. (1984) Meal initiation controlled by learned cues: basic behavioural properties. *Appetite*, 5, 147-158.

Weiss,J.M., Pohorecky,L.A., Salman,S. and Gruenthal,M. (1976) Attenuation of gastric lesions by psychological aspects of aggression in rats. *Journal of Comparative and Physiological Psychology*, 90, 252-259.

Williams,R.B. and Eichelman,B. (1971) Social setting: Influence on the physiological response to electric shock in the rat. *Science*, 174, 613-614.

Wise,R.A. (1987) Sensorimotor modulation and the variable action pattern (VAP): Toward a noncircular definition of drive and motivation. *Psychobiology*, 15, 7-20.

PLEASURE: THE ANSWER TO CONFLICTING MOTIVATIONS

Michel CABANAC
Département de physiologie
Faculté de médecine
Université Laval
Québec, GIK 7P4
Canada
email: CABANAC@LAVALVM1

Abstract

An experiment has been set up to explore the hypothesis according to which the solution of conflicts of motivations is reached by the trend to maximize pleasure. Subjects were placed in a situation of conflict where the pleasure of playing a videogame clashed with the increasing discomfort of a cold environment. The time lapse tolerated by each of the subjects could be predicted from the algebraic sum of the rating of displeasure aroused by the cold environment and the rating of pleasure aroused by the videogame, obtained in different sessions. This result supports the working hypothesis and perrmits the conclusion that pleasure is the common currency which allows tradeoffs among various motivations.

1. Models.

One may identify two types of models those describing the behavior of a given system and those describing the system itself. The first type of model is pragmatic. A good behavioral model is adequate enough when it replicates the behavior, and is able to predict future behavioral responses of the system. Yet this may be achieved without any knowledge of the intimate mechanisms that produce the behavior. In the same way as a given envelope function can be approximated with various summations of different functions, a model may ignore the inside of the black-box system whose behavior it nevertheless replicates adequately. The model may lump several functions into one, incorporate approximations, ignore some rare or extreme conditions but remain nevertheless a good model.

The second type of model is more ambitious and aims at theoretical description of the system. This type of model is more difficult to achieve because it includes the intimate laws of the system and because it implies the knowledge of the structures and functions of the constituent elements of the system. Eventually this second type also reaches the same goal as models of the first type.

In an enlightening chapter on instinct and motivation Epstein (1982) has recently reflected that there are behaviors which are simple reflexes in the Cartesian and Sherringtonian acceptation of the word, but there are also complex behaviors. The latter occur with the cooperative action of an endogenous component, an acquired component, and a reactive component. He stressed that «we need concepts that take account of the complexities of behaviors that are not reflexive». Theoretical models of complex behavior must include these three components. This implies the recognition and the incorporation of emergent properties and functions within the central nervous system, *i.e.* functions which cannot be predicted from the sum of properties of the elements of the central nervous system itself. Excellent theoretization on this point will be found in Toates (1986a).

Some theoretists are reluctant to admit the notion of emergence either because they believe that the properties of the whole CNS should be predicted eventually from the properties of its elements (Changeux, 1983), or because cognitive concepts are less phenomenological and thus less testable (Rachlin *et al.*, 1986). Yet emergent functions of the brain can be compared to the mathematical functions of a pocket calculator. It is far more simple to program in the calculator an additive or multiplicative function than to accumulate in memory the infinite number of possible combined additions and multiplications of all rational numbers. For the outside observer the result would be the same whether the calculator uses one method or the other to solve a problem. Yet, at a modest level of complexity, additive and multiplicative functions are emergent functions of the pocket calculator.

This present chapter aims at drawing the attention of model makers from the fields of robotics and artificial intelligence to the realm of cognitive psychology, and suggest that the best way to replicate natural complex behavior consists in including mentalistic concepts in their theoretical models. Among these mentalistic concepts pleasure is quite important to understand motivation at large (Cabanac, 1971), especially in the case of decision making. Pleasure has the advantage of involving Epstein's (1982) endogenous, acquired, and reactive three components.

2. Pleasure.

A basic postulate of Ethology is that animals rank their priorities at each instant and tend to select them so as to satisfy their most urgent motivation (Tinbergen 1950; Baerends, 1956). One shortcoming of the theories of the optimalization of behavior is that the mechanisms by which behavior is optimized, is never mentioned. In other words, these theories do not explain how the subject «decides». This is reasonable when the object studied is animal behavior with all the difficulties involved in exploring animal cognition. With humans, however, it is possible to go further than the mere measurement of behavior and to open the black box of cognitive motivation. One may explore quantitatively the relationship between the motivation thus analyzed, the behavior displayed by the subject, and the environmental parameters. This general method has been applied to the analysis of motivations leading to behaviors with a clear physiological aim such as thermoregulatory behavior, fatigue, and food intake (Cabanac, 1971). The results have demonstrated the paramount importance of sensory pleasure in the determinism of these behaviors.

One may question whether it is possible to extend this conclusion to domains other than biology and to less carnal motivations. The notion of «behavioral final common path» is especially enlightening from this perspective. Paraphrasing the image of the motoneuron final common path of all motor responses, McFarland and Sibly (1975) pointed out that behavior is also a final common path on which all motivations converge. This image incorporates all motivations into a unique category since behavior must operate in time-sharing to satisfy not only physiological motivation but also social, moral, esthetical, ludic motivations, etc. Indeed, it is often the case that behaviors are mutually exclusive: one cannot work and sleep at the same time. Therefore, the brain, responsible for the behavioral response, must rank priorities and determine tradeoffs in the decisions concerned with allocating time among competing behaviors. If it can be accepted that, at each instant, a subject responds to the most urgent motivation, then it can be expected that the brain operates this ranking by using a common currency.

3. Pleasure as the common currency.

The thesis presented here is that pleasure is this common currency and that the maximization of multidimensional pleasure, and the minimization of multidimensional displeasure, leads to the solution of motivational conflicts. This was verified in a situation of conflict where a biological motivation, cold discomfort, was pitted against a purely psychological one, the pleasure of playing a videogame.

The general principle of the experiment consisted in exposing each subject to three sessions: in one session (play) the subject played a videogame for one hour in a climatic chamber set at 25°C, and every five minutes was requested to rate on a magnitude estimation scale, the pleasure of playing the videogame; in the second session (cold) the ambient temperature was reduced progressively from 25°C to 7°C over one hour, and every five minutes the subject was requested to rate the pleasure or the displeasure aroused by ambient temperature. No anchor was given for the ratings; the only instruction received by the subjects was to use positive figures for pleasure and negative figures for displeasure. Pleasure is a state of mind and, as such, cannot be defined otherwise than tautologically but the subjects understood the word and did not report having problems in rating their pleasure.

In the third session (conflict) both videogame and cold ambient temperature were presented simultaneously and the only instruction received by the subjects was that they could interrupt the session whenever they wanted to. During all sessions, including the conflict session, heart rate, systolic and diastolic arterial pressure, tympanic temperature, skin temperature of the hand and the back were recorded every five minutes from the half-naked subjects. The sequence of three sessions combined six possibilities e.g. play - cold - conflict, play - conflict - cold, cold- play - conflict, etc. To suppress any possible influence on the group mean of the sequence of sessions, each of the six possible sequences was used in one female and one male subject for a total of 12 subjects. Informed consent was obtained after the nature and possible consequence of the study had been fully explained. After debriefing, no subject had guessed that the main aim of the experiment was to measure the duration tolerated during the conflict session. Most subjects believed that the aim was to compare their autonomic response in the cold with or without the videogame; the others believed that the aim was the measurement of their performance in a cold or a warm environment.

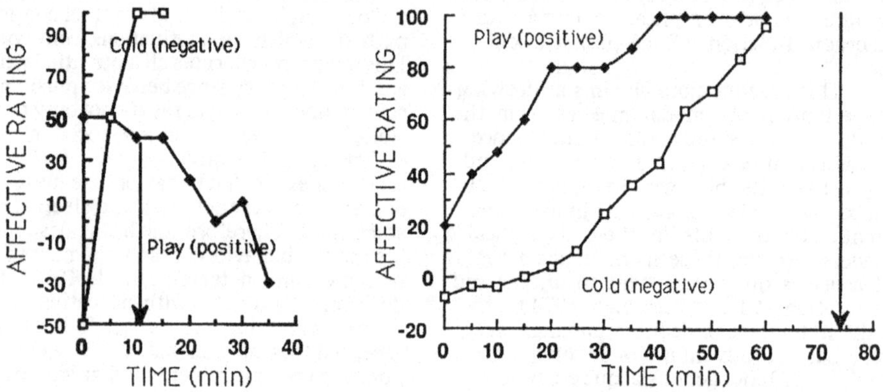

Figure 1. Results obtained on three separate sessions by the subjects who stayed the shortest (left) and longest (right) duration on the third session. The positive ratings of videogame (pleasure) were obtained in one session (♦). The ratings related to the cold environment (□) started positive (pleasure) then turned negative (displeasure) after a few minutes; they were obtained in another session. To make the comparison of both ratings more obvious and let the curves cross each other the cold ratings have been multiplied by -1 in the figure. On the third session, both videogame and cold environment were present and the arrows show the actual duration endured by the subjects (from Cabanac, 1989).

The main result of the experiment was the duration of the stay of the subjects in the climatic chamber in the conflict situation. This is shown by the arrows on figure 1 for the subjects with the shortest and longest durations in conflict sessions. In addition, this figure shows the ratings given to the pleasure of playing and to the displeasure of cold environment obtained in different sessions plotted against time. This figure gives the time course of positive ratings for the pleasure of playing and negative ratings for the displeasure of cold discomfort. None of the autonomic variables monitored correlated with ratings in such a way as to predict the duration of the conflict session. These variables may therefore be rejected as predictors of duration tolerated in the cold environment during the conflict session. The hypothesis under test states that at each instant subjects tend to maximize the algebraic sum of their pleasures. To hold true the hypothesis under test would imply that the subjects stayed in the climatic chamber during the conflict sessions as long as the pleasure of playing the videogame was stronger than the displeasure aroused by the decreasing ambient temperature. This can be verified first by the two examples of figure 1. It can be expected that the subjects would end the conflict session when the sum of the rating obtained in cold sessions plus that obtained in play sessions equalled zero, i.e. when the curves for the rating of cold displeasure crossed the curve for the rating of playing pleasure. For the subject shown on the left the arrow shows that actual duration was about six minutes longer than the time when both curves crossed each other. The subject shown on the right was a particular case: the ratings she gave during the 60 min sessions of play and cold did not cross each other. This was confirmed by her long stay in the conflict session: it can be seen that she stayed 74 minutes during the conflict session i.e. about 9 min beyond the extrapolated time of equality of the two curves. Thus, both subjects shown in figure 1 stayed somewhat longer than the predicted time, but the basic hypothesis remains valid since the error on actual duration was + 6 min and + 9 min out of a span of a theoretical duration extending from 5 min (left) to more than 60 min (right). It should be pointed out that the examples picked out for figure 1 were not the best in terms of matching theoretical and actual duration, but simply the shortest and longest durations in conflict sessions. Figure 2 shows the results of the group of 12 subjects. Actual duration, considered to be the dependent variable, is plotted against theoretical duration, considered to be the

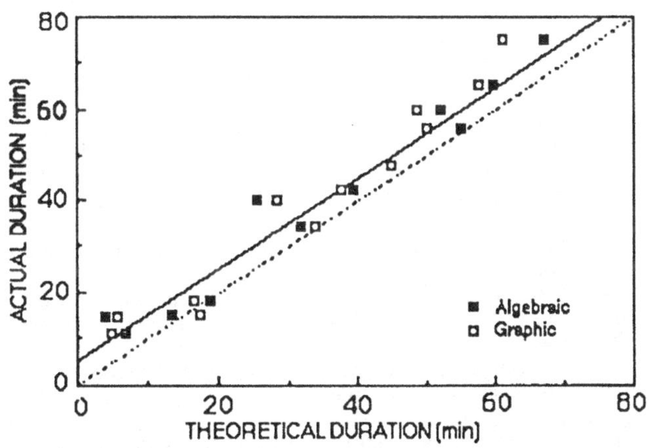

Figure 2) Actual duration endured by the 12 subjects, considered the dependent variable, plotted against theoretical duration, considered the independent variable. Theoretical duration was obtained through two different methods: graphic was the first crossing of the respective time courses of positive rating of the pleasure of playing, with negative rating of the displeasure of cold discomfort; algebraic was obtained in solving the family of 2 equations: y' = (rating play) and y = (rating cold) with y' = y. The dotted line gives the hypothetical case where actual duration equals theoretical. The solid line gives the regression of the actual duration plotted against the theoretical duration with the algebraic method (y = 0.993 x + 5.3, r = 0.977). The regression of the actual duration plotted against the theoretical duration with the graphic method is not represented. It can be seen that the actual duration for the group of subjects was the theoretical duration plus 5.3 min (from Cabanac, 1989).

independent variable. Theoretical duration was that predicted through two different methods:

i) graphic: theoretical duration obtained from the simple time course plot, as the time when positive rating was equal to negative rating as in figure 1;

ii) algebraic: theoretical duration obtained from the family of two equations fitted to the time courses of both ratings when their sum was equal to zero.

Both methods gave similar results which are plotted on figure 2 for the 12 subjects. The respective regression lines were: algebraic, y= 0.993 x + 5.29 with r= 0.977, and graphic, y = 1.093 x + 2.91 with r = 0.976. The coefficient of correlation as well as the slope was close to 1 whatever the method to obtain theoretical duration. From the ratings of pleasure and displeasure, it was therefore possible to predict the actual duration of the session in situation of conflict.

This result validates the hypothesis tested. It is therefore permitted to conclude that pleasure was the common currency of the motivations here pitted against each other. This extends to ludic motivation the conclusions made previously mainly on sensory pleasure and resulting behaviors with physiological aims such as temperature regulation (Cabanac and Leblanc, 1983), muscular work (Cabanac, 1986), or food intake (Cabanac and Ferber, 1987).

The intersect of the regression line with the y axis in figure 2 was y = 5.3 minutes and would be 2.9 minutes with graphic theoretical duration. This indicates that the subjects did not stop the session at the time when both motivations, to stay in the climatic chamber and to leave it, became equal but some minutes later. Such a delay necessary for the subjects to make up their minds, would tend to confirm the hypothesis that the subjects made their decision from the algebraic sum of pleasure/displeasure aroused in both dimensions, for, at the time of equal magnitude for both motivations, there was no more reason for a subject to remain in the climatic chamber, but there was none either to leave it. The few minutes of overstay can be understood as the time for the cold discomfort to become stronger than the pleasure of playing.

If maximization of the algebraic sum of pleasures aroused by behaviors, as different from each other as thermoregulatory behavior and playing a game, was the key to the conflict, it is likely that this result may be extrapolated to all motivations as a general law predicting behavior.

4. Animals also.

The central place of pleasure as motivating animal behavior has been defended long ago by Young (1959), and is now recognized in experimental (Waldbillig and O'Callaghan, 1980; Baldwin, 1985; Mehiel and Bolles, 1988; Bedard and Weingarten, 1989) and field (Hladik, 1977) observations, or from theoretical considerations (Toates, 1986b). The behavior displayed by the animals has been described also in terms of common currency. However the model produced to describe behavioral decisions related to lifetime fitness incorporated costs most of which had no biological interpretation (McNamara and Houston, 1986).

Is it possible to obtain evidence that animals will seek sensory pleasure just for the sake of it and even will be ready to trade off some amount of displeasure for it? In the obstruction method used by Warden (1931) rats could reach a bait at the cost of stepping on an unpleasant electrified mesh floor. In a more natural situation rats were trained to feed in a nest placed at one end of a 16 m zigzag alley (Cabanac and Johnson, 1983). The nest was provided with water and food in excess of the rats' needs. On the day of an experimental session the nest was heated and a palatable food such as shortbread was placed 16 m away, but this time the environment outside the warm nest was refrigerated down to -15°C, a temperature potentially lethal to rats. Invariably the rats ran to the cold feeder to obtain the highly palatable food. They faced the unpleasant and even painful ambient cold not out of necessity, since chow was provided at no cost in their warm nest, but for the pleasure of ingesting a palatable bait. The number of trips to the cold feeder was related to the palatability of the bait. When chow was placed in the cold feeder instead of shortbread the rats went to it once and did not return. Therefore the rats matched quantitatively the pleasure of food with the displeasure of enduring cold discomfort.

This shows that animal behavior seems to follow the same law as human behavior in situations of conflicting motivations involving decision making.

5. Conclusions.

The experimental method has been widely used before to explore cognition in situations of choice (Einhorn and Hogarth, 1981; Tversky and Kahneman, 1986). Most often the method consisted in using human subjects exposed to verbally couched hypothetical problems involving winning and losing, and in recording the subjects' choices. The experiment reported here was different in two ways: 1) The subjects reported their own pleasures or displeasures rather than reporting the result of a theoretical gambling; 2) these reports could be compared with the subjects' own behaviors. Yet the results confirm that the use of a mentalistic concept such as pleasure permits a good description and understanding of the mechanisms that lead to conation and to choice in a conflict of motivations. Animals also feel pleasure and displeasure and there is evidence that they are ready to trade off some amount of displeasure for the sake of increasing their pleasure. Models of human and animal behavior should therefore aim not only at describing behavior but also at incorporating this concept.

Acknowledgement.

The research reported in this chapter was supported by Conseil de la Recherche en Sciences Naturelles et en Génie, Canada.

References.

Baerends G. P., 1956. Aufbau des tierischen Verhaltens. Teil 10 (Lfg 7), VIII Mammalia, in W Kükenthal & T. Krumbach (eds), *Handbuch der Zoologie*, De Gruyter & Co, Berlin.

Baldwin B.A., 1985. Neural and hormonal mechanisms regulating food intake. *Proc. Nutr. Soc* . 44: 303-311.

Bedard M. and Weingarten H.P., 1989. Postabsorptive glucose decreases excitatory effects of taste on ingestion. *Amer. J. Physiol.* 256: R1142-R1147

Cabanac M., 1971. Physiological role of pleasure. *Science* 173: 1103-1107.

Cabanac M., 1986. Performance and perception at various combinations of treadmill speed and slope. *Physiol. Behav.* 38: 839-843.

Cabanac M., 1989. La maximisation du plaisir, réponse à un conflit de motivations. *C. R. Acad. Sci.* Paris 309: 397-402.

Cabanac M. and Ferber C., 1987. Pleasure and preference in a two-dimensional sensory space. *Appetite* 8: 15-28.

Cabanac M. and Johnson K.G., 1983. Analysis of a conflict between palatability and cold exposure in rats. *Physiol. Behav.* 31: 249-253.

Cabanac M. and LeBlanc J., 1983. Physiological conflict in humans: fatigue vs cold discomfort. *Am. J. Physiol* . 244: R621-R628.

Changeux J. P., 1983. *L'homme neuronal* Arthème Fayard, Paris.

Einhorn H. J. and Hogarth R. M., 1981. Behavioral decision theory: process of judgement and choice. *Ann. Rev. Psychol* . 32: 53-88.

Epstein A. N., 1982. Instinct and motivation as explanations for complex behavior, 25-58.*The physiological mechanisms of motivation* D. W. Pfaff (ed.) Springer-Verlag, New York.

Hladik C.M., 1977. A comparative study of feeding strategies of two sympatric species of leaf monkeys: *Presbytis senx Presbytis entellus,* 323-353*Primate ecology* T. H. Clutton-Brock (ed.) Academic Press, New York.

McFarland D. J. and Sibly R., 1975. M. The behavioural final common path. *Philos. Trans. Royal Soc.* (London) 270: 265-293.

McNamara J. M. and Houston A. I., 1986. The common currency for behavioral decisions. *Am. Naturalist* 127: 358-378.

Mehiel R. and Bolles R. C., 1988. Hedonic shift learning based on calories. *Bull. Psychon. Soc.* 26: 459-462.

Rachlin H., Logue A. W., Gibbon J. and Frankel M., 1986. Cognition and behavior in studies of choice. *Psychol. Rev* . 93: 33-45.

Tinbergen N., 1950. The hierarchical organization of mechanisms underlying instinctive behaviour. *Sympos. Soc. Exper. Biol* . 4: 305-312.

Toates F., 1986a. *Biological foundations of behaviour* . Open University Press, Milton Keynes.

Toates F., 1986b. *Motivational systems* . Cambridge University Press, Cambridge.

Tversky A. and Kahneman D., 1986. Rational choice and the framing of decisions. *J. Business* 59: S251-S278.

Waldbillig R. J. and O'Callaghan M., 1980. Hormones and hedonics, cholecystokinine and taste: A possible behavioral mechanism of action. *Physiol. Behav.* 25: 25-30.

Warden C. J., 1931. *Animal motivation, experimental studies on the albino rat.* Columbia University Press, New York.

Young P. T., 1959. The role of affective process in learning and motivation. *Psychol. Rev.* 66: 104-123.

MACHINE MOTIVATION

J. R. P. Halperin

Zoology Dept., Univ. of Toronto, Toronto, Can.,
fax: (416)-978-8532; email: janh@zoo.toronto.edu / uunet!attcan!utzoo!janh

Abstract

This paper presents a design for a sensory-motor interface to generate motivated behaviour and some capacity to learn, which may be useful in producing task-oriented autonomous robots. There is a single design for the basic sensory-motor interface. By varying sensor bias, parameter settings and type of sensory input one can generate many motivational phenomena observed in lower vertebrates. Chains of instinctive behaviours can be refined by the experience of whether a response to a particular stimulus is successful. The property of incentive motivation can be mimicked, as well as learning by punishment and drive-reduction learning. Throughout the discussion we attempt to point out some of the difficulties which might be encountered in designing bio-mimic autonomous robots.

There are three broad classes of behaviour which are generally referred to as motivated. In the first class, motivation is based on the reduction of physiological deficits. Hungers for calories, proteins and minerals are the obvious examples of these "drive-reduction" motivations. No deficit-correction is involved in the second category, where the motivations are "incentive"; however these incentive motivations involve stimuli which are inherently hedonic - either attractive or aversive. An example is the motivation to consume tasty non-caloric sugar substitutes like saccharin, which can reward behaviour quite effectively in hungry animals. Finally, there is motivation to perform sequences of displays or "fixed action patterns" with neither physiological deficit-correction nor obvious hedonic responses involved - for example the motivation of territorial animals to fight intruders.

This paper presents a model which was devised to simulate fighting behaviour in Siamese fighting fish (the last, least obvious class of motivation), and which turned out to model the other types of motivation as well. This model is so simple and explicit that it could be used as a design principle for building sensory-motor interfaces to control a range of motivated behaviours in autonomous robots.

The model has two basic elements: (1) a circuit diagram showing "neurons" connected by "synapses", some of which have variable strength (fig. 1), and (2) a connection strength change rule related to Hebb's rule, which controls the strengths of the synapses as a function of activity in the connected neurons (fig. 2).

We will describe the model in terms of an example rather than in abstracto, although a brief mathematical description is included for completeness at the end of the paper (fig. 4). A detailed mathematical model has been constructed (Halperin and Halperin, in prep.) and it forms the basis of a successful computer simulation (Halperin, Halperin, Rutherford and Dunham, in prep.). Since the fighting fish social behaviour which led to the development of the model would be difficult to visualize for anyone but ethologists, we will instead present the entirely hypothetical example of a scrap-collecting robot. The goal is to illustrate the principles and also to illustrate the problems which would have to be dealt with when using these ideas in a practical design context. Many of these problem are of the type which have been dealt with so successfully in the biological mimic robots built at MIT (e.g. Brooks and Flynn, 1989), and the success of these robots creates optimism that extensions to motivated bio-mimic robots can succeed.

The hypothetical example we will consider is a litter-collecting system for a park. There is a large, heavy collecting machine (or an expensive human to do the collecting) and a small army of

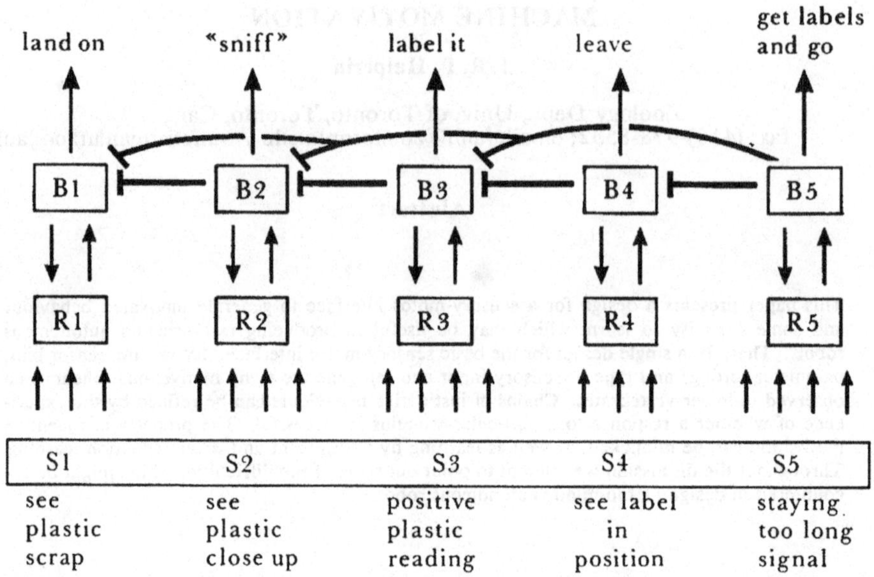

Figure 1. Basic wiring for the robot scrap-labelling interface, which permits the fine-tuning of the instinctive propensity to find plastic scraps and label them. Excitatory connections are arrows. Inhibitory connections are blunt-ended. Only the S to R connections have variable strength. Si represents the neuron pool feeding to Ri. In contrast with the R and B layers, these Si pools may overlap.

tiny robots all with one task - to recognize bits of litter and label them with miniaturized signal transmitters so the collector can easily and efficiently collect them. We will picture the robot as a mechanical fly which criss-crosses the park without bumping into trees, landing on plastic scrap and labelling it. The collector releases it, with many others, and they all zig-zag, half blown and half power-controlled, scanning for plastic.

The first step in designing a sensory-motor interface for our robot fly is to break its task into short sub-tasks. As shown in fig 1, for our robot fly these will be land, sniff, stick on the label and leave. There is an over-ride command to pick up the labels and leave if the robot has not successfully reached the initial "leave" stage within some pre-set time.

Figure 1 presents the interface circuit as it would be set up in the robot fly. There are three kinds of neurons. S neuron arrays are connected to unspecified sensory devices and can be thought of as responding to stimuli in the environment. They excite R neuron arrays which release behaviour by exciting B neuron arrays. The B neurons initiate the command to perform behaviour patterns. An important point is that each R neuron receives synapses from a pool of S neurons. It is these S to R synapses which have variable strength, and motivated learning occurs through the strength changes in these synapses.

Fixed inhibitory connections among B groups provide a somewhat winner-take-all situation in which usually only one behaviour at a time is activated. Normally, when a new action is triggered it will inhibit the previous one. At the same time, behaviour is stabilized by a positive feedback relationship between B and R. Once activated these neurons fire persistently until another B unit inhibits them, even though the stimulus pattern in S will generally disappear or be suppressed shortly after the behaviour is triggered. This simple circuit is an adequate interface to control a sequence of adaptive behaviours.

In the case of the robot fly, S1 is a subset of the neurons in a complex visual pattern-recognizing neural network device. The subset is chosen so that there are inherently likely to be many neurons in S1 active when the visual image contains features which we know will tend to identify plastic. We assume that there is some bias to recognize shininess, and a bias not to respond to leaf green. S2 must be biased to recognize features of plastic which can be detected close up. Since the park will contain many shiny leaves, the most important bias here might be a negative one, the tendency not to respond in the presence of the veined texture of leaves. The robot must have *some* adequately definitive sensing mechanism to confirm or reject the hypothesis that this object is plastic, at some stage of the process. We assume that there is a chemo-detection mechanism, which can distinguish between various types of plastic, either chemically, or by being integrated with the close-up visual pattern (really plastic + pattern x). There must be a criterion for deciding when the label is in place properly, in sensor output S4. And finally, S5 is a timing mechanism to "over-ride" the entire process if the robot has been at a single object for too long.

While the interface does not require a specific organization of the robot's sensory mechanism outputs we note that they could use the distributed, coarse coded representations typical of neural network simulators. We have not simulated sensing mechanisms, but merely define input patterns.

Assuming then that we have adequate sensing devices, so that output patterns in S1 can distinguish in principle between plastic scrap and wet leaves, we now have to train the connections from this S1 device to R1, the releaser for landing. We want to alter the strength of S to R connections so as to strengthen the response to real plastic scrap and weaken the response to patterns activated by objects which resemble plastic scrap but are not.

This is accomplished as follows: If during the robot's flight S1 is activated by some object in the sensor's field of vision, and there are currently strong enough connections that R1 then fires, the interface has hypothesized that the object below it is plastic scrap. R1 activates B1, which commands downward motion. As the robot lands, the visual pattern which activated S1 disappears, but the feedback between the B and R neurons maintains their activation. After landing, there will be a stable visual image available again. The plastic will have an un-veined texture, and therefore will activate the smooth-texture detectors of S2, which have biased connections to R2. Now R2 will activate B2, and the robot will sniff the object with its accurate plastic chemo-detector. In this case, inhibitory connections from B2 to B1 will shut off B1. Since R1 no longer has much S1 input, stopping B1 will stop R1 too. So the whole R1-B1 complex becomes silent.

From the point of view of the variable strength S1-R1 synapses which made the successfully confirmed hypothesis, the temporal firing pattern has been the strengthening pattern of fig. 2a: S1 on and off, R1 on and off somewhat later. Note that a delay occurs between the end of S1 activity and the end of R1's "co-activation", and an important feature of our strength change rule is that it permits strengthening to occur as a result of such *time-lagged* co-activity. Fig. 3 shows how the time lag constant, τ, should be set to ensure optimal strengthening in normal circumstances. The strengthening process occurs for any S1 neurons which had been recently activated by the plastic object.

To see how effectively this process can refine the robot's ability to recognize plastic, extend the picture by supposing that there are colour detection neurons in S1, which initially were all set with very weak connections to R1. Those colour detectors which were activated by this particular piece of plastic will have their connections to R1 strengthened, because they show the same co-activity with R1, and because *only* the temporal activity patterns control strength changes. The fact that these connections were so weak as to have negligible influence in pushing R1 above threshold is irrelevant. The connections strengthen, so when such an object is seen again, the colour will contribute more to the "hypothesis forming" or "recognition" process.

This is how the design allows already effective connections involved in performing "successful" identifications to self-reinforce, and permits weak connections used by other stimulus features to strengthen if there are repeated associations between these features and the successful triggering of the behaviour. Thus, the S1 which began with a bias to identify only shiny objects as plastic will continue to do so, and will gradually learn to identify them on the basis of colour also, if there are reliable colour cues available.

We would be pleased if the robot fly could increase its efficiency by learning not to land on shiny leaves, and this can happen as follows. Suppose that our robot lands on a shiny leaf. Now because leaves have veins the neuron pool S2 which represents the smooth-textured regions of the close-up stimulus is too small to activate R2. Thus B2 remains silent and does not inhibit B1. Now the B1-R1 feedback loop continues uninterrupted, while S1's representation of the shiny leaf seen from above has disappeared, Thus the firing pattern for S1 and R1 is as in fig. 2d, and so it causes weakening in these S1-R1 connections. As a result, the visual pattern in S1 caused by the shiny leaf will be less likely to activate R1 in the future, and we can say that the robot is learning not to land on this material, or alternatively that it is learning that it is not plastic.

Figure 2. Rules governing strength changes at excitatory synapses S --> R (fig. 1). The graphs are computer output from the simulation in Halperin, Halperin, Rutherford and Dunham (in prep.).
2a) Optimal strengthening, due to good correlation between firing of S and R. This is generally how perceptual learning is assumed to occur in "untutored" connectionist neural networks. Note the slight time lag between the firing of S and R. The general idea that time windows or time lags should be incorporated into synapses showing rewarded learning seems to have been first suggested by Sutton and Barto (1981).
2b) Weakening, because S fires and R doesn't - poor correlation.
2c) Strengthening, because once S fires, firing in R is correlated with it. That S did not cause R to fire is irrelevant. Only temporal association matters in strengthening the synapse. This rule is unusual because we propose that the initial poor correlation (S on, R off) is unimportant. In the biological context this property makes the surprising prediction that weak stimuli presented near the end of an after-discharge of behaviour should have their synapses to R strengthened, until they begin to elicit after-discharges themselves. We have confirmed this "backward conditioning" (in prep.).
2d) Weakening, because firing by R throughout the period after good correlation prevents consolidation of strengthening, replacing it with weakening.

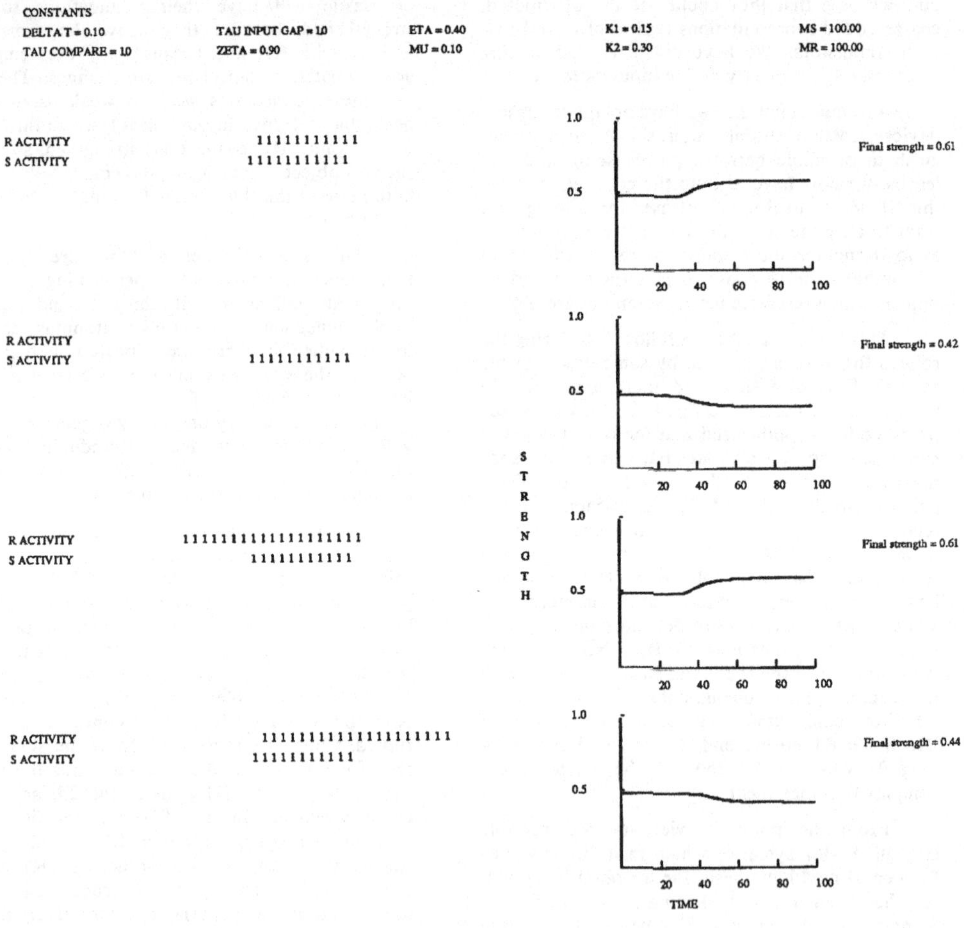

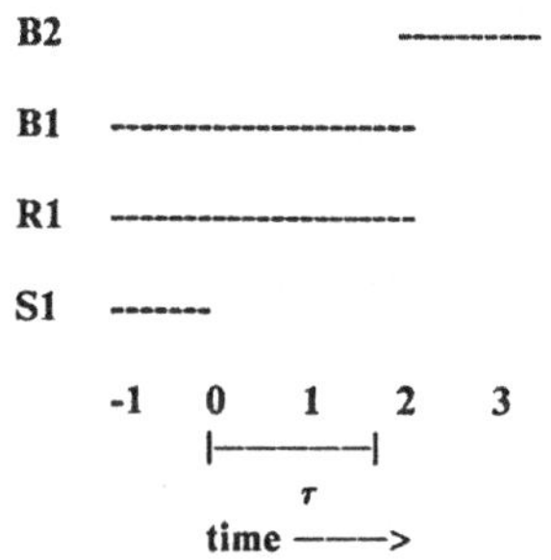

Figure 3. Choosing the parameter τ, as an example of setting parameters in this system, where almost all the parameters are time constants. Firing in neurons of an array is represented by dashes. S1 is a neuron which fires in response to an object, thereby "hypothesizing" that it is plastic and contributing to firing in R1. This releases landing via B1. At time 0 the visual input becomes so blurred by movement that S1 turns off. After landing is accomplished, at time 2s, the close-up recognition system activates B2. The action B2 quickly silences B1. So τ can be set at 2 seconds. In the task-controlling context τ can in general be set to the expected releaser offset time minus the expected stimulus offset time.

At this point we can note that while this strength change process is rather analogous to the Hebb rule (with a time lag) when there is a strengthening situation, the process which weakens connections if a mis-identification has been made is more unusual (but see Sinclair, 1981). It permits the machine to learn to correct erroneous identifications, and as we shall see now, together with the circuit design it allows information from a "bad" consequence to propogate backward down a chain of stimulus-response "instincts".

Consider, then, the case where a flaw in the label design has led to a situation in which some types of plastic scrap cannot be labelled. The robot will "see" the scrap, land, "sniff" it, and try to label it. The next correct stimulus situation is the sight of the label in place, recognized by S4. This triggers B4, which inhibits B3, the command for "labeling". If the scrap is of the type which can not be labelled successfully, stimulus situation S4 will not be achieved and B4 will not turn on to inhibit the B3-R3 complex. Therefore R3 will not turn off until the over-ride activates, long after S3 has become silent. Thus the temporal firing pattern for the S3 to R3 connections will be the same as in the S1 to R1 case when the shiny object is not plastic. Except for the initial period when S3 and R3 fired simultaneously, the pattern is S3 "off", R3 "on". This is the connection-weakening pattern shown in fig. 2d. Repeated weakening will lead to a situation in which the S3 pattern caused by the unlabellable plastic will not activate R3.

Now the robot fly will get to the point of sniffing the unlabellable scrap, but will then just sit there with its sniff command feedback loop running, until the over-ride timer activates. At this point the failure will propogate backwards to weaken the S2 to R2 connections in the same way. Once these are sufficiently weak the same process will finally cause the robot to learn not even to land on this type of plastic. In this way, the robot will have learned to ignore the unlabellable scrap, by propagating the information about the difficulty backwards down the chain of behaviours.

At this point we can see how to build in incentive motivations if we want them, by biasing the S array to respond to certain stimulus features. For example, suppose every second S neuron responded to silver. If, say, 6 very weak synapses firing simultaneously were sufficient to activate an R neuron, and there were 12 synapses to each R unit, with 6 of them being silver responders on average, then if the unlabellable plastic were very silvery, we could never eliminate the positive response to it. The robot fly would persistently land on it, and a psychologist studying it might be tempted to say that the silver shine per se was rewarding the fly, given that this is a system which usually responds to the consequences of actions.

Another type of "incentive" situation is one in which the "next" behaviour was always triggered and accomplished. Imagine a more sophisticated fly in which the first behaviour is a B1 allowing escape from birds by diving, and suppose that the fly recognizes birds through an S1 array biased to respond to rapidly looming objects associated with a certain shape. If the S2 which releases the B2 which shuts off B1 is the visual pattern associated with getting near the ground, then in a fly that stays near the ground this will always be activated quickly whenever diving occurs. This fly would always be rewarded for identifying a "bird" as a predator. Thus if some member of a seed eating and harmless bird species landed swiftly and suddenly near an inexperienced fly which mistook this action for an attack and dove, this fly would be automatically rewarded for its escape behaviour.

Such a fly could develop a bird phobia. Perhaps on another occasion when the fly encounters this species a bird will again happen to land suddenly and release escape. Now there will be more strong connections formed from S neurons responsive to the characteristic shape of this bird species up to the escape command. The fly may start to regularly avoid birds of this species even when they do not move in a "threatening" way. And the more it avoids, the more reinforcement it will get for "successfully" escaping. If the S array used distributed representations which allowed similar stimuli to activate overlapping neuron pools, one might develop a situation where the landing patterns of most birds activated enough S neurons to initiate escape. The robot fly would quickly learn to avoid all birds. We would say that it has an instinctive fear of birds, or that birds have a negative incentive motivational property for robot flies.

This brief discussion may suggest a few of the problems which would be encountered in trying to design bio-mimic robots for a real outdoor world, as well as showing that instinctive "hedonic" responses to stimuli would be very easy to produce, since they can simply be responses which are too strongly pre-wired to respond well to extinction, or responses which automatically self-reinforce.

We also remark that if two actions in sequence, B1 and B2 were potentially separated by a long time interval, then the timing parameters for the relevant S --> R connections would have to provide for a long time lag τ at the synapses between them. Again, for the sake of having an easily visualized example, consider designing the larger collector robot. There might be the risk of its encountering some unforeseen type of material which was corrosive, perhaps a type of plastic container with acid inside, that would not be detectable until a "trash compaction" manoeuvre was performed somewhat later. Say that action B1 was putting a material into the robot container. B2 would be the response to detecting something corrosive in the container. It might be a multiple activity, to back up and turn around, to empty contents, and simultaneously to notify base that there is a problem. One would arrange that there were normal synapses from S1 to R1 (the "put it in" releaser), but that S1 was a subset of S2 and had long τ synapses to R2. (There would also be normal synapses from the S2 damage detectors to the R2 emergency emptying reaction). If a particular S1 were followed τ seconds later by damage, this would activate the emergency reaction. When the emergency reaction was complete and B2 became silent, the S1 to R2 synapses would strengthen. The next time the machine encountered the dangerous material, S1's activation would tend to release action B2. The collection robot would back up, turn around, notify base, etc. As with biological systems, there is a possibility that it would even empty the harmless contents of its container even though the back up manoeuvre had prevented it from collecting the visually identified corrosive material. Unit S1 would also have developed strengthened connections to R1, since, in the short term, on the previous encounter the approach and collect routine apparently had been completed successfully. Therefore, one would have to ensure that the B2 unit had a strong inhibitory effect on the B1 unit, so that avoidance and leaving the area would dominate. It is a property of biological motivational systems of which a motivated bio-mimic robot designer would become aware, that learning by punishment is very different than learning by lack of reward.

We can now consider one final example of an application to robotics, the design of motivations which correct a deficit. Animal motivational systems like hunger are assumed to operate on drive-reduction principles, without actually being simple homeostatic systems. While the ideas to be presented in this section have never been tested in a biological system, they would be interesting to explore as a model for biological drive-reduction learning.

In this example, the R and B units are neuron arrays, if not in general so large as the S arrays. This feature permits considerable flexibility, and one application of this is to drive-reduction. The variable strength S to R connections have a fan-in fan-out arrangement so that each S unit sends connectors to many R neurons in a neighbourhood above it, and each R unit receives connections from many nearby S neurons. Information flow is thus topographically organized, with considerable redundancy. Exactly the same connection strength change rules apply as previously. The difference is that there is assumed to be some subset, Rd, of special deficit detection neurons within the R unit, which behave in a special way. These neurons are designed to receive input registering some deprivation state (e.g. low battery), and to continue to fire while the deficit continues. As the deficit increases, more and more neurons in the Rd group will be activated, since different neurons in Rd will have different thresholds.

Assume that the motor pattern commanded by the corresponding B unit is an approach toward whatever is being visually fixated at that moment, followed by an attempt to "plug in". If many Rd neurons were firing, (i. e., if the battery were low) the machine would begin to approach various objects in its environment. In a changable environment which provided new stimuli in the visual field, the robot would explore restlessly.

But considered from the viewpoint of the S to Rd connections, the situation is simple. Rd fires continuously, since these particular instances of approach have no effect on deprivation state. So although the firing pattern of S on-off, Rd on continuously could in principle lead to some weakening, here the synapses to Rd are not subjected to strength changes, because the Rd activity prevents consolidation. Therefore, connections to Rd are not affected by unsuccessful actions, that is, by actions which do not precede deficit-reduction.

Suppose the move forward is successful, - the machine plugs in and re-charges its batteries. Now the situation is completely different, because the deprivation state ends, Rd becomes silent, and strength change can occur. The relevant synapses are those connnecting whatever S neurons happened to be active just before the move forward, say S1, to the active Rd neurons. For these synapses the firing pattern is that shown in fig. 2c, which is a strengthening pattern. Thus, the next time the machine sees S1, the probability of Rd firing strongly enough to activate an approach will be greater than it was before the strengthening event. It is interesting to note that the prediction that firing pattern 2c should cause strengthening has been confirmed in our laboratory for fighting fish social behaviour, in a surprising instance of clear backward conditioning (Halperin and Dunham, in prep.).

If the successful approach led to only partial deficit-reduction, then only synapses from S1 to the subset of Rd neurons which became silent would occur. Thus the strengthening situation can be graded if there are deficit detectors with different thresholds.

We can now describe some of the corollaries of this design, occasionally using the example of human hunger drives, to make the usefulness of the arrangement more obvious, but also to emphasize limitations of directly mimicking biological systems. First, note that, if the strengthening event described above were to occur for our machine rather often, then seeing S1 would eventually be enough to release moving forward, even if the machine were not in a severe deprivation state at that moment. The machine would develop the tendency to plug in, if there were no strong competing tendency, before it became very energy deprived. The analogy with human hunger is suddenly feeling hungry upon sitting down at the table and smelling the types of food which have satisfied hunger in the past. This often leads to the performance of appetitive actions in the absence of strong physiological signals of deprivation, and assists in the maintainance of a surplus energy store (unfortunately).

Similarly, the threshold for performing the appetitive action would decrease with increasing deprivation. A machine which had a rather weak Sx to Rd connection might nevertheless be pushed over the threshold if Sx fired when Rd was already quite active. The analogy is that faint odors of food are enough to turn the head of a hungry person.

If the representations of stimuli have a degree of overlap proportional to their similarity, as is typical of connectionist systems, the machine will tend to generalize from success in one particular situation, and try re-charging in a novel but similar situation. The probability of doing this will be positively affected by both the degree of similarity and the current degree of deprivation. A rather strained human analogy is that if desserts have tended to just precede the shutting off of deprivation detectors in the past, a novel dessert will activate hunger if it looks not *too* unusual.

The design has some of the disadvantages, as well as advantages, of natural "homeostatic" systems such as the feeding system. It will have a tendency to over-correct, which would be a problem if over-charging were detrimental. This tendency would be exaggerated if the machine had experienced severe deficiency in the past. A once-starved machine would develop a tendency to approach energy sources. This tendency might be advantageous or not, depending on whether the environment still contained "dangerous" areas with no re-charge facilities. A full-charge detector, connected to a back-up and turn unit, would improve efficiency if time-wasting were an undesirable activity. Satiety detectors are present in most biological energy collection systems, which suggests that they are likely to improve the efficiency of systems based on this type of organization.

Successful deprivation-reducing situations will not tend to fade out of the connections' memories completely and be forgotten, even if there are effective satiety detectors. Say there is a

Figure 4. This shows two neurons and the variable synapse between them, with the essential elements of the mathematics which models the synaptic strength change processes assumed to occur at the sensory-motor interface as the neurons and their synapse process the information flowing through the system. Some sub-units use a time-weighted average of recently received information, so that they can respond to processes occuring in the recent past, as well as to curently arriving information. In particular, I(t), α(t) and C(t) use exponential smoothing.

Inputs are signals received by the neurons from pre-synaptic neurons or directly from sensors. Excitatory inputs are positive and inhibitory negative. A given synapse is excitatory-fixed or excitatory-variable, or inhibitory-fixed.

R and S are neurons, which receive their inputs, exponentially smooth them in I(t), fire if I(t) is above a threshold (which may be variable), and store an exponentially smoothed value of recent firing in α(t). The details of I(t) and the threshold function will be discussed in Halperin and Halperin, in prep., as they are not crucial for an understanding of the basic interface.

α(t) is a neuron's activity register. $\alpha'(t) = \varsigma(1-\alpha(t))$, if the neuron is on at t, $\alpha'(t) = -\eta\alpha(t)$, if the neuron is off at t.

C is the strength change command register. $C'(t) = \mu(A-C)$, where $A=A(\alpha_S,\alpha_R)$, and α_S is time-lagged by τ, and where μ is a constant such that $0<\mu<<\eta$.

ω(t) is the strength of the synapse, which is often referred to as the weight of the connection. It is changed by the change command in C(t), unless consolidation is blocked by high values in the activity registers.

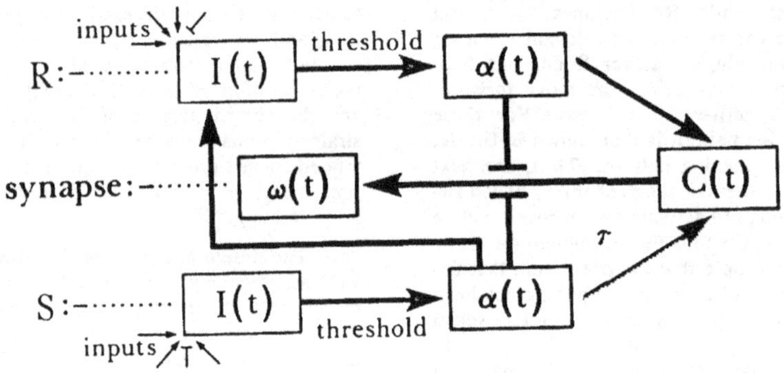

situation, S3, representing the "sight" of a particular plug which has often corrected energy deficits in the past. If S3 suddenly becomes incapable of satisfying energy needs (say the fuse for this plug has blown), the machine will not easily learn this. "Seeing" S3 will still cause firing in Rd even though the machine is not actually deprived. Furthermore, seeing S3 with or without plugging in to recharge will still be followed by the strengthening pattern caused by the end of firing in Rd as soon as S3 is no longer visible, because S3 now controls firing in Rd. So just as in the case where the machine sees a still-operational plug and plugs in despite not being deprived, seeing any "known" deprivation reducing plug will "remind" the machine of its experience and strengthen the connection.

If there is a real deprivation state, then Rd will simply stay on, so we have the firing pattern of fig. 2d. There is, however, no weakening; i.e. there is no negative re-inforcement to teach the machine that the plug is non-functional, because there is no neural silence for consolidation.

For deficit detecting situations, as before, the machine will be most likely to learn effectively if there is some bias in the objects to which it will respond. It will be most effective, for example, if there is some way of biasing the S neurons feeding into Rd so that they tend to respond to plugs. Otherwise, the machine will have to learn entirely by trial and error, after becoming so deprived that it goes about approaching objects at random. The machine, like a bird, would be at its most inefficient in the "fledgling" phase, when it was learning how to satisfy its own needs. Such an unbiased machine could have its performance markedly improved by a teacher who simply pointed it toward a plug when it was in an energy deprived state. After sufficient experience, the machine would almost never become energy deprived, if it were in an environment in which it could plug itself in often.

It may seem that this system is rather imperfect. But the biological analogue which the interface is designed to mimic is also far from perfect, and, like animals, such machines may be successful at performing adequately despite their limitations. The design features and the biological analogue suggest that either considerable biasing of the sensory devices, or a period of "social learning", would be necessary if it were the case that severe repercussions would result from initial failures to solve the deprivation-reduction puzzles facing the machine.

In conclusion, then, we have seen how the simple principles of wiring and connection strength change for a sensory-motor interface designed to model aggressive motivation in fighting fish can be used as design principles for a variety of robots with various types of motivational systems, flexibly adapting themselves to function in circumstances with unforeseen minor deviations from the environment envisaged by their designers.

References

Brooks, R. A. and Flynn, A. M. (1989) Fast, cheap and out of control: a robot invasion of the solar system. *J Brit. Interplanetary Soc.* 42: 478-485.

Hebb, D. O. (1949) *The Organization of Behaviour.* Wiley, N. Y.

Sinclair, J. D. (1981) *The Rest Principle: A Neurophysiological Theory of Behavior.* Erlbaum, Hillsdale, N. J.

Sutton, R. S. and Barto, A. G. (1981) Toward a modern theory of adaptive networks: Expectation and prediction. *Psychol. Rev.* 88: 135-171.

A Possibility for Implementing Curiosity and Boredom in Model-Building Neural Controllers

Jürgen Schmidhuber*
Institut für Informatik
Technische Universität München
Arcisstr. 21, 8000 München 2, Germany
schmidhu@tumult.informatik.tu-muenchen.de

Abstract

This paper introduces a framework for 'curious neural controllers' which employ an adaptive world model for goal directed on-line learning.

First an on-line reinforcement learning algorithm for autonomous 'animats' is described. The algorithm is based on two fully recurrent 'self-supervised' continually running networks which learn in parallel. One of the networks learns to represent a complete model of the environmental dynamics and is called the 'model network'. It provides complete 'credit assignment paths' into the past for the second network which controls the animats physical actions in a possibly reactive environment. The animats goal is to maximize cumulative reinforcement and minimize cumulative 'pain'.

The algorithm has properties which allow to implement something like *the desire to improve the model network's knowledge about the world.* This is related to *curiosity.* It is described how the particular algorithm (as well as similar model-building algorithms) may be augmented by dynamic *curiosity* and *boredom* in a natural manner. This may be done by introducing (delayed) reinforcement for actions that increase the model network's knowledge about the world. This in turn requires the model network *to model its own ignorance*, thus showing a rudimentary form of *self-introspective* behavior.

1. Introduction

In the sequel first an on-line algorithm for reinforcement learning in non-stationary reactive environments is described. The algorithm heavily relies on an *adaptive model of the environmental dynamics.* The main contribution of this paper (see the second section) is to demonstrate how the algorithm may be naturally augmented by *curiosity* and *boredom*, in order to improve the world model in an on-line manner.

Consider an 'animat' whose movements are controlled by the output units of a neural network, called the control network, which also receives the animat's sensory perception by means of its input units. The animat potentially is able to produce actions that may change the environmental input (external feedback caused by the 'reactive' environment). By means of recurrent connections in the network the animat is also potentially able to internally represent past events (internal feedback).

The animat sometimes experiences different types of reinforcement by means of so-called *reinforcement units* or *pain units* that become activated in moments of reinforcement or 'pain' (e.g. the experience of bumping against an obstacle with an extremity). The animat 's only goal is to minimize cumulative pain and maximize cumulative reinforcement. The animat is autonomous in the sense that no intelligent external teacher is required to provide additional goals or subgoals for it.

Reinforcement units and pain units are similar to other input units in the sense that they possess conventional outgoing connections to other units. However, unlike normal input units they can have *desired activation values* at every time. For the purpose of this paper we say that the desired activation of a pain unit is zero for all times, other reinforcement units may have positive desired values. In the sequel we assume a discrete time environment with 'time ticks'. At a given time the quantity to be minimized by the learning algorithm is $\sum_{t,i}(c_i - y_i(t))^2$ where $y_i(t)$ is the activation of the ith pain or reinforcement unit at time t, t ranges over all remaining time ticks still to come, and c_i is the desired activation of the ith reinforcement or pain unit for all times.

The reinforcement learning animat faces a very general spatio-temporal credit assignment task: No external teacher provides knowledge about e.g. desired outputs or 'episode boundaries' (externally defined temporal boundaries of training intervals). In the sequel it is demonstrated how the animat may employ a combination of two recurrent *self-supervised* learning networks in order to satisfy its goal.

Munro [2], Jordan [1], Werbos [12], Robinson and Fallside [6], and Nguyen and Widrow [4] used *'model networks'* for constructing a mapping from output actions

*This work was supported by a scholarship from SIEMENS AG

of a control network to their effects in in 'task space' [1]. The general system described below (an improved version of the variants described in [8] and [10]) also employs an adaptive model of the environmental dynamics for computing gradients of the control network's inputs with respect to the controller weights. The model network is trained to simulate the environment by making predictions about future inputs, including pain and reinforcement inputs. Training of the controller works as follows: Since we cannot propagate input errors (e.g. differences between actual pain signals and desired zero pain signals) 'through the environment', we propagate them through the combination of model network and controller. Only the controller weights change during this phase, the weights of the model network have to remain fixed. (Actually, we do not really propagate errors through the networks. We use an algorithm which is functionally equivalent to 'back-propagation through time' but uses only computations 'going forward in time'. So there is no need for storing past activations.)

Both the control network and the model network are fully recurrent. The algorithm differs from algorithms by other authors in at least some of the following issues: It aims at on-line learning and *locality in time*, it does not care for 'epoch-boundaries', it needs only reinforcement information for learning, it allows different kinds of reinforcement (or pain), and it allows both internal and external feedback with theoretically arbitrary time lags. Unlike with Robinson and Fallside's approach (which is the one that bears the most relationships to the one described here) 'credit assignment paths' are provided that lead from pain units back to output units back to all input units and so on. There are also credit assignment paths that lead from input units back to the input units themselves, and from there to the output units. The latter paths are important in the common case when the environment can change even if there were no recent output actions.

The discrete time algorithm below *concurrently* adjusts the fully recurrent model network and the fully recurrent control network. An on-line version of Robinson and Fallside's Infinite-Input-Duration learning algorithm for fully recurrent networks [5] (first implemented by Williams and Zipser [14]) is used for training both the model network and the combination of controller and model network. The algorithm is a particular instantiation of a more general form and is based on the logistic activation function for all non-input units.

In step 1 of the main loop of the algorithm actions in the external world are computed. Due to the internal feedback, these actions are based on previous inputs and outputs. For all new activations, the corresponding derivatives with respect to all controller weights are updated.

In step 2 actions are executed in the external world, and the effects of the current action and/or previous actions may become visible.

In step 3 the model network tries to predict these effects without seeing the new input. Again the relevant gradient information is computed.

In step 4 the model network is updated in order to better predict the input (including pain) for the controller. Finally, the weights of the control network are updated in order to minimize the cumulative differences between desired and actual activations of the pain and reinforcement units. Since the control network continues activation spreading based on the actual inputs instead of using the predictions of the model network, 'teacher forcing' [14] is used in the model network.

One can find various improvements of the systems described in [8] and [10]. For instance, the partial derivatives of the controller's inputs with respect to the controller's weights are approximated by the partial derivatives of the corresponding predictions generated by the model network. Furthermore, the model sees the last input and current output of the controller at the same time.

Notation (the reader may find it convenient to compare with [14]):

C is the set of all non-input units of the control network, A is the set of its output units, I is the set of its 'normal' input units, P is the set of its pain and reinforcement units, M is the set of all units of the model network, O is the set of its output units, $O_P \subset O$ is the set of all units that predict pain or reinforcement, W_M is the set of variables for the weights of the model network, W_C is the set of variables for the weights of the control network, $y_{k_{new}}$ is the variable for the updated activation of the kth unit from $M \cup C \cup I \cup P$, $y_{k_{old}}$ is the variable for the last value of $y_{k_{new}}$, w_{ij} is the variable for the weight of the directed connection from unit j to unit i, $p^k_{ij_{new}}$ is the variable which gives the current (approximated) value of $\frac{\partial y_{k_{new}}}{\partial w_{ij}}$, $p^k_{ij_{old}}$ is the variable which gives the last value of $p^k_{ij_{new}}$, if $k \in P$ then c_k is k's desired activation for all times, α_C is the learning rate for the control network, α_M is the learning rate for the model network.

$\mid I \cup P \mid = \mid O \mid$, $\mid O_P \mid = \mid P \mid$. If $k \in I \cup P$, then kpred is the unit from O which predicts k. Each unit from $I \cup P \cup A$ has one forward connection to each unit from $M \cup C$. Each unit from M is connected to each other unit from M. Each unit from C is connected to each other unit from C. Each weight of a connection leading to a unit in M is said to belong to W_M. Each weight of a connection leading to a unit in C is said to belong to W_C. Each weight $w_{ij} \in W_M$ needs p^k_{ij}-values for all $k \in M$. Each weight $w_{ij} \in W_C$ needs p^k_{ij}-values for all $k \in M \cup C \cup I \cup P$.

INITIALIZATION:
For all $w_{ij} \in W_M \cup W_C$:
begin $w_{ij} \leftarrow$ *random*,
for all possible k: $p^k_{ij_{old}} \leftarrow 0, p^k_{ij_{new}} \leftarrow 0$ *end.*
For all $k \in M \cup C : y_{k_{old}} \leftarrow 0, y_{k_{new}} \leftarrow 0$.
For all $k \in I \cup P$:
Set $y_{k_{old}}$ *by environmental perception,* $y_{k_{new}} \leftarrow 0$.

FOREVER REPEAT:
1. *For all* $i \in C : y_{i_{new}} \leftarrow \frac{1}{1+e^{-\sum_j w_{ij} y_{j_{old}}}}$.
   For all $w_{ij} \in W_C, k \in C$:
   $p^k_{ij_{new}} \leftarrow y_{k_{new}}(1 - y_{k_{new}})(\sum_l w_{kl} p^l_{ij_{old}} + \delta_{ik} y_{j_{old}})$
   For all $k \in C$:
   begin $y_{k_{old}} \leftarrow y_{k_{new}}$,
   for all $w_{ij} \in W_C : p^k_{ij_{old}} \leftarrow p^k_{ij_{new}}$ *end .*
2. *Execute all motoric actions based on activations of units in A. Update the environment.*
   For all $i \in I \cup P$:
   Set $y_{i_{new}}$ *by environmental perception.*
3. *For all* $i \in M : y_{i_{new}} \leftarrow \frac{1}{1+e^{-\sum_j w_{ij} y_{j_{old}}}}$.
   For all $w_{ij} \in W_M \cup W_C, k \in M$:
   $p^k_{ij_{new}} \leftarrow y_{k_{new}}(1 - y_{k_{new}})(\sum_l w_{kl} p^l_{ij_{old}} + \delta_{ik} y_{j_{old}})$.
   For all $k \in M$:
   begin $y_{k_{old}} \leftarrow y_{k_{new}}$,
   for all $w_{ij} \in W_C \cup W_M : p^k_{ij_{old}} \leftarrow p^k_{ij_{new}}$ *end.*
4. *For all* $w_{ij} \in W_M$:
   $w_{ij} \leftarrow w_{ij} + \alpha_M \sum_{k \in I \cup P}(y_{k_{new}} - y_{kpred_{old}}) p^{kpred}_{ij_{old}}$.
   For all $w_{ij} \in W_C$:
   $w_{ij} \leftarrow w_{ij} + \alpha_C \sum_{k \in P}(c_k - y_{k_{new}}) p^{kpred}_{ij_{old}}$.
   For all $k \in I \cup P$:
   begin $y_{k_{old}} \leftarrow y_{k_{new}}, y_{kpred_{old}} \leftarrow y_{k_{new}}$,
   for all $w_{ij} \in W_M : p^{kpred}_{ij_{old}} \leftarrow 0$,
   for all $w_{ij} \in W_C : p^k_{ij_{old}} \leftarrow p^{kpred}_{ij_{old}}$ *end.*

By employing probabilistic output units for C and by using 'gradient descent through random number generators' [13] we can introduce explicit explorative random search capabilities into the otherwise deterministic algorithm. In the context of the IID algorithm, this works as follows: A probabilistic output unit k consists of a conventional unit $k\mu$ which acts as a mean generator and a conventional unit $k\sigma$ which acts as a variance generator. At a given time, the probabilistic output $y_{k_{new}}$ is computed by

$$y_{k_{new}} = y_{k\mu_{new}} + z y_{k\sigma_{new}},$$

where z is distributed e.g. according to the normal distribution. The corresponding $p^k_{ij_{new}}$ have to be updated according to the following rule:

$$p^k_{ij_{new}} \leftarrow p^{k\mu}_{ij_{new}} + \frac{y_{k_{new}} - y_{k\mu_{new}}}{y_{k\sigma_{new}}} p^{k\sigma}_{ij_{new}}.$$

By performing more than one iteration of step 1 and step 3 at each time tick, one can adjust the algorithm to environments that change in a manner which is not predictable by semilinear operations (theoretically three additional iterations are sufficient for any environment).

The algorithm is local in time, but not in space. See [8] for a justification of certain deviations from 'pure gradient descent through time', and for a description of how the algorithm can be used for planning action sequences. See [7] and [9] for two quite different entirely local methods.

2. Implementing Dynamic Curiosity and Boredom

Only if the model network is a good predictor of the environmental dynamics we can expect the controller to converge. In the current section we motivate the introduction of the 'explicit desire to improve the world model' and show a possibility for implementing it in on-line model-building systems as the one described in the last section.

Many biological learning systems, particularily the more complex ones, show an interplay of goal-directed learning and explorative learning. In addition to certain permanent goals (like avoiding pain), goals are generated whose immediate purpose solely is to increase knowledge about the world. So far this interplay has not been addressed at all in the connectionist literature.

The explorative side of learning (related to something that usually is called *curiosity*) is not completely unsupervised, as it is sometimes assumed. Curiosity helps to learn how the world works, which in turn helps to satisfy certain goals. However, the goal-directedness of curiosity is less obvious than the goal-directedness of the algorithm described above (and of less general algorithms described in other papers on goal-directed learning).

Curiosity is related to what one already knows about the world. One gets curious as soon as one *believes that there is something that one does not know.* However, the goal of learning how the world works is dominated by other goals (like avoiding pain): One does not know *exactly* how it feels putting one's hand into the meat grinder. However, one does not *want* to know.

Since curiosity makes sense only for systems that can have *dynamic influence on what they learn*, and since curiosity aims at minimizing a dynamically changing value, namely, the degree of ignorance about something, it makes sense only in on-line learning situations where there is some sort of *dynamic attention.*

Thus the precondition of curiosity is something like our on-line learning algorithm described above. This algorithm builds a world model in order to use the world model for goal-directed learning of the controller. The controller's potential for *dynamic attention* is given by

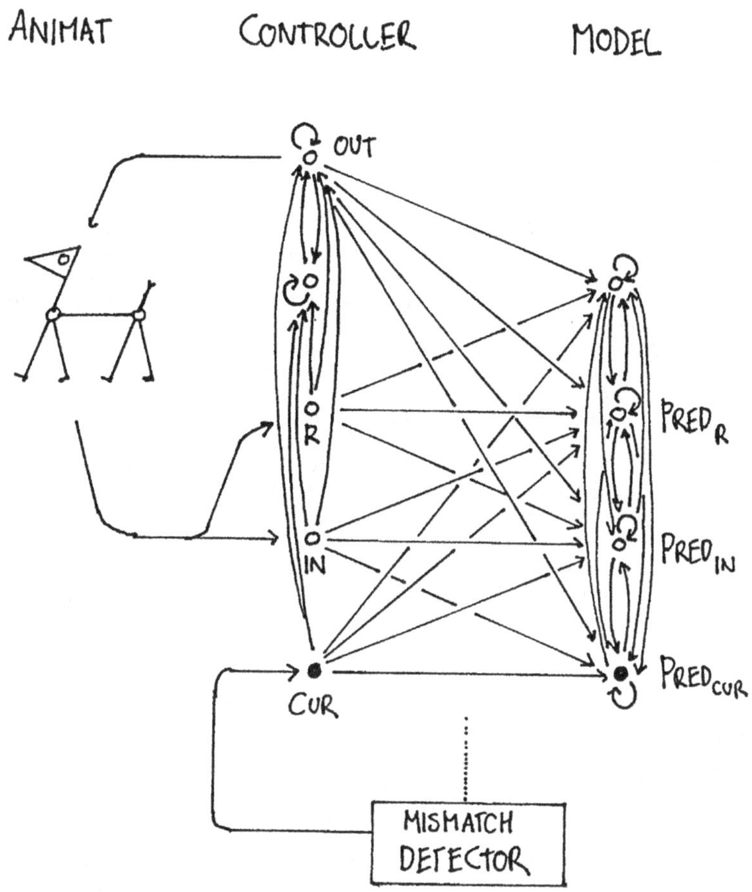

Figure 1. An animat controlled by a recurrent control network is shown. For simplicity, only one 'normal' input unit (IN), one 'normal' reinforcement input unit (R), one hidden unit and one output unit (OUT) are depicted. A recurrent model network is trained to emulate the environmental dynamics by predicting the control network's input ($PRED_{IN}$ and $PRED_R$).
An additional reinforcement unit CUR for the control network gets activated by 'ideal mismatches' between expectations of the model network and reality. The model network needs an additional output unit ($PRED_{CUR}$) for predicting CUR. It models its own ignorance, thus showing a rudimentary form of self-introspective behavior. The model network helps to encourage action sequences of the controller which lead to unfamiliar situations.

the external feedback. The world model adapts itself to whatever the controller focusses on (see [11] for an application of similar adaptive control techniques to the problem of learning selective attention). The direct goal of curiosity and boredom is to improve the world model. The indirect goal is to ease the learning of new goal-directed action sequences. The contribution of this section is to show one possibility for augmenting the algorithm by curiosity and by its counterpart, which is *boredom*.

The basic idea is simple: We introduce an additional reinforcement unit for the controller (see figure 1.). This unit, hereafter called the *curiosity unit*, gets activated by a process which at every time step measures the Euclidian distance between reality and prediction of the model network. The activation of the curiosity unit is a function of this distance. Its desired value is a positive number corresponding to the *ideal mismatch* between belief and reality. The effect of the algorithm described in the first section is that there is *positive reinforcement* whenever the model network *fails* to correctly predict the environment. Thus the usual credit assignment process for the controller encourages certain past actions in order to repeat situations similar to the mismatch situation.

As soon as the model network has learned to correctly predict the environment in former 'mismatch situations', actions leading to such situations automatically are dereinforced. This is because the activation of the curiosity unit goes back to zero. *Boredom* becomes associated with the corresponding situations.

The important point is: The same complex mechanism which is used for 'normal' goal-directed learning is used for implementing curiosity and boredom. There is no need *for devising a* separate *system which aims at improving the world model.*

The controller's credit assignment process is aimed at repeatedly entering situations where the model network's performance is not optimal. *It is important to observe that this process itself makes use of the model network!* The model network has to predict the activations of the curiosity unit. Thus the model network partly has to model its own ignorance, it has *to learn to know that it does not know* certain details.

What is the *ideal mismatch* mentioned above? In conventional AI the saying goes that a system can not learn something that it does not already *almost know*. If we want to adopt this view, then a consequence is that the function that translates mismatches into reinforcement is not a linear one. *Zero reinforcement should be given in case of perfect matches, high reinforcement should be given in case of 'near-misses', and low reinforcement again should be given in case of strong mismatches.* This corresponds to a notion from 'esthetic information theory' which tries to explain the feeling of 'beauty' by means of the quotient of 'subjective complexity' and 'subjective order' or the quotient of 'unfamiliarity' and 'familiarity' (measured in an information-theoretic manner). This quotient should achieve a certain ideal value. (See Nake [3] for an overview of approaches to formalizing 'esthetic information'. Interestingly, the number $\frac{1}{e}$ plays a significant role in at least some of these approaches.) However, at the moment the precise nature of a good mapping between (mis)matches and reinforcement is unclear and subject of ongoing research.

Currently some experimental research is going on in order to answer the following questions: What are useful learning rates (it is assumed that the model network should learn clearly faster than the controller)? What are useful relative strengths of pure goal-directed reinforcement and 'curiosity reinforcement'? And what are the properties of a good mapping from mismatches to reinforcement?

Although these questions are still open, in some preliminary experiments with a *linear* mapping from mismatches to reinforcement it already has been demonstrated that errors of the model network can be reduced by generating curiosity reinforcement in an on-line manner.

Concluding Remarks

The basic idea of implementing curiosity and boredom is not limited to the particular algorithm described in the first section. *Every* model-dependent on-line algorithm for learning goal directed behavior might be augmented by a similar implementation of 'the desire to improve the world model'. The basic motivation is: Instead of using some separate mechanism for improving the world model, we want to make use of the capabilities of the goal-directed learning algorithm itself.

The interesting side effect is: Since the learning algorithm depends on the model network, the model network has to make a prediction about its own current prediction capabilities. The *activations* of the model network are (partly) interpreted as a statement about the current *weights* of the model network. Note that this is already a rudimentary form of *self-introspective* behavior! The author believes that extensions of these rudimentary forms of introspective neural algorithms will be the key to learning systems which are much more sophisticated than the ones we know so far.

References

[1] M. I. Jordan. Supervised learning and systems with excess degrees of freedom. Technical Report COINS TR 88-27, Massachusetts Institute of Technology, 1988.

[2] P. W. Munro. A dual back-propagation scheme for scalar reinforcement learning. *Proceedings of the Ninth Annual Conference of the Cognitive Science Society, Seattle, WA*, pages 165–176, 1987.

[3] F. Nake. *Ästhetik als Informationsverarbeitung*. Springer, 1974.

[4] Nguyen and B. Widrow. The truck backer-upper: An example of self learning in neural networks. In *IEEE/INNS International Joint Conference on Neural Networks, Washington, D.C.*, volume 1, pages 357–364, 1989.

[5] A. J. Robinson and F. Fallside. Static and dynamic error propagation networks with application to speech coding. *Proceedings of Neural Information Processing Systems, American Institute of Physics*, 1987.

[6] T. Robinson and F. Fallside. Dynamic reinforcement driven error propagation networks with application to game playing. In *Proceedings of the 11th Conference of the Cognitive Science Society, Ann Arbor*, pages 836–843, 1989.

[7] J. H. Schmidhuber. A local learning algorithm for dynamic feedforward and recurrent networks. *Connection Science*, 1(4):403–412, 1990.

[8] J. H. Schmidhuber. An on-line algorithm for dynamic reinforcement learning and planning in reactive environments. In *Proc. IEEE/INNS International Joint Conference on Neural Networks, San Diego*, volume 2, pages 253–258, 1990.

[9] J. H. Schmidhuber. Recurrent networks adjusted by adaptive critics. In *Proc. IEEE/INNS International Joint Conference on Neural Networks, Washington, D. C.*, volume 1, pages 719–722, 1990.

[10] J. H. Schmidhuber. Reinforcement learning with interacting continually running fully recurrent networks. In *Proc. INNC International Neural Network Conference, Paris*, volume 2, pages 817–820, 1990.

[11] J. H. Schmidhuber and R. Huber. Learning to generate focus trajectories for attentive vision. Technical Report FKI-128-90, Institut für Informatik, Technische Universität München, 1990.

[12] P. J. Werbos. Backpropagation and neurocontrol: A review and prospectus. In *IEEE/INNS International Joint Conference on Neural Networks, Washington, D.C.*, volume 1, pages 209–216, 1989.

[13] R. J. Williams. On the use of backpropagation in associative reinforcement learning. In *IEEE International Conference on Neural Networks, San Diego*, volume 2, pages 263–270, 1988.

[14] R. J. Williams and D. Zipser. Experimental analysis of the real-time recurrent learning algorithm. *Connection Science*, 1(1):87–111, 1989.

ACTION SELECTION AND BEHAVIORAL SEQUENCES

Instinct as an Inductive Bias for Learning Behavioral Sequences

Lashon B. Booker*

Navy Center for Applied Research in Artificial Intelligence
Naval Research Laboratory, Code 5510
Washington, DC 20375-5000

Abstract

Ethologists have identified many ways that innate behavioral primitives and predispositions facilitate the learning of complex behaviors. This paper applies some of these insights to *classifier systems*. We show how certain built-in mechanisms for generating behavior in classifier systems provide an *inductive bias* that can be manipulated to improve learning. The effects of this bias on learning are illustrated by a classifier system that learns to solve a simple navigation task.

1. Introduction

Animals are born with a large repertoire of innate, coordinated patterns of muscle movement and behavior commonly referred to as *instincts*. Instincts are often thought of as inflexible, low-level "motor programs". However, ethologists have discovered that innate behaviors in fact often use learning as a strategy for filling in details that are too complex to specify completely in advance. While some instinctive behaviors may be rigid and stereotyped, many others are remarkably plastic.

The relationship between instinct and learning is not just relevant to ethologists however. Computational models of adaptive behavior can also benefit from understanding the important influences of prior structure on learning and behavior. Unfortunately, the role of prior structure and innate rules of behavior is ignored in most computational models of adaptive behavior. External reinforcement is usually viewed as the primary, if not the only, influence on learning.

This paper examines how prior structure or "instinct" can work together with reinforcement in *classifier systems*, a rule-based framework for studying adaptive behavior. The next section briefly reviews a few of the relationships between instinct and learning that are evident in animal behavior. In subsequent sections we describe the classifier system framework and show how it can be used to implement computational models of some of these relationships. In particular, we show how certain built-in mechanisms for generating behavior provide an *inductive bias* that can be manipulated to improve learning in classifier systems. The effects of this bias on learning are illustrated by a classifier system that learns to solve a simple navigation task.

* Author's current address is Artificial Intelligence Technical Center, The MITRE Corporation, 7525 Colshire Drive, McLean, VA 22102, booker@ai.mitre.org

2. Instinct and Learning

At one extreme, instinctive behaviors can be sufficiently preordained and inflexible that they proceed to completion automatically once they are triggered by an appropriate stimulus. These motor programs are called *fixed action patterns* and, once initiated, they often require little or no external feedback. A classic example of a fixed action pattern is the egg-rolling behavior of geese, which a goose will complete even if the egg is taken away (Tinbergen, 1951).

Other kinds of innate behavior patterns are rigidly programmed, yet exhibit a great deal of "run-time" flexibility (Gould, 1982). Examples include the construction of bird nests, beaver dams, and spider webs. These structures have fixed, species-specific characteristics, but the building behavior can adapt itself to a wide variety of both predictable and unpredictable contingencies in the immediate environment. Even more sophisticated examples of plasticity are evident in the way human infants learn to crawl and walk. One innate component of this behavior seems to be a goal-directed specification of what to learn — infants seems to have a built-in sense of which movements "feel right" — coupled with a motivational drive for repeated experimentation. Moreover, the learned movements show a further flexibility in the way they recalibrate themselves to accommodate the growth of the body. These are just a few examples of the many ways instinctive mechanisms and predispositions can facilitate the learning of complex behaviors.

It is clear that learning and instinct interact in a structured way, learning episodes being triggered and directed in particular contexts to achieve some specific, goal-oriented result. Ethologists use the phrase "programmed learning" to refer to the way instinct prescribes highly specific learning tasks:

> Learning, even in higher vertebrates, seems less a general quality of intelligence and more a specific, goal-oriented tool of instinct. Bouts of learning such as food-avoidance conditioning, imprinting, song learning, and so on, are specialized so as to focus on specific cues — releasers — during well-defined critical periods in particular contexts ... As a result animals know what in their busy and confusing world to learn and when, and what to do with the information once it has been acquired. Most learning, then, is as innate and preordained as the most rigid piece of instinctive behavior (Gould, 1982, p. 276).

However, instinct can guide learning in ways far more subtle than explicit programmed learning. This is especially evident in learned behavioral sequences which emerge under the direction of an implicit, innately specified goal. The ways in which innate pieces of behavior are orchestrated into complex behavioral sequences are not fully understood. It is clear, though, that more is involved than simply reinforcing chains of primitive motor programs. How can these more subtle relationships between learning and instinct be characterized? In particular, what role can innately set goals and predispositions play in computational models designed to learn sequences of behavior?

3. Adaptive Computational Models of Instinctive Behavior

As noted above, one important property of the way instinctive behavior is organized is the flexibility exhibited in the face of expected and unexpected contingencies in the environment. Achieving this kind of flexibility in a computational model of purposive behavior is a non-trivial challenge. At the level of individual units of behavior, much has been accomplished using computational implementations of what ethologists call an *innate releasing mechanism* (IRM). An IRM is an innate behavior pattern coupled with a schematic specification of the appropriate conditions to trigger the behavior (Tinbergen, 1951). The advantage of IRMs is that they can be implemented with simple connections from input to output, and they allow overall behavior to be directed by the relevant features of the immediate input configuration. This is an important opportunistic strategy for coping in real time with a complex and uncertain environment. As Simon (1969) has noted, complex behavioral sequences can sometimes emerge from the interaction of very simple mechanisms with a complex world.

The simple structure of an IRM is ideally suited to computational models that use condition/action rules to generate behavior. The structure is also convenient for implementing simple kinds of learning. Given an innately specified repertoire of actions, learning mechanisms can empirically discover associations between input and output that are adaptively significant. Behavioral sequences are not learned directly, but are an emergent property of the collection of learned input/output associations. Computational schemes that take advantage of this idea to simulate animal or robot behavior have been developed in adaptive frameworks like classifier systems (Wilson, 1985) and reactive systems (Whitehead & Ballard, 1989).

For complex behavioral sequences that require some explicit organization, it is somewhat more difficult to characterize the mechanisms that orchestrate behavior. Several metaphors have been proposed to help guide our intuition, including

- A hierarchy of controllers and command paths based on conventional principles of control theory (Arbib, 1972).
- A hierarchy of *instinctive centers*, coordinated by the principles of intra-level competition and inter-level subsumption of function (Tinbergen, 1951).
- A network of connections between sensors and effectors that directly encode various simple functional dependencies (Braitenberg, 1984).

There are relatively few computational models of adaptive behavior based on innate structures like these that have achieved significant learning of behavioral sequences. Successful approaches have relied on programmed learning to discover patterns in sensory or motivational input relevant to an individual unit of behavior. For example, a classifier system model has been developed for programmed learning in hierarchies of the kind described by Tinbergen (Booker, 1988).

Innate structures like those listed above are designed to directly control behavioral sequences. When the effects of instinct on a sequence are more indirect, it seems that learning must look beyond local relationships between input and output to a broader consideration of the overall behavioral context. The idea is to think of instinct as providing an abstract, perhaps implicit categorization of the current behavioral situation. The categorization is a predisposition about how to interpret what is going on that helps the more control-oriented instinctive mechanisms succeed. This kind of approach to behavior and action has recently been offered as an explanation of how abstract plans play a role in the moment-to-moment details of so-called *situated action* (Suchman, 1987):

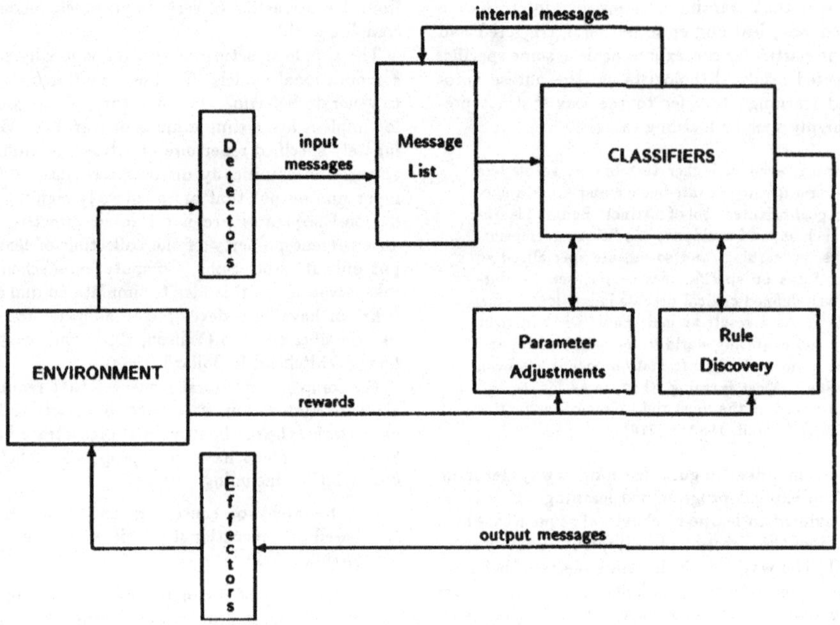

Figure 1: The basic classifier system architecture.

> The function of abstract representations is not to serve as specifications for the local interactions, but rather to orient or position us in a way that will allow us, through local interactions, to exploit some contingencies of our environment, and to avoid others. (p. 188)

If we view the more subtle aspects of instinct in this way as an abstract plan, then we must ask how such plans can be "hard-wired" and how they might facilitate the learning of behavioral sequences. The remainder of this paper takes a small step toward answering this question within the classifier system framework.

4. Classifier Systems

Classifier systems (Holland 1986) are parallel, message-passing, rule-based systems designed to avoid brittle behavior in "realistic" environments. An environment is considered realistic — from an information processing standpoint — if it is sufficiently complex and uncertain that it presents a perpetually novel stream of data to be categorized, predicted, and acted upon. In these environments, the correspondence between the unfolding complexity of the world and the representations being used by a problem solver cannot be taken for granted. Successful problem solving requires a pragmatic approach to discovering the structure and relevance of input categories. Moreover, it requires the capability to dynamically adjust representations and procedures to changing requirements without unnecessary disruption of existing capabilities. Classifier systems were designed to make this kind of flexibility possible.

The basic structure of a classifier system is illustrated in Figure 1. All information processing is achieved using binary strings of length k called *messages*. The set of messages to be processed on a given execution cycle is stored on a *message list*. Classifier systems process these messages by consulting a finite collection of condition/action rules called *classifiers*. Each classifier is a fixed-length string that has the same simple specification. The condition part of every classifier is a string in the alphabet $\{0,1,\#\}^k$. Conditions are matched against messages to determine which classifiers are eligible to be activated. A match requires that the 0's and 1's in the condition be identical to the values at corresponding positions of the message. The # symbol is a "don't care" place holder that allows general conditions to be specified. A probabilistic competition determines which classifiers are activated. Once a classifier is activated, the action part of the rule specifies a message to be generated. Some of these messages will initiate overt behavior at the output interface; others will be placed on the message list to be processed during the next execution cycle.

The typical classifier system also has two other basic elements: parameter adjustment mechanisms and rule discovery mechanisms. Classifier systems often maintain several statistics characterizing a classifier's past performance. The kinds of parameters typically maintained include measures of utility, relevance, consistency, etc. These parameters can be used in various ways to modify the behavior of the system. One very important use is in a *credit assignment* mechanism that evaluates each classifier and estimates how much it has contributed to successful problem solving behavior. This is especially important when overt feedback about performance is only available intermittently, after many rules have been activated over several execution cycles. The procedure most often used for credit assignment is the *bucket brigade* (Holland,1985).

The rule discovery mechanisms propose new rules to improve or expand upon current problem solving capabilities. The key requirement here is that the procedures generate *plausible* rules for functionally significant situations. Though plausibility can be difficult to define precisely, one robust way to proceed is to consider the constituent parts of rules as *building blocks* to be recombined in new ways (Holland, Holyoak, Nisbett, & Thagard, 1986). The average performance of existing rules that use a given part is a useful estimate of the "value" of that part. A new rule can be considered plausible to the extent that it is constructed from better than average parts. Most classifier systems use a *genetic algorithm* (Holland, 1975) for rule discovery, since those algorithms can efficiently manipulate large numbers of building blocks in the appropriate way.

This rather broad characterization of classifier systems leaves several questions unanswered about how the various components interact. Implementation decisions about the nature of those interactions cannot be made arbitrarily. See Booker (1988) for a review of some of the relevant implementation issues. Many variations on the basic themes discussed here are possible. A more detailed review of classifier systems can be found in Booker, Goldberg, & Holland (1989).

5. Classifier System Models of Instinctive Mechanisms

Previous work (Booker 1982, 1988) has shown how the classifier system framework can easily accommodate many of the hierarchically organized instinctive behaviors described by Tinbergen and other ethologists. In the ethologists view, an instinct is a hierarchy of mechanisms called *instinctive centers*, each of which has a releasing stimulus and an associated behavior. Every instinctive center has two potential sources of goal-directed control: the current motivational state, and other instinctive centers higher in the hierarchy. A center can only be activated to produce behavior when its control inputs and releasers are available. If several centers within a hierarchical level try to become active, the conflict is resolved by a winner-take-all competition based on strength of activation. If centers at more than one level are activated, control of overt behavior is subsumed by the center lowest in the hierarchy.

The rule discovery mechanisms in a classifier system can be used in a straightforward way to achieve programmed learning in these hierarchies. For example, Figure 2 shows a hierarchy of instinctive mechanisms that enable an animat to locate food and avoid noxious stimulation as it explores a simple environment. Stimulus signals indicating food and noxious objects serve as releasers that trigger the appropriate behavior — food-seeking or pain-avoidance — in the appropriate context. A classifier system architecture called GOFER has been developed that implements programmed learning in this hierarchy. GOFER learns rules that classify the objects experienced in the environment into categories that have affective significance. The affect associated with each category indicates the kind of behavior appropriate when an object is encountered. The representations GOFER learns can therefore be used as releasers to control food-seeking and pain-avoidance behavior. Details about the kind of conditioning needed for successful learning can be found in Booker (1988).

Though classifier systems lend themselves quite directly to programmed learning of behavior, they can also be used to study certain kinds of innately set goals and predispositions. The key to understanding how to do this is to examine the way coherent sequences of behavior emerge from local inductive mechanisms. There are two factors contributing to the coherence of behavioral sequences learned by a classifier system. The most obvious factor is credit assignment. The bucket brigade is a local credit assignment scheme that makes the estimated value of a rule proportional to the overt reward expected from the sequences in which the rule is activated. Sets of rules that play a key role in sequences leading to high rewards co-evolve and proliferate at the expense of rule combinations that are less successful. The genetic algorithm thus uses credit assignment information to generate selective pressures that weed out incoherent sequences. These selective pressures can be explicitly manipulated to focus learning and make rule discovery in classifier systems more efficient (Booker,

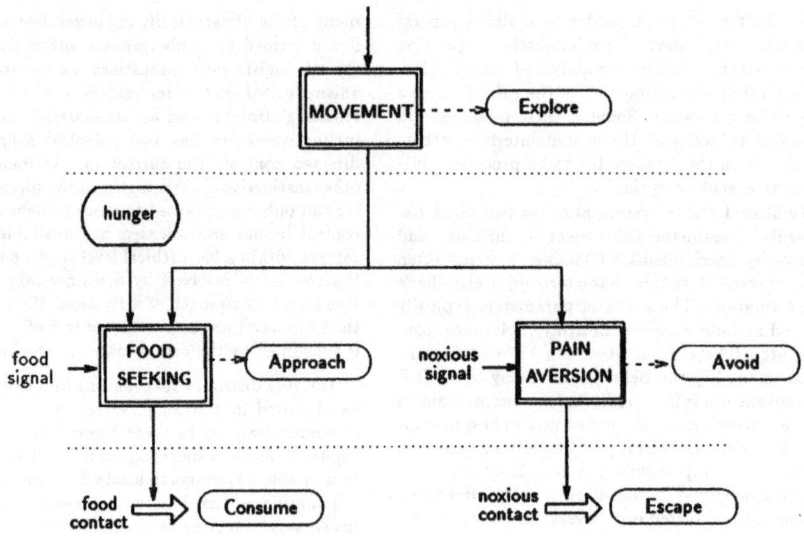

Figure 2: A hierarchy of instinctive centers controlling movement and feeding behavior in a simple animat. Classifier systems can be used to implement programmed learning in such hierarchies.

1989). This is one way to implement an *inductive bias* that influences what is learned by imposing a preference for particular kinds of inductive hypotheses.

A more subtle factor influencing the coherence of sequences is the way rules compete to become active. Selection can only operate with sequences that have been experienced, and the only sequences that get generated are those forged by competition among rules on each execution cycle. Rules compete probabilistically to become active, with individual probabilities determined by some built-in function of classifier parameters. The competition function is an innate criterion used to evaluate the available options for behavior. Since this criterion is usually based on credit assignment information, the behaviors most often selected via competition are those associated with the highest predicted reward. Consequently, the behavioral sequences that are generated tend to follow the "gradients" specified by successive predictions. The behavior remains relevant to the important contingencies in the environment because predictions are continually adjusted to reflect experience. Competition mechanisms therefore help to implement an "instinctive" preference for behavioral sequences that achieve goals. This preference is another kind of inductive bias that makes goal-related sequences easier to learn.

If we step back and take an ethologists view of how a classifier system behaves, it becomes clear that the criteria for manipulating selective pressure and the competition mechanisms are much like instincts. Each exerts an innate, context sensitive influence on learning to achieve a specific, goal-oriented result. This observation has two implications. First, it provides an explicit characterization — inductive bias — for some of the subtle effects of instinct on learning discussed earlier. This suggests that the machine learning literature on characterizing and manipulating inductive bias[1] may be relevant to some of the questions ethologists pose about the relationship between learning and instinct in animals. Second, it raises several questions about the role of competition in a classifier system. Once we think of competition as an instinctive mechanism, the traditional view of competition as a fixed, inflexible "operating principle" (Holland et al., 1986) immediately becomes suspect. It is natural to ask how well the mechanism is suited to the problems posed by a particular environment, if the mechanism should be rigid or flexible, if it should be organized as a single mechanism or a collection of several mechanisms, etc. To the extent that classifier system implementations have dealt with these issues at all, it is always in the form of

[1] See Gordon & Perlis (1989) for a review of this literature.

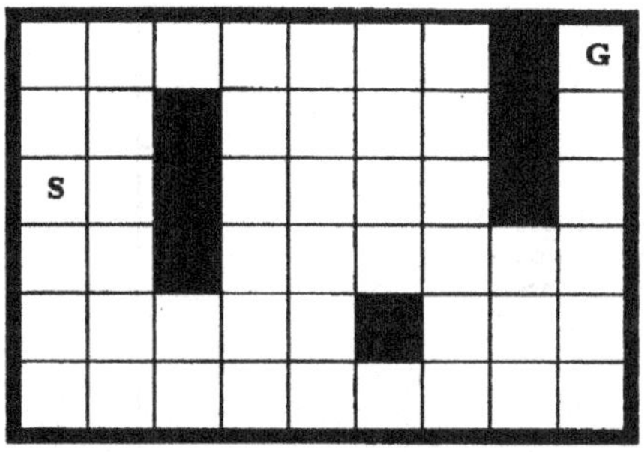

Figure 3: Sutton's (1990) navigation task. The starting location is labeled "S" and the goal is labeled "G".

a change to the competition function[2] as opposed to a change in the competition mechanism per se.

In order to show that there can be an advantage to taking a more ethologically motivated view of the competition mechanism in classifier systems, we now compare the effects of alternative mechanisms on performance in a simple navigation task.

6. A Simple Navigation Task

Navigation tasks provide a good challenge for studies of adaptive behavior in general and sequence learning in particular. The task proposed by Sutton (1990) is non-trivial yet easily simulated, so it is a good vehicle for a "proof of principle" exercise. The task is to find a path from a starting location to a goal in a simple two-dimensional space (see Figure 3). The space is organized into a 6 by 9 grid. Four actions are allowed at any location: UP, DOWN, RIGHT, LEFT. Any action that would lead to a forbidden location — those outside the space or in one of the shaded barriers — is ignored and leaves the position of the system unchanged. The optimum path from start to goal is only 14 steps long. However, a random walk from start to goal could easily take a few thousand steps. The task is therefore an adequate challenge for an adaptive system trying to learn behavioral sequences from experience.

The classifier system to be tested is a variant of the GOFER-1 system described by Booker (1989). GOFER-1 is a "single-step" classifier system architecture in which no internal messages are generated and every classifier proposes an overt action. The goal of learning in a system like GOFER-1 is to discover sets of rules that cover every input category the system encounters. In this sense, the system tries to build an "internal model" of its environment. This approach to reinforcement learning differs from that taken in other classifier systems, most of which focus only on optimal rules and behavior. One important innovation in GOFER-1 is its emphasis on *learning triggers*. A learning trigger is a mechanism for introducing local, context sensitive information about structure and performance into the rule discovery process. By emphasizing different inductive biases in different contexts, these triggers focus learning in a way that greatly enhances rule discovery in classifier systems.

A version of GOFER-1 containing 400 classifiers was tested on the navigation task using a simple reinforcement scheme. Whenever the system reached the goal state it was given an external reward of 1000. Every illegal action taken along the way resulted in an external reward of zero. No other external feedback about the task was available. On each time step, the system received a binary input message encoding the row and column indices of its current location.

In order to illustrate the inductive bias exerted by the competition mechanism, two different competition mechanisms were compared. The standard mechanism used estimates of expected rewards to select classifiers probabilistically as usual. The modified mechanism was inspired by the observation that it is sometimes advantageous to allow actions to persist (Selfridge, 1984). Accordingly, the standard mechanism was augmented with a rule that deterministically chooses a classifier that matches the current message if its action agrees with the last action taken. An action is eligible to persist in this manner until no matching classifiers advocate it, or an external reward is received. At that point the standard competition mechanism is used to select a new action. It is hypothesized that this predisposition for persistent action will facilitate sequence learning by shortening the chains of classifiers involved in credit assignment[3].

This hypothesis was tested by averaging the behavior of each mechanism over 100 experiments, each experiment consisting of 100 independent trials or traversals from start to goal. The comparative performance is summarized in Figure 4. As expected, the augmented mechanism for competition decisively outperforms the standard one. The differences are evident in the way the modified system avoids lengthy random walks in the early trials and finds near-optimal subsequences quite rapidly.

Note that overall performance approaches but does not reach the optimum level during the number of trials observed. That is because GOFER-1 learns a represen-

[2] For example, Wilson (1985) introduced a classifier parameter that estimated distance to food to bias competition so that his animat would prefer short paths to long paths.

[3] This is related to Holland's (1985) rationale for "bridging classifiers" that persist for several execution cycles and speed up the distribution of credit over long sequences.

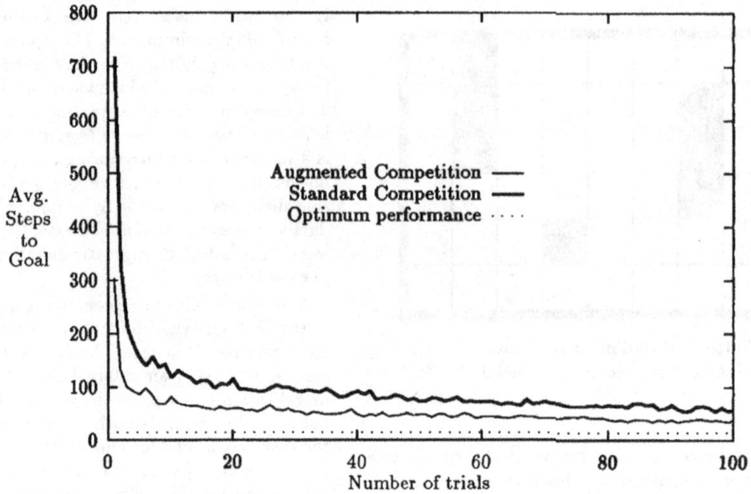

Figure 4: Learning curves showing improvement in the number of steps to the goal. Results are averaged over 100 experiments

tation of the environment that includes several paths to the goal in addition to the optimal one. Figure 5 shows a representative sample of the variety of paths discovered over several experiments.

Clearly, the instinctive bias for persistent action does not always generate optimal behavioral subsequences. It nevertheless is a useful preadaptation for discovering optimal sequences. Selective pressures eliminate the behaviors emerging from this bias that perform poorly. When the persistence bias does not apply, the standard competition mechanism generates a plausible new piece of behavior. The two instinctive mechanisms thereby work together to yield an overall instinctive influence on learning that, at least for this problem, is more effective than either mechanism acting alone.

7. Summary

The results given here are modest, but they are suggestive of what can be achieved when built-in classifier system mechanisms like competition are treated as instincts that have some goal-related function. By treating the competition mechanism as an instinct that biases learning, we explicitly acknowledge the role of prior structure in shaping adaptive behavior. In particular, we admit that prior structure need not be inflexible or "low-level". The stage is thus set for exploring how more sophisticated and flexible "instincts" might enhance the adaptive capabilities of classifier systems. In this regard, ethological studies of the organization of animal behavior can provide extremely valuable insights. A classifier system derived from a more balanced view of how structure and reinforcement shape behavior will not only be more plausible ethologically, it will potentially be more powerful as well.

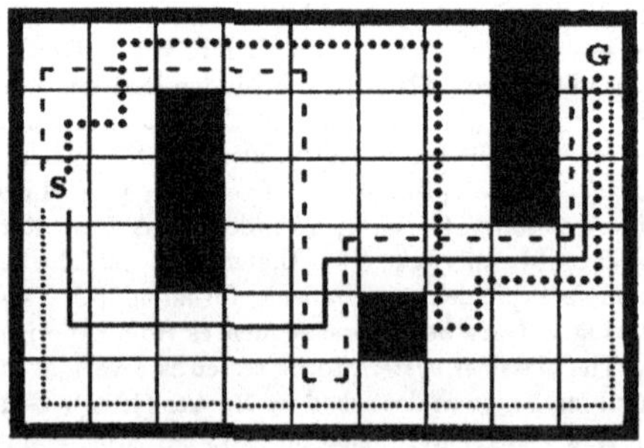

Figure 5: A representative sample of the paths to the goal learned during some of the experiments with the augmented competition mechanism.

† **Acknowledgements** I would like to thank Stewart Wilson for many helpful comments on an earlier draft of this paper.

References

Arbib, Michael A. (1972). *The metaphorical brain: An introduction to cybernetics as artificial intelligence and brain theory.* New York: Wiley.

Booker, Lashon B. (1982). *Intelligent behavior as an adaptation to the task environment.* Doctoral dissertation, Department of Computer and Communication Sciences, University of Michigan, Ann Arbor.

Booker, Lashon B. (1988). Classifier systems that learn internal world models. *Machine Learning, 3*, 161–192.

Booker, Lashon B. (1989). Triggered rule discovery in classifier systems. *Proceedings of the Third International Conference on Genetic Algorithms* (pp. 258–267). Fairfax, VA: Morgan Kaufmann.

Booker, L. B., Goldberg, D. E., & Holland, J. H. (1989). Classifier systems and genetic algorithms. *Artificial Intelligence, 40*, 235–282.

Braitenberg, Valentino (1984). *Vehicles: Experiments in synthetic psychology.* Cambridge, MA: MIT Press.

Gordon, Diana & Perlis, Donald (1989). Explicitly biased generalization. *Computational Intelligence, 5*, 67–81.

Gould, James J. (1982). *Ethology: The mechanisms and evolution of behavior.* New York: W. W. Norton.

Holland, John H. (1975). *Adaptation in natural and artificial systems.* Ann Arbor, MI: University of Michigan Press.

Holland, John H. (1985). Properties of the bucket brigade algorithm. *Proceedings of the First International Conference on Genetic Algorithms and Their Applications* (pp. 1–7). Pittsburgh, PA: Lawrence Erlbaum.

Holland, John H. (1986). Escaping brittleness: The possibilities of general-purpose learning algorithms applied to parallel rule-based systems. In R. Michalski, J. Carbonell, and T. Mitchell (Eds.), *Machine Learning: An Artificial Intelligence Approach* (Volume 2). Los Altos, CA: Morgan Kaufmann.

Holland, J. H., Holyoak, K. J., Nisbett, R. E., & Thagard, P. R. (1986). *Induction: Processes of inference, learning, and discovery.* Cambridge, MA: MIT Press.

Selfridge, Oliver G. (1984). Some themes and primitives in ill-defined systems. In O. Selfridge, E. Rissland, and M. Arbib (Eds.), *Adaptive Control of Ill-Defined Systems.* Proceedings of the NATO Advanced Research Institute on Adaptive Control of Ill Defined Systems held June 21–26, 1981 in Moretonhampstead, Devon, England. New York: Plenum

Simon, Herbert A. (1969). *The sciences of the artificial.* Cambridge, MA: MIT Press.

Suchman, Lucy A. (1987). *Plans and situated actions: The problem of human-machine communication.* New York: Cambridge University Press.

Sutton, Richard S. (1990). Integrated architectures for learning, planning, and reacting based on approximating dynamic programming. *Proceedings of the Seventh International Conference on Machine Learning* (pp. 216–224). Austin, TX: Morgan Kaufmann.

Tinbergen, N. (1951). *The study of instinct.* New York: Oxford University Press.

Whitehead, Steven D. & Ballard, Dana H. (1989). Reactive behavior, learning, and anticipation. *Proceedings of the 1989 NASA Conference on Space Telerobotics*, Pasadena, CA.

Wilson, Stewart W. (1985). Knowledge growth in an artificial animal. *Proceedings of the First International Conference on Genetic Algorithms and Their Applications* (pp. 16–23). Pittsburgh, PA: Lawrence Erlbaum.

A BOTTOM-UP MECHANISM FOR BEHAVIOR SELECTION IN AN ARTIFICIAL CREATURE

Pattie Maes

AI-Laboratory
Vrije Universiteit Brussel
Pleinlaan 2
B-1050 Brussels
Belgium
pattie@arti.vub.ac.be

AI-Laboratory
Massachusetts Institute of Technology
545 Technology Square
Cambridge, MA 02139
pattie@ai.mit.edu

Abstract.

In this paper we propose a mechanism for motivational competition and selection of behavior. One important characteristic of this mechanism is that the selection of behavior is modelled as an emergent property of a parallel process. This in contrast with mechanisms for behavior selection and motivational competition proposed earlier, which are based on a hierarchical, preprogrammed control structure. We show that selection of behavior can be modeled in a bottom-up way using an activation/inhibition dynamics among the different behaviors that can be selected. There is no weighing up of behaviors in a cognitive manner and neither are hierarchical or bureaucratic structures imposed. The paper elaborates upon the results we obtained with simulated creatures based on this mechanism. It draws parallels between characteristics observed in animal behavior and characteristics demonstrated by our artificial creatures. Examples are: displacement behavior, opportunistic behavior, fatigue, selective attention, and so on.

1. Introduction

This paper is concerned with the problem of behavior selection for an artificial creature. The context in which we discuss this problem is that of the behavior-based systems (Brooks, 1986) (Brooks, 1990), which embody a new philosophy for building artificial creatures, inspired by the field of Ethology (Mc Farland, 1981) and not unrelated to the Society of Mind theory (Minsky, 1986).

A creature is viewed as consisting of a set of behaviors. Examples of behaviors are: the feeding behavior, sleeping behavior, drinking behavior, etc. Only a few -- or often only one -- of these behaviors can be active at a time. However, a creature at every moment is probably motivated towards a variety of them. This means that there has to be some mechanism which decides which behavior "wins" and as such gets control over the "muscles" or actuators of the artificial creature.

In the case of simple animals, and also in the case of simple artificial creatures, the optimal strategy can be hard-wired respectively by nature (natural selection) or the programmer. The changes in behavior can be entirely preprogrammed, and selection of behavior is a matter of routine, showing very regular, rhythmic patterns. Although such a preprogrammed decision strategy may be useful for creatures living in a very stable and predictable environment, it does not suffice for creatures that have many jobs to do in an environment in which the opportunities to perform a job vary considerably (see Mc Farland, 1981, for the case of a natural creature) (see Maes, 1990b for the case of an artificial creature).

Complex creatures need a more flexible behavior selection mechanism which bases selection on the internal motivational state of the creature as well as on external circumstances. It is clear from observation of animal behavior that a change in the external environment may override the current behavior, for example, with some alarm response behavior. But animals also demonstrate changes in behavior without a change in the external situation, which suggests that behavior selection is also determined by internal motivation. E.g. a domestic hen, when presented with an egg may eat it on certain occasions or brood on it on other occasions (Mc Farland,

1981). The difference between these two behaviors is solely caused by the differences in motivation of being hungry or broody. Clearly a motivational competition is going on, which changes the priorities of different behaviors.

One of the insights of the first workshop on Artificial Life is that emergent functionality is an important organizing principle of living systems (Langton, 1989). One of the crucial aspects of "life" is the ability to flexibly and robustly react to a complex, dynamic environment. It has become clear that in complex and unpredictable environments, flexible and robust functionality is difficult to obtain in a "top-down", programmed, hierarchical architecture. A more succesful strategy is to design a functionality in a bottom-up fashion, by making it emerge as a global side-effect of some intensive, local interactions among the components that make up the organism.

This is exactly the strategy which we followed to implement the functionality of behavior selection. Instead of programming in advance a central control structure that regulates arbitration among behaviors, in our approach behavior selection emerges in a distributed fashion by parallel local interactions among the behaviors and between behaviors and the environment. A homeostatic process, or servo-control loop involving the behaviors of the creature and the environment is set up which converges towards the desired functionality. Its dynamics makes a selection of behavior emerge in response to the environmental conditions and the internal motivational state of the creature. As a result the behavior selection is more flexible, more robust, and more natural and fluid.

The following section presents the behavior selection mechanism we developed. It illustrates the algorithm with a particular simulated creature that bases its behavior selection on this algorithm. The third section discusses the results obtained with the algorithm. It draws some parallels between characteristics observed in animal behavior and characteristics demonstrated by our artificial creatures. Examples are displacement behavior, opportunistic behavior, fatigue, etc. Finally the last section discusses the advantages of a bottom-up mechanism for behavior selection as opposed to the more traditional top-down mechanisms that have been proposed. A more elaborate description of the algorithm and a mathematical model can be found in (Maes, 1990a) or (Maes, 1990b).

2. Mechanism

A creature is viewed as having a set of behaviors it can engage in. Every behavior has an associated "activation level", which is a real number. A behavior further also has a set of conditions (cf. stimuli), which have to be observed in order for the behavior to be "executable". Finally a behavior also has a threshold. An executable behavior whose activation level surpasses the threshold can be selected (become active). This means that a set of processes start running, which "realize" the behavior. Consider as an example the feeding behavior described in figure 1.

BEHAVIOR FEED
activation level: 78.3
conditions: (food-within-reach)
processes: <a set of processes that implement the feeding activity>

Fig. 1. An example behavior.

A creature also has a set of motivations, which have an associated number representing the strength of that motivation at a particular moment in time. Finally a creature has a set of perceptual conditions (or sensor readings) that it observes at a particular moment in its environment. Consider as an example, the creature illustrated in figure 2. It has ten different behaviors it can engage in: avoiding obstacles, exploring (random walk), fighting with another creature, going towards another creature, fleeing from a creature, eating (feeding), going towards food, drinking and sleeping. It has seven motivations: curiosity, lazyness, hunger, thirst, aggression, fear, and desire for safety. The strength of certain motivations is constant (e.g. safety, curiosity), while the strength of other motivations varies over time (e.g. moving makes the creature gradually more thirsty, hungry and sleepy, eating makes the creature more sleepy and decreases the hunger of the creature, and so on).

The different behaviors of a creature are linked in a network through "predecessor", "successor" and "conflicter" links. The idea is that these links among behaviors are either innate or learned through experience. There is a predecessor link from behavior A to behavior B (behavior A has behavior B as predecessor) if behavior B (through the actions it takes), makes certain conditions of behavior A come true. For example, 'exploring' may make the condition of the 'eating behavior' that food has to be within the reach of the creature, become true. Second, there is a matching successor link in the opposite direction for every predecessor link. Finally there is a conflicter link from behavior A to behavior B (behavior B is said to conflict with behavior A) if behavior B makes a certain condition of behavior A undone. For example, 'fleeing from another creature' may make the condition of 'fighting', that there is another creature nearby, untrue.

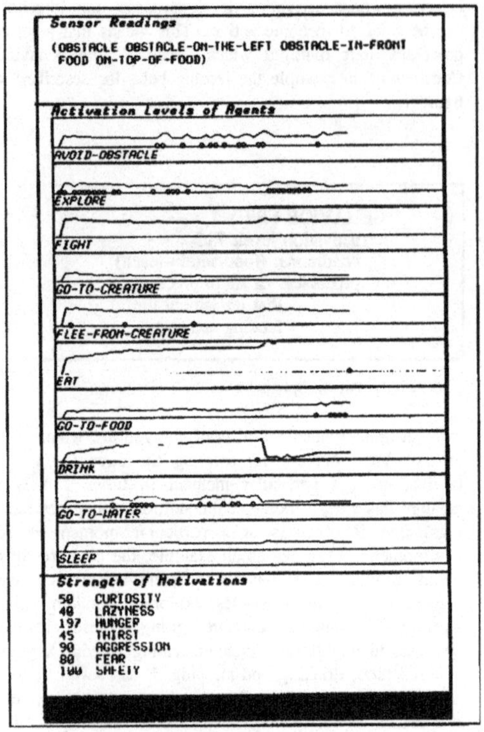

Fig. 2. The internal structure of an example creature.

Figure 3 illustrates the network for the particular creature studied in this paper.

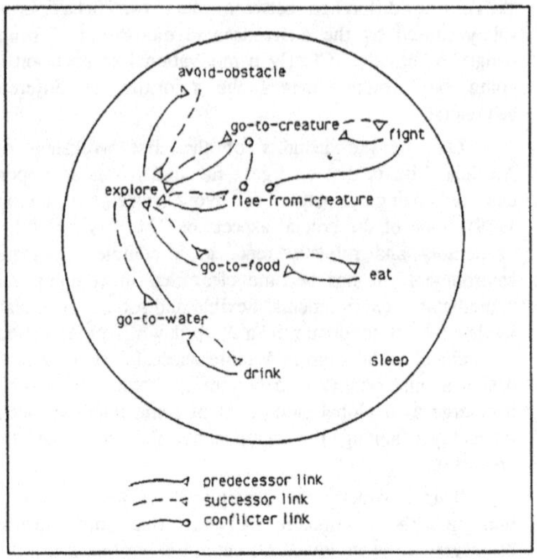

Fig. 3. The behaviors inside a creature are linked through predecessor, successor and conflicter links.

avoid-obstacle	safety
explore	curiosity
fight	aggression
flee-from-creature	fear
eat	hunger
sleep	lazyness
drink	thirst

Fig. 4. The consummatory behaviors of the creature and the motivations they are associated with.

Certain behaviors of a creature are associated with (one or more) motivations. These behaviors are called "consummatory behaviors" in the ethological literature (Mc Farland, 1981). For example, the 'feeding' behavior is associated with the motivation of obtaining food. The consummatory behaviors are the ones a creature really wants to indulge in when the associated motivation is high. In most cases, the execution of a consummatory behavior reduces the motivation it is associated with. For example, when executed, the 'feeding' behavior decreases the motivation for obtaining food. Figure 4 lists the consummatory behaviors of our example creature together with the motivations they are associated with.

Other behaviors are not directly associated with any motivation. These behaviors are called "appetitive behaviors". Examples are the behavior of 'going towards food' or 'going towards another creature'. Appetitive behaviors serve to make the conditions of a consummatory behavior come true. In the example, going towards food might make the condition of the 'feeding' behavior that there is food within the reach of the creature, come true. Or also 'exploring' the neighborhood might make the condition that another creature is observed nearby become true[1].

The intuitive idea behind the behavior selection mechanism is that behaviors use the links of the network to activate and inhibit each other (respectively increase and decrease each other's activation level), so that after some time the activation energy accumulates in the

[1] Certain behaviors are consummatory with respect to one motivation and appetitive with respect to another.

behavior that represents the "best" choice, given the current situation and motivational state of the creature. Once the activation level of a behavior reaches a certain threshold, it may be selected, and its processes start operating. The pattern of spreading activation among behaviors, as well as the input of new activation energy into the network is determined by the current situation and the motivational state of the creature. The input of new activation energy into the network of behaviors is defined as follows:

• *Activation by the Current Situation*
The currently observed perceptual conditions (sensor data) increase the activation level of the behaviors that have those conditions. For example, if the creature observes food within its reach, then the activation level of the 'eating' behavior is increased.

• *Activation by the Motivations*
The motivations of the creature increase the activation level of the consummatory behaviors that are associated with them with an amount which is relative to the strength of the motivation. For example, the activation level of 'eating' is increased with some amount proportional to the strength of the motivation 'hungry'. If more than one behavior is associated with a motivation, the amount given away by the motivation is distributed over the different behaviors.

These processes are continuous: there is a continual flow of activation energy towards the behaviors whose condition list partially matches the current situation and towards the modules that are associated with the motivations. If the situation or the motivational state changes (e.g. the perceived conditions change or the strengths of the motivations change), the input of activation energy automatically flows to other behaviors. Besides the impact on activation levels from the current situation and motivations, behaviors also activate and inhibit each other. Behaviors spread activation along their links as follows:

• *Activation of Successors*
An executable behavior spreads activation forward. It increases (by a fraction of its own activation level) the activation level of its successors. Intuitively, we want these successor behaviors to become more activated because they are "almost executable", since more of their conditions will be fulfilled after the behavior has become active. For example, if the 'going towards food' behavior is executable (i.e. food is perceived by the creature), this will increase the activation level of its successor, the 'eating' behavior, because 'going towards food' can make the condition that food has to be within reach, become true.

• *Activation of Predecessors*
A behavior that is not executable spreads activation backward. It increases (by a fraction of its own activation level) the activation level of its predecessors. Intuitively, a non-executable behavior spreads to the behaviors that can fulfill its conditions that are not yet true, so that the behavior itself may become executable afterwards. For example, if 'eating' is not executable, it will spread activation energy to 'going towards food' to encourage this behavior to become active (so that afterwards 'eating' can become active).

• *Inhibition of Conflicters*
Every behavior (executable or not) decreases (by a fraction of its own activation level) the activation level of its conflicters. Intuitively, every behavior will try to prevent a behavior that undoes one of its true conditions from becoming active. For example, the 'fighting' behavior will decrease the activation energy of its conflicter 'fleeing from another creature'.

The network in figure 5 illustrates the spreading activation process. The motivation 'hungry' is relatively high, while the motivation 'thirst' is relatively low (imagine that the creature has been drinking not so long ago, but has not fed for a while). There is water within reach of the creature, and food is being perceived further away from the creature (so the feeding behavior is not executable because its condition that food is within the reach of the creature is not fulfilled). The motivations increase the activation levels of the consummatory behaviors with a degree relative to their strength. This means that the activation level of 'feeding' will be increased more than that of 'drinking'. On the other hand, the current situation (the conditions perceived) will increase the activation level of 'drinking' and also of 'going towards food'. With respect to the spreading of activation within the network, 'feeding' will increase the activation level of 'going towards food' because feeding is not executable and 'going towards food' is a predecessor of 'feeding'. The other behaviors also spread activation towards predecessors or towards successors, or take away activation energy from their conflicters as indicated by the arrows.

The global algorithm performs a loop, in which at every timestep the following computation takes place over all of the behaviors:

• The impact of the current situation and motivations on the activation level of a behavior is computed.

• The way the behavior activates and inhibits related behaviors through its successor links, predecessor links and conflicter links is computed.

• A decay function ensures that the overall activation level remains within some boundary.

• The behavior that fulfills the following three conditions becomes active: (i) it has to be executable,

(ii) its level of activation has to surpass the threshold and (iii) it must have a higher activation level than all other behaviors that fulfill conditions (i) and (ii). When two behaviors fulfill these conditions (i.e., they are equally strong), one of them is chosen randomly. If none of the modules fulfills conditions (i) and (ii), the threshold is lowered by some percentage (e.g. 10). Once a behavior has been activated, its activation level is reset to 0 (but it may quickly increase again, since the spreading activation process goes on continuously).

In the example illustrated in figure 5, what happens is that the spreading activation process will go on long enough until one of the executable behaviors (sleep, drink, explore and go towards food) has accumulated enough activation energy to surpass the threshold. In this case, what happens is that because the creature is substantially more hungry than thirsty or sleepy or curious, 'going towards food' accumulates the most activation energy. This is the case if 'going towards food' receives more activation energy from on the one hand, the current situation, and on the other hand the 'eating' behavior (which in turn receives a lot of activation energy from 'hungry') than for example 'drinking' does, which receives energy from the current situation and also directly receives energy from the motivation 'thirst'.

So two "waves" of spreading activation happen in parallel. There is a wave of activation energy from the currently observed conditions forwards to the behaviors that match these conditions and on to the successors of these behaviors. And second, there is a wave of activation energy originating at the motivations, and flowing towards the consummatory behaviors for these motivations and further backwards to the predecessors of these behaviors. If a behavior contributes to several motivations or to several other behaviors (it is a predecessor of multiple behaviors), it receives activation energy from all of these sources. Also vice versa, the more conditions of a behavior are observed to be true, or the more executable predecessors a behavior has, the more activation energy it accumulates.

3. Results

We have evaluated the behavior selection mechanism presented above empirically by performing a wide series experiments in simulations and robots (Maes, 1989) (Maes, 1990a) (Maes, 1990b). This paper discusses the results obtained with an artificial simulated creature living in a 2-dimensional world. Figure 6 shows a bitmap of the simulation environment. The world is inhabited by

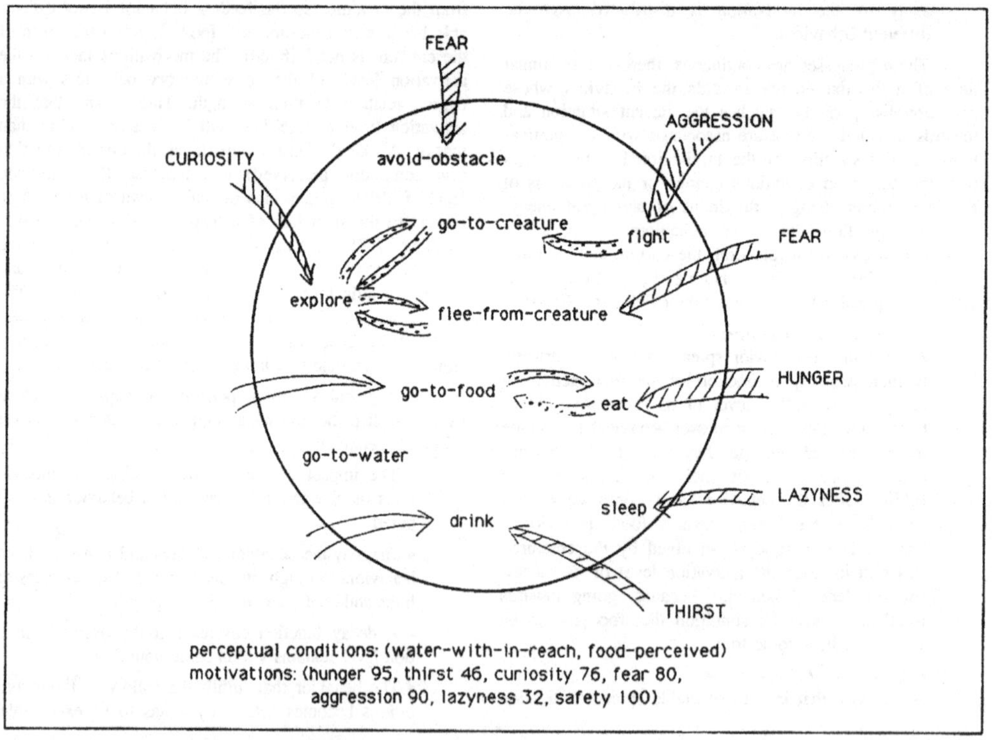

Fig. 5. A snapshot of the spreading activation process.

several identical creatures. One of the creature has a white cross on its back. This is the creature of which the internal state is shown on the right hand side of the interface. The upper window shows its current sensor readings. The window below that displays the different behaviors of the creature and their activation level over time. The little circles on the time axes show when a behavior is active. The lowest window displays the motivations and their current strength.

Creatures can sense each other, food (black circles), obstacles (the black boxes and the borders of the world), and water (gray stains). The sensors are noisy and return only local information. The information they return may be directional (e.g. 'food on the right').

The experimental set up allows a user to 'run' the system. In this mode time moves forward and all of the creatures 'live' and select behaviors on the basis of the algorithm discussed before. The right-hand windows show how the internal state of the white-crossed creature evolves. In this particular example the creature has already engaged in all sorts of activities except 'sleeping', 'fighting' and 'going towards another creature'. The latter two have not yet been selected because of the creature's motivational state: it is very fearful and not very aggressive.

Another thing one can do is to step through the evolution of the system. In this mode, it is possible to closely observe the spreading activation process and understand why a particular behavior is selected at a particular moment in time. at any moment the user can stop the simulation and change the world (add or remove food, obstacles and water) or the internal state of the creature (change the strengths of the motivations of the white-crossed creature). After that, the simulation can be resumed and the effects of these changes observed.

Using this simulation environment, we performed several experiments to test the properties of the behavior selection mechanism. The results were satisfactory in the sense that the creature chooses behaviors in an efficient and balanced way. The observed behavior selection is *flexible*. This is the case because the behavior selection process is completely "open". The environment as well as the motivations may change while a behavior is being selected. As a result, the external input/output as well as the internal activation/inhibition patterns will change to reflect the modified situation.

Even more, the external influence during spreading activation is so important that behaviors are only selected (structure is only formed) as long as the influence or or 'disturbance' from the environment and motivations is

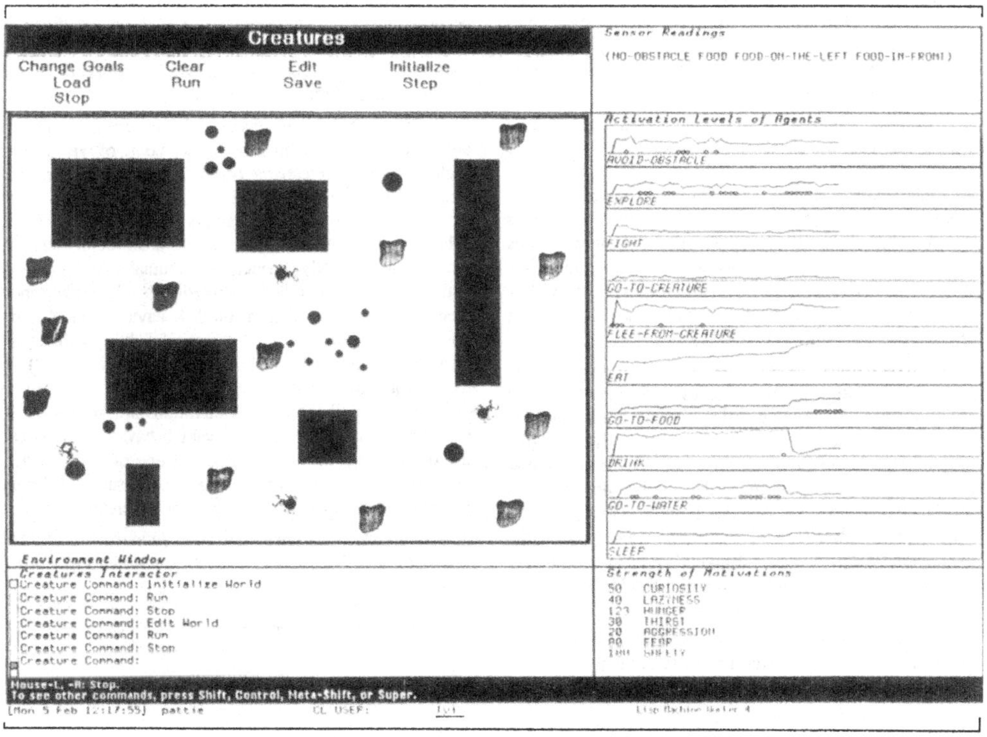

Fig. 6. The creatures in their 2-dimensional environment. The windows on the right show the internal state of the creature with a white cross on its back.

strong enough. I.e. when one decreases all of the motivational strengths to very low levels, the system might not receive enough new activation energy for a behavior to be selected. Another thing which might happen is that the creature becomes completely stimuli-driven (because that's where most of the new energy comes from).

The openness of the process ensures that changes in the world or in the creature itself are responded to *promptly* by the behavior selection. For example, if a creature is going towards some food and the user takes the particular food source away, the change in the creature's behavior will be almost immediate. On the other hand, the mechanism does implement some sort of "inertia". Because the activation levels are not reinitialised at 0 every timestep (or every time a behavior is selected), the past status of the world and the creature does have some influence on the activation levels and therefore on the behavior being selected.

The algorithm provides a numerical parameter which one can tune so as to make the behavior selection more or less sensitive to changes (more on the parameters to tune the characteristics of the behavior selection in Maes, 1990a and Maes, 1990b). In an environment in which the sensors are very noisy, there is some advantage to giving the system some inertia: when the food sensor at one timestep does not report any food being sensed, while it did so just before this, this might be due to some noise in the sensor (or the food might temporarily be out of sight). In this case, we do not want the creature to stop the behavior of going towards the food.

Another characteristic of the observed behavior selection is that it is *motivation-oriented.* If a behavior contributes to a motivation (i.e. there is a chain of successor links linking the behavior to the motivation), this behavior has more chances of being selected. A behavior that contributes to a stronger motivation has more chances of being selected than a behavior contributing to a less strong behavior. E.g. if the creature is more hungry than thirsty, the 'eating' behavior has more chances of being selected than the 'drinking' behavior.

The more motivations a behavior contributes to, the more its activation level increases and thus the more chances it has of becoming active. On the other hand, the 'further away' the behavior is from the motivations (the length of the chain of successor links), the lesser activation energy it receives from the motivations and the smaller its chances of becoming active. For example, since 'going towards food' is further away from the motivation 'hungry' than 'drinking' is away from the motivation 'thirst', 'drinking' might be selected, even if the creature is more hungry than thirsty.

The resulting behavior selection is also *situation-oriented.* The forward wave from the currently observed conditions forwards into the network implements some sort of *anticipation.* For example, if another creature is observed in the neighborhood, this will already activate the 'fighting' behavior via the 'go-towards-creature' behavior. Similar remarks as for motivation-orientedness hold for situation-orientedness. So, the more predecessor paths exist from the behavior to the observed conditions, the more chances the behavior has of becoming selected, and so on.

The fact that our algorithm allows both the motivations and the stimuli to play a role in the competition among behaviors represents an advantage with respect to behavior selection mechanisms proposed before, which have either been goal- (motivation) driven or data- (stimuli) driven. It allows the creature to show *opportunism.* Consider for example the world in figure 7. The creature has been chasing another creature for a while, in order to ultimately satisfy the 'aggression' motivation by activating the 'fighting' behavior. It happens to run over water while doing so, and even if the motivation for 'thirst' is lower than that for 'aggression', it uses this opportunity to drink some water.

Again there is a parameter which one can use to trade-off motivation-orientedness for situation-orientedness. The relative importance of the stimuli versus the motivations also varies by itself. When a lot of stimuli are observed to be true, the system will automatically become less situation-oriented and more motivation-oriented. Also vice versa, when all of the motivations have similar strength, the relative importance of the stimuli is much higher.

Some phenomena observed in our artificial creature are interesting from an ethological point of view. One such example is *irrelevant behavior*, also called *displacement behavior*[2] (Mc Farland, 1981), or the phenomenon that the most relevant behaviors eliminate each other. When two behaviors which are mutual conflicters are more or less equally strong, their mutual inhibition (they both decrease each other's activation level) might cancel them both out, so that a third behavior "wins". We observed this behavior in the case of fighting and fleeing from a creature. When both the 'aggression' and 'fear' motivation are very strong, and another creature is observed, and all the other motivations are very low, the creature might choose an irrelevant behavior, such as eating or sleeping, instead of choosing one of the important ones (with respect to the motivations) such as 'going towards the creature' or 'fleeing from the creature'.

Another such phenomenon is *fatigue* (Mc Farland, 1981), or the phenomenon that after a while a competing

[2] This is one of the small differences with the algorithm described in (Maes, 1990a) and (Maes, 1990b), which, in case of

motivation will take over, even when the stimuli and motivation for the current behavior have not changed. Since the activation level of a behavior is always reinitialised to 0 whenever it has been activated, and since the activation level of its competitors is not, one of these competitors might slowly accumulate more activation energy and finally become selected.

One more such phenomenon is that of *satiation* (Mc Farland, 1981). When an animal is feeding undisturbed from an abundant source of food, it will eventually stop feeding, and this before the state of zero hunger is reached. The same thing can be observed in our artificial creature: it does not eat till all the food of the source is gone, but until some other motivation becomes stronger (the 'hungry' motivation decreasing all the time) and makes the creature select a different behavior.

One final phenomenon we observed is that of *varying attention* (Mc Farland, 1981). This word is generally used by ethologists to denote the lower attention an animal pays to danger (e.g. predators), when it is in an extreme state of motivation, for example, when it is very hungry. Our simulated creature has to activate the 'avoid obstacles' behavior to avoid walking into the boundaries of the world. When one sets the 'hungry' motivation extremely high with respect to the 'safety' motivation, the creature sometimes runs into "the walls" of our simulated world. This is the case because the 'hungry' motivation increases the activation level of 'eating', which in its turn increases the activation level of 'going towards food' which in its turn increases the activation level of 'explore' in such a way that it is higher than that of 'avoid obstacles', even when an obstacle is being observed.

4. Discussion

In our algorithm, the control structure regulating when a particular behavior is activated is emergent: the dynamics of interaction between the behaviors themselves establishes the sequence of selected behaviors in a completely distributed way in response to environmental conditions and to the creature's motivational state. There are no global rules or programs that specify how the behavior selection functionality should be obtained. Instead, a dynamics takes place which forms structure in response to environmental conditions. Advantages of this bottom-up approach to the problem of behavior selection over the hierarchical, top-down, programmed approach are:

mutual conflict of modules, only makes the one with the higher activation level inhibit the other.

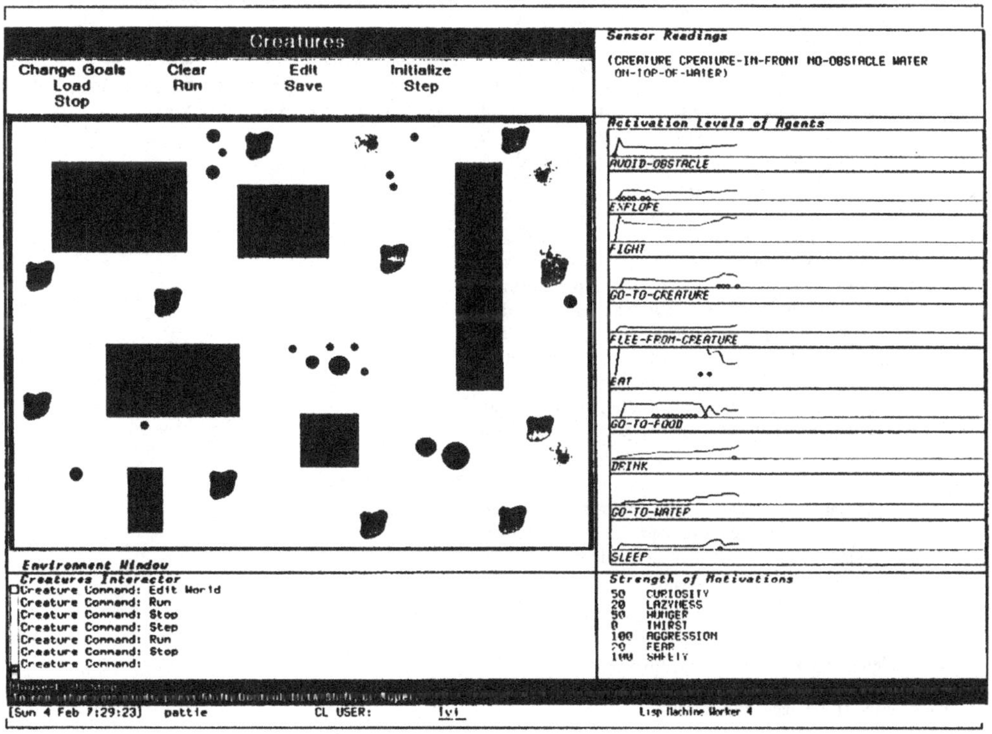

Fig. 7. An example of opportunistic behavior: the creature chooses to use the opportunity that it is on top of water to drink, while it is chasing another creature to fight with it. The strength of drink is rest to 0 (it was 40 before the drink behavior was selected).

• *The result is more more natural and fluid.*
This approach does not not require a categorization into discrete types of situations, which are responded to with a (radically) different behavior, as the hierarchical approach does. Instead, here the processes are continuous: there is a continual flow of activation energy towards the behaviors that partially match the current situation and towards the modules that are associated with the motivations. If the situation or the motivational state changes, the input of activation energy automatically flows to other behaviors. In addition, this change is a smooth one, as opposed to the very abrupt changes in response one finds in hierarchical systems.

• *The result is more flexible and reactive.*
In a hierarchical system, changes in the situation have to propagate from the top node of a hierarchy through all of the abstraction layers, down to the bottom components, where these changes should imply a real difference. In our creatures, the interactions with the environment happen directly in parallel through all of the bottom level components. Another disadvantage of hierarchical systems with respect to flexibility is that all of the possible situations have to be foreseen in advance, so that it can be programmed what behaviors the system should activate in each of them. In a system relying on emergent functionality the interactions with the environment happen directly in parallel through all of the bottom level components. As a result a much wider range of resulting activity can be obtained.

• *The result is more robust.*
None of the components (behaviors) of our creatures is more critical than another, which means that if one or more of the components break down, the system shows a smooth degradation of performance. One can, for example, add new behaviors or modify existing ones or even delete existing ones. The dynamics will automatically adapt to the new situation and still do whatever is in its possibilities. In hierarchical systems there are central, and thus more critical components which may cause the whole system to break down when they don't function properly.

Disadvantages of systems with emergent functionality are that the resulting behavior selection is less predictable, and related to this, that it is less understood how to obtain the desired global functionality.

Acknowledgements

The author wishes to thank Jan Torreele for writing the graphical simulator and Luc Steels for the introduction to the concept of emergent functionality.

This research is sponsored by the Belgian Ministry of Science (incentive programme on AI) through contract RFO/AI/10. The author is a fellow of the Belgian National Science Foundation.

References

Brooks, R. (1986) A Robust Layered Control System for a Mobile Robot. IEEE Journal of Robotics and Automation. Volume RA-2, Number 1.

Coderre, B. (1989) Modeling Behavior in Pet World. In: Artificial Life. Edited by C. Langton. Addison Wesley.

Langton, C. (1989) Artificial Life. In: Artificial Life. Edited by C. Langton. Addison Wesley.

Maes, P. (1989) The Dynamics of Action Selection. Proceedings of the IJCAI-89 conference, Detroit.

Maes, P. (1990a) How To Do the Right Thing. Connection Science Journal, special issue on Hybrid Systems. 1(3). February 1990. Also MIT AILAB Memo 1180.

Maes P. (1990b) Situated Agents can have Goals. Robotics and Autonomous Systems Journal. Special Issue on Designing Autonomous Agents. 6(1&2). Also to be published as a Bradford-MIT Press book.

Mc Farland, D. (1981) The Oxford Companion to Animal Behavior. Oxford University Press.

Minsky, M. (1986) The Society of the Mind. Simon and Schuster, New York, New York.

The Neural Basis of Behavioral Choice in an Artificial Insect

Randall D. Beer[1,2] and Hillel J. Chiel[2]
Departments of [1]Computer Engineering and Science, and [2]Biology
and the Center for Automation and Intelligent Systems Research
Case Western Reserve University
Cleveland, OH 44106 USA

Abstract

As its internal state and external environment continuously change, any truly autonomous agent must choose actions which are most appropriate to its immediate circumstances. This paper explores the idea that neurobiological design principles can be applied to the flexible control of autonomous agents. We describe a simulated insect whose behavior is controlled by an artificial nervous system. In particular, we focus on the neural basis of two different examples of behavioral choice in this artificial insect.

1 Introduction

One of the most fundamental problems faced by any agent, either natural or artificial, which must function autonomously in the real world is deciding what to do next. As both its external environment and internal state continuously change, an autonomous agent must constantly choose actions which lead to global behavior most appropriate for the current situation. Broadly speaking, the problem of behavioral choice encompasses the entire spectrum from minor adjustments of ongoing behavior to discrete switches between different behaviors. In addition, it involves the generation of groups of related behaviors with the appropriate timing and sequencing to accomplish specific objectives (McFarland, 1981, pp. 118-121).

How should the control system of an autonomous agent be organized to support such behavioral choice? Recently, there has been a trend toward more distributed approaches. For example, Brooks (1986) has been exploring the *subsumption architecture* for autonomous agent control. This architecture consists of layers of task-achieving behaviors each of which is implemented as a network of finite state machines augmented by timers and registers. Interactions between behaviors are handled by allowing machines in one layer to suppress interactions between machines in lower levels. In a similar vein, Maes (1989) has proposed an approach to action selection in which a collection of simple agents interact by passing activation along a variety of special-purpose links.

Our own approach is grounded in the study of the neuronal mechanisms underlying the behavior of simpler natural animals, a field known as *neuroethology* (Camhi, 1984). We have been exploring the idea that neural network control architectures for autonomous agents can be designed using principles drawn directly from biological nervous systems. This approach has the advantage that more direct interactions between biological and artificial mechanisms for autonomous behavior are possible. This style of modeling was first proposed by Braitenberg (1984). In this paper, we describe an artificial nervous system we have designed for controlling the behavior of a simulated insect. The design of this insect is based in part upon specific behaviors and neural circuits drawn from several natural animals. We focus here on two different examples of behavioral choice in this artificial insect.

2 The Artificial Insect Project

The artificial insect project is aimed at exploring the use of neuroethological principles to design artificial nervous systems for controlling the behavior of complete autonomous agents, an endeavor which we have termed *computational neuroethology* (Beer, 1990). We have developed a simulated insect, a simulated environment with which it must cope, and an artificial nervous system for controlling its behavior. The insect is capable of locomotion, wandering, edge-following, and feeding, as well as properly managing the interactions between its various behaviors. In order to understand behavioral choice in this artificial insect, it is essential to understand the details of its design, which we briefly review below.

The artificial insect is a two-dimensional abstraction of a biological insect (see Figure 5). Its body consists of a combined thorax/abdomen, a head, six legs and feet, two antennae, and a mouth. The feet may be either up or down. A leg can apply forces which move the body when its foot is down. The mouth can open and close. Both the mouth and the antennae contain tactile and chemical sensors. The insect also has a simple metabolism, in which an internal supply of energy is consumed at a fixed rate. The environment in which the insect is embedded can contain bricks, patches of food, and other insects. Bricks are immovable, and food patches emit an odor whose intensity is proportional to the size of the patch, and falls off as the inverse square of the distance from the patch's center.

The insect's nervous system is composed of networks of simplified model neurons which are similar, but not identical, to those that have been employed in artificial neural network research. A nerve cell is modelled as an RC circuit. The firing frequency of a model neuron is a saturating linear threshold function of its membrane potential. Connections between model neurons allow activity in one neuron to inject current into another one. A crucial distinguishing feature of our neural model is its inclusion of intrinsic currents which may be both voltage and time dependent. These intrinsic currents allow us to model some of the dynamical properties of nerve cells, and support the construction of model neurons with rhythmic activity or temporally extended responses to brief stimuli. Such properties appear to play an important role in the normal operation of neuronal circuits (Selverston, 1988).

The artificial insect's locomotion is controlled by a network of pacemaker neurons, which are neurons that intrinsically produce rhythmic bursts of activity (Beer *et al.*, 1989). Each pacemaker neuron is responsible for generating the swing and stance movements of a single leg. This basic pattern is shaped by feedback from leg position sensors and local leg reflexes. The individual leg circuits are coupled so as to produce statically stable gaits. The overall design of this locomotion circuitry is based on a neurobiological model of cockroach locomotion proposed by Pearson (1976). We have found that the artificial insect is capable of producing a continuous range of gaits similar to those observed in insects simply by varying the activity of a single command neuron.

Robustness is an essential substrate for generating behavior that can be modulated by ongoing events. In order to explore the operation and robustness of the locomotion controller, we undertook a series of lesion studies, i.e. examining the behavioral effects of selected removal of its component neurons and connections (Chiel and Beer, 1989). We found this circuit to be remarkably robust. The response of the model to certain lesions was quite similar to that observed in actual insects. In addition, we discovered unexpected subtleties in its operation. For example, we found that the relative importance of the central pattern generators and sensory feedback varied for different gaits, with the faster gaits primarily being generated centrally, and the slower gaits more dependent upon sensory feedback. The robustness of this locomotion controller allowed us to layer on additional neural circuitry for such exploratory behaviors as wandering, recoil, and edge-following.

One aspect of behavioral choice that goes beyond simple reactions to the external environment is the role played by internal drives (***motivational states***) which serve to direct an animal's responses toward the satisfaction of specific goals (McFarland, 1981, pp. 390-398). A hungry animal, for instance, generates behavior which is very different from a satiated one in the presence of food. Based upon work in the marine mollusc ***Aplysia*** (Chiel et al., 1990), we have designed neural circuitry which implements a feeding behavior for the artificial insect (Beer and Chiel, 1990). We have found that this feeding behavior shares many characteristics with the motivated behaviors of natural animals, including a build-up and decay of arousal (Kupfermann, 1974).

The complete artificial insect is capable of many different behaviors, each of which is controlled by its own neural circuitry. However, because many of these behaviors share the same body, they are potentially incompatible (McFarland, 1981, p. 305). For example, several of the behaviors involve turning the insect's body, possibly in opposite directions. Properly managing the interactions between its various behaviors is extremely important to the behavioral cohesiveness of the insect as it confronts its environment. Generally speaking, the consummatory phase of feeding should take precedence over the search for food (the appetitive phase). In addition, the appetitive phase of feeding should take precedence over edge-following, which should in turn take precedence over wandering. The artificial insect's behavior can therefore be arranged into the behavioral hierarchy illustrated schematically in Figure 1.

3 Neural Mechanisms of Behavioral Choice

The excitatory and inhibitory connections in Figure 1 are meant only to illustrate the interactions between the artificial insect's various behaviors. How can these interactions be implemented neurally? From the few instances of behavioral choice whose neuronal basis has been studied in biological organisms, there seems to be no single answer to this question. A variety of mechanisms are employed (Kristan *et al.*, 1989). In some cases, behavioral choice appears to be mediated by in-

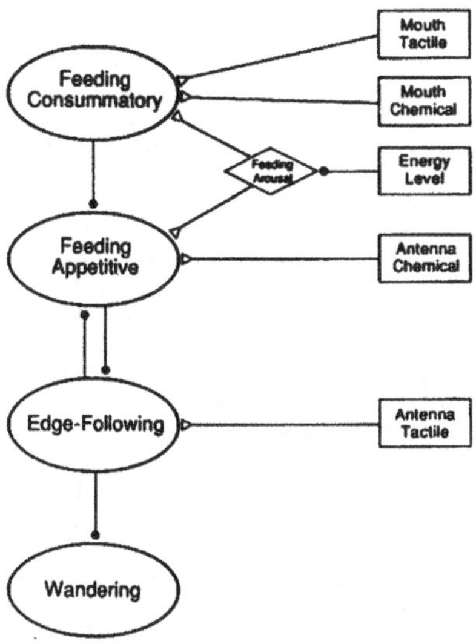

Figure 1: The behavioral organization of the artificial insect. Behaviors are shown as ellipses, boxes represent sensory stimuli, and the motivational state governing feeding is shown as a diamond.

hibition between the pattern generators for each behavior, e.g. the choice between walking and flight in the locust (Ramirez and Pearson, 1988). In other cases, such as the choice between forward and backward walking in crayfish (Kovac, 1974), the expression of a given pattern generator is varied in order to produce different behaviors. Finally, portions of the same pattern generator may be functionally rewired for different behaviors, e.g. the choice between withdrawal and swimming in the sea slug *Tritonia* in response to a tactile stimulus (Getting and Dekin, 1985). Combinations of all three mechanisms may be employed in some cases.

Indeed, Altman and Kien (1987; 1989) have argued that behavioral choice is not governed by discrete command centers which simply switch between different neural circuits. Rather, they suggest that decisions are made in a highly distributed fashion, by consensus among multiple centers potentially involving much of the nervous system. According to their view, the actual behavior expressed by an animal at any given point in time is actively constructed from the sum total of interactions among many interneuronal feedback loops, external sensory inputs, and internal physiological states.

The neural basis of behavioral choice in *P. computatrix* is similar in spirit to this view. Decisions are made by consensus among the individual controllers rather than by a centralized command center. However, these interactions are much simpler than those envisioned by Altman and Kien. In most cases, the required interactions between two behaviors can be directly implemented by explicit connections between the corresponding command neurons. For example, edge-following behavior can suppress wandering via direct inhibitory connections from a key neuron in the edge-following controller to a key neuron in the wandering controller. The neural implementation of other instances of behavioral choice in the artificial insect is more complex. In this paper, we will consider two such examples in some detail.

4 Recoil

A controller for a recoil response must be capable of dealing with several different contingencies. When the antennae glance off something to the side of the insect, it should briefly turn away from the object and continue walking. On the other hand, when the insect runs directly into an object, it should back up and turn before continuing on its way, in order to ensure that it will not repeatedly collide with the same object. Relative orientations intermediate between these two extremes should exhibit different proportions of forward and backward walking during the recoil response. The duration of the overall response should also be affected. A brief response is sufficient for a glancing contact, but a head-on collision requires an extended response. From the perspective of behavioral choice, the most interesting aspect of the recoil response is how it decides what pattern of response to generate for a particular contact.

These properties raise a number of issues for the design of a recoil controller. First, it must have some means for reversing the normal direction of walking. Second, it must be *multiphasic*, that is, it must be capable of generating both the forward and backward phases of the response in the appropriate sequence and with the appropriate timing. Third, the controller must exhibit *orientation sensitivity*. The durations of these two phases depend upon the relative orientation between the insect and the object contacted. In fact, if the angle of incidence is small enough, no backward walking occurs at all. Finally, the controller must exhibit some *hysteresis*. While the response is triggered by a brief sensory stimulus, it must run to completion even after contact has been lost.

Backward walking can be implemented by a controller very similar to that used for forward walking. The only real difference is that the roles of the motor neurons producing the swing and stance phases must be reversed and therefore some of the connections which implement these roles must be changed. Many exam-

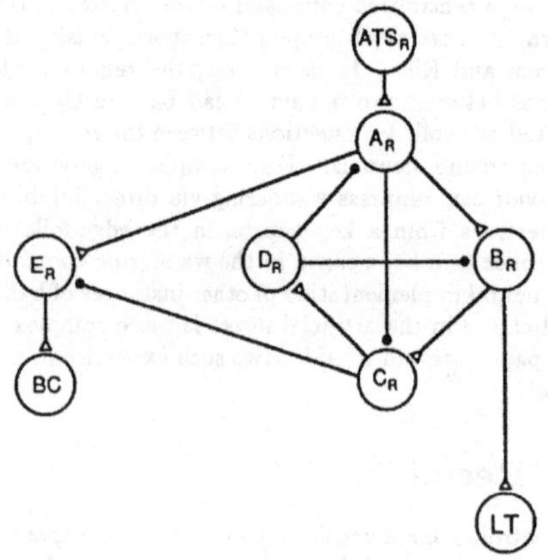

Figure 2: The recoil controller for the right antenna. Excitatory connections are shown as open triangles and inhibitory connections are shown as filled circles.

ples of such multifunctionality now exist in the invertebrate literature (e.g. Getting and Dekin, 1985). In fact, the crayfish appears to utilize the same pattern generator for both forward and backward walking, with only small modifications to vary the timing of muscle contractions (Kovac, 1974). In all of these cases, circuitry is shared among a number of related but distinct behaviors, with some mechanism for dynamically modifying the shared circuitry into the functionally distinct configurations required for each behavior.

The recoil response controller for the right antenna is shown in Figure 2. The controller itself consists of the antenna tactile sensor (ATS_R) and interneurons A_R, B_R, C_R, D_R, and E_R. BC is a command neuron which reconfigures the locomotion controller for backward walking. LT is the left turn neuron from the turning controller. There is another copy of this recoil response controller for the left antenna.

The antenna tactile sensor (ATS_R) generates a burst whose frequency is proportional to the relative orientation of the body and the tangent line of the object at the point of contact; the larger this angle, the higher the firing frequency. A head-on collision produces a high frequency burst and a glancing contact produces a low frequency one. This burst excites A_R, which is a *trigger cell*. Neuron A_R translates a brief burst of activity into an extended response whose characteristics depend upon those of the initial burst. Whenever external currents cause it to burst, an intrinsic current is triggered whose magnitude and duration is proportional to the membrane potential of the cell. This cell governs the duration of the entire recoil response. The other neurons B_R, C_R, and D_R govern the durations of the various phases of the response. These neurons have large time constants (on the order of tens to hundreds of milliseconds) and high thresholds.

This controller operates as follows. When the right antenna contacts an object, ATS_R generates a brief burst whose frequency is proportional to the contact angle. This brief burst produces an extended burst in the trigger neuron A_R whose frequency and duration are proportional to the activity of ATS_R. Activity in A_R excites E_R. If E_R goes above threshold, then it causes the insect to walk backward by exciting BC. Thus the threshold of E_R implements the decision whether or not to back up based upon the contact angle.

In addition, A_R excites B_R. After an initial delay due to its slightly larger time constant, B_R will begin to burst and cause a turn to the left by exciting LT. After a somewhat longer delay, B_R will cause C_R to burst, shutting off E_R. This removes any excitation that E_R may have been providing to BC, and allows the insect to continue walking forward. The time it takes C_R to go above threshold is also modulated by the inhibitory connection from A_R. The effect of this connection is to increase the delay before C_R bursts as a function of the activity of A_R. Therefore, the more head-on a collision is, the more backing up the insect will do.

C_R also excites D_R. After another delay, during which the insect is walking forward and still turning away from the contacted object, D_R will burst, inhibiting A_R and B_R and thus terminating the recoil response. For glancing contacts, the trigger current in A_R can terminate before the A_R, B_R, C_R, D_R loop completes. In this case, only a short turn away from the direction of contact with the object is generated. The response of an insect containing this controller is shown for three different angles of incidence in Figure 3. Note that the recoil controller generates rather different patterns of behavior based solely on the intensity of the initiating stimulus. This is similar to *Tritonia*, in which a single neural circuit generates either a brief withdrawal or an extended escape swim based only upon the intensity of the tactile stimulus (Getting and Dekin, 1985).

5 Behavioral Interaction

An important aspect of behavioral choice involves the selection of one behavior from a set of possible behaviors. This selection can be based upon both the environmental context and internal motivational state of an animal (McFarland, 1981, pp. 118-121). Perhaps the most interesting example of this in the artificial in-

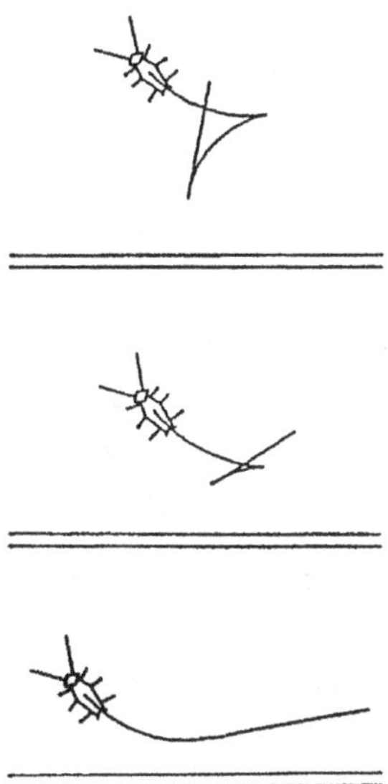

Figure 3: The recoil responses generated by the recoil controller for three different angles of incidence. (Bottom) A small angle of incidence produces only a turn away from the wall. (Middle) A larger angle of incidence results in a small amount of backing up as the insect turns away. (Top) A nearly head-on collision results in a great deal of backing up and turning away.

sect is the interaction between edge-following and feeding. Normally, the appetitive phase of feeding, in which the insect is following a chemical odor signal to food, takes precedence over edge-following in a hungry insect. However, if an obstacle stands between the insect and a food patch, a sensible strategy would be to temporarily invert this precedence, following the edge of the obstacle in the hopes of getting around it. In this case, the relationship between these two behaviors depends upon the context, as is illustrated in Figure 1 by a two-way connection between them. How can such context-dependent behavioral interactions be implemented in neural circuitry?

Clearly, the full answer to this question depends upon the design of the individual controllers. Unfortunately, space limitations prevent a complete characterization of the operation of the edge-following and feeding controllers. We will therefore provide only a brief sketch of the overall operation of these two controllers here. More details can be found in (Beer, 1990).

The edge-following controller is triggered whenever one of the insect's antennae contacts an obstacle, unless it is otherwise inhibited. Once active, the primary job of the edge-following controller is to maintain the insect's body in a nearly parallel orientation to the edge it is following. This is accomplished by using information from deflection sensors in the tips of the antennae to guide a turn of the appropriate magnitude either toward or away from the edge. The controller also exhibits a certain amount of hysteresis. If contact with an edge is lost, the insect will attempt to re-establish contact for a short period of time and then give up. In addition, there are actually two edge-following controllers, one for each antennae, and there are additional issues involved in handling the interaction between the two sides which we will not consider here.

The other controller we are concerned with here is that responsible for the appetitive phase of feeding. It's main goal is to get the insect to a food patch when it is in need of energy. It accomplishes this by integrating information from chemosensors in each antenna to direct a turn toward the direction of greatest odor strength. This turn is gated by information from an internal energy sensor, so that the insect only follows odor gradients to food when it is actually in need of energy.

Given this brief summary of the edge-following and feeding controllers, we now turn to the neural implementation of their interaction. The key issue for the artificial insect is whether or not an obstacle lies between the the insect and the food patch to which it is orienting. To a first approximation, this is equivalent to determining whether or not the direction of greatest odor strength lies on the same side of the body as the obstacle encountered. If it does, the insect should edge-follow around the obstacle. If it does not, the insect should ignore the obstacle and continue to orient to the food patch.

The neural circuitry underlying the interaction between edge-following and search operates as follows (Figure 4). In each edge-following controller, the F neuron is active for the duration of the response. Edge-following behavior on either side can be disabled by activity in the nonfollowing command neuron (NFC). The appetitive controller is designed to utilize the odor strength differential between the two antennae to guide the insect to a nearby food patch when it is in need of energy. The LOS and ROS neurons are driven by the chemical sensors in the antennae (not shown) and their activity represents the relative odor strength on the corresponding side of the body. These neurons turn the body in the direction of greatest odor strength by an amount proportional to the gradient by exciting the appropriate turn neuron (either LT or RT). Because the insect should only follow odor signals to food when it is actually in need of energy, the search command neuron (SC) gates the connections which are actually responsible for turning the insect via the "following odor motor

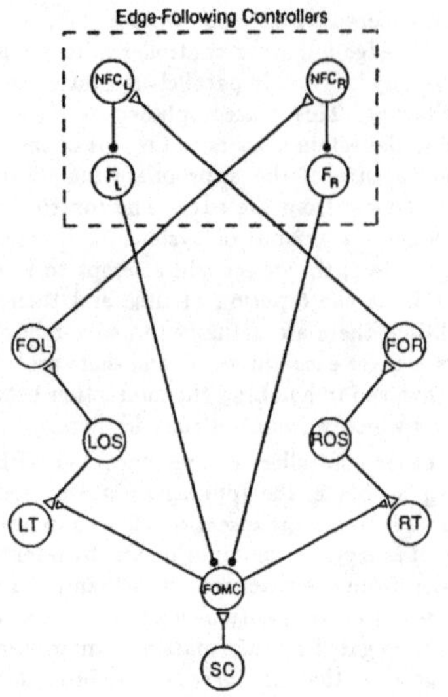

Figure 4: Neural circuitry mediating the interaction between edge-following and feeding. Only those portions of the appetitive and edge-following controllers which are actually involved in the interaction are shown.

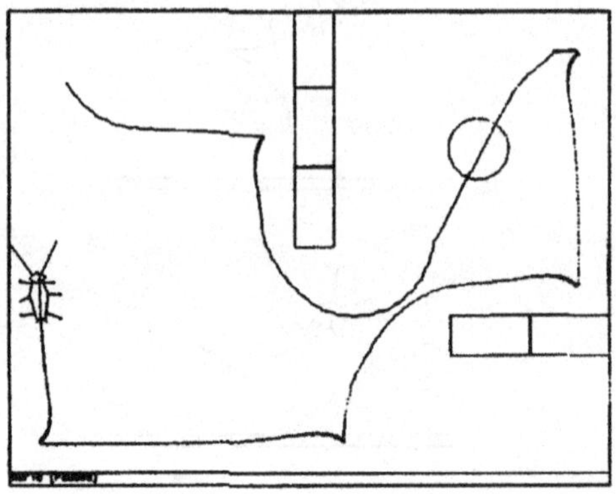

Figure 5: The path of the artificial insect as it traverses its environment.

command neuron" (*FOMC*). *SC* is active whenever the insect's energy level is low and it is not currently consuming food. The *FOL* and *FOR* neurons are active when the insect is currently following an odor gradient which is stronger to the left or right of its body, respectively.

When the insect is following an odor whose source lies to one side of its body, the cross-connections from *FOL* and *FOR* to NFC_R and NFC_L, respectively, prevent the insect from following the edge of any obstacle encountered on the opposite side of its body. However, if an obstacle is encountered on the same side of the body as the odor source, the edge-following controller must prevent the appetitive controller from following the odor gradient in order for the insect to follow the edge of the obstacle. This can be accomplished by inhibitory connections from each of the edge-following controllers to *FOMC*. Note, however, that the appetitive controller has only partially ceded control to the edge-following controller, because *SC* is not inhibited. If environmental conditions change (e.g. as the insect follows around the edge, the obstacle may no longer block its path to food), the appetitive controller can immediately regain control.

6 Discussion

We have described the modulation of a specific behavior and the interaction between two different behaviors in the artificial insect. How well does the system as a whole work to generate behavior which is globally appropriate? As an illustration of the capabilities of the complete artificial insect, which is controlled by a nervous system containing over 75 model neurons and over 150 connections, Figure 5 shows the path of the insect as it traverses its environment. Initially low on energy, the insect immediately orients to the circular food patch at the right. However, as it walks, it encounters the intervening wall. It edge-follows around this wall and then reorients to the food patch and feeds. After feeding, it exhibits periods of edge-following (which include the negotiation of several corners) and wandering.

The artificial insect illustrates some general principles about behavioral choice. As we stated at the outset, behavioral choice spans a continuum from minor adjustments to discrete switches between different behaviors. In the artificial insect, one can observe many different points along this continuum. For example, gait patterns may change as the insect changes its speed. In addition, the locomotion controller is sufficiently robust to adjust rapidly to small perturbations (e.g., changes in leg positions as the insect turns). Discrete changes within a particular behavior are illustrated by the recoil controller, which generates turns, and forward and backward walking patterns of different durations as a function of the event which triggers the response. The interactions we have described between feeding and edge following illustrate the other end of the continuum, a discrete switch between different behaviors. At the same time, this example illustrates the importance of external and internal contexts on the choices which are made. Thus, no choice between these behaviors is necessary if the insect is not "hungry", and the response to an obstacle when it is hungry is determined by the environmental context, i.e., whether or not the obstacle blocks access.

The artificial insect demonstrates that neuroethological principles can be applied to the design of artificial nervous systems for autonomous agents. More specifically, our results suggest that behavioral choice should be viewed as a global property of a nervous system, and is not, in general, localized to a command center or "switchboard" which directs discrete switches between different behaviors. Rather, it is an intrinsic part of the design of the nervous system and the interactions between its parts. We see the artificial insect project as a small step in the direction of constructing biologically-inspired distributed control systems for behavioral choice. We firmly believe, however, that it will be necessary to move to far more distributed mechanisms in order to achieve the richness and complexity of choice observed in natural animals. Clearly, further research into the neurobiological basis of behavioral choice could have a major impact on our ability to construct such distributed mechanisms.

Acknowledgements

We would like to thank Leon Sterling and Roy Ritzmann for their many useful suggestions and criticisms. This work was supported in part by a grant from the Cleveland Advanced Manufacturing Program. In addition, H.J.C. is grateful for support from the NSF, grant BNS-8810757.

References

Altman, J.S. and Kien, J. (1987). A model for decision making in the insect nervous system. In M.A. Ali (Ed.), *Nervous Systems in Invertebrates* (pp. 621-643). New York: Plenum Press.

Altman, J.S. and Kien, J. (1989). New models for motor control. *Neural Computation* **1**:173-183.

Beer, R.D. (1990). *Intelligence as Adaptive Behavior: An Experiment in Computational Neuroethology.* Academic Press.

Beer, R.D. and Chiel, H.J. (1990). Neural implementation of motivated behavior: Feeding in an artificial insect. In D.S. Touretzky (Ed.), *Advances in Neural Information Processing Systems 2* (pp. 44-51). San Mateo, CA: Morgan Kaufmann.

Beer, R.D., Chiel, H.J. and Sterling, L.S. (1989). Heterogeneous neural networks for adaptive behavior in dynamic environments. In D.S. Touretzky (Ed.), *Advances in Neural Information Processing Systems 1* (pp. 577-585). San Mateo, CA: Morgan Kaufmann.

Braitenberg, V. (1984). *Vehicles: Experiments in Synthetic Psychology.* Cambridge, MA: MIT Press.

Brooks, R.A. (1986). A robust layered control system for a mobile robot. *IEEE Journal of Robotics and Automation* **RA-2**(1):253-262.

Camhi, J.M. (1984). *Neuroethology: Nerve Cells and the Natural Behavior of Animals.* Sunderland, MA: Sinauer Associates.

Chiel, H.J. and Beer, R.D. (1989). A lesion study of a heterogeneous artificial neural network for hexapod locomotion. *Proceedings of the International Joint Conference on Neural Networks* (pp. 407-414).

Chiel, H.J., Weiss, K.R. and Kupfermann, I. (1990). Multiple roles of a histaminergic afferent neuron in the feeding behavior of *Aplysia. Trends in Neurosciences* **13**(6):223-227.

Getting, P.A. and Dekin, M.S. (1985). Tritonia swimming: A model system for integration within rhythmic motor systems. In A.I. Selverston (Ed.), *Model Neural Networks and Behavior* (pp. 3-20). New York: Plenum Press.

Kovac, M. (1974). Abdominal movements during backward walking in the crayfish. II. The neuronal basis. *J. Comp. Physiol.* **95**:71-94.

Kristan, W., Lockery, S., Wittenberg, G. and Cottrell, G.W. (1989). Behavioral choice - In theory and practice. In R. Durbin, C. Miall, and G. Mitchison (Eds.), *The Computing Neuron* (pp. 180-204). Addison-Wesley.

Kupfermann, I.J. (1974). Feeding behavior in *Aplysia*: A simple system for the study of motivation. *Behavioral Biology* **10**:1-26.

Maes, P. (1989). The dynamics of action selection. In *Proceedings of the Eleventh International Joint Conference on AI* (IJCAI 89) (pp. 991-997). Morgan Kaufmann Publishers.

McFarland, D. (1981). *The Oxford companion to animal behaviour.* Oxford University Press.

Pearson, K.G. (1976). The control of walking. *Scientific American* 235:72-86.

Ramirez, J.M. and Pearson, K.G. (1988). Generation of motor patterns for walking and flight in motoneurons supplying bifunctional muscles in the locust. *J. Neurobiology* 19:257-282.

Selverston, A.I. (1988). A consideration of invertebrate central pattern generators as computational data bases. *Neural Networks* 1(2):109-117.

An investigation of two mediation strategies suitable for behavioural control in animals and animats

Martin Snaith and **Owen Holland**

Technology Applications Group
Bolam's Mill
ALNWICK
NE66 1LN
United Kingdom
+ 44 665 604895 (phone)
+ 44 665 510624 (telefax)

Abstract

This paper gives details of recent experiments to determine the characteristics of two simple mediation strategies. We have chosen to compare two mutual inhibition strategies; the standard 'n-flop' model [Ludlow 76][Shackleford 89], and a scheme suggested by McFarland [McFarland 65]. To study the comparative features of each network we use TAG's 'non-spiking' neuron model (SNF block model [Snaith 89a][Snaith 89b][Holland & Snaith 90a]) both in simulation and on one of our Hilda series of mobile robots to show the mediation between simple 'fight' and 'flight' behaviours set up in conflict.

1. Introduction

A number of problems in animal behaviour centre around the situation where there are a number of stimuli present and a number of appropriate responses possible, only one of which can take place at any time. Which response (or behaviour) will occur? and when, and under what circumstances, will it be supplanted by another? In behavourial analysis variations in response have been attributed to set, fatigue, habituation, attention, displacement, and so on. At the physiological level, mechanisms have been proposed for mediating the interactions of a number of individual neurons, or a number of pools of neurons, each of which controls a behaviour, so that one or another is temporarily dominant.

This paper examines the consequences of making some very simple assumptions about the requirements which the mechanism must satisfy, and examines two possible implementations on paper, on the bench, and in a mobile robot.

2. Conflict and Resolution

We have already assured that the situation is one in which several stimuli are present. Let us further assume that each stimulus is linked to a separate behaviour, which is appropriate to the stimulus. For example, there may be food and water present, and behaviours for eating and drinking to which they are linked. The first step in considering any restrictions on the nature of any controlling mechanism might usefully be to imagine that no control mechanism is present. What will happen if the behaviours occur simultaneously? There appear to be three main possibilities:

a). The behaviours operate satisfactorily together eg: I can walk and chew gum at the same time.

b). The behaviours fail to operate satisfactorily together because they require effectors to carry out incompatible actions - eg: I cannot eat and roar at an enemy simultaneously.

c). The behaviours fail to operate satisfactorily together because the effects of one behaviour - eg: flight - make it impossible to continue with another - eg: mating.

Let us confine our attention to situations (b) and (c). While it is true that (c) may bring a sort of resolution to the conflict between the two behaviours. it is worth noting that there is no guarantee of this. An oscillation between the two behaviours might occur on an arbitrary time scale (flee for a second, mate for a second). This could also occur in (b), and might even in some circumstances be an effective strategy, amounting to time sharing [McFarland 73 [McFarland 74]. However, both the robot designor and the psychologist are forced at least to consider the attractions of an internal mechanism which permits only one behaviour to be produced by the robot or organism at any one time [Houston & Sumida 85] [Roper & Crossland 82].

The stimuli which are linked to the behaviours may be assumed to have some dimension corresponding to strength, and this dimension may reasonably be assumed to have something to do with the objective expected value to the robot or organism of indulging in the linked behaviour. For example, probability of success, cost of success, benefits of success, and cost of failure can be reduced to a composite measure, which could be monotonically related to a 'stimulus strength' dimension such as 'concentration of pheremones'. It will be possible to imagine this relationship existing for a great many stimuli and behaviours. If the discussion is limited to such cases, then the sensible strategy for the robot or organism to adopt is that of responding to the strongest stimulus, other things being equal. We emphasise that the robot or organism is not explicitly or implicitly computing the expected value of each behaviour, but rather that the strengths of certain stimuli which are clearly available, are good enough guides for a decision making strategy which approximates to choosing the behaviour with the highest expected value to the organism.

However, other things are not always equal. Some behaviours have expected values for relatively weak stimuli which completely overshadow the expected values for the strongest possible stimuli for other behaviours. The greatest possible discomfort from a flea will not lead to grooming if there is even the slightest smell of a lion around. A rough and ready characterisation of this dimension is as that of the importance of the behaviour. Other things being equal, the organism or robot should in reality produce the most important appropriate behaviour at any time.

We have now defined three requirements for a mechanism for controlling behaviour. In the most general terms they are:

1) Only one behaviour shall occur at any time.
2) Other things being equal, the strongest stimulus shall be responded to.
3) Other things being equal, the most important behaviour shall be produced.

The requirement in (1) can be satisfied by a simple scheme, using the idea that the 'successful' behaviour, or block, inhibits the others. By 'inhibit', we do not necessarily imply at this stage any involvement of neural inhibitory mechanisms, but rather something like 'the application of some signal to the mechanisms controlling the other behaviours to prevent them from implementing their output'. This is a different idea from that of there being some central 'channel' for producing behaviour which becomes 'unavailable' to the competing behaviours once 'captured' by the successful behaviour.

From considerations of symmetry, each behaviour (or unit) must be connected to each of the others by some connection with an 'inhibitory' effect. If this connection is direct, then there will be n(n-1) connections (assuming they are undirectional). Assume that each behaviour has a positive input related to the strength of the stimulus for that behaviour, and that the intensity of 'activity' of the behavioural block (or unit) represents the performance of the behaviour and supplies the inhibitory effect to competing behaviours. We straight away perceive that this is a mechanism involving a net positive feedback system - an active block inhibits its neighbours, reducing their inhibitory effects on itself. If each block receives a stimulus powerful enough to cause enough activity and consequent inhibition to completely inhibit the activity of the competing blocks, then the arrangement has n stable states, in each of which one block is active and the rest are silent. This is an n-flop; the two-block case is a flip-flop. [Shackleford 89]

This is surely the minimal implementation of requirement 1, but analysis shows that it exhibits a number of strikingly useful emergent properties. They are:
a) A block which is active can remain active at a lower input level than that at which it became active, This is <u>hysteresis.</u>
b) An active block, as long as it remains above the lower limit of the input to maintain it, gives the same performance as it would if no other blocks were present. This is <u>lack of distortion.</u>
c) If a new block becomes active, it will rapidly dominate the system if the input levels remain the same - that is, it will suppress activity in all other blocks. This is because its output must reduce the output of the presently dominant block, and hence must increase its own, further reducing the output of the dominant block - there is <u>clean switching between blocks.</u>
d) If the interconnections between blocks are asymmetrical, the capacity for different behaviours to dominate the net can be asymmetrical. This creates the possibility of a complex <u>dominance hierarchy</u> of behaviours.

Requirement (2) is satisfied, as long as the strongest stimulus is strong enough to overcome the inhibition produced by the currently dominant unit. However, the hysteresis effect means that a unit can retain 'control' of the system at an input level lower than that at which it acquired control, and may continue to do so in the presence of competing stimuli which are stronger, but not strong enough. This has obvious applications in the context of attention. An initial strong stimulus (such as a shout) can 'capture' the system, and a subsequent conversation at a normal volume can retain control until displaced by some other strong stimulus.

Requirement (3) is easily implemented. If the inhibitory connections to the block are weak, 'normal' intensity will be able to activate the block relatively easily. If the inhibitory connections from the block are strong, a relatively low level of activity will be able to suppress competing behaviours relatively easily. Taken together, these factors may be taken to define the importance of the behaviour.

Thus, the simple system for implementing (1) already includes mechanisms for (2) and (3), and more besides. However, it is at a disadvantage from the point of view of implementation, in that each new behaviour requires 2n connections, n from the existing blocks to itself, and n from itself to the existing blocks. Is there an equivalent system requiring fewer connections? The obvious thought is to take advantage of the fact that, when the mechanism is successful, only one block is active, and so if all blocks were connected to some distribution block, which provided inhibitory connections back to each block, it would be possible to distribute the correct amount of inhibition to each block. However, the active block must receive no inhibition for the scheme to work.

This can be arranged by making each block excite itself by an amount sufficient to cancel or balance out the inhibition it receives from itself when active.

This model is similar to one proposed by McFarland [McFarland 65]. If the units involved are linear and additive, and no saturation effects occur in the distribution block, then the two block McFarland model is exactly equivalent to the two-block standard n-flop model. However, when there are more than two blocks, there is a subtle difference between the two models. The dominance relationship between any pair of blocks in the n-flop model is independent of the relationship between either member of the pair and any other blocks. This is because each pair has its own pair of inhibitory interconnections, which may be of any strength whatsoever. However, in the McFarland model, the dominance relationship of a pair is defined by the set of connections of each unit to and from the distribution block, and so the contribution of a unit to this relationship is the same pair of parameters for each pair of which it is a member. This means that it may be possible to have a much more complex dominance hierarchy for a standard n-flop system than for a McFarland system. Whether or not this is a useful attribute remains to be determined.

The main advantage of the McFarland scheme is that it requires only 2n connections, and that a block can be added at a cost of only 2 connections - provided that the distribution block has enough spare capacity. The only cost disadvantage is that the distribution block - a sort of capital cost - must be provided. It is not possible yet to say which model is more appropriate for modelling and analysing animal behaviour. Further research into the properties of each model will be required, as will studies of dominance relationships between behaviours across species and preferably across phyla.

The remainder of this paper describes some tests of hardware implementations of the standard n-flop model and the basic McFarland model. These initial studies involved only two blocks. It should be borne in mind that some of the results may be peculiar to these particular implementations.

3. Description of the networks

The mediation networks are shown below as figures 1 & 2.

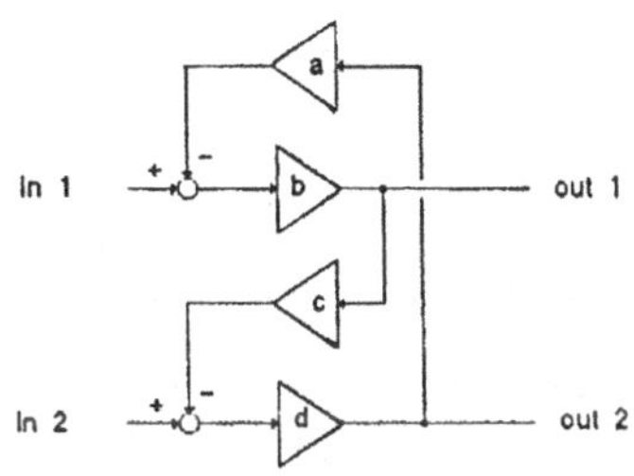

Figure 1: The n-dflop mediation pair. Shown above in a conventional summator and gain block diagram. And below in an SNF representation. The gain parameters are a thru d

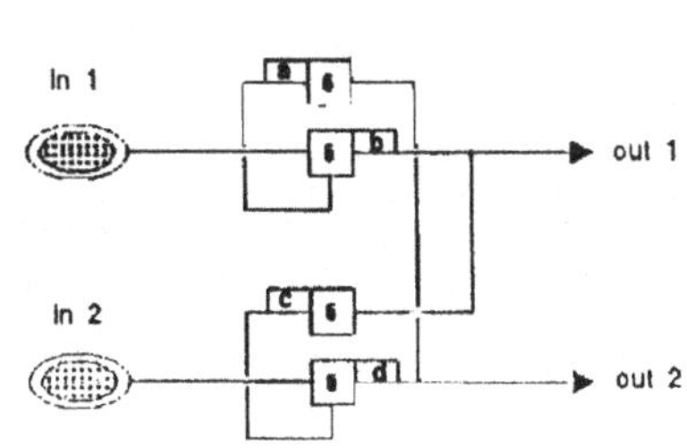

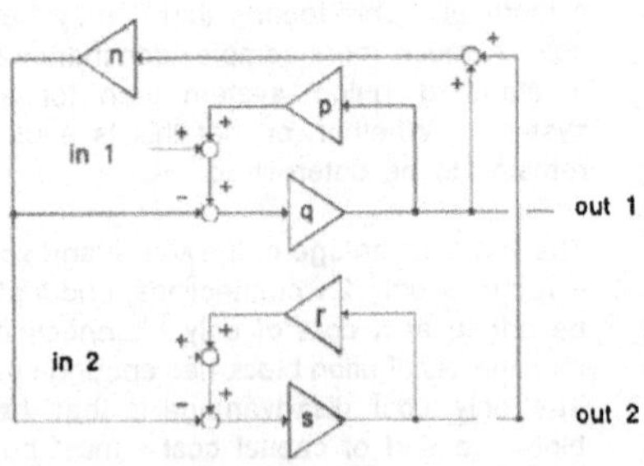

Figure 2: The McFarland mediation pair. Shown above in summator and gain block form and below in SNF. The gain parameters are n, p, q, r and s.

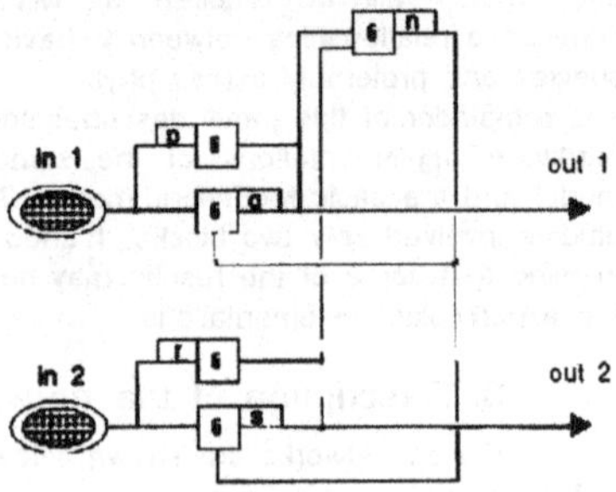

4. The SNF model

We have tested the validity of our models' approximations by implementing an equivalent network on an alternative pair of spiking and non-spiking models used by Ulrich Koch [Koch 90a]. This comparison showed strongly similar behaviour from both models. The main differences appear to involve the amounts of distortion introduced by the different implementations.

We are only using one of the six possible SNF blocks in this experiment [Snaith 89b]. The function of this block is described in figure 3. It simply takes the difference of input activity and inhibition and amplifies the result by a modulated gain parameter. (The modulation parameter is used to vary the characteristics of the network, especially the positive feedback and inhibition strengths; this allows us to search the parameter space of each particular configuration.)

Referring to figure 1, the reader will notice that some of the arcs in the SNF diagram have branching. The general rules for the construction of these diagrams are as follows: all arcs are mono-directional, from the activity output of the SNF to either input of another SNF; the activity input is a summing node on the long vertical side of the symbol, (in figure 2 this is the right-hand side for node 1a); the inhibition input is a summing node on the bottom, horizontal line of the symbol; the activity output is shown as the lower apex of the small vertical line on one side of the parameter rectangle, (in figure 1 this is the left-hand side for the block with parameter a).

5. The Benchtest

The networks were bench tested over the frequency range 2kHz to 0.01Hz using a sawtooth wave free-running between 10% and 80% activity with 1:1 mark space ratio for input 1 and a static input set at 60% activity for input 2. There was no apparent frequency dependence in either case.

The networks were first searched to find their 'best' configuration i.e. that which gave the outputs in a boolean complementary pair, and then the various feedback strengths were slowly modulated to search the parameter space.

The n-flop network showed a broader range of variation in hysteresis for a given level of transfer distortion, (see figures 4 & 5). Both networks degenerated to a distorted mixing of the input activies

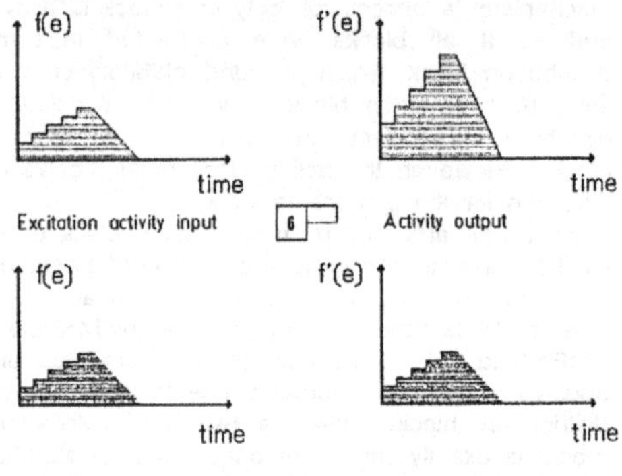

Figure 3: Defining the Class method for the Sub-Neural Function Class Generator. The diagram shows two cases: the lighter shaded in which the gain parameter is at 2; and the darker shaded in which the gain parameter is at 1. In both cases the inhibition is set to 0 and the gain parameter is unmodulated. Please note f(e) is some arbitrary function of the environment and in general Excitation_in = f(e).

A mathematical approximation of this SNF Class Method would be as follows:

Activity_out = (Gain_Parameter - Modulation_in) * (Excitation_in - Inhibition_in)

4a Definition of inputs. Input 1 is a free-running sawtooth wave with values from 10% of saturation to 80% of saturation. Input 2 is a fixed activity of 80%.

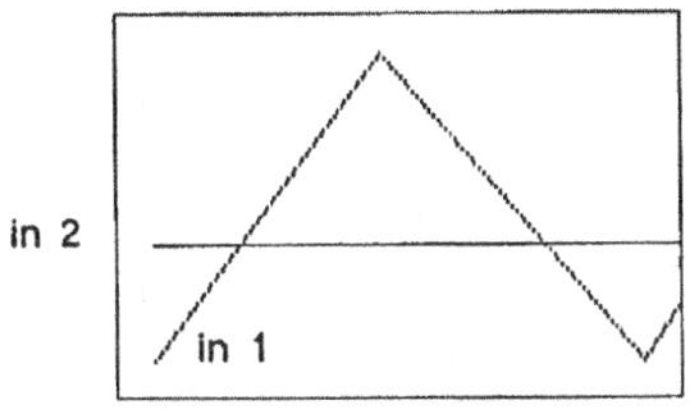

4b Typical transfer characteristic in 'best' configuration. Distortion to the stimulus input is below 1%. Gain parameters are a = 7, b = 9, c = 7, d = 9. (to about 10%).

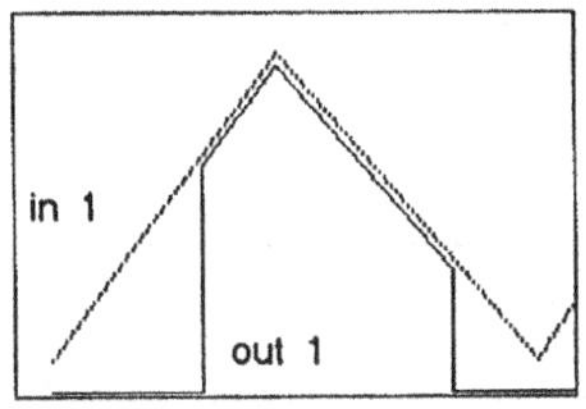

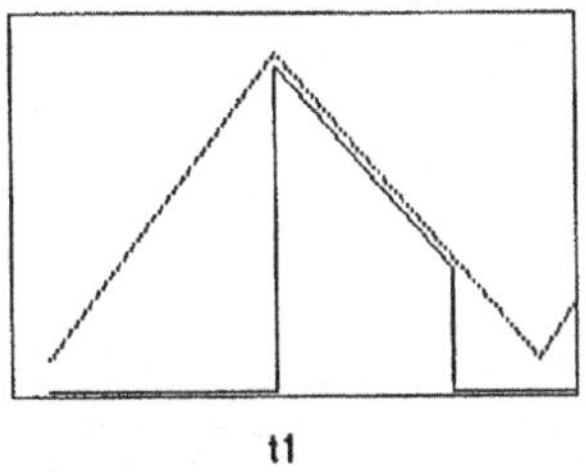

4c Full t1 hysteresis. It is possible to get to the peak of input activity before changing output characteristics

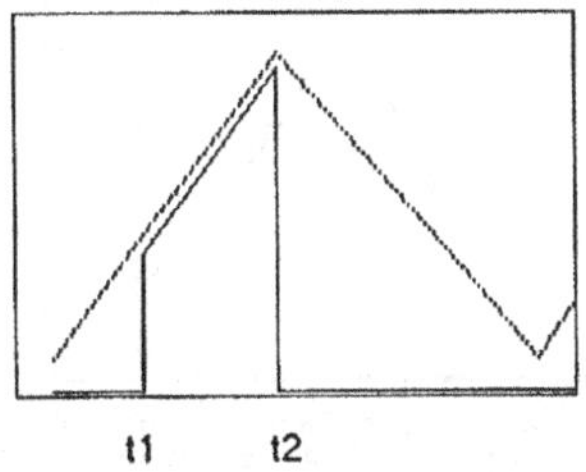

4d Full t2 hysteresis. This result is counter-intuitive and not within the first order approximations of the model. It may be due to a capacitive coupling between the stages of this implemetation, and as such may even be observable in natural systems.

Figure 4 (top of page) Showing transfer characteristics and hysteresis modulation for the n-flop pair. 4d gives details of the observed anomaly, which we are investigating further.

Figure 5 (bottom of page) Showing transfer characteristics and hysteresis modulation for the McFarland net.

5a Definition of inputs. The same inputs are applied to both nets. They are as defined in figure 4a. The frequency range used in the test is described in the text.

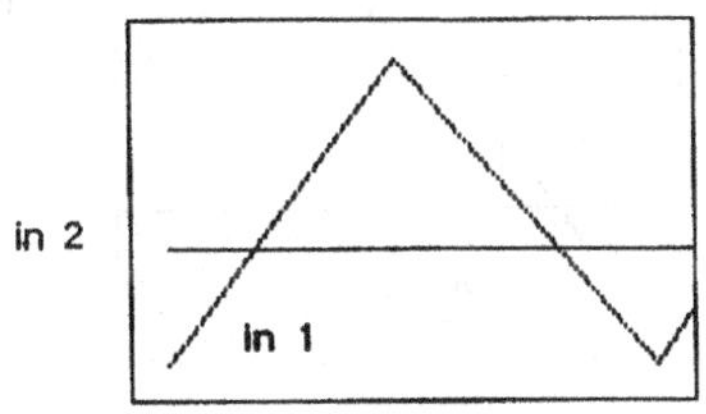

5b Typical transfer characteristic in 'best' configuration. Gain parameters are n = 6, p = 7, q = 9, r = 7, s = 9.

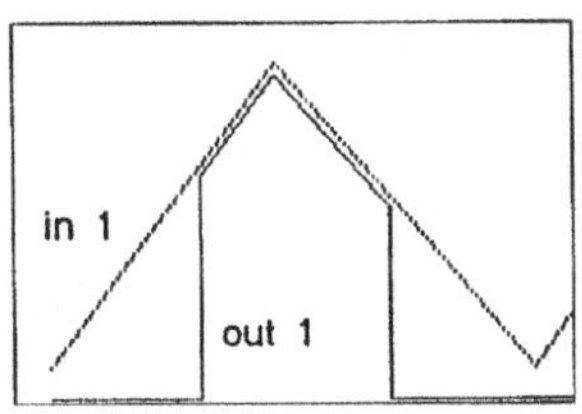

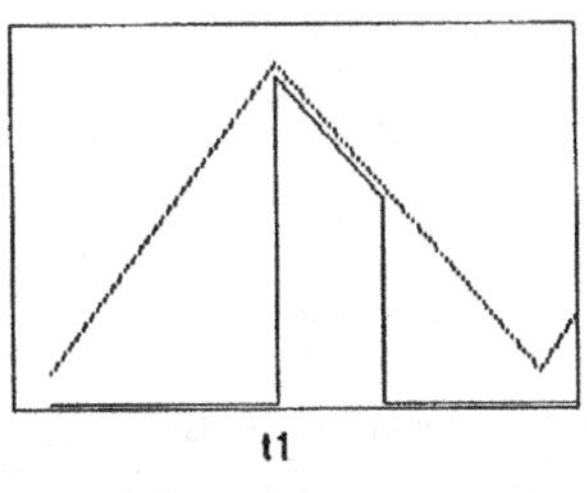

5c Full t1 hysteresis. Very similar ability to the n-flop net is observed in the McFarland net.

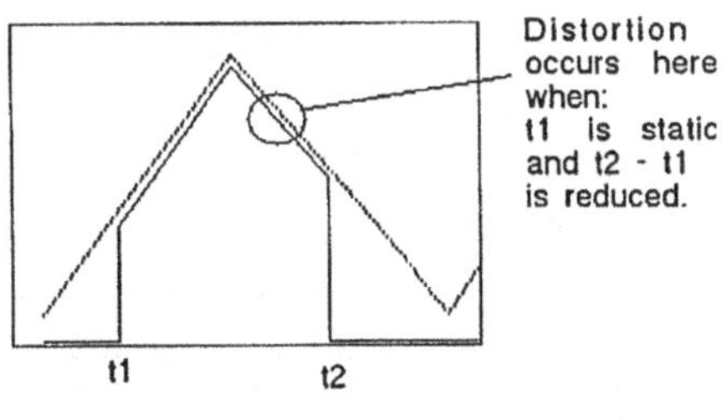

5d Full t2 hysteresis. The anomalous effect is not observed so strongly in the McFarland net. The falling edge distorts instead.

6a Defining the inputs. Input 1 is a free-running sawtooth wave with values from 10% of saturation to 80% of saturation. Input 2 is a fixed activity of 80%.

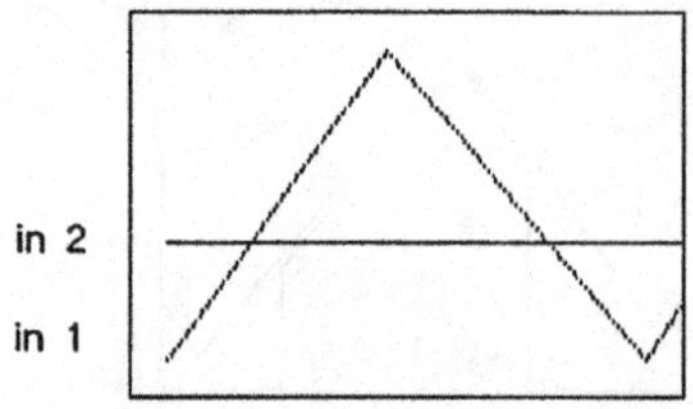

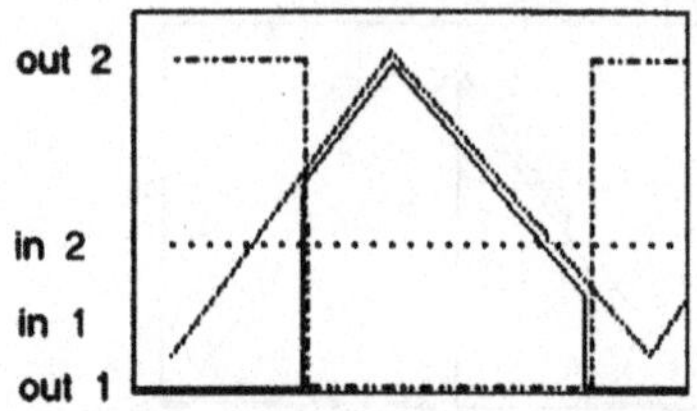

6b The two outputs compared in the 'best' configuration. Gain parameters are a = 7, b = 9, c= 7, d = 9. (to about 10%).

6c Degenerating the n-flop net by reducing gain parameter c. The two traces from output 1 are shown against the sawtooth applied to input 1.

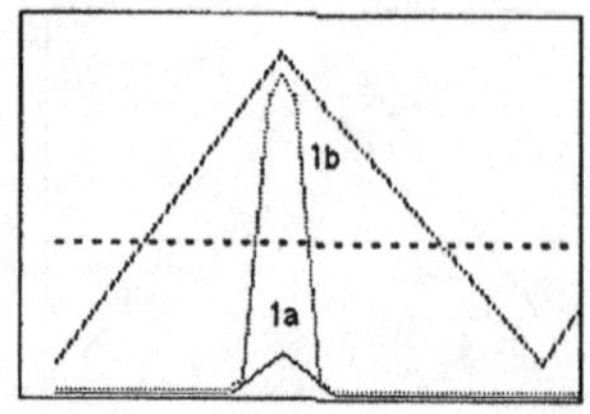

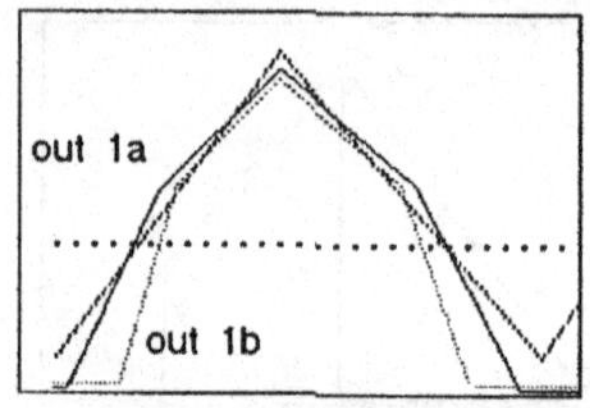

6d Degenerating the n-flop net by reducing gain parameter a. The original input waveform breaks through progressively as the inhibition is reduced.

Figure 6 (top of page.) Comparison of the outputs from the n-flop network. Degenerating the n-flop network by reducing the connection strengths. Trace 1a has degenerating parameter = 2.5, while trace 1b has degenerating parameter = 4.5. All other parameters are held at 'best' configuration.

Figure 7 (bottom of page) Comparison of the outputs from the McFarland network. Degenerating the McFarland network by reducing the connection strengths, both for overall inhibition and a single node's postive feedback.

7a Defining the inputs. These are indentical for those used on the n-flop degeneration shown in figure 6a.

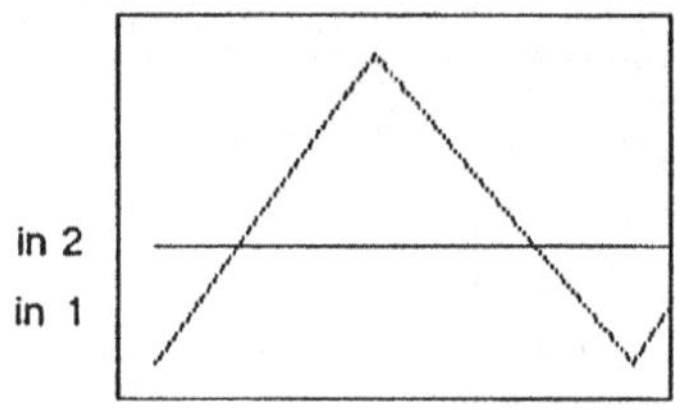

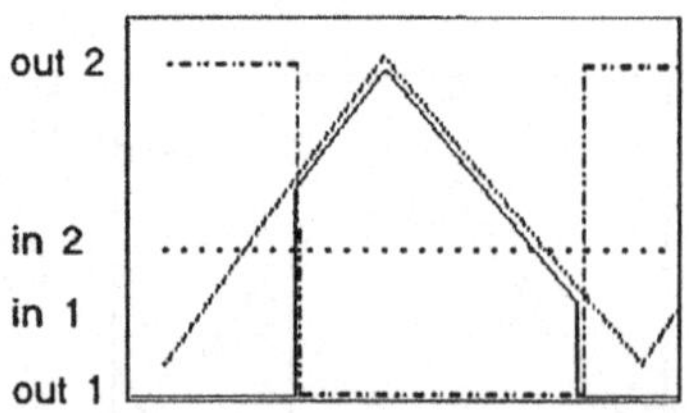

7b The two outputs compared in the 'best' configuration. Gain parameters are n = 6, p = 7, q = 9, r = 7, s = 9. (to about 10%).

7c Degenerating the McFarland net by reducing gain parameter n. Output 1 progressively saturates to 80% as the inhibition is reduced The traces are shown against input 1.

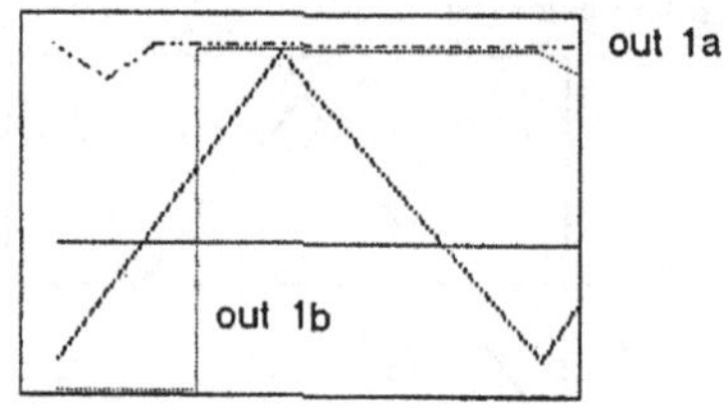

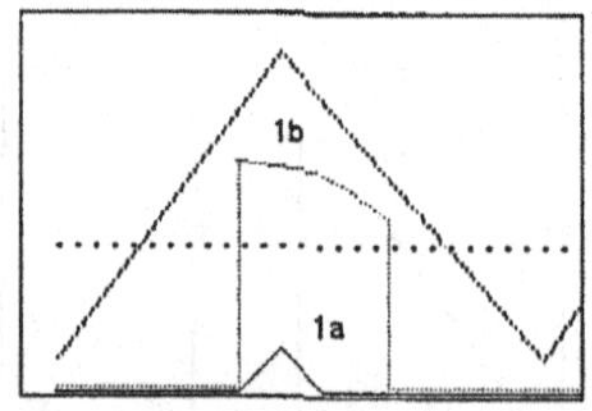

7d Degenerating the McFarland net by reducing gain parameter r. Output 1 traces are shown with the sawtooth applied to input 1. The output progressively decays as the positive feedback is reduced.

if the strength of connection was reduced too far, (see figures 6 & 7). The transition from degenerate to mediational modes of operation was quite rapid in both cases. A suggested quality factor for self training a network like this is to reduce the transfer distortion to a minimum whilst strictly enforcing either one output active OR the other but not both; however we have not as yet fully characterised the quality factors associated with these networks. The test of real mediation on a HILDA series autonomous mobile robot is available as an accompanying video tape.

6. Adaptation and architecture

We noted in the introduction that the n-flop model, the simplest mechanism for satisfying our first system requirement, also satisfies the other two, and also exhibits potentially useful emergent properties. The more economical McFarland implementation has some potential limitations, which are at the moment far from obvious. Two questions arise at this stage:

(i) How do these architectures relate to Brook's subsumption architecture [Brooks 86]?

(ii) What have any of the architectures to do with adaptive behaviour?

The answer to (i) is interesting. A subsumption architecture is what an n-flop architecture reduces to if the behaviours satisfy certain conditions. These conditions appear to be:

(a) Behaviours are ranked from 'least important' to 'most important' with no tied ranks

(b) Active behaviours send no inhibition to more important behaviours (to eliminate hysteresis) and sufficient inhibition to less important behaviours to inhibit them completely regardless of stimulus strength.

Note our requirement (b) means that a McFarland architecture cannot be made to implement a subsumption architecture, because under the McFarland scheme a block is constrained to treat and be treated alike by other blocks via the distribution block.

The answer to (ii) is not simple. None of the architectures is intrinsically adaptable in the sense that any of them change any aspects of their structures in response to time, circumstances, or experience. However the architectures register state in different degrees - the n-flop most, the McFarland to a lesser extent and the subsumption not at all. (The n-flop has a variable dominance hierarchy with hysteresis; the McFarland has a fixed dominance hierarchy with hysteresis and the subsumption has a fixed dominance hierachy without hysteresis). It is therefore possible to encode and produce greater apparent variability of response in the system with the greater capacity for storage of state. This variability may appear to be adaptive depending on the context. On the other hand it may be the result of an encoding produced by evolutionary adaptation, expressing itself not as apparent adaptive behaviour, but rather as appropriate behaviour within a context defined by the interplay between the succession of stimuli and emitted behaviours in a particular situation. It is only necessary to learn what you do not know; the n-flop architecture allows the encoding of more knowledge than either of the other schemes, and it is possible that the functional expression of this knowledge may give the false impression that some form of learning is taking place.

The main thrust of a reply to (ii) may be to answer a question with a question: what exactly would it be profitable to mean by 'adaptive behaviour'? Adaptation as we normally think of it takes place 'within behaviours'. However the present context asks us to think of adaptation manifesting itself in the balance 'between behaviours'. An analogy taken from the business world might be that the 'between behaviour' factor represents the robot's corporate strategy, whereas the 'within behaviour' factor represents departmental efficiency. Accumulated business wisdom says that it is more important to be doing the right thing at the right time with low efficiency than to be doing the wrong thing at the wrong time with high efficiency. An example from nature: the tree kangaroo is almost comically inefficient at coping with its chosen environment but from the point of view of strategy the rewards of the arboreal lifestyle are so great that the evolutionary 'decision' was and is undoubtedly correct. If it is accepted that the 'between behaviour' factor is important then it is necessary to formulate a model for a control mechanism in order to discover the point of action for adaptive change. (In the model considered in this paper, this point of action is at the level of the strengths of the inhibitory interconnections.) Adaptive change within the lifetime of the organism will be represented by short term or long term changes in these interconnection strengths. The effects of these changes will be to change the dominance hierarchies, and to alter the quantitative characteristics of hysteresis. The clean switching and lack of distortion are inherent properties of the models, and will be unaffected by adaptation; they are important in the context of adaptive behaviour only because they are desirable system properties which are preserved during and after adaptation.

Further work will be required to examine potential mechanisms for changing interconnection strength, and to investigate the results of such changes. Drive states, potentiation and accommodation (or habituation) are all obvious candidates for short term change, and some form of results-based learning for long term change. An encouraging indicator that variable dominance hierarchies may be an adequate expression of adaption is provided by recent developments in the theory of subsumption and manifold architectures concerning activation sets [Maes 90][Snaith 90a] A number of workers are now optimising their robots (or adapting them to their environment) by refining the activation sets of the individual behaviours; this paper offers a model which is simple and biologically plausible, and which offers an economical way of encoding the sort of information represented by activation sets.

7. Conclusion

Both the n-flop and McFarland models exhibit useful properties in the two-block situation. In addition to performing the required 1-of-2 selection, both models showed hysteresis, clean switching, substantial lack of distortion, and variable dominance. The n-flop model seemed to be more stable both from the point of view of set-up and performance, which would be expected as it is less complex than the McFarland model in the two-block implementation. Both models also exhibited the same counter-intuitive result (4d and 5d) of a block (that receiving the sawtooth input) switching at a higher input value than the switch-on value. This may be due to some sensitivity of the neural implementation to the sign of the rate of change of the difference between the inputs to the two blocks, or may be an emergent property of the models or of the implementation.

8. References

[Ludlow 76]
A.R.Ludlow, 'The Behaviour of a Model Animal.' Behaviour vol 58, pp 131-172.

[Shackleford 89]
J. Barry Shakleford, 'Neural Data Structures: Programming with Neurons.' Hewlett Packard Journal, June 1989 pp 69-78

[McFarland 65]
D.J.McFarland, 'Flow Graph Representation of Motivational Systems.' Brit. Jou. Math. & Stat. Psychology, vol 18 pp 24-43.

[Snaith 89a]
Snaith M.A. 'Neuro-Behavioural Control of Autonomous Mobile Robots.' Proc. IARP 1st workshop on Domestic Robotics, Newcastle Sept. 89 pp147-163.

[Snaith 89b]
Snaith M.A. 'Synthetic Neural Functions & Their Application to Autonomous Mobile Robots.' TAG internal paper Oct '89.

[Holland & Snaith 90a]
Holland O. & Snaith M.A. 'The use of Neural Modelling in the Control of Autonomous Mobile Robots.' Proc. INNC, Paris June '90.

[McFarland 73]
McFarland D.J. 'Time-Shared Feeding & Drinking.' Jou. Exptal. Psychology, vol 25, pp 48-61.

[McFarland 74]
McFarland D.J. 'Time-Sharing as a Behavioural Phenomenon.' Advances in the Study of Behaviour, vol 5, pp 201-225.

[Houston & Sumida 85]
Houston A. & Sumida B. 'A Positive Feedback Model for Switching Between two Activities.' Animal Behaviour, vol 33, pp 315-325.

[Roper & Crossland 82]
Roper T.J. & Crossland G. 'Mechanisms Underlying Eating - Drinking Transitions in Rats.' Animal Behaviour, vol 30, pp 602-614.

[Koch 90a]
Koch U. 'Simulation of Biological Neural Networks Using "Biological" Neurons.' Workshop on Biology, Cognition & Robotics, GMD, Sankt Augustin, June 90.

[Brooks 86]
Brooks R.A. 'A Robust Layered Control System for Mobile Robots.' IEEE Jou. Robotics and Automation RA-2, No. 1 1986 pp 14-23.

[Maes 90]
Maes P. 'Learning to Co-ordinate Behaviours.' Workshop on Biology, Cognition & Robotics, GMD, Sankt Augustin, June 90.

[Snaith 90a]
Snaith M.A. "A Virtual Field Uncertainty Recovery Strategy for Mobile Robot Navigation.' Workshop on Biology, Cognition & Robotics, GMD, Sankt Augustin, June 90.

Computer Simulation of an Animal Environment

Toby Tyrrell and John E. W. Mayhew
(e-mail: toby@aivru.uucp, jewm@aivru.uucp)
AIVRU, University of Sheffield.

Abstract

This paper describes a computer simulation of an animal environment which has been created as a tool for investigating the mechanisms behind 'behavioural choice' in animals. The simulated environment has been designed to provide sufficient complexity and realism for meaningful behavioural experiments.

1 The Problem

"All brains, even those of the tiniest insects, generate and control behaviour." [Albus, 81]. This basic task of animal brains can be split into three subtasks (e.g. [Brooks, 86]) as shown in Fig. 1:
1 - Sensing of the environment so as to be able to perceive what is going on at each moment in time (perception).
2 - Taking the interpretation of the environmental situation and using it to decide which of the animal's repertoire of behaviours is the most appropriate (behavioural choice).
3 - Transforming the chosen behaviour into a pattern of movements of parts of the body (motor control).

Our aim is to investigate the mechanisms underlying the second of the three stages outlined. This stage is perhaps the least well understood of the three due to the fact that the processes involved are internal and cannot be directly observed, only inferred from the resulting behaviour. The behavioural parts of a brain do not interact directly with the outside world, but rather through the interfacing systems of perception and motor control [Halliday, 83].

It should be noted that the use in this paper of the term 'behaviour' (e.g. in 'behavioural choice') refers to a pattern of actions such as eating, mating, avoiding predators, etc. This should be distinguished from the usual use of the term 'behaviour' in the AI literature, where it more often denotes the coordinated movement of a set of limbs (e.g. in order to grasp an object, navigate around an obstacle, etc). We want to look at high-level decision-making (e.g. should the animal obtain food from the nearby fruit bush or else flee from the predator that has just appeared in the distance). We do not want to address the problem of how the animal should move its limbs in order to pick the fruit and transfer it to its mouth or how the animal should best move its legs so as to be able to run away from the predator.

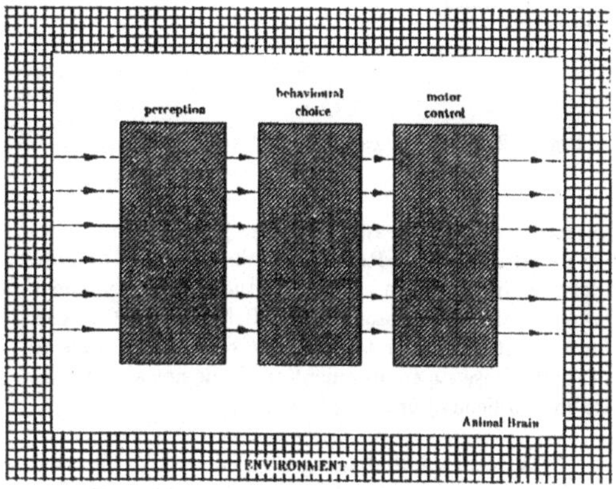

Figure 1: Three functions of an animal brain.

In short, we want to examine the second of the three functions of Fig. 1, while ignoring the other two as much as possible.

2 Why a Simulated Environment?

Given that we want to examine behavioural choice mechanisms, what is the best method of going about it?

One approach to examining behaviour has involved the building of robots which can navigate in a

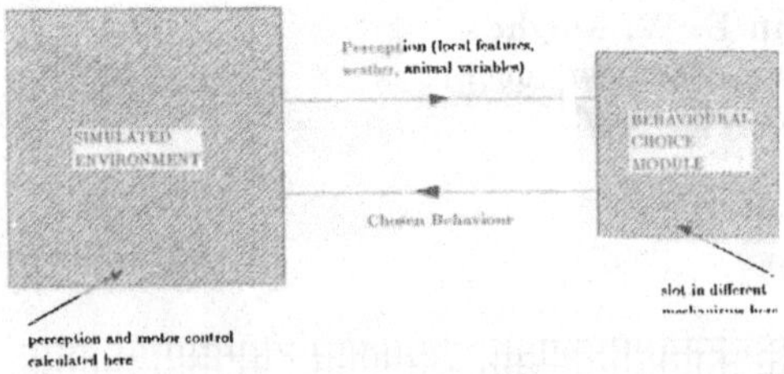

Figure 2: Advantages of a simulated environment.

selected environment and which are given a limited behavioural repertoire. Research of this kind not surprisingly requires a large amount of the time to be spent on perception and motor control. Behavioural strategies for a robot cannot be tested if the robot is not able to properly sense its environment or if it is not able to put its chosen behaviour into effect.

We decided that using a simulated environment would allow us to bypass the sensory perception problem (we just provide the animal with the relevant information about the environment), and also to bypass the motor control problem (we just calculate the effects of the animal's behaviours on the simulated environment without concerning ourselves as to how they are achieved). This would allow us to concentrate on behavioural issues. Fig. 2 illustrates the advantages of using a simulated environment. Time has to be spent on writing the programs to control the environment and to manage the interactions between the animal and the environment. Once this is finished though, it will be possible to just slot in different behavioural mechanisms and test them without altering any of the rest of the programming.

A major design issue concerned the degree of realism and complexity in the environment. We needed to create a balance between several factors. On the one hand we wanted to minimise the time spent on programming the environment and also to keep the complexity of the environment within manageable limits so that we could maintain a good understanding of its dynamics. On the other hand we wanted to make the environment fairly realistic so as to avoid the dangers of abstraction apparent in classic AI and other fields (see 'abstraction as a weapon' in [Brooks, 86]) and we also wanted to make the environment fairly complex so as to pose a difficult

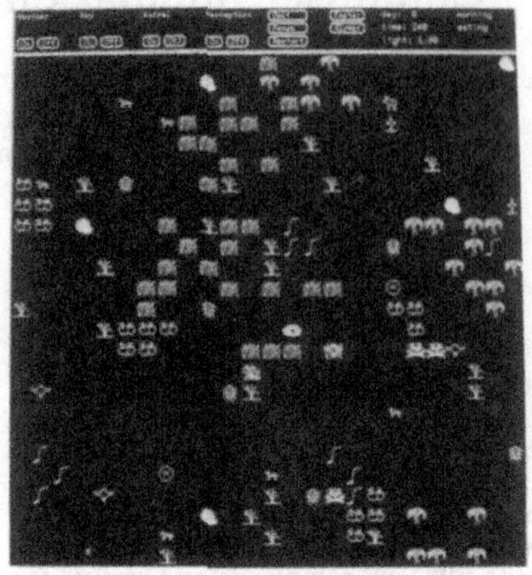

Figure 3: The simulated environment.

challenge to any behavioural strategy. Another factor was our desire to make the environment quite realistic so as not to prejudice the testbed by building into it our expectations of the solution. We decided to make the simulated environment as complex and realistic as the other constraints would allow.

3 Description of the Simulated Environment

The simulated environment consists of a large array of square regions, each of which corresponds to a small area which can contain one or more features (food, mate, water source, etc) to be described below. The environment is not static but changes over time as different features move, respond to the actions of an animal or respond to changes in the weather.

3.1 *Weather*

Weather is constant over the whole of the environment, and alternates between three weather states - sun, cloud and rain. The changes from one weather state to another are probabilistic. For example, if the weather state is currently rain then there might be an 80% chance of the state remaining as it is, an 18% chance of it changing to cloud, and a 2% chance of it changing to sun. Varying the probabilities of changes between weather states allows one to make adjustments to the average 'climate' in the environment and thus to alter the nature of the problems faced by the animal.

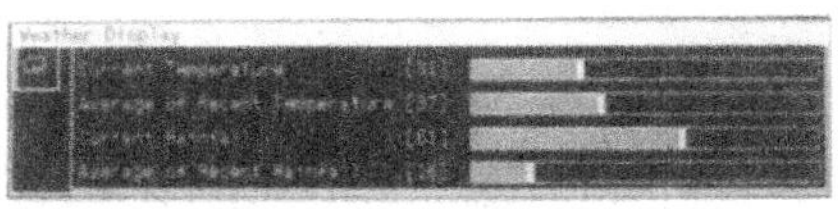

Figure 4: The weather display.

As well as the weather state there are two other weather variables - the current rainfall and the current temperature. These are linked to the weather state so that the current rainfall is zero unless the weather state is rain, and the current temperature is likely to be higher when the weather state is sun than when the weather state is cloud, and higher for cloud than rain. These two weather variables are also averaged over time to give measurements of the amount of rainfall and the average temperature in the recent past. The two averages are important in deciding the levels of the water sources and the instances of cereal type food (see below).

3.2 *Features*

There are several different types of feature that can be present in a square. Subject to a few constraints, all of the features are placed randomly around the environment, with some concentrated in clusters:

cereal type food These features correspond to foods such as grains and grasses. The value of these foods depends on rainfall and on whether they have been eaten recently. They can be toxic to the animal and damage its health.

fruit type food Similar to above but cyclic in nature with a 'ripening-decaying' cycle.

water source The levels of the water sources vary with recent rainfall, average recent temperature and on whether animals have been drinking from them recently. Some of these will be toxic.

predator type 1 These move across the environment and may chase and try and catch the animal if they catch sight of it. These predators will find it difficult to spot or chase the animal if it is in dense vegetation (see later features).

predator type 2 These features correspond to predatory birds such as hawks or owls. Similar to above except that they move more rapidly and can spot the animal from a longer distance away. These predators will find it difficult to perceive the animal if it remains completely motionless (i.e. 'freezes').

prey These features move slowly and randomly around the environment. They can be a source of food if caught by the animal.

irrelevant animal Groups of irrelevant animals occasionally traverse the environment and will ignore the modelled animal unless it is in the same square as them.

mate Potential mates appear every so often in the environment. Each mate can be receptive (i.e. 'on heat') or not receptive. If the potential mate is not receptive then it will not respond to courtship displays and will attack the animal if it tries to mate. For reasons given later (section 6.2) the animal is assumed to be male, and mates to be female.

den Provides protection from predators and the extremes of temperature. It will be advantageous for the animal to return to its den at night when its perception is limited and avoiding predators is difficult.

dangerous place This type of feature is representative of cliffs, marshes, etc, i.e. areas where the animal risks injury or death.

shade This type of feature allows the animal some protection from the extremes of temperature, as well as from detection by predators.

cover This feature represents dense vegetation in which the animal is very difficult to detect, and also very difficult to catch when being chased by predators.

landmark These are distinctive physical features which allow the animal to recognise positions, and help it to navigate successfully around the environment.

3.3 *Repertoire of Actions*

At each moment in time, the animal has a choice of several actions which it can perform [Kilmer et al., 69]. The animal's possible actions in the simulated environment are:

- resting
- sleeping (like resting but without perception)
- freezing (e.g. when bird type predator overhead)
- eating
- drinking
- displaying aggressively
- attacking

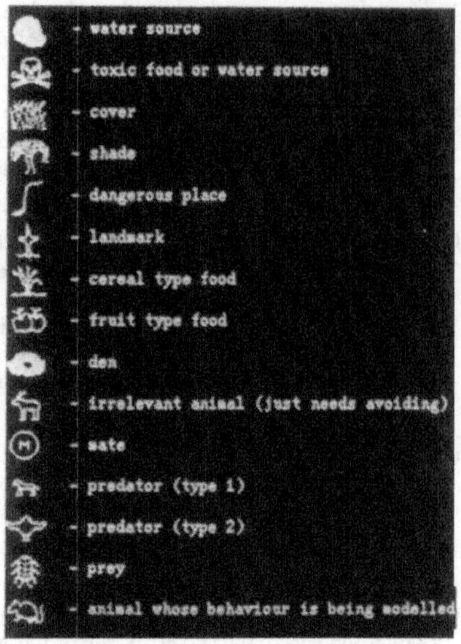

Figure 5: Key to environment features.

- courting
- mating
- moving
- moving rapidly (e.g. when trying to escape from a predator)
- looking around (scanning the local environment for predators, mates, etc.)
- looking in a specific direction

3.4 *Animal Variables*

In order to be able to use this artificial environment to look at the effectiveness of different behavioural strategies, some characterisation of the purpose of the strategies must be made, and this can then be used to measure how effectively each of the strategies achieves that purpose.

Animals may be considered as machines to enable the propagation of genes [Dawkins, 89]. For the purposes of this environment, it is assumed that the only way that an animal can help to propagate its genes is by reproduction. It is assumed that the animal knows nothing about, or doesn't come into contact with, its close relatives and that therefore it cannot further its genes by any sort of altruism to those carrying similar genes.

Therefore the purpose of the animal is to reproduce as often as possible during its lifetime. This obviously engenders a second goal which is to make the life of the animal as long as possible so that it has many chances to reproduce. But the final measure of how well an animal manages to exploit its environment (its 'genetic fitness') is the number of times it manages to reproduce.

As well as the variable 'genetic fitness', which is equal to the number of times the animal mates, there are four other animal variables. The first three of these are homeostatic variables (e.g [Toates, 80]), that is to say the animal needs to keep the values of these variables close to certain optimum values. The idea is not to maximise these variables, as is the case with genetic fitness. The three homeostatic variables are food, water and temperature.

Besides the three homeostatic variables, there is a fourth animal variable, 'health'. The value of this variable is reduced when the animal suffers injury by being attacked or by crossing a 'dangerous place' square. It is also reduced when the animal eats toxic food or drinks toxic water, and when the values of the homeostatic variables are far from their optimum value. As the value of the health variable approaches zero the animal becomes incapacitated and is forced to become less active. If the value of the health variable ever falls to zero, then the animal dies. As the age of the animal increases the value of the health variable drops and it becomes progressively harder for the animal to stay alive.

3.5 *Animal Perception*

In order to be able to choose the correct behaviour for each point of time in the environment, the animal has to be able to perceive the environment. In our system the animal has a limited range of perception. For each square within the animal's perceptual range, a probability of the animal's perception being correct is calculated. The probability is dependent on several factors:

1. how far away from the animal the square is;
2. what features are in the animal's square (for example the animal will find perception more difficult if it is in cover);
3. what features are in those squares that lie between the animal and the square that is being perceived;

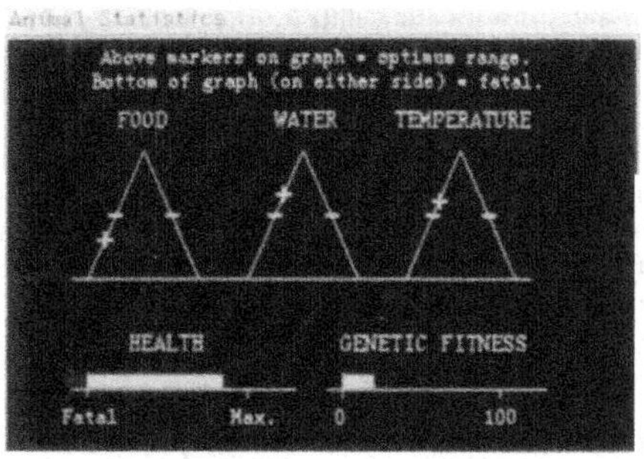

Figure 6: The animal variables display.

4. the action of the animal (if the animal is purposefully looking towards a square then it is more likely to perceive it correctly);
5. whether it is day or night.

3.6 *Behavioural Choice in the Environment*

The basics of the simulated environment have now been outlined. Next the relationship between the environment and the the behavioural choice of the animal will be considered.

Although a lot of programming is required for the controlling of the environment, hardly any of the information used for this controlling of the environment gets passed to the animal for use when making its behavioural choice. As can be seen in Fig. 2, the only information that gets passed concerns what the animal can see in its vicinity (whether correctly perceived or not), the state of the weather and the weather variables, and the values of the animal variables (food, water, temperature and health). There is a restriction on the information available to the behavioural choice module so that no information is given on, for instance, the lengths of the ripening cycles of the fruit type food sources, or the receptiveness of mates, but only information on the amount of food at a fruit source at certain instants and the response of a mate to courting displays. The only information that gets passed back from the behavioural choice module to the environment is the animal's choice of action at each moment in time. While there are restrictions on the information that gets passed between the environment and the behavioural choice mechanism, there are no restrictions on what the behavioural mechanism does with the information it receives (how much does it remember between timesteps, which information does it just ignore, etc).

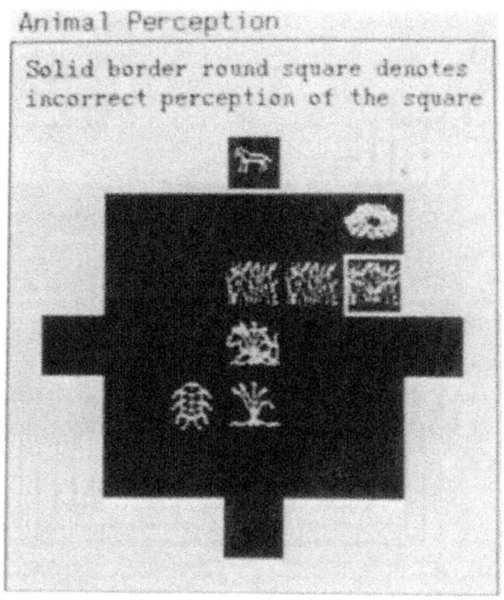

Figure 7: The animal perception display (with a predator type 1 incorrectly perceived to be in the top square and a predator type 2 incorrectly perceived to be in the square above and to the right of the animal).

4 Example 1 - A Blank Slate Adaptive System

The purpose of the two examples presented in this section and the next is to show that the development of a behavioural strategy for the simulated environment is not trivial and that the problems encountered mirror those faced with animals in the real world. The examples should also demonstrate that the simulated environment is a useful tool for highlighting difficulties in intrinsically plausible behavioural mechanisms. In presenting these naive examples, we are not trying to give a complete appraisal of the mechanisms, or to assert that they cannot be improved so as to overcome the difficulties outlined, but rather to show that the simulated environment does unearth the problems with the mechanisms. At the time of writing the programming is only just being finished and so the examples have not yet been fully evaluated in the simulated environment. However, some problems are already obvious.

For the first example we will look at an adaptive

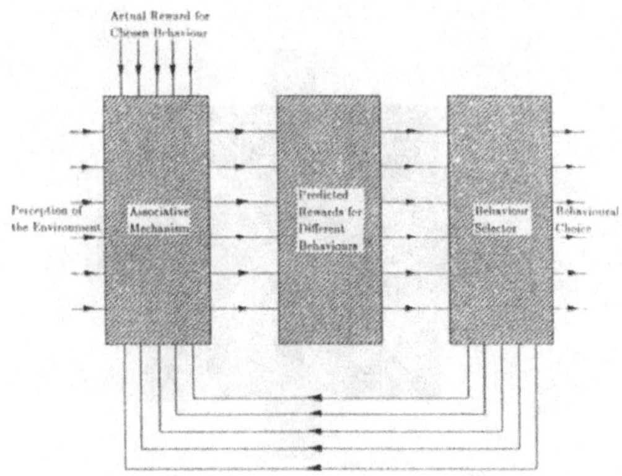

Figure 8: A simple blank slate adaptive system.

system that can learn from experience which actions are successful in which situations, but which starts off with no in-built knowledge (i.e. the system is initially a 'blank slate', or 'tabula rasa' with no predetermined propensities).

The simple, naive, adaptive mechanism of this example is shown in Fig. 8. This mechanism will start off by choosing random actions when presented with successive states of the environment. Whenever it carries out a certain action in a certain situation it makes associations between the action, the situation and the outcome. In this way the mechanism is able to build up evaluations for different actions in different situations. As these evaluations come to be formed, they can then be used by the action selector in order to choose the action with the highest predicted reward rather than just choosing randomly. By building up its knowledge of which actions work best in which situations, the animal is able to improve its performance over time. Two of the shortcomings of this mechanism for behavioural choice, both in the simulated environment and the real world, are as follows:

4.1 *Inappropriate Experimentation*

One problem with a system that contains no in-built knowledge is that it has to experiment in order to be able to learn about its environment. It has to test different actions in different situations to see whether they produce beneficial outcomes. In a real environment there are certain situations in which experimentation will have a high probability of resulting in serious injury or death. For instance, investigating cliffs for the possibility of finding food there, or investigating predators to see if they can be mated with will not be very conducive to survival.

A successful behavioural strategy will be one that allows experimentation with new responses, but which limits the experimentation so that it only occurs in non-critical situations. Some of the responses of the system will have to be hard-wired and inflexible.

4.2 *Inefficient Learning*

It has been shown that animals are prejudiced so as to make certain associations more readily than others. For instance, rats have been shown to be predisposed to associate visual and auditory stimuli (e.g. bright lights and loud noises) with electric shocks, and food stimuli with sickness. On the other hand the rats found it very difficult to associate visual/auditory stimuli with sickness and food with electric shocks.

In general, if a certain phenomenon in the real world (e.g. sickness) has a high probability of being due to a certain cause (e.g. eating a new type of food which is poisonous) then the animal's learning will be biased so that it is more likely to make the correct association.

An optimal behavioural strategy will make use of these regularities in the environment and should, for instance, be more likely to learn to associate a drop in temperature with entering shade than with eating a certain type of fruit.

5 Example 2 - the 'Simple Motivational Competitional Model'

The second example is what McFarland calls a 'simple competitional motivational model' [McFarland, 85]. In this model the animal experiences drives or tendencies for different behaviours. The intensity of a drive is an indication of the 'need' for that behaviour (for instance the intensity of the food drive will be proportional to the animal's food deficit). The behaviour selected is then that corresponding to the highest drive (see Fig. 9).

With this behavioural mechanism there will also be shortcomings. Some of these are:

5.1 *Dithering*

When the animal's needs for both water and food are approximately equal then with this simple behavioural strategy the animal will rapidly switch between the two

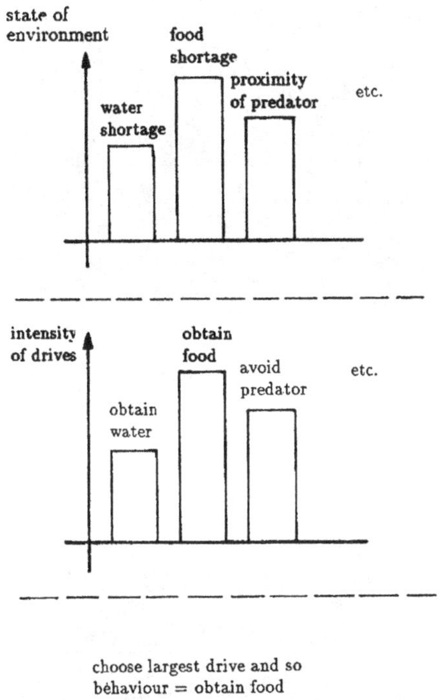

Figure 9: McFarland's 'simple motivational competitional model'.

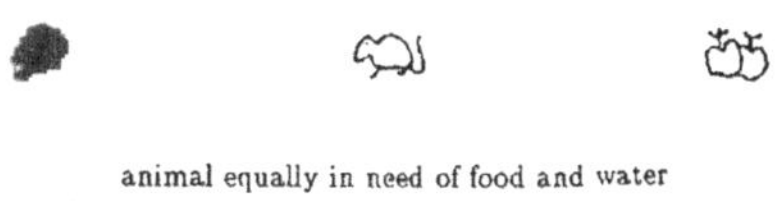

Figure 10: Dithering situation.

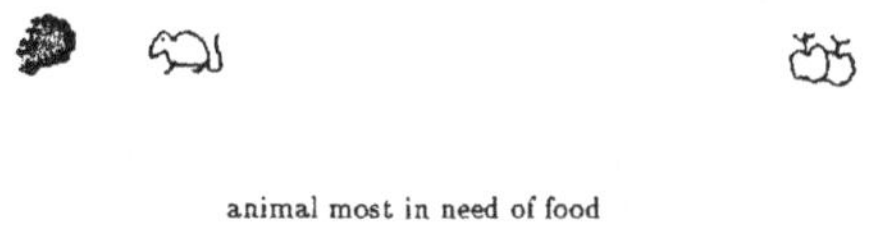

Figure 11: Opportunistic situation.

behaviours. Eating a small amount of food will decrease the food drive below the water drive, and the opposite will then occur when the animal travels to the water source. This will result in the animal travelling back and forth between the two sources, spending more time travelling than either eating or drinking.

5.2 *Opportunism*

There will be occasions when heading immediately for the most needed resource will entail passing up an easy opportunity to obtain a less important resource. For instance, if the animal is more in need of food than water, but the nearest water source is a lot closer than the nearest food source then it might pay the animal to visit the water source first and the food source later. If the animal visits the further away food source first, it can satisfy its hunger, but might then have to travel back to the water source in order to satisfy its thirst.

5.3 *Summation*

In this case it will be more efficient for the animal to visit the cluster of adjacent features rather than the single large food source which would best satisfy the animal's biggest drive.

5.4 *Local Conditions*

One problem faced in the 'design' of animals by the process of evolution is that the local environment for differ-

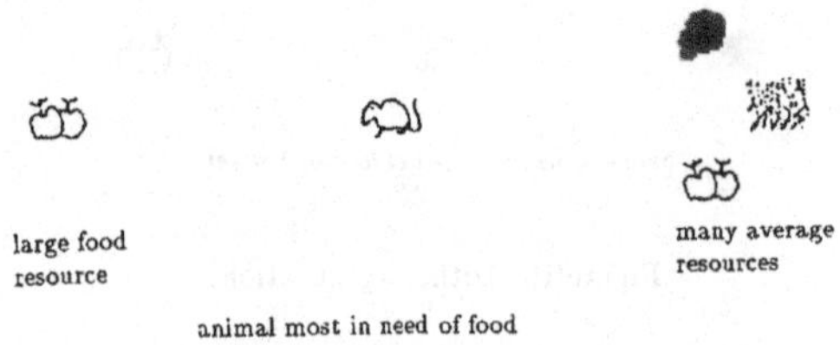

Figure 12: Summation situation

ent animals of the same species is not usually the same. In other words, animals of the same species will be spread about and will live in different locations. The different locations will not be identical and so will present different problems to the individual animals.

This is mirrored in the simulated environment by the use of a random number generator to calculate the positions of features. Each time the environment is generated, a different arrangement will result. These different arrangements of features provide different problems for the animal. For example, the environment in Fig. 13 has very little cover next to the food and water sources and so one of the most important aspects of the animal's strategy will be how to obtain food and water without laying itself open to predation. On the other hand, the environment in Fig. 14. has no water sources to the left of the animal's den and so the animal will need to learn to take this into account when deciding which areas of the environment to traverse.

Testing a behavioural strategy in the simulated environment will involve randomly generating a number of environments and measuring how well the animal performs in each of them.

The difference in local conditions (or successive simulated environments) means that the behavioural strategies have to be evolved (or designed) without any knowledge of the specific environment the individual animal will exist in. Therefore the best strategies will be those that are able to make use of knowledge gained about the local conditions to improve their effectiveness (i.e. those that are **adaptive**). The behavioural strategies that can improve with experience will be better able to exploit the local conditions.

6 Discussion

The two examples in the last two sections attempted to show that this simulated environment is able to re-

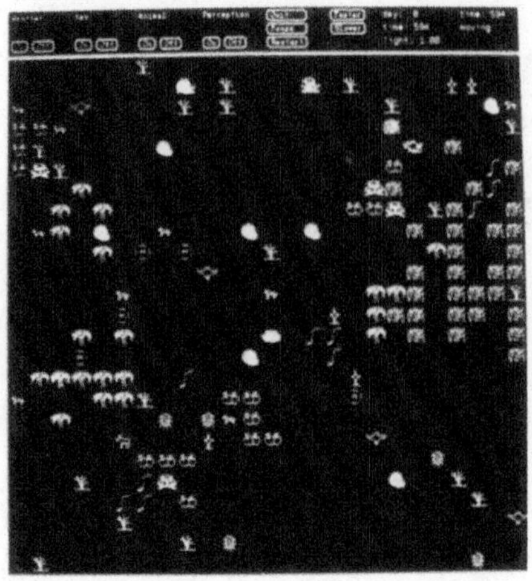

Figure 13: A randomly generated instance of the simulated environment.

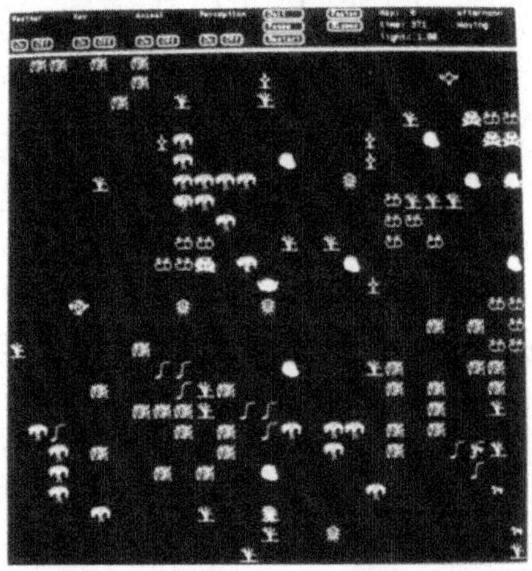

Figure 14: A randomly generated instance of the simulated environment.

produce the most important aspects of a real environment and that the problems encountered in building a behavioural choice mechanism for the simulated environment mirror those which had to be overcome by evolution in its design of real animals in real environments. Before concluding though that the simulated environment is a satisfactory 'testbed', it is necessary to examine and recognise some of the limitations of this approach.

6.1 *Separating The Three Stages*

One question that can be asked of the work in this paper is "Is it really possible to consider behavioural choice independently of perception and motor control, or are the three stages so interlinked that the chosen solution for one stage determines the solutions for the two other stages" So, for instance, it could be argued that the way in which the animal perceives the environment (does it mainly sense by smell or vision, what ability does it possess to classify/generalise perceptual inputs, etc.) crucially influences the form of the behavioural choice mechanism. Or, similarly, it could be argued that the way in which the animal executes its behaviours crucially influences the form of the behavioural choice mechanism.

While we accept that there is obviously some interdependence between the different systems, we believe that most of the problems that an animal's behavioural strategy has to be able to cope with exist independently of the way in which the animal senses the environment or carries out its chosen behaviours. In our work, we have assumed that an animal's perception serves just to provide the animal with information about the environment. Different perceptual systems will give the animal more or less reliable information about the environment, and will give the animal differing information about the environment. Also, we believe that the way in which an animal carries out particular behaviours (how many limbs it walks on, whether it picks food with its mouth or with its hands, etc.) does not seriously affect the nature of the behavioural choice problem.

Whatever the form of the perceptual and motor control systems, the animal still has to be able to weigh up the pros and cons of different actions, avoid dithering between different behaviours, use appropriate and efficient learning, etc.

6.2 *Social Interactions*

The main simplification that has been made in this work is to keep social interactions to a minimum. If the major part of an animal's environment consists of other animals of the same species then to accuratedly model the animal's environment, those other animals must also be correctly modelled. It is easier to correctly model physical features than other animals and so we chose an environment based on an animal which spends little time interacting with other animals of its kind. This also simplifies matters in that it allows us to use a simple function for genetic fitness (since a relatively solitary animal has little contact with close relatives).

The problem of modelling social interactions has lead us to assume that the animal is male (and so doesn't have to care for young), and also to ignore the possibility of transmission of successful behavioural strategies by cultural means.

While noting this drawback to the approach it is also worth noting that some animals (e.g. snakes, hedgehogs, tigers) do lead lives in which they contact others of their kind only very infrequently and in these cases our simulated environment will be fairly realistic. It is also worth noting that even when an animal relates frequently with others of its kind, it will still require a mechanism for choosing between different actions at each moment in time, even though the repertoire of actions will not be exactly the same (e.g. grooming, protection of others) and the results of actions will depend on other, less definite, factors such as the moods of other animals.

6.3 *Using a Grid of Squares*

Another simplification was to represent the environment as an array of square regions. Although this is not strictly necessary and it would have been possible to represent the environment as a continuous space (giving the features coordinates instead of tying them to squares), the programming for the environment would have taken much longer and the program would have run a lot more slowly.

7 Conclusions

The ultimate goal of the research this work is part of is to investigate mechanisms of behavioural choice, and to that end we wanted to create a testbed in which we could test and compare different mechanisms. Despite the simplifications that we have made, we believe we have satisfied that goal. We think that the simulated environment does sensibly reproduce the important aspects of an animal environment and that the stream of behavioural decisions the animal has to make (in response to its constantly changing environmental situation) are representative of the behavioural decisions animals need to make in the wild.

It is hoped that the use of the simulated environment will help us to gain useful insights into the requirements for successful behavioural choice mechanisms. Although the environment has not yet been used to test many different mechanisms (since the programming is only just being finished), some rather obvious conclusions can already be drawn. For instance, learning must be strictly constrained, some situation-action links must be hard-wired, the need for certain behaviours must be combined with the ease with which they can be carried out before deciding on whether to carry them out, etc. It is hoped that the continued use of the simulated environment will eventually lead to more, and deeper, insights.

Acknowledgements

Many thanks to members of AIVRU and the Psychology Department for helpful discussions. This work was funded by the Science and Engineering Research Council (grant no. 89310818).

References

[1] **Albus, James, S.** *Brains, Behaviours & Robotics,* BYTE Publications (1981).

[2] **Booker, Lashon B.** "Classifier Systems that Learn Internal World Models" *Machine Learning*, 3, 161-192, (1988).

[3] **Braitenberg, Valentino** *Vehicles, experiments in synthetic psychology,* MIT Press (1984).

[4] **Brookes, Rodney, A.** "A Robust Layered Control System for a Mobile Robot" *IEEE Journal of Robotics and Automation,* 2:14-23, (April 1986).

[5] **Coderre, Bill** "Modeling Behaviour in Petworld", chapter in *Artificial Life* (ed. Christopher G. Langton), Addison-Wesley (1989).

[6] **Dawkins, Richard** *The Selfish Gene* (2nd edition), Oxford University Press (1989).

[7] **Halliday, T. R.** "Motivation", chapter in *Causes and Effects* (eds Halliday, T. R. and Slater, P. J. B.) Blackwell Scientific Publications (1983).

[8] **Hinde, Robert A.** *Ethology: Its Nature and Relations to other Sciences,* Fontana Paperbacks (1982).

[9] **Kilmer, W. L., McCulloch, W. S., and Blum, J.** "A Model of the Vertebrate Central Command System" *Int. J. Man-Mach. Stud.*, 1, 279-309, (1969).

[10] **McFarland, David** *Animal Behaviour,* Pitman (1985).

[11] **Taylor, Charles E., Jefferson, David R., Turner, Scott R. and Goldman, Seth R.** "RAM: Artificial Life for the Exploration of Complex Systems", chapter in *Artificial Life* (ed. Christopher G. Langton), Addison-Wesley (1989).

[12] **Toates, F. M.** *Animal Behaviour - A Systems Approach,* Wiley (1980).

[13] **Travers, Michael** "Animal Construction Kits", chapter in *Artificial Life* (ed. Christopher G. Langton), Addison-Wesley (1989).

[14] **Wilson, Stewart W.** "Classifier Systems and the Animat Problem" *Machine Learning,* 2, 199-228, (1987).

ONTOGENY AND LEARNING

Ontogeny of Preferences in Guppies (Poecilia reticulata) Exposed to Food and Conspecifics

Patrick W. Colgan, T. Lynne Jamieson, Janice E. Frame, and J. Terry Smith

Queen's University at Kingston
Kingston, Ontario, Canada, K7L 3N6
Bitnet COLGANP@QUCDN
(613) 545-6144

Abstract

A statistical method for quantifying, summarizing, and evaluating information about the behaviour of a living system is illustrated using data from a study of the ontogenetic development of preferences in guppies (Poecilia reticulata). This method permits comparisons of groups of individuals in different treatments when the data on each individual consists of a long sequence of behavioural states together with the entry time into those states. This tool is readily adaptable to the study of a simulation of an organism, set of organisms, or the comparison of robotic models with living systems.

1. Introduction

The use of robotic systems to model living organisms has had varying degrees of success in expanding our knowledge of living systems. Generally, the exercise is fruitful when relevant aspects of the living system are modelled faithfully. It is important, therefore, to have tools for comparing the behaviour of robotic systems with the living systems they model.

We present here one example of such a tool: a statistical method for quantifying, summarizing, and evaluating information about the behaviour of a living system. We illustrate this statistical tool by a study of preferences in guppies (Poecilia reticulata) for food and conspecifics. The tool is readily adaptable to the study of a simulation of an organism or set of organisms, or to the comparison of robotic models with living systems.

Guppies (Poecilia reticulata) are an intensively studied species for many aspects of behaviour. Pilot studies in our laboratory indicated significant variation among both male and female guppies in their preferences for stimuli of food and conspecifics. Such individual variation has increasingly attracted the attention of behaviourists (e.g. for fish see Magurran, 1986; Gotceitas & Colgan, 1988) with respect to its basis and function. From an ontogenetic perspective, such variation raises questions concerning the role of social isolation in its development, sexual differences in this ontogeny, and the extent to which adult behaviour is predicted by juvenile behaviour.

Ontogenetic research on guppies has included studies on the development of stimulus preferences (Candland & Milne, 1966), avoidance of predators (Goodey & Liley, 1986), temperature preferences (Johansen & Cross, 1980), sexual activities (Liley & Wishlow, 1974), and the role of stress and rearing conditions (Pinckney & Anderson, 1967; Newton, 1982). Germane to the present work are the following findings from these studies. Isolated fish were less active than controls and that activity decreased over the duration of the observational trial (Newton, 1982). The isolates scored higher frequencies of social display and lower levels of sexual activity. Liley (1966) found that as experience was gained with females, previously isolated male guppies showed decreased levels of display. Pinckney and Anderson (1967) observed that group-reared fish spent a decreasing amount of time near the stimulus fish in contrast to the isolated fish that spent an increasing amount of time near the stimulus fish. The group-reared fish showed a preference for the stimulus fish of the same sex while the isolated fish showed no significant preference for either sex of stimulus fish.

All of these studies examined the behaviour of mature adult fish as the outcome of various manipulations. The objective of the present study was to monitor at frequent intervals the stimulus preferences and activity of guppies throughout the juvenile and early adult periods, and to determine the extent and ontogeny of individual differences in each sex. Individuals were raised as experimental fish in visual isolation, and as control fish in visual contact with each other. The resulting data on preferences were analyzed with both traditional statistical tools and a novel application of a time-series approach based on Zeger and Qaqish (1988). This approach provides a summary of the sequential data of each fish. Subsequent multivariate ANOVA applied to these summaries addresses the identification of individual differences, ontogenetic trends, and the effects of rearing and sex on the development of preferences.

2. Method

2.1 Fish and Tanks

Twenty guppies were randomly selected within a week of birth from a laboratory breeding stock and placed individually in a compartment of one of the two rearing tanks (Figure 1A). The tanks measured 44 cm long X 24 cm wide X 26 cm deep and were kept on a 16:8 L:D photoperiod at 21 ± 2° C. The partitions forming the compartments were transparent in the case of the control group and opaque in the case of the experimental group. The fish in the control group were placed in separate compartments for the purpose of individual identification.

The observational tank (Figure 1B), 41 cm long X 21 cm wide X 25 cm deep was divided into a waiting area 13 cm wide and an observation area containing a stimulus cylinder of clear plexiglass with a diameter of 14 cm trisected into three equal sectors containing floating Tetramin food, a mature male guppy, and a mature female guppy. The test fish could eat the food but was separated by plexiglass from the two stimulus fish which were used throughout the experiment. Outside of observational sessions the two stimulus fish were released in the tank. The test fish were fed daily with Tetramin food in the rearing tanks. On observation days the feeding followed the observational sessions.

2.2 Observations

Each fish was observed for 20 minutes every third or fourth day (i.e. about twice a week) between the hours of 1000 and 1600. It was transferred from its compartment in the rearing tank to the waiting area of the observational tank and permitted to settle for 10 minutes prior to the initiation of observation. Twenty-six observational sessions were made on each fish beginning in the second week of life and continuing until the 15th week of life. Since the development of female guppies to full maturity requires more time than for males a further 15 observational sessions were made on females only, starting in their 19th week of life and continuing until the 25th week.

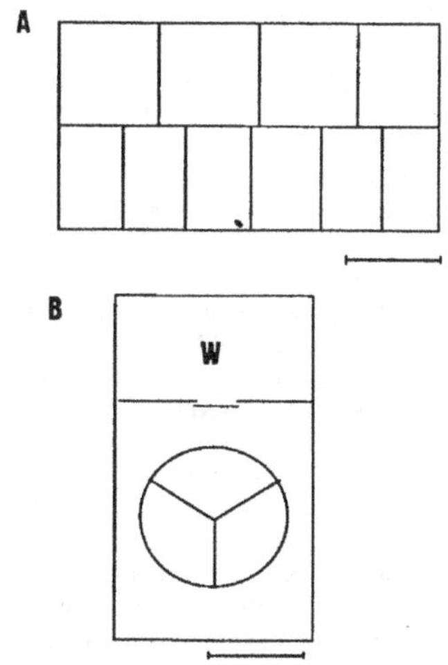

Figure 1. Top views of experimental tanks. A: Rearing tanks for the two groups. B: Observation tank, including the waiting area (W) and observations area with the stimulus cylinder. Scale = 10 cm.

During the week prior to the first observational session the test fish were transferred to the observational tank periodically in order to familiarize them with the procedure. In each observational session information was obtained on the location of the test fish in front of each sector and on its activities in that sector, including feeding and courtship. The time, measured in seconds from the beginning of the session, to the beginning of each behaviourial act was recorded on a Datamyte events recorder and transferred to an IBM mainframe. For each session with each fish, the sequence of act-change times was differenced to give a sequence of act durations. These became the raw material for the analysis of sequential patterning. Sample plots of the durations data for individual fish indicated that transformation by logarithms produced a more nearly normal distribution.

The statistical methodology developed in this paper addresses the problem of comparing groups of individuals under different treatments when the data on each individual consist of a long sequence of behaviour states together with the entry times into those states (hundreds of items of information). Traditional methodology typically treats sequences of behavioural states or, alternatively, sequences of state durations on an individual, and for purposes of comparing groups or taking covariates into account, is limited to relatively few items of information (five or ten, say) per individual. The strategy adopted here is to summarize the sequence information meaningfully in terms of a small number of items per individual and to compare groups in an appropriate way in terms of the derived summaries. The analysis thus proceeds in two stages.

At the first stage, the durations data for each fish were summarized by a set of parameter estimates based on a Markov regression model for the durations. At the second stage, between-individual and between-group comparisons were made in terms of these parameter estimates.

The first-stage analysis was based on a model (Zeger and Qaqish, 1988) for the conditional mean of the k-th duration as a linear function of the (k-1)-th duration, age of the fish, and effects for the conspecific and food stimuli. Coefficients of these "predictor variables" were regarded as subject-specific parameters summarizing the sequence behaviour of that individual. Parameter estimates and associated covariance matrices were generated by an iteratively reweighted least-squares algorithm using GLIM (Baker and Nelder,1978) using both the lognormal distribution and the gamma distribution to model the durations. These methods parallel the Yule-Walker methodology for conventional time series.

At the second stage, individual differences were assessed graphically in terms of the parameter estimates from stage one. Group comparisons were conducted with univariate and multivariate analyses of variance (ANOVA and MANOVA, respectively, using SAS), modified to account for the differing covariance matrices (assumed as estimated without error) across individual fish.

3. Results

3.1 Ontogeny

Fish in both groups grew well and behaved actively. Guppies cannot be externally sexed when young, and as the test fish matured the group-reared fish proved to consist of four males and six females while the isolated fish proved to consist of three males and seven females. Males showed mature colouration beginning between observational sessions 8 and 11. The test and stimulus fish interacted with each other with patterns typical of the species. Our analysis will be restricted to the pattern of time spent by the fish in each sector as a function of sex and rearing.

3.2 Sequential Patterning: First-Stage Analysis of Act Duration

The results of the Markov regression analysis of the durations are illustrated in Tables 1, 2, and 3 and Figure 2. The summary parameter estimates for one fish from the first stage of analysis are given in Table 1. Parameters representing a constant term, a term reflecting dependence on the preceding duration, the effects of the male and female stimuli (relative to that of the food), and the age of the fish are all significant. Table 2 compares the fit of models containing different sets of terms by an analysis of deviance (counterpart of analysis of variance for normally distributed data) for the same fish. Although the pseudo R^2 values are small, the terms reflecting the type of stimulus and the age of the fish are significant. The results given in Tables 1 and 2 are typical of all 20 fish.

Parameter	Estimate	Std. Error	P
Constant	4.641	.1573	<.001
Lagged duration	-0.062	.0366	NS
Male stimulus	-1.518	.1757	<.001
Female stimulus	-1.511	.1909	<.001
Age	-0.011	.0010	<.001

Table 1. Summary of parameter estimates for one fish.

3.3 Group Comparisons: the Second-Stage Analysis

Table 3 presents the results from the ANOVA's for each Markov regression parameter individually across sex and rearing conditions. The parameter estimates are shown graphically across sex and rearing condition in Figure 2. Note that the scales are different for different parameters. Males and females differ with respect to the dependence upon preceding duration and preference for the male

Model	G^2	Df	G^2	Df	R^2(%)	F	P
Independence, Null	1560	1795					
Auto,Null	1555	1794	5	1	0.32		
Auto,Stim	1532	1792	23	2	1.48	13.45	<.001
Auto, Stim + Age	1522	1791	10	1	0.64	11.76	<.001

Table 2. Analysis of deviance for one fish. (N=1561)
Auto=autoregression. Stim=stimulus. G^2 = scaled deviance.
F values are approximate.

Parameter	Sex Effect	Std. Error	P	Rearing Effect	Std. Error	P	R^2(%)
Constant	.313	.234	NS	-.076	.198	NS	10.0
Lagged duration	-.073	.032	<.025	.007	.028	NS	23.7
Male stim.	.329	.148	<.025	-.241	.126	**NS**	31.9
Female stim.	.183	.185	NS	-.347	.156	**<.025**	24.9
Age	-.004	.003	NS	.001	.002	**NS**	14.9

Table 3. Individual ANOVAs for summary parameter estimates stage one analysis.

stimulus over food, while grouped and isolated fish differ with respect to preference for the female stimulus over food.

Table 4 presents the significant results from the MANOVA comparing all the parameters simultaneously. The parameter estimates in the table are expressed as deviations from a baseline group of isolated females at age zero. There was an overall significant preference for males over food and for females over food, with no significant preference for males over females overall. Isolates showed the female-over-food preference significantly more strongly than did group-reared fish, the rearing effect being significantly stronger for males. There was a significant overall effect of age, mean duration decreasing about .005 for every unit increase of age. Males showed this effect significantly more strongly than females, this male-female difference being stronger in isolates than in group-reared fish.

3.4 Individual Differences in Sequential Patterning

Individual differences are depicted graphically in Figure 3 in terms of the parameters from the Markov regression analysis. The values represent contributions to the log mean duration for an individual fish. Again, note that the scales are different for different parameters, and that all three scales contain zero. Complete understanding of the significance of group differences requires knowledge of the standard errors attached to each of the parameter values, suppressed here in the interests of visual clarity.

Structurally, each panel of Figure 3 depicts visually the data for an unweighted two-way ANOVA for sex by rearing on the particular parameter depicted in the panel (not taking the weighting into account means that Figure 3 may not be wholly indicative of the results of Table 4). With regard to individual differences, we see that groups overlap considerably, and that within-group variation changes from one group to the next for a particular parameter estimate and, to a lesser extent, from one parameter estimate to another within a group (keeping in mind that scales for panels are determined by the total range of the 20 individuals for each panel separately).

Series Parameter	Manova Parameter	Estimate	Std. Error	P
Constant	Constant	3.634	0.123	.0001
Male vs Food	Constant	-0.585	0.116	.0001
	* Rearing	-0.328	0.169	.0554
Female vs Food	Constant	-0.455	0.117	.0002
	* Rearing	-0.473	0.166	.0055
	Sex x Rearing	0.596	0.348	.0900
Age	Constant	-0.005	0.001	.0001
	! Sex	-0.010	0.002	.0001
	Sex x Rearing	0.010	0.003	.0020

Table 4. Estimates from MANOVA (lognormal distribution, $R^2=97\%$)
[Significant or almost significant effects only]
* Group-reared vs isolates
! Male vs females

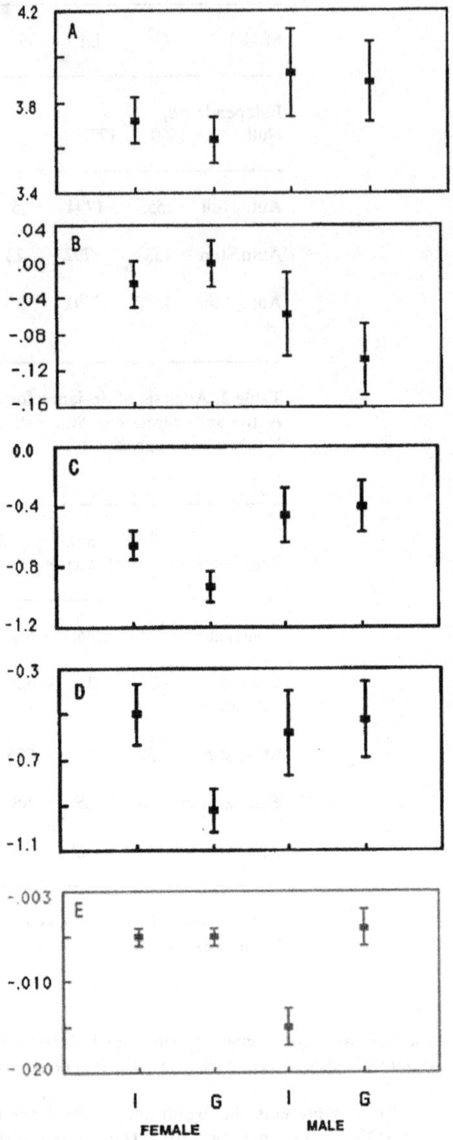

Figure 2. Summary of parameter estimates from the time-series model for isolated and group-reared male and female guppies. Data are means and 95% confidence limits. A: constant term, B - E are coefficients of: lagged duration, male stimulus, female stimulus, and age.

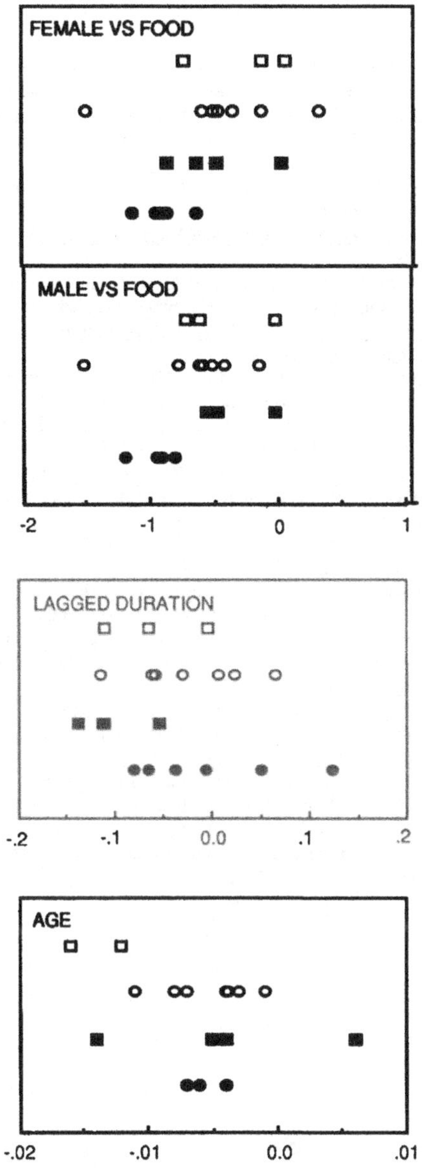

Figure 3. Estimates of parameters for individual fish from the time-series model. Data are solid=grouped, open=isolated, square=male, and circle=female.

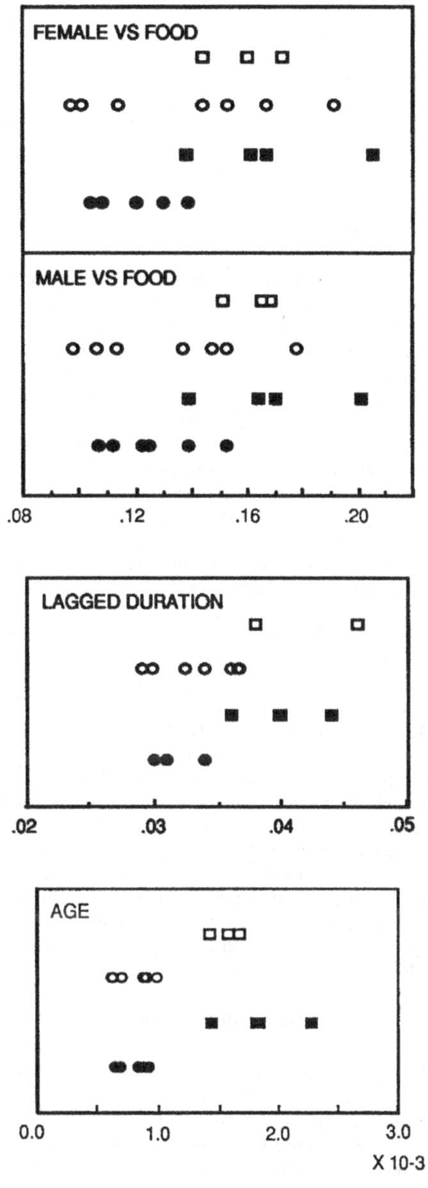

Figure 4 The associated standard errors of the time-series model for individual fish. Symbols as in figure 3.

Figure 4 presents information on within-individual variability as measured by standard errors of parameter estimates, the organization of the information corresponding exactly to Figure 3. While we expect standard errors to vary a lot more among individuals than mean values, we see that variability of these four summaries of behaviour sequences is quite heterogeneous across individuals.

4. Discussion

The data and analyses presented here indicate useful approaches which can be used to examine the differences and the effects of rearing and sex. The Markov regression analysis provides a more sensitive analytical tool than approaches which ignore the sequential properties of the data. For instance, lagged durations, while not varying significantly across sex or rearing condition (Table 4), varied significantly across individuals (Table 3). The use of MANOVA on the per-individual Markov regression parameters permits between-group comparison of the key descriptors of sequential patterning.

The results show that the temporal preferences of guppies for social and food stimuli change with age in a manner which is dependent upon the sex and rearing history of the fish and the type of stimulus. At a finer level of analysis, the sequential patterning of these preferences is also influenced by these factors as well as the duration of the preceding activity. Thus our data reflect more complex trends than those previously reported in the literature. This difference may be due to differences in experimental protocol. Alternatively, it may reflect the true richness which is to be found by examining the sequential patterning of individuals as they mature. Further studies are needed to reveal the causal and functional determinants of this patterning, including processes such as habituation and socialization.

More generally, the techniques employed in this paper can be applied to diverse behavioural situations, with data from natural or artificial systems.

Acknowledgements

This research was financially supported by an Operating Grant from the Natural Sciences and Engineering Research Council of Canada to Patrick Colgan.

Correspondence concerning this article should be addressed to Patrick Colgan, Biology, Queen's University at Kingston, Ontario K7L 3N6.

References

Baker, R.J., and Nelder, J.A., 1978. Generalized linear interactive modeling. Oxford: Royal Statistical Society.

Candland, D.K., and Milne, D., 1966. Species differences in approach-behaviour as a function of developmental environment. Animal Behaviour, 14, 539-545.

Goodey, W., and Liley, N.R., 1986. The influences of early experiences on escape behaviour in the guppy (Poecilia reticulata). Canadian Journal of Zoology, 64, 885-888.

Gotceitas, V., and Colgan, P., 1988. Individual variation in learning by foraging juvenile bluegill sunfish (Lepomis macrochirus). Journal of Comparative Psychology, 102, 294-299.

Johansen, P.H., and Cross, J.A., 1980. Effects of sexual maturation and steroid hormone treatment on the temperature preference of the guppy, Poecilia reticulata (Peters). Canadian Journal of Zoology, 58, 586-588.

Liley, N.R., and Wishlow, W., 1974. The interaction of endocrine and experiential factors in the regulation of sexual behaviour in the female guppy Poecilia reticulata. Behaviour, 48, 185-214.

Magurran, A.E., 1986. Individual differences in fish behaviour. In T.J. Pitcher (Editor) The behaviour of teleost fish (pp. 338-365). London: Croom Helm.

Newton, B.J., 1982. Early stress effects on growth and adult behaviour in Poecilia reticulata. Developmental Psychobiology, 15, 211-220.

Pinckney, G.A., and Anderson, L.E., 1967. Rearing conditions and sociability in Lebistes reticulatus. Psychonomic Science, 9, 591-592.

SAS Institute Inc., 1985. SAS User's Guide: Statistics, Version 5 Edition. Cary: SAS Institute Inc.

Zeger, S.L., and Qaqish, B., 1988. Markov regression models for time series: A quasi-likelihood approach. Biometrics, 44, 1019-1931.

SIMULATION STUDIES OF SONG LEARNING IN BIRDS

James M. Williams & P.J.B. Slater

Department of Biology & Preclinical Medicine,
University of St Andrews,
Fife KY16 9TS, U.K.

Email : ZORJW@uk.ac.st-andrews.sava
& ZOSPS@uk.ac.st-andrews.sava

Abstract

Song birds learn their songs, but the accuracy of the copying process varies. As a result the songs present in an area change with time.

Our evidence to date suggests that various aspects of chaffinch song in the wild can be accounted for by a simple random model in which individuals learn the songs that they sing from various adults. The distribution of songs between repertoires of different birds is best matched by simulations with a 15% copy-error rate. This rate of error, combined with a realistic mortality rate of 40%, also gives a good approximation to the changes in the song types present in a population with time.

Simulations have been used to examine the distribution of song types between individuals in a population. When the simple situation, where all birds had a single song type and four neighbours, from one of whom new birds copied their song was examined, small groups of birds sharing a song type were found, as in some dialect species. An extension of this approach to simulate variations in repertoire size or in numbers of neighbours has recently shown that both these factors can have a strong effect on the sharing of song types and their distribution in the population. If a bird chooses the commonest song type sung by its neighbours rather than one of them at random very large groups of birds can occur.

These simulations suggest that the complex distributions of song types often found in wild bird populations may result simply from random copying processes which are not always exact.

1. Introduction

Learning plays a role in the song development of all songbirds studied to date (see review by Slater 1989). In many cases the copying of song takes place from neighbours when young birds first set up their territories so that birds on adjacent territories tend to share songs, while the songs of those further apart are less similar. While the learning can be remarkably accurate, so that the songs of two individuals are often identical, there is good evidence from a number of studies that inaccuracies of copying may lead to new forms of song arising (e.g. Jenkins 1978, Slater & Ince 1979). These "cultural mutations" may be the reason why the songs present in a given area change with time, and why there are also differences in song between localities. Much of the geographical variation in song is complex, especially in cases where each individual has a repertoire of several different song types and these are not all learnt as a package from one other bird. However, in some species dialect areas have been described in which groups of birds share the same song type or types and are separated from each other by more or less sharp boundaries. Whether or not this mosaic pattern has any functional significance is a matter of a good deal of controversy (see Baker & Cunningham 1985).

2. Song in the Chaffinch

We have used computer simulation mainly to supplement our studies of song distribution in the chaffinch (*Fringilla coelebs*), a small European songbird. Its song has been extensively studied over many years, starting most notably with the work of Marler (1952) on song in the wild and the laboratory studies of Thorpe (1958) on song learning. The song consists of a series of phrases of repeated syllables (the trill) followed by a terminal flourish. It is usually about two and a half seconds long. Except in rare cases song is only produced by territorial males.

Each male has 1-6 song types and these are learnt in the first thirteen months of life, either after hatching in the summer or the next spring when the bird itself starts to sing (Slater & Ince 1982). All chaffinch songs share basic features but they differ in form both within the repertoire of a single bird and between birds, so that within a population a number of different song types are found. Some of these are shared by many birds, some are sung by only one bird. Slater *et al.* (1980) found that the frequency of song types in a population in Sussex fitted the distribution expected from a model of neutral alleles (Ewens 1972), and suggested that the songs in a bird's repertoire are learnt at random amongst those available and that different ones are often learnt from several different individuals. Since they found that songs are not shared more with neighbours than would be expected by chance, it is possible that some are learnt before territories are set up.

3. Chaffinch song simulations

3.1 Cultural Change

Slater *et al.* (1980) found that, in a wood containing 42 territories, a few common song types were sung by many birds and a lot of song types were sung by only one or two individuals. Thus 22 birds sang the commonest song, but there were 21 songs each sung by only one bird. A simulation of the rate of miscopying required to produce the distribution of sharing between repertoires observed produced a best match to the field data with a rate of 15%. The results were the means of three runs, each started with all birds singing a single song type and run for 100 'years', with a 40% annual mortality rate (data from the British Trust for Ornithology). A population of 121 songs was used, matching the total number of songs sung by the 42 birds in the wild sample. Running the program for 100 cycles ensured that the outcome was independent of the starting conditions.

Ince *et al.* (1980) went on to compare the 35 song types recorded in the same wood in 1978 with a sample of 22 that had been recorded there 18 years earlier. Only three types were similar enough to be considered as the same. Ten runs of a simulation based on these data were carried out using the 15% miscopying rate which had earlier been found to give the best fit to song distribution. These simulations predicted that a mean of 3.2 song types should be in common between the samples, a remarkably good fit to the data (Slater *et al.* 1980).

3.2 Dialects

The issue of whether song dialects exist and what their functions, if any, might be is a complex and controversial one. The main problem is one of definitions; the term "dialect" is used in several different senses. If clear dialects exist it should be possible to distinguish easily between individuals in different populations or subpopulations. Sharp dialect boundaries with such clear differences have only been described for a few species, and it is perhaps not surprising that these all have small repertoires of only one or two song types. Examples are the white-crowned sparrow (*Zonotrichia leucophrys*, Baptista 1975), European redwing (*Turdus iliacus*, Bjerke & Bjerke 1981) and corn bunting (*Emberiza calandra*, McGregor 1980, McGregor and Thompson 1988).

Some authors have argued that chaffinch song shows dialects (Metzmacher & Mairy 1972, Conrads 1986), but Slater *et al.* (1984) rejected this hypothesis as they found great variation within areas as well as some sharing of features between them. Such differences between areas are better referred to as geographic variation.

Goodfellow & Slater (1986) explored the possibility that dialects might arise as the result of random copying of song types. A 10 x 10 matrix was used, and the simulation run over 300 'years' with birds able to learn a single song type from one of up to four neighbours next to them. The simulation started with all birds singing the same song type. The mutation rate of 15% arrived at by Slater *et al.* (1980) and a mortality rate of 40% were used. If a new bird had no pre-existing neighbours to copy from it made up a new song type. A simple mosaic of shared songs resulted (see Figure 6 in Slater 1989). Groups of up to 20 birds sometimes emerged, but most birds shared their song with only a few neighbours. The size of the groups varied with the error rate used (see Figure 1 in Goodfellow and Slater 1986). When the simulation was started with the birds in one half of the matrix singing one song type, and those in the other half singing another, a stable border could be achieved provided error and mortality rates were low. The conclusion was that extremely accurate copying would be needed to produce the large numbers of birds sharing a song type described for white-crowned sparrows or corn buntings, but that the small song neighbourhoods of indigo buntings (*Passerina cyanea*) described by Payne *et al* (1981, 1988) could much more easily arise through the processes that were simulated.

Williams & Slater (1990) have recently further examined the possibility that dialects may form by random copying by investigating the effect of varying both the repertoire size of each bird (1 - 6), and the number of neighbours that it has to learn from (2, 4, or 8) (Figure 1). This is also a situation more relevant to birds, like the chaffinch, which have variable numbers of neighbours and usually have a small repertoire of song types. The simulation again used an annual mortality rate of 40% and a miscopying rate of 15%. It employed a 13 x 13 matrix and ran over 100 'years'. The runs started with each bird singing the same songs (labelled 1,2,..n).

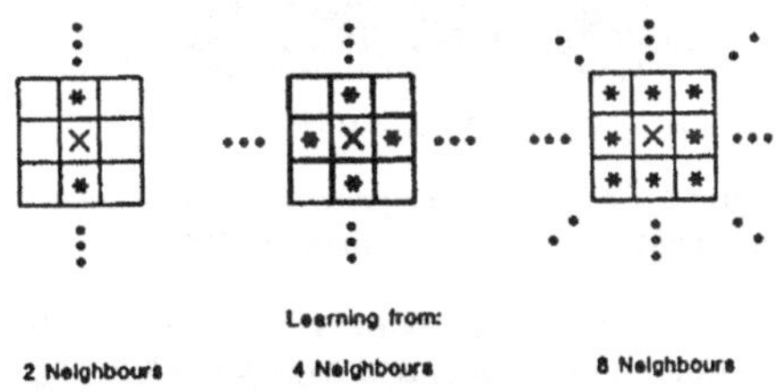

Figure 1. Locations of tutors from which birds can learn when learning from 2, 4, or 8 neighbours. A given cell (X) can learn from any of the cells marked (*), or, if these are 'dead', from the nearest extant cell in the directions indicated by the dots.

In contrast to the approach of Goodfellow & Slater, if a new bird had no extant neighbours to learn from, then it was allowed to learn from the nearest live neighbour in the same directions. This was limited by the edges of the matrix. Chaffinches appear to learn the different songs in their repertoire by copying each from a single (possibly different) individual (Slater & Ince 1979, Slater *et al.* 1980), so the learning process was managed by making a list of all the songs possessed by the neighbours and choosing at random from this. New birds could not learn from other new birds.

Both repertoire size and numbers of neighbours were found to have a strong effect upon song longevity and sharing (Table 1). Song longevity increased with increasing repertoire size, but decreased when a bird had more neighbours to learn from, possibly because the greater learning opportunities a bird has when it has more neighbours make any particular song less likely to be copied. This would affect rare songs more than those which are shared between several individuals. The maximum and mean number of birds sharing a song type increased both with increasing repertoire size and numbers of neighbours, but the proportion of the repertoire shared with a neighbour (measured using the Jaccard community similarity index) decreases with increasing repertoire size and numbers of neighbours.

	EFFECT OF INCREASING	
	REPERTOIRE SIZE	NUMBER OF NEIGHBOURS
Mean Song Longevity	Rise	Fall
Number with commonest song	Rise	Rise
Mean number with a song	No change	Rise
Jaccard Index	Fall	Fall
Fragmentation Index	No change	2 low, 4 & 8 higher

Table 1. Summary of results of simulations by Williams & Slater (1990) with varying numbers of neighbours and repertoire sizes.

The effect of the number of neighbours upon what is learnt may be important in the wild. In hedgerows birds are likely to have only two neighbours, but in areas of continuous habitat they may have many more possible tutors. This simulation makes a number of predictions which it may prove possible to test where habitats differ in structure in this way.

As far as dialects are concerned, the main finding of this simulation work was that the pattern of song sharing was extremely complex. This suggests that the existence of sharp dialect boundaries between groups of birds in which individuals have a small repertoire size is very unlikely, unless of course all the song types are learnt as a package, as appears to be the case in corn buntings (McGregor 1980).

4. Rules for Choosing Song Tutors

The simulations by Goodfellow & Slater and by Williams & Slater suggest that random learning of song types may lead to small groups of birds singing the same song as each other but would not explain cases where large numbers of birds share song features.

Using a modified version of the program described above, we have investigated the effect of different learning conditions on song sharing. The simulations reported in Williams & Slater (1990) used the entire repertoire of each neighbour in the list of possible songs to be learnt. Thus, with a repertoire size of three, and with four neighbours to learn from, the first song chosen was nominally one of twelve possibilities. However, if any of the songs are shared, then this changes the probability that this song will be learnt. If, for example, song 45 is in the repertoire of three neighbours, but song 67 is only in the repertoire of one, then song 45 has a three times greater probability of being learnt.

The modified program compared this unequal probability of learning a song type with the following alternatives:

(i) Each song has an equal probability of being learnt. This was managed by only adding songs to the list of those which could be learnt if they were not on it already. In this case, with a repertoire size of one, the program would make a random choice between songs 45 and 67, even though three neighbours sang the former and only one the latter.

(ii) Learning the rarest song present; in this case song 67 would be chosen.

(iii) Learning the commonest song; here song 45 would be chosen.

In situations (ii) or (iii), if there was no single rarest or commonest song, then the program made a random choice between the songs which were equally rare or common. To simplify the comparisons being made, all runs of this simulation were carried out with a repertoire size of one and with four neighbours. As before, a 13 x 13 matrix, 15% error rate and 40% mortality rate were used. The simulation was run for 300 years.

Ten runs for each of the learning situations were carried out and analysed with Minitab using one way ANOVAs. The major differences observed were between the situations involving learning of the rarest and commonest songs. There was little difference between the unequal and equal learning situations.

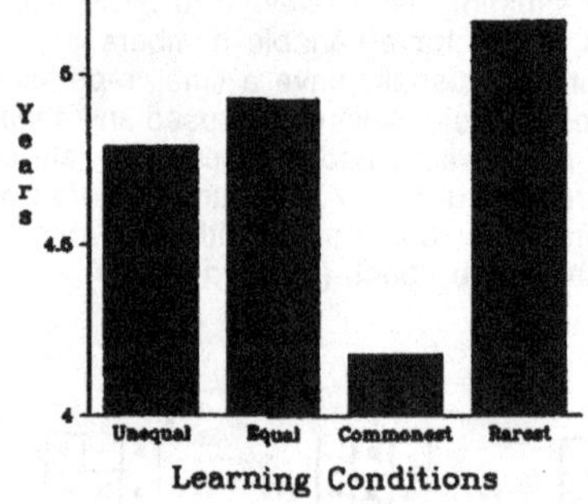

Figure 2. Mean song longevity under four learning conditions. Values are the mean of ten runs.

Mean song longevity is low when learning the commonest song, and high when learning the rarest (Figure 2). This is probably because it is difficult for a new song to spread into the simulation when the commonest is being learnt, but when the rarest is being learnt, just as a song is about to die out it becomes the rarest, and is therefore likely to be chosen for learning. The differences are significant ($F=94.04$ 3,36 d.f. $p<0.001$).

The maximum number of birds sharing a song type is much larger when the commonest song is learnt (Figure 3). The difference is less when the mean numbers shared are compared, but both results are significant: maximum shared $F=14.07$ 3,36 d.f. $p<0.001$; mean shared $F=6.94$ 3,36 d.f. $p=0.001$. In line with the pattern of sharing of songs, the proportion of the repertoire shared between neighbours, as measured by the

Jaccard index, is significantly greater when the commonest song is learnt ($F=32.24$ 3,36 d.f. $p<0.001$).

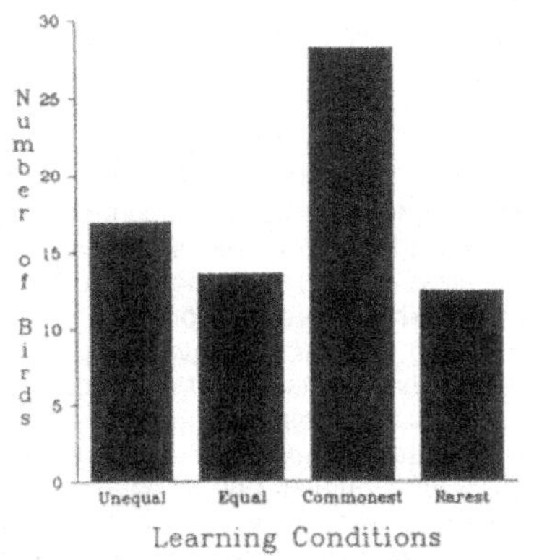

Figure 3. Number of birds with the commonest song type under four learning conditions. Values are the mean of ten runs.

Having found that large groups of birds sharing a single song type could be produced by learning the commonest song type the effect of varying the rate of miscopying was investigated. This is comparable to the work that Goodfellow & Slater did with random learning. A 20 x 20 matrix was used to increase the population of birds in anticipation of large groups at low mutation rates.

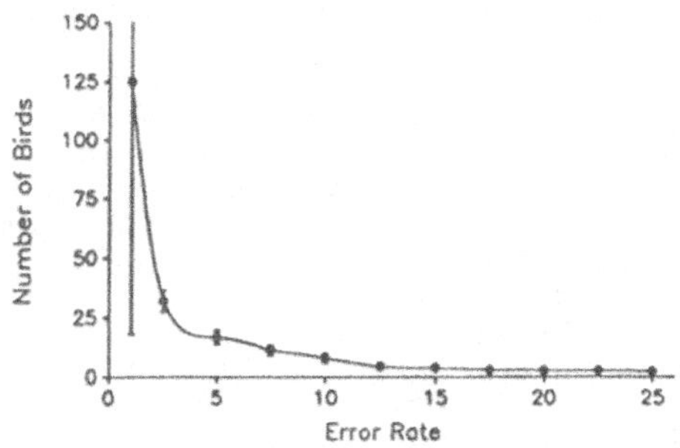

Figure 4. The effect of error rate on the mean number of birds sharing a song type when the commonest song is learnt. The values are means (± standard deviation) of ten runs.

Ten runs each for mutation rates of 1% and from 2.5% to 25% at 2.5% intervals were performed. When the commonest song is learnt and the matrix is seeded with all birds singing the same song, that song type can last for a very long time. Therefore, for this investigation of the effect of mutation rate each bird started with a different song type. Since song types had therefore to spread rather than just remain extant it did not matter if any of them persisted for the entire simulation. Nevertheless, runs were over 500 'years' to try as far as possible to ensure that an equilibrium had been reached. In fact, when the mutation rate was very low (1-5%), the song types at the end of the simulation were very often a selection of the initial ones.

With a very low mutation rate, very large groups of birds sharing the same song type can occur (Figure 4). The maximum number sharing actually approached the number of cells in the matrix for mutation rates of 1%. As suspected from the simulation of different learning conditions, the mean song longevity (which is calculated only for songs which arise during the simulation, not for any of the songs used to seed the matrix), is very low for very low mutation rates. This is probably because it is rare that new songs get established when the commonest song type is learnt.

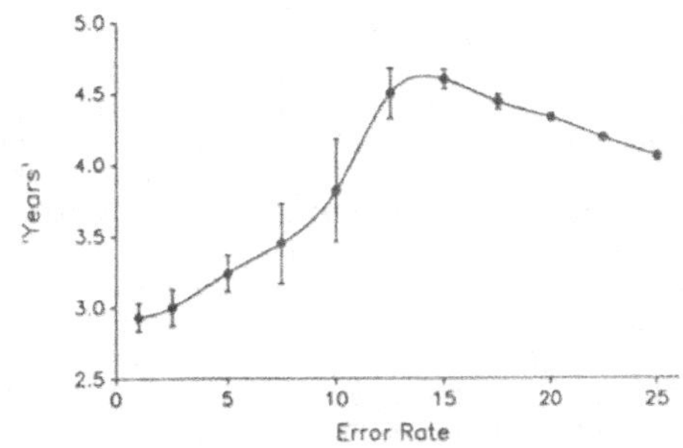

Figure 5. The effect of error rate on the mean song longevity when the commonest song is learnt. The values are means (± standard deviation) of ten runs.

The mean song longevity increases with increasing mutation rate (Figure 5), but decreases at rates greater than about 15%. This may be because at mutation rates above this it is difficult for a large group of birds singing the same song type to become established.

5. Discussion

Computer simulations have proved very useful in helping us to understand the processes likely to be involved in the distribution of songs in populations of wild birds. In the chaffinch, at most 15% of copying events result in errors and thereby introduce variation into the population of song types. In the simulations only small groups of birds are found to share songs when these are learnt at random from the neighbours surrounding them unless the accuracy of copying is assumed to be extremely high. Increasing the number of neighbours to learn from or the repertoire size results in more sharing, but not the large numbers sharing that are reported for some species. If the commonest song present amongst neighbours is learnt, rather than one being randomly chosen, then large groups can be formed when copying accuracy is high. Song longevity also depends on the learning conditions, on the number of neighbours available to learn from and on the repertoire sizes of the birds involved.

Our results (Williams & Slater 1990 and those reported here) suggest that the number of song types to which birds are exposed and whether or not they follow a simple rule such as 'learn the commonest song' can have a crucial effect upon the distribution of song sharing that is observed.

We can conclude that complex patterns of song sharing could result from relatively simple processes. It is therefore possible that Wiens (1982) is correct in suggesting that dialects may simply be by-products of vocal learning. Our new work on the conditions of learning also shows that sharp and stable dialect boundaries are possible if copying is very accurate, in the same way as was found by Goodfellow & Slater (1986). Given the controversy over the size and location of white-crowned sparrow dialects, it will be very interesting to discover just how accurate copying between individuals is in this species, and whether they do copy at random or not.

Another interesting data-set on which these simulations may shed light is that on the song sharing of Darwin's medium ground finches (*Geospiza fortis*) on the Galapagos Islands. Gibbs (1990) found the frequencies of different song types changed during the course of several years. The commonest type became progressively rarer while the other three became more common. It appeared that birds possessing the rarer types both survived longer and had greater breeding success. These results are perplexing, but they do suggest that, in this population, a rule such as 'learn the rarest song' might have some advantage. As we have shown here such a rule could have a dramatic effect on the distribution of songs within the population.

Much remains to be found out, and there is great diversity in the patterns of song sharing in the songbird species that have been studied to date. The data of Gibbs (1990) are exceptional in that they suggest that certain song types may be copied in preference to others. Random copying processes with a low error rate seem able to account for most of the results described for other species. While there may be a diversity of patterns of song distribution in wild populations, our simulations have shown that differences in features such as repertoire size, number of neighbours, accuracy of copying, and how the bird to be learnt from is chosen can all have an impact. The diversity and complexity of song distributions do not necessarily imply diverse and complex underlying mechanisms.

Acknowledgments

We thank colleagues in the Animal Behaviour Discussion Group at the University of St Andrews for constructive comments on this work. J.M. Williams was supported by a University of St Andrews Research Studentship.

References

Baker, M. C. & Cunningham, M. A. 1985. The biology of bird-song dialects. Behav. Brain Sci. **8**, 85-133.

Baptista, L. F. 1975. Song dialects and demes in sedentary populations of the white-crowned sparrow (*Zonotrichia leucophrys nuttalli*). Publ. Zool. Univ. Calif. **105**, 1-52.

Bjerke, T. K. & Bjerke, T. H. 1981. Song dialects in the redwing, *Turdus iliacus*. Ornis Scand. **12**, 40-50.

Conrads, K. 1986. Stabitat und Veranderungen eines Gesangsdialektes des Buchfinken (*Fringilla coelebs*) im Zeitraum von 1964/66 bis 1982/83 im Ostwestfalen. Ber. Naturwiss. Verein Bielefeld u. Umgegend **28**, 191-212.

Ewens, W. J. 1972. The sampling theory of selectively neutral alleles. Theor. Pop. Biol. **3**,

87-112.

Gibbs, H. L. 1990. Cultural evolution of male song types in Darwin's medium ground finches, *Geospiza fortis*. Anim. Behav. 39, 253-263.

Goodfellow, D. J. & Slater, P. J. B. 1986. A model of bird song dialects. Anim. Behav. **34**, 1579-1580.

Ince, S. A., Slater, P. J. B. & Weismann, C. 1980. Changes with time in the songs of a population of chaffinches. Condor **82**, 285-290.

Jenkins, P. F. 1978. Cultural transmission of song patterns and dialect development in a free-living bird population. Anim. Behav. **26**, 50-78.

Marler, P. 1952. Variations in the song of the chaffinch, *Fringilla coelebs*. Ibis 94, 458-472.

McGregor, P. K. 1980. Song dialects in the corn bunting (*Emberiza calandra*). Z. Tierpsychol. **54**, 285-297.

McGregor, P. K. & Thompson, D. B. A. 1988. Constancy and change in local dialects of the corn bunting. Ornis Scand. **19**, 153-159.

Metzmacher, M. & Mairy, F. 1972. Variations geographique de la figure finale du chant du Pinson des arbres *Fringilla coelebs coelebs* L. Gerfaut **62**, 215-243.

Payne, R. B., Payne, L. L. & Doehlert, S. M. 1988. Biological and cultural success of song memes in indigo buntings. Ecology 69, 104-117.

Payne, R. B., Thompson, W. L., Fiala, K. L. & Sweany, L. L. 1981. Local song traditions in indigo buntings: cultural transmission of behaviour patterns across generations. Behaviour **77**, 199-221.

Slater, P. J. B. 1989. Bird song learning: causes and consequences. Ethology Ecology & Evolution **1**, 19-46.

Slater, P. J. B., Clements, F. A. & Goodfellow, D. J. 1984. Local and regional variations in Chaffinch song and the question of dialects. Behaviour **88**, 76-97.

Slater, P. J. B. & Ince, S. A. 1979. Cultural evolution in chaffinch song. Behaviour **71**, 146-166.

Slater, P. J. B. & Ince, S. A. 1982. Song development in chaffinches: what is learnt and when?. Ibis **124**, 21-26.

Slater, P. J. B., Ince, S. A. & Colgan, P. W. 1980. Chaffinch song types: their frequencies in the population and distribution between the repertoires of different individuals. Behaviour **75**, 207-218.

Thorpe, W. H. 1958. The learning of song patterns by birds, with especial reference to the song of the chaffinch *Fringilla coelebs*. Ibis **100**, 535-570.

Wiens, J. A. 1982. Song pattern variation in the sage sparrow (*Amphispiza belli*): dialects or epiphenomena. Auk **99**, 208-229.

Williams, J.M. & Slater, P.J.B. 1990. Modelling bird song dialects: the influence of repertoire size and number of neighbours. J. Theor. Biol. (in press).

Reinforcement Learning Architectures for Animats

Richard S. Sutton
GTE Laboratories Incorporated
Waltham, MA 02254
sutton@gte.com

Abstract

In the first part of this paper I argue that the learning problem facing animats is essentially that which has been studied as the reinforcement learning problem—the learning of behavior by trial and error without an explicit teacher. A brief overview is presented of the development of reinforcement learning architectures over the past decade, with references to the literature.

The second part of this paper presents Dyna, a class of architectures based on reinforcement learning but which go beyond trial-and-error learning. Dyna architectures include a learned internal model of the world. By intermixing conventional trial and error with hypothetical trial and error using the world model, Dyna systems can plan and learn optimal behavior very rapidly. Results are shown for simple Dyna systems that learn from trial and error while they simultaneously learn a world model and use it to plan optimal action sequences. We also show that Dyna architectures are easy to adapt for use in changing environments.

1 Animats and the Reinforcement Learning Problem

What is an Animat? An animat is an adaptive system designed to operate in a tight, closed-loop interaction with its environment. An animat need not be a learning system, but often it is; some sense of adaptation of behavior to variations in the environment is required.

Figure 1 is a representation of the animat problem as I see it. On some short time cycle, the animat receives sensory information from the environment and chooses an action to send to the environment. In addition, the animat receives a special signal from the environment called the *reward.* Unlike the sensory information, which may be a large feature vector, or the action, which may also have many components, the reward is a single real-valued scalar, a number. The goal of adaptation is the maximization of the cumulative reward received over time.

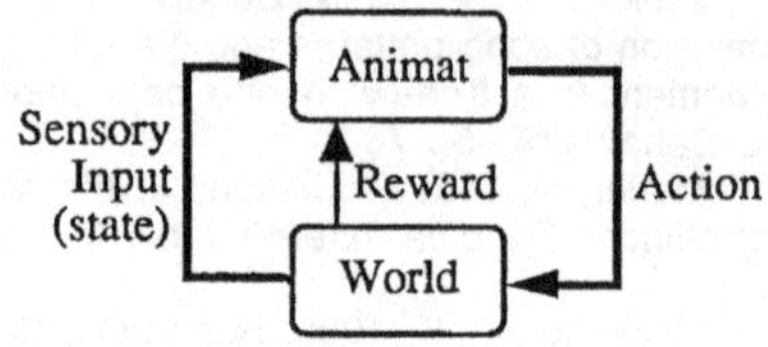

Figure 1. The Reinforcement Learning Problem facing an Animat. The goal is to maximize cumulative reward.

This formulation of the animat problem is the same as that widely used in the study of *reinforcement learning.* In fact, reinforcement learning systems can be defined as learning systems designed for and that perform well on the animat problem as described above. Informally, we define reinforcement learning as learning by trial and error from performance feedback—i.e., from feedback that evaluates the behavior generated by the animat but does not indicate correct behavior. In the next section we briefly survey reinforcement learning architectures.

One might object to the problem formulation in Figure 1 on the grounds that all possible goals have been reduced to a scalar reward. Although this appears limiting, in practice it has proved to be a useful way of structuring the problem. Some examples of goals formulated in this way are:

- Foraging: Reward is positive for finding food objects, negative for energetic motion, slightly negative for standing still.
- Pole-balancing (balancing a pole by applying forces to its base): The reward is zero while the pole is balanced, and then becomes -1 if the pole falls over or if the base moves too far out of bounds.
- Towers of Hanoi: Reward is positive for reaching the goal state.

- Recycling Robot: Reward is positive for dropping soda cans in the recycling bin, negative for bumping into things, more negative for bumping hard into things, most negative for being yelled at or for running down the battery, etc.
- Video Game Playing: One unit of reward for every point scored.

Another reason one might object to the problem formulation in Figure 1 is that the goal of learning is defined solely in terms of something (the reward) arising from the external environment, not from the animat itself. Often goals do concern the evolution of the the animat's own internal state, e.g., its energy reservoirs. Fortunately, it appears that most goals, perhaps all, can be put in the "external reward" form simply by redrawing the boundary between animat and environment. For example, if the goal concerns the animat's energy reservoirs, then these are considered part of the environment; if the goal concerns the positions of the animat's limbs, then these too can be considered part of the environment—the animat's boundary is drawn at the interface between the limbs and their control systems. Roughly speaking, things are considered part of the animat if they are completely, directly, and with certainty controllable; things are considered part of the environment if they are not. Since the goal is always something over which we have imperfect or uncertain control, it is placed outside the animat.

2 Overview of Reinforcement Learning Architectures

In this section we review the major steps in the development of reinforcement learning architectures over the last decade. These steps are illustrated by the four architectures shown in Figure 2.

All reinforcement learning involves the learning of a mapping from a representation of a situation or *state* to an appropriate *action* (or a probability distribution over actions) for that situation. This mapping is called the *policy*; it specifies what the animat will do in each situation at its current stage of learning. The simplest reinforcement learner that one might imagine, then, would consist only of a policy and a way of adjusting it based on reward, as shown in Figure 2a. Such architectures, in which the policy is the only modifiable data structure (and, indeed, the only structure at all) are here called *policy-only* architectures.

The policy can be implemented in any of a number of ways. It can be a connectionist neural network, or a symbolic learning structure such as a decision tree or lisp program, or a conventional set of statistics such as is used by maximum-likelihood or nearest-neighbor techniques. Any of these methods can be used—with varying advantages and disadvantages—to implement this and the other modifiable structures (in the other architectures) shown in Figure 2.

The learning algorithm for the policy must be of a slightly unusual type. Standard supervised learning methods such as backpropagation are not sufficient here, but must be modified, at least slightly, to take into account the fact that a "target" action is not directly available. Instead, a form of correlation must be done between the reward received and the actions taken by the animat, all with respect to the sensory input. Actions correlated with high reward have their probability of being repeated increased, while those correlated with low reward have their probability decreased. Examples of such algorithms for policy-only architectures can be found in (Farley & Clark, 1954; Widrow, Gupta & Maitra, 1973; Barto & Sutton, 1981; Barto & Anandan, 1985).

Policy-only architectures really only work well when it is clear a priori what constitutes a high reward and what constitutes a low one—for example, if all high rewards are positive and all low rewards are negative. Often, however, rewards are not distributed around a baseline of zero, but around some other, unknown value. Worse yet, the baseline may change from state to state. A low reward value in one state may be the highest attainable in another. To handle such variations, a baseline value must be learned that is a function of the state; the actual reward is then compared with the current state's baseline. This is what is done in *reinforcement-comparison* architectures (Figure 2b). As a baseline, these architectures usually use a *prediction* of the reward. The prediction error—the difference between predicted and actual reward—is used both as an enhanced, zero-balanced reward signal for adjusting the policy *and* as an ordinary error for learning the reward predictions (Figure 2b). A variety of reinforcement-comparison algorithms have been explored and compared (Barto, Sutton & Brouwer, 1981; Sutton, 1984; Williams & Peng, 1989; Williams, 1986).

Reinforcement comparison architectures are effective at optimizing immediate rewards, but not at optimizing total reward in the long run. The problem is that actions have two kinds of consequences—they affect the next reward *and* they affect the next state, but reinforcement-comparison architectures only take the first of these into account. Suppose an action produces high immediate reward but deposits the environment in a state from which only low reward can be obtained? In order to optimize long-term reward, these delayed affects of action must be taken into account.

The *adaptive heuristic critic (AHC)* architecture, shown in Figure 2c, was designed to take such delayed effects into account. The predictor of immediate reward has been replaced with a predictor of *return*, a measure of long-term cumulative reward. For any state x, the return is formally defined as the expected value of the sum of all future rewards, discounted by their delay, given that the systen starts in x:

$$return(x) = E\left\{\sum_{t=0}^{\infty}\gamma^t r_{t+1}\middle| x_0 = x\right\},$$

where γ, $0 \leq \gamma \leq 1$, is the discount rate determining how fast one's concern for delayed reward falls off with length of the delay.[1] The "return predictor" box

[1] This is analogous to the discount rate in economics—a

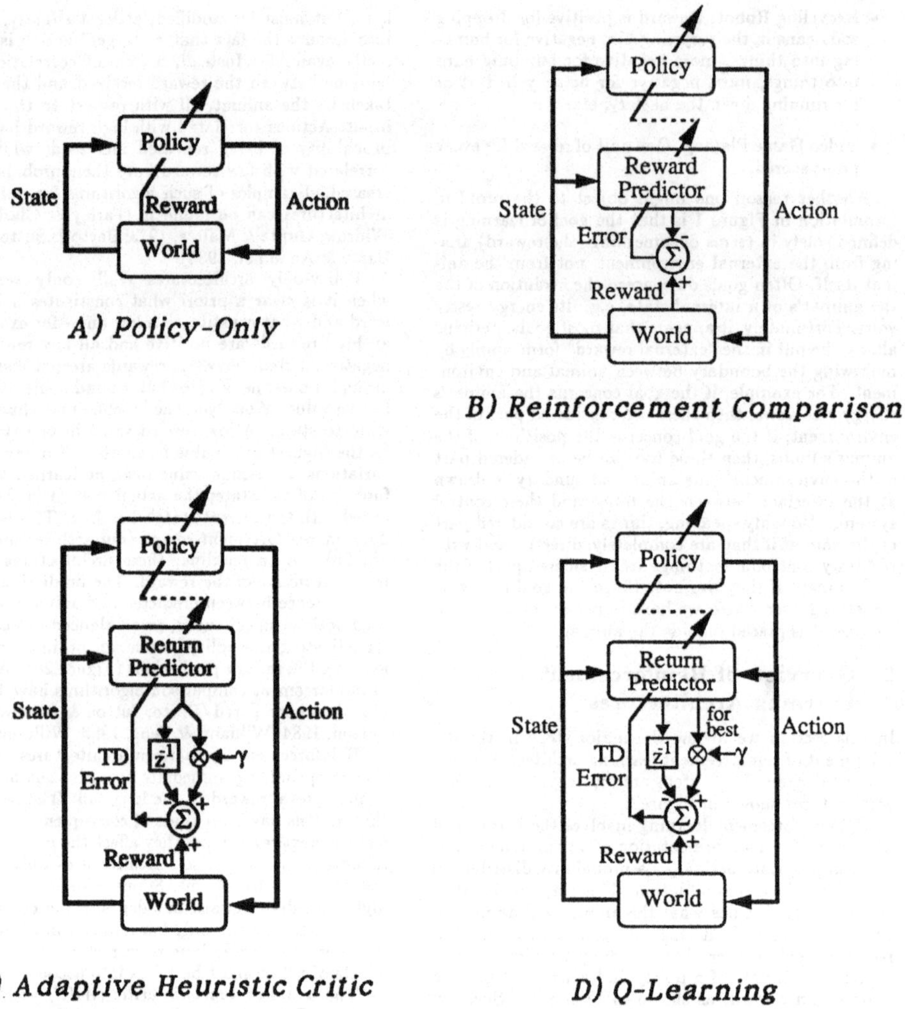

Figure 2. Overview of Reinforcement Learning Architectures. See text. The symbol $\boxed{z^{-1}}$ indicates a one time step delay, and the symbol ⊗ indicates multiplication. The line labeled "for best" in D is the prediction of return for the best action; the other output from the return predictor is the prediction of return for the action actually taken.

in Figure 2c comes to predict this return by virtue of the circuit shown below it for calculating a temporal-difference error (Sutton, 1988). In all other respects, the learning algorithm inside this box could be exactly the same as that used in the "reward predictor" box of Figure 2b. The AHC architecture has been used in a variety of learning control tasks (Sutton, 1984; Barto, Sutton & Anderson, 1983; Anderson, 1987; Barto, Sutton & Watkins, 1989).

Finally, Figure 2d shows the most recent reinforcement learning architecture, *Q-learning* (Watkins, 1989). The primary innovation here is that the predicted return is now a function of action as well as state. Formally, the return for a state x and action a is defined as

$$return(x,a) = E\left\{ \sum_{t=0}^{\infty} \gamma^t r_{t+1} \middle| x_0 = x, a_0 = a \right\}. \quad (1)$$

Two kinds of return are always predicted for the current state. One is the best predicted return for the state—the predicted return for the action with the highest predicted return. The other is the predicted return for the action actually selected. These two predictions are then combined according to the same circuit as used in the AHC architecture. In Q-learning, the policy could be a separate modifiable data structure as suggested by Figure 2d, but most often it is simply a dependent function of the return predictions. For example, the policy may be simply to pick the action that obtains the maximal return prediction. A particular Q-learning algorithm will be given in detail later.

3 Dyna Architectures

Reinforcement learning architectures are effective at trial-and-error learning, but no more. They can not do any of the things that are considered "cognitive," such as reasoning or planning. They do not learn an internal model of the world's dynamics, of what-causes-what, but only of what-to-do (policy) and how-well-am-I-doing (return predictions). This is an important limitation because potentially much more can be learned in the form of a world model than can be learned by trial and error; the reward signal is just a scalar, while the sensory input signal is a much richer potential source of training information. And what if the goal changes? Typically, a world model can remain relatively intact over goal changes and can assist in achieving the new goal, whereas policy and return predictions must be totally changed.

Dyna architectures are simple extensions of reinforcement learning architectures to include an internal world model (Sutton, 1990; Whitehead, 1989). The world model is defined as something that behaves like the world: given a state and an action it is supposed to output a prediction of the resultant reward and next state. If the world's state is observable, then it is straightforward to learn a world model using supervised learning methods and training examples taken

dollar today is worth more than a dollar tomorrow.

1. Observe the current state x and choose an action: $a \leftarrow Policy(x)$.
2. Send the action to the world and observe the resultant next state y and reward r.
3. Apply a reinforcement learning method to the experience x, y, a, and r.
4. Update the world model based on the experience x, y, a, and r.
5. Repeat the following steps k times:
   - 5.1 Select a hypothetical state x and hypothetical action a.
   - 5.2 Send x and a to the world model and obtain predictions of next state y and reward r.
   - 5.3 Apply a reinforcement learning method to the hypothetical experience x, y, a, and r.
6. Go to 1.

Figure 3. Main Loop of the Dyna Algorithm.

from actual interactions with the world. If the world is not observable, then it must be inferred estimated from the history of sensory input and action. Although there are a variety of algorithms for doing this, in the general case it is still an open problem. Here we assume for the time being that the state is observable.

In Dyna architectures the model is used as a direct replacement for the world in one of the reinforcement learning architectures shown in Figure 2. Reinforcement learning continues in the usual way, but, in addition, learning steps are also run using the model in place of the world, using predicted outcomes rather than actual ones. For each real experience with the world, many hypothetical experiences generated by the world model can also be processed and learned from. The cumulative effect of these hypothetical experiences is that the policy approaches the optimal policy given the current model; a form of planning has been achieved. The overall Dyna algorithm is given in Figure 3.

Figure 4 shows results for a Dyna architecture based on the AHC architecture, called *Dyna-AHC*. The task is to navigate through the maze shown in Figure 4 from the starting state "S" to the goal state "G". From each state there are four possible actions: UP, DOWN, RIGHT, and LEFT, which change the state accordingly, except where such a movement would take the system into a barrier (shaded state) or outside the maze, in which case the state is not changed. Reward is zero for all transitions except for those into the goal state, for which it is +1. The lower left portion of the figure shows learning curves for Dyna-AHC systems with $k = 100$, $k = 10$, and $k = 0$. The $k = 0$ case involves no hypothetical steps; this is a pure trial-and-error reinforcement-learning system. Although the length of path taken from start to goal falls dramatically for this case, it falls much *more* rapidly for the cases including hypothetical experiences, showing the benefit of planning using the learned world model. For $k = 100$, the optimal path was generally found and followed by the fourth trip from start to goal. This is extremely rapid learning.

Figure 5 shows why the Dyna-AHC system solved this problem so much faster than the pure reinforcement learning system. Shown are the policies found by the $k = 0$ and $k = 100$ systems half-way through the second trial. Without planning ($k = 0$), each trial adds only one additional step to the policy, and so only one step (the last) has been learned so far. With planning, the first trial also learned only one step, but here during the second trial an extensive policy has been developed that by the trial's end will reach back almost to the start state. By the end of the third or fourth trial a complete optimal policy will have been found and perfect performance attained.

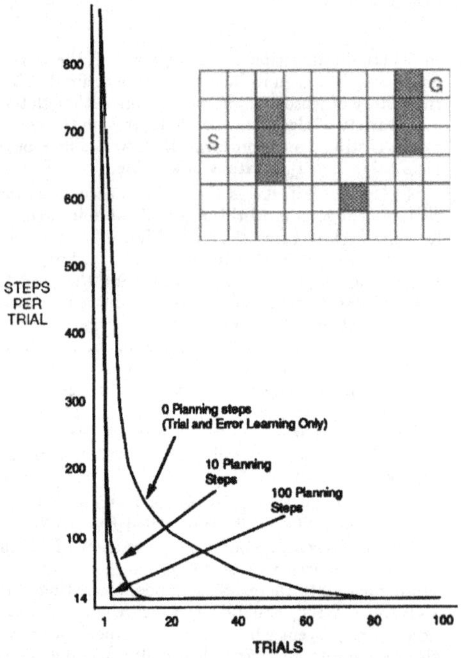

Figure 4. Learning Curves for Dyna-AHC Systems on a Simple Navigation Task. A trial is one trip from the start state "S" to the goal state "G". The more hypothetical experiences ("planning steps") using the world model, the faster an optimal path was found. These data are averages over 100 runs.

4 Problems of Changing Worlds

Suppose that, after a Dyna-AHC system has learned the optimal path from start to goal, a new barrier is added that blocks the optimal path. The Dyna-AHC system discussed above will run into the block and then try the newly ineffective action many hundreds of times. Eventually, the correct new path will be found, but the process is very slow. It seems inappropriately slow in that the system's world model is updated immediately. Even though the world model knows that the formerly good action is now poor, this is not reflected in the system's behavior for a long time. I call this the *blocking problem.*

Now consider a second sort of change in the environment. Suppose, after the optimal path has been learned, a barrier is removed that permits a shorter path from start to goal. The Dyna-AHC system is unable to take advantage of such a shortcut; it never wavers from the formerly optimal path and thus never discovers that the former obstacle is gone. I call this the *shortcut problem.* In seeking to improve the Dyna-AHC system to handle blocks, one might also seek to improve it to handle shortcuts. What is needed here is some way of continually testing the world model. In the next section we present a Dyna architecture based on Q-learning that handles both kinds of changes with little increase in complexity.

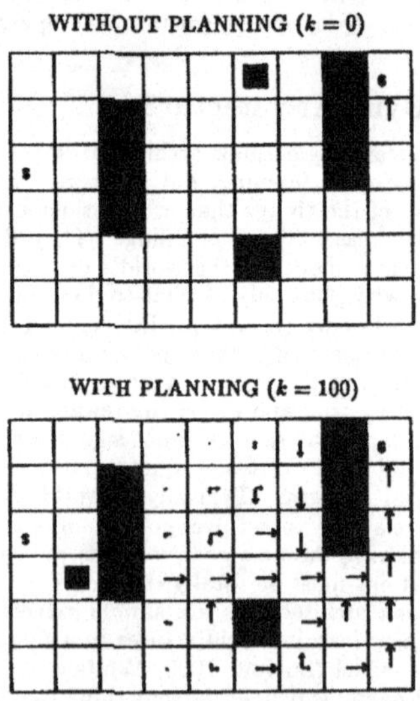

Figure 5. Policies Found by Planning and Non-Planning Dyna-AHC Systems by the Middle of the Second Trial. The black square indicates the current location of the Dyna system. The arrows indicate action probabilities (excess over the smallest) for each direction of movement.

5 Dyna-Q Architectures

In the rest of this paper we focus on Dyna architectures based on Q-learning. In this section we give a complete specification of two such Dyna-Q architectures, called *Dyna-Q-* and *Dyna-Q+*.

Recall that, in Q-learning, the return predictions are made on the basis of both a state and an action (Figure 2d). For any state x and action a, let Q_{xa} be the current prediction of $return(x, a)$; these are also called *Q-values*. Suppose the experience x, y, a, and r (state, next state, action, and reward) occurs. The simplest form of Q-learning is based on constructing from this the following training example:

$$Q_{xa} \quad \text{should be} \quad r + \gamma \max_{b \in A} Q_{yb}, \qquad (2)$$

where A is the set of actions and γ is the discount-rate parameter from (1). This training example could be passed to any appropriate supervised learning algorithm. In the results reported in the next section, the return predictor was implemented as a table. The complete update rule for this case is:

$$Q_{xa} \leftarrow Q_{xa} + \beta(r + \gamma \max_{b \in A} Q_{yb} - Q_{xa}),$$

where β is a positive learning-rate parameter. This is the only update rule in the version of Q-learning used here. The policy was to deterministically select the action with highest Q-value:

$$Policy(x) = \arg \max_{a \in A} Q_{xa}.$$

The Dyna-Q algorithms use Q-learning as their reinforcement learning method in Steps 3 and 5.3 of Figure 3. In the experiments presented in the next section, the starting states and actions for hypothetical experiences were selected uniformly randomly (Step 5.1). The model was implemented as a lookup table with an entry for each state-action pair. This completes the description of Dyna-Q-.

As we show below, Dyna-Q- is not able to solve the shortcut problem. To help on this problem, Dyna-Q+ uses an additional memory structure to keep track of the degree of uncertainty about each component of the model. For each state x and action a, a record is kept of the number of time steps n_{xa} that have elapsed since a was tried in x in a real experience. The square root $\sqrt{n_{xa}}$ is used as a measure of the uncertainty about Q_{xa}.[2] To encourage exploration, each state-action pair is then given an *exploration bonus* proportional to this uncertainty measure. For real experiences, the policy is to select the action a that maximizes $Q_{xa} + \epsilon\sqrt{n_{xa}}$, where ϵ is a small positive parameter. This method of encouraging variety is very similar to that used in Kaelbling's (1990) interval-estimation algorithm.

However, this approach alone does not take advantage of the planning capability of Dyna architectures. Suppose there is a state-action pair that has not been tested in a long time, but that is far from the currently preferred path, and thus extremely unlikely to be tried even with the exploration bonus discussed above. In a Dyna system, why not expect the system to *plan* an action sequence to go out and test the uncertain state-action pair? If there is genuine uncertainty, then there is potential benefit in going out and trying the action, and thus forming such a plan is simply rational behavior and should be done. It turns out that there is a simple way to do this in Dyna. The exploration bonus of $\epsilon\sqrt{n_{xa}}$ is used not only in the policy, but also in the update equation for the Q-values. That is (2) is replaced by:[3]

$$Q_{xa} \quad \text{should be} \quad r + \epsilon\sqrt{n_{xa}} + \gamma \max_{b \in A} Q_{yb}.$$

In addition, the system is permitted to hypothetically experience actions it has never before tried, so that the exploration bonus for not having tried them can be propagated back by the planning process. This can be done by starting the system with a non-empty initial model. In the experiments with Dyna systems reported below, actions that had never been tried were assumed to produce zero reward and leave the state unchanged. This completes the specification of Dyna-Q+.

6 Changing-World Experiments

Experiments were performed to test the ability of Dyna systems to solve blocking and shortcut problems. All three Dyna systems were used: Dyna-AHC, Dyna-Q-, and Dyna-Q+. All systems used $k = 10$. For the Dyna-AHC system, the other parameters were set as in the navigation experiment. For the Dyna-Q systems, they were set at $\beta = 0.5$, $\gamma = 0.9$, and $\epsilon = 0.001$. In all cases, the initial Q-values were zero.

The blocking experiment used the two mazes shown in the upper portion of Figure 6. Initially a short path from start to goal was available (first maze). After 1000 time steps, by which time the short path was usually well learned, that path was blocked and a longer path was opened (second maze). Performance under the new condition was measured for 2000 time steps. Average results over 50 runs are shown in Figure 6 for the three Dyna systems. The graph shows a *cumulative* record of the number of rewards received by the system up to each moment in time. In the first 1000 trials, all three Dyna systems found a short route to the goal, although the Dyna-Q+ system did so significantly faster than the other two. After the short path was blocked at 1000 steps, the graph for the Dyna-AHC system remains almost flat, indicating that it was unable to obtain further rewards. The Dyna-Q systems, on the other hand, clearly solved the blocking problem, reliably finding the alternate path after about 800 time steps.

[2]The use of the square root is heuristic but not arbitrary, as the standard deviation of stationary, cumulative random processes increases with the square root of the number of cumulating steps.

[3]Note that this differs from (2) only on hypothetical experiences, as $n_{xa} = 0$ on real experiences.

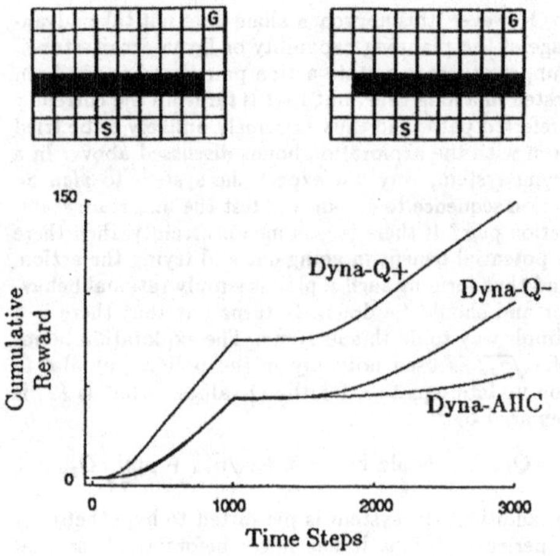

Figure 6. Average Performance of Dyna Systems on a Blocking Task. The left maze was used for the first 1000 time steps, the right maze for the last 2000. Shown is the cumulative reward received by a Dyna system at each time (e.g., a flat period is a period during which no reward was received).

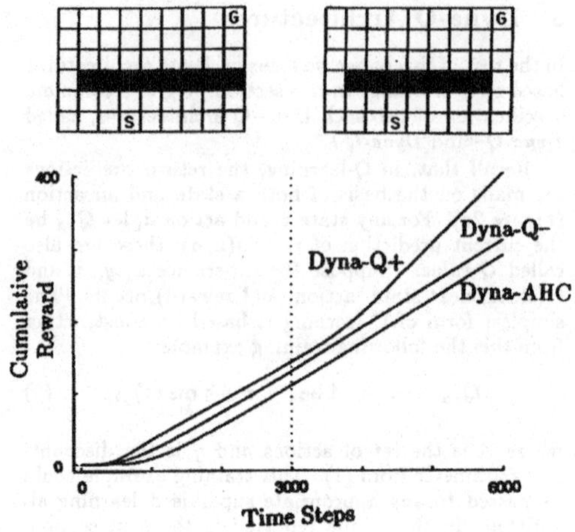

Figure 7. Average Performance of Dyna Systems on a Shortcut Task. The left maze was used for the first 3000 time steps, the right maze for the last 3000. Shown is the cumulative reward received by a Dyna system at each time (e.g., the slope corresponds to the *rate* at which reward was received).

The shortcut experiment began with only a long path available (first maze of Figure 7). After 3000 times steps all three Dyna systems had learned the long path, and then a shortcut was opened without interferring with the long path (second maze of Figure 7). The lower part of Figure 7 shows the results. The increase in the slope of the curve for the Dyna-Q+ system, while the others remain constant, indicates that it alone was able to find the shortcut. The Dyna-Q+ system also learned the original long route faster than the Dyna-Q– system, which in turn learned it faster than the Dyna-AHC system. However, the ability of the Dyna-Q+ system to find shortcuts does not come totally for free. Continually re-exploring the world means occasionally making suboptimal actions. If one looks closely at Figure 7, one can see that the Dyna-Q+ system actually acheives a slightly lower *rate* of reinforcement during the first 3000 steps. In a static environment, Dyna-Q+ will eventually perform worse than Dyna-Q–, whereas, in a changing environment, it will be far superior, as here. One possibility is to use a meta-level learning process to adjust the exploration parameter ϵ to match the degree of variability of the environment.

One strength of the Dyna approach is that it applies to stochastic problems as well as deterministic ones. We have explored this direction in recent work, but are not yet ready to present systematic results. The basic idea is to learn a model which predicts not a deterministc next state and next reward, but rather a probability distribution over next states and next rewards. In the simple cases we have explored, this reduces to counting the number of times each possible outcome has occurred. In hypothetical experiences, the expected value of the right of (2) is then estimated using the sample statistics. A slightly different exploration bonus is also needed. Promising preliminary results have so far been obtained for simple problems involving random autonomous agents and stochastic state transitions (e.g., action UP takes the system to the state above 80% of the time, and to a random neighboring state 20% of the time).

Further results are needed for a thorough comparison of Dyna-AHC and Dyna-Q architectures, but the results presented here suggest that it is easier to adapt Dyna-Q architectures to changing environments.

7 Limitations and Conclusions

The simple illustrations presented here are clearly limited in many ways. The state and action spaces are small and denumerable, permitting tables to be used for all learning processes and making it feasible for the entire state space to be explicitly explored. For large state spaces it is not practical to use tables or to visit all states; instead one must represent a limited amount of experience compactly and generalize from it. Both Dyna architectures are fully compatible with the use of a wide range of learning methods for doing this. For example, Lin (1990) has explored the use of Dyna architectures using backpropagation networks instead of tables.

We have also assumed that the Dyna systems have

explicit knowledge of the world's state. In general, states can not be known directly, but must be estimated from the pattern of past interaction with the world (Rivest & Schapire, 1987; Mozer and Bachrach, 1990). Dyna architectures can use state estimates constructed in any way, but will of course be limited by their quality and resolution. A promising area for future work is the combination of Dyna architectures with egocentric or "indexical-functional" state representations (Agre & Chapman, 1987; Whitehead, 1989).

Yet another limitation of the Dyna systems presented here is the trivial form of search control used. Search control in Dyna boils down to the decision of whether to consider hypothetical or real experiences, and of picking the order in which to consider hypothetical experiences. The tasks considered here are so small that search control is unimportant, and thus it was done trivially, but a wide variety of more sophisticated methods could be used. Particularly interesting is the possibility of using Dyna architectures at higher levels to make these decisions.

Finally, the examples presented here are limited in that reward is only non-zero upon termination of a path from start to goal. This makes the problem more like the kind of search problem typically studied in AI, but does not show the full generality of the framework, in which rewards may be received on any step and there need not even exist start or termination states.

Despite these limitations, the results presented here are significant. They show that the use of an internal model can dramatically speed trial-and-error learning processes even on simple problems. Moreover, they show how planning can be done with the incomplete, changing, and oftimes incorrect world models that are contructed through learning. Finally, they show how the functionality of planning can be obtained in a completely incremental manner, and how a planning process can be freely intermixed with execution and learning. I conclude that it is not necessary to choose between planning, reacting, and learning. These three can be integrated not only into one animat, but into a single algorithm, where each appears as a different facet of that algorithm.

Acknowledgments

The author gratefully acknowledges the extensive contributions to the ideas presented here by Andrew Barto, Chris Watkins and Steve Whitehead. I also wish to also thank the following people for ideas and discussions: Paul Werbos, Luis Almeida, Ron Williams, Glenn Iba, Leslie Kaelbling, John Vittal, Charles Anderson, Bernard Silver, Oliver Selfridge, Judy Franklin, Tom Dean and Chris Matheus.

References

Agre, P. E., & Chapman, D. (1987) Pengi: An implementation of a theory of activity. *Proceedings of AAAI-87*, 268–272.

Anderson, C. W. (1987) Strategy learning with multilayer connectionist representations. *Proceedings of the Fourth International Workshop on Machine Learning*, 103–114. Morgan Kaufmann, Irvine, CA.

Barto, A. G., & Anandan, P. (1985) Pattern recognizing stochastic learning automata. *IEEE Transactions on Systems, Man, and Cybernetics 15*, 360-375.

Barto, A. G., & Sutton. R. S. (1981) Landmark learning: An illustration of associative search. *Biological Cybernetics 42*, 1–8.

Barto, A. G., Sutton R. S., & Anderson, C. W. (1983) Neuronlike elements that can solve difficult learning control problems. *IEEE Transactions on Systems, Man, and Cybernetics 13*, 834–846.

Barto, A. G., Sutton, R. S., & Brouwer, P. S. (1981) Associative search network: A reinforcement learning associative memory. *Biological Cybernetics 40*, 201–211.

Barto, A. G., Sutton, R. S., & Watkins, C. J. C. H. (1989) Learning and sequential decision making. COINS Technical Report 89-95, Dept. of Computer and Information Science, University of Massachusetts, Amherst, MA 01003. Also to appear in *Learning and Computational Neuroscience*, M. Gabriel and J.W. Moore (Eds.), MIT Press, 1991.

Farley, B. G., & Clark, W. A. (1954) Simulation of self-organizing systems by digital computer. *I.R.E. Transactions on Inf. Theory 4*, 76–84.

Kaelbling, L. (1990) *Learning in Embedded Systems.* Stanford Computer Science Ph.D. Dissertation. Technical Report No. TR-90-04, Teleos Research, Palo Alto, CA.

Lin, L. (1990) Self-improving reactive agents: Case studies of reinforcement learning frameworks. *SAB-90.*

Mozer, M. C., & Bachrach, J. (1990) Discovering the structure of a reactive environment by exploration. In *Advances in Neural Information Processing Systems 2*, D. S. Touretzky, Ed. Morgan Kaufmann, San Mateo, CA. See also Technical Report CU-CS-451-89, Dept. of Computer Science, University of Colorado at Boulder 80309.

Rivest, R. L., & Schapire, R. E. (1987) A new approach to unsupervised learning in deterministic environments. *Proceedings of the Fourth International Workshop on Machine Learning*, 364–375. Morgan Kaufmann, Irvine, CA.

Sutton, R. S. (1984) Temporal credit assignment in reinforcement learning. Doctoral dissertation, Department of Computer and Information Science, University of Massachusetts, Amherst, MA 01003.

Sutton, R.S. (1988) Learning to predict by the methods of temporal differences. *Machine Learning 3*, 9–44.

Sutton, R. S. (1990) Integrated architectures for learning, planning, and reacting based on approximating dynamic programming. *Proceedings of the Seventh International Conference on Machine Learning*, 216–224, Morgan-Kaufmann.

Watkins, C. J. C. H. (1989) *Learning with Delayed Rewards.* Ph.D. dissertation, Cambridge University, Psychology Department.

Widrow, B., Gupta, N. K., & Maitra, S. (1973) Punish/reward: Learning with a critic in adaptive threshold systems. *IEEE Transactions on Systems, Man, and Cybernetics 5*, 455–465.

Whitehead, S. D. (1989) Scaling reinforcement learning systems. Technical Report 304, Dept. of Computer Science, University of Rochester, Rochester, NY 14627.

Williams, R. J., & Peng, J. (1989) Reinforcement learning algorithms as function optimizers. *Int. Joint Conference on Neural Networks.*

Williams, R. J. (1986) Reinforcement learning in connectionist networks: A mathematical analysis, University of California, San Diego Inst. for Cognitive Science Technical Report 8605.

Self-improving Reactive Agents: Case Studies of Reinforcement Learning Frameworks

Long-Ji Lin

School of Computer Science
Carnegie Mellon University
Pittsburgh, Pennsylvania 15213
e-mail: ljl@a.gp.cs.cmu.edu

Abstract

The purpose of this work is to investigate and evaluate different reinforcement learning frameworks using connectionist networks. I study four frameworks, which are adopted from the ideas developed in [Barto, Sutton & Watkins, 1989; Watkins, 1989; Sutton, 1990]. The four frameworks are based on two learning procedures: the Temporal Difference methods for solving the credit assignment problem, and the backpropagation algorithm for developing appropriate internal representations. Two of them also involve learning a world model and using it to speed learning. To evaluate their performance, I design a dynamic environment and implement different learning agents, using the different frameworks, to survive in it. The environment is nontrivial and nondeterministic. Surprisingly, all of the agents can learn to survive fairly well in a reasonable time frame. This paper describes the learning agents and their performance, and summarizes the learning algorithms and the lessons I learned from this study.

1. Introduction

Reinforcement learning is an interesting learning problem. It requires only a scalar reinforcement signal as a performance feedback from the environment. Reinforcement learning often involves two difficult subproblems. The first is called the *credit assignment problem*. Suppose the learning agent performs a sequence of actions and finally obtains certain outcomes. It must figure out how to assign credit or blame to each individual situation (or situation-action pair) to adjust its decision making and improve its performance. The second subproblem arises from the need to develop the appropriate internal representations required to achieve the target learning tasks. In the course of learning, both subproblems must be solved.

Several reinforcement learning frameworks or algorithms have been proposed in the literature (e.g., [Sutton, 1984; Williams, 1987; Barto, Sutton & Watkins, 1989; Watkins, 1989; Kaelbling, 1989; Sutton, 1990]). However, most have only been studied solving simple learning problems (e.g., [Anderson, 1989]). In addition, no serious comparison of different frameworks has been done. This work is thus intended to be a first step towards the investigation and evaluation of different reinforcement learning frameworks in solving nontrivial learning tasks. In particular, I am interested in reinforcement learning using connectionist networks.

In the paper I study four reinforcement learning frameworks, which are adopted from the ideas developed in [Barto, Sutton & Watkins, 1989; Watkins, 1989; Sutton, 1990]. All of these frameworks are based on two learning procedures: the Temporal Difference (TD) methods [Sutton, 1988] for solving the credit assignment problem and the error backpropagation algorithm [Rumelhart, et al., 1986a] for developing appropriate internal representations.

Generally speaking, reinforcement learning based solely on the TD methods is a slow process. In domains where reinforcements are sparse, the learning rate is slow, and if the cost of mistakes (e.g., physical damage) is also high, the agent would make more mistakes than allowed. A solution to these problems is to learn a world model, and practice with the model. This idea is embodied in two of the frameworks studied here.

My approach to evaluating different learning frameworks is to design a dynamic environment, implement learning agents to survive in it using different frameworks, and evaluate the performance of the agents. Four kinds of objects are involved in this environment: the agent, fixed food and obstacles, and moving enemies. Although survival in this environment is easy for humans, it is by no means trivial for knowledge-poor agents.

The remaining of this paper is organized as follows. Section 2 discusses the four learning frameworks. Section 3 describes the rules of the environment. Sections 4 and 5 present the implementation and performance of the learning agents. Section 6 assesses the merits of the agents. Finally, Section 7 concludes the paper by summarizing the lessons I learned from this study.

2. Reinforcement Learning Frameworks

Learning to survive in an unknown environment can be characterized as a kind of *reinforcement learning*. In reinforcement learning, the learning agent continually receives sensory inputs from the environment, selects and performs actions to affect the environment, and after each action, receives from the environment a scalar signal called *reinforcement*. The objective of learning is to construct an optimal action selection *policy* that maximizes the agent's performance. A natural measure of performance is the *discounted cumulative reinforcements* [Barto, Sutton & Watkins, 1989]:

$$V_t = \sum_{k=0}^{\infty} \gamma^k \cdot r_{t+k+1} \qquad (1)$$

where V_t is the discounted cumulative reinforcements starting from time t throughout the future, r_{t+1} is the reinforcement received after the transition from time t to t+1, and the *discount rate* γ, $0 \leq \gamma < 1$, adjusts the importance of long-term consequences of actions. For short, I use the term *utility* to refer to the discounted cumulative reinforcements. In the paper, both terms are used interchangeably.

In this section, four learning frameworks and the corresponding learning algorithms are described. Basically, the main idea behind the frameworks is to learn an *evaluation function* to predict the discounted cumulative reinforcements to be received. The evaluation function is represented using connectionist networks and is learned using a combination of the Temporal Difference (TD) methods [Sutton, 1988] and the error backpropagation algorithm [Rumelhart, et al., 1986a]. In essence, the TD methods compute the error (called *TD error*) between *temporally successive predictions*, while the backpropagation algorithm tries to minimize the error by changing the weights of the networks.

Before starting to discuss the learning frameworks, here I define two terms for later use:

- *eval(x)*: the discounted cumulative reinforcements to be received starting from world state *x*, or simply the utility of state *x*;
- *util(x,a)*: the discounted cumulative reinforcements to be received on the execution of action *a* in response to world state *x*; or simply the utility of the pair of state *x* and action *a*.

In a deterministic world, *util(x,a)* is equal to the immediate reinforcement *r* plus the utility of the next state y, discounted by γ:

$$util(x,a) = r + \gamma \, eval(y) \qquad (2)$$

Equation (2) can be generalized to a nondeterministic world by taking into consideration the probabilities of multiple outcomes.

2.1. Framework A

Framework A (Figure 2-1) was previously studied in [Sutton, 1984; Barto, Sutton & Watkins, 1989; Anderson, 1989]. It consists of three components: an evaluation network, a policy network and a stochastic action selector. In essence, the framework decomposes reinforcement learning into two subtasks. The first subtask is to construct, using the TD methods, an evaluation network that accurately estimates the utility of each world state. The second subtask is to adjust the policy to choose actions leading to high-utility states as measured by the evaluation network. Specifically, the policy network, which takes a world state as inputs, assigns to each possible action a value indicating the relative merit of performing that action in response to that world state. The policy network is adjusted so that actions resulting in higher utilities will be assigned with higher values. So, if such an optimal policy network is available, the best action can be obtained by applying the network to the given state and choosing the action with the highest merit value. To learn an optimal policy effectively, active exploration of different actions in response to the same situation is needed so that the relative merits of actions can be assessed. The stochastic action selector is used for this purpose.

In the course of learning, both the evaluation and the

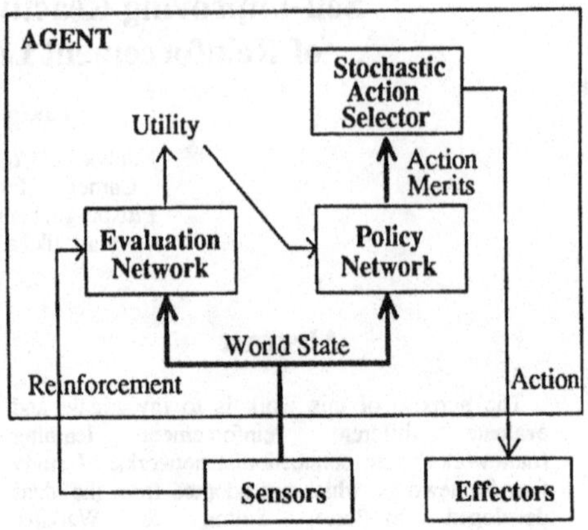

Figure 2-1: Framework A. The bold lines indicate a vector of signals, and the thinner lines a scalar signal.

---

1. $x \leftarrow$ current state; $e \leftarrow eval(x)$;
2. $a \leftarrow select(policy(x), T)$;
3. Perform action a;
   $(y,r) \leftarrow$ new state and reinforcement;
4. $e' \leftarrow r + \gamma * eval(y)$;
5. Adjust evaluation network by backpropagating TD error $(e' - e)$ through it with input x;
6. Adjust policy network by backpropagating error Δ through it with input x, where

$$\Delta_i = \begin{cases} e' - e, & \text{if } i = a \\ 0, & \text{otherwise} \end{cases}$$

7. goto 1.

Figure 2-2: The algorithm for Framework A

---

policy networks are adjusted incrementally. Figure 2-2 summarizes the learning algorithm, in which the simplest TD method is used. To use the TD methods to learn an evaluation function, the first step is to write down a recursive definition of the desired function. By definition, the utility of a state x is the immediate payoff r plus the utility of the next state y, discounted by γ.[1] So, if the prediction is accurate, Equation (3) should hold true:

$$eval(x)=r+\gamma \cdot eval(y) \qquad (3)$$

The TD error is the difference between the two sides of the equation, and the evaluation network is adjusted to reduce the error using the backpropagation algorithm.

The policy network is also adjusted (Step 6) according to the same TD error. The main idea is as follows: If $e' > e$, meaning that action a is found to be better than previously expected, the policy is modified to increase the merit of the action. On the other hand, if $e' < e$, the policy is modified to decrease the merit. The policy network is not updated with respect to actions other than a, since from a single experience we know nothing about the merits of the other actions.

In the course of learning, the stochastic action selector (i.e., the *select* function in the algorithm) chooses actions randomly according to a probability distribution determined by the outputs of the policy network-- actions favored by the policy network are more likely to be chosen. In this work, the probability of choosing action a_i is computed according to the following function:

$$Prob(a_i)=\frac{e^{x_i/T}}{\sum_k e^{x_k/T}} \qquad (4)$$

where x_i represents the input to the action selector corresponding to action a_i, and the *temperature* T adjusts the randomness of action selection.

2.2. Framework B

Framework B, which is also known as *Q-learning*, was developed by [Watkins, 1989]. Instead of learning two separate networks as Framework A does, Framework B only learns a single network called the *utility* network. The utility network is like the evaluation network of Framework A in that both are trained to predict the cumulative reinforcements. However, they differs in that for each action, the utility network maps a world state to the utility of each state-action pair, rather than the utility of the state (recall Equation (2) for the difference). So, if such an optimal utility network is available, the best action can be determined by applying the network to the given state and finding the action that results in the largest value (i.e., utility). Therefore, the utility network is also like the policy network of Framework A in that both are used to choose actions.

[1]See [Barto, Sutton & Watkins, 1989] for a detailed discussion.

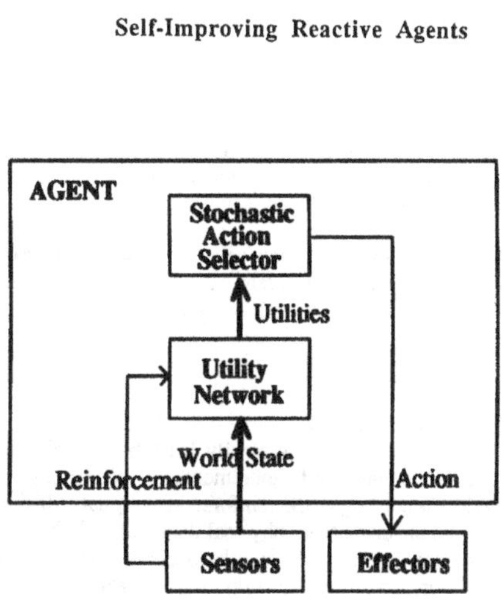

Figure 2-3: Framework B. The bold lines indicate a vector of signals, and the thinner lines a scalar signal.

1. $x \leftarrow$ current state;
   for each action i, $U_i \leftarrow util(x,i)$;
2. $a \leftarrow select(U,T)$;
3. Perform action a;
   $(y,r) \leftarrow$ new state and reinforcement;
4. $u' \leftarrow r + \gamma * Max_{k \in actions}\, util(y,k)$;
5. Adjust utility network by backpropagating error ΔU through it with input x, where

$$\Delta U_i = \begin{cases} u' - U_i, & \text{if } i=a \\ 0, & \text{otherwise} \end{cases}$$

6. goto 1.

Figure 2-4: The algorithm for Framework B

The connectionist learning algorithm for Framework B is depicted in Figure 2-4 and briefly described below. Generally speaking, the utility of an action a in response to a state x is equal to the immediate payoff r plus the best utility that can be obtained from the next state y, discounted by γ. So, if the prediction is correct, Equation (5) should hold true:

$$util(x,a)=r+\gamma \cdot Max_{k \in actions}\, util(y,k) \qquad (5)$$

The TD error for action a is the difference between the two sides of the equation, and the utility network is modified to reduce this error for action a. However, the network is not modified with respect to actions other than a, since from a single experience we know nothing about the utilities of the other actions.

The utility network described above has multiple outputs (one for each action). It could be replaced by multiple networks (one for each action) with a single output. The former implementation might be less desirable, because whenever the single utility network is modified with respect to an action, no matter whether it is desired or not, it is also modified with respect to the other actions as a result of shared hidden units between actions.[2]

Again, a stochastic action selector is needed to explore the consequences of different actions. The same action selector used by Framework A can be employed for this purpose.

2.3. Framework AM

Frameworks A and B typically learn by participating in the environment and incrementally modifying their behaviors according to the reinforcement signals. If the cost of making mistakes (e.g., physical damage) is high, learning agents based on the frameworks are subject to suffering. Furthermore, both frameworks mainly rely on the TD methods to propagate back credit or blame (i.e., reinforcements) to appropriate states and/or actions within action sequences. Generally speaking, this propagation process is slow particularly when the reinforcement signals are sparse. A solution to these problems is to use a world model. By using a world model, the agent can learn without participating in the real world, thus learning faster and committing less mistakes in the real world. Frameworks AM and BM to be discussed below are based on this idea.

Framework AM is very similar to Framework A; both consist of an evaluation network, a policy network, and an action selector. They only differ in that Framework AM also learns a world model and uses it to speed learning. This framework was studied in [Sutton, 1990]. In his work, a world model was merely used as a substitute of the real world, and the algorithm for the framework was exactly the same as the one in Figure 2-2 except that actions were sometimes taken with the real world and sometimes taken with a world model. Here I argue that a world model can be utilized in a more effective way, which is depicted in Figure 2-5 and discussed below.

The algorithm shown in Figure 2-5 is a modification to what is called *relaxation planning* in [Sutton, 1990]. In essence, with a world model, the evaluation and the policy networks can be effectively adjusted by looking ahead one step. Roughly speaking, the utility of a state x is equal to the immediate payoff r obtained from executing the best action a plus the discounted utility of the next state y:

$$eval(x) = Max_{a \in actions} r + \gamma \cdot eval(y) \qquad (6)$$

Thus, if a correct world model is available, the utility of any state can be more accurately estimated by looking ahead one step. During learning, the evaluation network is adjusted to reduce the difference between the two sides of the equation.

While the evaluation network is adjusted, the policy network can be updated as well. First we compute the average utility, μ, of a state x (Step 5) under the assumption

1. $x \leftarrow$ current state; $e \leftarrow eval(x)$;
2. Select promising actions S according to $policy(x)$;
3. If there is only one element in S, goto 8;
4. For $a \in S$ do
   4.a. Simulate action a;
   $(y, r) \leftarrow$ predicted new state and reinforcement;
   4.b. $E_a \leftarrow r + \gamma * eval(y)$;
5. $\mu \leftarrow \sum_{a \in S} Prob(a) \cdot E_a$; $max \leftarrow Max_{a \in S} E_a$;
6. Adjust evaluation network by backpropagating error $(max - e)$ through it with input x;
7. Adjust policy network by backpropagating error Δ through it with input x, where

$$\Delta_a = \begin{cases} E_a - \mu, & \text{if } a \in S \\ 0, & \text{otherwise} \end{cases}$$

8. exit.

Figure 2-5: The iteration of relaxation planning

that the current policy and the stochastic action selector will be used throughout the future. The *Prob* function is the one used to compute the probability distribution for selecting actions. In this work, it is Equation (4). Next, the policy network is modified to increase/decrease the merits of actions which are above/below the average (Step 7).

The algorithm of this framework is similar to that of Framework A, except that before the action selection step (Step 2 in Figure 2-2), relaxation planning may occur. To be efficient, relaxation planning is performed selectively (Steps 2 & 3 in Figure 2-5). For situations where the policy is very decisive about the best action, relaxation planning is not needed. If the policy cannot be very sure about which is the best action, relaxation planning is performed. In this way, at the beginning of learning, all actions are equally good and relaxation planning is performed a lot. As learning proceeds, relaxation planning is performed less and less often.

2.4. Framework BM

Framework BM is Framework B plus a world model, much as Framework AM is Framework A plus a world model. The algorithm is similar to that of Framework B, except that in Framework BM actions can be taken with a world model as well as with the real world. Figure 2-6 shows the algorithm for using a world model. In essence, for situations where the best action is obvious, no action simulation is done with the world model for efficiency reason. If there are several actions that seem to be promising, then these actions are tried with the model. The utility network is modified using a similar procedure as the one depicted in Figure 2-4.

[2]A similar argument can apply to the policy network of Framework A.

1. $x \leftarrow$ current state;
   for each action i, $U_i \leftarrow util(x,i)$;
2. Select promising actions S according to U;
3. If there is only one element in S, goto 6;
4. For $a \in S$ do
   4.a. Simulate action a;
   $(y,r) \leftarrow$ predicted new state and reinforcement;
   4.b. $U'_a \leftarrow r + \gamma * Max_{k \in actions}\, util(y,k)$;
5. Adjust utility network by backpropagating error ΔU through it with input x, where

$$\Delta U_a = \begin{cases} U'_a - U_a, & \text{if } a \in S \\ 0, & \text{otherwise} \end{cases}$$

6. exit.

Figure 2-6: The algorithm for Framework BM

3. The Dynamic Environment

The dynamical environment is a 25x25 cell world. A sample environment is shown in Figure 3-1. There are four kinds of objects in the environment: the agent ("I"), food ("$"), enemies ("E"), and obstacles ("O"). The outside of the world is considered as occupied by obstacles. The bottom of the figure is an energy indicator ("H"). At the start, the agent and four enemies are placed in their initial positions as shown in Figure 3-1, and fifteen pieces of food are randomly placed on unoccupied cells. On each move, the agent has four actions to choose from: walking up, down, left, or right to an adjacent cell. If the agent attempts to walk into obstacles, it will remain at the same position.

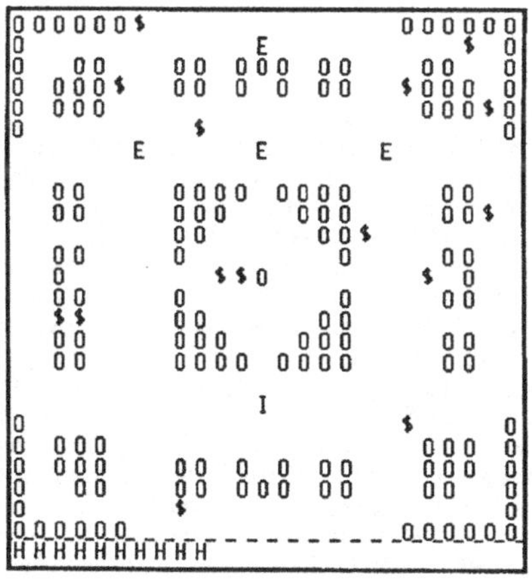

Figure 3-1: The dynamic environment

After the agent moves, each of the enemies is allowed to stay or move to an adjacent cell that is not occupied by obstacles. The speed of the enemies is about 80% of the full speed of the agent, so the agent is able to escape from chasing enemies. The behavior of the enemies is nondeterministic. However, there is a tendency that the enemies gradually move towards the agent, and the tendency becomes stronger when the agent is close.

The objective of the agent is to get as much food as possible while avoiding being caught by enemies. The play ends when the agent gets all of the food or dies. The agent can die by either colliding with the enemies or running out of energy. At the start of a new play, the agent is given 50 units of energy. Each piece of food provides the agent 15 units of energy, and each move costs the agent 1 unit. These three parameter values were empirically chosen so that survival in the environment would not be too easy or too difficult.

To make the learning task more interesting, the agent is allowed to see only a local area surrounding it. From the agent's point of view, the world is nondeterministic, not only because the enemies behave nondeterministically, but also because the world is only partially observable (so, what will be seen after a move is unpredictable). Although humans can avoid the enemies and get all of the food with little effort, survival in this environment is not trivial. To survive, an agent must learn to 1) approach food, 2) escape from enemies, 3) avoid obstacles, 4) identify and stay away from certain dangerous places (e.g., corridors) where it can be easily seized, and 5) seek for food when no food is within view.

4. The Learning Agents

In this section, I present the implementation of four learning agents, Agents A, B, AM, and BM, to survive in the environment. The agents are named after the learning frameworks on which they are based. In order to compare their performance, the agents use exactly the same reinforcement signals and sensory inputs as described below.

All of the connectionist networks are modified using a symmetrical version of the error backpropagation algorithm [Stornetta & Huberman, 1987]. In other words, I use 0.5 for representing boolean true and -0.5 for boolean false, and the squashing function used is a variation of the sigmoid function:

$$f(x) = \frac{1}{1+e^{-x}} - \frac{1}{2} \qquad (7)$$

4.1. The Reinforcement Signals

After each move, the learning agent could receive from the environment one of the following reinforcement signals.

- -1.0 if the agent dies
- 0.4 if the agent gets food

- -0.05 if the agent moves back or walks into obstacles
- 0.0 otherwise

Negative reinforcement is considered bad, and positive is good. It is important to choose the reinforcement values, because they determine what an optimal agent's behavior will look like. Using too large a reinforcement value for having food may direct the agent to neglect enemies in the face of food, resulting in death, while using too small a reinforcement value may result in slow learning. I tried a few different values for food reinforcement, and 0.4 gave roughly the best performance.

The agent gets a small penalty, -0.05, when it moves back to its previous position or attempts to walk into obstacles, because in general these actions are bad. In the earlier experiments, where this reinforcement was not used, the agent often moved back and forth when there were no interesting objects (i.e., food and enemies) around. The use of the reinforcement could instruct the agent to keep moving in a certain direction whenever no interesting objects are around, resulting in a behavior of seeking for food.

4.2. Input Representations

As described before, Agents A and AM consist of an evaluation network, a policy network, and a stochastic action selector, while Agents B and BM consist of a utility network and a stochastic action selector. The action selector is described in Section 2.1. This section describes the input representations of the three networks, while the next section describes the output representations.

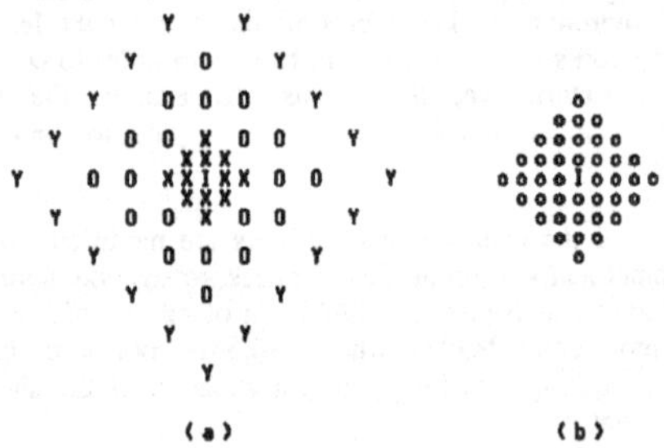

(a) (b)

Figure 4-1: The (a) food and (b) obstacle sensors

All of the three networks are structurally similar; each of them is a three-layer, feed-forward network, consisting of 145 input units and 1 output unit. (The number of hidden units is a parameter to be tuned for performance.) The networks are fully connected except that there are no connections between the input and output units. The input units of the networks can be divided into five groups: enemy map, food map, obstacle map, energy level, and the agent's previous heading. Each of the maps shows a certain kind of objects in a local region surrounding the agent and can be thought of as being obtained by an array of sensors fixed on the agent. Figure 4-1 shows the configuration of the food and obstacle sensor arrays.

The spatial positions of food are coarse-coded [Rumelhart, et al., 1986b], meaning that each food sensor may be activated by several nearby food objects. The food sensor array is composed of three different types of sensors, types "X", "O", and "Y". Different food sensor types have different resolution and different receptive fields-- "X", "O", and "Y" sensors can be activated by food at any of the five, nine, and thirteen nearby cells respectively. Through the use of multiple resolution and coarse coding techniques, the food positions are effectively coded without loss of critical information.

The enemy sensor array has a similar layout of the food sensor array, except that it consists of only "X" and "O" types of sensors. The obstacle sensors are organized differently; there is only one type of obstacle sensor, and each obstacle sensor is activated by an obstacle at the corresponding cell. The coarse coding technique is not used to encode obstacle positions, because it only works effectively when the features to be encoded are sparse [Rumelhart, et al., 1986b], and this is not the case for obstacles.

The agent's energy level is, again, coarse-coded by sixteen input units. In other words, each of the sixteen units represents a specific energy level and is activated when the agent's energy level is close to that specific level. Finally, the networks use five units to represent the agent's previous heading (up, down, left, right, and stop). The heading information allows the agents to learn that moving back is generally bad.

4.3. Output Representations

Each of the three networks has one output unit, with a value between -1 and +1. (Of course, it must be scaled to a value between -0.5 and 0.5 before the backpropagation algorithm applies.) The output of the evaluation network represents the estimated utility of the given input state. The single output of the policy network represents the merit of moving "up". The merits of moving in other directions can be computed using the same network by rotating the input signals, except those for the energy level, by 90, 180, and 270 degrees. Similarly, the single output of the utility network represents the utility of moving "up" in response to the input state. Again, the utilities of moving in other directions can be computed by rotating the input signals appropriately. By taking advantage of the action symmetry, the number of needed hidden units is reduced, and the learning rate was found to be significantly improved.

4.4. The World Model

Both Agents AM and BM learn a world model. The world model is intended to model the input-output behavior of the dynamic environment. More specifically, given an action to be performed, the world model is to predict what reinforcement signal will be received, where the food, enemies and obstacles will appear, and what the agent's energy level will be. In this nondeterministic environment, each action could have many possible outcomes. There are two alternatives to modeling the environment: the world model can generate either a list of outcomes associated with probabilities of happening or only the most likely outcome. The second alternative is adopted, because of its simplicity.

Since the food and obstacles do not move, their positions are quite easy to predict. So, to shorten simulation time without much simplying the model learning task, Agents AM and BM are only asked to learn a *reinforcement network* for predicting the reinforcement signal and an *enemy network* for predicting the enemy positions. The reinforcement network is structurally similar to the utility network of Agent B or BM. They differ in that the former estimates the reinforcement signal to be received immediately, while the latter estimates the discounted cumulative reinforcements to be received throughout the future.

The enemy network is also a three-layer, feed-forward network, consisting of 139 input units, 264 hidden units, and 99 output units. Unlike the other networks mentioned before, the enemy network is not fully connected and uses different inputs. Its inputs include the obstacle map and a high resolution, noncoarse-coded enemy map, which is more like the obstacle map than the enemy map mentioned before. (Note that the enemies' behavior is influenced by surrounding obstacles as well as the agent.) The outputs are also a high resolution enemy map, which is converted to a normal, coarse-coded enemy map by a hand-coded program. Since the enemies can only walk to an adjacent cell at each move, whether a pixel of the output enemy map will be activated is determined only by the nearby pixels of the input enemy and obstacle maps. Therefore, the enemy network does not need to be fully connected. Instead, each of the output units is indirectly connected to (a small number of) nearby input units through 3 hidden units or so.

The reinforcement and enemy networks are learned on-line just like the other networks. However, learning the two networks is a kind of *supervised learning*, and the encountered experiences (i.e., input-output patterns) can be saved (for a certain period of time) and repeatedly presented to the networks. Roughly speaking, the world model stops improving after being trained on the experiences from playing the first 50 environments (roughly 5000 patterns). Because the enemies are nondeterministic, predictions by such a well-trained model sometimes can be very wrong. For example, a nearby enemy might unexpectedly appear or disappear in the predicted map.

In addition to a learned model, a "perfect" model is provided. The perfect model is in fact the environment simulator which simulates the enemies' random behavior. Agents AM and BM can use either the perfect model or a learned model, but not both. Next section will present a performance comparison between using a perfect model and using a learned, potentially incorrect one.

5. Evaluation

For each learning agent, there are several parameters that can be tuned for performance. These parameters include the number of hidden units (H), the learning rate (η) and momentum factor (α) of the backpropagation algorithm, and the temperature (T) of the stochastic action selector. The discount factor γ is fixed to be 0.9, however. The parameter values used to generate the results below were 1) T=0.02 (a pretty low temperature) for all of the agents, 2) H=40, η=0.2, and α=0.9 for the evaluation network, 3) H=30, η=0.4, and α=0.9 for the policy network, and 4) H=30, η=0.3, and α=0.9 for the utility network. These values were manually chosen to give roughly the best performance for Agents A and B. No search was done for Agents AM and BM; they simply used the values found for Agents A and B. I have tried cooling temperatures, but I found no advantage of using them.

To evaluate the four learning agents, 700 training environments and 10 testing environments were randomly generated. The agents were allowed to learn only from playing the 700 training environments. Each time an agent played 10 training environments, it was tested on the 10 testing environments. The average number of food pieces obtained by the agent in the testing environments was then plotted versus the number of training environments that the agent had played so far. The obtained learning curves were not smooth, primarily due to the enemies' random behavior. For ease of comprehension, the curves are smoothed by averaging every three successive points (the first two points of the curves are left untouched, however). Figure 5-1 show the smoothed learning curves of the agents. Agents AM and BM each have two learning curves. One is obtained from using a perfect model, and the other from using a learned, potentially incorrect model. Although the results shown below were obtained with a particular set of training and testing environments, I did another experiment with a different set of environments and obtained similar results.

To summarize these learning curves:

- All of the agents learned quickly at the early stage of learning. They reached 80% of the asymptotic performance after playing less than 100 environments.
- Before learning, the agents can only get about 0.3 out of 15 food pieces (2%). After playing 250 environments, Agent A can get 10 pieces (67%) in average, Agent AM 11.5 pieces (77%), and Agents B and BM 13 pieces (87%).
- If we sort the agents by the asymptotic

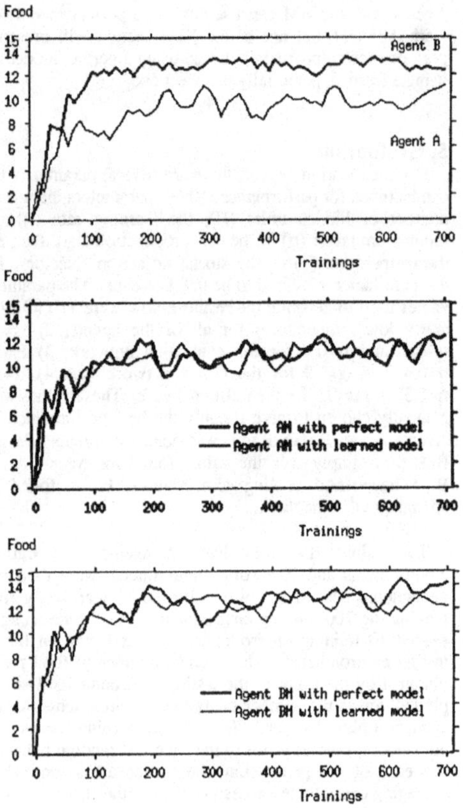

Figure 5-1: Learning curves of the agents

performance, the ordering is roughly: BM (perfect model) ≈ B ≈ BM (learned model) > AM (perfect model) ≈ AM (learned model) > A.

- If we sort the agents by the average learning rate in the first 50 environments, the ordering is roughly: BM (perfect model) ≈ BM (learned model) > AM (learned model) ≈ AM (perfect model) ≈ B > A.

Watching the agents playing, we can make the following observations:

- Agents B and BM were able to go get most food (including getting around obstacles) that were detected by the food sensors (the range of food sensors is 10 steps), while Agents A and AM were only able to go get food that is reachable within 6 and 8 steps or so, respectively.
- Agents B and BM learned good strategies for many rare and difficult situations, while Agent A did not. For instance, in the situations shown in Figure 5-2, Agent A would often move towards the food and get killed, while Agents B and BM would seldom make such mistakes. Agent AM was between Agents A and B.
- Even for handling common situations, Agents B and BM had better strategies than Agents A and AM. Agent AM was better than Agent A.
- Agents B and BM played very well. As a matter of fact, they got 14 or 15 pieces of food most of the time. They even learned how to get a collection of food efficiently, not simply take the closest one first. Of course they also made mistakes. Overlooking a potential danger is the most serious mistake. For example, being attracted by a piece of food, they might fall into a pocket formed by enemies and obstacles, and eventually get killed.

```
O O          O O
O $ I       $ O  I E      E $ I
O O E        O O
```

Figure 5-2: Difficult situations for Agent A

6. Discussion

Here is an interesting question: Why Agents B and BM performed better than Agents A and AM? My conjecture is as follows: Agents A and AM derive a policy from a learned evaluation function. In this fairly complex environment, the derivation process can be hardly perfect. On the other hand, Agents B and BM learn an evaluation function and use it directly for choosing actions. Thus, if all of the agents learn an evaluation function equally well, the performance of Agent A or AM would not be as good as that of Agent B or BM.

Another interesting question is why Agents B and BM performed equally well (in terms of asymptotic performance), while Agent AM was apparently better than Agent A? The reason is: What Agent BM gains from using a model is just additional experiences, while the relaxation planning provides Agent AM a more effective way to compute the evaluation function and to assess the relative merits of actions in response to the same situation-- this is not possible for Agent A to do. Because of this, we can expect that an agent using the model-based algorithm investigated in [Sutton, 1990] will be only as good as Agent A.

Not surprisingly, by using a world model, Agents AM and BM learned faster than their counterparts, A and B, even though the world model used was imperfect. However, if a sufficiently good world model is difficult to obtain, any advantage of using it could be negated. For example, Agent BM using a learned model performed slightly worse than Agent B.

In this work I only investigated how to use a world model to speed learning. There are other ways of using a model to improve performance that I did not study here. For example, we can use an evaluation function, a world model, and a game-tree search algorithm to select actions. In a fairly complex domain, an optimal policy is often unobtainable. By looking ahead a few steps, the non-optimality of a policy can be compensated, if a good world model is available.

7. Conclusion

I have studied four reinforcement learning frameworks. Frameworks A and AM learn an evaluation function of world states and construct a policy to keep themselves in good states as measured by the evaluation function, while Frameworks B and BM learn only an evaluation function of state-action pairs and use it to choose actions. Frameworks AM and BM also learn a world model and use it as another source for having experiences. I also have developed four learning agents, based on the frameworks, to survive in a dynamic environment. Through the study of the learning agents, I now have better understanding about the learning frameworks.

In general Frameworks B and BM work better than Frameworks A and AM. By using a world model, Frameworks AM and BM can learn faster than their counterparts, A and B, in spite of an imperfect model used. Framework AM not only learns faster but also performs better than Framework A. However, even if a perfect model is used, Framework BM can be only as good as Framework B in terms of asymptotic performance.

While it seems that all of the learning frameworks are promising, a good implementation is also important to building a successful learning agent. In the course of developing the learning agents, three techniques were found to be very helpful: 1) the coarse coding technique, 2) the use of concentric multiple resolution maps, which dramatically reduces the number of input units needed, and 3) taking advantage of the action symmetry (i.e., treating the four move actions as one by rotating input signals appropriately). Without using these techniques, the agents could not have been so successful.

Possible future work includes 1) adding complexity to the environment (e.g., moving food and pushable obstacles) to see the limitation of the learning agents, 2) using an evaluation function, a world model, and game-tree search techniques to select actions, 3) investigating other learning frameworks, and 4) applying the methodologies to real world tasks.

Acknowledgements

I would like to express my gratitude to Tom Mitchell and Jeffrey Schlimmer for their inspiration and advice on this work.

References

Anderson, C.W. Strategy learning with multilayer connectionist representations. In *Proceedings of the Fourth International Workshop on Machine Learning.* , 1989.

Barto, A.G., Sutton, R.S., and Watkins, C.J.C.H. *Learning and sequential decision making* (COINS Technical Report 89-95). Dept. of Computer and Information Science, University of Massachusetts, 1989.

Kaelbling, L.P. A Formal Framework for Learning in Embedded Systems. In *Proceedings of the Sixth International Workshop on Machine Learning.* 1989.

Rumelhart, D.E., Hinton, G.E., and Williams, R.J. Learning internal representations by error propagation. In *Rumelhart, D.E. and McClelland, J.L., editors, Parallel Distributed Processing: Explorations in the Microstructure of Cognition, vol.1.* Bradford Books/MIT press, Cambridge, MA, 1986.

Hinton, G.E., McClelland, J.L., and Rumelhart, D.E. Distributed representations. In *Rumelhart, D.E. and McClelland, J.L., editors, Parallel Distributed Processing: Explorations in the Microstructure of Cognition, vol.1.* Bradford Books/MIT press, Cambridge, MA, 1986.

Stornetta, W.S., and Huberman, B.A. An improved three-layer, back propagation algorithm. In *Proceedings of IEEE International Conference on Neural Networks, vol.2.* , 1987.

Sutton, R.S. *Temporal Credit Assignment in Reinforcement Learning.* Doctoral dissertation, Department of Computer and Information Science, University of Massachusetts, 1984.

Sutton, R.S. Learning to predict by the methods of temporal differences. In *Machine Learning, 3:9-44.* , 1988.

Sutton, R.S. First results with DYNA, an integrated architecture for learning, planning and reacting. In *Stanford Spring Symposium on Planning.* , 1990.

Watkins, C.J.C.H. *Learning with Delayed Rewards.* Doctoral dissertation, Psychology Department, Cambridge University, 1989.

Williams, R.J. A class of gradient-estimating algorithms for reinforcement learning in neural networks. In *Proceedings of the First Annual International Conference on Neural Networks, Vol II.* , 1987.

Exploring Adaptive Agency II: Simulating the Evolution of Associative Learning

Peter M. Todd **Geoffrey F. Miller**
Department of Psychology
Jordan Hall Building 420
Stanford University
Stanford, CA 94305
todd@psych.stanford.edu geoffrey@psych.stanford.edu

Abstract

We consider psychology as the study of *adaptive agency*, investigating the processes and mechanisms resulting in fitness-increasing behavior in the world. A central issue in psychology so construed becomes: what are the relations between the primary adaptive process of evolution by natural selection, and the adaptive processes psychologists call 'learning'? In particular, under what conditions would learning evolve? To explore this issue, we use genetic algorithms to simulate the evolution by natural selection of neural networks, which in turn control the behavior of simple creatures in virtual environments. We have developed what we consider the simplest possible environmental challenge in which unsupervised associative learning could prove adaptive: 'bootstrapping' the learned use of one highly accurate, but individually varying, sensory modality by another less accurate, but evolutionarily stable, modality. We have found a possibly quite general U-shaped curve relating the time (in number of generations) to evolve the use of unsupervised learning on the varying 'bootstrapped' modality, to the accuracy of perception in the stable modality which guides this learning. This U-shaped curve appears to represent a trade-off between the adaptive pressure to evolve learning (which peaks when perception accuracy in the stable guiding modality is at chance) and the ease of learning during a given lifespan (which peaks when this accuracy is perfect.)

1 Introduction

Natural selection has constructed animals' minds and behavioral repertoires for adaptive fit to the environmental problems they must face. As the study of such minds and behaviors, psychology should focus on the notion of *adaptive agency* -- the generation of action in the world in response to challenges to individual fitness. This framework encompasses many approaches, including (1) the elucidation of complex species-typical adaptations (as in human and animal experimental psychology and cognitive ethology), (2) the comparison of psychological adaptations across species and consideration of their phylogenetic origins (as in comparative psychology), and (3) the general exploration of the adaptive processes themselves that yield adaptive agency (e.g. by simulation methods, including those in the field of artificial life -- see Langton, 1989). In the current paper, we consider the phenomenon of 'learning' as an aspect of adaptive agency, by investigating via evolutionary simulations some conditions under which the ability to learn may prove adaptive and so spread through a population. The theoretical, historical, and methodological background for this work is presented more extensively in Miller and Todd (1990), and further extensions of this method applied to habituation and sensitization as adaptations to short-term environmental dynamics appear in Todd and Miller (in press).

Evolution as an adaptive process has itself undergone changes: "survival of the stable" probably preceded "survival of the fittest" (Dawkins, 1976). Evolution in the earth's early environment is likely to have selected for physical systems with relative stability in the shifting primordial soup. After stability came replication and metabolism: the ability to turn external material into copies and extensions of oneself. With the evolution of larger, more complex phenotypes came the opportunity to evolve behavior-generating systems that could produce innately programmed sequences of activity and movement. Sensory systems could then evolve to guide these behavior-generators more adaptively, based on sensitivity to particular environmental cues. Thus, blind activity based on preprogrammed behaviors may have preceded sensory-guided *reactivity*, that is, the ability to adaptively adjust to the current changing environment on a moment-by-moment basis. Only after these first two stages had evolved could a further adaptive process evolve -- 'learning,' defined as the ability to make long(ish)-term adaptive changes in behavior-generators in response to the environment.[1] In this hierarchy, learning emerges not as the primary adaptive force it is sometimes assumed to be, but rather as a tertiary one, following genotypic evolution and short-term environmental reactivity (see also Shepard, 1987, 1989). Once we re-conceptualize 'learning' as merely one process among several that generate adaptive agency, the questions we might ask about this

[1] By this definition, learning includes such processes as experience-guided development not commonly included in this category. For examples of such processes, see Knudsen, 1988, and Singer, 1984, 1988.

process begin to change as well.

Exploring adaptive agency by simulation thus requires simulating genotypic evolution, the generation of behavior, and the ability to change behavior generators by learning. Specifically, we use a genetic algorithm to evolve successive generations of a population of neural network architectures, which in turn control the behavior of simple creatures which can learn as they live in a simulated environment. This is an extension of earlier work exploring the use of genetic algorithms to design network architectures capable of learning specific input/output mapping tasks (Todd, 1988; Miller, Todd, & Hegde, 1989); here, the algorithm's measure of fitness depends not on learning an arbitrary task, but on behaving adaptively in the simulated environment.

Traditionally, researchers using simulation in this area have started by assuming learning as a primary adaptive process, and then asked how might evolution interact with, shape, and be shaped by learning? Examples include the work of Hinton and Nowlan (1987) and Belew (1989), who explicate the Baldwin effect in which learning "guides" evolution; and Stork and Keesing (1990) and Belew, McInerney, and Schraudolph (1990), who investigate the ways evolution and learning can combine to affect the initial structure of neural networks. But considering the hierarchy of adaptive processes spelled out above, the question *we* want to ask is, rather, given the already-powerful adaptive processes of genotypic evolution and environmental reactivity, under what conditions would the tertiary adaptive process of learning ever prove useful in terms of increasing individual fitness? Assuming *no* learning, when should learning evolve? Why should this extra level of adaptiveness exist? (For analysis of learning mechanisms as evolutionary adaptations, see Pulliam & Dunford, 1980; Dawkins, 1983; and Cosmides & Tooby, 1987.)

To answer such questions, 'learning' must be viewed not as a single monolithic process, but as the diverse set of distinct mechanisms, abilities, and dynamics that ethologists and psychologists have shown it to be (see Tinbergen, 1951; Lorenz, 1966; Hinde & Stevenson-Hinde, 1973; Bower & Hilgard, 1981; and Staddon, 1983). The adaptive (fitness-increasing) functions of each type of learning must be considered separately, with attention to the type of organism, econiche, and adaptive problem in question; this is the approach followed in the field of ecological learning theory (Davey, 1989), from which we take inspiration. In simulations, we must first specify some environment and what defines fitness in that environment, and then observe which adaptive processes are most important for maximizing individual fitness in "solving" the relevant environmental problems. Ackley and Littman's (1990) simulations go far to address these issues, showing the utility of evolving motivational systems to drive learning, but still his simulations either operate with learning or without, rather than setting up an environment whose adaptive problems allow the evolution of learning itself to be studied.

2 Adaptive Functions of 'Learning' Processes

Why should an organism allow environmental conditions to change how its behavior-generating mechanisms work (by 'learning'), rather than allowing natural selection to optimize those mechanisms (by 'hardwiring') just as its has optimized so many other physical adaptations (see Mayr, 1974; Staddon, 1983; Menzel, 1984)? We see two main adaptive functions for learning. First, and perhaps most importantly, 'learning' may serve to increase an organism's 'developmental leverage', allowing it to build a larger, more complex, more finely organized phenotype given a certain size genotype. To the extent that an organism can sense predictable environmental regularities during neural development, for example, it could use the resulting sensory activation patterns to guide the self-organization of its behavior-generating mechanisms (e.g. Singer, 1984). Learning may allow the genotype to 'store information in the environment' and let environmental regularities do much of the hard work of wiring up adaptive behavior-generators. The environmental regularities used in this way may take a rather abstract form. For example, parental 'imprinting' in birds (Lorenz, 1935/1970; Immelmann, 1984) can be viewed as a way of building a behavior-generator sensitive to the appearance of one's parent, based on the following environmental regularity: the first large moving thing one sees after hatching is very likely one's parent. Of course, the particular behavior-generator constructed by different birds for recognizing their parents will be different (the birds will 'learn' different parental images), but the species as a whole relies on the same environmental regularity to do the construction.

Second, 'learning' as sensitivity to dynamic changes in environmental conditions throughout life may allow organisms to adjust their behavior-generators faster than natural selection can track (see Belew, 1989). In this case, learning adapts phenotypes to ongoing changes and *particularities* of the environment, rather than building on environmental *regularities* during phenotype-construction. This is probably the adaptive function of learning most familiar to adult humans (e.g. learning a new restaurant location, or a new person's name), and therefore more commonly studied by psychologists, but it is likely to be less common than the phenotype-construction function throughout the animal kingdom (see Miller & Todd, 1990).

For our simulation studies of the evolution of learning, we attempted to devise the simplest, cleanest scenario in which learning could prove adaptive, focusing on the phenotype-construction function. If we make the genotype unable to know ahead-of time which one of two alternate econiches it will be born into, natural selection will be unable to choose the proper behavioral mechanisms to guide the phenotype's actions. Instead, it must select for the evolution of more general mechanisms that can flip into one of two states depending on which econiche the genotype finds itself in during phenotypic development. Thus, we settled on a kind of "imprinting" or parameter-setting based on the early environment as the simplest possible case in which learning, construed as adaptation to specific environmental regularities, could evolve.

3 Inclusive Fitness as the only Natural 'Supervisor' for Learning

Several previous efforts using genetic algorithms to evolve neural network architectures have evaluated architecture fitness by training the networks with a supervised learning procedure (i.e. one with an externally provided "target value" the network is to produce), typically back-propagation (Miller, Todd, & Hegde, 1989; Belew, McInerney, & Schraudolph, 1990; for a review, see Weiss, 1990). While supervised learning paradigms may be appropriate when engineering connectionist systems for particular industrial applications, they are problematic and perhaps misleading in scientific studies of adaptive agency. In particular, to be biologically plausible, the source of the "targets" must be justified. Organisms as whole functioning agents in real environments rarely receive patterns of information analogous to training signals in back-propagation (see Miller & Todd, 1990). Although the distinction between supervised and unsupervised learning procedures can be blurred, we have chosen to focus on the more defensible latter end of the spectrum, including self-organizing, associative, and simple feedback-based mechanisms. But even if we sidestep the issue of target-based training, the concept of feedback still raises problems.

Years of learning by "being taught" instill in us intuitions as to the usefulness of corrective feedback to promote learning. But such intuitions make it easy to overlook the fact that it is at least as difficult for natural selection to evolve the ability to perceive feedback signals from the environment to guide an organism's learning, as it is to evolve the perception of any other external stimuli. Consider for example the complexities involved in registering the information that one has just been rebuffed in a social exchange. Feedback signals cannot be assumed to be just somehow "provided" to an organism for it to use in adjusting its behavior. Instead, feedback systems must be understood as special sensory systems evolved to provide information to special learning mechanisms that in turn adaptively change the functioning of certain behavior generators. These feedback systems are likely to be comlex articulated systems themselves separate from the behavior generators; it is neurologically implausible that feed-forward signals are sent in one direction along axons while error (feedback) information is sent in the other. (Other sensory systems bias those behavior generators more directly and temporarily in response to ongoing environmental events -- reactivity, as we described earlier.) Feedback systems, whether motivational, emotional, volitional, or proprioceptive, evolve just like other aspects of adaptive agency -- by cumulative selection of incrementally better-adapted designs.

What an organism would like to have to guide its learning is direct information about how its inclusive fitness changes as a result of its behavior. But there is no such thing as an inclusive fitness transducer that can be used to supervise learning. Organisms must instead evolve to sense inclusive fitness indirectly, through whatever proximal sensory cues are reliably associated with increased fitness in their environment. Thus, natural selection itself is ultimately the only source of supervision for learning systems.

Natural selection's indirect supervision of learning leads to the complexity inherent in evolving learning systems. The learning scenario we present in this paper illustrates this point. To perform even the simplest kind of unsupervised associative (or correlational) learning we address, with real adaptive consequences through behavior rather than just forming perceptual categories, the following building blocks are required: (A) one input modality that has evolved to sense an environmental cue that has been reliably associated with a change in inclusive fitness; (B) another input modality that has evolved to sense a different cue which has not been reliably associated with a change in fitness over evolutionary time, but which can have such an association for some time during the organism's lifespan; (C) an output modality (effector) capable of generating some behavior that has fitness consequences related to the environmental situation sensed by A; (D) a connection between input A and output C with an evolved behavioral valence indicating whether or not that environmental cue should activate or inhibit that behavior (e.g. should there be a positive or negative weight from A to C?); and (E) a connection between input B and output C, whose valence can be built up by association with the activation of C caused by A. In the simulations to follow, we have limited the maximum processing capabilities of the creatures as closely as possible to these building blocks, as we show in section 5.

4 A Simple Scenario for the Evolution of Unsupervised Learning

To investigate conditions under which learning might evolve, we wanted to start with the simplest simulated world we could devise where learning could prove adaptive, and then see if it would spread through a population of simulated creatures behaving in that world. We explain the scenario by outlining a biological metaphor that identifies the structure of the econiche and the nature of the adaptive problem. The actual simulation methods used to instantiate this metaphor are presented in the next section.

Our scenario can be imagined as an underwater realm, in which parent creatures produce gametes that float around randomly, possibly undergoing sexual recombination with other gametes, until they fall to a particular feeding patch, where they hatch into offspring. The offspring then attach to the substrate, grow as sessile (stationary) feeders, and attempt to ingest the appropriate kinds of substances that float past. The offspring live for a fixed lifespan, and just before death they produce some number of gametes of their own, based on the amount of food they've eaten (which thus determines their fitness). These creatures thus never interact with each other (only their gametes may interact genetically), and at each moment in their life they merely decide whether or not to eat the particular substance currently in their vicinity.

The floating substances come in two forms, food and poison, that appear equally often (same base rate) and at random. Eating food *increases* the creature's internal energy store (by an arbitrary 10 points) and, since the average number of gametes released is proportional to lifetime energy accumulation, it likewise increases potential fitness. Eating poison *decreases* the creature's energy by the same amount (10 points), representing the metabolic activity necessary to counteract its effects, and thus reduces lifetime fitness. If a creature decides not to eat at a particular time, its energy (and hence its fitness) remains unchanged. Once a creature's lifespan ends, the average number of gametes released (and thus average fitness) is proportional to the amount of food eaten minus the amount of poison eaten. Thus, those creatures with "brains" (simulated by neural networks) that allow them to eat food and avoid poison will save up more energy and have more offspring, so that their beneficial brain-designs will spread through the population.

Food and poison can be distinguished by olfactory and visual features. Food is sweet and poison is sour across the entire world, forever. However, turbulence in the watery environment may carry odors in such a way that a creature receives more sour odorants than sweet odorants even though the nearest thing is really food. Turbulence thus decreases the accuracy of the smell sense, so food might not always *smell* sweet, at a given instant for a given creature, and poison might not always smell sour. We define a smell-sense accuracy for a given world (based on the degree of turbulence in the water) which determines the percentage of time each substance has its proper smell for the creatures. For example, 75% smell-accuracy means food smells sweet to each creature a random 3/4's of the time, and sour the other 1/4 of the time, and vice-versa for poison.

Food and poison also have associated colors: red or green. But this world is divided up into different food patches, and in half of these patches, food is always red and poison always green; in the other half, food is green and poison is red. Thus, during each sessile creature's lifetime, food will always be one color and poison the other. But the mapping from color to object will vary between individual creatures, depending on which kind of food patch they happen to land in at birth. The creatures' visual sense, though, unlike their olfactory sense, is completely accurate -- unaffected by the turbulence -- so that food and poison always appear their true color for that patch.[2]

To summarize, then, each creature in this world lives a fixed lifespan, eating or ignoring food and poison, and amassing energy which helps determine its eventual number of offspring in the next generation. Food smells sweet and poison sour across all creatures, but with some error rate determined by the smell-sense accuracy. Food and poison each have characteristic fixed colors *within* one creature's life, but the meaning of each color varies *between* creatures, food being red for some and green for others. Thus natural selection can 'predict' the association between smell and substance, but not between color and substance -- this will be the task for learning.

[2] The following results should generally hold even if the color-sense is merely *more* accurate than the smell-sense, not absolutely accurate. Postulating absolute accuracy in color sensing simply makes analysis of creature performance and interpretation of results easier.

5 The Genetic Algorithm for Evolving Neural Networks

To simulate the evolution of learning in the scenario just described, we use a relatively standard form of Holland's (1975) genetic algorithm, combined with a simple "developmental" scheme which translates genotypes into neural network architecture phenotypes. Each network, representing an individual creature, is evaluated in the simulated world over several time-steps (e.g. 200) representing the creature's lifespan. During each time-step, the network receives sensory input based on the current environmental cues available to the creature, processes that input according to its architecture and current weights, generates behavior based on the activation of its output units, and changes its connection weights using an unsupervised correlational learning rule (e.g. Hebbian association). The effects of the creature's network-generated behavior on the world and its own fitness are then registered, and the next time-step begins.

The genotypes of all initial populations are randomly-generated bit-strings. A strong genetic specification scheme (as defined by Miller, Todd, & Hegde, 1989) interprets each genotype as a *connectivity constraint matrix* that directly specifies the nature of each unit and connection in the network architecture. (Given the extremely simple design and small size of the networks investigated, more powerful development schemes seemed inappropriate.)

For the current scenario, we use the simplest network architecture we can: three units connected in a feedforward manner (unit 1 to units 2 and 3, and unit 2 to 3 alone). This compact genotype severely constrains the amount of neural "machinery" evolution has to play with in any given creature. However, this size is sufficient to incorporate the five building blocks necessary for simple associative learning as described in section 3, since all we need are two input units (A and B), an output unit (C), and connections between them (D and E).

To fill in this architectural structure, genotypes can specify six possible types of units: four possible types of input unit, a hidden unit, or an output unit. Different input unit types can sense sweet smell, sour smell, red color, or green color, being activated if their corresponding stimulus is present; hidden units compute the standard logistic function of their summed inputs; and output (motor) units generate eating behavior if turned on over some threshold level, or indicate abstinence if not. Genotypes also specify feedforward connections between and biases on the units (no recurrence is allowed in the simulations discussed here). Connections and biases can have positive or negative weights of four possible magnitudes (0, 1, 2, 3); and each is either genetically fixed or learnable via a simple Hebbian rule: $\Delta w_{ab} = lr * A_a * A_b$, where the change in connection strength Δw_{ab} from unit a to unit b is equal to some learning rate lr times the activation A_a of unit a and the activation A_b of unit b. Hidden and output unit activations may range from -1 to $+1$, while input activations may range from 0 to 1. With unit types specified by 4 bits (the extra bits helping to protect unit-type genes against mutation and crossover), initial biases and weights by 4 bits, and learnability by one bit, the genotype length for a 3 unit network is $(4*3) + (1*3 + 3*3)*(4 + 1) = 72$.

Grefenstette's (1987) standard genetic operators are used in the genetic algorithm: fitness-proportionate reproduction (with range-based fitness scaling), two-point crossover (p = 0.6 per offspring), and point mutation (p = 0.005 per genotype bit). Genes are ordered within genotypes by having all weights *to* a particular unit placed consecutively. Thus crossover tends to preserve associated groups of incoming connections to a given unit, but crossover points can occur anywhere within the genotype. Each generation contained 100 individuals for all runs reported here.

6 Some Poor Designs in Initial Populations

The problem facing our creatures is to eat food and avoid eating poison, when one sensory modality is noisy but evolutionarily stable, and the other sensory modality is accurate but evolutionarily variable. In this section we consider how well various poorly-adapted designs that might be randomly generated in our initial populations could do in this scenario. We expect these maladapted creatures to be quickly weeded out, but they are worth discussing because their defects illuminate important aspects of our watery world.

The simplest failure occurs in those creatures who do not even have any genetically specified motor unit (building block C) -- they would be unable to eat, so they'd always have zero fitness. Even worse, though, would be a creature with a smell unit (building block A), a motor unit (C), and a fixed connection between them (D) that has the wrong 'behavioral valence', i.e. that has the wrong fixed weight. For instance, if a creature's nervous system contains a sweet smell unit connected by a fixed *negative* weight to the motor (eating) unit, it will tend to eat poison and avoid food -- even worse for its fitness than not eating at all.

Another design that has a somewhat more subtle flaw would be a network with a smell unit (A), a color unit (B), and a motor unit (C), but with only learnable connections (each with zero initial strength) between these units. If such a creature found itself in a green-food patch, for example, it could gradually build up a Hebbian association between green color and sweet smell, that is, between units A and B, because the activity on those two units would be highly correlated. But it would be unable to build up the appropriate positive or negative weights from either of these input units to the motor unit (i.e., the necessary connections D and E would never be properly strengthened), because there would be nothing turning on and off the motor (eating) unit appropriately in response to food, and thus no association of sensation to food would be achievable. In other words, the creature would never be able to attach an adaptive behavioral valence to either color or smell. Such a creature would have found the statistical regularities in its world without being able to map them onto actual behavior. Thus, its average fitness would be zero.[3]

[3] This raises a general problem for unsupervised learning algorithms that allow only the development of self-organizing perceptual maps or the extraction of statistical regularities in input vectors (e.g. Linsker, 1988, 1990). Although the resulting systems will in a sense 'know' the structure of the world they inhabit, they will have no way of translating that knowledge into appropriate behavior and thus into increased fitness.

7 Evolving Better Networks

As the genetic operators process genotypes again and again through many generations, recombination and mutation will sometimes produce better designs. One reasonable design we can expect consists of a sweet smell unit (A) connected to a motor unit (C) by a fixed positive weight (D), with the other unit having only zero-strength incoming and outgoing connections and so being disregarded.[4] A creature equipped with this nervous system will depend purely on smell to decide when to eat, with its behavioral accuracy constrained by the inherent noisiness of the smell sense in its world (i.e. by amount of turbulence). Although this design will sometimes make mistakes (ignoring food that smells sour because of the turbulence -- a 'miss' -- and eating poison that smells sweet -- a 'false alarm'), still on average it will eat more food than poison. Thus these creatures' energy, and number of offspring, will be higher than if they were just eating randomly. We call this design the *color-blind eater*, or, more fancifully, ***Chromignarus simplex***, and it is shown in the top portion of Figure 1.

The best design though is an elaboration on the color-blind eater chassis. Creatures of this type gain an additional adaptive edge by including a red or green color unit (B), along with a learnable connection (E) to the eating unit. This creature design we name ***Chromodiscipulus simplex***, color-learning eaters, as shown at the bottom of Figure 1.

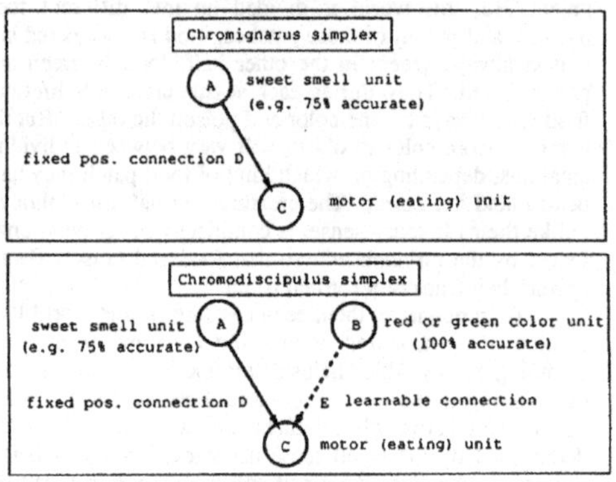

Figure 1. The network designs of ***Chromignarus*** (without color learning) and ***Chromodiscipulus*** (with color learning), showing the corresponding building blocks (A-E).

With this design, an appropriately excitatory or inhibitory link (E) between the color unit (B) and the eating unit (C) will be built up over successive time-steps in the creature's life by the Hebbian associative weight-change

[4] The sweet-smell sensing unit can always be replaced by a sour-smell sensing unit, with the appropriate reversal of connection strength to the eating unit. Since mating is unrestricted in our simulations, though, we do not tend to see the evolution of populations containing both subspecies simultaneously, because their interbreeding would tend to disrupt the linkage between smell-sense type and connection-valence.

mechanism. This will happen as follows: The sweet-smell unit (A) will, via its positive connection (D), preferentially turn on the eating unit (C) in the presence of food; the eating unit will have a bias so that it stays off otherwise (i.e. for poison). Now suppose the creature also has a red color-sensing unit (B), and food for this creature's patch is red. Then the red unit and the eating unit typically will be firing in unison in the presence of food, causing the weight between them (E) to increase. Eventually, this weight will be large enough that the red unit alone can cause the eating unit to come on, *regardless* of what the sweet unit says to do -- the creature has now learned that red means food. And since red, for this creature, *always* means food, the creature can rely solely on this visual cue, rather than the inaccurate smell cue, and always choose to eat properly, thereby increasing its fitness further. On the other hand, if food was green for this creature, the red unit and the eating unit would mostly come on separately, rather than in unison, and the weight between them would be decreased and eventually become inhibitory; the red color unit would thus serve to tell the creature when *not* to eat, and once again increase its fitness. So in either case, learning has given the creature a fitness bonus, and thereby proved adaptive.

8 Initial Simulation Results

We initially ran this simulation with a world where smell accuracy was 75%. The average population fitnesses of eight runs over 750 generations in this world are plotted in Figure 2. The populations tend to start with a collection of the sorts of poor designs discussed in section 6, with an average fitness of about zero. But very rapidly, the *Chromignarus* design evolves and spreads through the population, causing a leap in average fitness (to a value of about 2). All of the runs exhibit the rapid spread of this design's strong fixed positive connection between a sweet-smell unit and an eating unit within about the first 100 generations. *Chromignarus* usually rules the turbulent seascape for many generations, resulting in the first population plateau (at average fitness value 2) seen along each curve in Figure 2. After this plateau is reached, evolution typically takes a while to add a color unit with a learnable connection to the eating unit onto the *Chromignarus* design.[5] But once this addition occurs, and *Chromodiscipulus* appears, it rapidly takes over the population, increasing the average population fitness to a new plateau close to ceiling level. The switch to *Chromodiscipulus* happens quite quickly (over only a few tens of generations), as the step-like rise between the two plateau-levels on the curves in Figure 2 shows; but for each population, this switch occurs at different times.

[5] Note that the particular color (red or green) sensed by this unit is *not* universal throughout the population, since as described in the previous section either color can be used equally well to guide the learned behavior.

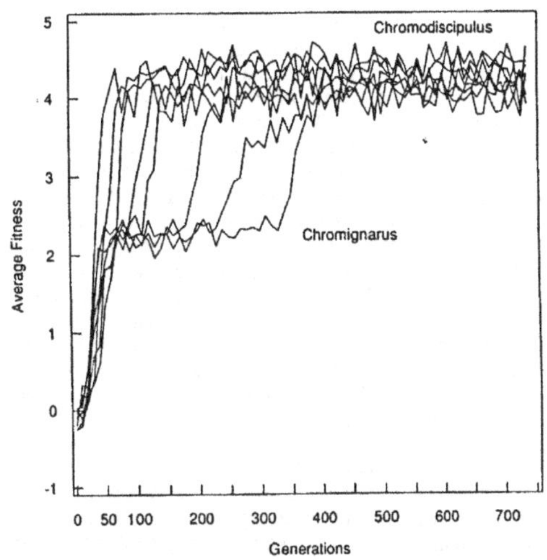

Figure 2. Average population fitness for several evolutionary runs with 75% smell-sense accuracy. *Chromignarus* predominates at the first fitness plateau, then *Chromodiscipulus* takes over at the second, with the evolution of color learning.

The fact that the creatures evolved to use learning indicates that our world was properly constructed with a behavioral challenge for which learning would prove adaptive. As such, witnessing the evolution of learning in our simulations is a confirmation of our approach, but is not in itself surprising. We continued our investigation of the conditions under which learning could evolve with the more interesting question of how *quickly* learning would evolve, given various smell accuracies in different worlds. By tracking population average fitness values, it is possible to tell when the use of learning has spread through the population, as we showed in Figure 2. Initially, the average fitness quickly rises to a plateau where the fixed sweet-smell to eating unit connection is present in most of the creatures (the *Chromignarus* species). After remaining fairly level at this average fitness value for possibly many more generations, the average population fitness again jumps, indicating that the learnable connection from a color unit has penetrated the population (the *Chromodiscipulus* species). After this jump, the average fitness again levels out around its final highest value. Thus recording when fitness jumps occur can tell us when the different creature designs predominate in the population.

9 Analysis of the Adaptive Problem at Different Smell Accuracies

No matter what network design has evolved, adaptiveness in this scenario depends critically on smell accuracy. If the smell-sense is 50% accurate (chance level), so that food smells sweet half the time and sour the other half, then no useful information can be gained from any smell-sense units. Can the creatures still evolve to behave adaptively? No, because there would be no evolutionarily stable information for the system to build upon. The color-blind eater design with a fixed connection from smell to eating will have an average fitness of zero in such a world, since there's no systematically helpful information for it to access. Any color-sensing creature designs with only a fixed connection from a color unit to the motor unit will also have zero average fitness across the two types of color patches in the world: in one patch, a given such design would do perfectly (always eating food and avoiding poison), but in the other kind of patch, it would be wired up exactly the wrong way and do terribly, always eating poison and avoiding food, so that the genotype's average performance would be nil. Nor would a learnable connection from a color unit to an eating unit be of any help: there would be no way of bootstrapping the learning by using a statistically dependable association between smell and eating.

If smell accuracy is 100%, the color-blind eater design, *Chromignarus*, will take over the population and no design could possibly do better. With the association between smell and substance type both evolutionarily stable and completely accurate, nothing more is required for maximally adaptive behavior. Food will always smell sweet and poison always sour, so the creatures should evolve with just two utilized units, a sweet-sensing unit and an eating unit, connected by a strong positive weight (and both with zero or negative biases) -- exactly the *Chromignarus* design. This setup is sufficient to always make the right decision: eat when there's a sweet smell, since it's guaranteed to be food, and don't eat when there's a sour smell, from poison. No additional color information is needed.

But we now wondered what would happen for smell accuracies *between* 50% and 100% -- would learning evolve for all of these accuracies, and if so, how quickly? We proceeded to run many populations for accuracies in this range to explore these questions. To determine whether or not in fact learning evolved in each population, we used the average fitness plateau-monitoring method mentioned in the previous section.

The easiest way to discover when a population has evolved networks that learn color associations adaptively is to compare population average fitness at each generation with the average fitness that would be expected if the creatures did not learn to use color information. That is, if the creatures are on average doing better than creatures without color learning typically could do, then a large number of the creatures have most likely (barring large random effects) evolved to learn to use color. Computing expected fitness without color learning for different levels of smell accuracy is relatively easy.

In general, for a creature with a sweet-smell unit linked to an eating unit by a fixed positive connection, expected fitness without color learning is just $E(F) = P(f)*F_f*P(s|f) + P(p)*F_p*P(s|p)$, where $P(f)$ is the probability of food, F_f is the fitness benefit of eating food, $P(s|f)$ is the probability of food smelling sweet (a sweet smell given the presence of food), that is, the smell accuracy, $P(p)$ is the probability of poison, F_p is the fitness decrement from eating poison, and $P(s|p)$ is the probability of poison smelling sweet. Since $P(f) = P(p) =$ 0.5 (food and poison appear equally often), $F_f = +10$ points, $F_p = -10$ points, and $P(s|p) = (1 - P(s|f))$, expected fitness varies simply with smell accuracy. For example, if smell accuracy is 100%, then $E(F) = (0.5)(+10)(1.0) + (0.5)(-10)(0.0) = 5$. For 90% accuracy, we get $E(F) = (0.5)(+10)(0.9) + (0.5)(-10)(0.1) = 4$. Likewise, smell accuracies of 80%, 70%, 60%, and 50% result, respectively, in expected fitnesses of 3, 2, 1, and 0, without color learning.

Calculating expected fitness *given* color learning is more complex, depending not only on smell accuracy (which determines ease of learning), but on learning rate and number of life-steps. Given any inaccuracy in the smell-sense, creatures will start out performing relatively poorly, until the connection-strength from their color-sensing unit to the eating unit has been built up (to override the imperfect smell-sense). Thus their expected fitness even with color learning will generally be lower than the maximum achievable with perfect behavior, here 5 (the expected fitness computed above for 100% smell accuracy). Longer lifespans tend to raise the expected fitness ceiling by making the period of improper behavior before complete learning relatively shorter. Various learning rates tend to affect the expected fitness in a non-monotonic way, with a trade-off between noisy weight oscillation given high learning rates and slow weight build-up given low ones. (Adding 'momentum' to the Hebbian learning rule would tend to smooth the biased random walk build-up of weights, speed up learning, and result in higher expected fitness.) Thus, many factors combine in still-uncharacterized ways to determine the expected average population fitness once color learning has evolved; however, these considerations are unnecessary for our present goal of simply determining *when* learning has evolved in the population.

10 Parametric Exploration of the Learning Scenario

For each of 17 smell accuracies between 50% and 100%[6] we ran 20 populations of 100 individuals for 1500 generations each. Figure 3 shows how many generations it took each population to make each of the two jumps to new fitness-plateaus. These two jumps correspond to the widespread appearance of *Chromignarus* (without color learning) -- indicated by asterisks -- and of *Chromodiscipulus* (with color learning) -- indicated by bullets. The bottom curve shows the average number of generations taken to evolve *Chromignarus* across the 20 runs at each accuracy level, and the top curve indicates the analogous average generations to evolve *Chromodiscipulus*, and thus learning itself.[7]

---

[6] 15 approximately evenly spaced accuracy levels from 52% to 92% inclusive.

[7] If *Chromodiscipulus* had not evolved by the 1500-generation simulation limit for a particular run, we simply took that limit as the evolution time for color learning. Thus, for accuracies that tend to produce many 'failed evolutions', the average we plotted underestimates the actual evolution time.

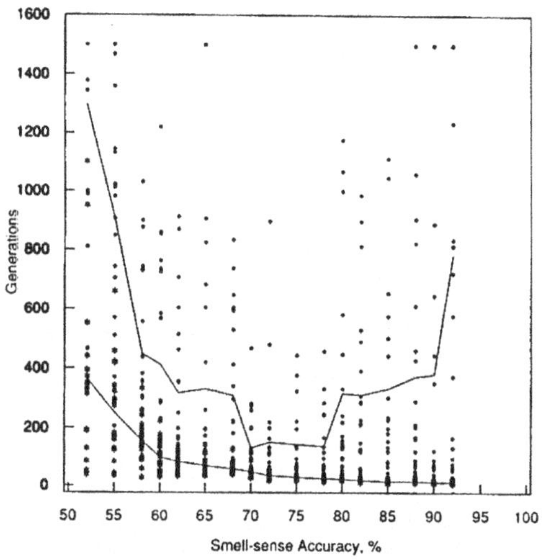

Figure 3. Generations to evolve *Chromignarus* (without color learning) and *Chromodiscipulus* (with learning) plotted against smell-sense accuracy. Asterisks indicate time to evolve *Chromignarus* for each of 20 runs at each of 17 smell-sense accuracy levels; the dotted line indicates average time across the 20 runs at each smell accuracy. Bullets indicate time to evolve *Chromodiscipulus* for the same 20 runs at each accuracy level; the solid line indicates average time across the 20 runs at each accuracy. Note that average time to evolve *Chromignarus* decreases monotonically as smell-sense accuracy increases, but average time to evolve *Chromodiscipulus* follows a U-shaped Learning Evolution Curve.

As indicated in section 8, the fixed smell connection (the Chromignarus design) evolves rapidly, in less than 100 generations for most accuracy levels. The greater accuracy of smell, the more quickly the fixed smell connection spreads, because the adaptive advantage to be gained from evolving it -- that is, the adaptive pressure -- increases. At 90% accuracy, for example, the fixed connection will allow the creatures to get 90% of their decisions right, and thus will be a very useful trait. At 60% accuracy, though, the connection only gives the creatures a 10% increase over chance performance. In this case the connection is not nearly so useful, and will not give those creatures who possess it much of a fitness edge. When smell accuracy is lower, natural selection has a harder time distinguishing creatures with the appropriate fixed connection from creatures without. As a result, lower accuracy leads to slower evolution of *Chromignarus.* In the limit, the generations taken to evolve *Chromignarus* should asymptote to infinity when accuracy is 50%. The lower (dotted) curve in Figure 3 shows some suggestion of this tendency, but our 1500-generation limit leads to underestimating average time to evolve *Chromignarus* at the lower accuracies.

More interesting is the effect of smell accuracy on time taken to evolve color learning (in *Chromodiscipulus*). Here we found an unexpected U-shaped relationship: color learning evolved most quickly for smell accuracies around 75%, and took longer and longer for accuracies diverging on either side of that middle range (as shown by the upper, solid, curve in Figure 3). For example, it takes about three times longer for color learning to evolve with 60% *or* 90% smell accuracy, than with 75% accuracy. For both 50% accuracy (chance) and 100% accuracy, time to evolve color learning appears to asymptote to infinity (as suggested by the analysis in section 9). In further simulations, we have found that this U-shape is quite robust over a variety of population sizes and creature lifespans.

11 Theoretical Interpretation of the U-Shaped Learning Evolution Function

This U-shaped learning-evolution function may be a quite general pattern across certain adaptive learning scenarios. Such a relationship was not anticipated when we devised the learning scenario, so we have since endeavored to explore its significance both theoretically and empirically, by considering the factors thought to underlie it, and by manipulating those factors and observing the resulting evolutionary patterns. For expository convenience, the U-shaped relation between smell accuracy and time taken to evolve color learning will hereafter be referred to as the *learning evolution curve*, or LEC.

We view the U shape as emerging indirectly from a trade-off between the phylogenetic adaptive pressure to evolve learning (during species-wide evolution), and the ontogenetic ease of learning (during each individual creature lifetime). These two forces have roughly mirror-image effects on the time it will take to evolve learning as we move away from middle levels of smell-sense accuracy (75%), each being high (infinite in fact) at one end of the 50%-100% accuracy scale, and low at the other. They combine to produce the symmetrical U shape we see in Figure 3.

Consider first the phylogenetic adaptive pressures for learning. As outlined previously, *Chromignarus* performs poorly at low smell accuracies. Evolving color learning in such cases would add significantly to this creature's fitness, by overcoming its error-prone, color-blind, smell-guided behavior. But the higher the smell-sense accuracy, the less color learning will be able to add to the average fitness, because the smell-sense alone will yield increasingly adaptive behavior. Thus, the phylogenetic adaptive pressure to evolve color learning peaks when smell accuracy is closest to chance (50%), and decreases as accuracy improves, ending up at zero when smell accuracy is perfect (100%).

Now consider the ontogenetic problems associated with actually doing color learning. Although *Chromignarus* in a low smell accuracy world would benefit greatly from color learning, that very inaccuracy makes learning difficult. The changes in the weight from the color unit to the motor unit are dependent on the correlation between the motor unit's activity, controlled by the sweet-smell unit, and the activity of the particular color-sensing unit; the lower the smell accuracy, the smaller this correlation will be, and the smaller the correlation, the longer it takes for the weight changes to build up in the appropriate direction. In fact, at low accuracies, *Chromodiscipulus*' finite lifespan may elapse before it has time to build up a weight from its color unit to its motor unit that is strong enough to begin to allow color information to override the inaccurate smell information. But conversely, at high smell accuracies, the correlation between the eating activity and the color will be large, since the motor unit will come on (via its sweet-smell unit connection) almost exclusively in the presence of food, which is always one particular color. As a result, the weight changes on the learnable color connection will add up quickly in the proper direction. Thus, the ease of learning color-guided behavior peaks when smell accuracy is highest (100%), and becomes increasingly more difficult as accuracy reaches chance (50%).

All else being equal, when the adaptive pressure to evolve learning is high, the number of generations required to evolve learning should be low. Similarly, all else being equal, when the ease of learning is high, the number of generations required to evolve learning should be low. The mapping from each of these two factors in isolation to the time required for evolution thus could be a reciprocal function. If this is the case, then since the adaptive pressure to evolve color learning is zero when smell accuracy is perfect (100%), the time taken to evolve color learning (based on this first factor alone) should be infinite. At the other end of the accuracy scale, the adaptive pressure peaks, and so the amount this factor contributes to the time to evolve learning will be small. Conversely, since ease of learning is zero (i.e. learning is impossible) when smell accuracy is at chance (50%), the time taken to evolve color learning (based on this second factor alone) will also be infinite. And at 100% accuracy, where ease of learning peaks, the amount *this* factor contributes to learning evolution time will also be small. When the two factors combine, whether additively or multiplicatively, the factor indicating infinite time to evolve learning will win out. This creates the two arms of the U shape in the LEC which asymptote to infinity, at 50% and 100% accuracies. In the middle ranges of the smell accuracy, though, the two factors contribute relatively small times to the evolution of learning, and their combined effect is also relatively small. This gives the bottom of the U, at around 75% smell accuracy.[8]

To spell this out a bit further, at low smell accuracies, the difficulty of gaining benefit from learning offsets the (large potential) benefit itself, leading to increased time for learning to evolve. At high smell accuracies, in contrast, there is little adaptive pressure to evolve color learning, although color learning would be easy to perform. Natural selection cannot distinguish *Chromignarus* from *Chromodiscipulus* if they are both doing almost perfectly; this 'ceiling effect' will keep *Chromodiscipulus* from proliferating, so again color learning will take a long time to evolve.

But for middle smell accuracies, color learning is relatively adaptive and relatively easy. Color learning gives a significant fitness increase over using smell alone, and learning can occur fairly quickly, since the eating unit comes on rather more often to food than to poison. Mid-level smell accuracy represents a happy medium between phylogenetic adaptive pressure and ontogenetic ease of learning, leading to the rapid evolution of color learning and its spread through the population.

[8] Given our learning rate and lifespan, the two evolution-time factors happened to combine in such a way that they yielded a roughly symmetric LEC about 75% accuracy. The LEC could undoubtedly be skewed in either direction by changing the learning rate and/or lifespan, which would affect ease of learning, and hence the relative contributions of the two factors.

12 Planned Extensions and Future Research

The implications of the U-shaped learning evolution curve have yet to be fleshed out. Just how general it is across parameter variations, and how it changes with manipulation of learning rates, lifespans, genetic operators and representations, and informational structure in the environment all need to be explored. More importantly, though, we seek evidence of this type of relation in actual evolutionary processes in terrestrial environments. Such empirical verification of our results will be the truest sign of their significance.

In further work, our theoretical motivation will continue to be the exploration of adaptive agency and interactions among adaptive processes; our methodological strategy will continue to focus on the search for simple, elegant scenarios that reveal unanticipated, potentially general patterns and dynamics underlying adaptive agency. We intend to focus on the evolution of learning for a variety of adaptive purposes, beginning with issues of foraging and energy recruitment. This work will serve to provide further support for the emerging models of ecological learning theory. But we also hope that, through careful and thorough simulation of the processes embodied in adaptive agency, a general theory concerning the interaction of adaptive processes might emerge -- a concrete understanding of the interactions among adaptive pressures, cue structures in different environments, genetic representations and operators, developmental mechanisms, learning, information-processing, and behavior-generation.

Acknowledgements

This research was partially supported by National Science Foundation Graduate Fellowships to both authors. Any opinions, findings, conclusions, or recommendations expressed in this publication are those of the authors and do not necessarily reflect the views of the National Science Foundation.

We wish to thank Dave Rumelhart, Roger Shepard, the Evolutionary Psychology Group at the Center for Advanced Study in the Behavioral Sciences, and a population of other colleagues who have influenced and guided this research along the way.

References

Ackley, D.H., & Littman, M.S. (1990). Learning from natural selection in an artificial environment. In *Proceedings of the International Joint Conference on Neural Networks* (pp. 189-193). Hillsdale, NJ: Lawrence Erlbaum Associates.

Belew, R.K. (1989). *Evolution, learning, and culture: Computational metaphors for adaptive algorithms* (Tech. Rep. No. CS89-156). San Diego: University of California, Computer Science and Engineering Department.

Belew, R.K., McInerney, J., & Schraudolph, N.N. (1990). *Evolving networks: Using the genetic algorithm with connectionist learning* (Tech. Rep. No. CS90-174). San Diego: University of California, Computer Science and Engineering Department.

Bower, G.H., & Hilgard, E.R. (1981). *Theories of learning* (5th ed.). Englewood Cliffs, NJ: Prentice-Hall.

Cosmides, L., & Tooby, J. (1987). From evolution to behavior: Evolutionary psychology as the missing link. In J. Dupre (Ed.), *The latest on the best: Essays on evolution and optimality*. Cambridge, MA: MIT Press/Bradford Books.

Davey, G. (1989). *Ecological learning theory*. London: Routledge.

Dawkins, R. (1976). *The selfish gene*. Oxford, UK: Oxford University Press.

Dawkins, R. (1983). Universal Darwinism. In D.S. Bendall (Ed.), *Evolution from molecules to men*. Cambridge, UK: Cambridge University Press.

Grefenstette, J.J. (1987). *A user's guide to GENESIS* (Tech. Rep.). Washington, DC: Naval Research Laboratory, Navy Center for Applied Research in Artificial Intelligence.

Hinde, R.A., & Stevenson-Hinde, J. (1973). *Constraints on learning*. London: Academic Press.

Hinton, G.E., & Nowlan, S.J. (1987). How learning guides evolution. *Complex Systems, 1*, 495-502.

Holland, J. H. (1975). *Adaptation in natural and artificial systems*. Ann Arbor, MI: University of Michigan Press.

Immelmann, K. (1984). The natural history of bird learning. In P. Marler & H.S. Terrace (Eds.), *The biology of learning*. Berlin: Springer-Verlag.

Knudsen, E.I. (1988). Sensitive and critical periods in the development of sound localization. In S.S. Easter, Jr., K.F. Barald, & B.M. Carlson (Eds.), *From message to mind: Directions in developmental neurobiology*. Sunderland, MA: Sinauer Associates.

Langton, C.G. (Ed.). (1989). *Artificial life*. New York: Addison-Wesley.

Linsker, R. (1988). Self-organization in a perceptual network. *Computer, 21*, 105-117.

Linsker, R. (1990). Perceptual neural organization: Some approaches based on network models and information theory. *Annual Review of Neuroscience, 13*, 257-281.

Lorenz, K.Z. (1970). Der Krumpans in der Umwelt des Vogels. In K.Z. Lorenz (Ed.), *Studies in animal and human behaviour, Vol. 1* (pp. 101-258) (R. Martin, Trans.). Cambridge, MA: Harvard University Press. (Reprinted from *Journal of Ornithology*, 1935, *83*, 137-213)

Lorenz, K.Z. (1966). *Evolution and modification of behavior*. Chicago: University of Chicago Press.

Mayr, E. (1974). Behavior programs and evolutionary strategies. *American Scientist, 62*, 650-659.

Menzel, R. (1984). Biology of invertebrate learning (Group Report). In P. Marler & H.S. Terrace (Eds.), *The biology of learning*. Berlin: Springer-Verlag.

Miller, G.F., & Todd, P.M. (1990). Exploring adaptive agency I: Theory and methods for simulating the evolution of learning. In D.S. Touretzky, J.L. Elman, T.J. Sejnowski, & G.E. Hinton (Eds.), *Proceedings of the 1990 Connectionist Models Summer School*. San Mateo, CA: Morgan Kaufmann.

Miller, G.F., Todd, P.M., & Hegde, S.U. (1989). Designing neural networks using genetic algorithms. In J.D. Schaffer (Ed.), *Proceedings of the Third International Conference on Genetic Algorithms* (pp. 379-384). San Mateo, CA: Morgan Kaufmann.

Pulliam, H.R., & Dunford, C. (1980). *Programmed to learn*. New York: Columbia University Press.

Shepard, R.N. (1987). Evolution of a mesh between principles of the mind and regularities of the world. In J. Dupre (Ed.), *The latest on the best: Essays on evolution and optimality*. Cambridge, MA: MIT Press/Bradford Books.

Shepard, R.N. (1989). Internal representation of universal regularities: A challenge for connectionism. In L. Nadel, L.A. Cooper, P. Culicover, & R. M. Harnish (Eds.), *Neural connections and mental computation*. Cambridge, MA: MIT Press/Bradford Books.

Singer, W. (1984). Learning to see: Mechanisms in experience-dependent development. In P. Marler & H.S. Terrace (Eds.), *The biology of learning*. Berlin: Springer-Verlag.

Singer, W. (1988). Ontogenetic self-organization and learning. In J.L. McGaugh (Ed.), *Brain organization and memory: Cells, systems, and circuits*. New York: Oxford University Press.

Staddon, J.E.R. (1983). *Adaptive behavior and learning*. Cambridge, UK: Cambridge University Press.

Stork, D.G., & Keesing, R.C. (1990). *Evolution of initial weights for neural networks*. Talk presented to the Stanford PDP Research Group, Stanford, CA.

Tinbergen, N. (1951). *The study of instinct*. Oxford: Clarendon Press.

Todd, P.M., & Miller, G.F. (in press). Exploring adaptive agency III: Simulating the evolution of habituation and sensitization. In *Proceedings of the First International Conference on Parallel Problem Solving from Nature*. Berlin: Springer-Verlag.

Todd, P.M. (1988). *Evolutionary methods for connectionist architectures*. Psychology Department, Stanford University. Unpublished manuscript.

Weiss, G. (1990). *Combining neural and evolutionary learning: Aspects and approaches* (Tech. Rep. No. FKI-132-90). Munich: Technical University of Munich.

Lookahead Planning and Latent Learning in a Classifier System

Rick L. Riolo
University of Michigan
Ann Arbor, MI 48109 USA
rick_riolo@um.cc.umich.edu

Abstract

Classifier systems (CSs) have been used to simulate and describe the behavior of adaptive organisms, animats, and robots. However, classifier system implementations to date have all been reactive systems, which use simple S-R rules and which base their learning algorithms on trial-and-error reinforcement techniques similar to the Hullian Law of Effect. While these systems have exhibited interesting behavior and good adaptive capacity, they cannot do other types of learning which require having explicit internal models of the external world, e.g., using complex plans as humans do, or doing "latent learning" of the type observed in rats. This paper describes a classifier system that is able to learn and use internal models both to greatly decrease the time to learn general sequential decision tasks and to enable the system to exhibit latent learning.

1 Introduction

Classifier systems (CSs) have been used to understand, through metaphor and simulation, the behavior of adaptive organisms and robots from animats ([Holland and Reitman, 1978], [Booker, 1982], [Wilson, 1985]) to rabbits [Holyoak *et al.*, 1990] to humans [Holland *et al.*, 1986]. However, all CSs implemented to date have been *reactive* systems, i.e., they store and use knowledge as rules of the form "If the situation is X, do A" (where X may describe a set of world states). In the terms of animal psychology, the system stores S-R associations [Walker, 1987]. These systems all have used a trial and error learning technique, the ***bucket brigade algorithm (BBA)***,[1] to assign priorities to rules; the priorities determine which rules will fire and what the system will do in a given situation. The BBA is reminicent of the Hullian "Law of Effect," i.e., rules active when reward is received (from the environment or from subsequently active rules) have their priorities modified in proportion to the reward. (They use a more sophisticated learning algorithm, the ***genetic algorithm***, to form generalizations over the space of situations, i.e., to form concepts.) These CSs use only a very simple ***model*** of the world, in which a rule's priority in effect ***predicts*** the reward expected if that rule is fired. Despite using such simple representational and learning techniques, CSs have shown surprisingly interesting behavior when controlling animats which must learn and adapt to survive in simple enviroments; they have also been used to solve concept learning [Wilson, 1987a], dynamic control [Goldberg, 1988], and sequential decision problems ([Grefenstette, 1988],[Booker, 1989]).

A second type of learning involves building more complex models which not only predict rewards, but also ***predict world states***.[2] These models can be implemented as rules of the form "If the situation is X, and I do A, then expect situation Y," i.e., S-R-S associations. Systems can create and update these models continuously, even when no rewards are being received, by predicting the (non-reward) outcomes of actions and then modifying the model when the predictions are incorrect. That is, rather than just using the usually infrequent feedback provided by rewards or punishments to build simple S-R models, systems can exploit the flood of non-reward experiences they have to build much more complete models of the world. Predictions of expected states then can be integrated with motivations and predictions of rewards to choose actions that lead to goals. Using internal models enables systems to reduce the number of trials required to learn tasks ([Sutton, 1990], [Whitehead and Ballard, 1989]). Further, a model enables a system to easily integrate newly acquired knowledge about the world or about changes in goals or motivations ([Holland *et al.*, 1986], [Dickinson, 1980]). Internal models also have been used to simulate Piagetian cognitive development during infancy ([Drescher, 1986], [Drescher, 1989]).

[1] The bucket brigade algorithm is a temporal difference method [Sutton, 1988].

[2] Unless stated otherwise "internal model" and "model" will refer to this more complex type of model.

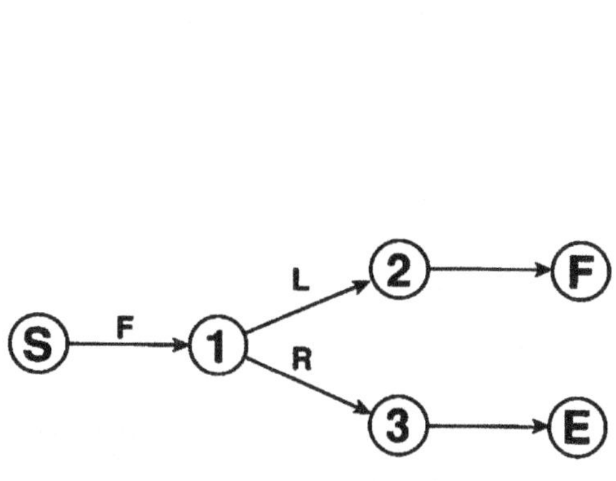

Figure 1: A simple maze for latent learning experiments.

Internal models are the sine qua non of the ability to formulate and execute complex plans to achieve goals, as done by humans and other higher animals. The use of internal models can also be observed in lower animals, e.g., rats. One of the first demonstrations of the use of models in animals was the "latent learning" experiments done by Tolman and his students (for reviews, see [Mackintosh, 1974] or [Walker, 1987]). A basic latent learning experiment proceeds as follows (ignoring the necessary control details): In a pre-reward phase, satiated rats are allowed a number of trials in the maze shown in Figure 1, where a trial consists of being placed in the start box S, making a choice to go left or right, and then being removed from one of the end boxes, F or E, depending on the choice made. (there is no food in either end box). Next the rats are not fed for 24 hours, placed *directly* in one of the end boxes with food in it, and allowed to eat for one minute. Then the rat is placed in box S and the direction the rat chooses on the *first* post-eating trial is recorded, i.e., toward the box with the food or not. Typical results are: (1) Rats not given pre-reward trials in the maze choose to go toward the box with food 50% of the time, as expected if they guess with no information; (2) Rats allowed pre-reward trials but with identical end boxes again go toward the box with the food 50% of the time; (3) But rats allowed pre-reward trials with boxes that are easily distinguishable for a rat (e.g., one has white walls and the other has dark)[3], choose the path to the box with food 90% of the time, even though that choice of turn-left (or turn-right) has never led to a food reward!

There are three things to note about this experiment. First, in the pre-reward phase the rats clearly learned the structure of the maze without explicit rewards, contrary to any simple version of the Law of Effect. Second, and more important, what they learned were S-R-S type associations, e.g., "If in Start, and turn Left, predict the Black walls." And third, the rats immediately were able to attain the food (goal) when they learned of its association with the Black walls box; in effect, they performed a kind of *lookahead search* of their model of the maze, and used the value found associated with box F to bias the choice to go left from the start box. In other words, they used a "practical inference" [Dickinson, 1980] to integrate the information about the structure of the maze (learned in pre-reward training) with the information about the location of the food (learned when put in F and allowed to eat), and to use that integrated information to make a goal-directed choice. This type of inference could not be made by a system that stores and uses only S-R (reactive) rules, e.g., any of the CSs referenced above.

In this paper I will describe CFSC2, a CS which is able to learn and use internal models both to greatly decrease the time to learn general sequential decision tasks and to enable the system to exhibit latent learning of the type observed in rats. Section 2 describes the basic architecture of CSs. Section 3 describes CFSC2, including how internal models can be implemented and learned. Section 4 presents results using the CFSC2 system, and Section 5 discusses these results, the limits of the implementation used, and directions for future work.

2 Classifier System Architecture

This section describes the basic architecture and operation of "Holland" (or "Michigan") classifier systems. For a more detailed description of this and other types of CSs, see [Booker *et al.*, 1989] or [Goldberg, 1988].

Classifier systems are parallel, message-passing rule-based systems with four basic parts:

- A *message list*, which acts as a "blackboard" for communications and short term memory. Messages are fixed length binary strings (say of length 32). Typically the message list has a fixed size.
- A *classifier list*, which consists of condition-action rules called *classifiers*.[4] The condition and action parts are strings of the same length as the messages, built from the alphabet $\{0, 1, \#\}$. In the condition part, the # acts as a wildcard in the match process. A classifier's condition part is *satisfied* when its condition is matched by one or more messages on the message list. A satisfied classifier may produce one message. The message is produced by using the action part as a "mask" applied to one of the messages that match its condition; i.e., where the action has a 1 or 0, that value appears in the new message, and where the action has a #, the corresponding value from the matching message is "passed-through" to the new message. Each classifier also has an associated *strength*, which is related to its usefulness in attaining rewards for the system.[5]

[3] The obvious controls have been done, e.g., to control for rats bias to go to dark spaces over light.

[4] Classifiers can have one or more conditions, but in this paper I will discuss only one-condition rules. In fact, CFSC2 uses two-condition rules, but by creating rules with identical conditions, they can be treated as one condition rules.

[5] Some systems also associate a *specificity* value with each rule, which is a measure of the generality of the classifier's condition.

- An *input interface*, which provides the CS with information about its environment, by placing one or more *detector* messages on the message list.
- An output interface, which provides a way for the CS to communicate with or change its environment, by interpreting (some of) the messages posted by rules as commands to the interface.

Note that *learning* CSs also must have a *reward* mechanism, as found in all reinforcement learning systems, providing feedback from the environment in the form of a scalar value.

Detector messages typically start with some fixed bits (e.g. "001") which identify, or *tag*, the message as from the detectors. Similarly, the effectors will interpret only messages that start with some other tag (e.g. "111") as commands to the interface. For example, a rule like:

001011##/111100##

will respond to detector messages like 00101101 (which may mean there is a large object to the right) and produce an effector message 11110001 (which may be interpreted as the command "turn left"). Tags are a general way for a CS to establish communications and control, i.e., a semantics, since only rules that match the tag bits will respond to a given message and only rules that post messages with the appropriate tags can control the interface or activate other rules. Tags will play a critical role in representing models in CSs(see Section 3.1).

To see how a CS interacts with a task domain, consider the maze task in Figure 1. This maze could be simulated as a finite state world in which each box is a state and the paths between boxes are transitions. For such a world, the CS's input interface could post one detector message that indicates which state the system is in. Classifiers could produce effector messages that specify available actions, e.g., move left, move forward, eat, etc. For each trial, the system is started in state S and it moves to one of the end states F or E; the system is then returned to S for another trial. The task is to move from S to F and eat the food there, in which case the system receives a positive reward (e.g., 1000).

In such a task domain, a CS is run by repeatedly executing the steps shown in Figure 2. When a competition is run to determine which classifiers are to be activated, the probability that a given (satisfied) classifier i will win is:

$$Prob(\ i\ wins\ ) = B_i(t) / \sum_j B_j(t)$$

where $B_i(t)$, the *bid* of classifier i at step t, is:

$$B_i(t) = k * S_i(t) * \psi_i(t)$$

k is a small constant (e.g., 0.1), which acts as a "risk factor," and $S_i(t)$ is the classifier's strength at step t.

1. Add the detector message indicating the current state to the message list.
2. Compare all messages to all classifiers and record matches.
3. Generate new messges by activating satisfied classifiers. If activating all the satisfied classifiers would produce more messages than will fit on the message list, a competition is run to determine which classifiers are to be activated. Classifiers are chosen probabilistically, without replacement, until the message list is full. The probability that a given classifier is activated is proportional to its *bid*.
4. Process the new messages through the output interface, resolving conflicts and selecting one response, r, to be used for the current time step.
5. Apply the *bucket brigade algorithm*, to redistribute strength from the environment to the system and from classifiers to other classifiers.
6. Apply *discovery algorithms*, to create new classifiers and remove classifiers that have not been useful.
7. Replace the contents of the message list with the new messages, and return to step 1.

Figure 2: The basic CS execution cycle.

$\psi_i(t)$ is the relative *support* for classifier i at t, which is a measure of the number of messages matching the classifier and the amount bid to post those messages. Support provides a way to bias activity towards rules that respond to many messages, even if the rule has low strength. That is, rules that match the current situation well can win the competition even if they have a strength lower than other bidders. (For a further discussion of support, see [Holland *et al.*, 1986].)

Note that the output interface of a CS may have to resolve conflicts, e.g., when one classifier produces a message that means "turn left" and another produces a message that means "turn right." Basically, when there are conflicts the response r is chosen probabilistically, with the probability that $r = r'$ equal to:

$$\sum_{m'} B_{m'}(t) / \sum_m B_m(t)$$

where m' ranges over the effector messages that set r to r', m ranges over all effector messages, and $B_{m'}(t)$ (or m) is the bid of the classifier that posted message m' (or m). The winning value r' is used to determine which transition is executed.[6] Once an effector value is chosen, all messages that are inconsistent with that setting are deleted from the new message list.

There are two ways CSs can learn, one a short term and one a longer term learning process. In the short

[6] In this and other tasks described in this paper, the results of responses are deterministic.

term, the strength of existing rules can be adjusted so that rules that lead to good performance are more likely to win the competition. Adjusting strengths is done by the *bucket brigade algorithm (BBA)*. In the longer term, new rules can be added to the system, and rules that are not useful can be removed. Creating and removing rules is done by the *rule discovery algorithms*.

The BBA is basically a type of temporal difference method [Sutton, 1988] for updating rule strengths.[7] When a classifier posts a message, its strength is reduced by its bid. When a reward is received, *all* classifiers that posted effector messages (after any effector conflicts are resolved) have the payoff added to their strengths. Thus when the activation of a classifier tends to be *directly* associated with a high reward from the environment, that classifier's strength is increased. The BBA also redistributes strength from classifiers to other classifiers. In particular, when a classifier posts a message, the amount it bid (or a share of it) is added to the strength of the classifiers that made it possible for that classifier to become active. In summary, the strength $S_i(t+1)$ of a classifier i at $t+1$ is:

$$S_i(t+1) = S_i(t) + R(t) + P_i(t) - B_i(t)$$

where $S_i(t)$ is the strength of classifier i at t, $R(t)$ is the reward from the environment at step t, $P_i(t)$ is the sum of all payments to classifier i from classifiers that matched messages produced by i during the previous step, and $B_i(t)$ is the bid of i. It can be shown that a classifier's strength reaches a fixed-point when the amount it receives at t is equal to the amount it pays.

Most classifier systems use a number of rule discovery operators, but the most novel is the *genetic algorithm (GA)*. Essentially the GA creates new rules by combining parts of rules that have high strength; typically these new rules then replace rules with lower strengths. (Thus the GA will work only if the BBA allocates strength properly.) The GA will not be used in the results presented in this paper, so it will not be described further.

3 Lookahead in Classifier Systems

For a system to do lookahead, it must have:

- A model of the environment, including a way to represent states and transitions between them.
- A way to represent the value (expected rewards) of actions and/or states.
- A mechanism to do lookahead planning, i.e., a way to consider possible future states, actions and rewards and to use that information to influence the choice available at the current state.

[7] The BBA is also related to the Q-Learning techniques described by [Watkins, 1989].

Traditional CSs like those described in the previous section do represent the current state (as rules matching detector messages) and the value of actions from that state (as the strengths of those rules). However, those systems were not able to do lookhead because (1) they do not represent transistions (i.e., explicit S-R-S associations) and (2) they choose an action every cycle (and so allow no time for lookahead). Further, because those systems use only one value, strength, to indicate the worth of a rule and to calculate its probability of being active when matched, they do not have a way to distinguish the *prescriptive value* of a rule (what reward can be expected if it is fired) from the *predictive accuracy* of a rule (how well the rule predicts what state the system will be in next). Thus in a latent learning experiment, the rules that represent the "latent" model of the world, before rewards are received, would have no strength and so would be removed by rule discovery operators like the GA.

In the following subsections I will describe CFSC2, a CS that can do lookahead. CFSC2 was implemented by making just a few changes to the traditional CS architecture. The key differences/extensions required are: (1) Allowing the system to execute more than one cycle per action, so that it can examine its model before acting. (2) Using rule discovery operators that create rules that use tags in a coordinated way to represent a model of the world. (3) Using two "strength" values (rather than one) to distinguish the long term value of a rule from a short term, context dependent value determined by the current lookahead process. (4) Using a third "strength" to represent the predictive accuracy of the rule, and adding a mechanism to update predictive strengths.

3.1 Representing an Internal Model

This section describes how CFSC2 uses tags to represent internal models of its environment. A *tag* is any set of bits in messages and rules that by convention have a certain meaning. For example, detector messages all have a prefix of "001" to indicate they are produced by the input interface; the rest of the bits may carry information about the attributes of the current state. Similary, effectors only respond to messages that have a prefix of "111"; thus any message that starts with that prefix means "emit some response *r*" (where other bits in the message determine exactly which response is emitted).

One way to represent a model of the world is in terms of states and transitions between states.[8] In CSs, the classifiers are the only form of long term memory, so the states and transitions of a model must be implemented as rules. The messages that rules match and post can be

[8] Note that in any but toy worlds, most or all model states will actually represent *many* world states. Thus model states can also be thought of as *events*, and the model will be a homomorphic or quasimorphic image of the world [Holland et al., 1986]. Because there are few states in the tasks described here, model states are one-to-one with world states.

Bits	Symbol	Meaning
00	d	detector message
01	e	event (state) message
10	a	association message
11	t	transition message

Figure 3: Four message types and tag symbols.

(1)	$d^*.0.S_S/e^*.0.S_S$	(5)	$e\#.0.S_1/t\#.R.S_3$
(2)	$e\#.0.S_S/t\#.F.S_1$	(6)	$t\#.\#.S_2/e.0.S_2$
(3)	$t\#.\#.S_1/e.0.S_1$	(7)	$e\#.0.S_2/t\#.F.S_F$
(4)	$e\#.0.S_1/t\#.L.S_2$	(8)	$t\#.\#.S_F/e.0.S_F$

Figure 4: Rules to model part of the world in Figure 1. The actions abbreviations are R (Right), L (Left), and F (Forward).

thought of as activation of parts of a model, i.e. a short term memory.

All messages and rule condition/action parts will be treated as having three parts: (1) three bits for a type-tag, (2) two bits for an action, and (3) the remainning bits for event attribute values. For clarity, rules and messages are shown in symbolic form rather than as uninterpreted binary and ternary strings, and periods are used to distinguish the three parts. Actions will be shown as Right, Left, and so on. The event attribute values will be shown as the corresponding state symbols S_1, S_2, etc. The "#" will still be used as a "match-any" character in rule conditions, and to indicate "pass-through" in rule actions. Two of the type-tag bits will be used to specify four types of messages, as shown in Figure 3. The third bit in the type-tag is used to designate messages that describe "reality,", i.e. the current situation and action choices. If the bit is 1, an asterisk will appear after the type-tag symbol. For example, a detector message is "$d^*.0.S_1$" and an effector message is "$t^*.Up.S_2$" (recall that dectector and effector message tags are "001" and "111", respectively, so both are "real"). The middle part of the detector message is just zeros, since no action is specified; only "t" messages will have non-zero actions.

Simple models can be implemented with three types of rules:

1. $d^*.0.S_1/e^*.0.S_1$ Rules. When matched this rule-type represents the activation of of a model state by the detection of some event (state) in the world. The "d^*/e^*" rules transform detector messages into event messages, thus allowing an arbitrary encoding of world events into model representations. The message shown is the simple case (used throughout this paper) in which the model representation is identical to the (detected) attributes. Thus when the detector message $d^*.0.S_1$ is posted, this rule will be matched and it may post a message which means "the current event is S_1".

2. $e\#.0.S_1/t\#.Right.S_2$ Rules. The "$e\#/t\#$" rule-type is used to represent S-R-S transitions. The rule shown means "if the event is S_1, then a system response of Right will lead to event S_2". Note that the # in the type-tag of the condition means this rule can match either $e^*.0.S_1$ ("the *current* event is S_1") or $e.0.S_1$ ("the *hypothetical* event is S_1"). Further, because the # in the action part "passes through" the "reality" bit value (the asterisk or lack of one), if the matching message is $e^*.0.S_1$, then the message posted is $t^*.Right.S_2$, which not only *predicts* the result of the action "Right", it also *prescribes* that action to the output interface (since it is an effector message, i.e., is starts with the bits "111"). On the other hand, if the matching message is $e.0.S_1$, the message produced will be $t.Right.S_2$, which simply predicts the result (S_2) of going right. Thus one transition rule can be used to represent either a real (currently possible) transition or a hypothetical one, depending on the matching message.

3. $t\#.\#.S_2/e.0.S_2$ Rules. The "$t\#/e$" rule-type is used to represent activation of model states by model transitions. This rules means "If *any* transition, real or hypothetical, predicts event S_2, then post a hypothetical-event message for event S_2". This type of rule is necessary (a) to turn-off the reality bit, rather than have it passed through to all activated states and transitions, and (b) to have a single rule (and the message it posts) represent the activation of a hypothetical state, no matter how many transitions there are to it.

To see how a set of these type of rules can represent a model of the world as well as activation of parts of it, consider the world shown in Figure 1 and the rules in Figure 4, which model a part of it. Assume the current state is S_S and that the system will not actually respond (and so not change the world state) for several cycles. On the first step the input interface will add $d^*.0.S_S$ to the message list, and rule 1 will fire. On the next step, rule 1 will post $e^*.0.S_S$ and the detector will add its message again, so rules 1 and 2 will fire. On the next step, rule 3 also will fire (matching the message from rule 2). At this point the system has activated the representation of the current state S_S, its one transition, and the (predicted) state S_1. Next the system also will activate the (hypothetical) transitions from state S_1 to states S_2 and S_3 (rules 4 and 5). By continuing in this fashion, the CS can use a kind of parallel spreading activation process to activate as much of its model as it can, until it chooses a response (or its message list fills).

1. Add the detector message to the message list.
2. Record all matches and calculate bids.
3. Generate new messsges, using a bid competition if necessary.
4. Add the bid of each $t^*.r.S$ message to the support for response r. If the support for any r exceeds θ_{se}, or if more than C_{max} cycles have passed without acting, pick an action (probabilistically) and execute it.
5. Update rule strengths.
6. Apply *discovery algorithms*, to create new classifiers and (prehaps) remove classifiers that have not been useful.
7. Replace the contents of the message list with the new messages, and return to step 1.

Figure 5: The CFSC2 execution cycle.

3.2 Using an Internal Model

This section describes how CFSC2 uses the activated model described above to bias its current choice of a response. Figure 5 shows the basic execution cycle of the CFSC2 system.

The CFSC2 execution cycle is very similar to that described in Section 2. One key difference is that the system does not act every cycle step. Instead, it acts only when the support for some response exceeds a threshold, θ_{se}, or when a maximum number of cycles, C_{max}, has passed without acting. In short, CFSC2 is able to post messages and so activate its internal model until it has a good idea about what to do, or until it reaches a time-out.

Another key difference is that each rule has three strength values associated with it (rather than one):

- S_v. This is the long term value of the rule. S_v is analogous (in meaning and rate of change) to the single strength value used in traditional CSs. S_v is updated by a temporal differnce method so it reflects the rewards obtained in the past when the rule (and its successors) were fired:

$$S_v(t+1) = \delta_v S_v(t) + (1-\delta_v)[R(t) + (\gamma_v P_v(t))]$$

  $0 \leq \delta_v \leq 1$ is a constant, $R(t)$ is reward from the environment at t, $0 \leq \gamma_v \leq 1$ is a discount factor on future rewards, and $P_v(t)$ is the "S_v payment" received from rules that use the message it produced at $t-1$. In CFSC2, the $P_v(t)$ to rule i is just the maximum S_v over all rules that use any message produced by i at $t-1$. For transition rules ($e\#/t\#$), S_v is updated *only* when the system actually makes the response suggested by the rule. For other rule types, S_v is updated every step they post messages.

- S_t. This is a rapidly changing, context dependent value of a rule. S_t is updated *every* step, whether the rule is active or not *and* whether the system acts or not. In particular,

$$S_t(t+1) = max \begin{cases} \delta_t S_t(t) + (1-\delta_t)[R(t) + (\gamma_t P_t(t))] \\ \lambda_{vt} * S_v(t+1) \end{cases}$$

  $0 \leq \delta_t \leq 1$ is a constant, $R(t)$ is reward from the environment, $0 \leq \gamma_t \leq 1$ is a discount factor, and $P_t(t)$ is the "S_t payment" received, i.e., the maximum S_t over all rules that used messages it produced at $t-1$. Thus when a rule is active, its S_t rapidly comes to reflect the maximum S_v found in the rules it activates (directly or indirectly). When it is inactive, its S_t tends to $\lambda_{vt} * S_v$.

- S_p. This value (from 0.0 to 1.0) reflects the predictive accuracy of transition rules (S_p is always 1 for other rule types). For $e\#/t\#$ rules,

$$S_p(t+1) = \delta_p * S_p(t) + (1-\delta_p) * P_p$$

  $0 \leq \delta_p \leq 1$ is a constant and P_p is 1 if the rule's prediction is correct and zero otherwise. In the tasks reported here, a prediction is correct if the attribute bits S_i in the posted $t^*.r.S_i$ message match the attribute bits S_j in some $e^*.0.S_j$ message. Thus a rule will have an S_p approaching 1.0 if it accurately predicts which *model* state will be active next.

A third key difference between CFSC2 and earlier systems is the way bids are calculated. In particular, $B_i(t)$, the bid of rule i at t is:

$$B_i(t) = S_p(t) * S_t(t) * \psi_i(t)/max_j(\psi_j(t))$$

where S_p and S_t are defined above, and the third term is the support for i at t, $\psi_i(t)$, relative to the maximum support over all bidders j at t. In CFSC2, $\psi_i(t) = \gamma_I * \sum I_m$, where m ranges over the messages that match i at t, I_m is the *intensity* of message m and $0 \leq \gamma_I \leq 1$ is a discount factor. In general, I_m is the bid of the classifier that posted m; for detector messages, I_m is set to 10 times the average $S_t * S_p$, i.e., 10 times a (low) estimate of the average bid expected).

The combination of (a) using tags to link rules representing the model and (b) the strength updating algorithms mentioned above enable CFSC2 to utilize lookahead to bias its current choice toward higher-payoff futures. For example, consider the rules in Figure 4, a (partial) model of the world in Figure 1. If the system has reached state F and been rewarded there, then rule 8 will have a high S_v (and so a high S_t). If the system is returned to state S, then a "spreading activation" of rules and messages will take place, in effect doing a lookahead search through the model. The lookahead will continue until the chain of rules $2 \rightarrow 3 \rightarrow 4 \rightarrow 6 \rightarrow 7 \rightarrow 8$ are all active. As a result of the strength updating algorithms described above, S_t will begin to flow back from rule 8 through the other rules in that chain until it gets to rule

2 (and 1). As the S_t of rule 2 rises, its bid increases, which in turn means the support for action *Forward* increases until the threshold θ_{se} is reached, at which point the system will respond and go to state 1. Because rule 4 is in the chain of strengthed rules, whereas rule 5 is not, the bid of 4 will be greater than that of 5 (assuming equal support and S_p). As a result of the disparities in bids, the CS will (probably) choose the left path instead of the right, thereby receiving the food reward.

Note that a rule's S_v stores information about the results of actual experiences using that rule, not information obtained from hypothetical, lookahead processes. On the other hand, when a rule is active, its S_t is generated by the lookahead process described above, so its S_t indicates the maximum value found in that part of the model activated (directly or indirectly) by that rule. Thus S_t is much more variable than S_v, because it reflects a value generated by a dynamic and *context dependent* lookahead process. That is, other factors (e.g., new messages about the world or about motivations) can alter the S_t by changing which rules are matched and activated, even when the underlying model has not changed. The implications of these differences between S_v and S_t are discussed further in Section 5.

In summary, considering just the S_t and S_p terms[9] a rule's bid is basically the (discounted) *expected value* of that part of the model the rule represents *and* all the parts of the model that have been activated as successors in the current lookahead process. Because the bids of rules determine the outcome of the competition to post messages to a limited-size message list, the lookahead process will be biased toward higher expected payoff paths. In addition, because bids determine the outcome of the competition at the effector interface, the lookahead process will result in following paths that lead to higher payoff sequences.

3.3 Learning a Model

For CFSC2, learning a model involves two processes: first, discovering rules to represent the model states and transitions, and second, assigning proper S_v and S_p values to those rules and transitions.

Consider the second process first. As CFSC2 actually picks actions and completes trials, the S_v and S_p values of rules are updated as described in Section 3.2. Over many *real* actions and trials, S_p comes to reflect the predictive accuracy of rules and S_v comes to reflect the value of transitions and states. One implication of this type of learning is that after many real actions and trials, little or no lookahead will be required to make a choice. The reason is that since S_t is set to be the maximum of (a) the S_t values backed up by the lookahead process and (b) $\lambda_{vt} * S_v$, when S_v increases, the value $S_t = \lambda_{vt} * S_v$ will immediately indicate which action is best.

[9]There is not space to discuss the support term in detail.

CFSC2 uses two rule discovery operators to create a model of its world. Both are *triggered* by a "surprise" to the system, i.e., some event occuring that is not predicted by the system's model. Both use only local, current information (e.g., messages on the current or last message list, the most recent action taken, etc.) to create new rules of the three types described above. In the current version of CFSC2, rules are simply added to the classifier list; none are removed.

The first operator is a kind of *Cover Detector Message* operator [Robertson and Riolo, 1988], which is triggered when there is a detector message $d^*.0.S_i$ that is not matched by any rule.[10] This operator creates two new rules. One is a rule "$d^*.0.S_i/e^*.0.S_i$", which when executed represents the activation of the event S_i by information from the input interface. The second is the rule "$t\#.\#.S_i/e.0.S_i$", which represents the activation of event S_i by transitions from other model states.

The second rule discovery operator, *Create Transition*, is triggered when no rule predicts the occurence of an event. It creates a new transition rule to predict that event. In particular, after the system has acted and a new "$e^*.0.S_i$" message appears on the message list (i.e., at the onset of activity of model state S_i), if no rule has predicted it (i.e., there is no message "$t^*.r.S_i$", where r is any action), then a new rule is created that would have predicted the event. The new rule is "$e\#.0.S_j/t\#.r'.S_i$", where S_j is the attribute part from some $e^*.0.S_j$ message on the old message list (i.e., a message representing the current model state before the action), r' is the action just taken, and S_i is the attribute part from the unpredicted $e^*.0.S_i$ message. To the extent that event S_i actually follows S_j after action r', this new rule will accurately represent a transition found in the world.

To encourage the system to try novel actions, a bias was added to support action values not supported by any rule. In particular, when the system takes an action because it reached the maximum number of cycles without acting, C_{max}, one non-supported action (chosen randomly) is given a support value of θ_{se}. A response is then chosen probabilistically in the normal way. If the unsupported action is selected, there will be no "$t^*.r.S_i$" message predicting the post-action event, so the Create Transition operator will be triggered to create a new rule that will predict the results of taking the novel action.

4 Results

This section descibes how well the CFSC2 system defined in the previous section performs on three tasks. Note that no systematic effort was made to optimize parameter settings. Except where noted, results are the average of 10 runs. The values used for all experiments were $\delta_v = \delta_p = 0.9, \gamma_v = \gamma_t = \gamma_I = 0.9$, and $\lambda_{vt} = 0.1$.

[10]Wilson [Wilson, 1985] used a similar operator, which he called the "Create" operator.

The maximum payoff was 1000. The amount of lookahead is controlled by (a) the message list size, M, (b) the S_t update parameter δ_t, (c) the action support threshold, θ_{se}, or (d) the maximum steps allowed before an action is taken, C_{max}. Except where specified, to use lookahead $M = 63$, $\delta_t = 0.7$, $\theta_{se} = 100$, and $C_{max} = 20$. To prevent the use of information from lookahead, $C_{max} = 2$ and $S_t(t+1)$ is just $\lambda_{vt} * S_v(t+1)$. In all cases, the minimum number of steps per action, C_{min}, was 2, to allow time for rule discovery algorithms to work.

The first task for CFSC2 was the simple latent learning experiment described in Section 1 (and Figure 1). The system was given 50 pre-reward trials, during which the rule discovery operators created rules that implement a model of the maze. (During this phase the system chose to go to F 50% of the time.) Then it was put in state F and rewarded 10 times for "eating," so that the S_v and S_t of the rule representing F increased dramatically. Finally the system was put in state S, and its choice at state 1 on the first post-food trial was measured. This experiment was repeated 20 times for each condition studied. The following table shows the results:

Lookahead Cycles	% to Food
16 (No-Food)	52
2	45
8	80
16	95

As expected, when no food was in F, or when the time allowed for lookahead was too small, the system chose to go to F about 50% of the time. However, as more cycles were allowed for lookahead, the system chose to go to F far above random guessing, at rates comparable to those observed with rats. Thus CFSC2 is able to create a model of the maze "latently", without reward or change in behavior, which it can then use to make a good choice when reward is introduced at some distal state in the model.

The second task is a simple sequential decision task, "Gref3", which has been studied using other classifier systems ([Booker, 1989], [Grefenstette, 1988]). This world has 288 states, arranged in nine rows of 32 states each. The system is started in one of the states in the first row (picked at random). From each state, the system can go forward, right or left to a state in the next row (the states in each rows are laid out like a torus). The only rewards are for entering states in the ninth row, after which the system is returned to the first row and a new trial is started. The rewards range from 0 to 1000, in a bimodal distribution over the ninth row. The maximum reward per trial is 1000; random performance yields 250.

Note that in the Gref3 task (and the next as well), the CS is learning the structure of the world (the transitions) and the value of the goal states (the rewards) from the start of the experiment, and its performance is being measured continuously. This appears to be somewhat different from the latent learning task, in which the experimental design introduced three phases, i.e., (1) learn the structure of the world, (2) learn that a state has value associated with it, and (3) integrate that knowledge to make a choice (the measured performance). Of course these phases are somewhat artificial, in that the CS is learning the value of states in the first phase, it just is the case that the values are all zero. Similarly, the CS is trying to improve performance (receive more rewards) in all phases, it just has no way to do so until the third (test) phase.

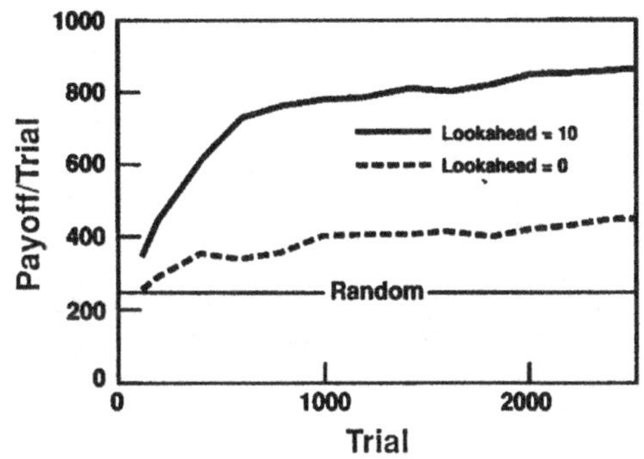

Figure 6: Performance in the Gref3 sequential descision task.

Figure 6 shows the performance (average reward per trial over 200 trials) of the CFSC2 system with lookahead ($C_{max} = 10$) and without. Lookahead clearly leads to faster learning of this task. After 2500 trials, with lookahead the average performance (for 7 runs) is 830 per trial, with one run getting up to 930. Without lookahead, the average is only about 450 (it slowly increases to 510 after 5000 trials, the longest tried). Despite the fact that $C_{max} = 10$, CFSC2 used only three times as many total cycles with lookahead as without. The reason is that after many trials, S_v is increasing further and further back from the goal row, so that less and less lookahead is needed to make a decision (i.e., reach θ_{se}).

For a comparison, after 25,000 trials Booker's Gofer system reached 850 per trial (using 200 rules), and Grefenstette's Rudi system reached 715 (using 3200 rules, divided into 50 "chromosomes" of 64 rules each). Thus CFSC2 appears to solve this task much faster than either Gofer or Rudi. However, to be fair it must be pointed out that both Gofer and Rudi in fact have a harder task, since they have far fewer rules available and so must generalize across states, whereas CFSC2 treats each state as unique (by using about 1400 rules).

Figure 7 shows a simple navigation task like those studied in [Sutton, 1990]. There are 30 states in a 5 by 6

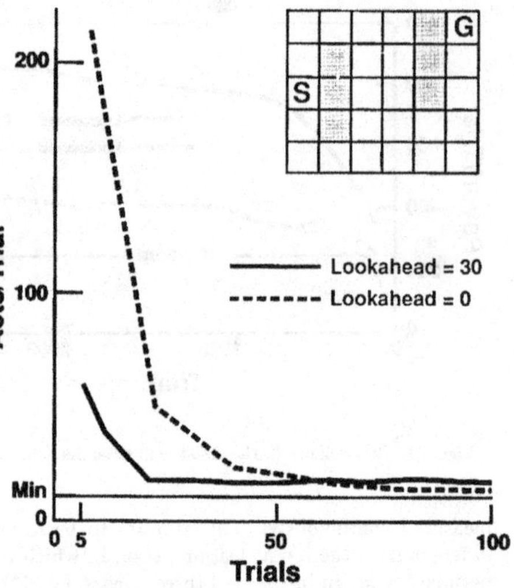

Figure 7: Performance in a maze with and without lookahead.

array. The system can produce effector messages to move Up, Down, Right or Left. If the system tries to move into a block (shaded areas) or one of the outer edges, it remains in the same state. For each trial, the system starts in state S and must move to state G. On entry to G, it receives a reward (1000), remains two steps, and then is returned to S to start the next trial. The minimum number of actions per trial is 11; random action choice yields 415 actions per trial.

Unlike the Gref3 task, in this task the system can take actions which leave it in the same state, which return it to the previous state, or which result in cycling between small numbers of states. In order to improve performance in this world, a bias against actions that would return the system to the same or the previous state was introduced. In particular, the support for such actions was multiplied by 0.01, which made it quite unlikely the system would choose those actions. In addition, to help CFSC2 focus its "attention" on the current situation, the support, $\psi_i(t)$, for each rule i that matches a message with the "real" bit on (i.e, a message with an $*$ in the type-tag) was multiplied by 20; this increases that rule's bid, and so helps insure rules representing the current situation remain active when matched.

Figure 7 shows the performace of the CFSC2 system with lookahead ($C_{max} = 30$) and without. (Each point is the average number of actions per trial over the most recent 10 trials.) CFSC2 clearly learns much faster with lookahead than without. Even by five trials, performance is about eight times better than random; by 20 trials with lookahead the system has reached its best performance, about 17 acts per trial. Without lookahead, the inital improvement is slower and it takes about 60-80 trials to reach its best, about 13 actions per trial. As with the Gref3 task, the total cycles with lookahead is only about 2.5 times the total without.

Note that in the limit performance without lookahead is better than with lookahead. The basic reason is that in CFSC2, the probability of winning a competition to post a message or pick an action is in direct proportion to the rules' bids. With lookahead, bids are determined primarily by the S_t values. As the S_t values come to reflect the expected payoffs of alternative routes, the bids for very similar paths are not very different (the discount factor γ_t is 0.9). Thus there is a good chance that rules that suggest routes that are slightly longer than the minimum will win the competition and control the system's behavior. However with no lookahead, the bids are determined by the S_v values. Since there are relatively few trials, there are relatively few actions on any given path, so most rule's S_v values are far from their equilibrium. Thus there is a kind of "founder effect"[11] in which rules that suggest the shortest paths rapidly get higher S_v values, and continue to control behavior, never allowing the alternative rules to increase their S_v.

Similar experiments have been run with the larger (54 state) worlds used with the Dyna-Q system [Sutton, 1990]. Again lookahead significantly improves performance. And while the results with CFSC2 are not quite as good as with Sutton's Dyna-Q, they are comparable, with lookahead reaching its best performance in about 30 or 40 trials (using $C_{max} = 40$, which is $k = 20$ for Dyna-Q). See the next section for a further comparison of Dyna-Q and CFSC2.

5 Discussion

The results presented above for the CFSC2 system show that with just a few modifications, CSs can learn and use internal models of the world. Building and using a model of the world greatly decreases the number of trials required to solve sequential decision tasks, at the expense of more lookahead planning. This can be especially important when "trials" (or mistakes) in the real world are expensive, as is often the case for real robots and animals. The ability to learn models without explicit rewards is also required to explain "latent learning" in rats, and more complex planning observed in humans and other higher animals. In summary, these results clearly show the importance of using the *non-reward* feedback a system gets as part of the continuous "what-follows-what" experiences it has as it interacts with its environment.

The lookahead techniques used by CFSC2 are similar

[11] Similar founder effects have been observed in other classifier systems, e.g., [Wilson, 1985] and [Booker, 1989].

to those used by the Dyna-Q system [Sutton, 1990].[12] Dyna-Q uses simple tables to represent the states and transitions of the model; it uses a kind of fixed-length, random search to do lookahead, and it updates its one "strength" parameter as a result of that lookahead. In the work described here, CFSC2 basically uses classifiers to implement tables. However, the use of classifiers should provide a direct way for CFSC2 to be extended to more complex worlds not easily represented in tables (see below). In addition, CFSC2 uses a kind of *directed spreading activation* mechanism for its lookahead search, starting from the current state and biased toward higher expected value paths, which is terminated when enough information is gathered to make a good decision. CFSC2's lookahead process only updates the S_t value, which decays rapidly, so that in effect CFSC2 forgets some of its lookahead results. While the decay of S_t may not be useful in some applications, it is one way to explain some observations of "irrelevant incentives" learning in rats (see below).

Another CS-based system to do lookahead is the ARS system described by [Whitehead and Ballard, 1989]. Their system (not implemented at that time) did not use the standard CS language (built on $1, 0, \#$), and it did not have to learn the model. However, they did suggest having a variable length of time allowed for the lookahead process (their termination conditions were different from those used in CFSC2).

One critical limitation in all these systems is that they implement their internal models as tables of states and transitions. In any real world task, e.g., those facing robots or animals trying to survive in uncertain, complex environments, a system typically is provided with extensive amounts of data from its input interface, much or most of it irrelevant. Such a system therefore must deal with the "perpetual novelty" problem, because it never sees exactly the same input twice. To do that, the system must generalize, creating model states that represent many equivalent world states by ignoring many or most input values—in effect creating "concepts" or "symbols." CSs have long been used to solve concept formation tasks ([Booker, 1982],[Wilson, 1987a],[Riolo, 1989]); in short, CSs have used the *genetic algorithm* to discover generalizations that enable them to use just a few rules to model complex worlds with many states. Thus one future line of research is to incorporate the GA into the CFSC2 system, using it to to build more complex models in such a way that the lookahead processes described in this paper can still be used.

Complex "real world" tasks also will bring forward two other problems, prediction verification and lookahead capacity. CFSC2 uses a simple match/no-match function to decide if a rule accurately predicts the next event. For complex worlds with many irrelevant features, a more sophisticated *partial verification* procedure will have to be developed. It may also be necessary to use a more complicated syntax to express predictions. Note, however, that classifier conditions already express partial matching functions [Booker, 1982], so there may be simple ways to use classifiers to verify predictions.

As for lookahead capacity, first note that in models like that used by CFSC2, there can be an exponential growth in the number of active states and transitions with each action considered. This is controlled by the size of the message list, which limits the depth of lookahead possible (in a given number of cycles). In situations where many "primitive actions" are necessary to achieve goals, some way to hierarchically organize rules in "subroutines" may be necessary to speed learning and allocation of credit [Wilson, 1987b]. One way to approach this problem in CFSC2 is make it possible for rules to do *multiple-step predictions.*

Finally, one other limitation of CFSC2 (and the other systems mentioned) is that they are able to handle only one goal (motivation) at one time. Once they have assigned an S_v value to rules as a result of solving one task (e.g., find food), it will be difficult for them to use the same model to attain some other goal (e.g., find water). Using the same knowledge structure for many different goals is a clear way to transfer information across tasks and so reduce learning time and storage costs. Such behavior can be observed not only in humans but also in rats, as seen in "irrelevant incentives" experiments (for reviews, see [Mackintosh, 1974] or [Walker, 1987]). While the separation of the S_v and S_t values in CFSC2 may make it possible to simulate some behavior, it seems likely that more complicated rules (e.g., multiple condition rules with some conditions sensitive to current goals) will be needed for full use of the same models for different goals.

Acknowledgements

I would like to thank the BACH group (Arthur Burks, Bob Axelrod, Michael Cohen, John Holland and Carl Simon) for comments and suggestions made during the course of this work. The CFSC2 system is an implementation of a combination of ideas on lookahead in classifier systems, starting with two independent proposals (one by Holland and one by me), and then modified as a result of discussions of the BACH group over the past year. I would also like to thank Rich Sutton for aquainting me with his own work on Dyna-Q, the work of Watkins on Q-learning, and that of Whitehead and Ballard on adaptive reactive systems. Thanks also to Stewart Wilson for helpful comments on a draft of this paper. This work was supported in part by National Science Foundation grant

[12] There are other differences between CFSC2 and Dyna-Q not discussed here, e.g., the way they encourage "curiosity."

IRI 86-10225 and by a University of Michigan Kellog Presidential Initiative Fund grant to the BACH group.

References

[Booker, 1982] Booker, L.B. "Intelligent Behavior as an Adaption to the Task Environment." Ph.D. Dissertation. University of Michigan.

[Booker, 1989] Booker, L.B. "Triggered Rule Discovery in Classifier Systems." *Proc. Third Internat. Conf. On Genetic Alogirithms,* 265-274. J.D.Schaeffer (ed).

[Booker *et al.*, 1989] Booker, L.B., Goldberg, D.E. and Holland, J.H. "Classifier Systems and Genetic Algorithms." *Artificial Intelligence, 40,* 235-282.

[Drescher, 1986] Drescher, Gary L. "Genetic AI: Translating Piaget into Lisp." MIT A.I. Memo No. 890.

[Drescher, 1989] Drescher, Gary L. "Made-Up Minds: a Constructivist Approach to Artificial Intelligence." Ph.D. Thesis, Dept. Elec. Eng. and Computer Science, MIT.

[Dickinson, 1980] Dickinson, A. *Contemporary Animal Learning Theory.* Cambridge University Press.

[Goldberg, 1988] Goldberg, D. E. *Genetic Algorithms in Search, Optimization, and Machine Learning.* Addison-Wesley, Reading MA.

[Grefenstette, 1988] Grefenstette, J.J. "Credit Assignment in Rule Discovery Systems based on Genetic Algorithms." *Machine Learning, 3,* 225-245.

[Holland, 1986] Holland, J.H. "Escaping Brittleness: The Possibilities of General-Purpose Learning Algorithms Applied to Parallel Rule-Based Systems." In *Machine Learning: An Artificial Intelligence Approach, Volume II,* Michalski, R. S., *et al.* (Eds). Morgan Kaufman Publishers, Inc.

[Holland and Reitman, 1978] Holland, J.H. and Reitman, J. S. "Cognitive Systems based on Adaptive Algorithms." In *Pattern Directed Inference Systems,* D.A. Watermann and F. Hayes-Roth (eds). Academic Press, NY.

[Holland *et al.*, 1986] Holland, J.H., Holyoak, K.J., Nisbett, R.E. and P.A. Thagard. *Induction. Processes of Inference, Learning, and Discovery.* The MIT Press.

[Holland, 1990] Holland, J.H. "Concerning the Emergence of Tag–Mediated Lookahead in Classifier Systems." Physica D, in press.

[Holyoak *et al.*, 1990] Holyoak, K.J., Koh, K. and Nisbett, R.E. "A Theory of Conditioning: Inductive Learning within Rule-Based Default Hierarchies." *Psych. Review,* in press.

[Mackintosh, 1974] Mackintosh, N.J. *The Psychology of Animal Learning.* Academic Press, London.

[Riolo, 1988] Riolo, R.L. *Empirical Studies of Default Hierarchies and Sequences of Rules in Learning Classifier Systems.* Ph.D. Dissertation, University of Michigan.

[Riolo, 1989] Riolo, R.L. "The Emergence of Default Hierarchies in Learning Classifier Systems." *Proc. Third Internat. Conf. On Genetic Alogirithms,* 322-326. J.D.Schaeffer (ed).

[Robertson and Riolo, 1988] Robertson, G.G. and Riolo, R.L. "A Tale of Two Classifier Systems." *Machine Learning, 3,* 139-159.

[Sutton, 1988] Sutton, R.S. "Learning to Predict by the Methids of Temporal Differences." *Machine Learning, 3,* 9-44.

[Sutton, 1990] Sutton, R.S. "Integrated Architectures for Learning, Planning, and Reacting Based on Approximating Dynamic Programming." *Proceedings of the Seventh International Conference on Machine Learning.*

[Walker, 1987] Walker, S. *Animal Learning. An Introduction.* Routledge and Kegan Paul, New York.

[Watkins, 1989] Watkins, Christopher J.C.H. "Learning from Delayed Rewards." Ph.D. Thesis, King's College.

[Whitehead and Ballard, 1989] Whitehead, S.D. and Ballard, D.H. "Reactive Behavior, Learning, and Anticipation." *Proceedings of 1989 NASA Conf. on Space Telerobotics.*

[Wilson, 1985] Wilson, S.W. "Knowledge Growth in an Artificial Animal." *Proc. First Internat. Conf. On Genetic Alogirithms,* 16-23. J.J. Greffenstette (ed).

[Wilson, 1987a] Wilson, S.W. "Classifier Systems and the Animat Problem." *Machine Learning, 2,* 199-228.

[Wilson, 1987b] Wilson, S.W. "Hierarchical Credit Allocation in a Classifier System." In *Genetic Algorithms and Simulated Annealing,* Lawrence, David (Ed). Morgan Kaufman Publishers, Inc., Los Altos, CA.

Learning Hill-Climbing Functions as a Strategy for Generating Behaviors in a Mobile Robot

David Pierce
Department of Computer Sciences
University of Texas at Austin
Austin, TX 78712
(dmpierce@cs.utexas.edu)

Benjamin Kuipers
Department of Computer Sciences
University of Texas at Austin
Austin, TX 78712
(kuipers@cs.utexas.edu)

Abstract

We consider the problem of a robot with uninterpreted sensors and effectors which must learn, in an unknown environment, behaviors (i.e., sequences of actions) which can be taken to achieve a given goal. This general problem involves a learning agent interacting with a reactive environment: the agent produces actions that affect the environment and in turn receives sensory feedback from the environment. The agent must learn, through experimentation, behaviors that consistently achieve the goal. The difficulty lies in the fact that the robot does not know *a priori* what its sensors mean, nor what effects its motor apparatus has on the world.

We propose a method by which the robot may analyze its sensory information in order to derive (when possible) a function defined in terms of the sensory data which is maximized at the goal and which is suitable for hill-climbing. Given this function, the robot solves its problem by learning a behavior that maximizes the function thereby resulting in motion to the goal.

1 The credit assignment problem

The learning problem addressed in this paper is illustrated in Figure 1. The learning agent, which we are calling a "critter," receives sensory input (vector s) from the world and acts on the world via motor outputs (represented by a, the action vector). In addition, the critter has access to a reward signal, r, by which it knows when it has achieved its goal. (In the experiments discussed later, the reward signal is incorporated into the sense vector for simplicity.) The critter's task is to learn a behavior which reliably achieves the goal. This behavior is a sequence of actions (most likely dependent on the concomitant sequence of sense vectors) which takes the critter from its present state to the goal state. The problem is difficult because the reward signal does not provide feedback for every action. The critter only knows that it has done the right thing when it stumbles onto the goal and is rewarded. It is then faced with the credit assignment problem – the problem of deciding which actions led to the goal.

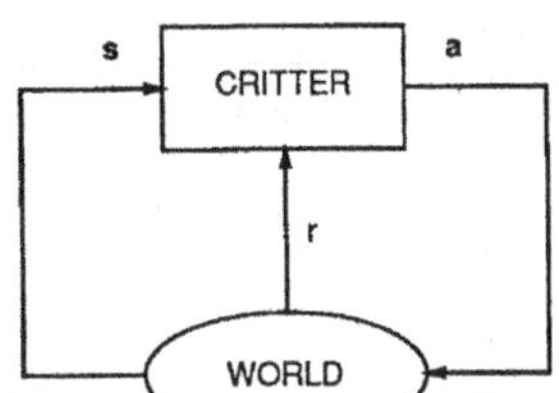

Figure 1: The general problem is for the learning agent, the "critter," to learn sequences of actions which produce rewards. The critter is rewarded when it is in a goal state.

2 A solution

In this paper, we propose the following solution to this problem:

1. Derive a function defined in terms of the sense vector (which is itself a function of the state of the world) such that this function is maximized at the goal state and is suitable for hill-climbing. It may in some cases be impossible to find such a function, in which cases, the method fails.
2. Learn a behavior that does gradient ascent on this hill-climbing function.

The problem explored in this paper can be viewed as the problem of learning a hill-climbing function to replace an *a priori* function where the latter is not appropriate for gradient ascent. In general, this problem can be described as follows: There is some function, given *a*

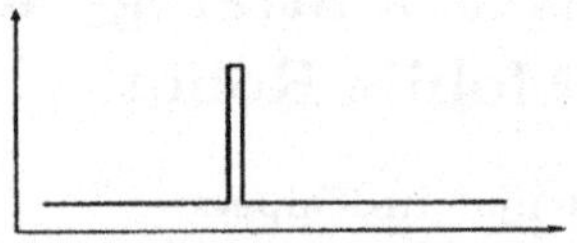

Figure 2: A goal function tells when the goal is achieved, but might not tell how to get to the goal.

priori, which defines a goal. This function is inappropriate for hill-climbing because it is almost everywhere zero and thus gives almost no information when the critter is not at the goal. Such a function is illustrated in Figure 2. The horizontal axis represents the space of possible states of the world or some subset of this space. For example, this space may be the set of possible positions of a robot in a room. In such a case, there may be only one position or a small set of positions which defines the goal state. The problem is that the goal function tells the critter when it is at the goal but does not tell the critter how to get to the goal if it is not already at the goal.

The aim of the research described here is to replace the *a priori* function with one which is maximized at the goal and which is suitable for hill-climbing. Such a function is illustrated in Figure 3. Ideally, this function will be differentiable and its gradient will everywhere point roughly toward the goal.

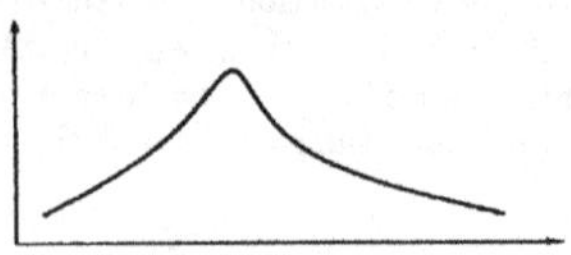

Figure 3: This goal function tells not only when the goal is achieved, but how to get to the goal as well, namely, by performing hill-climbing.

The ultimate goal of this research is to demonstrate how a robot can experiment with its sensors and effectors, in an unknown world, in order to learn the *distinctiveness measures* and control strategies that are the foundation of the Spatial Semantic Hierarchy of Kuipers and Byun (1988).

2.1 Overview

In the next section, we pose a particular instance of the general learning problem described in Figure 1. In Section 4, we develop a language of functions and operators useful for analyzing sensory data observed by a critter exploring its environment. In Section 5, we use the language to define a hill-climbing function suitable for the particular problem. In Section 6, this definition is used by a critter to learn hill-climbing functions for finding food in two different environments – one with one food source and one with two food sources. In Section 7, we demonstrate the usefulness of the learned hill-climbing functions by showing a number of ways in which the critter can use them to actually move to the food.

3 A particular problem: food-finding

In Sections 1 and 2, a general problem was posed along with a suggested approach to solving the problem. We now demonstrate this approach with a particular problem, the "food-finding problem."

Here is the setup of the learning problem: The critter is placed in a rectangular world with no other objects except a food source (see Figure 4). The critter has the

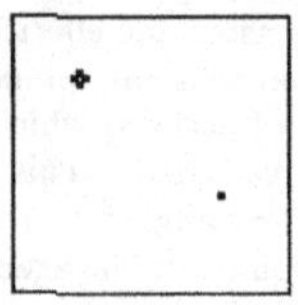

Figure 4: The critter world with one food source. The critter is represented by a small square. The food is represented by the diamond in the upper-left (northwest) corner.

following sensors, organized into a sense vector, **s**, of 21 components. Each sense component's value lies between −1.0 and 1.0.

At-food: This component is equal to 1.0 if the critter is at the food and is equal to 0.0 otherwise. This serves as the reward signal or goal function. The critter is at the food if it is inside the 7x7 square centered at the food source.

Smell: This is equal to 1.0 if the critter is at the center of a food source. It decreases with distance from the nearest source of food.

Position: This is a pair, (x,y), which gives the absolute position of the critter (but of course the critter doesn't know that this pair gives its position).

Sonar: This is a 16-component vector, $(s_0, s_1, ..., s_{15})$, whose components give the ranges to the nearest objects (food or walls in this case) in each of sixteen directions. These have a maximum value of 1.0 which they take on when the nearest object is beyond a certain distance away.

Compass: This gives the absolute orientation of the critter, with respect to east. A value of 1 corresponds to 180 degrees, -1 corresponds to -180 degrees.

The motor apparatus is

Turn-and-advance: This action function takes two parameters. The first determines an angle for the critter to turn. The second determines the amount that the critter advances in this new direction. The maximum turn is 45 degrees. The maximum distance moved is five grid locations.

The world is simulated as a 50 by 50 grid. A food source is placed (arbitrarily) somewhere in the northwest corner. It is represented as a solid diamond covering four grid locations. The fact that it is solid means that the critter cannot move into those four locations and that the critter's sonar can sense the food. For the purpose of the *at-food* sensor, however, the food is effectively a 7x7 square and thus occupies almost one fiftieth of the entire area of the world.

The problem posed to the critter is to learn a hill-climbing function defined in terms of the sense vector which is maximized when at the food. We must be careful to make explicit what is and is not known by the critter initially. The critter knows that the first sense component is *at-food* (the reward signal) and that its goal is to find a hill-climbing function such that maximizing this function brings the critter to the food (or, more correctly, makes the first sense component become 1). The critter knows nothing else about the sensorimotor apparatus.

In order to learn the hill-climbing function, the critter wanders around its world for 10,000 time steps, taking a random action at each step. During this time it explores its world including the area with the food source. In the next section, we will see how the hill-climbing function is learned.

We should emphasize that this particular problem is just an example. The setup of the experiment could have been different. In the current implementation of the critter world simulator, a number of different sensory and motor systems are defined. For example, the *translate* motor apparatus provides translational motion in which the critter's orientation never changes. When a critter is created, it may be given any combination of sensory and motor systems. The critter's task is to do as well as it can with whatever sensorimotor apparatus it is given.

3.1 A solution using a "goal sense vector"

Before describing the learning process more formally, let us consider the problem and its solution informally. The critter should learn that the smell sensor should be maximized. It should also learn that the x and y position sensors take on certain values at the food and that the food may be obtained by making the current x and y values match these values. The critter should also learn that the sonar and compass information is not as useful for finding the food (i.e., it is hard to know how to move based on compass and sonar information). What we would like the critter to learn, then, is that x and y should have certain values (e.g., about 0.2 and 0.8, respectively), that the smell should be near 1.0, and that the other sense components are unimportant. A hill-climbing function can then be proposed which is maximized when the current x, y, and *smell* components match their targets.

The learning process is outlined below:

Input: The critter's learning algorithm is given a sense vector history (a sequence of sense vectors, one per time step) plus the knowledge that the first component of the sense vectors corresponds to the reward signal *at-food*. The critter knows that when this sensor is active, it is at the food. (This sense thus takes the place of the reward signal in the description of the general problem.)

Output: The algorithm generates a function which is maximized when the critter is at the food and which can be effectively used for hill-climbing. It is not sufficient for the hill-climbing function to be maximized at the food. This is true for the sense component *at-food* itself, but *at-food* is not a good hill-climbing function because it is not a smooth function of position. We would like the function to be a continuous, differentiable function of position.

Method: The hill-climbing function is implemented as a similarity metric between a target sense vector and the current sense vector. This target vector is taken to be the average of the sense vectors seen while at the food source. With this similarity metric, different components of the target vector have different weights or "importances" where the weight is a measure of that component's usefulness. In the setup used in this experiment, for example, we expect the critter to learn that it is more important for the smell and position components to match their targets than it is for the compass and sonar components to match their targets. Thus, as a result of learning, the smell and position components should have large weights and the other components should have small weights.

The *target vector* and *importance vector* together comprise the *goal sense vector*. The goal sense vector can thus be viewed as a vector of target values where each target value has an associated importance. Given **u**, the importance vector, **g**, the target vector, and the current sense vector, **s**, the hill-climbing function, d, is taken to be the following:

$$d = -\mathbf{u} \cdot |\mathbf{g} - \mathbf{s}| = -\sum u_c(|g_c - s_c|)$$

Here, c ranges over the components of the vectors. This is the dot product of the importance vector and an absolute difference vector. Since we have already defined the target vector, the problem reduces to finding the importance vector (and then showing that this hill-climbing function can actually be used to achieve the goal).

The importance vector is computed by analyzing a history of sense vectors. In this experiment, we used histories 10,000 steps in length to ensure that the critter would sufficiently explore its world. From such a history one can calculate the measures E, C, R, S, I, and P, which we will define in Section 5. From these measures, one can define the importance vector, **u**. Before defining these measures, we introduce a number of definitions that will provide the language with which to describe the measures.

4 Tools for analyzing sensory histories

The raw data used in this analysis is the *history*. In general, to record all potentially useful information, we would want a history to contain all the senses and actions seen and taken by the critter. However, in the present work, we only need the history to keep track of sensory information.

We can think of a history as an abstract object which contains sensory information for a set of time instances. We will represent this object in different ways depending on the context. The actual representation is not important as long as it contains all the information we need.

The history may be represented as a function of type[1] HIST $= \mathcal{Z}_0^+ \rightarrow [-1,1]^{ns}$:

$$h(t) = \mathbf{s}^t = (s_0^t, s_1^t, \ldots s_{ns-1}^t).$$

Here, h is a function whose value at t is the sense vector seen by the critter at time t. The history may also be represented as a function of type $(\mathcal{Z}_0^+ \times [0 \ldots ns-1] \rightarrow [-1,1])$ where $h(t,c)$ gives the value of component c at time t. Thus, for example, $h(t,0)$ is the value of the *at-food* or reward sense at time t. Another way to represent the history is as a set of $(ns+1)$-tuples where the first component is the time and the remaining components comprise the sense vector.

In analyzing a sense component, we have found the frequency distribution to be a helpful tool – it abstracts away time and reduces a very large history to an object of manageable size. The distribution of a sense component, c, partitions the range of c (i.e., $[-1,1]$) into a number of subintervals (which are called "class intervals" in some statistics texts, and which we will be calling "boxes") and associates with each subinterval the number of times in the history that the value of component c lay in that subinterval. If nb is the number of boxes in the distribution, then a distribution, p, may be represented as a function of type DIST $= [0 \ldots (nb-1)] \rightarrow \mathcal{R}_0^+$. We allow $p(i)$ to have real values as well as integral values. We will want, for example, to speak of relative frequency distributions where we scale the distribution so that $\sum_i p(i) = 1$.

[1] Some notation: $\mathcal{Z}_0^+$ is the set of nonnegative integers. $[a,b]$ is the set of reals between a and b inclusive. $[-1,1]^{ns}$ is the set of ns-component vectors whose components are real valued and lie between -1 and 1 inclusive. $\mathcal{R}_0^+$ is the set of nonnegative reals. $[0 \ldots n]$ is the set of integers between 0 and n inclusive.

We introduce the operator $\mathcal{D}$ which maps histories into distributions. Thus $p_c = \mathcal{D}_c\{h\}$ is the distribution for component c. The operator $\mathcal{D}$ is parameterized by nb, the number of boxes in the partition, as well as by c. In our experiments, $nb = 100$. The operator $\mathcal{D}$ is of type HIST $\rightarrow$ DIST.

We define $\mathcal{D}_c$ as follows:

$$\mathcal{D}_c\{h\}(i) = \textstyle\sum_t b_i(h(t,c))$$
$$\text{where } b_i(x) = \begin{cases} 1, & \text{if } x \in box_i \\ 0, & \text{otherwise} \end{cases}$$

Here, box_i is the ith subinterval in the distribution's partition of the interval $[-1,1]$. The summation is taken over all values of t for which h is defined.

The frequency distribution, as we will see, captures a good deal of useful information about sense components. This amount of information may be reduced further, and made more manageable, by applying elementary statistics functions to the distributions. Thus we define the mean operator, μ:

$$\mu\{p\} = \frac{\sum_i \nu(i)p(i)}{\sum_i p(i)}$$

where $\nu(i)$ is the class midpoint of the ith class interval box_i. For example, if $nb = 100$, then $\{\nu(i), 0 \leq i < 100\} = \{-0.99, -0.97, \ldots, 0.99\}$. If we are given the history upon which the distribution of p is based, then we may calculate the mean more accurately from the history itself.

We similarly define the operator $\sigma\{p\}$ which gives (approximately) the standard deviation of distribution p:

$$\sigma\{p\} = \left[\frac{\sum_i(\nu(i)-\mu)^2 p(i)}{\sum_i p(i)}\right]^{1/2}$$

where μ is the mean of distribution p. Again, if the original history is available, the standard deviation may be computed more accurately from the history.

There are two more operators defined on histories which have proven useful and which are thus included in our language. The first is the "filter" operator, $\mathcal{F}_P$. This operator is parameterized by a predicate P. In describing the operation of $\mathcal{F}_P$, it is best to think of histories as sets of $(t,\mathbf{s})$ tuples. The value of $\mathcal{F}_P\{h\}$ is the subset of h consisting of all the elements in h whose sense vectors satisfy predicate P. Another way to describe the operation $\mathcal{F}_P$ is given below:

$$\mathcal{F}_P\{h\}(t) = \begin{cases} h(t), & \text{if } P(h(t)) \\ \textit{undefined}, & \text{otherwise} \end{cases}$$

Another useful operator is the operator ∂_t which can be viewed as taking the derivative of a complete history with respect to time. Thus

$$\partial_t\{h\}(t) = h(t) - h(t-1), \qquad t > 0$$

A Language for Analyzing Sense Histories			
Two Basic Data Types			
History (HIST):	h	:	$\mathcal{Z}_0^+ \to [-1,1]^{ns}$
Dist'n (DIST):	p	:	$[0\ldots(nb-1)] \to \mathcal{R}_0^+$
Operators on Histories			
Time Derivative:	∂_t	:	HIST $\to$ HIST
Filter:	$\mathcal{F}_P$	:	HIST $\to$ HIST
Dist'n Operator:	$\mathcal{D}_c$	:	HIST $\to$ DIST
Operators on Distributions			
Mean:	μ	:	DIST $\to \mathcal{R}$
Standard Dev.:	σ	:	DIST $\to \mathcal{R}_0^+$
Division:	$\mathcal{Q}$	:	(DIST $\times$ DIST) $\to$ DIST
Miscellaneous Functions			
Box midpoint:	ν	:	$[0\ldots(nb-1)] \to [-1,1]$
In-box:	b_i	:	$[-1,1] \to \{0,1\}$
Threshold:	th_l	:	$\mathcal{Z} \to \{0,1\}$

Figure 5: Here is a concise description of the language used for analyzing sense histories. The basic data types are the history and the distribution. A history maps time into the space of sense vectors. A distribution for a sense component, c, is a mapping from a partition (i.e., a set of subintervals) of the range of c into frequencies. These operators can be used to define measures used in the analysis of the sensorimotor apparatus. In this paper, we use the operators to define measures which rate sense components according to their usefulness as components in a hill-climbing function.

Remember that $h(t)$ is the sense vector that appears in the history at time t.

We define the division operator which computes the pointwise quotient of two distributions:

$$\mathcal{Q}\{p_1, p_2\} = p_1/p_2 = p$$
$$\text{where } p(i) = \begin{cases} p_1(i)/p_2(i), & \text{if } p_2(i) \neq 0 \\ 0, & \text{otherwise} \end{cases}$$

A function which will later prove useful is the threshold function, th:

$$th_l(x) = \begin{cases} 1, & \text{if } x \geq l \\ 0, & \text{otherwise} \end{cases}$$

To summarize, the objects, operators, and functions introduced in this section are listed in Figure 5 with their types. These operators, though few in number, are very expressive as they are based on powerful mathematical and statistical concepts. The computations involved in calculating them are temporally local which reduces memory requirements and makes them implementable in connectionist-style architectures. In the next section, we define a set of measures in terms of these operators which are useful in analyzing sensory data to determine which sense components may be used in defining a hill-climbing function.

5 Rating the sense components

Recall that our present goal is to define an importance vector **u** based on an analysis of a sense history. In this section, we define a number of measures which can be used to rate a sense component's usefulness. These measures are illustrated using a particular history obtained from the critter world simulator.

5.1 Extent

Sense components like *at-food*, which take on discrete values, are not good for hill-climbing. The purpose of the measure E is to eliminate discrete sense components based on an analysis of the history. The measure E will allow us to distinguish between components with sparse ranges (i.e., with a small set of possible values) and components with dense ranges.

To eliminate components like *at-food* from consideration in the goal sense vector, we require that a component's range cover a good percentage of the interval $[-1,1]$. For example, the x position sense component takes on values between 0 and 1 and thus its range would have one half the extent of the total interval $[-1,1]$. The extent of the *at-food* component would be very small (or zero, depending on how the extent is defined). With these examples as motivation, we define the measure E in terms of the sense components' distributions of values.

The measure E for sense component c, given history h, is defined below:

$$\begin{aligned} \text{let} \quad p_c &= \mathcal{D}_c\{h\} \\ th_{10}(x) &= \begin{cases} 1, & \text{if } x \geq 10 \\ 0, & \text{otherwise} \end{cases} \\ \text{then} \quad E(c,h) &= \tfrac{1}{nb} \textstyle\sum_i th_{10}(p_c(i)) \end{aligned}$$

A component's extent is defined in terms of the number of "full" boxes, i.e., those with at least 10 samples in them.[2] The extent is the number of full boxes divided by the total number of boxes, nb. Thus, for *at-food*, the extent is $2/100 = 0.02$. For x, it is $50/100 = 0.5$.

The importance of a component, for the purposes of defining hill-climbing functions, is defined to be 0 if the extent is less than a certain number. In this experiment, this number was 0.25. The value of this threshold is not critical – the point of the extent measure is to weed out sense components with discrete ranges. For a variable which takes on only a few discrete values, the extent should be much less than it is for a continuous variable.

5.2 Continuity and Range

A component useful for hill-climbing should be continuous in the sense that small actions should result in small changes in the component's value. Thus we propose the

[2]The value of 10 was chosen rather arbitrarily.

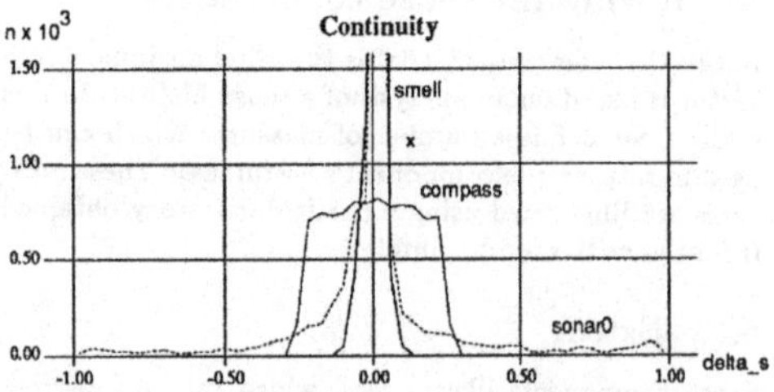

Figure 6: This graph shows the distributions of differences between consecutive sense vectors for the components *smell*, x, $sonar_0$, and *compass*. The measure C, for *continuity*, is equal to the standard deviation of this distribution. The smaller the value of C, the better, for use in a goal sense vector. This is because we do not want factors contributing to a hill-climbing function to fluctuate drastically with changes in position or orientation. Notice that the distributions for *smell* and x are much narrower than the ones for $sonar_0$ and *compass*.

measure C, for *continuity*, which will give an indication of whether or not a sense component is continuous. The smaller the value of C, the better. C, for a component, is defined as the standard deviation of the distribution of differences in consecutive values of that component.

Here is the definition of C using the time derivative operator:

$$C(c, h) = \sigma\{\mathcal{D}_c\{\partial_t\{h\}\}\}$$

For the position and smell sensors, this measure will be small because the critter's movement doesn't cause rapid fluctuations in these components. Sonar sensors, on the other hand, will have fairly large C-values since even a small turn action can result in a large change in a sonar sense's value. Figure 6 gives the distributions of differences for *smell*, x, $sonar_0$, and the *compass*.

The importance of a component is taken to be inversely proportional to the ratio C/R where R, the "range," is a term used for normalization. R is defined as the standard deviation of the component's overall distribution as opposed to the distribution of differences.

$$R(c, h) = \sigma\{\mathcal{D}_c\{h\}\}$$

The idea here is that it makes more sense to compare the measure C/R (rather than just C) across sensory modalities. For example, if a sense component has a very small range, it will also have a small C value, even though it may not be good for hill-climbing.

5.3 Shape

The previous measures (extent and continuity) do not depend on the goal for which a hill-climbing function is sought. In this experiment, we do not just want to know what components might be used for hill-climbing

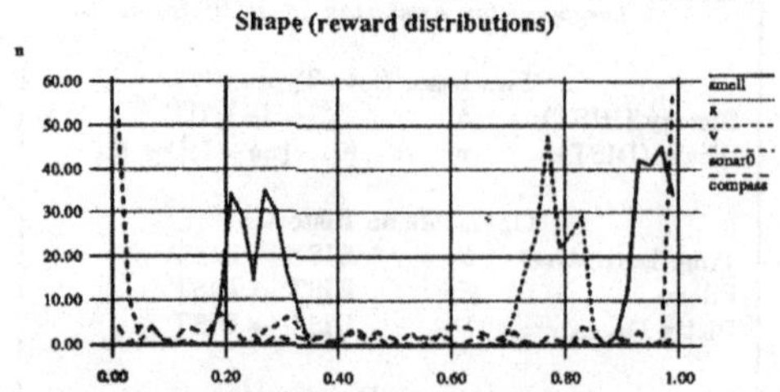

Figure 7: This graph shows the reward distributions for sense components *smell*, x, y, $sonar_0$, and *compass*. Notice that *smell*, x, and y all have well-defined intervals of values that they assume when at the food. Thus they all have small values of S, (for "shape" of the reward distribution), which is taken to be the standard deviation of the distribution. As with measure C, smaller values of S are preferred. We can see from this graph that the *compass* can take on the whole gamut of values while the critter is at the food. The same is true for $sonar_0$. (There is a peak at 0.0 for $sonar_0$ because the critter often reaches the food by running head-on into it.)

in general, we want to know which are useful for the particular task of finding food.

Smell, x, and y are good components because they all take on values from a small interval when the critter is at the food. We want the value of the component to be a good predictor of whether or not the critter is at the food. For example, *smell* is a good predictor: if the *smell* is near 1.0, then the critter is near the food. Conversely, if the *smell* is not near 1.0, that the critter is not close to the food.

One measure motivated by this idea is S, for shape, which is defined in terms of the component's distribution of values that it assumes when the critter is at the food. The measure S is defined as the standard deviation of this "reward distribution:"

$$\begin{array}{lrcl} \text{let} & P & \equiv & (h(t,0) = 1) \\ \text{then} & S(c,h) & = & \sigma\{\mathcal{D}_c\{\mathcal{F}_P\{h\}\}\} \end{array}$$

The predicate P holds when the critter is at the food. $\mathcal{F}_P$ thus contains only the sense vectors seen when the critter was at the food. As with C, this term is normalized by dividing by R. The importance will be inversely proportional to S/R. Figure 7 shows the reward distributions of *smell*, x, y, $sonar_0$, and *compass*.

5.4 Mutual Information

This measure is taken from information theory (see Jones (1979) for an introduction to the subject). Given two random variables x and y, the "mutual information" $MI(x, y)$ gives a measure of how much information one variable gives about the other. Thus the measure

$MI(r, c)$ (where r is the reward or *at-food* sense component) will tell us how much information variable c gives about the value of *at-food*. In order to define MI, we first introduce the information theoretic concept of entropy.

In information theory, the entropy of a random variable is a measure intended to formalize the intuitive notion of uncertainty. If the value of a variable is perfectly certain, then its entropy is zero. If a variable takes on n different equally probable values, then its entropy is maximal and is equal to $\log n$.

For the variable *at-food* which has only two values, the entropy is defined as

$$H(r) = -p_r \log p_r - (1 - p_r) \log(1 - p_r)$$

where we let p_r be the probability that *at-food* $= 1$. (Here, r (for reward) stands for the *at-food* sense component variable.)

We are interested in the change in the uncertainty of variable r given knowledge of the value of component c. If knowing the value of c decreases the uncertainty of r, then we can say that variable c gives information about r. In information theoretic terms, we are interested in the *mutual information* $MI(r, c)$ between variables r and c which is equal to the difference between the *self entropy* $H(r)$ for variable r and the *conditional entropy* $H(r|c)$.

We now derive an expression for the conditional entropy $H(r|c)$. Consider the distributions below, where $n = \sum_i \mathcal{D}_c\{h\}(i)$ is the total number of samples in the distribution $\mathcal{D}_c\{h\}$.

$$\begin{aligned} p_c &= \tfrac{1}{n}\mathcal{D}_c\{h\} \\ p_{r\wedge c} &= \tfrac{1}{n}\mathcal{D}_c\{\mathcal{F}_{(r=1)}\{h\}\} \\ p_{r|c} &= p_{r\wedge c}/p_c \end{aligned}$$

The value $p_{r|c}(i)$, being the ratio of the number of sense vectors in box_i seen at the food to the total number of sense vectors in box_i, gives an estimate of the probability that $r = 1$ given that $c \in \text{box}_i$. Associated with this probability is the entropy $\text{ent}(p_{r|c}(i))$ where

$$\text{ent}(p) = -p \log p - (1 - p) \log(1 - p).$$

What we want is a measure which takes into account the entropies for all of the boxes in the distribution. This measure is the conditional entropy $H(r|c)$ which is the weighted sum of the $\text{ent}(p_{r|c}(i))$ terms where the weight is the probability $p_c(i)$:

$$H(r|c) = \sum_i p_c(i) \text{ent}(p_{r|c}(i))$$

To reiterate, this conditional entropy term is a measure of the uncertainty in variable r given knowledge of c. The mutual information is given by

$$MI(r, c) = H(r) - H(r|c)$$

the difference between the self entropy of variable r and the conditional entropy of r given c. $MI(r, c)$ is at least zero, and is equal to zero when the variables are statistically independent.

Finally, we define the measure I:

$$I(c, h) = MI(r, c)$$

with $MI(r, c)$ as defined above.

5.5 Conditional probability

Another measure directly related to a component's ability to predict whether the critter is at the food is the conditional probability that the critter is at the food given that that component's value is near its target value. Recall that the target value is the mean value of the component's reward distribution (i.e., the distribution of values that the component assumes while the critter is at the food). The measure P will allow us to determine whether this target value is useful with respect to the goal of being at the food.

The conditional probability or expected reward, P, is calculated approximately using a component's overall distribution and its reward distribution. The distributions, remember, divide the interval $[-1, 1]$ into 100 subintervals or boxes. One of these, the "target box," contains the component's target value. In this calculation, we count the number of samples in the target box and two boxes on each side of it. For conciseness, call these five boxes the "target region." The approximation used for P, is the ratio of the number of samples, n_r, counted in the target region for the reward distribution to the number of samples, n, counted in the target region for the overall distribution. This is an estimate of the likelihood that the critter is at the food given that the value of the component under consideration lies in its target region. Figure 8 shows the ratios of the *at-food* distribution to the overall distribution for components *smell*, x, and y. From the figure, it should be clear that *smell* has a much higher P-value than does x or y.

Below is the formal definition of P:

$$\begin{aligned} \text{let} \quad P &\equiv (h(t, 0) = 1) \\ p_c &= \mathcal{D}_c\{h\} \\ p_{r\wedge c} &= \mathcal{D}_c\{\mathcal{F}_P\{h\}\} \\ tb &= \nu^{-1}(\mu\{p_{r\wedge c}\}) \\ &\quad \text{(the index of the target box)} \\ n_r &= \textstyle\sum_{tb-2}^{tb+2} p_{r\wedge c}(i) \\ n &= \textstyle\sum_{tb-2}^{tb+2} p_c(i) \\ \text{then} \quad P(c, h) &= n_r/n \end{aligned}$$

Here, $\nu^{-1}(x)$ gives the value of i such that $x \in \text{box}_i$.

5.6 Importance

The set of measures described in the preceding sections should illustrate the descriptive power of the language of

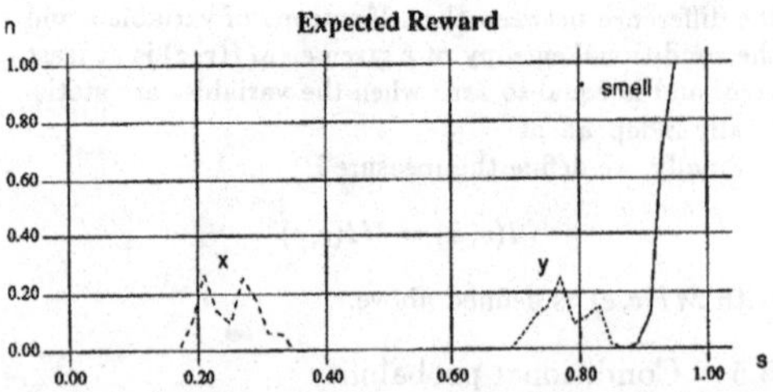

Figure 8: This shows an approximation of the expected reward (i.e., the probability that the critter is at the food source) given any particular sense value for the components *smell*, x, and y. These distributions are obtained by point-wise division of the *at-food* distribution and the overall distribution. The target values for these components (calculated as the means of the *at-food* distributions) are 0.95, 0.26, and 0.79 respectively. From the graph, it is clear that when the *smell* is near its target, there is a high probability of reward. When x or y is near its target, there is roughly a 15 percent chance of reward. For the other components (not shown), the expected reward is much less.

operators defined on sense histories. The set of measures described is certainly not exhaustive, but it is a useful set as the rest of the paper will argue. Given these measures, we now define the importance measure in a straightforward way. The (unnormalized) importance of a component is defined to be

$$u = \begin{cases} 0, & \text{if } E < 0.25 \\ (R/C) \cdot (R/S) \cdot I \cdot P, & \text{otherwise} \end{cases}$$

The importance vector, **u** resulting from these calculations is normalized by dividing each component by the maximum importance value. This gives the importance vector to be used with the target vector in calculating the hill-climbing function. Remember that the equation for the hill-climbing function, d, is

$$d = -\mathbf{u} \cdot |\mathbf{g} - \mathbf{s}| = -\sum u_c(|g_c - s_c|)$$

where **s** is the current sense vector and **g** is the target vector.

The definitions of this section are summarized in Figure 9.

6 An experiment

In this section, we apply the preceding definition of the importance vector to histories generated by a critter wandering randomly in a simulated environment with one food source. Intuitively, we expect the importance vector to assert that the components x, y, and *smell* are important and that the rest are not for the purpose of defining a goal sense vector.

In this experiment, the critter explores a world with one food source (see Figure 4). Its task is to learn a hill-climbing function which can be used to move to the food.

Below are the statistics after 10k steps of wandering. The number of samples is the length of the history. The "filter samples" are those seen when the critter is at the food (i.e., when the *at-food* component is equal to one).

```
Number of samples        = 10000
Number of filter samples = 176
```

Name	Index	Mean	R	C	S	E	P	I	GSV	U
at-food	0	0.02	0.13	0.08	0.00	0.02	1.00	0.128	1.00	0.00
smell	1	0.52	0.25	0.02	0.02	0.50	0.63	0.116	0.95	1.00
x	2	0.51	0.32	0.02	0.04	0.50	0.18	0.058	0.26	0.14
y	3	0.63	0.32	0.02	0.03	0.50	0.14	0.054	0.78	0.13
sonar0	4	0.42	0.41	0.24	0.43	0.50	0.02	0.008	0.46	0.00
sonar1	5	0.45	0.42	0.25	0.43	0.50	0.02	0.007	0.52	0.00
sonar2	6	0.50	0.42	0.25	0.42	0.50	0.02	0.007	0.53	0.00
sonar3	7	0.55	0.41	0.25	0.42	0.50	0.02	0.011	0.51	0.00
sonar4	8	0.59	0.40	0.25	0.40	0.50	0.03	0.012	0.53	0.00
sonar5	9	0.63	0.39	0.26	0.39	0.50	0.04	0.013	0.51	0.00
sonar6	10	0.66	0.37	0.26	0.37	0.50	0.04	0.017	0.52	0.00
sonar7	11	0.68	0.36	0.26	0.36	0.50	0.04	0.019	0.48	0.00
sonar8	12	0.67	0.36	0.25	0.34	0.50	0.06	0.020	0.47	0.00
sonar9	13	0.65	0.37	0.25	0.32	0.50	0.06	0.020	0.44	0.00
sonar10	14	0.61	0.39	0.25	0.31	0.50	0.07	0.019	0.38	0.00
sonar11	15	0.56	0.40	0.25	0.31	0.50	0.08	0.017	0.35	0.00
sonar12	16	0.50	0.41	0.25	0.33	0.50	0.07	0.015	0.34	0.00
sonar13	17	0.45	0.41	0.24	0.36	0.50	0.06	0.013	0.36	0.00
sonar14	18	0.42	0.41	0.24	0.40	0.50	0.05	0.011	0.41	0.00
sonar15	19	0.40	0.41	0.23	0.42	0.50	0.05	0.010	0.43	0.00
compass	20	0.06	0.57	0.47	0.44	1.00	0.04	0.012	0.14	0.00

Here, Mean is the overall mean for all samples in the 10k-step history. The measures R, C, S, E, P, and I were defined in section 5. GSV (for goal sense vector), is the vector of target values. U (for usefulness) is the vector of importance values. The goal sense vector has the targets *smell*= 0.95, $x = 0.26$, $y = 0.78$. The other components have all been eliminated from consideration, that is, their importances are very small. Notice that *smell* has the highest importance rating and that x and y both have importances less than a sixth that of *smell* indicating that the smell sense is significantly better for hill-climbing.

Let us analyze why the eliminated components (those with very small importance values) received the importance ratings they did. The *at-food* component is eliminated because its extent is too small. The sonar sensors all fare poorly on measures C, S, P, and I. C is large meaning that the sonar sensors fluctuate significantly with small motions. S is large meaning that there is a wide range of values that the sensors take on when at the goal. P is small meaning that the target values are not good indicators of the presence of food. I is small indicating that the sonar values do not give much information about the value of the reward variable. Similarly, the *compass* fares poorly on measures C, S, P, and I.

A second experiment was run in which a second food source was placed near the bottom middle of the world.

Measures for Rating Sense Components as Hill-Climbing Functions			
Extent:	$E(c,h)$	=	$\frac{1}{nb}\sum_i th_{10}(\mathcal{D}_c\{h\}(i))$
Continuity:	$C(c,h)$	=	$\sigma\{\mathcal{D}_c\{\partial_t\{h\}\}\}$
Range:	$R(c,h)$	=	$\sigma\{\mathcal{D}_c\{h\}\}$
Shape:	$S(c,h)$	=	$\sigma\{\mathcal{D}_c\{\mathcal{F}_{r=1}\{h\}\}\}$
Mutual Information:	$I(c,h)$	=	(see text)
Probability of reward:	$P(c,h)$	=	(see text)
	The Goal Sense Vector		
Target:	$g(c,h)$	=	$\mu\{\mathcal{D}_c\{\mathcal{F}_{r=1}\{h\}\}\}$
Importance:	$u(c,h)$	=	$(R/C)\cdot(R/S)\cdot I\cdot P$, if $E \geq 0.25$
Hill-climbing function:	d	=	$-\sum u_i(\lvert g_i - s_i\rvert)$

Figure 9: These measures (described in detail in the text) are used to determine, for each sense component, its "importance" or usefulness as a component of a hill-climbing function. The hill-climbing function used for finding the food is a function on the current sense vector, and is defined using the target and importance vectors.

In that experiment, only the smell sense was labeled as being important.

7 Using goal sense vectors to find food

Once the critter has learned a goal sense vector, it needs to learn the corresponding behavior, i.e., it needs to learn to maximize the derived hill-climbing function and move to the food. In this section, we briefly describe a number of methods for using the goal sense vectors.

7.1 A brute-force method

The first method is a very general, brute-force approach which does not require additional learning. The hill-climbing algorithm involves randomly choosing an action and executing it as long as it results in an increase in the hill-climbing function. When the hill-climbing function decreases, a new random action is taken. A trace of a typical path taken by the critter in finding the food is given in Figure 10b. This algorithm works, but it is not very efficient. The critter makes a lot of bad moves and wanders around a good deal before achieving its goal. The brute force hill-climbing method does not work very well with the ***turn-and-advance*** motor apparatus whose actions result in circular paths. With most actions, the critter makes only small progress toward the food before turning away. For the purposes of comparison, Figure 10a shows the path taken by a critter which wanders randomly until it stumbles onto the food.

7.2 Learning reactive behaviors

Another method for implementing a hill-climbing behavior is to use a (**sense** $\rightarrow$ **action**) mapping whereby, at each step, the critter computes the action to take by applying the mapping function to the sensory input. This mapping is learned using connectionist techniques, such that the resulting reactive behavior solves the problem of maximizing the hill-climbing function. The subject of learning such behaviors is beyond the scope of this paper, but we do briefly discuss one example of such a learning method here. In Figure 10c, we show the path taken by a critter that has learned an appropriate reactive behavior. In these examples, the critter takes 10,000 random actions to learn a goal sense vector and then takes another 10,000 more random actions to train the reactive behavior. The figures show traces of paths made after both learning phases.

The learning method used by this critter is straightforward. The critter learns a linear mapping from sense vectors to actions vectors: Suppose that during learning the critter sees sense vector **s** and responds with random action **a**. The critter learns by reinforcing the association (**s** $\rightarrow$ **a**) using the delta rule[3] to modify the weight matrix that determines the (**sense** $\rightarrow$ **action**) mapping. The amount of reinforcement is proportional to the immediate reward and to future rewards where the contribution of future rewards falls off exponentially with time. The reward here is taken to be the increase in the hill-climbing function (and can therefore be negative).

7.3 Exploiting motor knowledge

The brute force hill-climbing method has the advantage that it works without additional learning. All that is needed is the hill-climbing function itself. The method can be greatly improved if the critter has knowledge of turns and travels. (In other work, we are exploring methods of diagnosing turn and travel actions given an uninterpreted sensorimotor apparatus.) Given such knowledge, the critter can travel forward as long as the

[3]See McClelland (1988, p. 86) for a good discussion of the delta rule.

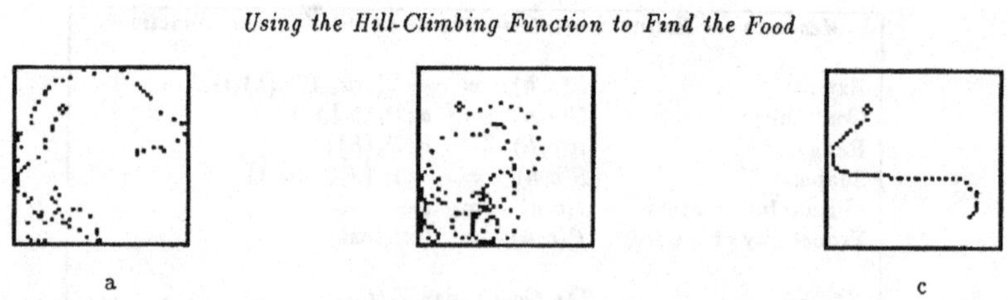

Figure 10: These figures show traces of paths taken by the critter while trying to find the food. In Figure a, the critter takes random actions producing a random walk that ends when it happens onto the food. In Figure b, the critter knows the hill-climbing function and uses a simple minded method to perform hill-climbing (see the text for a description). In Figure c, the critter has learned a reactive behavior. It uses its sensory input to determine what action to take at each step.

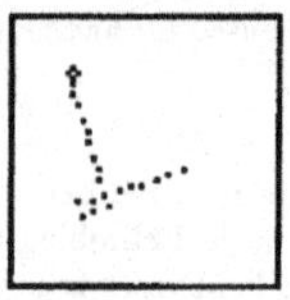

Figure 11: A sequence of pure turn and travel actions can be used to move efficiently to the goal if the critter knows how to make such actions.

Figure 12: The critter learns an improved reactive behavior when it has a *food-beacon* sensor giving it the direction to the food.

hill-climbing function increases and then take a 90 degree turn. In Figure 11, we show a typical path taken by a critter that uses such a scheme for hill-climbing.

7.4 Exploiting additional sensory knowledge

Given additional sensory information, the critter can learn improved behaviors. The critter learns to exploit the sensory information it is given. In Figure 12, we see a typical path taken, after learning (as in Section 7.2), by a critter that has a *food-beacon* sensor in addition to the rest of the sensors. The *food-beacon* sensor has 12 binary components corresponding to 12 different directions. Exactly two components have value 1 corresponding to the two directions between which the food source lies. Using this sense, the critter can learn to turn toward the food and then advance to it. Incidentally, the critter assigns importance values of zero to the food beacon sensors.

The hill-climbing methods have been tried for the two-food-source case as well, with similar results. Which food source the critter ends up at depends on the critter's initial position and orientation.

8 Summary

We have addressed a particular version of the credit-assignment problem, namely that faced by an agent, with uninterpreted sensors and effectors, learning goal-directed behaviors in a reactive environment. Our solution is to exploit the sensory information given to the learning agent in order to replace a goal function with a hill-climbing function.

To that end, we have developed a language of operators defined on sense histories which has enabled us to analyze the agent's sensory apparatus in order to discover which senses might be used as components in a hill-climbing function. A hill-climbing function, d, was defined in the language and shown to be useful in the context of two experiments in which the goal was to find sources of food.

Acknowledgements

This work has taken place in the Qualitative Reasoning Group at the Artificial Intelligence Laboratory, The University of Texas at Austin. Research of the Qualitative Reasoning Group is supported in part by NSF grants IRI-8602665, IRI-8905494, and IRI-8904454, by NASA grant NAG 2-507, and by the Texas Advanced Research Program under grant no. 003658175.

References

[1] Jones, D.S. (1979). *Elementary Information Theory.* Oxford: Clarendon Press.

[2] Kuipers, Benjamin, & Byun, Yung-tai (1988). A robust, qualitative method for robot spatial learning. *Proc. AAAI-88,* St. Paul/Minneapolis, pp 774-779.

[3] McClelland, James L., and Rumelhart, David E. (1988) *Explorations in Parallel Distributed Processing.* Cambridge: MIT Press.

Some Parallels Between Associative Learning and Object Classification

David R.Shanks

MRC Applied Psychology Unit
15 Chaucer Road
Cambridge CB2 2EF
England

Abstract

Certain parallels are explored between the mechanism of associative learning in animals - and its description by contemporary learning theories such as that of Rescorla and Wagner (1972) - and object classification, which can be construed as the problem of learning to associate visual features or micro-fearures with a category. A specific associative learning theory of classification is presented. While *prototype* effects can be easily accommodated within associative theories of classification, *exemplar* effects appear to be fatal for such accounts, since no explicit representations of exemplars are stored in associative networks. An experiment is reported which attempted to see whether the model would be able to reproduce exemplar effects. Surprisingly, it could. In examining why this was the case, a new interpretation of exemplar effects emerged.

1. Introduction

Some years ago, it was very fashionable to assume that people represented categories of visual objects in terms of their *prototypes*. As a result of learning that a large number of different visual stimuli all belong to a certain category - such as the category *dog* - people were assumed to have extracted a prototype representing the central tendency of the stimuli on a variety of feature dimensions.

Figure 1 (left-hand panel) illustrates how this abstraction is supposed to work. Imagine a number of exemplars (marked by X's) of a category which vary on, say, two dimensions. For instance, the exemplars could be dogs varying in colour and size. According to prototype theories, what is actually extracted and mentally represented of this category is the central tendency of the exemplars within the feature space. The prototype (marked by a dot) has a value on each dimension corresponding to the modal value of the actual exemplars on that dimension.

What happens when a new stimulus is presented for classification ? According to prototype theories, the similarity of the test stimulus to the prototype is determined - this being just the inverse of the distance between them - and the stimulus is classified as being a member of this category if its similarity to the *dog* prototype is greater than its similarity to any other category prototypes. The greater the similarity to the prototype, the faster, more confident, or more accurate the classification decision.

The crucial commitment, then, of prototype theories is that during the process of classification, no representations of specific training exemplars play a role in classification. It is only the abstracted prototype which determines the course of classification. Of course, representations of specific training exemplars clearly do exist: thus each of us has memory traces of specific faces that we are familiar with - our mother's face, for example. Prototype theories maintain, however, that when a new physical stimulus appears before me and I decide that it is a face, the trace that exists in memory of my mother's face plays no role in this process. It is only the prototypical face that I have abstracted from all the faces that I have ever experienced that determines my classification decision.

Now, there is one piece of empirical evidence which appears to provide strong encouragement for the prototype view, and this is the fact that subjects often respond more accurately, or with greater confidence, to the prototype of a category than they do to the specific training exemplars, even though they may never have seen the prototype before. A clear example of this occurs in an experiment by Knapp and Anderson (1984). They generated training stimuli, which were dot patterns, by distorting a particular prototype pattern. Subjects saw a number of distortions of the prototype, and learned correctly to classify all of these patterns. When they were subsequently tested either with new distortions, the original training patterns, or with the prototype itself, they responded most accurately to the prototype (I shall call this the *prototype* effect). This was most evident when the number of different training stimuli used had been large. Of course, this result is precisely what one would expect on a prototype account, since the prototype pattern corresponds to what the

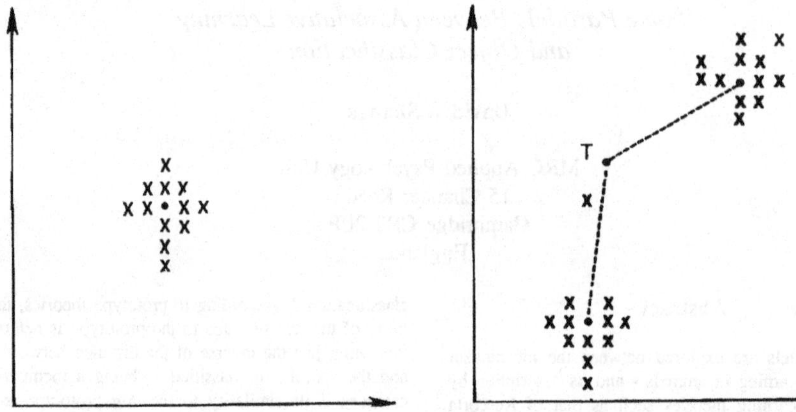

Figure 1. Left-hand panel: the prototype (marked by a dot) corresponds to the central tendency of the training exemplars (the X's) on the relevant feature dimensions. Right-hand panel: arrangement of stimuli in which a test stimulus (T) is equidistant in the space from the prototypes (marked by dots) of two categories, but highly similar to an exemplar of one of the categories. The categories are represented by sets of training exemplars (the X's).

subjects would actually have abstracted from the training stimuli.

2. A Connectionist Model of Object Classification

All of this is highly congenial to associative or connectionist accounts of object classification, which in general are prototype theories. These accounts propose that low-level visual processing yields a description of an object which can be characterized as a set of features or micro-features, which in turn forms the input to a network of processing elements. The output of this network is a classification decision. Recently, several authors have argued that certain aspects of human classification data can be accounted for by such theories (e.g. Anderson & Mozer, 1981; Estes, Campbell, Hatsopoulos & Hurwitz, 1989; Gluck & Bower, 1988; Hintzman, 1986; Knapp & Anderson, 1984; Shanks, 1990). I shall briefly describe in detail how such a network can classify stimuli.

Consider a single-layer feedforward network consisting of a set of input units, each of which is connected to a set of output units. An appropriate learning algorithm for adjusting the weights in the network is the "least-mean-squares" (LMS) rule of Widrow and Hoff (1960), also called the "delta rule", which is formally equivalent to the Rescorla and Wagner (1972) theory of associative learning in animals. What the theory proposes is that each feature has independent weights (or associative strengths) connecting it to each category, in just the same way that a conditioned stimulus (CS) is associated with an unconditioned stimulus (US).

Although associative learning and object classification have many surface similarities, it has only recently become apparent that a common mechanism might be at work in the two situations. Thus Gluck and Bower (1988) have drawn attention to several parallels. More importantly, there have been demonstrations of common phenomena occurring in the two situations. Shanks (1990) has shown that a crucial phenomenon of associative learning, namely sensitivity to contingency, occurs not only in Pavlovian and instrumental conditioning, but also in human classification as well.

Suppose that the subject experiences a series of trials on which exemplars are presented, each with its appropriate category. What we want to know is how the features become associated with the various catgories. On each trial, the weight connecting each feature that is present with the category occurring on that trial is incremented according to the equation:

$$\Delta W_{fi} = \alpha_f . \beta_i .(\lambda - \sum_f W_{fi}) \qquad (1)$$

where ΔW_{fi} is the weight change of a particular feature for category i, α_f is a measure of the salience of the feature, β_i is a measure of the salience of the category occurring on that trial, λ is the asymptote of each weight (usually 1.0), and $\sum_f W_{fi}$ is the sum of the weights on the connections from all the features present on that trial to the category that actually occurs. At the same time, the weights connecting each feature that is present to each

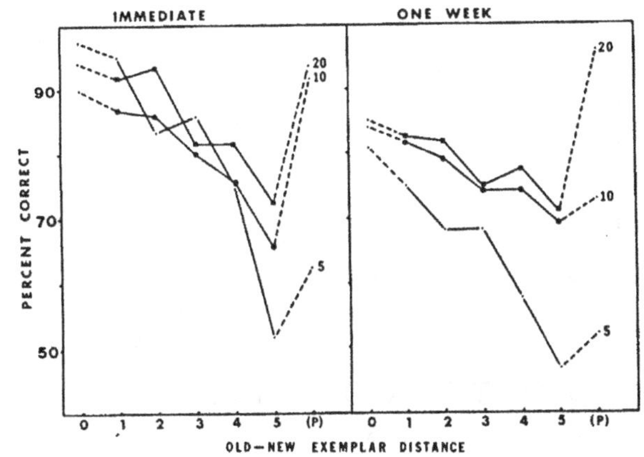

Figure 2. Results of Experiment 2 of Homa, Sterling and Trepel (1981). Mean classification accuracy on new test stimuli as a function of old-new similarity, category size, and time of test.

category that does *not* occur on that trial are reduced according to the equation:

$$\Delta W_{fi} = -\alpha_f . \beta_i . \sum_f W_{fi} \qquad (2)$$

where $\sum_f W_{fi}$ now refers to the sum of the weights of the features present for the category not occurring on that trial.

At the end of a series of training trials, if we present the subject with a test stimulus possessing certain features, it is assumed that the probability of classifying the stimulus as a member of category j will simply be related to the output produced on the category j output node divided by the total output across all output units:

$$P(j) = \frac{\sum_f W_{fj}}{\sum_f W_{fj} + \sum_{i \neq j} \sum_f W_{fi}} \qquad (3)$$

Simple though it is, this classification network has no difficulty reproducing prototype effects. The reason for this is not hard to see. Essentially, when the prototype is used as an input pattern, the input units which have the largest weights for the category will all be switched on, and so a larger output will occur at the appropriate category node than for any other pattern. The matrix of weights the network acquires during training represents an abstraction from the training exemplars corresponding to a prototype. An illustration of how the theory reproduces prototype effects will be provided shortly.

Unfortunately, though, there is a further finding from classification studies which has been universally taken to be incompatible with the sort of prototype theories I have described. The right-hand panel of Figure 1 illustrates a situation in which a test stimulus is equidistant from two prototypes. According to prototype theory, then, subjects should be indifferent about classifying this pattern between the two categories. However, the pattern is particularly similar to a specific training pattern from one category, but not from the other. What has been found, in experiments using this procedure, is that subjects are biased in such circumstances: they tend to classify the test pattern as belonging to the category for which it is similar to a training pattern (I shall call this the *exemplar* effect). For example, Homa, Sterling and Trepel (1981) trained subjects to classify geometrical patterns which were distortions of prototypes. The categories consisted of either 5, 10 or 20 different training stimuli. In the test stage, which took place either immediately or one week after training, the subjects had to classify 5 new patterns which were all equidistant from the prototype, but which got progressively less similar to a specific training stimulus. Their results are shown in Figure 2. As the old-new exemplar distance was increased, classification accuracy deteriorated. This effect appears to indicate that similarity to training exemplars is important in classification, and seems to be quite incompatible with prototype theories.

As a result of such demonstrations, it is now widely acknowledged that specific exemplars do play a role in object classification. In fact some authors, such as Medin and Schaffer (1978) and Nosofsky (1989) have gone to the extreme of assuming that classification depends solely on a comparison of the test stimulus with the traces

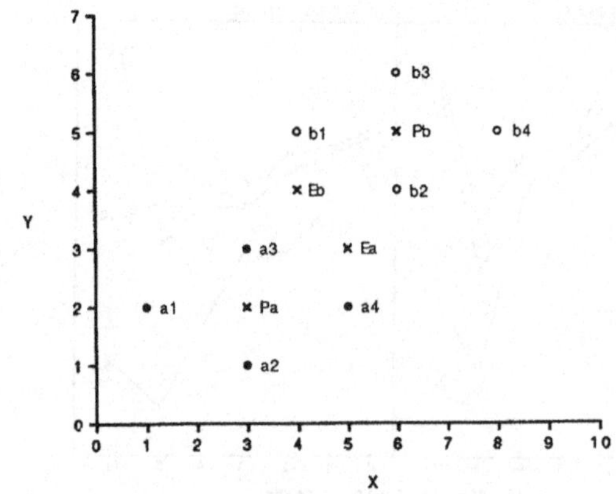

Figure 3. Arrangement of stimuli for the experiment. Patterns a1, a2, a3 and a4 are the stimuli for category A, patterns b1, b2, b3 and b4 for category B. The prototypes are patterns Pa and Pb. Pattern Ea is equidistant from the two prototypes, but more similar to a category A exemplar. Pattern Eb is more similar to a category B exemplar.

of the training stimuli, and have abandoned the notion of prototype abstraction altogether. It might appear that this sort of view would be unable to explain prototype effects, but in fact this is not so (see Nosofsky, 1989).

The clear implication of this is that associative theories of classification should be unable to reproduce exemplar effects. This is because the matrix of weights in a network does not preserve any information at all about the training exemplars. The matrix is simply an aggregate of information provided by the training stimuli, but those stimuli could never be recovered from the matrix. If it is therefore the case that connectionist models of classification are unable to reproduce exemplar effects, then these theories will have to be abandoned as serious contenders for explaining how organisms classify stimuli. I now report a simulation experiment which attempted to test this prediction.

3. Experiment

Figure 3 illustrates the locations within a 2-dimensional space of the set of stimuli that were used in the experiment. Four stimuli (a1, a2, a3 and a4) were members of category A and four (b1, b2, b3 and b4) of category B. The prototypes, corresponding to the modal values of the exemplars, are marked Pa and Pb, and the critical test stimuli, equidistant from the prototypes, are marked Ea and Eb. Pattern Ea is more similar to a category A exemplar than to any category B exemplar, since it differs from pattern a4 by only two features in the coding scheme shown in Table 1, but differs by 4 features from the nearest category B exemplar (b2). The converse is true for pattern Eb.

The stimuli were arranged in such a way as to ensure that the two categories would be equally difficult to learn. If this were not the case, then the model would produce exemplar effects simply in virtue of the fact that the weights most relevant to the easier category would be greater than those most relevant to the harder of the categories. Homa et al.'s (1981) also tried to arrange for their categories to be equally difficult to learn.

Each of the stimuli was coded as a 24-bit binary number which was then presented to a network containing 24 input units and two output units. The input units were divided into 14 that coded the co-ordinate of the stimulus on the x-axis, and 10 that coded the y-axis co-ordinate. The patterns corresponding to each value were as shown in the table on the next page. Thus, pattern a1, with co-ordinates (1,2) was coded by the vector 000000011111110000101111 (row 1 from the table of x-axis co-ordinates, row 2 from the y co-ordinates), which was then presented to the 24 input units. Zero indicates that the unit was turned off, 1 that it was turned on.

This scheme has the essential characteristic of yielding a gradation of feature overlap as one moves along a dimension. Thus, a stimulus that has an x co-ordinate of, say, 4 has more features in common with a stimulus with a co-ordinate of 5 than it does with one with a co-ordinate of 6. Such a stimulus gradient is obviously a characteristic of most natural featural dimensions, and needs to be captured by any coding scheme that is adopted.

The network was presented with 48 learning trials. On each trial, one of the 8 training stimuli (a1, a2, a3, a4, b1, b2, b3 or b4) was randomly chosen as the input pattern. At the end of the trial, the weights were adjusted in

x Co-ordinate	Input Unit (x-axis)													
	1	2	3	4	5	6	7	8	9	10	11	12	13	14
1	0	0	0	0	0	0	0	1	1	1	1	1	1	1
2	0	0	0	0	0	0	1	0	1	1	1	1	1	1
3	0	0	0	0	0	1	1	0	0	1	1	1	1	1
4	0	0	0	0	1	1	1	0	0	0	1	1	1	1
5	0	0	0	1	1	1	1	0	0	0	0	1	1	1
6	0	0	1	1	1	1	1	0	0	0	0	0	1	1
7	0	1	1	1	1	1	1	0	0	0	0	0	0	1
8	1	1	1	1	1	1	1	0	0	0	0	0	0	0

y Co-ordinate	Input Unit (y-axis)									
	15	16	17	18	19	20	21	22	23	24
1	0	0	0	0	0	1	1	1	1	1
2	0	0	0	0	1	0	1	1	1	1
3	0	0	0	1	1	0	0	1	1	1
4	0	0	1	1	1	0	0	0	1	1
5	0	1	1	1	1	0	0	0	0	1
6	1	1	1	1	1	0	0	0	0	0

Table 1. Coding scheme for the stimuli used in the experiment. The x co-ordinate of each stimulus was coded by 14 bits, and the y co-ordinate by a further 10.

accordance with equations (1) and (2). After these training trials, there were 4 further test trials, with the prototype patterns Pa and Pb and the critical stimuli Ea and Eb. During this test stage, the weights on the connections remained fixed. The results presented below are averaged across 10 complete runs of the above procedure.

4. Results

Figure 4 shows the results of the experiment. The y-axis measures the output from equation (3), that is, the output for category A divided by the total output for categories A and B. Large values correspond to classifications of category A, small values to category B, and values close to 0.5 represent indifference.

First, the network had no difficulty learning to classify the training stimuli. Figure 4 gives the mean output on the last training trial for each of these patterns. Large outputs are produced for the A patterns and small outputs for the B patterns. Next, the actual prototypes were very accurately classified, even though they had never been seen by the network during the training phase. The mean output for Pa was 0.97, while for the A training exemplars the mean at the end of training was only 0.91. For category B, the figures were 0.03 for pattern Pb and 0.11 for the category B training patterns. Thus we have a demonstration that prototype effects seen in human categorization studies (e.g. Knapp & Anderson, 1984) can be reproduced by the associative theory.

The most important results are the classifications of the stimuli Ea and Eb. Surprisingly, the network responded differently to these two stimuli, classifying Ea as a member of category A and Eb as a member of category B.

5. Discussion

Why does the network - which is an implementation, after all, of a sort of prototype account - differentially classify two stimuli equidistant from the prototypes ? I shall argue that the reason for this is straightforward: although pattern Ea is equidistant in the experimenter-described space from prototypes Pa and Pb, in fact these distances are not equal in the network's "psychological" space. The reason for this is not difficult to see. The y-dimension much more readily discriminates categories A and B than does the x-dimension. Thus we might expect the network to give more weight to the y-dimension than

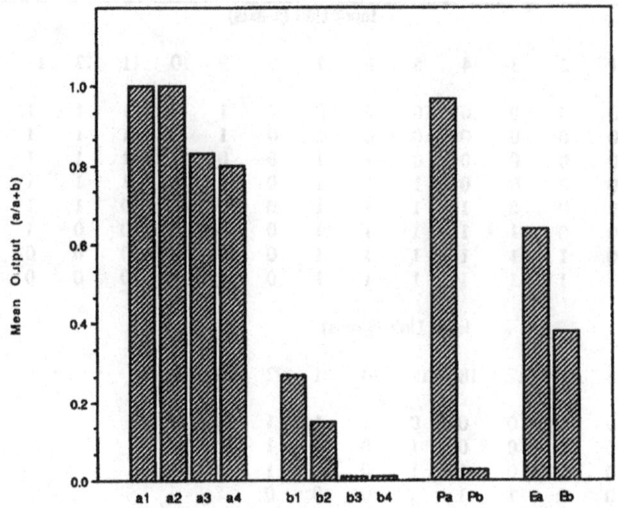

Figure 4. Results of the experiment. Stimuli a1, a2 a3 and a4 were the training stimuli for category A, stimuli b1, b2, b3 and b4 the training stimuli for category B. Pa and Pb were the prototypes for the two categories, and Ea and Eb the test patterns equidistant from the two prototypes.

to the x-dimension, and this is exactly what happens. The mean of the unsigned weights on the connections between the y-dimension input units and the output units was 0.083. For the x-dimension units the mean was only 0.049, that is, only 59% of the weight given to the y-dimension. Such differential weightings to different dimensions have been found in actual subjects' classification data by Nosofsky (1989).

It might appear, therefore, that the differential response to patterns Ea and Eb had little to do with the distances of these stimuli from the prototypes, but was instead simply due to the fact that the arrangement of the training patterns led to differential weights being given to the two dimensions. From this, an adversary of prototype theory might argue that the simulation was not correctly designed - what is needed is a demonstration that a connectionist model can produce exemplar effects even when the dimensions are given equal weight. The results of the present simulation give no reason to believe that this will be possible. But I shall argue that, in principle, it will always be the case that the dimensions in exemplar-effect experiments are given unequal weights.

In order to try to set up a situation in which the dimensions *are* given equal weights, we basically have to arrange that the stimuli have equivalent projections onto the various dimensions. Figure 5 shows the only two fully general ways of achieving this, and at the same time having categories that are equally easy to learn. In both cases the stimuli are symmetrically arranged around a diagonal. This ensures that the projections onto the x- and y-axes are identical, and hence that the two dimensions will achieve equal weights.

But the problem is that in neither of these cases is it possible to define a stimulus that is equidistant from the two prototypes, yet more similar to a training exemplar of one category than of the other. Inspection of the figures shows that the symmetry constraint will make this impossible. Therefore, I conclude that the following chain of reasoning must be valid: (1) It is impossible to set up a situation in which the dimensions are guaranteed to be given equal weights, while at the same time arranging for there to be a test stimulus which is both equidistant from the prototypes and at the same time more similar to a training pattern from one category. (2) Therefore, published demonstrations of exemplar effects, such as that by Homa, Sterling and Trepel (1981), must necessarily have used dimensions which the subjects would not weight equally. (3) Therefore, demonstrations of exemplar effects must necessarily be open to the interpretation that the effects do not emerge because of the specific exemplars, but because the space has become psychologically distorted, as happened in the simulation. Although this argument derives from consideration only of two-dimensional stimuli, I believe that it will generalize to multidimensional situations too.

We can conclude, then, that exemplar effects present no particular difficulty for associative theories of classification.

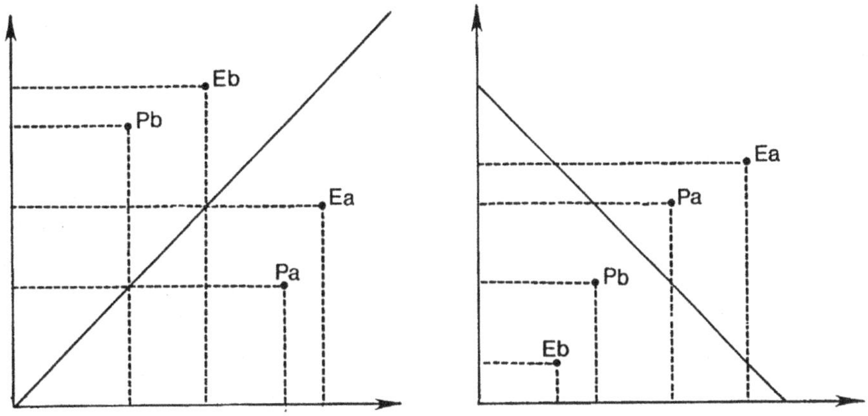

Figure 5. Possible arrangements of stimuli which arrange for the weights given to the dimensions to be equal. Pa and Pb refer to the prototypes for two categories, A and B, while Ea and Eb are typical training patterns from the categories. In each case, symmetry ensures that the categories will be equally difficult to learn.

Acknowledgements

I thank Ian McLaren for his many helpful comments.

References

Anderson,J.A. & Mozer,M.C. (1981). Categorization and selective neurons. In Hinton,G.E. & Anderson,J.A. (Eds.), *Parallel Models of Associative Memory*. Hillsdale, N.J.: Lawrence Erlbaum.

Estes,W.K., Campbell,J.A., Hatsopoulos,N. & Hurwitz,J.B. (1989). Base-rate effects in category learning: A comparison of parallel network and memory storage-retrieval models. *Journal of Experimental Psychology: Learning, Memory and Cognition, 15*, 556-571.

Gluck,M.A. & Bower,G.H. (1988). From conditioning to category learning: An adaptive network model. *Journal of Experimental Psychology: General, 117*, 227-247.

Hintzman,D.L. (1986). "Schema abstraction" in a multiple-trace memory model. *Psychological Review, 93*, 411-428.

Homa,D. Sterling,S. & Trepel,L. (1981). Limitations of exemplar-based generalization and the abstraction of categorical information. *Journal of Experimental Psychology: Human Learning and Memory, 7*, 418-439.

Knapp,A.G. & Anderson,J.A. (1984). Theory of categorization based on distributed memory storage. *Journal of Experimental Psychology: Learning, Memory and Cognition, 10*, 616-637.

Medin,D.L. & Schaffer,M.M. (1978). Context theory of classification learning. *Psychological Review, 85*, 207-238.

Nosofsky,R.M. (1989). Further tests of an exemplar-similarity approach to relating identification and categorization. *Perception and Psychophysics, 45*, 279-290.

Rescorla,R.A. & Wagner,A.R. (1972). A theory of Pavlovian conditioning: Variations in the effectiveness of reinforcement and nonreinforcement. In Black,A.H. & Prokasy,W.F. (Eds.), *Classical Conditioning II: Current Theory and Research.*. New York: Appleton-Century-Crofts.

Shanks,D.R. (1990). Connectionism and the learning of probabilistic concepts. *Quarterly Journal of Experimental Psychology, 42A*, 209-237.

Widrow,B. & Hoff,M.E. (1960). Adaptive switching circuits. *Institute of Radio Engineers, Western Electronic Show and Convention, Convention Record, Part 4*, 96-104.

COLLECTIVE BEHAVIORS

Task differentiation in Polistes wasp colonies : a model for self-organizing groups of robots

Guy THERAULAZ [*], Simon GOSS [**]
Jacques GERVET [*] and Jean-Louis DENEUBOURG [**]

[*] CNRS - UPR 38, 31 chemin Joseph Aiguier, 13402 Marseille cedex 09, France
[**] Unit of Theoretical Behavioural Ecology, CP 231, Université Libre de Bruxelles, 1050 Bruxelles, Belgium

Abstract

In this paper we described a simplified model for a case of functional self-organization. It deals with the emergence of a particular form of task assignment and parallel hierarchical organization within a social group which depend basically on the interactions occurring between individuals and with their immediate local surroundings. The task organization within the colony appeared to be a distributed function which does not require the presence of an individualized central organizer. We discussed how such elementary processes could potentially be applied in the coordination and self-organization of groups of interacting robots with simple local computational properties to perform a wide range of tasks.

Introduction

In eusocial insect societies, all the individuals must cooperate to perform a certain number of tasks the nature of which depends on the internal needs of the colony, as well as on the particular environmental conditions. At any time each individual can act and interact either with other individuals or with their environment and thus causes changes in the state of the group. The group is nevertheless the focal point for a **stable, self-regulated organization** of individual behaviors. The study of the processes leading to **the emergence of a stable collective order** in insect societies has recently emphasized the importance of individual interaction dynamics (Deneubourg et al., 1987; Deneubourg and Goss, 1989; Goss et al., 1990; Beckers et al., 1990). This research has demonstrated that quite simple elementary rules of individual behavior often make it possible for the society to create surprisingly complicated patterns and to make efficient decisions when certain types of external constraints are encountered.

Our biological study examined the processes involved in **task assignment** in primitive *Polistes* wasp colonies (Theraulaz et al., 1990 a, b and c). *Polistes dominulus* is a species which is to be found in temperate, northerly regions. These wasps had two advantages from the point of view of our study, since they build their nest with no envelope and their colonies contain a relatively small number of individuals ($n \approx 20$), which makes it possible to observe all the members of a colony over a period of time. These primitively eusocial species have little individual differentiation and no morphological differences between castes or predetermined control of activities depending on age or on any other known physiological predetermination. The integration and coordination of individual activities therefore depend largely on the interactions which take place among the members of these societies and on the immediate relationships between a society and its environment.

A two fold **morphogenetic process** thus occurs within the colony :

• Each individual acquires its own **behavioral profile** which is characterized by all the observable stable behavioral items in which it takes part. All the profiles can be described by a reduced number of **behavioral forms** to which the various individual profiles belong (see Theraulaz et coll., 1990 a).

• In a society at a given moment in time, the whole set of individual behavioral profiles does not constitute a random sample of all the possible profiles; they constitute a **profile configuration** which can be defined by the proportion of individuals belonging to each of the behavioral forms.

The way this configuration is controlled by internal and external constraints on the colony constitutes the task assignment process. The model presented here aims at describing the task assignment process in a hierarchically structured society.

Results

1. Characteristics of individual interactions

Two types of interactions were found to control individual behavior in *Polistes dominulus* Christ colonies:

1.1. a *hierarchical type* of interaction which can be observed during dominance scenes. The hierarchy, which is initially set up by fighting, or sometimes just by the preliminaries to fighting, results in complementary poses being adopted within any pair of individuals which come into contact : the one performs an act of domination, and the other an act of submission. The constancy of these relationships among individuals are part of a larger **linear hierarchy** ($\alpha > \beta > \gamma > ... > \omega$) extending throughout the society as a whole. Each individual can thus be characterized by its social rank, which indicates the number of individuals towards which it must behave submissively. The animal of rank 1 is called by definition α individual.

Figure 1 gives the ratio between the number of dominance acts performed and the total number of hierarchical interactions (dominances + subordinations) depending on the hierarchical rank. This ratio expresses the probability that an individual will dominate during a hierarchical interaction. This probability decreases according to the individual's hierarchical rank. The relationship between an individual's hierarchical rank and its propensity to perform acts of domination towards other members of the colony is also shown (see figure 1). This is expressed for an individual of a given rank by the ratio between the number of acts of domination it performs to the total time spent at the nest by all the lower-ranking individuals.

In general, the more subordinate the individual is, the less likely it is to show dominance behavior. It can also be seen that the higher the probability that an individual will dominate in a hierarchical interaction, the greater its propensity will be to frequently dominate its subordinates. Another factor which is concomitant with the hierarchical rank is **the degree of mobility** on the comb. Figure 2 shows the variations in the time spent walking on the nest during the accession to a given rank from a lower one. It demonstrates that the α individual developed hyperactive behavior on the nest, while individuals which had just moved up to ranks 2 to 12 also showed a constant but much lower increase in activity.

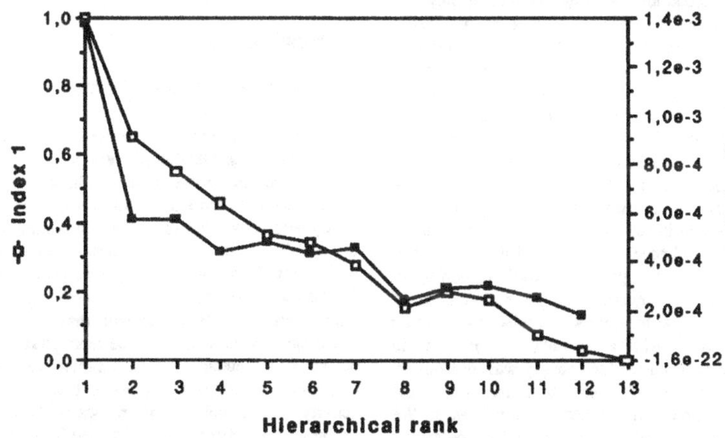

Fig. 1

Index 1 : Dominance Index :

$$\frac{\textbf{Dominances}}{\textbf{Dominances + Subordinations}}$$

Index 2 : **Relative dominance rate on the nest :**

$$\frac{\textbf{Number of dominance acts per time unit at the nest}}{\sum \textbf{time spent at the nest by subordinate individuals}}$$

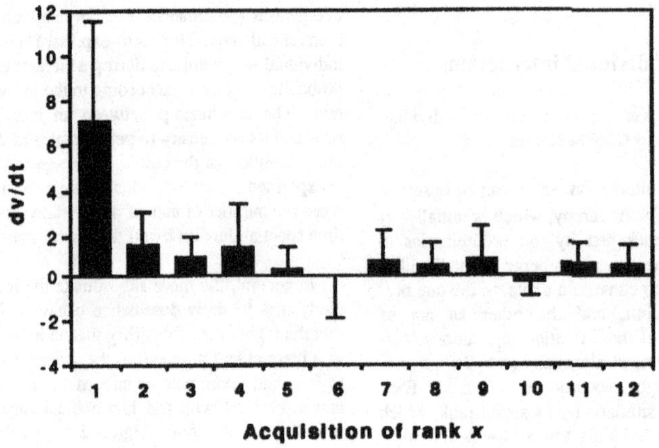

Fig. 2

1.2. A *trophic type* of interaction controlling relationships between individuals and the environment, particularly with the brood, which in turn triggers some types of activities towards the brood, such as brood-tending and/or collecting prey and the subsequent feeding of larvae. We have shown that a **differential reactivity** exists between the main characteristic behavioral profile types in a colony and the various brood components, that is the eggs, small and large larvae, and pupae (Theraulaz et al., 1990 b and c). For exemple, some individuals characterized by a foraging profile have a low response threshold to stimulations from large larvae. Other types of profiles reacted on the contrary to an increase in the number of large larvae by intensifying their brood-tending activities, but took no part in foraging. The increase in the number of large larvae, and hence in the needs of the colony, had no effect on the other types of profiles. In addition, during experiments in which we investigated the replacement of foragers, it was found that foragers were replaced by a small number of descendents with a tendency to specialize in foraging. In addition, when the food supply was insufficient, the intensity of the larval stimulation increased, and consequently the number of foragers also increased. We attributed this process to a self-facilitating process controlling foraging for prey activity. This type of process has also been observed in other social insects, particularly among foragers in *Cataglyphis bicolor* (Wehner et al., 1983) and *Neoponera apicalis* ants (Deneubourg et al., 1987).

We hypothesized that first, the animal reinforces its probability of dominating through the number of times it is successful in hierarchical interactions and each subordination has the opposite effect. Secondly, there is a parallel self-facilitating process which causes the increase in the amount of time spent foraging by individuals which have already foraged previously.

2. Model description

2.1. Local specification of system elements

The model proposed here describes **task assignment and parallel colony organization processes** without establishing any direct causal links between the two. In order to account for the various parallel events occurring in a colony, we investigated the dynamic properties of systems consisting of small number of **mobile units and possessing local computational properties.** In many ways, eusocial insects societies can in fact be taken to consist of several relatively simple units interacting with each other and with their immediate local surroundings. So, each wasp can be represented by a unit having the ability to interact both with similar units and with some characteristics of the environment. Our model describes the activity of a colony of n wasps where dominance-subordination relationships exist, and integrates two local action rules. Each individual i is characterized by a variable force F_i which intervenes in random encounters within the nest, and by a reponse threshold σ_i to larval stimulation from a specific zone of the nest : that occupied by the larval brood.

At the beginning of each stimulation, the n individuals are identical. At each time interval, the order of the n individuals' passage in the algorithm is selected at random.

2.1.1 Hierarchical interaction

During movements within the nest, two individuals i and j occasionally meet (i.e. occupy nearby positions within the space mesh) (see Fig. 3). These two individuals then interact (1), and the outcome of the interaction (expressed here by the probability of success of individual j) is determined by the Fermi function of the force variables F_i and F_j :

$$\mathcal{F}(F_i,F_j) = \frac{1}{1 + e^{\eta(F_i - F_j)}}$$

2.1.2. Foraging Activity

The system describes a colony where wasps are confronted with a single foraging task (hunting and carrying food) and brood care. The brood occupies a specific portion of the nest space, thus creating a lack of spatial homogeneity, and is characterized by an overall variable D which expresses the total larval demand for food and care. The brood has its own way of functioning which expresses the evolution of the degree of hunger satisfaction over time. Thus, at every time interval p, the overall value of the demand D increases by a quantity d, expressing an increase in the overall larval demand.

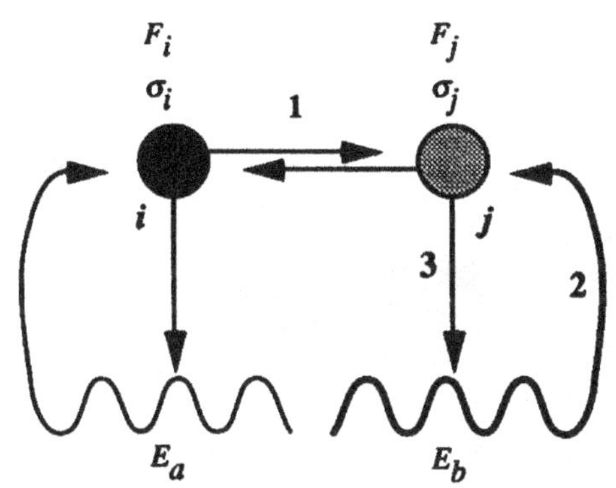

Fig. 3 : Individual local action rules in Polistes

1. : Hierarchical interaction between i and j individuals ; 2. : Stimulation capable of triggering foraging ; 3. : Foraging activity : this one decreases larval demand ; E_a,E_b : Social demand in two brood zones $E_a \ll E_b$; F_i, F_j. : Force of i and j individuals ; σ_i, σ_j. : Foraging threshold of i and j individuals.

Coefficient η is used to modulate the Fermi function by regulating the extent to which the outcome of the interaction is predetermined ($\eta \rightarrow \infty$ corresponds to an all or nothing response). Although from a biological point of view the choice of the Fermi function was arbitrary, we used it because it is a classic exemple of function with two parameters, which easily allow us to modulate both the threshold and the deterministic aspect of the response. The dominant animal's force F thus increases by ε, while the subordinated animal's force decreases by the same ε. The parameter ε is **an intrinsic coefficient of reinforcement** which is common to all the individuals. During stimulation, the force variable F evolves within the interval $[F_{min}, F_{max}]$, where F_{min} and F_{max} are the minimal and maximum values of F.

Updating the demand to time period t is shown by the following formula :

$$D_{(t)} = D_{(t-1)} + d_{(t-1)} - W_{(t-1)},$$

where W is the work performed during the time period t-1 by all the individuals in the colony. At each interval, and as long as the individual considered is on the brood area (i.e. in the presence of stimulations), we tested whether the individual j would or would not forage for food.

The outcome (i.e. the probability of leaving to forage for food) was determined using the Fermi function with the two following variables D_t and σ_j, demand at time t and the individual j's response threshold :

$$\mathcal{F}(\sigma_j, D_t) = \frac{1}{1 + e^{\eta(\sigma_j - D_t)}}$$

therefore (see Fig. 3) when some types of stimulation (E_b) emanating from the brood (2) trigger a foraging or feeding activity (3) in an animal j which is located nearby, the response threshold of this individual σ_j to the same stimulations decreases by a quantity ξ, which thus increases the probability of the animal responding in a similar way when it once again comes into contact with these stimulations. The parameter **ξ is an intrinsic learning coefficient** which is common to all individuals in the colony. However, when an individual does not forage during the time interval t, the value of its foraging threshold increases by a quantity ϕ which constitutes **a forgetting coefficient** common to all the elements in the system. During simulation, the variable

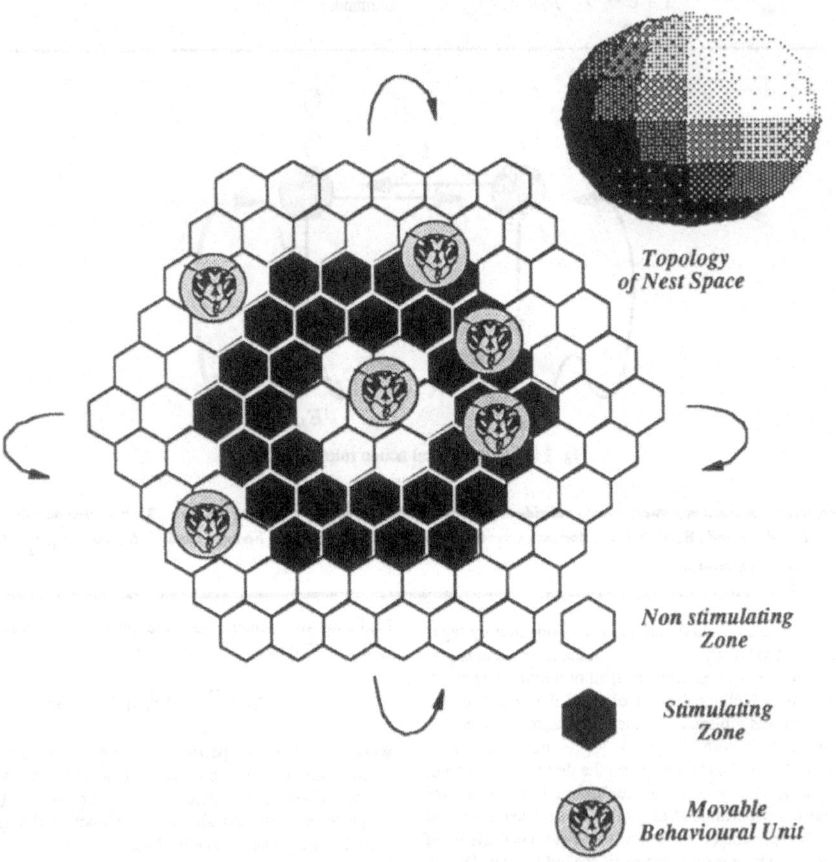

Fig. 4 : Structuration of the nest space

evolves within the interval (σ_{min}, σ_{max}), where σ_{min} and σ_{max} are the minimum and maximum values of σ.

2.2. Structuration of nest space

The model proposed here takes **spatial differentiation** into account. In fact, the space in which the units move in our system is conventionally made up of two non-stimulating zones separated by a zone where larval stimulations act directly (see figure 4). Each unit will thus be either subjected or not to larval stimulation depending on its position within this space. The nest has an ovoïd shape.

2.3. Movement on the Nest

The speed of individual i's movements on the nest is directly proportional to the force F_i of this individual. The more force it has, the greater the amount of distance it can cover per time interval, and the more it will tend to interact with other individuals in the colony, depending on the individual force of the latter. However, individuals which are foraging for food during a time period t_f cannot engage in any hierarchical interaction.

3. Overall system behavior and comparison with biological data

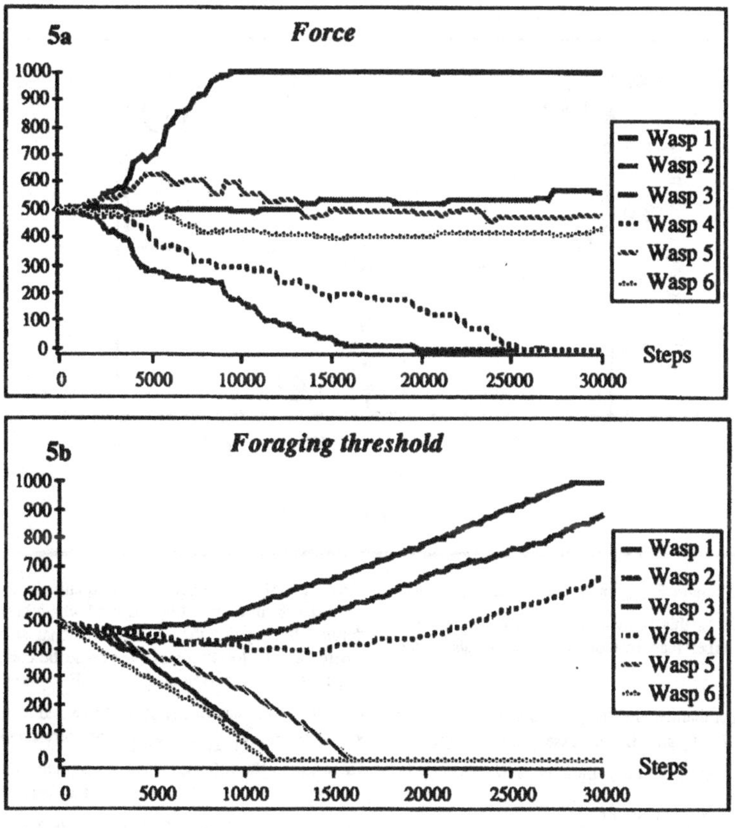

Fig. 5

3.1. Combined changes in force and foraging threshold

If we now consider how the system as a whole behaves, it can be seen that after a certain period of time, it will evolve towards one of its **attractors**, namely towards **a stable configuration** formed by the individual properties of the constituent units. For a colony with n = 6 individuals : figures 5a and 5b show the changes in the force and foraging threshold variables for each individual. Each individual has the same characteristics at the outset. All the simulations performed made it possible to isolate 3 main types of pattern in the development of individual characteristics.

3.1.1. A single individual, here I_1, stands out from the others quite early since its force increases gradually and regularly up to the maximum F_{max} value. At the same time, the foraging threshold of this individual tends to increase and always remains higher than that of other individuals in the colony.

3.1.2. Concomitantly, the foraging thresholds of other individuals, here I_3, I_5 and I_6, decrease until reaching the minimum value. These individuals thus specialize early on in foraging activities and provide all the colony's needs. They all three also maintain a high level of force.

3.1.3. Finally, other individuals (I_2 and I_4) maintain a high foraging threshold and rarely participate in foraging for food, while their force decreases rapidly. They remain at the nest and are often subjected to frequent hierarchical interactions with individual α.

The combined action of the two processes we used to characterize individual behavior generates a stable hierarchy among the individuals, at the top of which α individual never leaves the nest over which it performs a considerable amount of dominance activity, while at the same time its response threshold to larval stimulations increases. The process automatically leads to the specialization of some individuals for the tasks of foraging and feeding the larvae in the nest.

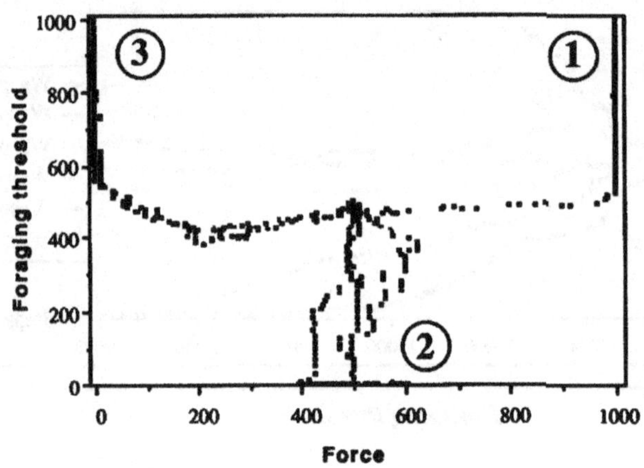

Fig. 6

3.2. Study of states space

Here, a system composed of a relatively low number of elements or units (n ≤ 13) was used where behavior could be defined by a point in a two-dimensional states space, and where each of the two variables could be characterized by a minimal and a maximum value. If F_{min} and σ_{min} are the respective minimal values of the force and foraging threshold variables, and F_{max} and σ_{max} the corresponding maximum values, then **the internal state** of each element in the system can be characterized by a point in the plane $[(F_{min}, F_{max}), (\sigma_{min}, \sigma_{max})]$. If the combined behavioral changes in the units in a system with n = 6 elements are represented in this plane, the graph in figure 6 is obtained. Here, the development over time of the six elements from their undifferentiated common origin ($F_i = 500$, $\sigma_i = 500$) up to the final equilibrium state of the system (after 30000 time intervals) is shown.

The individual trajectories converge towards 3 zones in the states space which constitute privileged **attractive configurations** :

1). one unit can be singled out by a maximum strength and a very high foraging threshold; by analogy with what we have observed in colonies, we have called this **unit** α.

2). other units, however, maintain a high force, with a minimal foraging threshold. These units are **specialists** and perform all the foraging tasks.

3). finally, some units are characterized by inverse properties, i.e. a minimum or very low force and a very high foraging threshold. It should be noted that these units also move very slowly across the nest and spend most of their time at the periphery of the nest. (nb. a high foraging threshold is equivalent to a low probability of foraging).

All the simulations performed show that with very different initial force and foraging threshold values, and with different parameters ε, ξ and ϕ, the final evolutionary state of the system converges towards a configuration characterized by the existence of 3 domains in the space of individual states. Only the relative proportions of individuals in either categories 2 or 3 vary; category 1 is always characterized by one single unit. This demonstrates the stability and reliability of the final attractive configuration in terms of the reinforcement values ε and ξ which intervene during hierarchical and trophic interactions.

When this configuration is compared with that observed *in vivo* using descriptors considered to be equivalent, the three zones of accretion can be distinguished which orient the distribution, except that there is a continuum between zones 2 and 3. Figure 7 shows the relationship which exists between :

1) on the one hand : 1/average foraging time per hour, which yields an index of the response threshold to larval stimulations,

2) on the other hand : the number of dominance acts performed/total number of dominance-subordination interactions indicates the probability of dominating during a hierarchical interaction, which is taken as being representative of the force value.

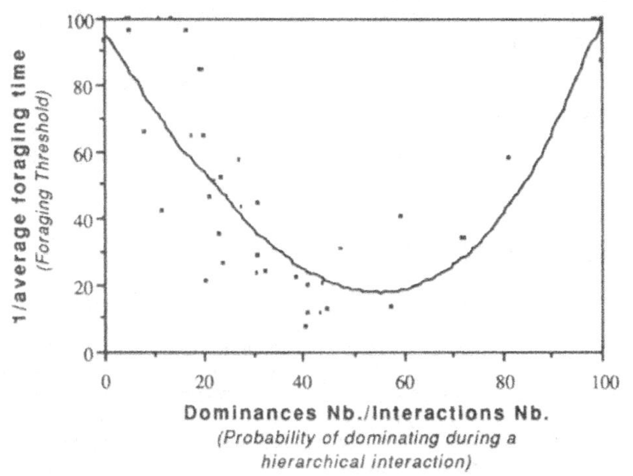

Fig. 7

The average values of these two indexes are given for each of the hierarchical ranks in 4 different societies. If behavioral forms appearing in the colony defined not by the 2 descriptors, but by 31 behavioral items are examined, it can be found that individuals integrated into the colonies show 3 main behavioral profiles : the single α individual characterized by the form 1 profile : it remains on the nest where it carries out a high level of dominance activity and activity in general; the foragers (form 2 profile) and finally inactive individuals located at the edge of the nest and often dominated, characterized by the form 3 profile (see Theraulaz et al., 1990 b). At a very general level, this model can thus reflect the **process of behavioral morphogenesis** at work in an insect colony.

Discussion and perspectives

We have presented here a deliberately simplified model which might be called a **functional self-organization** model: it involves the emergence of a certain form of task distribution within a social group generated fundamentally by interactions occurring both between individuals and with their immediate environment. In fact, **a collective pattern is generated** by individual entities which receive purely local information. Each individual enjoys autonomous behaviour and it is the history of its interactions with the other colony members and with the environment which shapes the individual's own intrinsic characteristics. Two comments can be made on the model's functional level:

- The task organization in the group can be said to be **a distributed function** which does not require the presence of an individualized central organizer. However, all the elements do not play an identical role but differences in function emerge with time which did not necessarily exist before.

- Group function is generated by a double feed-back system controlling individual behavior :

. **Positive feedbacks** rapidly amplify small random fluctuations among the competing units, which have morphogenetic effects. In the case of the force value, direct competition among individuals leads to the hyperactivity of a single element up to the biologically (i.e. formally) possible limit. The more indirect nature of the competitive interaction between the foragers explains why elements can reach the limit value in variable numbers depending on the values given to the model's parameters (this number is particularly a function of the intensity of larval demand and foraging time).

. A double system of **negative feed-back** regulates the stability of performances and thus has morphostatic effects. Apart from the α individual, the force value of other elements stabilizes around two other values which constitute the function attractors: any random increase (or decrease) in force causes an increase (or decrease) in encounters with stronger subjects which reestablishes the initial level. The cyclical relationship between brood satisfaction and foraging activity maintains a regular level of social demand which regulates the number of foragers, i.e. the elements with a minimum foraging threshold.

These rules suffice to compensate for the risks associated with the disappearance of the α individual or that of the foragers and to quickly reproduce an organization similar to that which was fortuitously destroyed (Theraulaz et coll., 1990 d). Thus simple and strictly local rules of environmental response allow for the appearance, stabilization and reestablishment of collective forms of activity when interactions exist between elements. In addition, these structures, by interacting with the environment, generate different types of collective functions from the same elementary behavior algorithms. This type of model combines simplicity, reliability and adaptability and can be used for problems of coordination of the performance of complex tasks by groups of robots with simple reshaping properties .

This model can in fact be extended to **multi-task systems** by determining a response threshold to the different tasks introduced. In the same way, more complex relationships can also be established between response thresholds, and the development of the spatial organization of the environment can be studied as a structuring source for the system, where both mutually contribute to specifying each other.

Acknowledgments

We wish to thank Dr. Jessica BLANC for revising the english version. This study was supported by a doctoral grant from the French Ministry of Scientific Research.

References

BECKERS R., DENEUBOURG J.L., GOSS S. and J.M. PASTEELS (1990). Collective decision making through food recruitment. *Ins. Soc.* (in press).

GOSS S., BECKERS R., DENEUBOURG J.L., ARON S. and J.M. PASTEELS (1990). How trail laying and trail following can solve foraging problems for ant colonies, in **Behavioural Mechanisms of Food Selection**, R.N. HUGHES ed. *NATO ASI Series*, Vol. G 20, pp. 661-678. Springer-Verlag, Berlin, Heidelberg.

DENEUBOURG J.L., GOSS S., PASTEELS J., FRESNEAU D. and J.P. LACHAUD (1987). Self-organization mecanisms in ant societies (II) : learning in foraging and division of labour, in **From Individual to Collective Behaviour in Social Insects**, J.M. PASTEELS & J.L. DENEUBOURG eds. *Experientia Supplementum* vol. **54**, pp. 177 - 196. Birkhäuser Verlag, Basel.

DENEUBOURG J.L. and S. GOSS (1989). Collective patterns and decision-making. *Ethology Ecology and Evolution* 1, pp. 295 - 311.

THERAULAZ G., M. PRATTE and J. GERVET (1990 a). Behavioural profiles in *Polistes dominulus* (Christ) wasp societies : a quantitative study. *Behaviour* **113**, pp. 223-250.

THERAULAZ G., J. GERVET, B. THON, M. PRATTE and S. SEMENOFF (1990 b). The dynamics of colony organization in the primitively eusocial wasp *Polistes dominulus* (Christ) *Ethology* (in press).

THERAULAZ G., J. GERVET and S. SEMENOFF (1990 c). Social regulation of foraging activities in *Polistes dominulus* Christ wasp colonies : a systemic approach to behavioural organization. *Behaviour* (in press).

THERAULAZ G., GOSS S., GERVET J. and J.L. DENEUBOURG (1990 d). Self-organization of behavioural profiles and task assignment based on local individual rules in the eusocial wasp *Polistes dominulus* Christ, in **Social insects and the environment,** *proceedings of the symposium Models and Theoretical Approches to the Study of Social Insects*, 11th International Congress of IUSSI 1990, Bangalore, pp. 535 - 537. Oxford & IBH Publishing Co. Pvt. Ltd, New Delhi, Bombay.

WEHNER R., HARKNESS R.D. and P. SCHMID-HEMPEL (1983). **Foraging strategies in individually searching ants *Cataglyphis bicolor*.** G. Fisher Verlag, Stuttgart.

THE DYNAMICS OF COLLECTIVE SORTING ROBOT - LIKE ANTS AND ANT - LIKE ROBOTS

J.L. Deneubourg[1], S. Goss[1], , N. Franks[2], A. Sendova-Franks[2], C. Detrain[1] and L. Chrétien[1]

[1] Unit of Behavioural Ecology, CP 231
Université Libre de Bruxelles, 1050 Bruxelles

[2] School of Biological Sciences
University of Bath, Bath BA2 7AY

Abstract

A distributed sorting algorithm, inspired by how ant colonies sort their brood, is presented for use by robot teams. The robots move randomly, do not communicate, have no hierarchical organisation, have no global representation, can only perceive objects just in front of them, but can distinguish between objects of two or more types with a certain degree of error. The probability that they pick up or put down an object is modulated as a function of how many of the same objects they have met in the recent past. This generates a positive feed-back that is sufficient to coordinate the robots' activity, resulting in their sorting the objects into common clusters. While less efficient than a hierarchically controlled sorting, this decentralised organisation offers the advantages of simplicity, flexibility and robustness.

1. Introduction

What is the common point between a shopkeeper and an ant colony? Each of these organisms is able to sort similar but different objects. When one examines an ant nest it is clear that neither the workers, the brood nor the food are randomly distributed. For example the eggs are arranged in a pile next to a pile of larvae and a further pile of cocoons, or else the three categories are placed in entirely different parts of the nest. The same is true in a shop. There is, however, an essential difference. The shopkeeper decides where he is going to put his different goods, and if he has assistants he tells them where to place what. Ants work in parallel but do not, as far as we can tell, have the capacity to communicate like the shopkeeper, nor do they have a hierarchical organisation whereby one individual makes the necessary decisions and the others follow. Nevertheless, if you tip the contents of a nest out onto a surface, very rapidly the workers will gather the brood into a place of shelter and then sort it into different piles as before.

This article describes a simple behavioural algorithm, to be followed by each worker, that generates a sorting process. Sorting is achieved without requiring either external heterogeneities (e.g. temperature or humidity), hierarchical decision-making, communication between the individuals or any global representation of the environment. We also stress that the ants/robots have only very local information about the environment and a very short-term memory, and furthermore move randomly, no oriented movement being necessary. They can't see far off nor move directly towards objects or piles of objects.

Our aim in this article is not to prove that the model proposed is actually how the ants behave, but to show that such an algorithm both works and could be used by a team of robots. Inspired from our knowldege of the importance of functional self-organisation or distributed intelligence in ant colonies (Deneubourg, 1977; Deneubourg et al., 1984, 1986, 1987; Deneubourg and Goss, 1989; Goss et al., 1990; Aron et al., 1990), our idea presents a working illustration of how such a distributed system can have practical applications in robotics, in accordance with ideas developed by ourselves (e.g. Deneubourg et al., 1984, 1990; Deneubourg and Goss, 1989), and others (e g. Beni, 1988; Brooks and Flynn, 1989; Sandini and Dario, 1989; Fukuda and Kawauchi, 1989; Brooks et al., 1990; Steels, 1990). The environment for a real-time multi-robotic demonstration of this algorithm (and others related to ant-like foraging behaviour) is under preparation at MIT, directed by Maja Mataric and Rodney Brooks.

Firstly we describe a Monte-Carlo simulation of this collective behaviour, secondly we present the continuous mathematical model and its steady state analysis.

2. Monte Carlo Model

The model is based on the following principle. The ALRs or RLAs (ant-like robots or robot-like ants) move only randomly. When they come across an object the probability of picking it up is all the greater the more the object is isolated, i.e. the less the number of similar objects there are in the immediate neighbourhood. When carrying the object an ALR's probability of putting it down is all the greater as there are more of the same in the immediate neighbourhood. Either of these two rules is sufficient to form separate clusters of the two object types, but both rules together act much faster.

Isolated objects are picked up. When a small, albeit loose cluster appears by chance, it "encourages" passing carriers to add their load to the cluster thus increasing its "attractivity". With this positive feed-back mechanism the clusters grow, "absorbing" isolated objects and the smaller clusters through the action of the carriers. As clusters of each object type "attract" and fill up the nearby space with essentially the same type, this effectively isolates any object of another type in the immediate vicinity, thus making it more likely to be picked up by an ALR. Sorting is the consequence of this clustering and crowding out behaviour.

The environment is a square network of points. At time zero a number of ALRs and objects of type A and B are placed at random in the network, only one object and/or one ALR being allowed at each point. At each time step, and in random order, the ALRs move randomly north, south, east or west, although they cannot move into the wall around the space or into a point already occupied by an ALR. When an ALR moves onto a point containing an object it decides whether or not to pick it up. The less objects of the same sort there are in the immediate enviroment, the greater the probability that it will pick it up, as given by the following function:

$$p\ (\text{pick up}) = (k^+ / (k^+ + f))^2$$

where f is an estimation of the fraction of nearby points occupied by objects of the same type, and k^+ is a constant. The probability thus decreases with f, from 1 (when f=0), to 1/4 (when $f = k^+$), and less as f tends to 1.

An ant could (probably) estimate f by the strength of an odour associated with each brood type or else by tactile investigation, sight being less important as such sorting is usually done underground. A robot would need a rather sophisticated visual, chemical or other sensory system to do the same, and so we propose the following sampling-based estimation which, while less precise, has the advantage of being much simpler and more easily implemented.

Each ALR has a short-term memory of m steps, that records what it met in each of the last m time steps. Thus at t=10, a memory of length 10 could hold the string 00AB0AA0B0, indicating that during the previous ten time steps the robot met 3 objects of type A and 2 of type B, the other points having been empty. f_A would be equal to 3/10, and f_B to 2/10.

As the robot walks randomly, this sampling provides a rough estimation of the density of the two sorts of objects in the immediate neighbourhood. Note that similar sampling techniques are known or suspected to exist in ants (e.g. Lumsden and Hölldobler, 1983) and could possibly be at work in the way they decide to pick up or put down a larva.

As the memory is ten steps long, at the eleventh step the robot would forget what it met at the first step and add what it met at the eleventh step. In this example the string could become 0AB0AA0B0A if it next encountered a type A object.

Whatever its decision, the ALR then carries on its random walk. If it has picked up an object, then at each step that it finds itself in an unoccupied point it decides whether or not to put the object down. The more objects of the same sort there are in the immediate environment, the greater the probability that it will do so, as given by the following function:

$$p\ (\text{put down}) = (f / (k^- + f))^2$$

where f is as before, and k^- is a constant. The probability thus increases with f, from 0 (when f=0), to 1/4 ($f = k^-$), and more as f tends to 1.

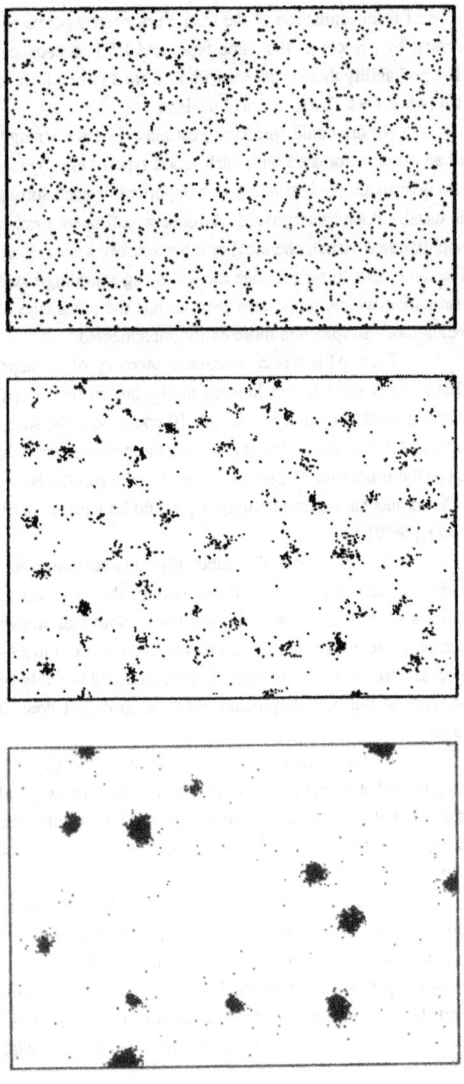

Fig. 1. Clustering after 1, 100000 and 2000000 steps. 100 ALRs, 1500 objects, $k^{+}=0.1$, $k^{-}=0.3$, m=50, e=0, space=290x200 points. Small evenly spaced clusters rapidly form, and later merge into fewer larger clusters.

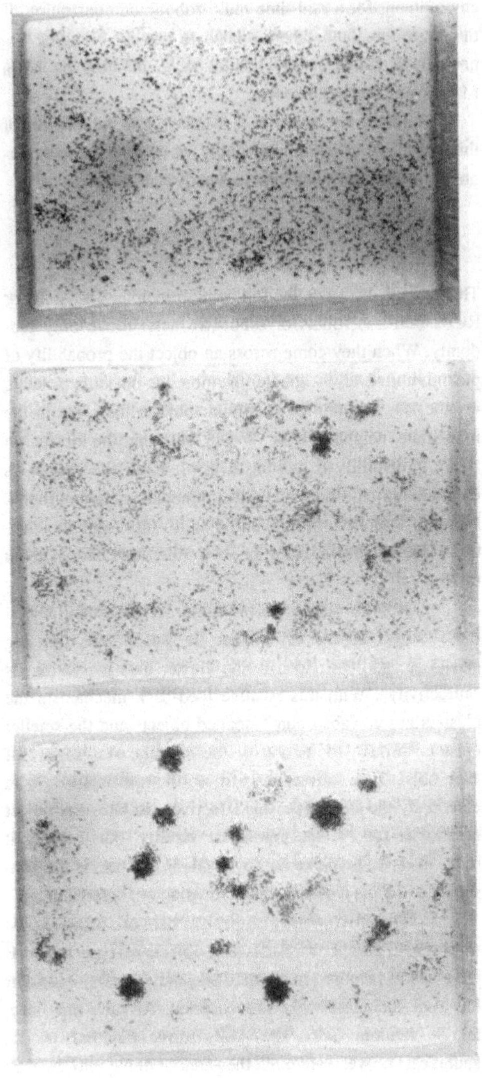

Fig. 2. Clustering in a colony of *Pheidole pallidula*. 4000 corpses were placed on a 50x50cm arena, and photos taken at time 0, 20 and 68 hrs. Small evenly spaced clusters rapidly form, and later merge into fewer larger clusters.

3. Monte Carlo Simulations and Comparable Experiments With Ants

Fig. 1 shows how randomly distributed objects of one type are rapidly gathered into small and regularly spaced clusters, which over a longer period of time gradually merge into a smaller number of larger clusters. The clusters are constantly having elements removed and added, and therefore tend to drift about slowly. When two clusters meet they fuse.

Fig. 2 shows a similar process in a colony of *Pheidole pallidula*. When ants die, workers carry the corpse out of the nest, and in laboratory conditions place them in a pile, a behaviour common to many ant species. In this experiment, a large number of ant corpses were spread out on an arena. Very quickly the workers (or robot-like ants) gathered them into a number of small clusters, which after a long period of time merge into one or two large clusters (the experiment shown in fig. 2 did not run long enough for this last stage to be seen).

Fig. 3 shows how randomly distributed objects of type A and B are rapidly sorted into small clusters of each type, which again over a longer period of time gradually merge into a smaller number of larger clusters of each type.

The parameter values used have been selected more or less arbitrarily, helped by the fact that the model sorts efficiently within a wide range of parameter values. A more formal analysis of the influence of the parameter values will be performed elsewhere on the continuous model described below. It is nevertheless clear that a very long memory length prevents effective sorting or clustering as it gives the ALRs the equivalent of a perceptive radius as large as the space they operate in, and so they could not distinguish between an isolated object and one in a local cluster.

An interesting variant of the model introduces some overlapping, or imperfection of discrimination, between the two sorts of objects, in the following sense. An unloaded ALR has met an object of type A and must thus calculate f for A, or f_A, being the number of objects A met in the last m steps divided by m. However one can introduce some confusion between A and B by calculating f_A as the number of objects of type A plus a fraction, e (the error rate), of objects of type B, the sum being divided by m. In the example above, 00AB0AA0B0, f_A would become (3 + 2e)/10. This is equivalent to the ALRs making a certain

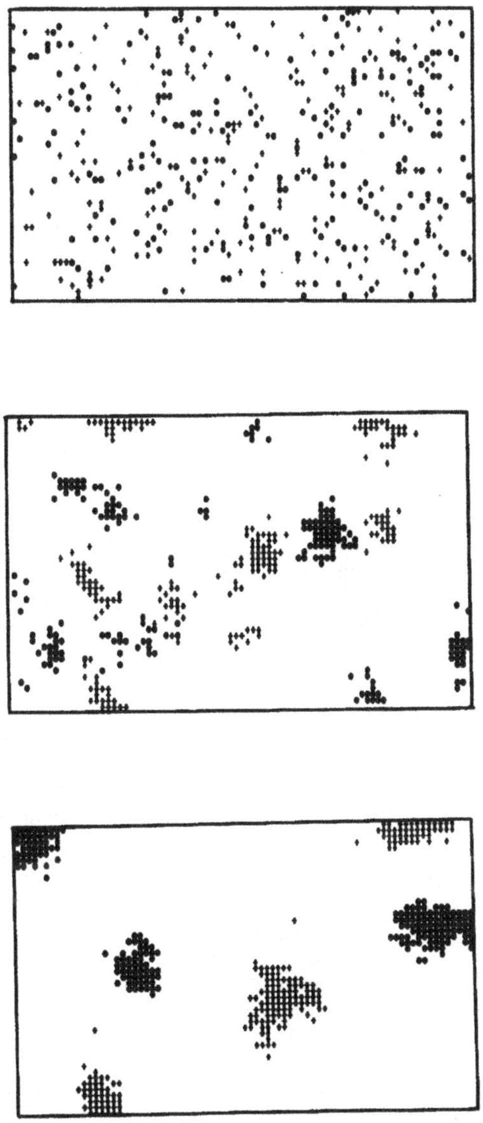

Fig. 3. Sorting after 1, 60000 and 570000 steps. 20 ALRs, 200 o and 200 + objects, k^+=0.1, k^-=0.3, m=15,e=0, space=80x49 points. Small evenly spaced clusters of each object type rapidly form, and later merge into fewer larger clusters with a high degree of sorting.

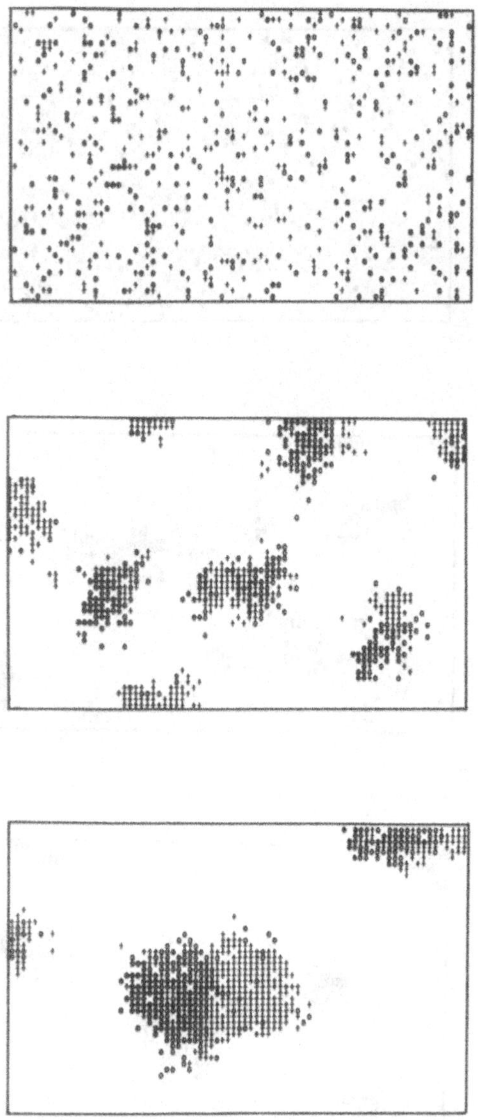

Fig. 4. Sorting after 1, 225000 and 1660000 steps. 20 ALRs, 300 o and 300 + objects, $k^+=0.1$, $k^-=0.3$, m=15, e=0.2, space=80x49 points. Small evenly spaced clusters, containing both types of object but placed adjacently, rapidly form, and later merge into fewer larger clusters, with a high degree of sorting.

Fig. 5. Sorting 1h, 1h40m and 72h after a nest of *Leptothorax unifasciatus* has been mixed up. The larvae are rapidly gathered into a number of small clusters, and later merge into one large cluster with the small larvae (centre left) clearly separated from the large larvae (centre and top and bottom left).

number of mistakes in their identification of the two types of object.

As on might expect, with a large error rate ($e>0.3$), the objects are clustered indiscriminately and not sorted. With a small error rate ($e<0.1$), the sorting is more or less as efficient as with no error (fig.3). With an intermediate error rate, the objects are efficiently sorted, but into overlapping clusters whose dispersion is thus reduced (see fig. 4).

Figure 5 shows a similar process in an ant colony. A small microscope-slide nest of *Leptothorax unifasciatus* is tipped out onto an area. The workers rapidly bring the larvae back into the nest (5a), placing them in small piles. At this point the sorting is not very strong. Rapidly the piles merge (5b), and eventually one big pile is formed (5c) with the small larvae clearly separated from the large larvae.

4. The Continuous Model

(1) $\delta_t O_i = b_i(1-\Sigma O_i/e)R_i - aO_iR_0/b_i$

(2) $\delta_t R_i = D\nabla^2 R_i - \delta_t O_i$

(3) $\delta_t R_0 = D\nabla^2 R_0 + \Sigma\delta_t O_i$

(4) $b_i = (b + b'\Sigma\int_{-\infty}^{\infty} g_{ij}(r,z)O_j(z)dz)^n$

with:

O_i the number of objects of type i
R_i the number of ALRs carrying an object i
R_0 the number of ALRs not carrying an object
b_i the probability of putting down an object i
a/b_i the probability of picking up an object i
e no of empty places per unit surface
g_{ij} the distance-interaction function between objects i and j
D, a, b and b' being constants, and $n>=1$.

The continuous model functions more or less exactly as the Monte Carlo model. Equation (1) states that the number of objects i added, at a given point and time, is given by the product between the number of ALRs carrying an object i, the probability of putting down an object i and the fraction of room left at that point and time. The number of objects removed is a product of the number of empty ALRs and the probability of picking up an object i.

Equations (2) and (3) state that the robots diffuse randomly, and that empty and carrying ALRs are "created" by the putting down and picking up process. Equation (4) expresses the probability of picking up (the inverse of putting down) as a distance-dependant function of the number of objects of different types both in the immediate neighbourhood and, to a lesser extent, further away. The distance-interaction function expresses an instantaneous distance-dependant measure of the number of each object type, and the interaction between the different types of object. Note that in the Monte Carlo simulations the corresponding n was equal to 2.

The complete stability analysis will be published elsewhere, but it is already clear that, depending on the parameter values, the model exhibits either only a homogeneous solution or else this homogeneous distribution is unstable and the model reaches an inhomogeneous distribution, i.e. sorting, these inhomogeneties being like a Turing instability.

With one object type, at low density there is no clustering. As the density increases, clustering appears, with a wavelength dependent on the different parameter values, notably the distance-interaction function and the size of the nest.

With two object types the system has roughly the same behaviour, with respect to the size and number of clusters. What is of interest is of course the relative position of each object type. With no error, i.e. no distance-interaction between the different object types, each cluster contains only one object class. With a small distance-interaction between the different object types, the clusters contain mostly one sort of object, but are positioned adjacently. If the density is, however, very low, the clusters appearing contain both object types, sorting appearing progressively as the density increases.

5. Conclusions

Whereas we stated that the ALRs (and the RLAs) do not communicate, one might be tempted to say that they communicate in a very indirect manner, via their effect on the environment. In other words each ALR is influenced by its own and the other ALRs' past actions in moving objects. By this means they give the appearance of coordinating their activity to form clusters, in the sense that because many different ALRs add objects of type A to a particular cluster of type A, they appear to have agreed to form a cluster of type A at this point, whereas we know that no such consensus exists.

By the same logic it is evident that one solitary ALR can also form clusters and sort by interacting with its own past actions, although it would act for example ten times as slowly as a group of ten ALRs.

Another example of how such indirect "communication" can organise the activity of a group of non-communicating agents can be seen in the way the foragers of some species of ants set up individual and non-overlapping foraging territories (Deneubourg et al., 1987). Individual foragers learn progressively to return to the area where they find food. When an individual becomes specialised to an area then by its activity it reduces the amount of food there. Other individuals passing through that zone will have less chance of finding food and so will be less encouraged to return to that zone by their own learning process. The same principle can also be used to allocate different tasks dynamically among the members of a group (see also Theraulaz et al., this volume).

Returning to the sorting context, it is clear that a hierarchical system, wherein either a human supervisor or an alpha-robot decides exactly where to put which type of object, would sort more efficiently. However such a robot would no longer be ant-like, and would require a capacity for the analysis of how many types of object there are in the environment and a means of communicating its decision to other robots. All the robots would require at least a rudimentary map of the environment to transport the objects they find to the pre-arranged locations, or else would need to home towards a beacon placed by some means at each location. Furthermore, any fluctuation in the environment could make the original decision inappropriate.

The ALR system, while less efficient, requires no supervision and is capable of operating in a wide range of environments without specific programming, and with a large number of object types. The ALRs are only capable of perceiving an object at the point they occupy, and have no long range perception either for objects, piles of objects or homing beacons.

This simplicity makes the ALRs cheaper and more robust, and the lack of any hierarchy prevents fatal breakdowns, no one unit being essential. Even if a number of ALRs broke down this would only slow and not prevent the sorting. Again even if the ALRs frequently mistake one object for another, sorting is unimpaired. Indeed, rather than simply tolerating a certain degree of error, it can even be desirable to deliberately add error where none or little exist. We have seen how a small error in discriminating objects can lead to the ALRs placing the pile or piles of the two sorts of objects adjacently. Another example in the social insect world shows how a certain degree of error in following nestmates' trails to food sources allows the colony to exploit spatial heterogeneities in the food distribution more efficiently (Deneubourg et al., 1984). Put simply, if they followed the trails too exactly they would never find new sources nearby the one they were guided to. Error can thus be more creative than inefficient, and room should be allowed for it (see also the different works of Charles Darwin).

Overall, the system's simplicity, flexibility, error tolerance and reliability largely compensate for their lower efficiency. This is a general characteristic of systems in which the collective behaviour of a group of autonomous agents is emergent rather than explicitly programmed, and is surely one of the reasons for the 100 million year long evolutionary triumph of social insects. There are many circumstances in which a robot team could be profitably organised in a similar fashion.

Acknowledgements

This work was supported in part by the Belgian Program for Inter-University Attraction Poles, the Belgian Fonds National de Recherche Scientifique, NATO Scientific Affairs Division Grant 034/88, the Venture Research Unit of British Petroleum and the Ernst och Victor Hasselblads Foundation. We would like to thank Stefano Focardi for his helpful scepticism.

References

Aron S., Deneubourg J.L., Goss S. and Pasteels J.M. (1990) Functional self-organisation illustrated by inter-nest traffic in ants: the case of the Argentine ant. In: W. Alt and G. Hoffman (eds.) Biological Motion, Lecture Notes in Biomathematics, Springer Verlag, Heidelberg.

Beni G. (1988) The concept of cellular robotic systems. IEEE International Symposium on Intelligent Control, Arlington, VA, August 24-26.

Brooks R. and Flynn A. (1989) Robot beings. IEEE International Workshop on Intelligent Robots and Systems (IROS '90), Tokyo, Japan, Sept 4-6.

Brooks R.A., Maes P., Mataric M.J. and More G. (1990) Lunar base construction robots. Proc. Japan-USA symposium on Flexible Automation (in press).

Deneubourg J.L. (1977) Application de l'ordre par fluctuations à la description de certaines étapes de construction du nid chez les termites. Insectes Sociaux 24: 117-130.

Deneubourg J.L., Pasteels J.M and Verhaeghe J.-C. (1984) Quand l'erreur alimente l'imagination d'une société: le cas des fourmis. Nouvelles de la Science et des Technologies 2: 47-52.

Deneubourg J.L., Aron S., Goss S., Pasteels J.M and Duerinck G. (1986) Random behaviour, amplification processes and number of participants: how they contribute to the foraging properties of ants. In: Los Alamos Conference on Evolution, Games and Learning, Physica 22D: 176-186.

Deneubourg J.L., Fresneau D., Goss S., Lachaud J.-P. and Pasteels J.M (1987) Self-organisation mechanisms in ant societies (II): learning during foraging and division of labor. In: J.M. Pasteels and J.L. Deneubourg (eds.) From Individual Characteristics to Collective Organisation in Social Insects. Experientia Supplementum 54, 177-196, Birkhaüser, Bâle.

Deneubourg J.L. and Goss S. (1989) Collective patterns and decision making. Ecol. Ethol. Evol. 1: 295-311.

Deneubourg J.L., Goss S., Sandini G., Ferrari F. and Dario P. (1990) Self-organizing collection and transport of objects in unpredictable environments. Proc. Japan-USA symposium on Flexible Automation (in press).

Fukuda T. and Kawauchi Y. (1989) Cellular Robotics, Construction of Complicated Systems from Simple Functions. In: P. Dario, G. Sandini and P. Aebisher (eds.) Robots and Biological Systems. Proc. NATO ARW, Il Ciocco, Italy.

Goss S., Deneubourg J.L., Aron S., Beckers R. and Pasteels J.M. (1990) How trail laying and trail following can solve foraging problems for ant colonies. In : R.N. Hughes (ed.) Behavioural mechanisms of food selection: 661-678. NATO ASI Series, Vol G 20, Springer Verlag, Heidelberg.

Lumsden C.J. and Hölldobler B. Ritualized combat and intercolony communication in ants. J. theor. Biol. 100: 81-98.

Sandini G. and Dario P. (1989) Sensing strategies in cellular robotic systems. In: T. Kanade, F. Groen and L. Hertzberger (eds.) Proc. IAS 2, 937-942.

Steels L. Cooperation between distributed agents through self-organisation. In: Y. Demazeau and J.-P. Müller (eds.) Decentralized AI, 175-196, Elsevier, Holland.

EVOLUTION OF BEHAVIOR

EVOLUTION AND CO-EVOLUTION OF COMPUTER PROGRAMS TO CONTROL INDEPENDENTLY-ACTING AGENTS

John R. Koza
Computer Science Department
Stanford University
Stanford, CA 94305 USA
Koza@Sunburn.Stanford.Edu
415-941-0336

ABSTRACT

This paper describes the recently developed "genetic programming" paradigm which genetically breeds populations of computer programs to solve problems. In genetic programming, the individuals in the population are hierarchical computer programs of various sizes and shapes. This paper also extends the genetic programming paradigm to a "co-evolution" algorithm which operates simultaneously on two populations of independently-acting hierarchical computer programs of various sizes and shapes.

1. INTRODUCTION AND OVERVIEW

This paper describes the recently developed "genetic programming" paradigm which genetically breeds populations of computer programs to solve problems. In genetic programming, the individuals in the population are hierarchical compositions of functions and arguments of various sizes and shapes. Each of these individual computer programs is evaluated for its fitness in handling the problem environment and a simulated evolutionary process is driven by this measure of fitness.

This paper also extends the genetic programming paradigm to a "co-evolution" algorithm which operates simultaneously on two (or more) populations of independently-acting hierarchical computer programs of various sizes and shapes. In co-evolution, each population acts as the environment for the other population. In particular, each individual of the first population is evaluated for "relative fitness" by testing it against each individual in the second population, and, simultaneously, each individual in the second population is evaluated for "relative fitness" by testing it against each individual in the first population. Over a period of many generations, individuals with high "absolute fitness" tend to evolve as the two populations mutually bootstrap each other to increasingly high levels of fitness.

In this paper, the genetic programming paradigm is illustrated with three problems. The first problem involves genetically breeding a population of computer programs to allow an "artificial ant" to traverse an irregular trail. The second problem involves genetically breeding a minimax control strategy in a differential game with an independently-acting pursuer and evader. The third problem illustrates the "co-evolution" and involves genetically breeding an optimal strategy for a player of a simple discrete two-person game represented by a game tree in extensive form.

2. BACKGROUND ON GENETIC ALGORITHMS

Genetic algorithms are highly parallel mathematical algorithms that transform populations of individual mathematical objects (typically fixed-length binary character strings) into new populations using operations patterned after (1) natural genetic operations such as sexual recombination (crossover) and (2) fitness proportionate reproduction (Darwinian survival of the fittest). Genetic algorithms begin with an initial population of individuals (typically randomly generated) and then iteratively (1) evaluate the individuals in the population for fitness with respect to the problem environment and (2) perform genetic operations on various individuals in the population to produce a new population. John Holland of the University of Michigan presented the pioneering formulation of genetic algorithms for fixed-length character strings in *Adaptation in Natural and Artificial Systems* (Holland 1975). Holland established, among other things, that the genetic algorithm is a mathematically near optimal approach to adaptation in that it maximizes expected overall average payoff when the adaptive process is viewed as a multi-armed slot machine problem requiring an optimal allocation of future trials given currently available information. Recent work in genetic algorithms and genetic classifier systems can be surveyed in Goldberg (1989), Davis (1987), and Schaffer (1989).

3. BACKGROUND ON GENETIC PROGRAMMING PARADIGM

Representation is a key issue in genetic algorithm work because genetic algorithms directly manipulate the coded representation of the problem and because the representation scheme can severely limit the window by which the system observes its world. Fixed length character strings present difficulties for some problems — particularly problems in artificial intelligence where the desired solution is hierarchical and where the size and shape of the solution is unknown in advance. The need for more powerful representations has been recognized for some time (De Jong 1985, 1988).

The structure of the individual mathematical objects that are manipulated by the genetic algorithm can be more complex than the fixed length character strings. Smith (1980) departed from the early fixed-length character strings by introducing variable length strings, including strings whose elements were if-then rules (rather than single characters). Holland's introduction of the classifier system (1986) continued the trend towards increasing the complexity of the structures undergoing adaptation. The classifier system is a cognitive architecture into which the genetic algorithm is embedded so as to allow adaptive modification of a population of string-based if-then rules (whose condition and action parts are fixed length binary strings). Applications of classifier systems are discussed in Wilson (1987a, 1987b, 1988).

In addition, we have recently shown that entire computer programs can be genetically bred to solve problems in a variety of different areas of artificial intelligence, machine learning, and symbolic processing (Koza 1989, 1990a). In this recently developed "genetic programming" paradigm, the individuals in the population are compositions of functions and terminals appropriate to the particular problem domain. The set of functions used typically includes arithmetic operations, mathematical functions, conditional logical operations, and domain-specific functions. Each function in the function set must be well defined for any element in the range of every other function in the set. The set of terminals used typically includes inputs (sensors) appropriate to the problem domain and various constants. The search space is the hyperspace of all possible compositions of functions that can be recursively composed of the available functions and terminals. The symbolic expressions (S-expressions) of the LISP programming language are an especially convenient way to create and manipulate the compositions of functions and terminals described above. These S-expressions in LISP correspond directly to the "parse tree" that is internally created by most compilers.

The basic genetic operations for the genetic programming paradigm are fitness proportionate reproduction and crossover (recombination). Fitness proportionate reproduction is the basic engine of Darwinian reproduction and survival of the fittest and operates for genetic programming paradigms in the same way as it does for conventional genetic algorithms. The crossover operation for genetic programming paradigms is a sexual operation that operates on two parental LISP S-expressions and produces two offspring S-expressions using parts of each parent. In particular, the crossover operation creates new offspring S-expressions by exchanging sub-trees (i.e. sub-lists) between the two parents. Because entire sub-trees are swapped, this genetic crossover (recombination) operation produces syntactically and semantically valid LISP S-expressions as offspring regardless of which point is selected in either parent.

For example, consider the parental LISP S-expression:

```
(OR (NOT D1) (AND D0 D1))
```

And, consider the second parental S-expression below:

```
(OR (OR D1 (NOT D0))
    (AND (NOT D0) (NOT D1)))
```

These two LISP S-expressions can be depicted graphically as rooted, point-labeled trees with ordered branches. Assume that the points of both trees are numbered in a depth-first way starting at the left. Suppose that the second point (out of 6 points of the first parent) is randomly selected as the crossover point for the first parent and that the sixth point (out of 10 points of the second parent) is randomly selected as the crossover point of the second parent. The crossover points are therefore the NOT in the first parent and the AND in the second parent.

The two parental LISP S-expressions are shown below:

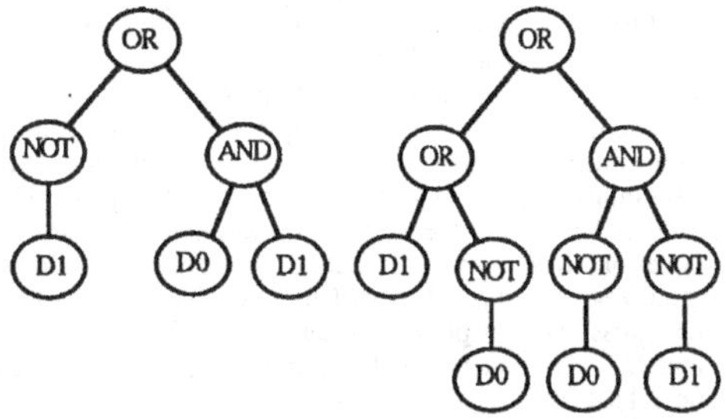

The two crossover fragments are two sub-trees shown below:

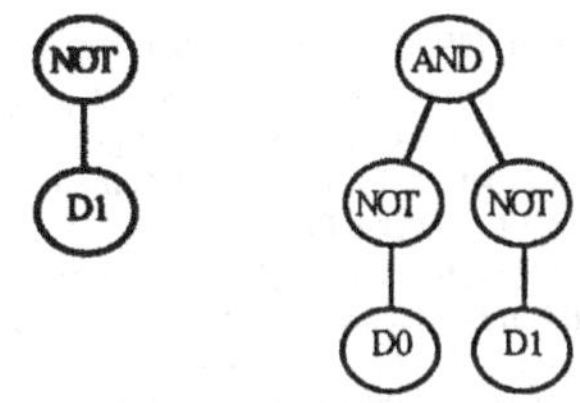

These two crossover fragments correspond to the bold, underlined sub-expressions (sub-lists) in the two parental LISP S-expressions shown above. The two offspring resulting from crossover are shown below.

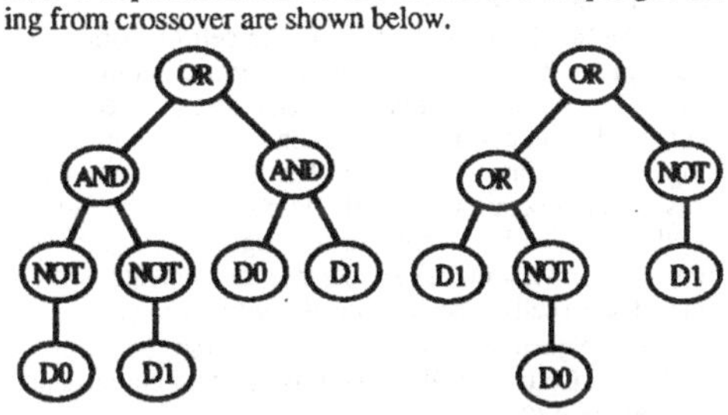

Note that the first offspring above is a perfect solution for the exclusive-or function, namely

```
(OR (AND (NOT D0) (NOT D1)) (AND D0 D1)).
```

Details on this algorithm can be found in Koza (1990a).

This new genetic algorithm paradigm has been successfully applied (Koza 1989, 1990a) to example problems in several different areas, including (1) machine learning of functions (e.g. learning the Boolean 11-multiplexer function, (2) planning (e.g. developing a robotic action sequence that can stack an arbitrary initial configuration of blocks into a specified order), (3) automatic programming (e.g. discovering a computational procedure for solving pairs of linear equations, solving quadratic equations for complex roots, and discovering trigonometric identities), (4) sequence induction (e.g. inducing a recursive computational procedure for the Fibonacci and the Hofstadter sequences), (5) pattern recognition (e.g. translation-invariant recognition of a simple one-dimensional shape in a linear retina), (6) optimal control (e.g. centering a cart and balancing a broom on a moving cart in minimal time by applying a "bang bang" force to the cart) (Koza and Keane 1990a, Koza and Keane 1990b), (7) symbolic "data to function" regression, symbolic "data to function" integration, and symbolic "data to function" differentiation, (8) symbolic solution to functional equations (including differential equations with initial conditions, integral equations, and general functional equations), (9) empirical discovery (e.g. rediscovering Kepler's Third Law, rediscovering the well-known econometric "exchange equation" MV = PQ from actual time series data for the money supply, the velocity of money, the price level, and the gross national product of an economy) (Koza 1990b), and (10) simultaneous architectural design and training of neural networks.

4. THE "ARTIFICIAL ANT"

Jefferson et. al. (1990) devised a planning task involving an "artificial ant" attempting to traverse an irregular trail and successfully used the conventional string-based genetic algorithm to discover a finite state automaton enabling the "artificial ant" to traverse the trail.

The setting for the problem is a square 32 by 32 toroidal grid in the plane. The "Santa Fe" trail is a winding trail of stones in 89 of the 1024 cells. The trail has single missing stones, double missing stones, single missing stones at some corners, double missing stones at other corners (knight moves), and triple missing stones at other corners (long knight moves).

The "artificial ant" begins in the cell identified by the coordinates (0,0) and is facing in a particular direction (i.e. east). The artificial ant has a sensor that can see only the single adjacent cell in the direction the ant is currently facing. At each time step, the ant has the capacity to execute any of four operations, namely, to move forward (advance) in the direction it is facing, to turn right (and not move), to turn left (and not move), or to sense the contents of the single adjacent cell in the direction the ant is facing.

The objective of the ant is to traverse the entire trail. Jefferson, Collins et. al. limited the ant to a certain number of time steps (200).

Jefferson, Collins et. al. started by assuming that the finite automaton necessary to solve the problem would have 32 or fewer states. They then represented an individual in their population of automata by a binary string representing the state transition diagram (and its initial state) of the individual automaton. The ant's sensory input at each time step was coded as one bit and the output at each time step was coded as two bits (representing the three actions). The next state of the automaton was coded with 5 bits. The complete behavior of an automaton was thus specified with a genome consisting of a binary string with 453 bits (64 substrings of length 7 representing the state transitions plus 5 additional bits representing the initial state of the automaton). They then processed a population of 65,536 individual bit strings of length 453 on a Connection Machine™ using a genetic algorithm using crossover and mutation operating on a selected (relatively small) fraction of the population. After 200 generations in a particular run (taking about 10 hours on the Connection Machine), Jefferson, Collins et. al. reported that a single individual in the population emerged which attained

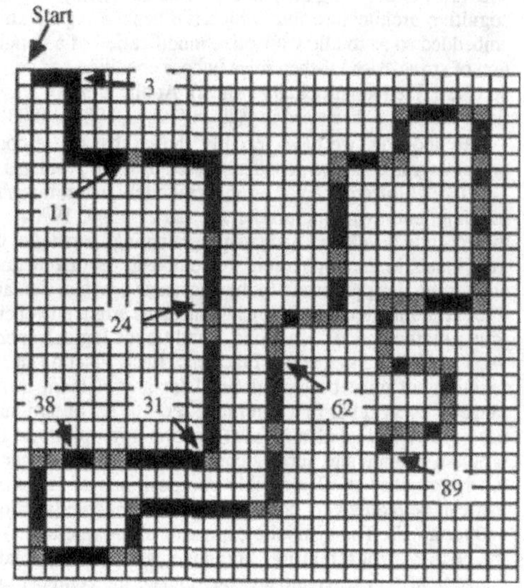

a perfect score of 89 stones. As it happened, this single individual completed the task in exactly 200 operations.

In our approach to this task using the genetic programming paradigm, we used the function set consisting of the functions F = {IF-SENSOR, PROGN}. The IF-SENSOR function has two arguments and evaluates the first argument if the ant's sensor senses a stone or, otherwise, evaluates the second argument. The PROGN function is the LISP connective function that sequentially evaluates its arguments as individual steps in a "program." The terminal set was T = {ADVANCE, TURN-RIGHT, TURN-LEFT}. These three terminals are actually functions with no arguments. They operate via their side effects on the ant's state (i.e. the ant's position on the grid or the ant's facing direction). IF-SENSOR, ADVANCE, TURN-RIGHT, and TURN-LEFT correspond directly to the operators defined and used by Jefferson, Collins et. al. We allowed 400 time steps before timing out.

The initial generation (generation 0) consisted of randomly generated individual S-expressions recursively created using the available functions and available terminals of the problem. For this problem (and each of the other problems described in this paper), each new generation was created from the preceding generation by applying the fitness proportionate reproduction operation to 10% of the population and by applying the crossover operation to 90% of the population (with reselection allowed). The selection of crossover points in the population was biased 90% towards internal (function) points of the tree and 10% towards external (terminal) points of the tree. Mutation was not used. For practical reasons, a limit of 4 was placed on the depth of initial random S-expressions and a limit of 15 was placed on the depth of S-expressions created by crossover.

In one run, an individual LISP S-expression scoring 89 out of 89 emerged on the 7th generation, namely,

```
(IF-SENSOR (ADVANCE)
```

```
(PROGN (TURN-RIGHT)
       (IF-SENSOR (ADVANCE) (TURN-LEFT))
       (PROGN (TURN-LEFT)
              (IF-SENSOR (ADVANCE)
                (TURN-RIGHT))
              (ADVANCE)))).
```

This plan is graphically depicted in Figure 1.

This individual LISP S-expression is the solution to the problem. In particular, this plan moves the ant forward if a stone is sensed. Otherwise it turns right and then moves the ant forward if a stone is sensed but turns left (returning to its original orientation) if no stone is sensed. Then it turns left and moves forward if a stone is sensed but turns right (returning to its original orientation) if no stone is sensed. If the ant originally did not sense a stone, the ant moves forward unconditionally as its fifth operation. Note that there is no testing of the backwards directions.

We can undertake to measure the number of individuals that need to be processed by a genetic algorithm to produce a desired result with a certain probability, say 99%. Suppose, for example, a particular run of a genetic algorithm produces the desired result with only a probability of success p_s after a specified choice (perhaps arbitrary and non-optimal) of number of generations N_{gen} and population of size N. Suppose also that we are seeking to achieve the desired result with a probability of, say, $z = 1 - \varepsilon = 99\%$. Then, the number K of independent runs required is

$$K = \frac{\log(1-z)}{\log(1-p_s)} = \frac{\log \varepsilon}{\log(1-p_s)}, \text{ where } \varepsilon = 1-z.$$

For example, we ran 111 runs of the Artificial Ant problem with a population size of 1000 and 51 generations. We found that the probability of success p_s on a particular single run was 43%. With this probability of success, 8 independent runs are required to assure a 99% probability of solving the problem on at least one of the 8 runs. That is, it is sufficient to process 408,000 individuals. This requires about 6 hours of computing time on the Texas Instruments Explorer II+™ workstation.

The graph below shows that the probability of success p_s of a run is 67% for a population size of 2000 with 51 generations. The graph also shows that the probability of success of a run is 81% for a population of 4000 with 51 generations. A population of 2000 requires 4 independent runs (i.e. 408,000 individuals are sufficient) to achieve the desired 99% probability. A population of 4000 requires 3 independent runs (i.e. 612,000 individuals are sufficient).

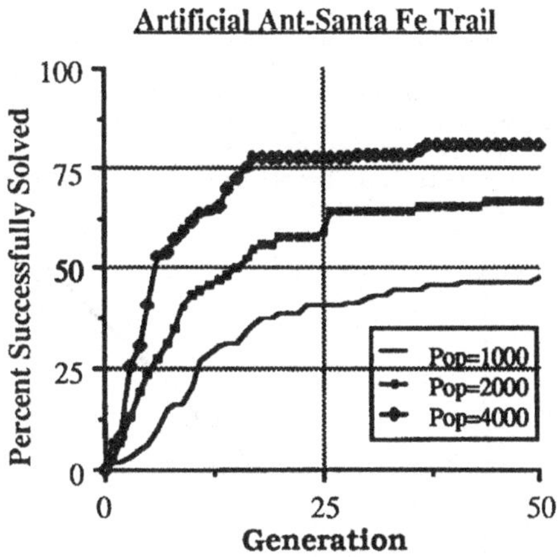

5. DIFFERENTIAL PURSUIT GAME

As a second illustration of the genetic programming paradigm, we consider a differential pursuer-evader game. Our objective is to find an optimal strategy for one player when the environment (fitness function) consists of an optimal opponent.

In a game, there are two or more independently-acting players who make choices (moves) and receive a payoff based on the choices they make. The differential "game of simple pursuit" is a two-person, competitive, zero-sum, simultaneous-moving, complete-information game in which a fast pursuing player P is trying to capture a slower evading player E. The "choices" available to a player at a given moment consist of choosing a direction (angle) in which to travel. In the simple game of pursuit, the players travel in a plane and both players may instantaneously change direction. Each player travels at a constant speed, and the pursuing player's speed w_p (1.0) is greater than the evading

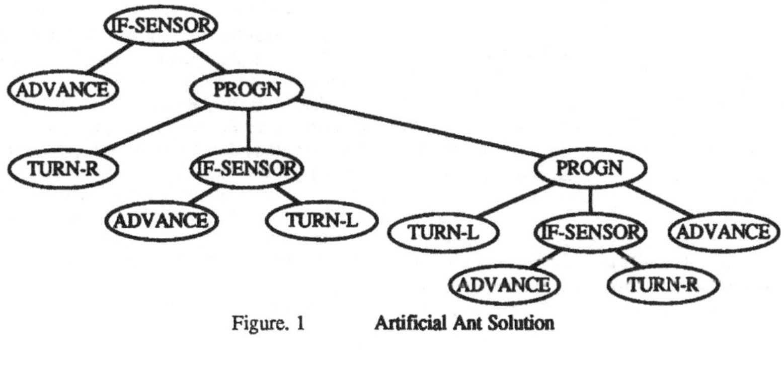

Figure. 1 **Artificial Ant Solution**

player's speed w_e (0.67).

The state variables of the game are x_p, y_p, x_e, and y_e representing the coordinate positions (x_p, y_p) and (x_e, y_e) of the pursuer P and evader E in the plane.

At each time step, both players know the position (state variables) of both players. The choice for each player is to select a value of their control variable (i.e. the angular direction in which to travel). The pursuer's control variable is the angle ϕ (from 0 to 2π radians) and the evader's control variable is the angle ψ. The players choose their respective control variable simultaneously.

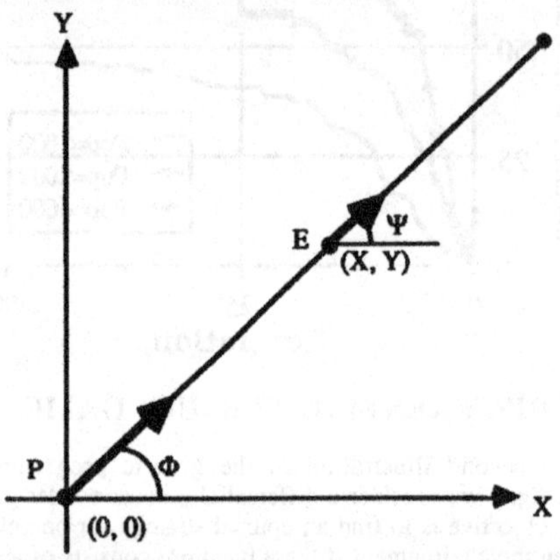

The analysis of this game can be simplified by reducing the number of state variables from four to two (Isaacs 1965). This state reduction is accomplished by simply viewing the pursuer P as being at the origin point (0,0) of a new coordinate system at all times and then viewing the evader E as being at position (x, y) in this new coordinate system. The two numbers x and y representing the position (x, y) of the evader E thus become the two "reduced" state variables of the game. Whenever the pursuer P travels in a particular direction, the coordinate system is immediately adjusted so that the pursuer is repositioned back to the origin (0, 0) and then appropriately adjusting the position (x, y) of the evader to reflect the travel of the pursuer.

The state transition equations for the evader E follow:

$$x(t+1) = x(t) + w_e \mathrm{Cos}\ \psi - w_p \mathrm{Cos}\ \phi$$

$$y(t+1) = y(t) + w_e \mathrm{Sin}\ \psi - w_p \mathrm{Cos}\ \phi.$$

In order to develop optimal playing strategies, we use a set of random environmental starting condition cases consisting of N_e (= 10) starting positions (x_i, y_i) for the evader E. Each starting value of x_i and y_i is between -5.0 and +5.0. The two players may travel anywhere in the plane. We regard the pursuer P as having captured the evader E when the pursuer gets to within some specified small distance ε= 0.5 of the evader E.

The payoff for a given player is measured by time. The payoff for the pursuer P is the total time it takes to capture the evader E over all of the environmental cases. The pursuer tries to minimize the time to capture. The payoff for the evader E is the total time of survival for E. The evader tries to maximize this time of survival.

A maximum "time out" time (100 time steps) is established so that if a particular pursuer strategy has not made the capture within that amount of time, that maximum time becomes the payoff for that particular environmental case and that particular strategy.

The problem is to find the strategy for choosing the control variable of the pursuer so as to minimize the total time to capture for any set of environmental cases when playing against an optimal evader.

For this particular simple game, the best strategy for the pursuer P at any given time step is to chase the evader E in the direction of the straight line currently connecting the pursuer to the evader. And, for this particular simple game, the best strategy for the evader E is to race away from the pursuer in the direction of the straight line connecting the pursuer to the evader.

If the evader chooses some action other than the strategy of racing away from the pursuer in the direction of the straight line connecting the pursuer to the evader (as shown below), the evader will survive for less time than if he follows his best strategy. If the evader initially chooses a suboptimal direction and then belatedly chooses the optimal direction, his time of survival is less than if he had chosen the optimal direction from the beginning.

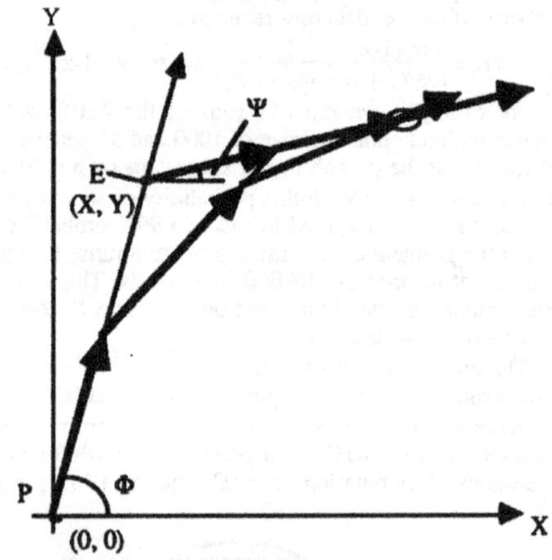

The situation is symmetric in that if the pursuer does not chase after the evader E along the straight line, he fails to minimize the time to capture.

The "value of the game" is the payoff (time) such that, no matter what the evader does, the evader cannot hold out for longer than this amount of time, and, if the evader does anything other than direct fleeing, his survival time is a shorter amount of time. Conversely, no matter what the pursuer does, the pursuer P cannot capture an optimal evader E in less than that amount of time. And, if the pursuer does anything other than direct pursuit, the evader can remain at large for a longer amount of time.

The genetic programming paradigm can be used to solve

the differential game of simple pursuit by genetically evolving a population of strategies for the pursuing individuals over a number of generations.

The genetic programming paradigm is especially well suited to solving this kind of problem because the solution takes the form of a mathematical expression whose size and shape may not be known in advance.

The terminal set is T = {X, Y, R}. The two state variables X and Y represent the position of the evader E in the plane in a "reduced" coordinate system where the pursuer is always positioned (or repositioned) to the origin. The terminal R is an ephemeral random constant. Each occurrence of an ephemeral random constant in an individual in the initial random population is assigned a different random floating point value (between -1.0 and +1.0). These ephemeral random constants give the algorithm the building blocks with which to construct the particular numeric constants that may be needed in the solution. Additional details are found in Koza (1990a).

The function set for this problem can be a set of arithmetic and mathematical operations such as addition, subtraction, multiplication, division (using the operation % which returns a zero when division by zero is attempted), and the exponential function EXP. Thus, the function set is F = {+, -, *, %, EXP}.

If the population of individuals represents pursuers and we are attempting to genetically breed an optimal pursuing individual, the environment for this "genetic algorithm" consists of an optimal evading individual. The optimal evader travels with the established constant evader speed w_e in the angular direction specified by the Arctangent function.

The genetic programming paradigm is successful in genetically breeding optimal pursuing individuals. As one progresses from generation to generation, the population of pursuing individuals typically improves. After several generations, the best pursuing individuals in the population can capture the evader in a small fraction (perhaps 2, 3, or 4) of the 10 environmental cases within a certain amount of time. Then, after additional generations, the population improves and the best pursuing individuals in the population can capture the evader in a larger fraction (perhaps 4, 5, or 6) of the 10 environmental cases within a shorter amount of time. Often, these partially effective pursuers are effective in some identifiable fraction of the plane or at some identifiable range of distances, but ineffective in other parts of the plane or at other distances. However, as more and more generations are run, the population of pursuing individuals typically continues to improve.

In one run, a pursuer strategy emerged in the 17th generation which correctly handled all 10 of the environmental cases. This S-expression is shown below.

```
(% (- (% (+ (* 2.0 Y) -0.066) -0.365)
      (% Y -0.124))
   (+ (EXP X) Y -0.579))
```

This S-expression is depicted graphically below:

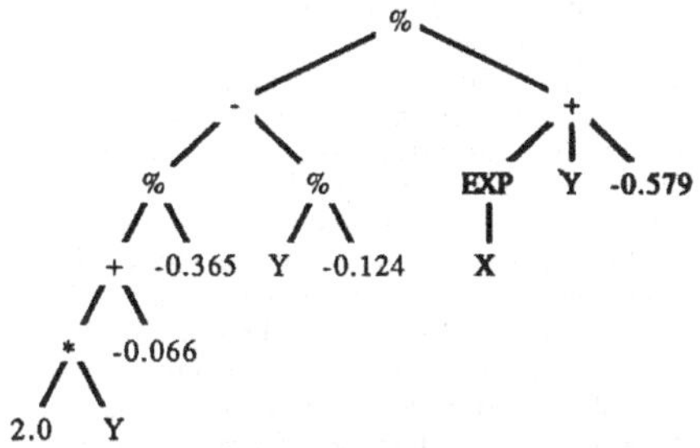

This S-expression is equivalent to

$$\frac{\dfrac{2y - 0.066}{-0.365} + \dfrac{y}{0.124}}{e^x + y - 0.579}$$

which in turn is equivalent to

$$\frac{2.58y + 1.81}{e^x + y - 0.579}$$

When this apparently optimal pursuing individual is retested against a much larger set of environmental cases (i.e. 1000), we then find that it also successfully handles 100% of the environmental cases. Thus, this S-expression is an optimal solution to the problem. It is also, as a result, an excellent approximation to the Arctangent function.

For example, we ran 111 runs of the differential pursuer-evader game problem with a population size of 500 and found that the probability of success p_s was 55% after 51 generations (see graph below). With a probability of success p_s of 55%, 6 independent runs are required to assure a 99% probability of solving the problem. That is, 153,000 individuals must be processed.

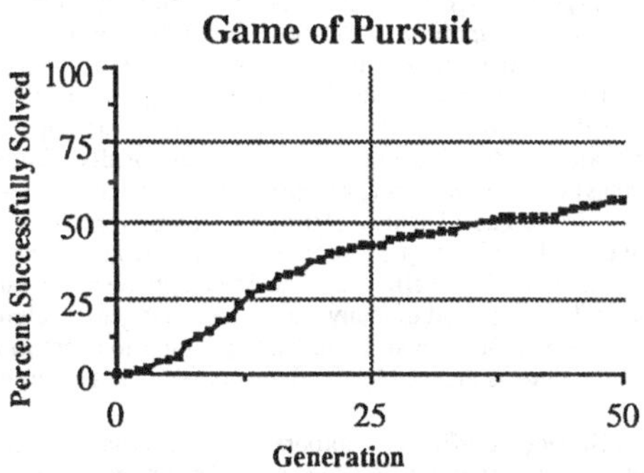

An optimal evader has been similarly evolved using an optimal pursuer (i.e. the Arctangent strategy).

6. CO-EVOLUTION IN NATURE

The evolutionary process in nature is often described as if one population of individuals is trying to adapt to a fixed environment. This description is, however, only a first order approximation to the actual situation. The environment actually consists of both the physical environment (which is usually relatively unchanging) as well as other independently-acting biological populations of individuals which are simultaneously trying to adapt to "their" environment. The actions of each of these other independently-acting biological populations (species) usually affect all the others. In other words, the environment of a given species includes all the other biological species that contemporaneously occupy the physical environment and which are simultaneously trying to survive. In biology, the term "co-evolution" is sometimes used to reflect the fact that all species are simultaneously co-evolving in a given physical environment.

A biological example presented by Holland illustrates the point (1990). A given species of plant may be faced with an environment containing insects that like to eat it. To defend against its predators (and increase its probability of survival in the environment), the plant may, over a period of time, evolve a tough exterior that makes it difficult for the insect to eat it. But, over a period of time, the insect may evolve a stronger jaw so that that the insect population can continue to feed on the plant (and increase its probability of survival in the environment). Then, over an additional period of time, the plant may evolve a poison to help defend itself further against the insects. But, then again, over a period of time, the insect may evolve a digestive enzyme that negates the effect of the poison so that the insect population can continue to feed on the plant.

In effect, both the plant and the insects get better and better at their respective defensive and offensive roles in this "biological arms race". Each species changes in response to the actions of the other.

7. BACKGROUND ON CO-EVOLUTION AND GENETIC ALGORITHMS

In the "genetic algorithm," described by John Holland in his pioneering *Adaptation in Natural and Artificial Systems* (1975), a population of individuals attempts to adapt to a fixed "environment." In the basic genetic algorithm as described by Holland in 1975, the individuals in the population are fixed-length character strings (typically binary strings) that are encoded to represent some problem in some way. In the basic "genetic algorithm", the performance of the individuals in the population is measured using a fitness measure which is, in effect, the "environment" for the population. Over a period of many generations, the genetic algorithm causes the individuals in the population to adapt in a direction that is dictated by the fitness measure (its environment).

Holland (1990) has incorporated co-evolution and genetic algorithms in his ECHO system for exploring the co-evolution of artificial organisms described by fixed-length character strings (chromosomes) in a "miniature world." In ECHO, there is a single population of artificial organisms. The environment of each organism includes all other organisms.

Miller (1988, 1989) has used co-evolution in a genetic algorithm to evolve a finite automaton as the strategy for playing the Repeated Prisoner's Dilemma game. Miller's population consisted of strings (chromosomes) of 148 binary digits to represent finite automata with 16 states. Each string in the population represented a complete strategy by which to play the game. That is, it specified what move the player was to make for any sequence of moves by the other player. Miller then used co-evolution to evolve strategies. Miller's co-evolutionary approach contrasts with Alexrod's (1987) solution to the repeated prisoner's dilemma using genetic algorithms. Axelrod measured performance of a particular strategy with a fixed weighted mix of the strategy's results against eight superior computer programs submitted in an international programming tournament for the prisoner's dilemma. A best strategy for one player (represented as a 70 bit string with a 3-move look-back) was then evolved with the weighted mix of eight opposing computer programs serving as the environment. Hillis (1990) used co-evolution in genetic algorithms to solve optimization problems.

8. CO-EVOLUTION AND THE GENETIC PROGRAMMING PARADIGM

In the "hierarchical co-evolution algorithm," there are two populations of individuals. The environment for the first population consists of the second population. And, conversely, the environment for the second population consists of the first population.

The co-evolutionary process typically starts with both populations being highly unfit (when measured by an absolute fitness measure). Then, the first population tries to adapt to the "environment" created by the second population. Simultaneously, the second population tries to adapt to the "environment" created by the first population.

This process is carried out by testing the performance of each individual in the first population against each individual (or a sampling of individuals) from the second population. We call this performance the "relative fitness" of an individual because it represents the performance of one individual in one population relative to the environment consisting of the entire second population. Then, each individual in the second population is tested against each individual (or a sampling of individuals) from the first population.

Note that this measurement of relative fitness for an individual in co-evolution is not an absolute measure of fitness against an optimal opponent, but merely relative measure when the individual is tested against the current opposing population. If one population contains boxers who only throw left punches, then an individual whose defensive repertoire contains only defenses against left punches will have high relative fitness. But, this individual will have only mediocre absolute fitness when tested against an opponent who knows how to throw both left punches and right punches (the optimal opponent).

Even though both initial populations are initially highly unfit (both relatively and absolutely), the virtually inevitable variation of an initial random population will mean that some individuals have slightly better relative fitness than others. That means that some individuals in each population have somewhat better performance than others in dealing with the current opposing population.

The operation of fitness proportionate reproduction (based on the Darwinian principle of "survival and reproduction of the fittest) can then be applied to each population using the relative fitness of each individual currently in each population. In addition, the operation of genetic recombination (crossover) can also be applied to a pair of parents, at least one of which is selected based on its relative fitness.

Over a period of time, both populations of individuals will tend to "co-evolve" and to rise to higher levels of performance as measured in terms of absolute fitness. Both populations do this without the aid of any externally supplied absolute fitness measure serving as the environment. In the limiting case, both populations of individuals can evolve to a level of performance that equals the absolute optimal fitness. Thus, the hierarchical co-evolution algorithm is a self-organizing, mutually-bootstrapping process that is driven only by relative fitness (and not by absolute fitness).

Co-evolution is likely to be especially important in game theory because one almost never has *a priori* access to a minimax strategy for either player. One therefore encounters a "chicken and egg" situation. In trying to develop a minimax strategy for the first player, one does not have the advantage of having a minimax second player against which to test candidate strategies. In checkers or chess, for example, it is difficult for a new player to learn to play well if he does not have the advantage of playing against a highly competent player.

9. CO-EVOLUTION OF A GAME STRATEGY

We now illustrate the "hierarchical co-evolution algorithm" to discover minimax strategies for both players simultaneously in a simple discrete game represented by a game tree in extensive form.

In the hierarchical co-evolution algorithm, we do not have access to the optimal opponent to train the population. Instead, our objective is to breed two populations simultaneously. Both populations start as random compositions of the available functions and arguments. The entire second population serves as the "environment" for testing the performance of each particular individual in the first population. And, at the same time, the entire first population serves as the environment for testing the performance of each particular individual in the second population. The algorithm does not have access to the "absolute fitness" measure provided by an optimal opponent. In other words, only relative fitness is used to drive the co-evolution algorithm.

Consider the following simple discrete game whose game tree is presented in extensive form in Figure 2. Each internal point of this tree is labeled with the player who must move. Each line is labeled with the choice (either L or R) made by the moving player. Each endpoint of the tree is labeled with the payoff (to player X).

This game is a two-person, competitive, zero-sum game in which the players make alternating moves. On each move, a player can choose to go L (left) or R (right). After player X has made three moves and player O has made two moves, player X receives (and player O pays out) the payoff shown at the particular endpoint of the game tree (1 of 32).

Since this game is a game of complete information, each player has access to complete information about his opponent's previous moves (and his own previous moves). This historical information is contained in five variables XM1 (X's move 1), OM1 (O's move 1), XM2 (X's move 2), OM2 (O's move 2), and XM3 (X's move 3). These five variables each assume one of three possible values: L (left), R (right), or U (undefined). A variable is undefined prior to the time when the move to which it refers has been made. Thus, at the beginning of the game, all five variables are undefined. The particular variables that are defined and undefined indicate the point to which play has progressed during the play of the game. For example, if both players have moved once, XM1 and OM1 are defined (as either L or R) but the other three variables (XM2, OM2, and XM3) are undefined (have the value U).

A strategy for a particular player in a game specifies which choice that player is to make for every possible situation that may arise for that player. In particular, in this game, a strategy for player X must specify his first move if he happens to be at the beginning of the game. A strategy for player X must also specify his second move if player O

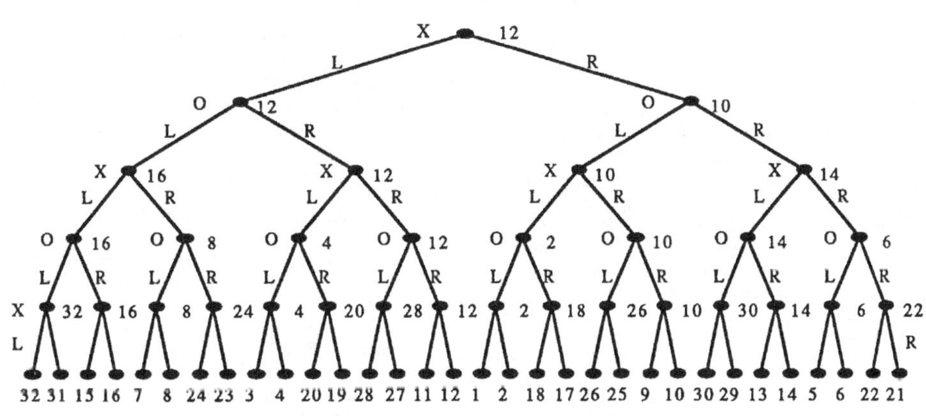

Figure 2 Game Tree with Payoffs

has already made one move and it must specify his third move if player O has already made two moves. Since Player X moves first, player X's first move is not conditioned on any previous move. But, player X's second move will depend on Player O's first move (i.e. OM1) and, in general, it will also depend on his own first move (XM1). Similarly, player X's third move will depend on player O's first two moves and, in general, his own first two moves. Similarly, a strategy for player O must specify what choice player O is to make for every possible situation that may arise for player O. A strategy here is a computer program (i.e. S-expression) whose inputs are the relevant historical variables and whose output is a move (L or R) for the player involved. Thus, the set of terminals is T = {L, R}.

Four testing functions CXM1, COM1, CXM2, and COM2 provide the facility to test each of the historical variables that are relevant to deciding upon a player's move. Each of these functions is a specialized form of the CASE function in LISP. For example, function CXM1 has three arguments and evaluates it first argument if XM1 (X's move 1) is undefined, evaluates its second argument if XM1 is L (Left), and evaluates its third argument if XM1 is R (Right). Functions CXM2, COM1, and COM2 are similarly defined. Thus, the function set for this problem is F = {CXM1, COM1, CXM2, COM2}. Each of these functions takes three arguments.

Our goal is to simultaneously co-evolve strategies for both players of this game.

In co-evolution, the relative fitness of a particular strategy for a particular player in a game is the average of the payoffs received when that strategy is played against the entire population of opposing strategies.

The absolute fitness of a particular strategy for a particular player in a game is the payoff received when that strategy is played against the minimax strategy for the opponent. Note that when we compute the absolute fitness of an X strategy for our descriptive purposes here, we test the X strategy against 4 possible combinations of O moves — that is, O's choice of L or R for his moves 1 and 2. When we compute the absolute fitness of an O strategy, we test it against 8 possible combinations of X moves — that is, X's choice of L or R for his moves 1, 2, and 3. Note that this testing of 4 or 8 combinations does not occur in the computation for relative fitness. When the two minimax strategies are played against each other, the payoff is 12, which is the value of this game. A minimax strategy takes advantage of non-minimax play by the other player.

As previously mentioned, the co-evolution algorithm does not use the minimax strategy of the opponent in any way. We use it in this paper for descriptive purposes. The co-evolution algorithm uses only relative fitness.

In one run (with population size 300), the individual strategy for player X in the initial random generation (generation 0) with the best relative fitness was

```
(COM1 L (COM2 (CXM1 (CXM2 R (CXM2 R R R) (CXM2 R
L R)) L (CXM2 L R (COM2 R R R))) (COM1 R (COM2
(CXM2 L R L) (COM2 R L L) R) (COM2 (COM1 R R L)
(CXM1 R L R) (CXM1 R L L))) (CXM1 (COM2 (CXM1 R
L L) (CXM2 R R L) R) R (COM2 L R (CXM1 L L L))))
R).
```

This simplifies to

```
(COM1 L (COM2 L L R) R).
```

This individual has relative fitness of 10.08.

The individual in the initial random population (generation 0) for player O with the best relative fitness was an equally complex expression. It simplifies to

```
(CXM2 R (CXM1 # L R) (CXM1 # R L)).
```

Note that, in simplifying this strategy, we inserted the symbol # to indicate that the situation involved can never arise.This individual has relative fitness of 7.57.

Neither the best X individual nor the best O individual from generation 0 reached maximal absolute fitness.

Note that the values of relative fitness for the relative best X individual and the relative best O individual from generation 0 (i.e. 10.08 and the 7.57) are each computed by averaging the payoff from the interaction of the individual involved with all 300 individual strategies in the current opposing population.

In generation 1, the individual strategy for player X with the best relative fitness had relative fitness of 11.28. This individual X strategy is still not a minimax strategy. It does not have the maximal absolute fitness.

In generation 1, the best individual O strategy attained relative fitness of 7.18. It is shown below:

```
(CXM2 (CXM1 R R L) (CXM2 L L (CXM2 R L R)) R).
```

This individual O strategy simplifies to

```
(CXM2 (CXM1 # R L) L R).
```

Although the co-evolution algorithm does not know it, this best single individual O strategy for generation 1 is a minimax strategy for player O. It has maximal absolute fitness in this game. This one O individual was the first such O individual to attain this level of performance during this run. If it were played against the minimax X strategy, it would score 12 (i.e. the value of this game).

Between generations 2 and 14, the number of individuals in the O population reaching maximal absolute fitness was 2, 7, 17, 28, 35, 40, 50, 64, 73, 83, 93, 98, and 107, respectively. That is, the minimax O strategy began to dominate the O population.

In generation 14, the individual strategy for player X with the best relative fitness had relative fitness of 18.11. Although the co-evolution algorithm does not know it, this one individual scored maximal absolute fitness. This one X individual was the first such X individual to attain this level of performance during this run. If it were played against the minimax O strategy, it would score 12 (i.e. the value of this game). This individual was

```
(COM2 (COM1 L L (CXM1 R R R)) L (CXM1 (COM1 L L
(CXM1 R R R)) (CXM2 L R R) R)).
```

This individual X strategy simplifies to

```
(COM2 (COM1 L L R) L R).
```

Between generations 15 and 29, the number of individuals in the X population reaching maximal absolute fitness was 3, 4, 8, 11, 10, 9, 13, 21, 24, 29, 43, 32, 52, 48, and 50, respectively. That is, the minimax X strategy began to dominate the X population. Meanwhile, the O population became even more dominated by the O minimax strategy.

By generation 38, the number of O individuals in the population reaching maximal absolute fitness reached 188 (almost two thirds of the population) and the number of X individuals reaching maximal absolute fitness reached 74 (about a quarter). That is, by generation 38, the minimax strategies for both players were dominant.

Interestingly, these 74 X individuals had relative fitness of 19.11 and these 188 O individuals had relative fitness of 10.47. Neither of these values equals 12 because the other population is not fully converged to its minimax strategy.

In summary, we have seen the discovery, via mutual bootstrapping, of the minimax strategies for both players in this game. This mutually bootstrapping process found the minimax strategies for both players without using knowledge of the minimax strategy (i.e. any *a priori* knowledge of the game) for either player.

10. CONCLUSIONS

We used the genetic programming paradigm to discover a plan allowing an "artificial ant" to traverse an irregular trail and to discover the minimax strategy for a pursuer in the differential game of simple pursuit. We then extended the genetic programming paradigm so that two populations of individuals are simultaneously co-evolved wherein each population serves as the environment to guide the evolution of the other population. We then illustrated the hierarchical co-evolution algorithm by simultaneously breeding an optimal game-playing strategy for both players of a discrete game in extensive form.

ACKNOWLEDGMENTS

James P. Rice of the Knowledge Systems Laboratory at Stanford University made numerous contributions in connection with the computer programming of the above.

REFERENCES

Axelrod, R. "The evolution of strategies in the iterated prisoner's dilemma." In *Genetic Algorithms and Simulated Annealing* , edited by L. Davis. London: Pittman 1987.

Davis, L. (editor) *Genetic Algorithms and Simulated Annealing* London: Pittman 1987.

De Jong, Kenneth A. Genetic algorithms: A 10 year perspective. *Proceedings of an International Conference on Genetic Algorithms and Their Applications.* Hillsdale, NJ: Lawrence Erlbaum Associates 1985.

De Jong, Kenneth A. Learning with genetic algorithms: an overview. *Machine Learning*, 3(2), 121-138, 1988.

Goldberg, D. E. *Genetic Algorithms in Search, Optimization, and Machine Learning.* Reading, MA: Addison-Wesley 1989.

Hillis, W. Daniel. "Co-Evolving Parasites Improve Simulated Evolution as an Optimization Procedure." In *Emergent Computation: Self-organizing, Collective, and Cooperative Computing Networks.* edited by S. Forrest. Cambridge, MA: MIT Press 1990 (to appear).

Holland, J. H. *Adaptation in Natural and Artificial Systems.* Ann Arbor, MI: University of Michigan Press 1975.

Holland, John H. Escaping brittleness: The possibilities of general purpose learning algorithms applied to parallel rule-based systems. In Michalski, Ryszard S., Carbonell, Jaime G. and Mitchell, Tom M. Machine Learning: An Artificial Intelligence Approach, Volume II. P. 593-623. Los Altos, CA: Morgan Kaufman 1986.

Holland, J. H. "ECHO: Explorations of Evolution in a Minature World." In *Proceedings of the Second Conference on Artificial Life.* edited by C. G. Langton, and J. D. Farmer, J. Doyne. Redwood City, CA; Addison-Wesley 1990. In press.

Isaacs, Rufus.. *Differential Games.* New York: John Wiley 1965.

Jefferson, David, Collins, Rob, et. al. "The Genesys System: Evolution as a Theme in Artificial Life." In *Proceedings of Second Conference on Artificial Life,* edited by C. G. Langton and D Farmer. Redwood City, CA: Addison-Wesley. 1990. In Press.

Koza, John R. "Hierarchical Genetic Algorithms Operating on Populations of Computer Programs." In *Proceedings of the 11th International Joint Conference on Artificial Intelligence (IJCAI).* San Mateo: Morgan Kaufman 1989.

Koza, John R. *Genetic Programming: A Paradigm for Genetically Breeding Populations of Computer Programs to Solve Problems.* Stanford University Computer Science Department Technical Report STAN-CS-90-1314. June 1990. 1990a.

Koza, John R. "A Genetic Approach to Econometric Modeling." Sixth World Congress of the Econometric Society. Barcelona, Spain. August 27, 1990. 1990b.

Koza, John R. and Keane, Martin A. "Cart Centering and Broom Balancing by Genetically Breeding Populations of Control Strategy Programs." In *Proceedings of International Joint Conference on Neural Networks, Washington, January 15-19, 1990.* Volume I. Hillsdale, NJ: Lawrence Erlbaum 1990.

Koza, John R. and Keane, Martin A. "Genetic Breeding of Non-Linear Optimal Control Strategies for Broom Balancing." In *Proceedings of the Ninth International Conference on Analysis and Optimization of Systems. Antibes, June, 1990.* Berlin: Springer-Verlag, 1990.

Miller, J. H. "The Evolution of Automata in the Repeated Prisoner's Dilemma." In *Two Essays on the Economics of Imperfect Information.* PhD dissertation, Department of Economics, University of Michigan, 1988.

Miller, J. H. *The Co-evolution of Automata in the Repeated Prisoner's Dilemma.* Sante Fe Institute Report 89-003. 1989.

Schaffer , J. D. (editor) *Proceedings of the Third International Conference on Genetic Algorithms.* San Mateo, Ca: Morgan Kaufmann Publishers Inc. 1989.

Smith, Steven F. *A Learning System Based on Genetic Adaptive Algorithms.* PhD dissertation. Pittsburgh: University of Pittsburgh 1980.

Wilson, Stewart. W. Classifier Systems and the animat problem. *Machine Learning*, 3(2), 199-228, 1987a.

Wilson, Stewart. W. Hierarchical credit allocation in a classifier system. *Proceedings of the Tenth International Joint Conference on Artificial Intelligence*, 217-220, 1987b.

Wilson, Stewart W. Bid competition and specificity reconsidered. *Journal of Complex Systems.* 2(6), 705-723, 1988.

The evolution of information gathering: operational constraints

Cynthia Kurtz
Dept. of Ecology & Evolution
State University of New York at Stony Brook
Stony Brook, NY 11794 USA
e-mail: kurtz@sbbiovm.bitnet
ckurtz@sbbiovm.bitnet

Abstract

I present a scheme for partitioning information used in decision making. Three types of information are recognized: internal information, or an individual's internal state; external information, or environmental factors; and relational information, or rules for predicting transformations of internal state. A genetic simulation model is described which tracks the evolution of alleles for high and low information access in each information type in a population with density dependence. Stable polymorphisms result. Interactions between the three genes are explored. The relevance of the model to foraging situations is discussed.

1. Introduction

A major criticism of classical optimal foraging models is that they assume complete information (Stephens & Krebs 1986). The problem of incomplete information has received considerable attention in the past twenty years, and the basic prey and patch models have been expanded to consider incomplete information in prey recognition (Houston et al. 1980, Getty & Krebs 1985), patch sampling (McNamara 1982, Lima 1984, Bernstein et al. 1988), and tracking a changing environment (Stephens 1987, Shettleworth 1988). In each of these cases the information studied is "about" the environment; that is, animals sense the states of pertinent environmental parameters.

Is environmental information the only sort of information an animal needs to make its decisions? How might one partition information in a way that is useful in thinking about the evolution of behavior?

Let us begin with a simplified look at the process of survival and reproduction. Any individual in a population can be described at a given time by an internal state which largely determines its current reproductive potential. That state will be a complex of many factors, including stored energy, health, fertility, and (in a sexual population) attractiveness to mates. The lifetime of any individual can be thought of as a time series of internal state transformations. Transformation will be influenced by the individual's own behavior and by factors in the environment which influence that individual, and will be governed by a set of rules which we may collectively describe as a transforming function.

From this basic scheme, there appear three types of information to which an individual making decisions may have access. First, the individual may be aware of its own state. Call this internal information. Second, it may be aware of environmental conditions, or external information. Third, it may be aware of the form of the transforming function, or relational information.

Now consider foraging models with these three information types in mind. The classical models assume complete external and relational information; but since they are static, or state-independent, ignore internal state. Their refinements deal with deficiencies in external information only. Dynamic optimization (Mangel & Clark 1988) considers state-dependent decision making; but it assumes complete internal information. Finally, studies of rules of thumb (Janetos & Cole 1981, Green 1984) implicitly consider reductions in relational information, mainly as satisficing strategies (Simon 1956) due to functional constraints, but assume complete internal and external information.

This paper presents a model that examines the interaction of internal, external and relational information genes in an evolving population. I chose simulation methods to model this situation for two reasons. First, including enough processes to generate information of each type makes the model sufficiently complex to prohibit an easy analytical analysis. Second, I was interested in the influence of population-level effects on the evolution of these genes. Also of significance is the fact that no costs of gathering information are included in the model. This was done to discover whether any inherent constraints might exist in the evolution of information gathering which the addition of costs might obscure.

2. The Model

2.1. Energy Gain and Reproduction

This is a three-locus, haploid population genetic model, with one gene for each type of information. Each locus has one high- and one low-information allele. The model environment consists of three resource patches in which individuals forage. Mean qualities of the three sites are such that a lone forager will receive 194 (good), 100 (medium), or 3.4 (bad) units of energy in one time frame of foraging (Figure 1). Each patch is replenished in each time frame with a quantity of energy randomly selected from a normal distribution with variance one-tenth of the mean. Every 25 time frames, patch identities are randomly reassigned, so that the mean quality of each patch fluctuates between the three means. Three hundred individuals are placed randomly in the patches at time(t) = 1, and forage until they reach an age of 5000. (The age of each individual is randomly assigned at the start, as is their patch residence time and time of last reproduction; this allows the simulation to begin as though it were already in motion.) Reproduction and premature death occur probabilistically, with probabilities dependent on energy states. Reproduction is asexual, and mutation does not occur.

Each individual's energy state at time t, X(i,t), changes according to the following function:

$$X(i,t) = X(i,t-1) + \triangle X(j,t) - C(m) - C(r) - C(v)$$

where $\triangle X(j,t)$ is the amount of energy any individual consumes given its choice of feeding site j; C(m) is a unit maintenance cost; C(r) is a conditional reproduction cost; and C(v) is a movement cost subtracted when an individual switches sites. The function that determines $\triangle X(j,t)$ is

$$\triangle X(j,t) = \frac{T}{N(j,t) + K/Q(j,t)},$$

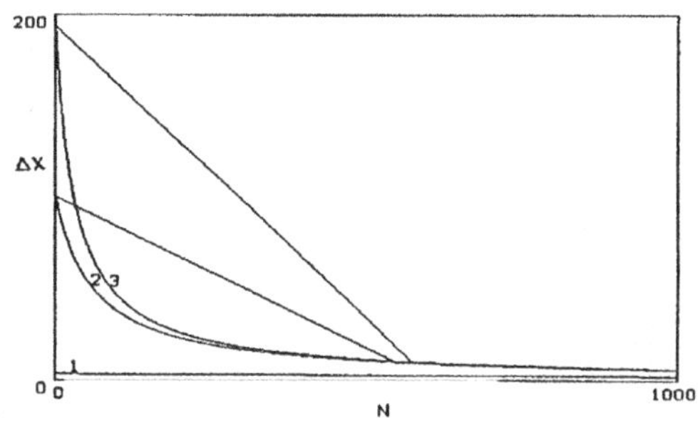

Figure 1. Effect of the relational gene: density dependence functions and linear approximations for each of three sites.

where N(j,t) is the number of individuals in patch j at time t, Q(j,t) is the quality of patch j at time t, T is a constant regulating the maximum energy available to an individual foraging alone, and K determines the steepness of the curve (see Figure 1). This is what I have above called the transforming function.

Each individual may reproduce only in every tenth time frame, as determined by its age; and its probability of doing so is calculated by the function

$$R[X(i,t)] = 1 \quad \exp[a(X(i,t) \quad b)]$$

where a determines the steepness of the function and b is an energy threshold below which individuals do not reproduce (see Figure 2). Reproduction occurs when a random number selected from a uniform distribution falls below R[X(i,t)]. Death is also a probabilistic function of energy: D[X(i,t)] = -exp[a(X(i,t))].

Each individual has the opportunity to switch sites in every third time frame (again determined by its age). A site is selected by calculating the expected energy gain from feeding at each site, translating these values into effects on reproductive probability, and choosing the site whose effect is greatest. Travel is instantaneous.

2.2. Information-Gathering Genes

High- and low-information alleles for the three genes work as follows. To assess internal information, each individual keeps an estimate of its own energy level. Individuals with the internal high-information allele have an accurate estimate; E[X(i,t)] = X(i,t). Individuals with low internal information have an estimate that can take one of two values:

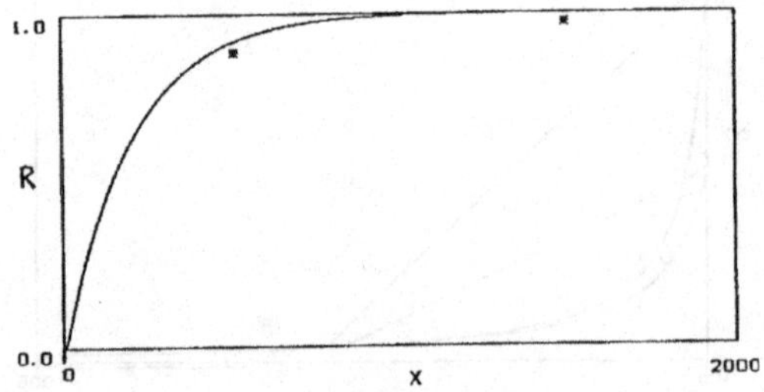

Figure 2. Reproduction probability as a function of energy state. Asterisks mark low-information estimates of internal state.

$$E[X(i,t)] = 1/4\ X^* \text{ if } X(i,t) < 1/2\ X^* \quad \text{or} \quad 3/4\ X^* \text{ if } X(i,t) > 1/2\ X^*$$

where $X^* = [X(R=0.99) - X(R=0.0)]$, or the difference between those values of $X(i,t)$ at which reproduction probabilities are 0.99 and 0.0. Thus the probable range of energy values is divided in half, and one quarter value or the other is selected (see Figure 2). This is roughly analogous to a perception of hunger versus satiation.

To assess external information, individuals update estimates of site qualities and densities periodically. Individuals with the high-information external allele receive accurate information on all sites every fifth time frame. Low-information individuals update their information every thirtieth time frame. Thus the high-information individual tracks changes in the environment with six times more precision than the low-information type. These updates are provided without cost and without having to visit sites. Estimates are updated according to a one-step memory function

$$E[Q(j,t)] = mQ(j,t) \quad (1-m)E[Q(j,t-1)],$$

where m is a memory constant which defines the weighting of past and present estimates. This constant was held at 0.5 for all simulations.

Finally, relational information defines how an individual perceives the function relating patch quality and density to per capita consumption (the transforming function). Individuals with the high-information relational allele use the function itself to calculate expected energy gain at each site. Individuals with the low-information allele use a linear approximation to the curve (see Figure 1),

Table 1

means and standard deviations of high-information allele frequencies for each gene

INT	0.536 ± 0.265
EXT	0.797 ± 0.096
REL	0.524 ± 0.071

Means collected from 190 sample periods in last 4500 time frames of simulation, and results of 18 runs (with starting allele frequencies of 0.25, 0.50 or 0.75) combined. Low standard deviation implies equilibrium. Total n for each line 3420.

$$E[\triangle X(j)] = T \quad \frac{\triangle X(j,1) - C(m)^*}{1 - N(j,\wedge Xi=C(m))} \quad E[N(j)]$$

To calculate the effect of the expected energy gain at each site on reproductive probability, all individuals use the true exponential function.

3. Results

I report here six sets of three simulations, each with a different seed value for the random number generator. The three runs within each set began high-information allele frequencies for each gene at 0.5, 0.75 and 0.25 to test for attraction to a stable equilibrium. A run length of 5000 time frames allowed the allele frequencies to reach equilibria. Table 1 shows the means and standard deviations of allele frequencies for the combined results, and Figure 3 shows a representative run. Since all runs were considered together, a small standard deviation shows that the allele frequencies tended to return to a particular equilibrium value whatever their starting values. The internal high-information allele remained at whatever frequency it was started; the external high-information allele reached an equilibrium at 0.797; and the relational high-information allele stabilized at 0.524.

The interaction of the three genes was examined in a series of genetic environment runs (Table 2). Each gene's activity was noted in four situations: 1) fixation of high-information alleles for both remaining genes, 2)

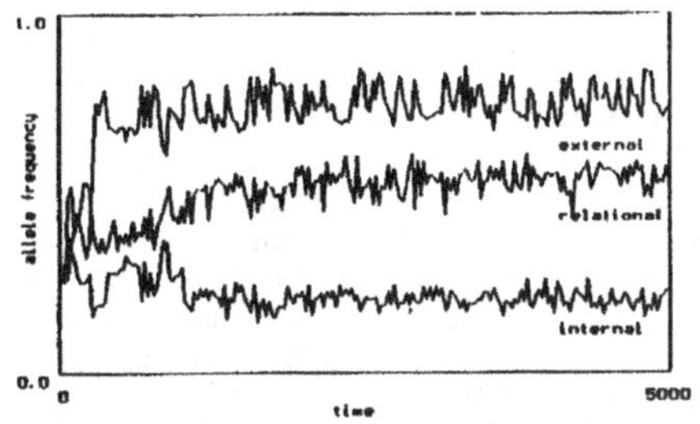

Figure 3. High-information allele frequencies for internal, external and relational genes. Sample run with all frequencies starting at 0.25.

fixation of low-information alleles for both remaining genes, 3) and 4) fixation of the high-information allele of one gene and the low-information allele of the other. Again allele frequencies were started at 0.25, 0.50 and 0.75 and the results of the three runs considered together.

Table 2

Mean frequencies of high-information alleles for each gene in genetic environment runs

	INT	EXT	REL
a	0.546 ± 0.351	1.0	1.0
	0.416 ± 0.307	0.0	0.0
	0.520 ± 0.282	0.0	1.0
	0.505 ± 0.049	1.0	0.0
b	1.0	0.529 ± 0.078	1.0
	0.0	0.610 ± 0.101	0.0
	0.0	0.520 ± 0.076	1.0
	1.0	0.605 ± 0.102	0.0
c	1.0	1.0	0.513 ± 0.058
	0.0	0.0	0.674 ± 0.194
	0.0	1.0	0.509 ± 0.059
	1.0	0.0	0.583 ± 0.125

Means collected from 190 sample periods in last 3600 time frames of simulation and results from three runs (with starting frequencies of 0.25, 0.5 and 0.75) combined. Low standard deviation implies equilibrium. Total n for each line 570.

4. Discussion

4.1. Main Simulations

Table 1 shows that two of the three genes reach a stable polymorphism of high- and low-information types. At first glance, this result seems counter-intuitive; one would expect information-gathering ability modeled without cost to fixate rapidly. The explanation lies in the over-concentration of high-information individuals at high-quality sites. As the frequency of the high-information allele increases, the proportion of individuals correctly assessing site quality increases. These individuals move in relative unison and therefore suffer from overcrowding; for although their estimates of site attendances are good, they cannot predict the concurrent movements of other individuals. The uninformed decisions made by low-information individuals actually produce a competitive advantage in this situation, since the two sites not deemed "best" at any time are sparsely populated. Thus the high-information allele frequencies decline until their descent is checked by the selective advantage (under non-crowded conditions) of choosing the best site.

The internal gene's neutrality in the first set of simulations is explained by the process of decision making used by model individuals. The function relating energy state to fitness is monotonically increasing; therefore the difference between correctly estimating one's point on the energy axis and choosing one of the two mid-points is not in the ranking of the site effects, but in the perceived magnitude of differences between the sites. Knowing one's true energy state will only be selectively advantageous when it is of value to know the difference between site effects.

One such situation is when costs of movement are high. In deciding to switch sites, an individual is comparing the effect of staying put with that of moving, and finding the difference greater than the cost of movement. A high movement cost should make the internal gene more responsive to selection. The value of internal information is thus affected by the cost of travel.

4.2. Genetic Environment Runs

The effect of the relational gene on the external gene is as follows: when R = 1, E goes to 0.5, but when R = 0, E goes to 0.6 (see Table 2, part b). The error : correct choice ratio of individuals with high external information is greater in a genetic environment of high relational information than in one of low relational information. Therefore decisions made by high external information types will tend to concur more often, and a lower proportion will produce excessive crowding. This suggests an adaptive argument for the use of rules of thumb; they may soften the detrimental effect of identical information access by all individuals.

The external gene in turn affects the relational gene. The latter is neutral when E = 0, but reaches an equilibrium when E = 1 (see Table 2, part c). This shows that external information is the most crucial of the three types, since without it decision rules are arbitrary. Sampling the environment, then, is much more important to the decision-making process than is any other step.

The internal gene reaches an equilibrium in only one genetic environment: fixation of the external high-information allele and the relational low-information allele (E = 1, R = 0) (see Figure 2, part a). In this situation, all foragers use a linear approximation to the density-dependence function (see Figure 1). Therefore, perceived differences between site effects are greater than true differences except in the short range of the graph where N is small. This heightened sensitivity to site differences brings out the small selective effect of the internal gene and causes high internal information types to experience the same crowding situation experienced by high-information individuals of the other genes in other milieus. This phenomenon does not occur in the environment of E = 0, R = 0, because as we have seen above, the usefulness of the relational gene depends on good external information.

4.3. General Discussion

A critical assumption of any simulation model comes from its repetitive nature, which forces a discrete representation of time. In this model, individuals move simultaneously from site to site, and are not allowed to abort a movement upon arriving. Once they have chosen a site, they must stay and eat, no matter how many others have arrived with them. This situation essentially creates the equilibria seen here. Further development of the model will extend the time of travel to allow an individual to modify its decision as it "nears" a site.

However, the present model does have implications for some real situations. Four situations in particular suggest themselves. 1) If travel time between sites is high compared to the feeding time at any site, it may not be wise to abort movement, especially if the quality and attendance of a site cannot be surmised from afar. This may occur, for example, in the case of desert animals which travel long distances in search of water (Smith 1980: 305). 2) If a high cost is associated with travel, an individual may be forced to feed at a site rather than move on because its energy budget has dropped below some critical threshold. Some birds (Welty 1982: 112) and insects (Borror et al. 1981: 71) are known to expend tremendous amounts of energy in flight. 3) If the resource which exists at the sites is ephemeral and must be collected within some fixed interval of time, it may be disadvantageous to use up time in running between sites rather than traveling to one site and remaining there whatever the reward. This is actually a situation where time _is_ divided into discrete units in nature; the units depend on the appearance and disappearance of the resource. The most obvious case of this situation is during a breeding season, the resource being available breeding sites; if one is too choosy one may not obtain a site at all. Density-dependence here takes the form of crowding at prime breeding areas, which is known to reduce offspring survival (Welty 1982: 324). 4) Finally, if variation in site quality is such that there infrequently occur concentrations of resources so high that density-dependence ceases to be a factor, it may be better not to base decisions strictly on the number of individuals at a site when one arrives, and instead stay at the site until an estimate of its quality is completed. Fruit trees especially are known to produce biomass much higher than the mean in some seasons (Smith 1980: 565).

The appearance of these equilibria in allele frequencies also depends heavily on the fact that no intermediate information levels exist in the model population. No individual has a sampling period of 15, or a partial linear approximation. If alleles for intermediate levels of information access were introduced by mutation, these would reach fixation. Thus incomplete information is the best solution in this situation. However, if canalization in the sensory and integrative organs reduces the likelihood of intermediates, an equilibrium between discrete types might result.

Regardless of the applicability of the results presented here, they suggest operational constraints on the gathering of information which do not rely on cost. This challenges current intuition about information and presents an important point: that this is not a conclusion immediately obvious without the consideration of interaction between individuals, and that attention to the emergent properties of groups may be necessary to adequately explain the evolution of many behavioral traits.

Acknowledgements

C. Janson and J. Labate provided criticism of an earlier draft. Contribution no. 769 in Ecology & Evolution from the State University of New York at Stony Brook.

References

Bernstein, C., A. Kacelnik and J.R. Krebs. 1988. Individual decisions and the distribution of predators in a patch environment. J. Anim. Ecol. 57:1007-1026.

Borror, D.J., D.M. DeLong and C.A. Triplehorn. 1981. An Introduction to the Study of Insects. CBS College Publishing, New York.

Getty, T. and J.R. Krebs. 1985. Lagging partial preferences for cryptic prey: a signal detection analysis of great tit foraging. Am. Nat. 125:39-60.

Green, R.F. 1984. Stopping rules for optimal foragers. Am. Nat. 123:30-40.

Houston, A.I., J.R. Krebs, and J.T. Erichsen. 1980. Optimal prey choice and discrimination time in the great tit (Parus major L.). Behav. Ecol. Sociobiol. 6:169-175.

Janetos, A.C. and B.J. Cole. 1981. Imperfectly optimal animals. Behav. Ecol. Sociobiol. 9:203-210.

Lima, S.L. 1984. Downy woodpecker foraging behavior: efficient sampling in simple stochastic environments. Ecology 65:166-174.

Mangel, M. and C.W. Clark. 1988. Dynamic Modeling in Behavioral Ecology. Princeton University Press, Princeton NJ.

McNamara, J.M. 1982. Optimal patch use in a stochastic environment. Theor. Pop. Biol. 21:269-288.

Shettleworth, S.J., J.R. Krebs, D.W. Stephens and J. Gibbon. 1988. Tracking a fluctuating environment: a study of sampling. Anim. Behav. 36:87-105.

Simon, H.A. 1956. Rational choice and the structure of the environment. Psychol. Rev. 63:129-138.

Smith, R.L. 1980. Ecology and Field Biology. Harper & Row, New York.

Stephens, D.W. 1987. On economically tracking a variable environment. Theor. Pop. Biol. 32:15-25.

Stephens, D.W. and J.R. Krebs. 1986. Foraging Theory. Princeton University Press, Princeton, New Jersey.

Welty, J.C. 1982. The Life of Birds. CBS College Publishing, New York.

Representations for Artificial Organisms

Robert J. Collins and David R. Jefferson
rjc@cs.ucla.edu jefferso@cs.ucla.edu
Department of Computer Science
University of California, Los Angeles
Los Angeles, CA 90024

Abstract

We are interested in simulations of biological evolution, i.e. simulations of populations of organisms over many generations living in a complex and dynamic environment. Our simulations are microanalytic, meaning that each individual organism and gene is separately represented, and the biologically significant events in an organism's life are all separately simulated in detail. Although we have been successful with simple models, we have encountered fundamental difficulties when scaling up the complexity of the organisms and the complexity of behaviors we expect of them. These difficulties all lead to a single question: What is an appropriate representation for an organism, i.e. what is an appropriate programming paradigm in which to express the complex behavior of organisms, and how should such programs be encoded into strings so that genetic algorithms will be successful over them?

The project that brought these issues to the surface is a complex evolutionary simulation called AntFarm, in which we are attempting to evolve cooperative foraging behavior in a population of colonies of artificial "ants." In this paper we survey a number of candidate representations for organisms, that we have considered for AntFarm, all of which have been used in the past for simple evolution models. We show that none of the representations are well-suited for AntFarm. From their inadequacies we abstract a number of principles that we believe are necessary for successful evolution of complex artificial life. Finding a representation that has all of the properties we identify is still an open problem.

1 Introduction

The simulation of evolving populations of artificial organisms is very important in the study of ecological, adaptive, and evolutionary systems whose dynamics are too complex to study analytically or experimentally [11]. In this paper we consider simulating organisms that live and reproduce in relatively complex environments, with many sensors (external and internal), and many possible actions at each moment. In addition the organisms possess some amount of internal memory, allowing their behavior to be history sensitive. In the course of its life each organism is born, makes thousands of decisions (eat, move, mate, etc.), and eventually dies. The reproductive success of a particular organism is affected by its behavior throughout its lifetime.

Our simulations are *microanalytic*, meaning that each individual organism and gene is separately represented, and the biologically significant events in an organism's life are all separately simulated in detail. Each organism in the evolving population is separately represented as a program. Each organism's life is represented as a process, a detailed sequence of events including its birth, its interaction with a dynamic environment, its competition with other organisms, its mating and reproduction (if any), and its death.

The structural representation of an organism consists of the following parts:

- interpreter: used to execute organism behavior functions (programs);
- phenotype: the behavior function (program), that maps sensory inputs and memories into memories and effector outputs;
- genotype: a bitstring that encodes the behavior function;
- development function: the mapping that decodes the genotype to produce the phenotype.

In all of our experiments, the development function is fixed for all organisms and for all time; it is not subject to evolution. The genotype, of course, differs from animal to animal, but is static throughout the life of the organism; it is the genetic material used in reproduction. At the time of reproduction, recombination and mutation operators are applied to a pair of parent genotypes to produce an offspring genotype. The phenotype of an organism, determined initially by the genotype and development function, can in principle change during an organism's lifetime if there is some provision for learning; however, our simulations do not attempt this, and we presume that an organism's phenotype is also static throughout its lifetime.

The behavior functions of the artificial organisms are evolved with a genetic algorithm. The behavior function (which converts an organism's sensory inputs to motor outputs) is encoded as a bit-string, and the genetic algorithm is applied to the population of bit-strings. The genetic algorithm translates each string into the corresponding behavior function, and places the resulting organism in the environment in order to determine its relative "fitness." The reproductive success of each organism is a function of this relative fitness.

2 AntFarm

We are building a simulation of an evolving population of ant colonies, called AntFarm [2]. Each colony is composed of a number of ants that are genetically related (sisters). Sexual reproduction occurs at the level of *colonies*, not at the level of individual ants. Two successful colonies reproduce sexually by crossing the genomes of representatives of each colony according to simple genetic operators. All of the colonies in a population live in a single shared environment so that they can directly compete with one another for food scattered in the environment. Competition among colonies is local, meaning that a successful colony (with more offspring colonies) is one whose food gathering success ranks high among its neighbors; but it need not rank high compared to the entire population, nor compared to any absolute standard. Because the ants in a single colony are closely related, it is evolutionarily advantageous for them to cooperate with each other in food gathering; and because competition and mating are local, it is to the advantage of each colony to frustrate the food gathering efforts of neighboring colonies (if it can do so at small cost to itself) so that its relative local ranking increases.

Each individual ant (Figure 1) has a number of internal state sensors and local environment sensors, providing information concerning the the presence of nearby food, nests, and pheromones (chemicals used by ants for communication). It also has a sensor to detect whether it is carrying food, and sensors that input random noise. We provide each ant with a *compass*, which always points in the direction of its nest. (We have decided not to simultaneously evolve solutions to both the foraging and navigation tasks—real ants typically use fairly complex methods to find their way back to the nest.) An ant can walk, drop pheromones, and can pick up, carry, or drop food. Pheromones diffuse in the environment, and eventually disappear. AntFarm runs on the Connection Machine [5], a massively parallel supercomputer with up to 65,536 processors.

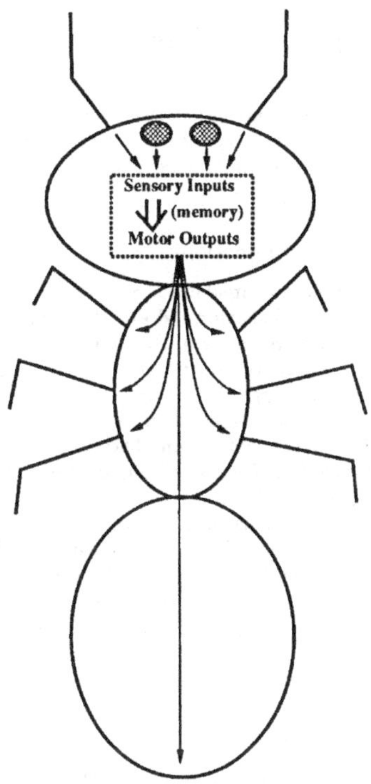

Figure 1: The internals of an AntFarm ant. The box represents the behavior function. The behavior function takes sensory inputs and its internal state (memory), and produces motor outputs. Among other things, the motor outputs control locomotion and the deposition of pheromones (from the tip of the abdomen).

The AntFarm simulation is intended for study of a number of interesting issues, including

- the computational feasibility of simulating the evolution at the scale of tens of thousands of organisms for hundreds of generations in a complex environment with thousands of bits of genome per organism;
- the evolution of communication and cooperation among closely related individuals; and
- the evolution of competitive strategies (among neighboring colonies).

In AntFarm all ants have the same sensory and effector apparatus; they differ only in their behavior functions, the heritable function that maps sensory inputs to motor outputs. For our purposes, the behavior function can be considered to be the phenotype of the organism.

3 Representations

In this section, we consider a number of organism program representations for use as behavior functions in AntFarm. Although all of these representations have been used successfully in simple simulations, none are satisfactory for AntFarm.

Parameterized Functions

RAM is a powerful simulator shell that is used for modeling population behavior and evolution [11]. In RAM, the organism representation is a parameterized Lisp [10] function in which the programmer defines the possible behaviors of the organism. The exact behavior that an individual expresses depends on its environment and the parameters to its Lisp function. These parameters are the behavior function of the organism, and the Lisp function is the organism interpreter.

The main problems with the representation of organisms as parameterized functions is that most of the behavior of the organism is specified in the interpreter, rather than in the behavior function—most of the behavior is not under genetic control. The space of potential behavior functions we wish to consider in AntFarm cannot be expressed as a set of parameters.

Lisp S-Expressions

Koza describes a technique for evolving computer programs that are represented as Lisp symbolic expressions (S-expressions) [8]. An S-expression has the form

$$(\text{function operand}_1\ \text{operand}_2 \ldots \text{operand}_N)$$

where *function* is to be applied to the N operands. The set of possible functions is F, and each of the operands is either an S-expression or one of the set of terminal symbols T. Examples of functions that might be elements of F include IF, AND, OR, NOT, $<$, $>$, $+$, $-$, $*$, etc. The genetic operators are recombination and mutation, which are defined such that legal S-expression syntax is maintained (the S-expression is not encoded as a linear bit-string). A given S-expression is a behavior function, and the organism interpreter is constructed to implement the primitive functions in F.

Although Koza has successfully applied the evolution of S-expressions to a wide variety of problems, this representation is not appropriate for use in AntFarm. In particular, this technique requires that a set T of terminals and a set F of functions—the basic building blocks of the behavior function—be specified. The only heuristic for choosing these sets that Koza provides is that these sets must be "sufficient" for the particular problem, and the functions must be chosen and defined so that any composition of the functions in F is valid for any value that any operand might assume. In order to construct F and T, we must decide what sorts of computations the behavior function needs to perform. Like the use of parameterized functions (above), the S-expression representation requires us to bias and limit the possible outcomes of the evolutionary process by specifying task-specific information in the organism interpreter. The AntFarm task is sufficiently complex that this is a critical flaw.

Deterministic Finite State Automata

Tracker is a Connection Machine program that evolves large populations of simple organisms [1][6]. Each Tracker organism has its own two dimensional environment containing a noisy, broken trail. The "fitness" of a Tracker animal is based on the amount of the trail that it can follow during its lifetime, so trail-following behavior evolves. The Tracker organism is very simple, receiving one bit of input that indicates the presence or absence of a trail marker in the location ahead of the organism. It can choose one of four options at each time step: move forward, turn left (in place), turn right (in place), or do nothing. The organisms each have five bits of internal memory. The populations sizes in the Tracker experiments range from 8192 to 262,144 organisms. Organisms are represented by finite state automata (FSA) or artificial neural networks (ANN), encoded in a 460-bit chromosome.

The FSA Tracker organisms use a deterministic input/output transducer finite state automata (Figure 2). Conceptually, the automata consists of a table with four columns and an initial state. In the FSA Tracker organism representation, the genotype consists of the concatenation of the bit representation of the initial state and the rows of the FSA table. The rows of the table are placed in a canonical order, so only two of the columns are explicitly part of the genotype. The size of the FSA behavior function is

$$(S+O)2^{I+S}+S$$

bits, where I is the number of bits of input and S is the number of bits of state (memory), and O is the number of output bits (encoding possible actions). In Tracker, $I=1$, $S=5$, and $O=2$. During each time step, the FSA interpreter takes the sensory input and the previous internal state, and produces a motor output and a new internal state.

Although the FSA representation is fine for Tracker-sized problems, the exponential rate of scaling on the number of inputs and amount of internal memory makes finite state automata unsuitable for AntFarm.

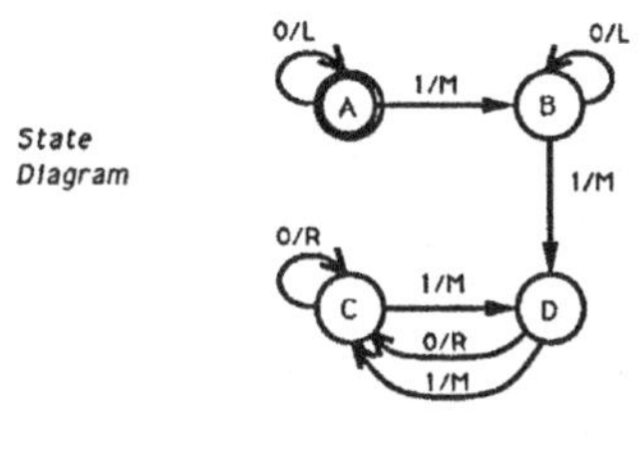

State Transition Table

Old State	Input	New State	Move
00	0	00	01
00	1	01	11
01	0	01	01
01	1	11	11
10	0	10	10
10	1	11	11
11	0	10	10
11	1	10	11

Genome 0000010111010111111010111110011011

Figure 2: The finite state machine and encoding that controls the Tracker organism's behavior. The state diagram is our conceptual model of the machine. The state diagram can be encoded in a four column state transition table. Since the rows in this table are placed in a canonical order, only two of the rows need to be represented in the genome. The genome consists of the concatenation of the bit representations of the initial state and the rows of the table. Tracker uses a thirty-two state machine (a four state machine is shown here). To read the diagram, consider the arc connecting state A and state B (labeled 1/M). When the automata is in state A, and the input is "1", the automata will move forward ("M") and make a transition to state B.

Primitive Rule-Based Organisms

Rule systems can be viewed as a variation on finite state automata (e.g. classifier systems [3]). In an FSA representation, each row of the table corresponds to the action and new state based on the inputs and current state. There is a row for every possible combination of inputs and current state (resulting in unfortunate scaling properties). Rule systems are an attempt to avoid the explosion in representation size; they do not represent actions for every possible situation.

A rule-based behavior function consists of K rules, each rule having four parts: inputs, current state (memory), actions, and new state (memory). The interpreter uses the rule that has an (input, current state) pair that most closely matches the current situation to determine the behavior (action) and new state. The alphabet of the (input, current state) pair may include a "don't care" symbol, which will match any of the other symbols.

The genotype of the rule system representation consists of the concatenation of the bit representation of the K rules and the initial state. The size of the representation is roughly

$$(I + S + O + 2S)K + S$$

bits, where I is the number of bits of input, O is the number of output bits, S is the number of bits of memory, and K is the number of rules. The main problem for the rule representation seems to be evolvability of appropriate behaviors from low-level sensory inputs. Consider the problem of following a gradient of pheromones in AntFarm, where the inputs are the nine pheromone levels in a 3 x 3 neighborhood of locations. This simple task will require a large number of rules, because there is no way to express relations among the various inputs, except by enumerating the possibilities. To make the rule system expressive enough, K will have to be made large, making the representation too big to be practical. Or, the rules would have to allow symbolic expressions, and we would have the same problem as with S-expressions.

Artificial Neural Networks

The ANN representation used in the Tracker simulations (described above) consists of three layers: input, hidden, and output (Figure 3). The input layer is fully connected to both the hidden and output layers. The hidden layer consists of five units (the five bits of memory) and is fully connected to both itself (recurrent connections) and the output layer. The activations of the input layer are set based on the sensory input from the environment, and the output layer specifies the desired motor action. The behavior function consists of the weights, thresholds, and initial activations of the network. A generic ANN interpreter is used to execute the behavior function.

In the ANN Tracker organism representation, the genotype consists of the concatenation of the bit representation of the weights, thresholds, and initial activations of the network. The size of the representation is

$$W(IO + IH + H^2 + HO) + (T + A)H$$

bits, where I is the number of inputs, O is the number of outputs, H is the number of hidden units (bits of memory), W is the number of bits per weight, T is the number of bits per threshold, and A is the number of bits per initial hidden unit activation. This rate of scaling seems reasonable for our target organisms. Unfortunately, the next section demonstrates that the Tracker-style ANN is unable to evolve the behaviors required for foraging in AntFarm.

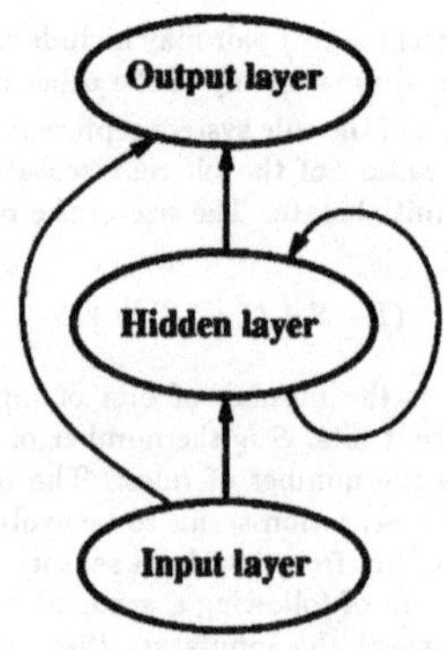

Figure 3: The Tracker ANN, showing the architecture of the neural network that controls the behavior of the Tracker organisms. The network consists of three layers. Each arc in the figure indicates that the layers are fully connected. The inputs are fully connected to the hidden layer and the output layer. The hidden layer is fully connected to itself (recurrent connections) and the output layer.

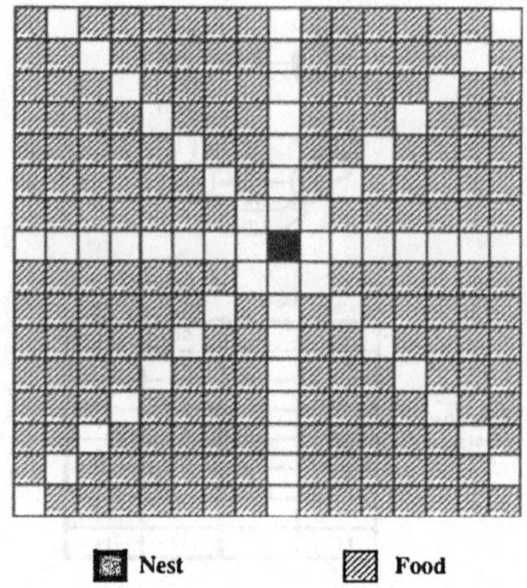

Figure 4: The food distribution in the experiments. The nest is the black square in the center. One unit of food is placed in each location in the environment, except on the axes and diagonals that are even with the nest.

4 Artificial Neural Networks vs. AntFarm

Of the representations we have surveyed, only neural networks seem to show any promise for use in AntFarm: parameterized functions and S-expressions require problem-specific information in the interpreter whereas FSAs and rule-based systems do not scale. In this section, we empirically compare both recurrent and feed-forward ANN representations to three foraging benchmarks and a hand-crafted ANN-based behavior function.

To perform this study, we have made a number of simplifications to the AntFarm model. Rather than all colonies foraging in a large, shared environment, each colony is assigned to a separate 16 x 16 environment, which is shared only by the ants within the colony (conceptually, the separate environments are still part of a common environment, since local colony mating is still used). In addition, the initial food distribution for each colony in each generation is always the same: one unit of food in each location, except for locations on a straight (horizontal, vertical, or diagonal) line with the nest (Figure 4). We chose this food distribution because foraging that only requires walking in a straight line from the nest would involve only a few neurons.

The population consists of 16,384 colonies, each of which contains four identical ants (initially located in its nest). Each run is 500 generations long, each lasting 50 time steps (i.e. each ant can move 50 times). An ant can sense the presence or absence of the pheromone, food, and the nest in a 3 x 3 square (centered at its current location), along with the direction towards the nest (encoded as four bits), a bit indicating whether it is carrying food, and four bits of random input (a total of 36 inputs per time step). In each time step, an ant can choose to drop some of the pheromone, move in any of the eight directions (encoded as four bits), drop food, and pick up food (a total of seven outputs). The genetic operators are random crossover at a rate of 0.0001 per bit and point mutations (bit flips) at a rate of 0.0001 per bit.

How can we tell how well a population is foraging? It is clearly impossible for four ants to carry all 196 units of food in the environment to the nest in only 50 time steps, so how much food can we expect them to recover? We have hand-coded three simple behavior functions (in C++) to serve as foraging benchmarks (Table 1). BM1 forages using only a random walk. BM2 searches for food with a random walk (ignoring the food sensors) and carries food to the nest by following the compass (the inputs that always point to the nest). BM3 improves on BM2 by using the the food sensors while searching. These benchmarks provide an absolute measure of foraging efficiency.

We have implemented three different ANN-based behavior functions: a Tracker-style three layer recurrent network (ANN1), a feed-forward network with five hidden layers (ANN2), and a hand-crafted ANN behavior function made up of two Tracker-style recurrent networks (ANN3). We suspect that a successful forager will be-

BM	Search/Transport	Mean	Max
1	random/random	1.07	6
2	random/compass	15.07	21
3	random+food/compass	20.82	25

Table 1: The foraging task is treated as two separate tasks: search for food and transport of the food back to the nest. Random indicates that only the random inputs are used, compass indicates that the inputs pointing the way to the nest are used, and food indicates that the food sensors are used. Mean and Max refer to the amount of food recovered in the population of 16,384 colonies.

have quite differently based on whether or not it is carrying food, so ANN3 invokes one of the networks when the ant is carrying food and the other when it is not. Comparisons of performance between the other two and this dual-network behavior function will help us determine how successfully the networks have evolved this type of behavior.

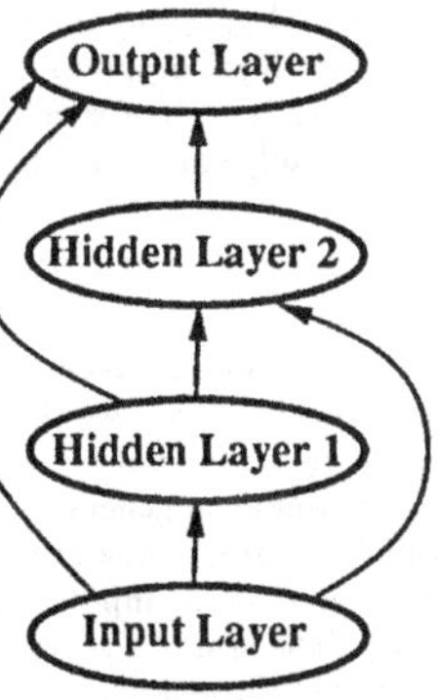

Figure 5: The architecture of ANN2, a feed-forward neural network. Each layer is fully connected to all "forward" layers. Each arc in the figure indicates that the layers are fully connected. ANN2 has five hidden layers (only 2 are shown here).

Each of these behavior functions is encoded in a bit-string of approximately 8000 bits (Table 2). In all three ANN behavior functions, weights are encoded as 3-bit one's-complement integers and the activations of the hidden units are initialized to 0 at the beginning of the generation. The architecture of the ANN1 and ANN3 networks is shown in Figure 3. ANN1 has a recurrent hidden layer consisting of 32 units (32-bits of memory). ANN2 consists of seven layers, each layer fully connected with all forward layers (Figure 5). There are five hidden layers, each with eight units. This is a feed-forward network, so it has no memory. The two networks in ANN3 each have an 18 unit recurrent hidden layer. Although it has a total of 36 recurrent units, ANN3 only has the equivalent of 18 bits of memory because at the beginning of each time step, the activations of the hidden layer of the network that was invoked on the previous time step are copied into the network that will be run this time.

Function	Connections	Memory	Chromosome
ANN1	2652	32	7956
ANN2	2612	0	7836
ANN3	2(1325)=2650	18	7950

Table 2: A summary of the ANN behavior functions, including the number of connections, the number of bits of memory, and the number of bits in the chromosome.

The results are summarized in Figure 6. The dual-network representation (ANN3) out-performs all of the other representations, foraging nearly as effectively as BM3, which uses both the food and compass sensors perfectly. It is not surprising that ANN3 performs the best, since it has a great deal of information about the foraging task built into the representation. The other two representations perform rather badly—they are not good at changing their behavior based on whether or not they are carrying food, although the recurrent network (ANN1) does better than the feed forward network (ANN2).

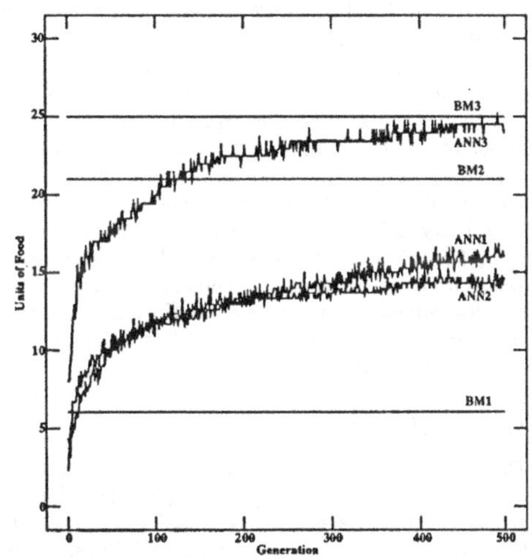

Figure 6: The maximum units of food brought back to the nest in a population of 16,384 colonies across 500 generations. Each curve is the average of three runs, differing only in the initial random seed. These simulations required 6 days of Connection Machine computation (on 8K processors).

5 Properties of Representations

In the previous section, we have considered a number of candidate representations for AntFarm organisms. We have assessed their strengths and weaknesses, and none are well–suited for AntFarm. From their inadequacies we abstract a number of properties that we believe are necessary for successful evolution of complex artificial life. We list the properties here, and then discuss them in detail. A good representation for organisms in microanalytic simulations of evolution should have the following properties:

1. (approximate) *closure* of the set of legal genotypes under the action genetic operators;
2. *smoothness* of the phenotype under the action genetic operators, i.e. the behavior function should tend to change smoothly as the genotype is changed by mutation and recombination;
3. the ability to *scale* to large behavior functions, i.e. those that can handle large amounts of input and output data;
4. the ability to *evolve* phenotypes that exhibit *both continuous and discrete behaviors* as a function of their inputs; and
5. a *uniform computational model*, i.e. the programming paradigm in which the behavior functions are expressed should not contain features that include any kind of explicit or implicit knowledge of the environment, nor bias toward a particular evolutionary trajectory.

Closure (1) and smoothness (2) are properties of the development function and the genetic operators. Scaling (3), the ability to evolve continuous and discrete behavior (4), and uniformity (5) are all properties of the computational model of the behavior function. Smoothness and the evolvability of continuous/discrete behavior are very different constraints. The smoothness property refers to the effect of changes in the *genotype* on the overall behavior function of the organism, while continuous/discontinuous behavior refers to the effect of changes in the *sensory inputs* on subsequent behavior during the lifetime of the organism.

Syntactic Closure

The property of syntactic closure constrains the development function, the encoding of the phenotype into the genotype. To be syntactically closed (approximately), the genetic operators must always, (or almost always) produce syntactically legal genotypes when applied to other syntactically legal genotypes, where the function that maps the genotype to phenotype defines the legal syntax. An evolutionary system will not work if a single mutation or recombination is likely to transform a genotype that encodes a perfectly good behavior function program into a program that is syntactically illegal.

Smoothness

The smoothness property requires that most changes to the genotype due to the application of genetic operators result in small changes in the phenotype. For example, a mutation in an ant should usually have a small affect on its foraging algorithm. Of course, it need not *always* cause a small change; some small changes will be fatal, and a few small changes may cause profound but beneficial effects. Still, evolution cannot work if the phenotype space is not relatively smooth as a function of genotype. The encoding and mapping functions should be smooth not only under mutation, but also under recombination, implying that functionally related genes should usually be inherited as a unit (strongly linked). The smoothness property has the effect of requiring the "adaptive landscape" [7] to be correlated with respect to the genetic operators. Evolution can successfully searches the space of possible organisms only in correlated adaptive landscapes.

Smoothness is an extension of the closure property. Not only does it require a legal program to be (usually) transformed into another legal program by the genetic operators, but it requires it to be (usually) transformed into one that is semantically similar to the original.

Scaling

The scaling properties of a representation are of extreme practical importance, because we must be able to simulate the evolution of large populations (tens of thousands) of colonies for thousands of generations. Scaling refers to the rate at which the size of the representation grows as a function of the number of inputs, outputs, and bits of internal memory. The size of the representation includes the number of bits in the genotype, the number of bits required to store the behavior function, the amount of time to translate from the genotype to the behavior function, and the amount of time to run a set of inputs through the behavior function to produce the outputs. We are interested in organisms with dozens or hundreds of inputs, outputs, and bits of internal memory.

Continuous/Discrete Behavior

We characterize behaviors as either continuous or discrete (although many behaviors are intermediate between these two extremes). Roughly speaking, a behavior function is producing continuous behavior when a small change in the inputs to the function results in a small change in the outputs, and discrete behavior when a small change in the inputs results in a large change in the outputs. We believe that in our target class of environments and organisms, a combination of both continuous and discrete behaviors will generally be necessary.

Consider the AntFarm task. Foraging ants seem to have two modes of behavior: (1) search and (2) transport and recruitment. While an ant searches for food, small changes in the local pheromone levels should result in small changes in its behavior (continuous). On the other hand, its behavior should change completely based on whether or not it is carrying food. When an ant is carrying food, it should walk directly home (possibly leaving a pheromone trail if the pile of food is big) and deposit the food in the nest. When an ant is not carrying food, it should search for food, perhaps by doing a pseudo-random walk or following a pheromone trail if one is encountered. The behavior of the ant must change completely based on a single bit of input (a very small part of the sensory inputs), an example of discrete behavior. Most or perhaps all complex behaviors will involve the combination of different modes of behavior, and thus require the evolution of discrete behavior. Within each mode, the behavior is continuous, but the transition from one mode to another involves only a small part of the total set of inputs.

Computational Uniformity

The final property is that the computational model of the representation should be uniform. In particular, it must be able to describe all desired behavior functions *without designing features of the problem or possible solutions into the representation.* For example, through all the years of neural network research, the area of network design has remained a "black art" [9][4][12]—each new application requires a new design, and no rigorous design principles exist. This is a situation that we must avoid.

One of our main assumptions about the environment and organisms is that their interactions are too complex for us to completely understand, so we cannot easily construct a good behavior function. If we are trying to shed light on a biologically motivated hypothesis, the results will be invalid if we bias the organisms toward (or away from) some evolutionary path. Even if we are simply interested in engineering an organism for survival in a particular environment, we will almost certainly do a bad job of guiding the representation toward good solutions by designing a special-purpose representation. The computational model must not have knowledge of the problem or solutions embedded in it.

6 Discussion

Although many organism-based evolutionary simulations have been run, most of the problems and models have been very simple. While many interesting questions can be answered by evolution of simple organisms in simple environments, there are many which cannot (e.g. the evolution of communication/cooperation). There is a wide variety of animal representations that are appropriate for simple simulations (e.g. parameterized functions, Lisp S-expressions, finite state automata, fully connected neural networks, rule systems, etc.). Unfortunately, none of these representations scales to the environment/organism complexity that is necessary for our investigations.

We have identified five properties that an organism representation must embody: (1) syntactic closure of the genotype under the genetic operators, (2) smoothness of the behavior function under the genetic operators, (3) scalable to a large number of inputs and outputs, (4) able to evolve both continuous and discrete behaviors, and (5) the computational model of the representation must be uniform. Finding a representation that has all of these properties is still an open problem.

The most difficult obstacle seems to be a result of the tradeoff between the size of the representation and its ability to evolve both continuous and discrete behaviors. A very large representation like a finite state automaton can easily express both continuous and discrete behaviors. A compact representation like a fully connected artificial neural network has difficulty expressing discrete behavior (where a small change in the inputs leads to a large change in behavior). The smaller representation must "generalize;" small changes in the inputs are smoothed away, making the evolution of discrete behavior unlikely.

Acknowledgments

We thank Chris Langton and Doyne Farmer for their valuable comments and ideas about this issue. This work is supported in part by W. M. Keck Foundation grant number W880615, University of California Los Alamos National Laboratory award number CNLS/89-427, and University of California Los Alamos National Laboratory award number UC-90-4-A-88.

References

[1] Robert J. Collins. Tracker: Preliminary results. (unpublished), 1990.

[2] Robert J. Collins and David R. Jefferson. AntFarm: A progress report. In Christopher Langton, J. Doyne Farmer, Steen Rasmussen, and Charles Taylor, editors, *Artificial Life II*. Addison-Wesley Publishing Company, (submitted, in press).

[3] David E. Goldberg. *Genetic Algorithms in Search, Optimization and Machine Learning.* Addison-Wesley Publishing Company, Inc., Reading, Massachusetts, 1989.

[4] Steven Alex Harp, Tariq Samad, and Aloke Guha. Towards the genetic synthesis of neural networks. In *Proceedings of the Third International Conference on Genetic Algorithms*, June 1989.

[5] W. Daniel Hillis. *The Connection Machine.* The MIT Press, Cambridge, Massachusetts, 1985.

[6] David Jefferson, Robert Collins, Claus Cooper, Michael Dyer, Margot Flowers, Richard Korf, Charles Taylor, and Alan Wang. The Genesys System: Evolution as a theme in artificial life. In Christopher Langton, J. Doyne Farmer, Steen Rasmussen, and Charles Taylor, editors, *Artificial Life II.* Addison-Wesley Publishing Company, (submitted, in press).

[7] Stuart Kauffman and Simon Levin. Towards a general theory of adaptive walks on rugged landscapes. *Journal of Theoretical Biology*, 128:11–45, 1987.

[8] John R. Koza. Genetic programming: A paradigm for genetically breeding populations of computer programs to solve problems. Technical report, Department of Computer Science, Stanford University, 1990.

[9] Geoffrey F. Miller, Peter M. Todd, and Shailesh U. Hegde. Designing neural networks using genetic algorithms. In *Proceedings of the Third International Conference on Genetic Algorithms*, June 1989.

[10] Guy L. Steele Jr. *Common LISP: The Language.* Digital Press, 1984.

[11] Charles E. Taylor, David R. Jefferson, Scott R. Turner, and Seth R. Goldman. RAM: Artificial life for the exploration of complex biological systems. In Christopher G. Langton, editor, *Artificial Life*, pages 275–295. Santa Fe Institute, Addison-Wesley Publishing Company, Inc., 1989.

[12] Gerhard Weiss. Combining neural and evolutionary learning: Aspects and approaches. Technical report, Institute für Informatik, Technische Universität München, 1990.

Evolving organisms that can reach for objects

Federico Cecconi Domenico Parisi

Institute of Psychology, C.N.R
15, Viale Marx - 00137 Rome
Tel.: 0039-6-8894596
Fax: 0039-6-824737
E-mail: cecconi at irmkant.bitnet
domenico at irmkant.bitnet

Abstract

An evolutionary method based on selective reproduction and random mutation was used to evolve neural networks that control two types of simple organisms which can reach for objects using their 2-segment arm. One kind of organism does not move and can only capture an object if it is at reaching distance; the other can displace itself and therefore it first approaches an object and then captures it. Individual learning during lifetime to predict changes in the position of an object or of the hand relative to the organism's body helps in the evolution of the object reaching capacity, although inheritance of the weight matrix is strictly Darwinian. Finally, a more sofisticated fitness criterion which penalizes arm movements causes the more complex organism to move its arm only when an object is at reaching distance.

1. Introduction

Our purpose in the present paper is to describe an attempt at evolving simple simulated organisms that have the capacity to reach for objects using their single 2-segment arm. We will describe two such organisms. One does not move and can only reach for objects if the objects are located at reaching distance from the organism. The second organism can displace itself in space and therefore can approach objects and, when they are at reaching distance, it can reach for them with its arm.

The methodology we have used to develop such organisms is an evolutionary technique based on selective reproduction and random mutation (Holland, 1975; Goldberg, 1989). (For other approaches to developing similar behaviors see Jordan, 1989; Booker, 1988; Patarnello and Carnevali, 1989.) An initial population of organisms each randomly different from all others is created. Each organism lives in its individual environment which contains a number of objects. At the end of their life they are rank ordered based on their performance on the object reaching tasks. There is no learning of this task during life but performances vary because of chance. Only the best performing organisms are allowed to reproduce by generating a number of copies of themselves while the others estinguish without leaving any offspring. A small amount of random variation is added to the copies so that some offspring will result in a better individual and some in a less good individual than their common parent. However, selective reproduction will insure that a better offspring is more likely to reproduce than a less good one. The net result of this evolutionary process is that the capacity to reach for objects gradually increases across a number of generations.

Another purpose of this research is to examine how the performance that emerges evolutionarily can be controlled and shaped by appropriately manipulating the fitness criterion, that is, the criterion in terms of which individuals are rank ordered and which therefore dictates who will reproduce and who won't. In the second simulation with the organism which both displaces itself in space and moves its arm, we will show that the two capacities can be separately controlled by an evolutionary mechanism that uses a sufficiently sophisticated fitness criterion.

Evolution is one mechanism for the emergence of capacities in organisms. Another is individual learning that requires an organism to modify itself during its life based on its experience. We have run our simulations in two versions. In one version the organisms were only subject to evolutionary learning of the object reaching capacity as already described. In another version, in addition to being selected for their reaching capacity, the organisms were also individually taught during their life another capacity. This was the capacity to predict how the sensory information from objects or from the organism's own hand would change as a function of a planned displacement of the body or of a planned movement of the arm. This capacity was taught using the backpropagation algorithm (Rumelhart, Hinton, and Williams 1986). Our purpose was to investigate the interaction between evolution and individual learning and more specifically to determine if individual learning can have a beneficial effect on evolution even in the absence of inheritance of acquired characteristics.

The nervous system of our organisms is modeled by a neural network, i.e. by a set of units and interconnections among units with a defined architecture. The networks are of the feedforward type with connections that go from an input layer to an intermediate hidden layer and from there to an output layer. The organisms live in environments that have a specified structure. At least part of the input codes sensory information from the environment and part of the output codes motor actions. An algorithm simulates the environment with its structure (e.g. where objects are located). The algorithm computes the sensory information from the environment to the organism and executes the motor actions of the organism, which may change the sensory information in the next activity cycle.

2. Experiment A: A fixed organism which can move its arm to reach for objects

Our first organism sits at the center of a semicircular environment which contains an average of 20 randomly distributed objects. The organism has an arm formed by two segments and it must learn to move its arm so that the arm's endpoint (= hand) comes to be located at the same location as one of the objects (Figure 1).

Figure 1

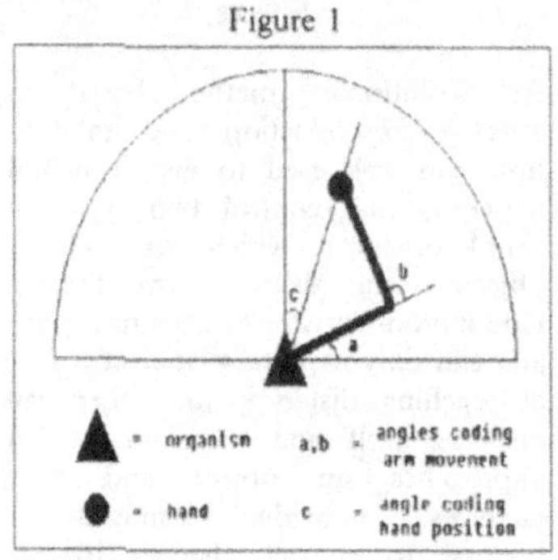

All the objects are located within reaching distance of the organism's arm. The nervous system of the organism is a neural network with 6 input units, 7 hidden units, and 4 output units (Figure 2 [1]).

---

[1] The 6 input units receive activation patterns that code the following information:
(a) 2 input units represent the current position of the hand. The position of the hand is coded as the distance of the hand from the body of the organism and the angle of the hand with the organism's fixed facing direction (North).
(b) 2 input units represent the position of the object which is currently closest to the hand. The position of this nearest object is coded in the same manner as the hand (distance and angle from the organism, as (a) above).
(c) 2 input units represent the currently planned movement of the arm, that is, the movement which has been decided by the organism in the preceding cycle but has not been executed yet. The role of this will become clear when we will talk about the simulation with individual learning of the prediction ability (see below).
The 4 output units code the following:

Figure 2

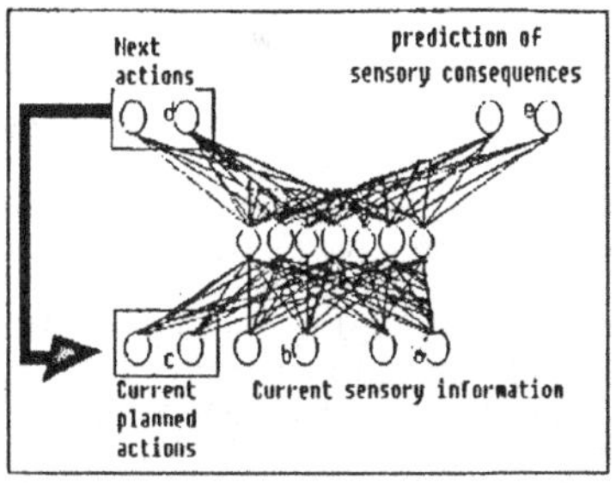

Arm movements are coded as the two angles that must added to or subtracted from the current position of the arm's two segments. Each segment can move a maximum of 30 degrees. Angles are normalized from 0 to 1, with 0 = 30 degrees left, 1 = 30 degrees right, and .5 = no movement.

In each cycle the organism is informed about the current position of its hand and of the object which is closest to the hand. Furthemore, the organism knows which arm movement it is going to do. Activation spreads through the network yielding two results: a decision on the next arm movement and a prediction on the new position of the hand that will be caused by the actual execution of the currently planned arm movement. Between one cycle and the next the algorithm executes the currently planned arm movement, actually displacing the arm in the environment, and computes the new position of the hand resulting from this movement. The actual new position of the hand is used as the input coding the current position of the hand in the next cycle and, in the simulation with prediction learning, also as teaching input for correcting the predicted hand position as generated by the network.

We create an initial generation of 100 networks with an identical architecture (as described in (a)-(e) above) but individually different randomly generated sets of weights on the networks' 70 connections (6x7x4). Each network lives in 5 different environments with 3000 arm movements in each environment, for a total of 15,000 arm movements (= lifetime). Each environment contains an average of 20 randomly located objects. In Simulation 1 there is no learning during an organism's life, hence no change in the network's connection weights. The network automatically generates predictions on the next position of the hand because of the network's architecture but these predictions are not corrected using the teaching input on the actual new positions of the hand as computed by the algorithm.

At the end of its life each network is assigned a fitness measure based on the total number of objects that have been reached by the organism's hand. Networks will vary on this measure because for purely chance reasons some networks will have a matrix of connections weights which generates arm movements pushing the hand toward a sensed object whereas other networks have weights that do not have this tendency. The best 20 networks reproduce, that is, 5 copies of their weight matrices are created and the resulting 100 (5x20) individuals constitute the next generation. An average of 5 randomly selected weights in these matrices are slightly changed by adding a quantity randomly varying between + 1.0 and -1.0 to them.

The evolutionary process goes on for 65 generations. The resulting evolution of the capacity to reach for objects is charted in Figure 3 which shows how the capacity changes across the 65 generations both for the average and the peak individual in each generation.

(d) 2 output units represent the next arm movement, which becomes the currently planned arm movement in the next cycle (see (c) above).

(e) 2 output units represent a prediction concerning the next position of the hand that will result when the currently planned arm movement will be executed. This prediction is coded in terms of distance and angle of the hand (see (a) above).

Figure 3

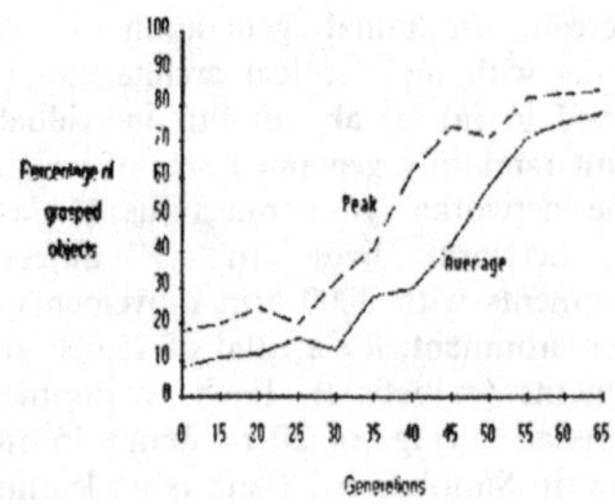

As it is clear from this Figure, a capacity to reach for objects can evolve in simulated organisms as a consequence of selective reproduction and random mutation. Figure 4 shows a set of 30 sequential arm movements generated by an organism belonging to the first generation compared with a set of 30 movements generated by an organism belonging to the last generation. The first organism reaches only two objects with its hand while the second organism reaches seven objects.

Figure 4

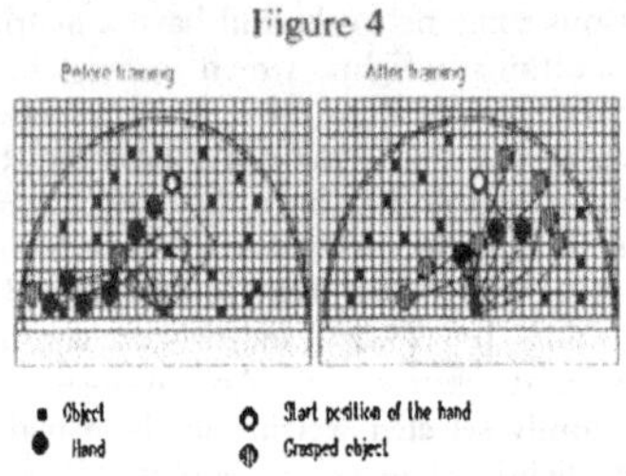

In Simulation 2 our purpose was to examine how the evolution of the object reaching capacity interacted with the individual learning during life of the capacity to predict how the hand's position changed with arm movements. The simulation was identical to the preceding one with the exception that now the actual new positions of the hand resulting from the organism's arm movements were used as teaching input to the network to be compared with the predictions generated by the network. The backpropagation procedure was applied to backpropagate error and change consequently the weights on the connections. Notice that this resulted not only in a change in the predictions (which became progressively better predictions) but also in a change in the arm movements themselves since the "lower" weights, those connecting the input units to the hidden units, were changed by the backpropagation procedure, and these weights are in part responsible for the arm movements generated by the network. (Compare the network architecture in Figure 2.) As a matter of fact, there are changes in the arm movements generated by a network which is learning to predict the sensory changes resulting from these movements as compared with the arm movements generated by a network which is not taught how to predict these changes.

Figure 5, reprinted here from another paper (Parisi, Cecconi, and Nolfi, 1990), shows the amplitude of the angles that correspond to arm movements generated by a network in the first and in the second case. The abscissa shows a sequence of 50 arm movements and the ordinate the amplitude of the two angles for the two segments of the arm summed together with the same sign. As this Figure demonstrates, the arm movements become less sweeping and irregular if the network that generates them is trying to learn to predict their sensory consequences.

Figure 5

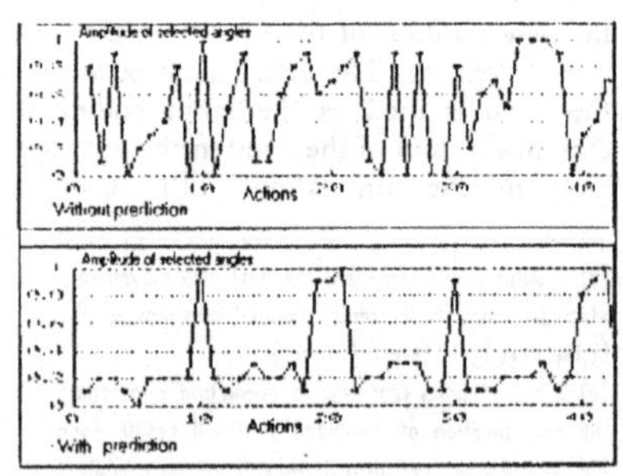

Aside from such prediction learning during

their life, the networks in this second simulation reproduce in the same way as those of the first simulation. It must be stressed that inheritance is rigorously Darwinian, that is, a parent hands down to its offspring the unmodified weight matrix it has inherited at birth from its parent and there is no transmission of the weight changes that have occurred during life as a consequence of prediction learning.

This notwithstanding, not only individual networks learn during their life to predict where the hand will end up as a consequence of a given arm movement but this learning appears to have a beneficial effect on the evolution of the object reaching capacity across generations. Figure 6 shows the evolution of the object reaching capacity in the organisms that do learn to predict compared with those that do not learn (same data of Figure 3). As can be seen, the evolutionary performance of the organisms in Simulation 2 is better than that of the organisms of Simulation 1 considering both the average and the best individual in each generation.

Figure 6

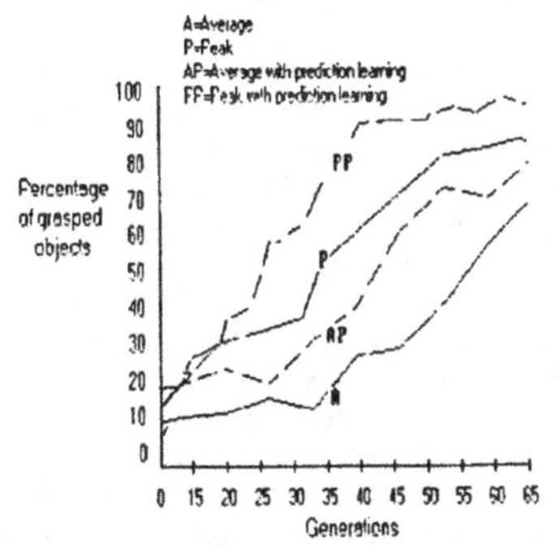

In Nolfi, Elman, and Parisi (1990) we have explained this beneficial effect of learning on evolution as due to the fact that learning allows the evolutionary mechanism to explore the region surrounding a given point representing an individual network in the high-dimensional space of weight matrices. Since networks which reproduce are selected on the basis of their performance as modified by learning at the end of life, networks with better surrounding regions (i.e., regions containing higher performance levels than the performance level of the point they surround) are more likely to be selected than networks with less good surrounding regions. When mutations are introduced in the offspring of a particular reproducing network, these mutations result in the creation of weight matrices lying in the region surrounding the point representing the parent network. And these offspring networks are more likely to exhibit a better performance than their parent if parents with good surrounding regions have been selected for reproduction. Hence, by allowing the evolutionary mechanism to take into account not only the fitness of individual networks but also the regions surrounding individual networks in the weight matrix space, learning can help evolution, as observed in our results, even if there is no inheritance of learned modifications. (For a related treatment of the effect of learning on evolution, see Hinton and Nolan, 1987.)

3. Experiment B: An organism that displaces itself in space and moves its arm

The first organism we have described is very limited in its object reaching possibilities because it cannot move. A consequence of its immobility is that only objects that are within reaching distance from the organism and its arm can be reached. The new organism that we will describe in this section overcomes this limitation. This organism has two different types of motor behaviors in its repertoire: it can displace itself in space and it can move its arm to reach objects. The kind of behavior that we expect it to acquire is first to approach distant objects and then, when these objects are at reaching distance, to move its arm to reach them with the arm's endpoint.

We have studied in previous experiments organisms that learn to displace themselves in space in such a way that they succeed in approaching distant objects (Nolfi, Elman, and Parisi, 1990; Parisi, Cecconi, and Nolfi, 1990). In these simulations the fitness criterion is the number of objects that an individual organism succeeds in approaching with its own body, i.e. the organism's body must coincide in space with an object. In the present experiment the fitness criterion remains that of the previous simulation: it is the arm's endpoint (hand) that must coincide in space with an object. However, since objects can be located outside reaching distance from the organism, the organism's task is to first approach an object and then operate its arm to reach it.

The network architecture for this new organism is necessarily more complex than the architecture of the previous organism. There are two kinds of movements, of the body and of the arm, and since we were interested in the role of prediction in the evolutionary emergence of the appropriate behavior as in the preceding experiment, the network receives input information on both the currently planned body movement and the currently planned arm movement and generates predictions on the new positions of the closest object and of the hand with respect to the organism's body as the planned movements of the body and of the arm are actually executed. The new network includes 8 input units, 8 output units, and 7 hidden units (see Figure 7 [2)])

2)The 8 input units represent the following information:
a) 2 units represent the current position of the hand (distance and angle with respect to the organism's facing direction
(b) 2 units represent the current position of the object which is closest to the hand (distance and angle as above
(c) 2 units represent the currently planned arm movement (two angles to be added or subtracted to the current position of the two arm segments)
(d) 2 units represent the currently planned body movement (see below).
The 8 output units are subdivided as follows:
(e) 2 units for the predicted hand position
(f) 2 units for the predicted object position
(g) 2 units for the next arm movement
(h) 2 units for the next body movement

Figure 7

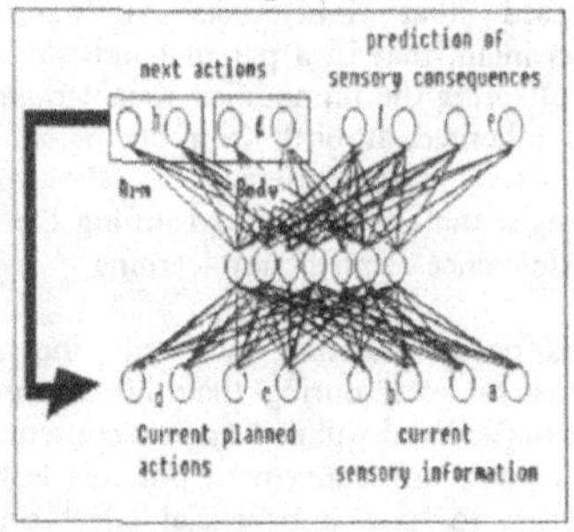

A body movement is composed of a turning of the body and of a movement forward in the facing direction after the body has been turned. The body can be turned either to the right or to the left for a maximum of 30 degrees. These values are normalized from 0 to 1, with 0 = 30 degrees left, 1 = 30 degrees right, and .5 = no turning. The forward movement can be 0 to 5 steps forward, also normalized from 0 to 1. Of the two units coding body movement one codes the turning of the body and the other its movement forward.

Arm movements are coded as in the previous organism, with a maximum displacement of each segment of the arm of 30 degrees left or right, normalized from 0 to 1 (0 = 30 degrees left; 1 = 30 degrees righ, .5 = no displacement).

The environment of these organisms is a square matrix of 80x80 cells, with 1 to 5% of these cells occupied by objects. (Each object occupies a single cell.) The environment has "walls" in the sense that if a body movement or an arm movement attempts to go outside the square matrix, it is simply stopped at the "wall". Every organism lives in 5 different environments each with a different random distribution of objects. An organism executes 50 actions (body plus arm movements) for each of 20 "epochs" in an environment, each epoch being initiated by a new random starting position of the organism in the environment. Hence, the total number of actions in a lifetime is 5,000.

We have conducted two different simulations in order to demonstrate how manipulating the fitness criterion can be a powerful means for controlling the kind of behavior that emerges evolutionarily. In the first simulation the fitness criterion was simply the total number of objects that are reached by the arm of an individual organism during the organism's life. In these conditions organisms learn to approach distant objects and when an object is at reaching distance one or two arm movements are sufficient to grasp the object (the arm's endpoint is in the same spatial location as the object). However, an undesired feature of the organisms' behavior in this first simulation is that while approaching a distant object the organism would pointlessly move its arm. This appears to be an unnatural behavior since the object is not at reaching distance yet and therefore these arm movements are useless. A more appropriate behavior would be to keep the arm quiet while approaching an object and then move the arm as soon the object is brought at reaching distance. This is what we have attempted to obtain in the second simulation. In this simulation a more complex fitness criterion was used. We wanted the arm movements to have a cost from the point of view of reproductive success, in such a manner that arm movements were generated by an organism only when they had a chance to increase fitness but not in other occasions. Hence, the fitness formula was total number of objects grasped minus a small fixed quantity multiplied by the number of arm movements executed by the organism during its life. (More precisely, each degree of an executed arm movement had a unitary cost which was subtracted from the fitness of individual organisms.) By making arm movements costly in terms of fitness, we wanted a more interesting kind of behavior to emerge across generations: a typical organism of the last generations should approach a distant object while keeping its arm quiet and, then, when the object is at reaching distance, one or two movements of the arm should bring the arm's endpoint to where the object is located.

We have attempted to use the same evolutionary algorithm of the preceding experiment, with the best 20 individuals in each generation of 100 organisms reproducing themselves by generating 5 offspring each. However, the task to be learned by the new organisms is much more complex than that of the organisms of the first experiment. As a consequence, the increase in fitness which is obtained after a number of generations is quite limited.

Given the unsatisfactory results obtained with the simple deterministic evolutionary algorithm used in Experiment A, we constructed a more complex stochastic algorithm which is more similar to Holland's genetic algorithms (Holland, 1975). The new algorithm selects the reproducing individuals on the basis of a probability distribution which is a function of the organisms' fitnesses. Furthermore, each reproducing individual generates a single copy of itself which, after mutation, replaces an existing individual in the population. The individual to be replaced is selected from the population on the basis of a probability distribution which is an inverse function of the organisms' fitnesses. With this new evolutionary algorithm a more substantial increase of fitness across generation was obtained. The results for the simulations with and without prediction learning during life are shown in Figure 8. As with the other simpler organisms, the evolution of the object reaching capacity is enhanced if during life individual organisms learn to predict how object position and hand position change when the organism moves its body and its arm.

Figure 8

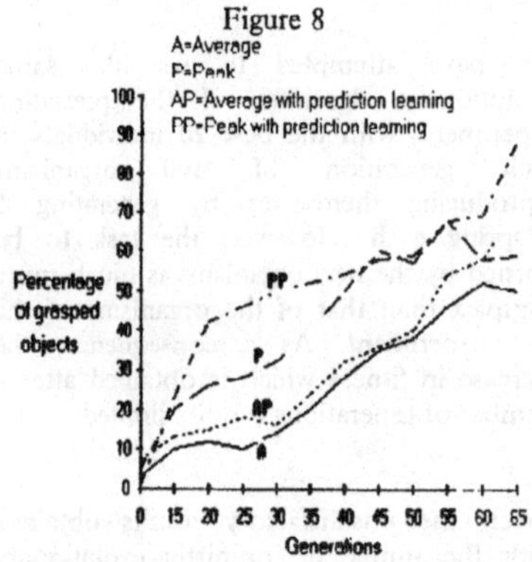

The addition of a cost for arm movements to the fitness criterion resulted in a rather interesting kind of behavior. If arm movements had no cost and networks were selected only for number of objects grasped during life, organisms typically approached a distant object while at the same time agitating their arm without much purpose since the object could not be reached by the arm. However, when networks were also selected for making a more judicious use of their arms, organisms would approach a distant object with their arm kept quiet and only when the object was at reaching distance the arm was moved to capture the object. Figure 9 shows a sequence of positions of an organism's hand in the first simulation and in the second.

Figure 9

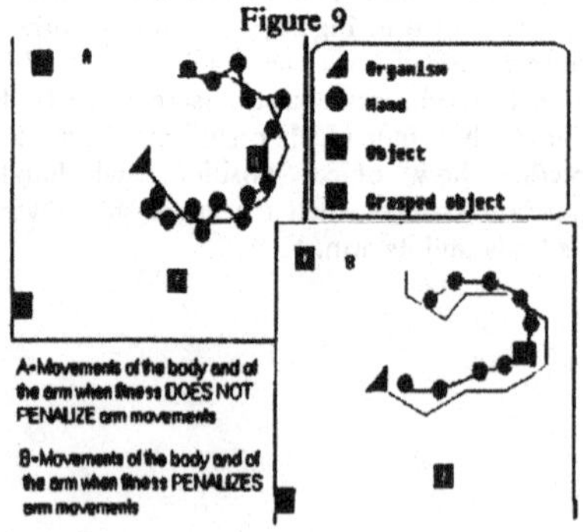

4. Conclusion

We have shown that evolutionary methods can be used successfully to develop simulated organisms that possess the capacity to reach for objects in their environments. We have succeeded in developing a simpler organism which does not displace itself in space but can move its arm to capture objects situated at reaching distance and a more complex organism which both approaches objects not at reaching distance and then operates its arm to capture them. We have also shown that the evolution of such capacities is enhanced if individual organisms learn during their life to predict the sensory consequences of their movements, even if the inheritance mechanism is strictly Darwinian and there is no Lamarckian inheritance of acquired traits.

Finally, using our more complex organism we have demonstrated that variations in the fitness criterion can determine interesting differences in the behavior that evolves. More specifically, we have succeeded in eliminating pointless arm movements when there is no object at reaching distance by adding to the fitness criterion a cost for arm movements. In these circumstances, our more complex organisms exhibit the rather sensible behavior of keeping their arm quiet while approaching objects not at reaching distance, and then operating successfully their arm to capture an object when it has been brought to reaching distance.

5. References

Booker, L.B. Classifier systems that learn world models. Machine Learning, 1988, 3, 161-192.

Goldberg, D.E. Genetic Algorithms in search, optimization, and machine learning. Reading, Mass.: Addison-Wesley, 1989

Hinton, G.E. and Nolan, S.J. How learning guides evolution. Complex Systems, 1987, 1, 485-502.

Holland, J.J. Adaptation in natural and artificial systems. Ann Arbor, Mich.:

University of Michigan Press, 1976.

Jordan, M.I. Supervised learning and systems with excess degrees of freedom. In D. Touretsky, G. E. Hinton, and T.J. Sejnowski (eds.), Proceedings of the 1988 Summer School on Connectionist Models. Palo Alto, Cal.: Morgan Kaufmann, 1989, 62-75.

Nolfi, S. Elman, J.L. and Parisi, D. Learning and evolution in neural networks. Center for Research in Language, University of California, San Diego, 1990.

Parisi, D., Cecconi, F. and Nolfi, S. ECONETS: Neural networks that learn in an environment. Network, 1990, 2, 1-21.

Patarnello, S. and Carnevali, P. A neural network model to simulate a conditioning experiment. International Journal of Neural Systems, 1989, 1, 47-53.

Rumelhart, D.E., Hinton, G.E. and Williams, R.J. Learning internal representations by error propagation. In D.E. Rumelhart and J.L. McClelland (eds.), Parallel distributed processing, Vol.1 Foundations. Cambridge, Mass.: MIT Press, 1986.

The Artificial Evolution of Behaviour

Inman Harvey
School of Cognitive and Computing Sciences, University of Sussex
Brighton BN1 9QH, England
email: inmanh@cogs.susx.ac.uk

Abstract

Darwinian evolution produces intelligent behaviour without a designer, and this can be used to evolve behaviour in simulated organisms. The problems associated with using genetic algorithms to evolve programs in a conventional language, and to evolve the architecture of connectionist networks, are compared. For an open-ended space of network architectures, a developmental language to model the production of a network from the genotype is necessary, as is a new theoretical analysis of genetic algorithms not limited to genotypes of a fixed length.

1 Introduction

Artificial Intelligence arose as a field of study from the belief that intelligent human behaviour could be formalised, and hence could be mechanised. Problem solving must be done according to rules, so this approach went; put the rules in a machine, and we will have an intelligent machine. The greatest successes have been, understandably, in just those fields of human intelligence where the problems can be formally defined, e.g. chess-playing, expert systems in simple domains; even here, success has been limited. Connectionism (McClelland and Rumelhart 1987) uses a different approach; investigating network models which are based on a very simplified model of the brain, namely large numbers of simple processing nodes with many wires connecting them, passing activations throughout the network. A major insight to come from this approach is that the behaviour of the whole can look as though it is obeying explicitly programmed rules, even though one can see that this is just an emergent property of the underlying mechanisms.

The conventional A.I. approach tries to design into a program intelligent behaviour; the connectionist tries to design networks that will produce intelligent or adaptive behaviour. Yet Darwinian evolution shows us that there can be intelligent behaviour without a designer.

Hence as one can consider the intelligence and adaptability of humans and other animals to be an emergent property of their evolutionary history in their environment, one can also consider the possibility of the emergence of intelligent and adaptive behaviour in simulated organisms, in simulated environments, with some form of evolutionary algorithm.

Approaches on these lines have been the subject of much recent interest, and conferences on similar themes have been held under the title of 'Artificial Life' in Santa Fe in 1987 and 1990 (Langton 1989, Langton *et al.* 1991), and with the titles 'Evolution, Games and Learning' and 'Emergent Computation' by the Center for Nonlinear Studies, Los Alamos, in 1985 and 1989 (Farmer *et al.* 1986, Forrest 1990).

This paper begins with an introduction to the most developed work on evolutionary algorithms, Genetic Algorithms (GAs). There follows a discussion of the problems involved in using an evolutionary approach to developing conventionally structured programs, such as might be required by a conventional, symbol-processing, A.I. approach to simulating behaviour; a program evolution system is described that has a novel evaluation function which sidesteps one of these problems, at the expense of others. Classifier systems are mentioned, leading to a discussion of recent work on applying evolutionary techniques to developing connectionist networks with architectures appropriate for the cognitive task they are facing. Finally, the necessity is stressed for a new theoretical backing to GAs, and the underlying pitfalls of evolving computational systems are looked at.

2 Genetic Algorithms

GAs were developed initially by John Holland in the 1960's (Holland 1975) as a form of search technique modeled on Darwinian evolution. The most accessible introduction is by Goldberg (1989); other sources are Davis (1987), and the Proceedings of the first three GA conferences (Grefenstette 1985, Grefenstette 1987, Schaffer 1989).

Given a search problem, with a multi-dimensional space of possible solutions, a 'genetic-code' representation is chosen such that each point in the search space is represented by a string of symbols, the genotype. A number of initial random genotypes are produced, typically by a random number generator, which form the initial population. Each of the corresponding points in the search space (which can be considered as representing the corresponding phenotypes) is evaluated by the appropriate evaluation function which gives higher scores to those nearest the solution sought, the 'fitter' ones.

The next generation of points is generated from the present population by selection and reproduction. The selection process is based on the scores of the present population, such that the fitter genotypes contribute more to the reproductive pool; typically this is also done probabilistically.

From the reproductive pool of selected genotypes, a new set of genotypes for the next generation is derived, using genetic operators such as crossover and mutation. The crossover operator works by taking parent genotypes in pairs, selecting a crossover point somewhere at random along the length of the genotype, and taking the left-hand section of one parent, as far as the crossover, and joining it to the right-hand section of the other parent; so the offspring inherits genetic material from both parents. The mutation operator just changes at random a very small proportion of the symbols on the offspring genotype to some other valid symbol.

The new population so derived has thus inherited genetic material selectively, but probabilistically, from the parent generation. New search points in the search space have been generated, exploiting the information from the performance of the parents, and improved performance can be expected from the new population. The cycle of selection and reproduction can then be repeated.

Variations on these genetic operators can be used, and there are decisions to be made on suitable sizes of populations, and the rate at which mutation and other operators are applied. In the choice of selection techniques, a recurring problem is keeping a balance between enough selective pressure for continued improvement, against too much selective pressure which leads to premature convergence, with loss of the diversity needed to escape from local optima. Subject to these caveats, however, GAs are a very robust search technique that can operate successfully in many varied, and indeed varying, search domains, and some of these properties can be demonstrated theoretically.

An early theoretical result by Holland (1975) was the Schema Theorem which demonstrates that (subject to certain conditions) schemata, which are the 'building blocks' in his representation, of above average fitness will receive exponentially increasing numbers of trials in successive generations — in other words improvement will take place. Holland also demonstrated the implicit parallelism of GAs; in a population of size P, the number of these schemata being processed each generation is of order P cubed — in other words, it is an efficient algorithm. Just for once the combinatorial explosion is on our side. These results remain the theoretical underpinning to applications of GAs. The conditions under which they apply are frequently ignored; until more general results are found this is perilous.

Typically GAs are used on well-defined problems where the search space is so large that other approaches are computationally infeasible. The problems the user faces are: what choice of genetic coding to make; what variations on the standard genetic operators and on the selection system for reproductive opportunity to make; and how to set the balance of selective pressure.

3 Evolution of simple programs

With the evolution of simulated organisms in a simulated environment, the search space is that of appropriate behaviours. Assuming that there are sensory inputs, and motor outputs, the symbol-processing A.I. approach would be to code these in symbolic form, and evolve programs which produced outputs appropriate to their given inputs.

An early evolutionary programming attempt used small finite-state machines (Fogel *et al.* 1966). From a single parent machine a single offspring was derived by mutation of the parent's state transition table; the better of the two at a sequential symbol prediction task was chosen. No crossover operator was used to gain the power of recombination, and the result was little more than random search, which is only feasible in a small search space.

Moving to programming languages, GAs deal with populations represented by strings, and a program in most languages is a string of symbols. However the dependence on order in the string is so critical that the simple insertion or deletion of a punctuation mark can completely destroy the meaning of a program, and more often than not produce a syntax error. Since the usual genetic operators are disastrous for both the syntax and the semantics of traditional programming languages, emphasis has been placed on less traditional languages with simpler forms. A simplified form of Lisp was used by Fujiki (Fujiki and Dickinson 1987) to produce programs for the strategy for the iterated prisoners' dilemma. Cramer (1985) used a subset of the simple algorithmic language PL, a pseudo-assembly language, to attempt to produce two-input, single-output multiplication functions.

Cramer attempts to preserve any potentially useful but fragile code in the population from the ruthless genetic operators; the program is divided into blocks which are exchanged as a whole by the crossover operator, and mutation is restricted to 'non-catastrophic' parts of the program. The real problem, however, is that in attempting to evolve a program that performs multiplication, the evaluation function displays what Minsky has termed the *mesa* phenomenon (as quoted in Conrad 1988); peaks are separated by untraversable flatlands, and there are no intermediate values between the very few programs that perform the required function, and all the rest that are useless. An attempt was made, with only limited success, to hand-craft an evaluation function which gave partial credit to functions that did something 'multiplication-like'. These difficulties indicate that perhaps there are not any schemata in this representation which are manipulated by the genetic operators in such a way that Holland's Schema Theorem holds true.

Recently a new approach to the evolution of programs as applied to a wide-ranging variety of problems has been put forward by Koza (1990). The programs are given in the form of LISP S-expressions, which can be depicted as rooted point-labeled trees with ordered branches. The problem of 'illegal offspring' is met by having the primary genetic operator of crossover, or recombination, swap complete sub-trees between the parents, which necessarily results in syntactically correct offspring. The twin problem of how to give partial credit to a less-than-perfect program is met by computing, for each possible set of inputs (or, with a large input space, each set from a sample of all possible), the distance in the output space between the output given by the program, and the correct output; the fitness function for this program is then based on the sum of all these distances for all the cases considered.

Koza gives examples of the application of his genetic programming paradigm to fields ranging from learning a function given a small number of examples, through solving equations to neural network design. The best-developed approach to date, discussion of whether it still has drawbacks will follow in section 7.

4 Alien mathematicians

In order to investigate an evaluation function which is intended to value, without hand-crafting, 'interesting' programs of increasing difficulty — where 'interesting' would include a two-input one-output multiplication function — I have been working with a primitive assembly language suitable for a simple register machine. Such a language can be Turing equivalent; whereas a Turing machine has an initially fixed and finite number of internal states, but unlimited memory capacity in the form of an endless tape, in a register machine this is traded off for a finite size memory, a fixed set of registers; but each register can hold an integer of any size. It can be shown (Minsky 1967) that only two registers are necessary for Turing-equivalence.

The register contents are limited to non-negative integers, and only a small set of primitive assembly language instructions are necessary, although more than the bare minimum can be used. A program is a list of instructions which are carried out sequentially, except where the instructions themselves require a jump to a different part of the program. Such assembly language instructions include (where A is the label of any register)

ZERO	A	set contents of register A to zero
INC	A	increase contents by 1
DEC	A	decrease contents by 1, minimum value zero
LD	A	set contents of register A to that of one particular register, the accumulator
ST	A	set the contents of the accumulator equal to those of register A
JPZ	B	test the contents of the accumulator; if they are zero, jump to a different part of the program specified by B, otherwise carry on to the next instruction normally.

This test-and-branch instruction is fundamentally necessary to the Turing-equivalence of such a language, and can be implemented in various ways. One such is to have numbered labels in the program, which are not themselves instructions, but define the places such jumps should aim for. An advantage of this register machine language is that any program, however manipulated by genetic operators, is syntactically correct

By the addition of an instruction of the form JPSZ B, reading 'jump to a sub-routine labeled B if the accumulator is zero', subroutines which can be called several times can be included in the program. The labels which appear in the program string are also treated as the end of the previous block of the program, and hence as 'RETURN' instructions which finish a sub-routine, and direct the execution of the program back to where the sub-routine was originally called from. Furthermore, if these markers are treated as points in the genotype where the genetic operator of crossover is more likely to occur, this will increase the likelihood of whole sub-routines, which may be useful building blocks in a program, being transmitted complete through successive generations.

A possible program string could now look like this:

```
ZERO A , INC D , JPZ F , LABEL R , DEC B ,
LABEL S , JPSZ R , INC B , ...
```

where the `LABELS` are preferential points for crossovers. A new genetic operator that is biologically plausible becomes possible with this representation, namely translocation; two points in the genotype are chosen at random (though with a preference for the labeled points), and the section of code between them is removed and re-inserted at a third, similarly randomly chosen, point. This operator allows the possibility of two sections of code which are originally far apart on the genotype, and hence almost certain to be separated by the crossover operator, to be brought closer together, and hence more likely to be inherited as a group. (This is the desired effect of the 'inversion' genetic operator, which is frequently mentioned in the literature, but rarely implemented with any success.)

This puts us in the position of a having a simple register machine language, where arbitrary strings of the symbols of the language are syntactically correct. We have three genetic operators, crossover, mutation and translocation, which allow generation of new strings of differing sizes by taking crossover points at different places in each of the parent genotypes. Through preferential cutting for the crossover and translocation operators at the labels in the program, these allow for the formation of building blocks that can be passed to following generations.

To evolve programs that take two inputs and return one output, the convention used initialises the values of two specific registers to the inputs; and at the end of the program (as signalled by the program pointer either reaching the end of the list of instructions, or reaching a label marker signalling a 'RETURN' instruction when it is not within a sub-routine with somewhere to return to), the value of one specific register is taken as the result of the program. In general, of course, of an initial population of randomly generated programs, although all will be syntactically correct, some may well have infinite loops. In the running of each program a step-counter with a finite limit rejects these.

The evaluation function is designed to produce increasingly 'interesting' programs, and to reject 'meaningless' random programs. It attempts the impossible, in being an evaluation function for Pure Mathematicians, and is loosely based on those attempts at inter-galactic communication, where the space-probe is sent off with contemporary cultural artefacts, and some mathematical formulae which are thought to be timeless and recognizable by any alien civilization. Hence two separate populations of (initially random) programs are maintained on separate 'planets'. For each of a series of trials, two inputs are randomly chosen, and these same inputs are given to all the programs on both 'planets'. Each member of the first planet scores a point for each member of the other planet which agrees with its output, and vice versa; the total scores over the series of trials is the evaluation.

Needless to say, most of the initial programs on each planet are random junk which produce an output of zero regardless of the inputs; since there will be near unanimity on both planets, these will be highly scored. If the top scorer on a planet exceeds a threshold, it is classified as an 'interesting' program and put into a catalogue (associated with that planet) of 'interesting' programs. Thereafter, during the next scoring phase, all generated offspring are penalised if they agree (in input-to-output relationships) with any one member of the catalogue over a series of trials with random inputs. Hence the second generation will not contain any programs which blindly output zero, as such a program will in practice be the first member of the catalogue.

Reproduction takes place separately on each planet, so that no genetic material is common to both planets, except by chance; the members of the catalogue associated with each planet can also contribute genetic material to the offspring with a certain probability. The separation of the planets helps to prevent genetic drift, where if the scoring had been done on the basis of matching with the same population, any program which through a random fluctuation had above-average representation in the population would quickly take over. Because of the wide variation in scores, rank-based allocation to the reproductive pools is used, with the lowest scorers having a fixed fraction of the allocation of the highest scorers. As the catalogue increases in size, the evaluation function is effectively bootstrapping itself up from nothing, using the Red Queen phenomenon where Alice (in Wonderland) has to run ever-faster to stay in the same place; there is no hand-crafting of an evaluation function to give partial credit en route to a desired end. In the same way, in the real world, there was no hand-crafting to give partial credit to jelly-fish and monkeys while moving towards that pinnacle of evolution which some (humans) claim humans stand on. Indeed, there is no global evaluation function at all in the real world, except for the negative one of non-survival, or non-reproduction; from which, in a particular context only, one can derive *a posteriori* some locally valid evaluation function.

The catalogue is prevented from being filled with just any trivial program by the requirement of agreement of input-output relationships with a significant number of members of a different population with a completely separate gene pool; for a population of programs of a given length, a simple input-output relationship can be realised in many more ways than a more complex such relationship. In runs of this system made to date, the initial programs to start filling the catalogue are ones that return a fixed number regardless of inputs; ones that return one input, or the other; and then ones that return first simple and then more complex functions of one input or the other. As examples of these programs enter the catalogue, they become potential building blocks as part of the genetic pool for future programs. The catalogue builds up, aiming to produce at some relatively early stage programs that output simple functions of the two inputs, such as their sum, or their product.

The *mesa* phenomenon has hence been by-passed, at the expense of having an evaluation function which is no longer aiming at a fixed target. It is not seriously suggested that the design of pocket calculators should be undertaken on these lines, but the implementation of this system allows assessment of just how appropriate a relatively conventional programming language is for evolutionary development. Further discussion of theoretical issues comes in section 7.

There is a useful discussion of the issues involved in using GAs to search program spaces in (De Jong 1987), where he concludes that currently the production system languages as developed for classifier systems are the most suitable.

5 Classifier Systems

Classifier systems were put forward by Holland (1975) as a family of languages which were designed for GAs to manipulate. These are production system languages, in which rules are strings with a left-hand side which represents the 'if' condition, and a right-hand side which asserts the consequent action. Given (for instance) a binary alphabet {0, 1} with the addition of the '#' wildcard or 'don't-care' symbol, a rule 01#00# → 110110 would insert a new string or 'message' 110110 into the environment if and only if any of the four messages 010000, 010001, 011000, 011001 were already in the environment. A program can consist of a set of these rules whose left hand sides are in parallel monitoring the environment, and effecting changes to it. Such rules are tailor-made for sensible manipulation by the usual genetic operators of crossover and mutation, and classifier systems have been much developed and are the subject of a considerable literature (Goldberg 1989, Grefenstette 1985, Grefenstette 1987, Schaffer 1989).

In the 'Michigan' style of classifier systems, individual rules are members of the population, and the complete population forms the rule set which is tested against the environment; a GA for producing new rules is but one small part of a complex system, in which a 'bucket brigade' plays a prominent role in apportioning of credit from the relatively few external payoffs to the rules which contributed to them. The different 'Pitt' approach, in which each member of the population is a complete rule set, is shown for instance in the LS-1 system of Smith (1980).

While classifier systems seem to be a formal mechanism very far removed from the network models of connectionists, there is a mapping from classifier systems to Boolean networks (Miller and Forrest 1989). Boolean networks consist of a set of nodes, each of which has a possible state of 0 or 1, and an associated fixed Boolean function. There are directed connections between nodes, and at discrete time-steps each node updates its state to the Boolean function of its inputs received from other nodes. By mapping all possible messages of the classifier system to nodes; the presence of any message in the environment to the firing, or being in state 1, of the associated node; and mapping the classifier rules into the appropriate connections and Boolean functions in the network, one can preserve the functional behaviour of the classifier system. Where there are strengths associated with each classifier, these can be mapped into a probabilistic function (or 'synaptic weighting') for the Boolean functions that implement the classifier.

6 Connectionist networks

The connectionist paradigm is based on seeing interesting, perhaps 'intelligent' behaviour coming from a black box inside which there is merely a large network of nodes with weighted interconnections, with rules for propagating activations through the network and altering the weights. An analogy obviously exists with the complex networks we see in the brain, but with our present inadequate knowledge of the behaviour of neurons it would be foolish to identify the connectionist's nodes with neurons. Connectionist models should be seen as simple examples of a particular style of computation — perhaps similar to that the brain uses — which is very different from the von Neumann style.

Genetic algorithms have been used as a technique for setting the weights within a connectionist network, as an alternative to, for instance, back-propagation; indeed, some success has been reported (Whitley 1989, Montana and Davis 1989). However this setting of weights is for learning in an individual, with a given network architecture; GAs should be able to search through the space of possible architectures for the more appropriate ones for the task in hand.

There has been some discussion on the GA-list (GA-1989) about whether using GAs to design networks might not be just adding an outer loop, of testing over generations, around an inner loop of learning which is already computationally expensive. It is suggested that the computational effort involved results in geological time-scales. Would it not be more efficient, it is suggested, if one is going to put such vast amounts of computational resources into producing a network, to concentrate the resources on running the inner loop of learning, for longer, and abandoning the outer loop altogether? Harp's response to this (GA- 1989) is to agree that this might well be wise

> if your sole objective is a high precision trained backprop network to address a single application problem ... [but] If you are interested, for example, in finding networks that can learn new problems rapidly, you should probably be resigned to the costs of training a few networks.

Approaches to evolving connectionist network architectures that have been taken include those in (Kerszberg and Bergman 1988, Harp *et al.* 1989, Miller *et al.* 1989, Muhlenbein and Kindermann 1989); see also references in (Rudnick 1990). All these have used some form of genetic algorithm to search through a pre-defined finite space of possible network architectures. In other words, at a more or less sophisticated level, the basic architecture has been defined with some parameters left as variables, and the GA has been used to tweak the parameters to optimal values.

To move to an open-ended space of possible architectures, rather than adjusting parameters in a finite space, a genetic coding able to produce a network from a genotype of indefinite length must be used; and this in turn raises theoretical issues dealt with in the next section. Since in this case there can be no direct mapping from each pre-defined section of the genotype to a parameter of the network, the process of translating the genotype into the network cannot all be done in parallel, and hence at least part of the translation must be a sequential process. This is in contrast to standard GA practice, where in functional terms the translation from genotype to phenotype could well be done in parallel, and the genotype is in a sense a simple description of the salient characteristics of the phenotype.

Once the genotype is interpreted sequentially, the production of the phenotype is now a developmental process, where the sequence of interpretation is of significance. So it is necessary to find some appropriate developmental language, such that the genotype determines the developmental process, which results in a network. It should be noted that in the real world, the appropriate parts of the DNA in each cell are translated sequentially; and the development of the form of the body, including the brain, from the population of cells which constitute it, is itself a sequential process of immense and little-understood complexity. The information in the DNA is too small by many orders of magnitude to uniquely specify each connection in the brain, and hence there must be some form of modularity in this specification. One presumes that there are building blocks, for instance layers with a certain pattern of projection for the connectivity between them, that are useful in many different parts of the brain; rather than each occurrence being individually coded for in the genotype, the coding must be equivalent to such a building block being coded for in perhaps just one place, and this information used many times over in the development of the brain structure.

7 Theoretical Issues

One issue that has been raised in the discussion of both the evolution of conventional programs, and the evolution of networks, is that of the theoretical basis of GAs as extended to varying length genotypes. The theoretical backbone of conventional GAs is Holland's Schema Theorem, which explicitly defines a type of building block, the schema, and proves that the conventional genetic operators will produce exponentially increasing numbers of any particular schema that is associated with above-average fitness. The problem then is reduced to finding a coding for the genotype such that schemata do indeed correspond to independent properties in the real world; building blocks that are associated with above- (or below-) average fitness. With genotypes of varying length, however, this analysis is no longer valid.

The alternative way of looking at the Schema Theorem is that, in a high-dimensional search space, the schema corresponds to a hyperplane which may be associated with above average fitness; the GA then concentrates the search towards the intersections of these hyperplanes. With genotypes of varying lengths, however, we no longer have a search space of known dimensionality.

Smith (1980) with his LS-1 system uses genotypes, of rule sets, of varying length. He suggests that for a hyperplane analysis an alternative 'attribute space' representation should be used. For one of the particular domains LS-1 works in, analysis would show, he suggests, that there are a finite number k of distinct attributes, relative to the task at hand. The search can be seen as one for combinations of these attributes yielding high performance, and hence as optimization in a k-dimensional space, where each dimension represents the presence or absence of a particular attribute. A hyperplane analysis can then be applied to the structures represented as points in this k-dimensional space, provided that it can be shown that the genetic operators applied to the genotype tends both to explore new untested hyperplanes in the attribute space, and to exploit previously tested hyperplanes by generating new instances of them; but unfortunately this last condition need not be true in the general case.

A completely different approach to the analysis of the conditions under which GAs will move a population towards optimal fitness is suggested by the concept of protein space, originally introduced in (Maynard Smith 1970). This space has a point for each possible example of a genotype, in his case the coding for a protein, and a neighbourhood metric which gives all those other points which can be reached by a single mutation from a given point. If there are N neighbours to a given point, and f is the fraction of these neighbours which are of equal or greater fitness, then the threshold condition for evolution is that fN should be greater than or equal to one.

Kauffman (Kauffman and Levin 1987, Kauffman 1990) has extended this model to produce a general theory of adaptive walks on rugged fitness landscapes (landscapes where the fitness function is represented by the height). The expected number of local optima, and the length of time taken to reach local optima through hill-climbing steps can be derived, depending on how correlated (smooth) or uncorrelated (rugged) the fitness landscape is. This theory is fairly well developed for one-step mutations as a means of generating a walk through such a fitness landscape. Traditionally in GAs, however, mutation is seen as a genetic operator of minor significance compared to crossover, and a full extension to the other genetic operators would be required in order to produce an alternative to Holland's Schema Theorem valid for genotypes of varying lengths.

The above issue is the same for both evolving conventional programs of varying length, and for evolving networks with genotypes of varying length. A second theoretical issue is whether there are inherent advantages or flaws in either of these styles of computation when an evolutionary algorithm is applied to them. In an attack on the evolvability of structured programs, Conrad (1988a, 1988b) argues that for a program to satisfy Maynard Smith's threshold condition for evolution it must in general have neighbours reachable by a single structural change, i.e. mutation, which are at a minimum halting programs. Yet as a corollary to the unsolvability of the halting problem it is in general impossible to ensure that this is so, although it may be possible in special cases.

A 'structurally nonprogrammable' system, such as a connectionist network, however, is gradually transformable. It has, for instance, been shown that a feedforward network can learn arbitrary classifications using algorithms that either extend the network forward, in the sense of adding new layers, one at a time until the classification is learnt; or extending a network with a fixed number of layers laterally with the same objective (Frean 1989, Honavar and Uhr 1989). To build a connectionist network as a virtual machine on top of a conventionally programmed computer does not alter the fact that the virtual machine may be suitable for evolutionary development whereas the underlying real machine is not — the mutations of structure are at the virtual machine level only. The price paid for this, however, is the computational inefficiency of simulating one type of computation with another.

Koza's (1990) approach, as outlined in section 3 above, seems to have some success notwithstanding being based on a conventional programming language. In the majority of the examples he gives the functions used within the LISP S-expressions are limited so as to only produce primitive recursive expressions; these do not have the power of Turing-equivalence, and hence the halting problem never arises. Some of his examples, such as block-stacking problems, or finding prime numbers, do have something like a DO UNTIL operator, which gives them the power of partial recursion, i.e. Turing-equivalence.

Where these programs that potentially may never halt are used, a 'time-out' system is brought into play to place limits on the number of iterations carried out. The point is made that in natural evolution, unlike simulations on a serial computer, all organisms are effectively evaluated in parallel; and the infeasibility of any one individual does not stop the continued propagation of the others. In the real world, the time taken to compute a decision is necessarily one of the factors to take account of in the evaluation of that decision. Should we strive for Turing-equivalence in modeling our intelligence, or should we recognise that we are no more than finite state automata?

Nevertheless, where Koza claims evolutionary success with his partial recursive expressions, this would seem contrary to the arguments put forward above for the intrinsic non-evolvability of structured programs. However the real test would be to see how his approach scales up for more than toy-size problems.

8 Conclusions

In Artificial Intelligence, it is always a good idea when in doubt to look again at the only uncontroversial examples of intelligent and adaptive behaviour, those provided by the real world. That these were produced by evolution rather than by design was Darwin's great insight. That a similar approach can be used in developing artificial examples of intelligence and adaptive behaviour has only been appreciated relatively recently.

Genetic algorithms provide an evolutionary technique which has been used to attempt to evolve programs, and to evolve connectionist networks. The former run into the '*mesa* problem', where there are no intermediate values between the very few fit programs and the majority that are useless; though an outline has been given of an evaluation function which avoids this problem, at the expense of no longer pointing in the direction of a fixed target. Work done so far on evolving connectionist networks has been restricted to searching through a pre-defined fixed space of possible architectures. To extend this to an open-ended space of possible networks, a developmental language which allows the production of a network from a genotype will be necessary.

To deal with genotypes of varying length, the theoretical underpinning of Holland's Schema Theorem is inadequate, and a new analysis is necessary. The theoretical consideration of what is necessary for a system to be amenable to evolution suggests that connectionist networks are inherently more suitable than conventional programming languages.

Acknowledgements

This work has been supported by a grant from the SERC, and I acknowledge many useful and enjoyable discussions with Phil Husbands and other colleagues at Sussex.

References

[Conrad 1988a] M. Conrad. The Price of Programmability. In R. Herken, editor, *The Universal Turing Machine: A Half-Century Survey*, pages 285–307. Oxford University Press.

[Conrad 1988b] M. Conrad. Prolegomena to Evolutionary Programming. In M. Kochen and H. M. Hastings, editors, *Advances in Cognitive Science: Steps Toward Convergence*, pages 150–168, Boulder, Colorado, 1988. AAAS Selected Symposia Series, Westview Press.

[Cramer 1985] N. L. Cramer. A Representation for the Adaptive Generation of Simple Sequential Programs. In J.J. Grefenstette, editor, *Proceedings of an International Conference on Genetic Algorithms and Their Applications*, Hillsdale, N.J., 1985. Lawrence Erlbaum Associates.

[Davis 1987] L. Davis, editor. *Genetic Algorithms and Simulated Annealing.* Pitman Press, London, 1987.

[De Jong 1987] Kenneth De Jong. On Using Genetic Algorithms to search Program Spaces. In J. J. Grefenstette, editor, *Genetic Algorithms and their Applications: Proceedings of the Second International Conference on Genetic Algorithms*, Hillsdale, N.J., 1987. Lawrence Erlbaum Associates.

[Farmer *et al.* 1986] J.D. Farmer, A. Lapedes, N. Packard, and B. Wendroff, editors. *Evolution, Games and Learning: Models for Adaptation in Machines and Nature. Proceedings of the 5th Annual International Conference of the Center for Nonlinear Studies, Los Alamos, N.M.*, May 1985. Physica D, v.22, 1986.

[Fogel *et al.* 1966] L.J. Fogel, A.J. Owens, and M.J. Walsh. *Artificial Intelligence through Simulated Evolution.* John Wiley, New York, 1966.

[Forrest 1990] S. Forrest, editor. *Emergent Computation: Self-organizing, Collective and Cooperative Phenomena in Natural and Artificial Computing Networks. Proceedings of the 9th Annual International Conference of the Center for Nonlinear Studies, Los Alamos, N.M.*, 1989. Physica D, v.42, June 1990.

[Frean 1989] Marcus Frean. The Upstart Algorithm : A Method for Constructing and Training Feed-forward Neural Networks. Technical Report 89/469, Physics Department, Edinburgh University, 1989.

[Fujiki and Dickinson 1987] Cory Fujiki and John Dickinson. Using the Genetic Algorithm to Generate Lisp Source Code to Solve the Prisoner's Dilemma. In J. J. Grefenstette, editor, *Genetic Algorithms and their Applications: Proceedings of the Second International Conference on Genetic Algorithms*, Hillsdale, N.J., 1987. Lawrence Erlbaum Associates.

[GA- 1989] Ga-list. email bulletin board, from GA-List-Request@AIC.NRL.NAVY.MIL, 1989.

[Goldberg 1989] David E. Goldberg. *Genetic Algorithms in Search, Optimization and Machine Learning.* Addison-Wesley, Reading, Massachusetts, USA, 1989.

[Grefenstette 1985] J. J. Grefenstette, editor. *Proceedings of an International Conference on Genetic Algorithms and Their Applications*, Hillsdale, N.J., 1985. Lawrence Erlbaum Associates.

[Grefenstette 1987] J. J. Grefenstette, editor. *Genetic Algorithms and their Applications: Proceedings of the Second International Conference on Genetic Algorithms*, Hillsdale, N.J., 1987. Lawrence Erlbaum Associates.

[Harp *et al.* 1989] Steven A. Harp, T. Samad, and A. Guha. Towards the Genetic Synthesis of Neural Networks. In J. D. Schaffer, editor, *Proceedings of the Third International Conference on Genetic Algorithms*, San Mateo, California, 1989. Morgan Kaufmann.

[Holland 1975] John Holland. *Adaptation in Natural and Artificial Systems.* University of Michigan Press, Ann Arbor, USA, 1975.

[Honavar and Uhr 1989] V. Honavar and L. Uhr. A Network of Neuron-like Units that Learns to Perceive by Generation as well as Reweighting of its Links. In *Proceedings of the 1988 Connectionist Models Summer School*, pages 472–484, San Mateo, California, 1989. Morgan Kaufmann.

[Kauffman and Levin 1987] Stuart Kauffman and Simon Levin. Towards a General Theory of Adaptive Walks on Rugged Landscapes. *Journal of Theoretical Biology*, 128:11–45, 1987.

[Kauffman 1990] S. A. Kauffman. *Origins of Order: Self-Organization and Selection in Evolution.* Oxford University Press, 1990.

[Kerszberg and Bergman 1988] Michel Kerszberg and Aviv Bergman. The Evolution of Data Processing Abilities in Competing Automata. In Rodney M. J. Cotterill, editor, *Computer Simulation in Brain Science*, pages 249–259. Cambridge University Press.

[Koza 1990] John R. Koza. Genetic Programming: A Paradigm for Genetically Breeding Populations of Computer Programs to Solve Problems. Technical Report STAN-CS-90-1314, Department of Computer Science, Stanford University, 1990.

[Langton *et al.* 1991] C. G. Langton, J. D. Farmer, S. Rasmussen, and C. Taylor, editors. *Artificial Life II: Proceedings Volume of Santa Fe Conference Feb. 1990.* Addison Wesley: volume XI in the series of the Santa Fe Institute Studies in the Sciences of Complexity, 1991.

[Langton 1989] C. G. Langton, editor. *Proceedings of the Interdisciplinary Workshop on the Synthesis and Simulation of Living Systems, Los Alamos, NM, Sept, 1987.* Addison-Wesley: volume VI in the series of the Santa Fe Institute Studies in the Sciences of Complexity, 1989.

[Maynard Smith 1970] John Maynard Smith. Natural Selection and the Concept of a Protein Space. *Nature*, 225:563–564, 1970.

[McClelland and Rumelhart 1987] J. L. McClelland and D. E. Rumelhart, editors. *Explorations in Parallel Distributed Processing.* MIT Press/Bradford Books, Cambridge Massachusetts, 1987.

[Miller and Forrest 1989] John H. Miller and Stephanie Forrest. The Dynamical Behavior of Classifier Systems. In J. D. Schaffer, editor, *Proceedings of the Third International Conference on Genetic Algorithms*, San Mateo, California, 1989. Morgan Kaufmann.

[Miller *et al.* 1989] Geoffrey F. Miller, P. M. Todd, and S. U. Hegde. Designing Neural Networks using Genetic Algorithms. In J. D. Schaffer, editor, *Proceedings of the Third International Conference on Genetic Algorithms*, San Mateo, California, 1989. Morgan Kaufmann.

[Minsky 1967] Marvin Minsky. *Computation: Finite and Infinite Machines.* Prentice-Hall, 1967.

[Montana and Davis 1989] David J. Montana and Lawrence Davis. Training Feedforward Neural Networks using Genetic Algorithms. In *Proceedings of the 11th International Joint Conference on Artificial Intelligence*, pages 762–767, 1989.

[Muhlenbein and Kindermann 1989] H. Muhlenbein and J. Kindermann. The Dynamics of Evolution and Learning - towards Genetic Neural Networks. In R. Pfeifer, Z. Schreter, F. Fogelman-Soulie, and L. Steels, editors, *Connectionism in Perspective*, pages 173–197. Elsevier Science Publishers B.V. (North-Holland).

[Rudnick 1990] M. Rudnick. A Bibliography of the Intersection of Genetic Search and Artificial Neural Networks. Technical Report CS/E 90-001, Oregon Graduate Institute, Dept. of Computer Science and Engineering, 1990.

[Schaffer 1989] J. D. Schaffer, editor. *Proceedings of the Third International Conference on Genetic Algorithms*, San Mateo, California, 1989. Morgan Kaufmann.

[Smith 1980] Stephen F. Smith. *A Learning System based on Genetic Adaptive Algorithms.* PhD thesis, Department of Computer Science, University of Pittsburgh, USA, 1980.

[Whitley 1989] D. Whitley. Applying Genetic Algorithms to Neural Network Learning. In *Proceedings of the 7th Conference for the Study of Artificial Intelligence and Simulated Behaviour, Sussex, England*, London, 1989. Pitman Publishing.

The Rise of Interaction

Intrinsic simulation modeling of the onset of interacting behaviour

Fabio De Luigi

Town Municipality of Ferrara, C.E.D, 44100 Ferrara, Italy

Vittorio Maniezzo

MP-AI Project Dipartimento di Elettronica
Politecnico di Milano, 20133 Milano, Italy

E-mail: MANI%ipmel1.infn.it@iboinfn.bitnet

Abstract

A simple model is presented, aimed at capturing the essential traits of the interacting behaviour of adapting species. The modelling paradigm followed is centered on ***intrinsic adaptation***, with no explicit fitness function, and is implemented using a probabilistic cellular automaton. Some simulation results are shown, regarding the outset of a predator species and prey/predator population dynamics, also with respect to environmental structural changes.

1. Introduction

Complex systems science aims at capturing the fundamental characteristic of adaptive systems composed of multitudes of interacting entities. The fundamental concept around which all its approaches hinge is that of *self-organization* [Nicolis, Prigogine, 1977], that is the emergence of organized behaviour in systems which were not designed explicitly to manage entities at the level of the outputs. A common feature of all such models in fact regards the input specification and the system description, which are defined at an aggregation level quite far - both in terms of object structuring and of characteristic time scale from that of the output of interest.

This approach allows simulation of realities unapproachable with traditional analytic techniques. A paradigmatic example is the evolution of an ecosystem where several species coexist. In this situation every species affects with its presence the environment it lives in, consisting both in the world and in its other occupants: its survival probability, along with its fitness to the environment, should therefore continuously be readjusted. In the paper we present a model where fitnesses are only implicitly dealt with, in that they emerge from environment self-organizing evolution and can be computed only *a posteriori*. Given the description of an environment and an initial uniform population, we simulate the adaptation of the population to the environment, along with the possible outset of new species.

The paper is organized as follows: in section 2 we introduce the simulation approach followed in our research along with the essential features of our model, in section 3 we describe the methodology followed to introduce interaction among the basic individuals, in section 4 we give a more detailed description of our implemented system and propose some simulation result. Finally, in section 5, we briefly outline our current activities regarding extensions of the model.

2. Modeling evolution: intrinsic simulation

The approach to simulation followed in this research tries to capture through a reductionistic/ aggregationistic process, that we call *intrinsic simulation*, the modifications on the environment caused by the presence of living entities, the alterations of this modification caused by the evolution of the entities and the effect of the evolution of entity species (thus their changes and possible onset or extinction) over the overall environment and therefore over the evolving species themselves.

Biological modeling, and evolution modeling with it, is usually performed using a set of differential equations identified at the level of population over a set of - usually phenotypical - parameters. A fitness is assigned to each population and evolution tries to maximize this fitness, that is to move the point characterizing the considered population in the space of parameters onto the top of a peak of the landscape defined by the differential equation system. This has been called *extrinsic adaptation modeling* [Packard, 1988].

Intrinsic adaptation on the other hand does not use any a priori fitness function and lets every organism evolve on the basis of the external environment and of the other organisms interacting with it. This reductionistic shift of the modeling focus from population to individuals characterizes *intrinsic simulation*.

The modeling approach we follow, originally presented in [Packard, 1988], aims at capturing the essential features of the processes of interest, namely *adaptation* and *evolution*, while remaining as simple as possible. The result is an ecosystem which does not exhaustively mimic any real-world situation, but for the features of interest (even though a parallel with chemotactic bacteria has been advanced). We willstudy *a posteriori* populations fitnesses on the basis of individual survivals and breeding capacity with respect to environment status and changes.

The environment, the world where our entities (in the following referred as *bugs*) live is an array having toroidal structure, whose cells can be occupied by *food* or by *bugs*. The food is represented by a discrete, upper-bounded, cell-specific variable: it is incremented according to some predefined feeding policies (which can be freely changed also during the simulation) and decremented by the bugs.

The bugs are individual organisms, characterized by their *genetic material*, capable of moving, eating and reproducing. Eating occurs when a bug occupies a cell with positive food level. The bugs then decrease this level by an amount depending upon their genotype, and thereby increase their internal energy level.

The bugs' energy is diminished primarily by their movements: each bug has a certain *sight* over its neighbouring cells and moves toward the nearest one containing food; when no food is acknowledged, it moves randomly. Energy is consumed in accordance to the length of the performed movement. When a bug's energy level reaches zero, it dies.

Reproduction occurs when a bug's energy level reaches a bug-specific threshold. When a bug reproduces, it splits into a number of offsprings, each with a fixed amount of energy, and its energy level is proportionally decreased. Initially, all the generated bugs occupy the same position.

Two basic computational mechanisms underlie the simulated world evolution: a *cellular automaton* and an adapted *genetic algorithm*.

The cellular automaton [von Neumann, 1966; Wolfram 1984] determines the evolution of the world. At each time-step the state of every cell is in fact determined from that of the nearest ones, on the basis of two-dimensional locality, according to evolutionary rules, determining the bugs' movements. More precisely, at each time-step the feeding policy is actuated and the cells containing bugs along with their neighbouring ones determine the movement of the bugs, that is the state of these cells themselves at the next time-step. For each moved bug, eating occurs, thus potentially preventing those that were on the same starting cell to perform the same movement. The actions performed by the bugs, that is eating, moving and reproducing (described later), are all activated on a local basis and this ensures congruence with the cellular automata paradigm. World evolution is still diachronic, but it can easily become synchronic by expliciting interaction among bugs even at this level.

The genetic algorithm[1] [Holland, 1975] underlies the reproduction of the bugs. It is at the moment implemented in a simplified version, retaining only an adapted reproduction and the mutation operators. It works on a population constituted by all the bugs of the world, each described by the string of its genes (thus by non-binary allelic values). Reproduction - here meant as a GA operator, not as the bugs breeding process - is very simplified. The standard operator involves a population with a fixed number of individuals: for each individual a fitness is computed, then the individual repruduction probability is defined as the ratio between its own fitness and the sum of the fitnesses of all the individuals. In our model there is no fitness function and the population has a variable number number of individuals. The standard operator has thus been modified, in our algorithm all zero-energy bugs are assigned a zero probability to be cloned in the next generation (that is, in the world at the next time-step), all the others are assigned a probability of one. Cross-over could provide the basic mechanism for introducing sexual dimorphism, diploidity and dominance could also be of interest. Nevertheless due to our intent of studying the simplest model showing evolutionary behaviour, these possibilities have been for the moment neglected.

Evolution occurs with the action of the mutation operator. This is a standard GA mutation, modifying an allele into another, which has been constrained to be "not too different" from the initial one. Behavioural variations follow from this genetic mutation, thus identifying genotypes with phenotypes. It is specified how individual bugs can change from one generation to the next, but evolution may be seen only by observing the changes of the entire population with time. Bug subpopulations (or *species*) are defined as parts (actually disjoint hypercubes) of the multi-dimensional space defined by the genes: whenever a bug has allelic values comprised within certain ranges, it belongs to a certain subpopulation. At the beginning of the simulation bugs of just one subpopulation are present, during the simulation others are generated by mutation.

3. Interaction among entities

Bugs interact when, at the end of a movement, one occupies a cell already inhabited by others. The interaction takes place between the newcomer and each one, in turn, of the old occupants. The basic evolutionary mechanism can give rise to different interacting behaviours. These can be seen as resulting from a mapping (the *interaction function*):

$$f:\chi^n \times \chi^n \rightarrow \mathbf{B}$$

[1] Genetic Algorithms are a stochastic computational device that maintain a *population* of entities (called *chromosomes*) and update it through gentics-inspired *operators*. The most important operators are called *reproduction*, *mutation* and crossover.

where each χ^n stands for the chromosomic material of the interacting bugs, and **B** is a set of possible behaviours. To ensure computational tractability, a default behaviour of "do nothing" is implemented, and only few circumstances prescribing special behaviours are made explicit. The interaction function is statically determined at the beginning of the simulation, and prescribes a set of interaction rules, to be followed by facing bugs. At present, the rules are defined on the basis of only one gene, but the extension to more structured activation patterns is straightforward.

Some of the implemented rules are presented in figure 1.

Two bug species are possible, with respect to aggressivity. The first is characterized by a zero in the second gene and corresponds to prey bugs: they can eat the food dispensed on the world and represent a social species, in that prey bugs can coexist in the same cell without any energy loss. Their possible dispersion is an indirect consequence of the depletion of the food of a cell caused by many eating bugs. When the food is finished the bugs will move toward the nearest cells containing food: if the first moved bugs exhaust the food content of the nearest cells, the last ones will be forced to move toward progressively far off ones.

The second possible species has a one in the second gene and correspond to predator bugs: they can only eat prey bugs and represent a non-social species. The table of fig. 1 shows how the eventuality of a predator getting in a cell already occupied by another predator would lead to a "fight" among them, resulting in a loss of energy for both of them. Moreover, the weakest of the two will be forced to move away.

At present, the bugs can only see their appropriate nourishment and decide their movements only on this basis. This was chosen in order not to put too much intelligence into the bugs, anyway in the 5th section a current development of the model is presented, which overcomes this limitation.

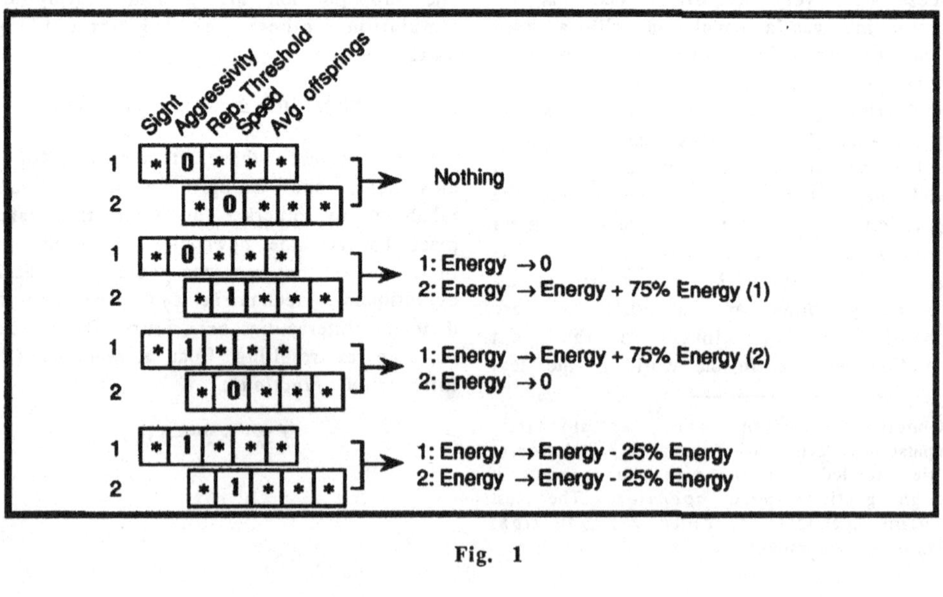

Fig. 1

Since no explicit dependency, or constraint, is placed between the value of the aggressivity gene and any of the values of the other genes, there is no correlation between aggressivity and any other phenotypical feature (e.g. predators are not heavier, thus more energy-consuming). Exhibited correlations among them are only due to natural selection disjunctively operating on each of them and leading to a coherent result.

4. Model implementation

The general criteria followed in the implementation of the introduced model, as well as during model design, has been the minimality of the developed system. We tried to introduce as few features as possible, in order to simulate the behaviours of interest.

The world is implemented as a square array, all the procedure dealing with it consider the upper row as contiguous to lowest one, and the leftmost as contiguous to the rightmost, in order to avoid boundary biases. The resulting structure has a toroidal geometry, which has been preferred to a spherical one (which could have been implemented by viewing every cell of the upper/lower row as contiguous to every other of the same row) for computational reasons.

Every cell can host food and/or bugs. Food is placed from the external, according to some predefined, run-time swappable feeding policies. The policy followed has a deep impact on the evolution, both from the viewpoint of the number of individuals and from that of their genotypic variety. Two parametrized policies have been followed: a uniform probabilistic distribution and a localized one. The latter corresponds, in suitably large matrices, to the definition of small "islands" which could evolve differently when differently fed.

Bugs are characterized by their alleles. These are bounded integer values taken on by record variables corresponding to the genes. We use five genes:

- *aggressivity*, already introduced;
- *sight*: specifies the maximal distance from the bug's position at which nourishment can be acknowledged by the bug itself;
- *speed*: specifies the maximal distance that can be covered by a bug in a single movement (i.e. during one simulation step). This value is upper bounded by the bug's sight allele;
- *Reproduction threshold*: determines the energy level required for reproduction;
- *Average offsprings*: determines the average number of offsprings generated by the bug when reproducing. The actual number is determined stochastically, on a range with boundary distance proportional to the average value.

At each simulation time-step, each bug looks around for food and moves toward the nearest cell containing it, possibly the same one on which it already was. If no food is in sight, a random movement is performed. The see-and-move action is actually implemented as a rule of the cellular automaton: the state of each cell is determined by those of the cell itself and of the neighbouring ones at the previous time-step. Neighbourhood is defined on a two-dimensional basis with a radius given by the maximum value possible for the sight allele. Bugs with different genotypes identify different cell states, thus lead to the application of different rules. A bias to this functioning is introduced by the serial nature of automaton evolution, which causes the cell states to be updated on the basis of an already partially updated matrix. While this drawback could be countered (also on a computer with traditional architecture), we don't feel that the potential benefits in terms of simulation results is worth the computational overload caused by the parallelization.

While automaton rules are specified only at the level of the individuals, results are of interest at the level of populations.

A population is defined as a set of bugs having allelic values included within particular ranges. The initial population that we use consists of bugs with identical genetic material, as is shown in fig.2 with respect to the values of speed and reproduction threshold genes (the last one actually takes values on the range 10-20, it has been normalized to allow graphical processing).

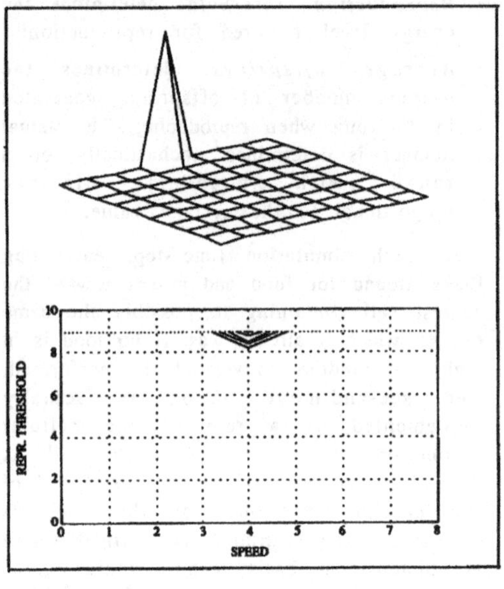

Fig.2

The genetic algorithm, and specifically its mutation operator, causes with low probability - a variation of the values of some gene of some bugs at each time-step. This causes evolution, observable in the population time-scale, which has characteristic times that are orders of magnitudes longer than those of individual mutations. The same population of fig. 2, after 1000 steps, led to the configuration of fig.3, where it is shown how the initial population evolved in a bimodal one, that is, it gave rise to two subpopulations identifiable with opportune ranges of the allelic values (in this case of the single reproducton threshold gene).

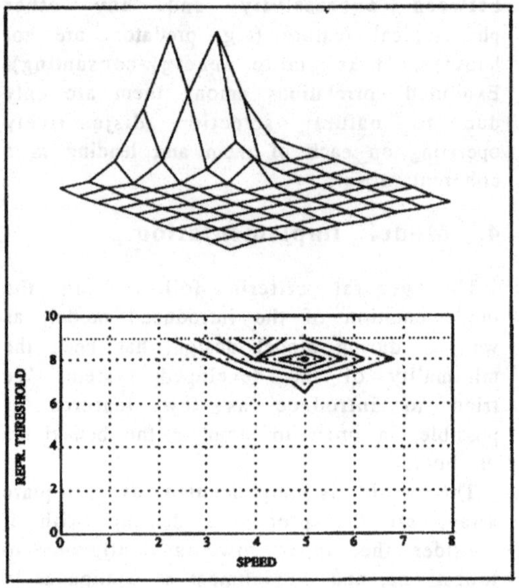

Fig.3

In contrast to most analytical population modeling techniques, this model accounts for *spatial dependent* populations: the organisms behave according to spatially local rules and the overall system evolution can be illustrated by a direct visualization of the world state. Food and the various populations are given different colours and each time the matrix is updated, it can be shown on the screen.

A sample run is presented in figure 4.

In order to quantify the visualized populaton distributions we defined a *distribution index* applied to bug populations. For each population, typically the prey and the predator ones, we compute the ratio α between the number of prey/predator bugs having another bug of the same population in its immediate proximity and the total number of bugs of that same population currently present on the world.

This is an a-posteriori measure of the degree of clusterization of the population.

The spatial evolution is reflected in the course of this index, and shows how the bugs are thickly clustered in the initial transitory phase and scatter when the food becomes more difficult to get.

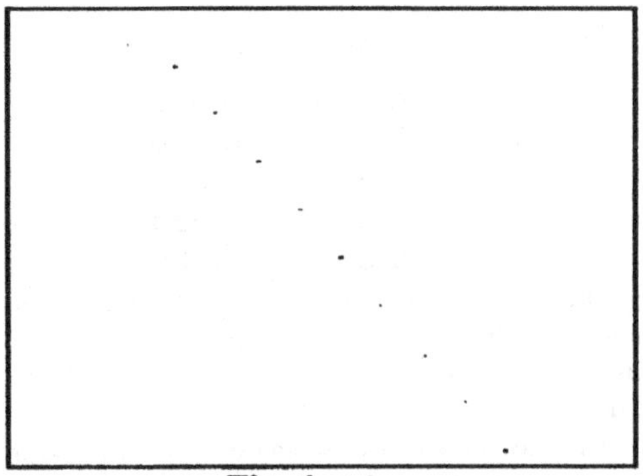

Fig 4 a

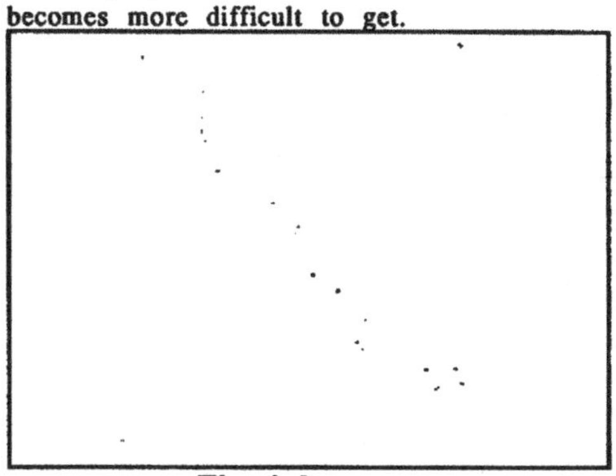

Fig 4 b

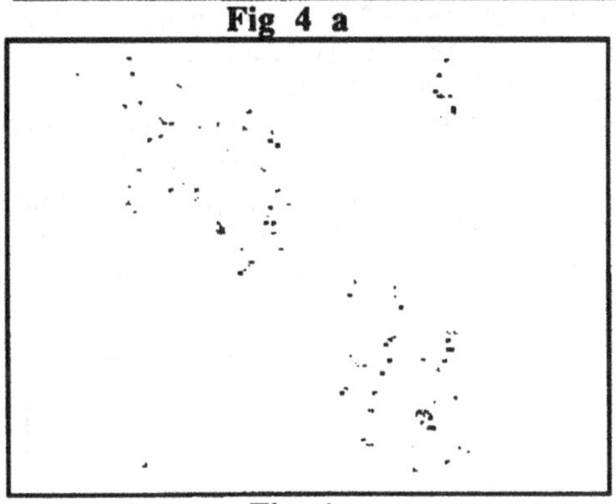

Fig 4 c

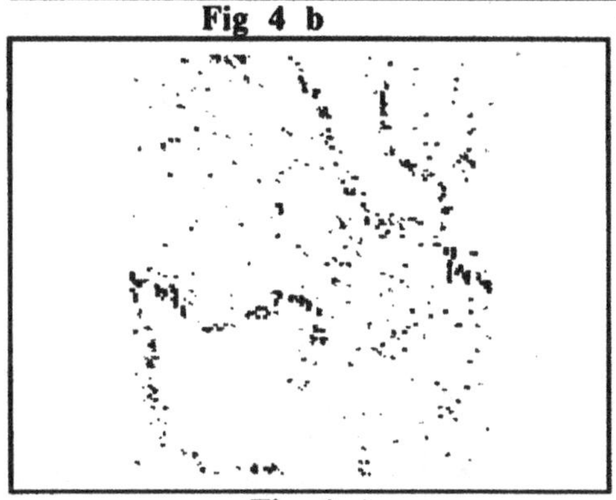

Fig 4 d

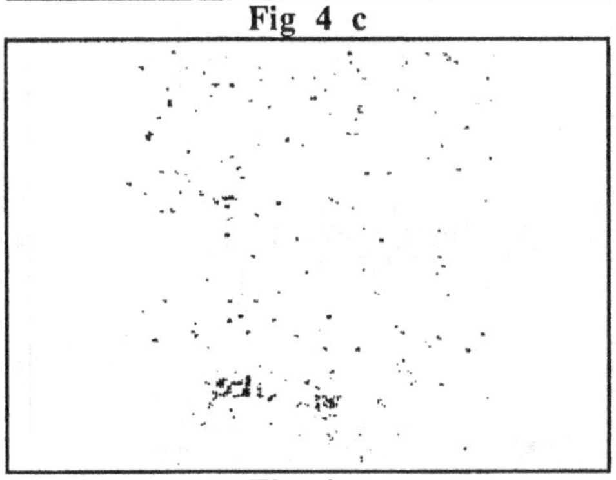

Fig 4 e

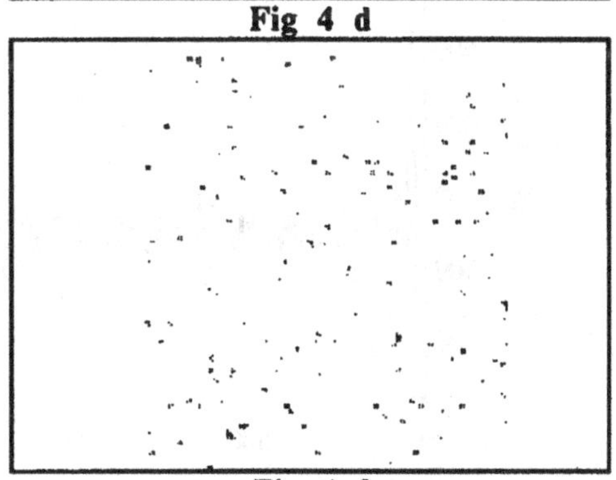

Fig 4 f

It is interesting to notice the different behaviour of the predators: after a longer initial transitory phase, due to the delay of their outset and then to their presence only in the world region where they originated, they converge to an index value significantly smaller than

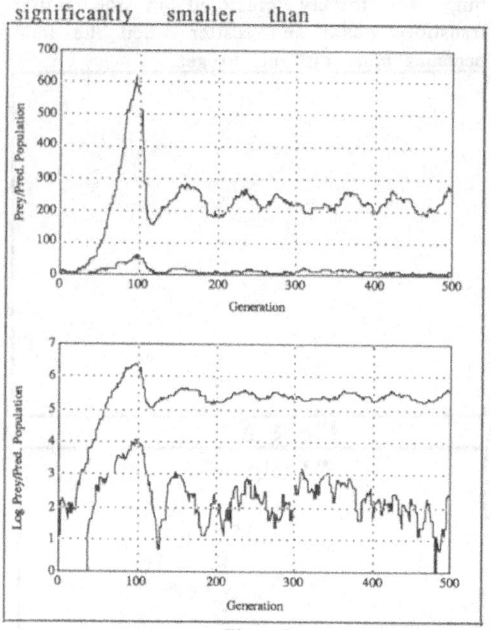

Fig. 5

that of prey. The following table summarizes the results of the computation for a distance of 2 (i.e. 25 cells centered on each bug) and for the first 200 generations.

Generation	α_{Prey}	$\alpha_{Predators}$
1	0.000	0.000
50	0.531	0.000
100	0.679	1.000
150	0.536	0.533
200	0.597	0.428

The last computed values are subject to some oscillation, expecially for prey, but are basically maintained during the rest of the simulation.

This diversity of behaviour is also seen for the dynamics of the two populations. Fig. 5 shows their total amount and their logarithms and displays how predator population is almost independent from limited variations of prey number.

Moreover, the predator/prey ratio keeps around 0.05, and is considerably higher than mutation probability for the aggressivity gene.

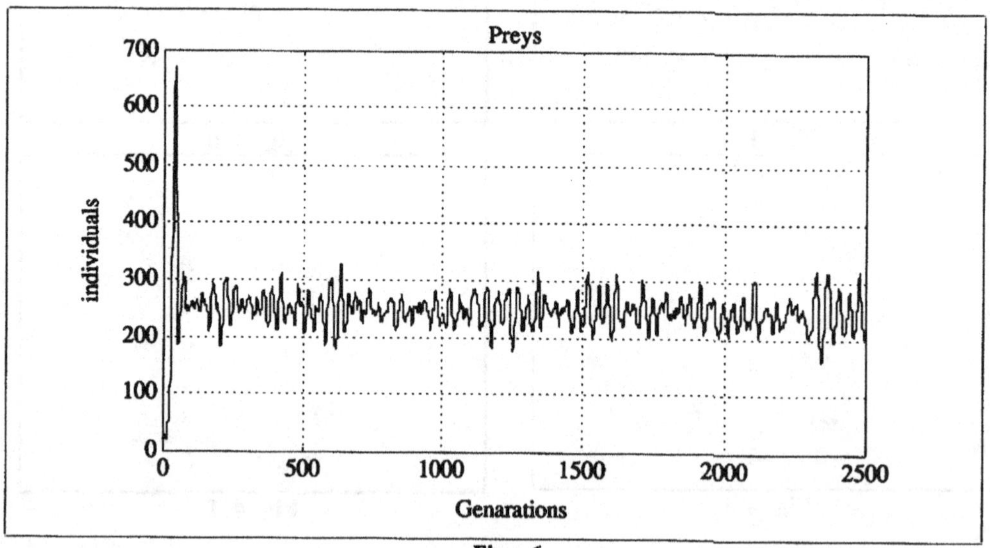

Fig. 6

Data regarding prey population suggest a cyclic equilibrium. To ascertain this, here are the results of a longer simulation, along with the fast fourier transform of the results, without considering the initial transitory phase. Fig. 6 displays these outcomes.

Fig.7 shows a plot of the prey population data on itself, with time-axis shifted of 50 generation and with the initial point placed at the end of the initial transition. The complexity of the population dynamics is evident. Data regarding predators lead to analogous results.

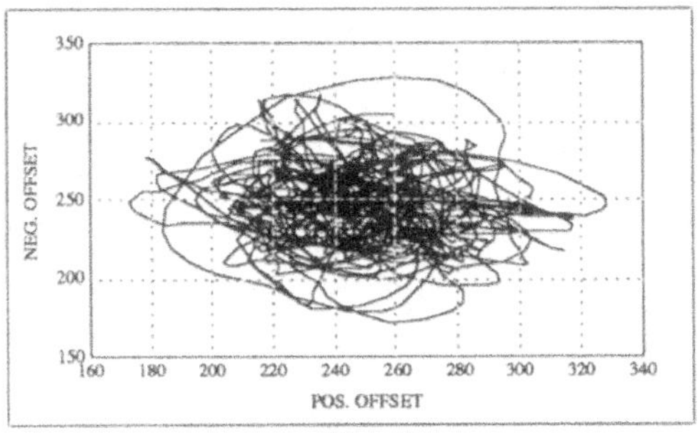

Fig. 7

Finally, in Fig. 8 we present the result of a simulation where we changed the feeding policy at run-time, switching from a uniform distribution with a probability of 2% for each world cell to receive a food unit, to a probability of 4%. The induced transition and the new final conditions are evident. The latter comprise higher average population levels for both prey and predators and larger elongations for oscillations in predation.

5. Conclusions and ongoing work

We presented a self-organizing model meant to capture the essential traits of interacting species, with special reference to prey-predator relationship. While associating ourselves with a recently proposed paradigm for ecological modelling [Packard, 1988], we showed how the introduction of a single dychotomic gene in the strings describing bug individuals to the GA underlying our system allows the identification of strongly differentiated

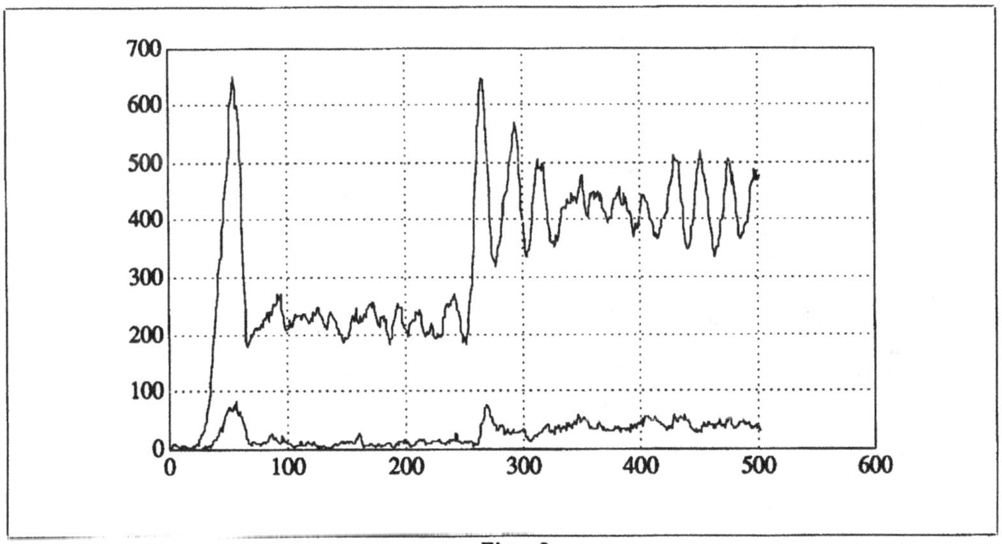

Fig. 8

behaviour for the two bug populations so identified. This is obtained preserving the minimality of the resulting model, which is one of the objectives of our approach. More specifically, we showed how our model accounts for the evolution of species and the onset of new ones, how distal features are explicitely handeled and accessible both through a direct visualization of the world state and through a distribution index that we defined in order to quantify the distribution of the various species on the available space. Moreover we showed how the dynamics of the different species is a complex one, and how they are strongly dependent upon the feeding policy used.

Our current goal is to eliminate the external determination of the interactions, allowing the system itself to evolve configuring species which exhibit particular interacting traits as a result to natural selection. This will be achieved by integrating a classifier system [Booker et al., 1989] into our program, which will devise the behavioural rules of interest. It is expected that - without changing the genetic material - different interacting behaviours will emerge, including sociality, aggressivity and others.

References

[Booker et al., 1989] Booker L.B., Goldberg D.E., Holland J.H.: "Classifier systems and genetic algorithms", Artificial Intelligence, 40, 1989

[Holland, 1975] Holland J.: "Adaptation in natural and artificial systems" Univ. of Michigan Press, 1975

[Nicolis, Prigogine, 1977] Nicolis G., Prigogine I.: "Self-organization in non-equilibrium systems", John Wiley, 1977

[Packard, 1988] Packard N.H.: "Intrinsic adaptation in a simple model for evolution" in C. Langton (ed.): "Artificial Life" Addison Wesley, 1988

[von Neumann, 1966] von Neumann J.: "Theory of self-reproducing automata" (Burks A.W. ed.), Univ. of Illinois press, 1966

[Wolfram, 1984] Wolfram S.: "Universality and complexity in cellular automata" Physica, 22D, 1984

THE EVOLUTION OF BEHAVIOR: SOME EXPERIMENTS.

JAN PAREDIS

RIKS
Postbus 463
NL-6200 AL Maastricht,
The Netherlands

ABSTRACT.

This paper investigates the evolutionary development of (problem solving) behavior. Through evolution, artificial animals learn to survive in a given world.

We use layered neural networks (NNs) as the substrate on which evolutionary learning operates. The fault tolerance of neural networks allows for a genotype / phenotype distinction which maintains the variation in the genetic pool. Furthermore, we define building blocks which take into account the functionality of the NNs.

The result of the algorithm can be inspected at two levels. First, there is the behavior of the individual animals. A description of their behavior is obtained through the induction of decision trees which describe the functionality of the NN. Second, the behavior of the population as a whole can be described. The distribution of the animals over the world often provides an analogical representation of a problem solution.

Keywords: autonomous agents, evolutionary learning, genetic algorithms, inductive learning, machine learning, neural networks.

1. Introduction.

Natural evolution continues to intrigue mankind. Particularly, the complexity to which it leads often surprises us. Examples of this complexity are abundant. Animals in a prey - predator relation, for example, develop complex defensive and offensive behavior, such as camouflage and imitation. Insect colonies (e.g colonies of ants, bees, wasps, termites etc.) are another instance of the complexity to which evolution can give rise. Such complexity is obvious if one examines the nests termites build or the social organisation within insect colonies [Wilson 85]. Many other examples can be found in [Dawkins 86] or [Tinbergen 65], amongst others.

Computational methodologies based on the evolutionary metaphor have been developed for a wide-range of problems, such as search, optimization and machine learning. One of the most notable methodologies are genetic algorithms (a good overview of GAs can be found in [Goldberg 89]). Another, related, methodology is evolutionary learning, which searches through a space of behaviors using the principles of variation and selection. In contrast with GAs, evolutionary learning does not require an explicit, domain specific fitness measure. We only need to specify the characteristics of the environment (e.g. the amount of food present) and the effects of an animal's actions on itself and on the environment.

In this paper we propose an evolutionary framework in which successive generations of animals learn to improve their chance of survival in a given environment. Or, in other words, successive generations adaptively develop behavior (such as: look for food, avoid predators ...) in correspondence with their natural needs. In order to be successful, an animal has to find an answer to the following question: When should I perform which actions? This question is answered through evolutionary learning on NNs. Our system learns only through evolution, no life-time learning mechanism is incorporated. We should stress that our primary goal is to obtain adequate (problem solving) behavior. The biological relevance of the algorithm is not our main concern.

The structure of this text is as follows: section 2 describes the worlds we work with, section 3 proposes the animal architecture. The fourth section explains our evolutionary learning mechanism. It is followed by experimental results. Finally, we discuss our planned future research and we conclude.

All the results discussed here have been implemented. Figures 6, 7, 9 and 10 are actually taken from our interface after overnight runs.

2. Problem Description.

Let us first look at the worlds we use, and the animals populating them. Figures 1 and 2 depict two simple artificial worlds . They are both planar and consist of 15 by 15 locations. The locations on the edge of these worlds are on fire. Four locations in world1 (W1) contain food, and two locations contain drinking-water. We also use a world W0 which is similar to W1 except that it does not contain any water. World2 has no water but four food sources move around in it. This food moves clockwise along the perimeter of a square.

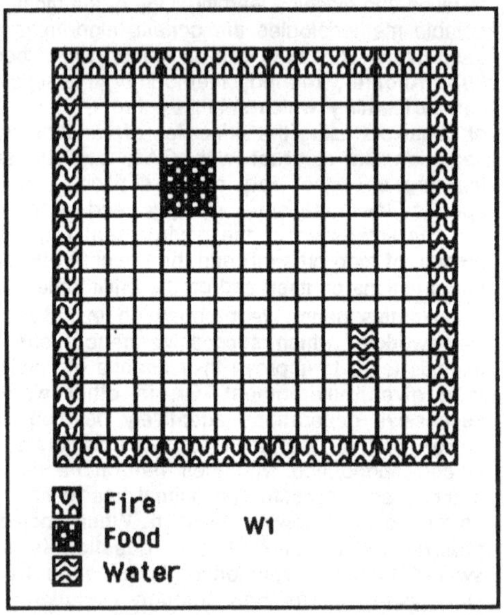

Figure 1: World1.

An animal eats or drinks when it is at a location which contains respectively food or water. This results in the storage of food or water in internal stores. Each move of an animal decreases the contents of these internal storages by a fixed amount. In W0 and W2 an animal dies when the contents of its food store drops below a given level. In W1 the contents of both the internal food store and the internal water store have to remain above a threshold. In all three worlds the animals have to avoid the locations on fire, otherwise they burn to death.

We should note that in the worlds described here the food sources do not get depleted. Or in other words, the quantity of food and water present at a location is not changed through eating or drinking.

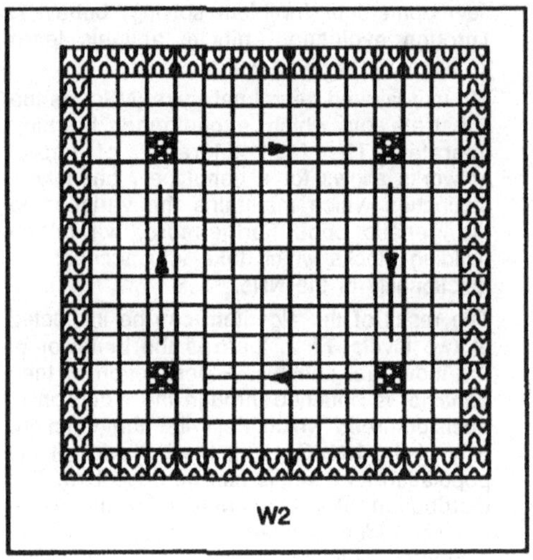

Figure 2: World2.

At discrete time intervals the animals synchronously move to one of their eight neighbouring locations. These eight moves constitute the possible actions of the animals. The problem we want to solve can now be described more precisely: from every position in the world the animal has to learn how to survive. The following goals are necessary conditions for survival : avoid starvation (W0, W1, W2), avoid dying of thirst (W1), and avoid burning to death (W0, W1, W2). In the next section, we put forward an architecture which enables animals to develop behavior in correspondence with their natural needs.

3. Animal Architecture.

An animal consists of two parts (see figure 3). Its *local state* represents data about the animal itself. It describes, for example, the animal's internal food reserve, current heading, position in the world, age ... . Secondly, a neural network implements the behavior of an animal. This net maps the animal's perception of the world, and its local state, to actions. In turn, these actions possibly change the world or the animal's local state. Moving, for example, changes an animal's local state: the contents of its internal storages decrease. Moving, also changes the state of the world: the presence of animals at the locations changes.

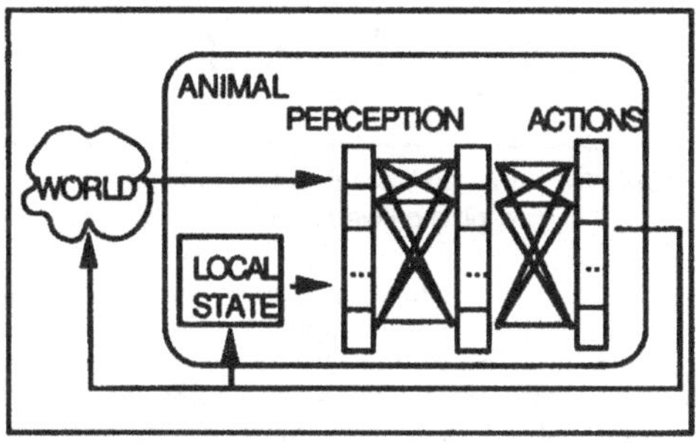

Figure 3: Animal Architecture.

The layered feed-forward networks which we use consist of: an input layer, one hidden layer and an output layer. Lippmann [Lippmann87] has shown that any partitioning of arbitrary input spaces can be learned by networks with two hidden layers. Hence, for more complex tasks no more than two hidden layers are needed.

Each node of our network is connected to all nodes in the next layer. The standard sigmoidal threshold is used and all weights are in the closed interval [-1,1]. In worlds W0 and W2 the input layer contains only two inputs: the x and y position of the animal in the world. In addition to their position, animals living in W2 use the level of their internal food and water reserve as inputs. This way the animal's behavior is not only determined by its position but also by its hunger and thirst: hungry animals should move towards the food, whereas thirsty animals should look for drinking water. In all three worlds we use one hidden layer consisting of 10 neurons. Each output node represents one of the eight possible directions in which the animal can move. These moves are mutually exclusive, i.e. only one of them can be chosen at a time.

At each point of time the following steps are executed: 1) fill in the input layer, 2) calculate the activation level of the output nodes through forward propagation, 3) The animal executes the move associated with the output neuron with *highest activity level*. After this move the animals living in W0 and W2 eat, those in W1 eat and drink - when there is food or water at their location.

The learning goal can now be stated as follows: determine the weights of the NNs such that they provide a good mapping from perception and local state to actions. Even for simple network architectures, this weight space constitutes a large, high dimensional search space. The animals living in W0, for example, consist of 118 weights (including the threshold biases). Those in W1 define a 138 dimensional search space.

4. Evolutionary Learning.

The animal's neural network is the substrate on which evolutionary learning operates. The initial population consists of animals whose networks have random weights. These weights are the "genes" of the animal. Just as in nature, the fitness of an animal is determined by its ability to find food, to survive predators and to avoid all other obstacles to adulthood and reproduction. Children inherit their weights from their parents through crossover. As in GAs, this operator is followed by a low probability mutation operator.

This section deals with two questions: 1) How do we implement inheritance?, and 2) What are the advantages of using NNs as an evolutionary substrate?

4.1. Building Blocks for Neural Networks.

In order to be successful, the sets of weights which are responsible for good performance on a (sub)task should have a high probability of being passed on to the next generations. GAs similarly try to combine good partial solutions into a good global solution.

Let us first look at some alternative decompositions of our "behavioral NNs" into subproblems. One decomposition partitions the set of weights into subsets containing all the weights involved in the mapping of one input node (e.g.

internal food reserve) to one output node (e.g. a direction the animal has to move in). These subsets contain weights which are clearly functionally related. By analogy with GAs [Holland et al. 86] such a set of functionally related elements is called a Building Block (BB). Figure four depicts the BB described above, which we call a Fan out Fan in BB (FOFI BB). It shows the links whose weights determine the mapping between an input (I) and an output node (O).

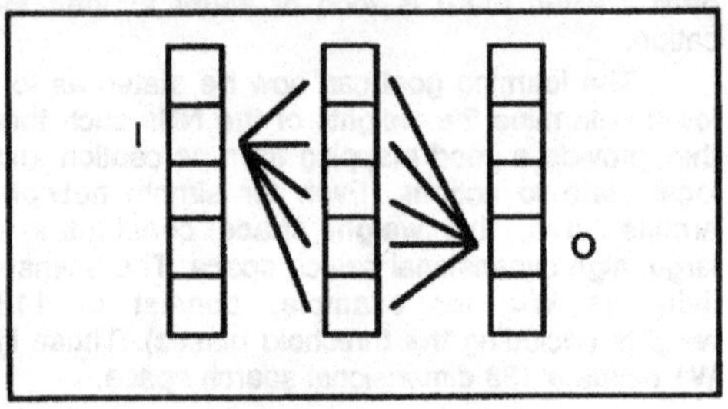

Figure 4: The FOFI Building Block.

When, during reproduction, a child inherits a "good" BB from one of its parents (e.g. the mapping from hunger to a direction) and another good one from the other parent (mapping thirst and direction, for example) then it might have a selective advantage. In this case, it has a high chance to reproduce and to spread its BBs through the population. After a while good BBs occur in large numbers throughout the population.

The FOFI BBs described above have an important disadvantage: they become large when used for NNs with more than one hidden layer. They also overlap to a large extent: all connections between two hidden layers belong to all FOFI BBs. Such large BBs typically cause premature convergence: once a good BB has been found, reproduction quickly dries out the genetic variation.

For these reasons we use a smaller BB: the path BB (see figure five). It consists of a random path between an input / output node pair. The reproduction using the path BB proceeds in three steps: 1) during reproduction the child initially copies all its weights from one of its parents, 2) from each input node a random path to a random output node is chosen, 3) all the child's weights on this path are overwritten with weights from the other parent.

Newborn animals always replace dead ones. This way, the population size is held constant. Furthermore, only the 25 % oldest animals reproduce. This gradually increasing age of sexual maturity ensures that animals are tested longer and longer before they reproduce. At first, relatively inexperienced animals reproduce, at later stages the reproducing animals are more sophisticated. From this 25% percentile, parents are chosen with a probability proportional to the level of their internal stores. Hence, the parents don't have to be located near to each other in order to reproduce.

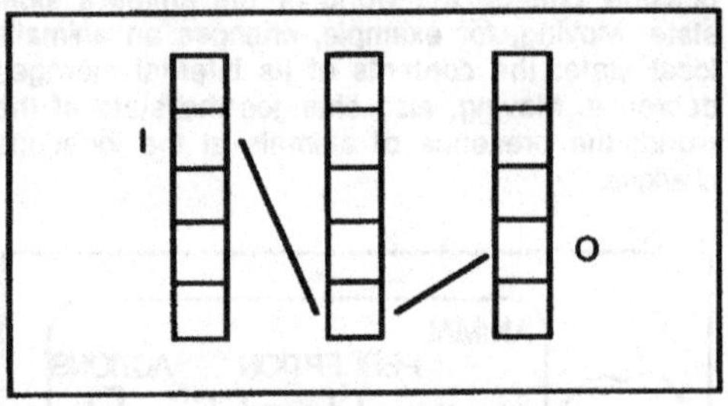

Figure 5: The path BB only partly defines the mapping between nodes I and O.

Section 5 provides examples of the efficiency of the search algorithm based on the principles explained above. This efficiency critically depends on the level of abstraction (i.e. the size of the BBs used) at which the search is done. The right level of abstraction nicely balances exploration and exploitation. Search at too low a level (e.g. at the level of individual weights) results in slow convergence. Search at too high a level prematurely converges. We are currently setting up experiments to investigate this trade off in more detail.

4.2. Suitability of NNs as an Evolutionary Substrate.

It is well known that changing a few weights of a neural network does not necessarily impair its functioning. Due to this *fault tolerance*, one or more regions in the weight space (= the space of all NNs with a given architecture) represent the same input-output mapping. Or, in other words, the same mapping can be represented by NNs with different weights.

It is their fault tolerance which makes NNs a good evolutionary substrate because various genotypes give rise to the same phenotype. Here, the weight settings of a NN constitute the genotype, whereas the behavior of the animal is its phenotype. Evolutionary learning typically operates through a process of genotype variation and phenotype selection. Hence, the behavior of an animal determines its success, not its genes. Due to the fault tolerance of NNs, the genetic pool (i.e. variation of the weights in the population) is kept relatively large during the entire simulation. The principal source for keeping up this variety is the mutation operator. It changes a child's weights without necessarily changing its behavior. Even a population in which all animals belong to one phenotype typically consists of a wide variety of genotypes. Such a large genetic pool enables rapid adaptation in response to changes in the environment.

5. Experimental Results.

The results of our system can be inspected at two levels. Firstly, there are the *micro results* which describe the behavior of individual animals. Secondly, one can examine the behavior of the population as a whole (= the *macro result*).

5.1 Micro Solutions.

Figures 6 and 7 show typical micro solutions for W0 and W1. The triangles indicate the direction in which the animal moves from a given location.

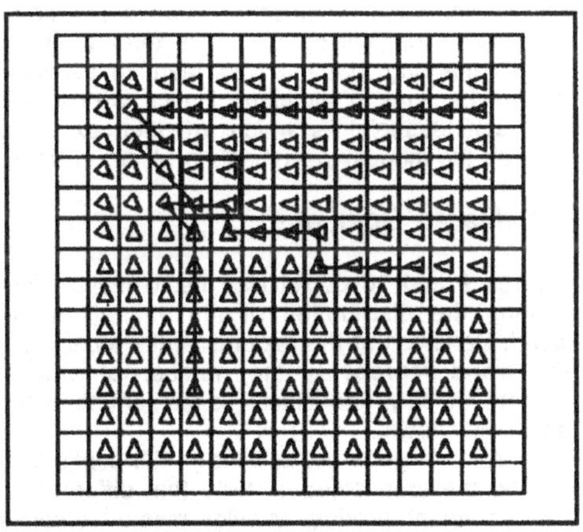

Figure 6: A micro result for W0. This animal finds the food from every location. Three paths are shown.

In W0 the animals learn to move towards the food where they end up on a cyclic path. In this example, old (i.e. good) animals are periodically relocated to a random position in the world. Hence, the animals which know how to move from every position in the world have a clear selective advantage.

The micro level solutions for W1 are more difficult to depict: the behavior is not only a function of the position in the world but also of the animal's hunger and thirst: a hungry animal moves towards the food even when near to the water. Figure 7 shows a typical path between food and water as obtained by our algorithm.

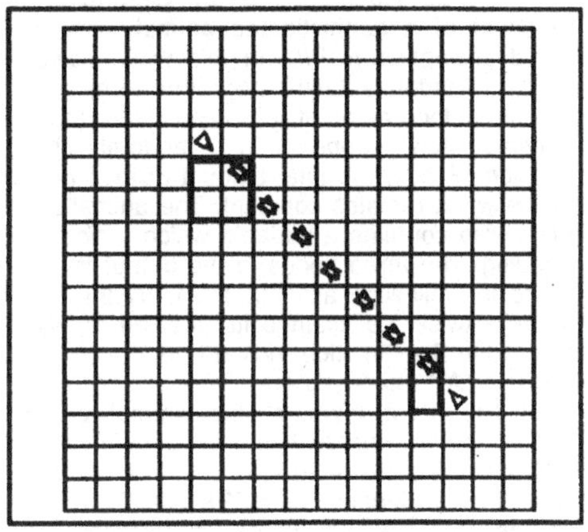

Figure 7: A path between food and water learned by an animal in W1.

On average, it takes about 10000 time steps to obtain an animal in W0 which knows how to go to the food from every location. Animals in W1 which have learned a path between the food and the water appear after about 5000 time steps. Note that at each time step only a few children are born.

The graphical representation used above, does not provide a complete and clear description of the mapping from the state of the world and an animal's local state into actions. This text proposes the use of inductive learning techniques to "explain" the mapping encoded by an animal's NN. The approach we use is called TDIDT, which stands for top-down induction of decision trees [Quinlan 86]. From the class of TDIDT algorithms we chose to use the NPPA algorithm [Talmon 86] because it learns to classify concepts with *real-valued* attributes. It learns such a classification from a given training set of pre-classified examples. Here, the examples consist of randomly filled input layers of the animal one wants to study. The classes represent the animal's actions. The class of an example is determined by forward propagation through the animal's NN starting from the input layer represented by the example. Hence, the class of an example symbolically represents the action associated with the output node with highest activity level. From these labelled examples NPPA constructs a binary decision tree. Leaves of such a tree are class names, the branching points test an attribute-value with respect to a threshold value.

Let us illustrate this by describing the behavior of a sophisticated animal living in a world analogous to W0. In this world however, food is randomly distributed. Furthermore, the amount of food eaten at a location disappears, but it immediately reappears at a randomly chosen location. Hence, the total amount of food present in the world remains constant. The animal's local state also contains a variable which indicates its heading. The three nodes in the output layer activate the following actions: 1) TURN the animal 45° clockwise, i.e. change its heading, 2) MOVE forward in the direction indicated by the heading, and 3) STAY where you are.

The input layer of the animal contains values for: its energy level, i.e. the level of its internal food store (= ENERGY), the food present at its present location (= HERE-FOOD), the food at the location the animal is heading to (= FUTURE-FOOD), and the presence of fire at that same location (= FUTURE-FIRE). Figure 8 represents part of the decision tree of an animal which survives in such a world. We see that NPPA comes up with a tree in which the most discriminating feature - the presence of fire at the future location (which is a binary value) - is the first splitting criterion.

The three terminal nodes depicted in figure 8 represent the following rules: 1) If the location I am heading to is not on fire but it contains only a small amount of food (< 0.3) and at my current location there is more than 0.19 food then I stay where I am. Or, in other words, it is not worthwhile to waste energy by moving forward, there is enough food here. 2) If there is however, more than 0.3 food at the future location and my energy level is high enough then I move in the direction indicated by my heading 3) If the location I am heading to is on fire then I turn 45°.

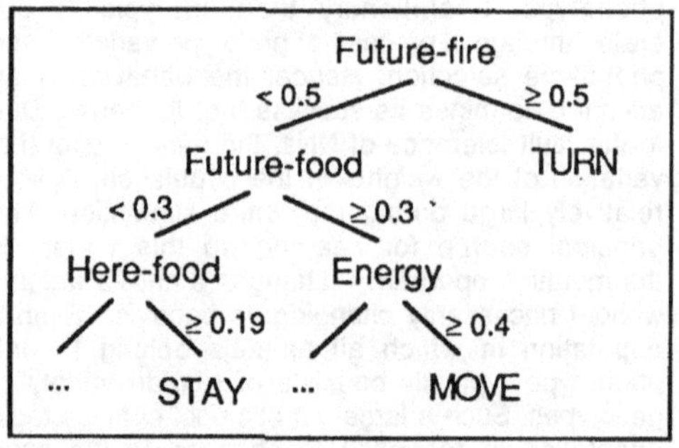

Figure 8: Part of a decision tree.

Such a collection of rules gives us insight in the solutions found by the evolutionary learning algorithm. They allow us to inspect solutions to complex non-linear worlds which might be hard to find through mathematical analysis.

The main advantage of the approach sketched above is that it combines the advantages of both NNs and TDIDT systems. NNs are known to be good pattern associators. They are however black boxes: they cannot provide a good explanation of this association. The representation obtained by TDIDT systems is understood more easily.
The learning task itself, i.e. the association of actions with situations, cannot be achieved by the TDIDT approach because it requires pre-classified examples [Wilson 87]. It is through evolutionary learning that the NNs learn to associate the good actions to situations. Hence, we use NNs as label associators which are then explained through the inductive synthesis of decision trees.

5.2. Macro Solutions.

Figures 9 and 10 depict two macro maps representing the distribution of the animals in the world. Darkness of the textures at the locations in the world represents the number of animals present. If only one animal is present then that animal is shown.

As time goes by, animals in W0 concentrate around the food source. Note that different behaviors might produce similar macro solutions. All animals in W0 walking on short circular paths nearby the food give rise to macro maps which look similar to the one depicted in Figure 9. Hence, macro solutions show some degree of robustness with respect to the micro solutions they are composed of. The macro solution of W1 consists of a path of dark textures between food and water.

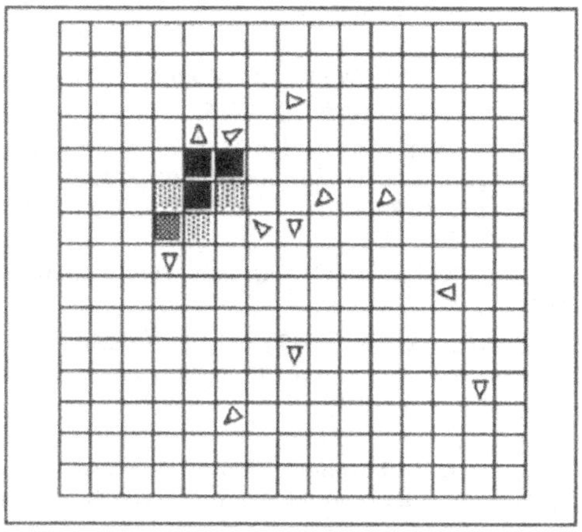

Figure 9: A macro solution for W0.

It is also interesting to see how changes in the world can lead to different micro as well as macro results. Figure 9 depicts a macro solution for W2 in which all animals remain close to the trajectory of the food. If however, the amount of food in the four moving objects is smaller, then the animals will have to move synchronously with the food. In this case, the macro solution consists of four locations at which the animals are concentrated.

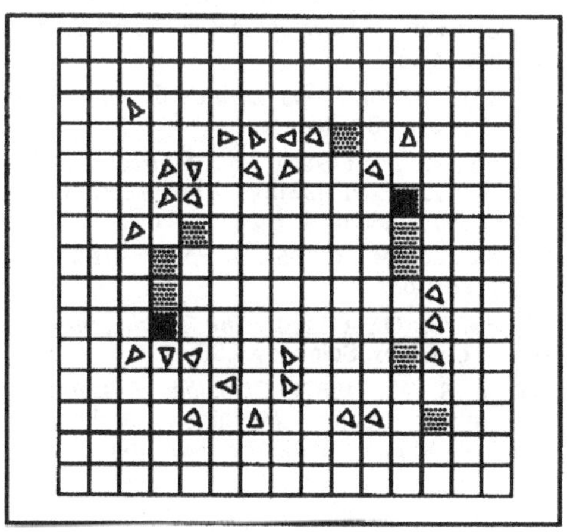

Figure 10: A macro solution for W2.

Macro solutions as the ones depicted here are typically obtained after about 15000 (W0) and 20000 (W2) time steps. Less pronounced macro maps appear earlier. These results, of course, heavily depend on the "tightness" of the world, which is, for example, determined by the amount of food present in the world.

The macro solutions pictorially represent solutions to various problems (find food, find shortest path ...). Such representations which bear a structural similarity to what is represented are called analogical representations [Sloman 84]. They implicitly represent properties (e.g. is-a-square) and relations (e.g left-of). These implicit properties and relations can be deduced directly from the analogical representation, instead of from a symbolical representation. [Steels 88] describes the use of analogical representations for problem solving.

6. Future Research.

Future research will investigate additional "driving forces" towards complexity. Let us first make the distinction between biological and system parameters.

System parameters define the evolutionary search. Examples of system parameters, and the values we used for them are: mutation rate (15% probability that one weight of a child is changed), population size (50), selection of parents ... Most of these system parameters were chosen once and held constant for all the worlds.

Energy consumption (i.e. the decrease of the internal reserves) is one example of a biological parameter. Such biological parameters model the world and the effects of an animal's actions. They strongly determine the selective pressure and its direction. Or in other words, they implicitly define the fitness landscape. Changing biological parameters might change the micro and/or the macro results. A gradual increase of an animal's energy consumption, for example, pushes the animals towards more efficient use of their resources. It forces the animals in W0 to find a short path to the food. Unfortunately, such gradually changing biological parameters bring with them some difficult questions, such as: At what rate can these parameters be changed? If they are changed too slowly then convergence will be slow. Changing the too quickly does not allow the necessary time to adapt to the new situation. In such a case, one risks losing all the genetic material found so far.

Nature uses more elegant mechanisms for gradually obtaining more complex behavior. One of these is called the *evolutionary arms race* [Dawkins 86]. In nature, there is strong evolutionary pressure for preys to defend themselves better (e.g. by running quicker, growing bigger shields, camouflage ...) in response to which the predators develop better attacking strategies (e.g. stronger claws, better eye-sight ...). More efficient predators provide, on their turn, selective pressure on the prey which has to "respond" with more efficient behavior as well. Such a stepwise arms race leads towards increased complexity. It also ensures that the gradual increase of selective pressure is in step with the complexity of the population. The use of such arms races will be investigated in future research.

7. Conclusion.

The objective of this paper was to put forward an architecture which allows for the evolutionary development of behavior.

We argued that NNs are a good evolutionary substrate because of their fault tolerance. This tolerance leads to a genotype / phenotype distinction which, through the variation in the genetic pool, allows for rapid adaptation. Next we defined building blocks which take into account the functionality of NNs. We discussed two types of building blocks which reside at different locations on the exploration - exploitation spectrum.

Finally, we discussed the two levels at which the results can be inspected: the behavior of the individual animals and the behavior of the population as a whole. We also showed how the functionality of NNs can be explained through the inductive synthesis of decision trees. These trees give us insight in the behavior of the animals which survive in a given world.

Acknowledgements.

The expertise of NNs and/or GAs of Hans Henseler, Bernard Manderick, and Piet Spiessens enabled me to advance quickly with this research. I am indebted to Tanja van Rij for developing the interface, and to Floor Verdenius for making his NPPA implementation available to me. Eric Postma, and Ronald Westra provided useful comments.

References.

Dawkins, R., (1986), *The Blind Watchmaker*, W.W. Norton, New York.

Goldberg, D.E., (1989), *Genetic Algorithms in Search, Optimization & Machine Learning*, Addison-Wesley, Reading, Mass.

D.E., Holland, J.H., Holyoak, K.J., Nisbett, R.E., & Thagard, P.R. (1986), *Induction: Processes of Inference, Learning and Discovery*, Cambridge: MIT Press.

Langton, G.L., (1989), *Artificial Life*, Addison-Wesley.

Lippmann, R.P., (1987), *An Introduction to Computing with Neural Nets*, IEEE ASSP Magazine, April 1987.

Moyson, F., Manderick, B., (1988), *The Collective Behavior of Ants: An Example of Self-Organisation in Massive Parallelism*, Proc. of the AAAI Symposium on Parallel Models of Intelligence, Stanford University.

Packard, N.H., (1989), *Intrinsic Adaptation in a Simple Model for Evolution*, in [Langton 89].

Paredis, J., (1989), *Learning the Behavior of Dynamical Systems from Examples*, Proc. 6th International Workshop on Machine Learning, Ithaca, Morgan Kaufmann.

Quinlan, J.R., (1986), Induction of Decision Trees. Machine Learning 1, Kluwer Academic Publishers, Boston, MA

Sloman, A., (1984), *Why We need many Knowledge Representation Formalisms*, in "Research and Development in Expert Systems" (Bramer ed.).

Steels, L., (1988), *Steps towards Common Sense* , Proc. ECAI-88, Pitman.

Talmon, J.L., (1986), *A Multiclass Non-parametric Partitioning Algorithm*, Pattern Recognition Letters 4, North-Holland.

Tinbergen, N., (1965), *Animal Behaviour*, Time-Life Books.

Wilson, E. O., (1985), *The Sociogenesis of Insect Colonies*, Science, 288.

Wilson, S. W., (1987), *Classifier Systems and the Animat Problem*, Machine Learning 2: p. 199-228, Kluwer Academic Publishers.

A Von Neumann Approach to a Genotype Expression in a Neural Animat

Dan Wood
ThinkAlong Software, Inc., PO Box 359, Brownsville California 95919 USA
thinkalong@applelink.apple.com (916) 675-2478

Abstract

This paper describes an approach to evolution upon an artificial insect existing in a simulated two-dimensional world, comprised of a "spreading-activation" neural network. Study of the difficulty of hand-building neural networks has resulted in a genetic expression, based on the notion of the Von Neumann computer architecture and the biological principles of DNA, which incorporates both operations and data in a simulated DNA strand. Work in progress indicates that by embodying behavior parameters and actual neural connections in a genotypic "language," and by expressing that phenotypically as an animat, the computer simulation appears able to evolve a better species of animat through mutations upon the genotype.

Introduction

A domain of "artificial life" explores limited computer simulations of animal behavior in an artificial environment. This class of simulated animal is coming to be known as the *animat*, as first coined by Stewart Wilson in [Wilson, 1985]. Research on the notion of an artificial insect has proceeded in many diverse headings (see for instance [Travers, 1988], [Maes, 1990], [Park, 1988], [Wilson, 1987]). The implementation described here is a direct extension of Jack Park's animat as described in [Park, 1988].

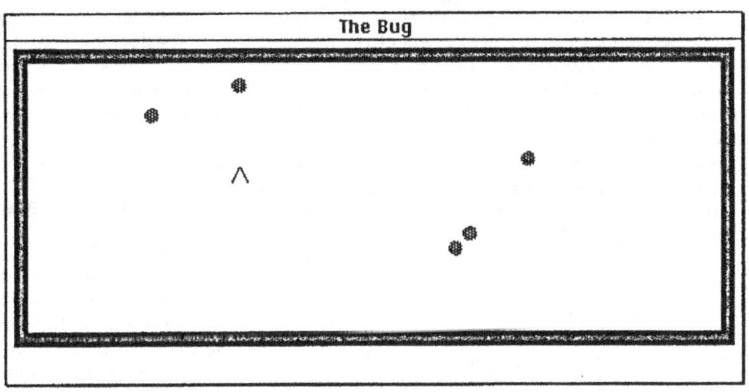

Figure 1

The existing animat system upon which this program has extended utilizes a spreading-activation [Collins and Loftus, 1975, Anderson, 1983] neural network which was originally conceived as an exercise to ascertain the capabilities of a neural net in *controlling* some sort of process. This neural implementation has been applied not only to the animat described here, but to process control in the manufacturing domain [Park, et al., in prep.], part of a scientific discovery system [Wood and Park, in press], and to the study of the Piagetian development of an infant brain [Wood, manuscript].

The "wetware" implementation controlling this animat consists of a few dozen "neuron" nodes, each connected to several other nodes with varying levels of positive (excitory) or negative (inhibitory) strength. Sensors (input of sight, pain, hunger, taste, and satiation) stimulate neurons which in turn dissipate activation to other neurons; muscle neurons that have reached a certain threshold of activation will cause appropriate routines to execute (to cause the bug to move, turn, eat, or move randomly). Additionally, the system applies a constant decay, to keep the overall activation steady, and random noise injected into the activation levels of all neurons. The system does not employ learning methods during the "lifetime" of the animat; this project studies only the non-plastic neurons found in the lower animals.

The animat's environment, implemented on a Macintosh II, is a bound arena (figure 1) with morsels of food scattered about, represented by dark particles. The bug, represented by an arrow facing the direction of its movement, can move and look straight ahead as well as exactly 45 degrees in either direction. At this stage of development, the population is a single bug rather than a group of competing or cooperating insects.

The animat, initially, has been equipped with a network that will allow it to move about and find food, though it behaves rather obtusely. For instance, it might be "looking" at a piece of food some distance away but fail to build up enough activation to cause it to start moving forward before it turns away. The problem may be due to the value of some global system parameter, such as the maximum strength that one neuron can spread to another, or it may be due to a shortcoming in the specific neural connections, such as the sense of hunger lacking a strong enough connection to drive the animat to move forward. Specific combinations of hand-tuned connections over the development of this system have resulted in different levels of acceptability in behavior.

The Problem

Fine-tuning a neural network—both the individual connections and the system parameters—can be a very tedious chore, since changing one value often results in undesired side-effects. That is, the *fitness landscape* [Kauffman, 1989]—the distribution of fitness over the space of phenotypes—is not easy to understand by manual tuning. It is anticipated that such an obstacle will increase in complexity as the size and complexity of the applied network grows. This difficulty leads to the need to study another approach to building such neural networks. A likeness of biological evolution, analogous to what Holland [Holland, 1975, see also Goldberg, 1989] calls the genetic algorithm, is an obvious candidate for dealing with this problem.

In this section I argue that there are fundamentally two approaches to the problem of representing the evolutionary process in a computer. Genetic simulation has been explored by Holland [1975, 1986] in his classifier system, in which the classifier is a direct expression of the phenotype *itself*. Mutations are performed upon the strands of three-state codes which represent the left- and right-hand sides of classifier rules. The classifier approach has been applied to animats in [Wilson, 1986].

The other approach to simulated evolution is what Dress [1990] calls "In-Silico Gene Expression"—a method for expressing genes as a phenotype. Mutation in this class of system is performed upon the genetic representation; the phenotype is created based upon the genotype. This paradigm of evolution tends to match the biological strategy, in which mutations take place only at the genetic level and are manifested in the phenotype. This approach has been modelled, for instance, in a statistically-built frequency-coded neural bug in [Dress, 1989], and in [Dawkins, 1986] as a visual expression of a static creature. The gene expression approach will be discussed in this paper as a model for a explicitly expressing an analog neural animat.

As a sidebar, an interesting contrast can be drawn between the classifier and neural approaches to the animat problem. In the classifier paradigm, a population represents a collection of classifier rules, the *components* of an animat, competing for survival. Mutation that takes place upon a single classifier could potentially improve its chances for survival as a component of a single animat. Under the other approach, mutation takes place upon the genes that are expressed as an entire animal; competition would thus need to take place among a society of animats. This notion is a useful topic for further study.

Gene-Building Instructions

Given the approach I have chosen to explore, it is thus necessary to invent a mapping from genotype to phenotype. The metaphor chosen here is that the genes can represent *instructions* for building the neural connections and for setting up the global parameters of the network as well. Discussed below, then, are some tenets and strategies that I believe to be important to the evolution of a neural network.

The notion of "DNA as program" was introduced in the early days of AI research as a means of attempting evolution of machine language programs [Lenat, 1983a]. This approach was not fruitful, due to the fact that empirically, gradual hill-climbing on the fitness landscape did not seem possible. This problem equates to Lenat's work on Eurisko, when it was found that the LISP code in AM, the program's predecessor, was too complex for useful mutations [Lenat, 1983b].

The domain of constructing a neural network out of "instructions" is a different matter. Syntax is limited, and the animat may not change its behavior significantly in spite of changed instructions. Understanding the suitability of this approach to evolving a better organism is the goal of this work.

The Data Space of a Gene Expression

Choosing a representation for an individual gene has been based on the principle of mixing instructions (opcodes) for constructing the neural network with its data serving as the parameters to the construction operations. This notion parallels the Von Neumann Architecture used in most computers. A gene in this implementation represents a single byte or word (8 or

Figure 2

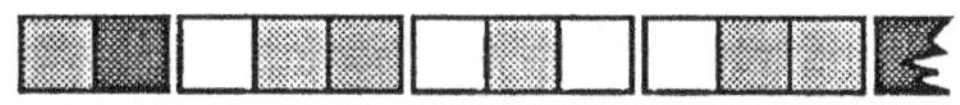

Figure 3

16 bits) which most microprocessors today use as the basic *width* of their native instruction and data space. For contrast, the width of a single point of biological DNA is much smaller; the four bases of A, C, G, and T are equivalent to only two bits.

For this system, I have chosen 4 bits—representing the numbers 0 through 15—as the basic unit size for an individual gene. Chains of genes, expressing the nucleotides, can consist of both instructions and data, freely mixed together in a single strand of DNA-like instructions. Sixteen is a logical base size for a microcomputer, since a number from 0 to 15 is comprised of exactly half of a byte. When a gene represents an instruction, a 16-opcode implementation allows room for several basic operations, while still allowing for several "unused" operations as argued for below. As a point in the data space, four-bit genes can represent a simple value or be chained to represent a larger binary number, as would be needed to refer to individual neurons. For instance, a chain of two genes can represent an eight-bit number, ranging up to 255.

Von Neumann Mutation

Uncertainty in the results of a mutation is, arguably, an important aspect of a system which performs mutations. A change merely to the value of a neural strength or to a numerical parameter in the neural network may produce an interesting change in behavior, but the construction of the animat, when compared to its predecessor, will not likely be surprising. If the mutations are performed on the building *instructions*, though, the result could indeed be unexpected.

Mutation, not surprisingly, is blind to the semantics of the individual gene. The value of the given point could be changed from a valid opcode to another opcode or to an invalid opcode. This is indeed a possibility in the present implementation, given the notion of *variable-length instructions* common to machine languages. Invalid opcodes could be made valid again, with the trailing points in the strand given *new meanings*. Even if one unused point is changed, its sense could result in new behaviors after further mutations take place.

If point mutation takes place which changes an instruction opcode, the meaning of the rest of the chain, for several units of length, may alter drastically. This phenomenon is familiar to machine-language programmers: haphazardly changing one byte in a sequence of instructions can severely affect the machine's interpretation of that instruction and the instructions following in memory. It may be several instructions, either single-opcode or unknown, before the alignment is as it was before.

Abstractly, the effects of a point mutation with a variable-length instruction stream of DNA can be visualized in this simple example. In figure 2, each entire instruction (consisting of an operation and data) is grouped together.

A mutation of the first (leftmost) instruction from white to medium-gray changes not only the first instruction but the next five instructions as well, substituting them with four new instructions (figure 3).

This potential results of opcode mutation elicits the need for a wide range of single-opcode instructions (including unknown "no-operation" instructions), to assist in returning the flow of instructions back to "normal" within a few points on the strand. Without such instructions, a single mutation could change the entire meaning of a chain for hundreds of points. Such a mutation might be *too* drastic. The genetic instruction set presented below, then, leaves a number of operations (out of sixteen possible) unused.

Spread-out Instructions

The system that I have studied, of encoding a representation of a neural network in a genome, plays to the notion of spreading out the instructions over a strand that is longer than would be minimally needed to describe the network. With a number of unused (uninterpretable) instructions, redundancies, or modifications to connections made in instructions that could have already been interpreted in the building of the phenotype, a mutation upon the genotype could result in a wide range of behaviors.

As an example, take a possible instruction, say one of the instructions interpreted first, that tells the system to connect one neuron to another with a given strength. A mutation upon that instruction could invalidate the entire connection, or simply change one of the parameters. If a later instruction performed the same (or similar) connection, then the original mutation might not have any effect at all. A later instruction could, alternately, merely *modify* the original connection.

The notion of incremental or modificational building instructions seems particularly appealing for planned future versions of the animat simulator in which reproduction will take place through sexual recombination (rather than asexually as is done now). Merging genes from two slightly different animats, where roughly one half of one animat combines with the other half of another's genes. This could allow for some "interaction" between the genetic instructions as realized in one animat with the instructions from another. This phenomenon might be paralleled in the human biological domain, for example, in which dark skin appears to be a combination of several genes [Winkless and Browning, 1975].

The applicability of this conjecture of dispersing genes throughout the genotype to result in some particular characteristic in the phenotype is an open point of study; later examination of the recombination process should provide insight into the benefit of this view.

Some Desirable Instructions

The operations that are needed for this particular implementation can be divided into two categories: setting a system parameter (such as a global neural threshold) and connecting one neuron to another.

System parameters are important so that a particular neural network structure can be exercised with different global system parameters. In a particular network implementation, such as the one developed by hand, the performance of the animat depends heavily on the system parameters being set properly. Unfortunately, it's not always easy to determine the optimal settings by hand. For example, the bug may be likely to satisfy its goal of eating before it starves if the settings encourage rapid muscle firing, bringing it to its food quickly. On the other hand, the bug may simply thrash, where a more "relaxed" setting might allow the bug to more methodically seek out and move towards its food.

The definition of the actual neural network is the other purpose of the genotypic instructions. Certain neural connections in the animat may be crucial to its succeeding (such as the connection between the desire to eat and the action of moving), while other connections and neurons may not necessarily be useful in satisfying the fitness test. The animat's performance could easily be affected by the presence, absence, or particular linked strength of any neuron.

With this background in mind, it is now appropriate to introduce an instruction set upon which this implementation is based.

Set a parameter to a given value.

The following point is the global parameter number (0 through 15). The next point represents a number, say, 1/17th of the value. Thus, 0 becomes 0 (minimum value); 15 becomes 255 (maximum value).

Increase a parameter.

The following point is the global parameter number. This instruction will increase the value of the parameter by a constant amount, such as by multiplying by 1.5.

Decrease a parameter.

The following point is the global parameter number. This decreases the parameter, such as by multiplying by 0.67.

Connect from one neuron to another.

The following two *pairs* of points are neuron numbers. Thus, a neuron is represented by two points, and can thus represent a number up to 16^2=256, an adequate value for this simple model. The final point becomes a strength value, 0 representing -100% up to 15 representing +100%.

Select a formula.

In the case where certain internal algorithms (such as the accumulation method used in spreading activation) can use different methods, this will set the one to use. For example, this operation could set a decaying function to be either linear or logarithmic. The operation is followed by a point representing the formula type and then another point representing the formula to use. The meanings of the formulas types or formulas is completely symbolic, and is set in the program.

Increase all strengths from a given neuron.

The following pair of points represents a neuron from which the connections to all other neurons are increased in strength. Currently, this is set to multiply the values by 120%.

Decrease all strengths from a given neuron.

Similar to above, this decreases the outbound neural strengths for a given neuron. Currently, the value is 80%.

Randomly connect one neuron to another. (This uses no other points as data.)

Randomly disconnect one neuron from another. (This also uses no other points.)

Discussion

As currently implemented, this set of instructions is capable of describing the construction of the spreading-activation network used in the animat. Though not the *minimum* instruction set needed for building a network (so as to allow for partial mutations of features), the instruction set is small enough that the probability of the phenotype-building "interpreter" encountering an unused instruction is fairly high, allowing the effects of a single mutation to stay fairly local.

Work on this form of evolutionary animat (using a genetic representation of the current hand-tuned network and parameters) is in progress, for now utilizing a simple fitness function of survival (ability to eat before starving) and only a single animat population. The genetic strategy is this: if the results of a mutated animat are superior to its predecessor, it is kept and allowed to "produce offspring." If not, the genes revert back to that of the parent and a new mutated offspring is generated. Results to date are promising; an overnight run of the program produces about a half a dozen new generations that are considered "superior" to the previous generation based upon the current fitness function. Current work involves studying the results of the offspring to ascertain which connection instructions and parameter settings appear to be favored for the system, and to provide mechanisms for the animat to evolve further.

Implementing a multiple-animat society, to explore the issue of neural evolution using sexual reproduction and multiple populations, is now starting. Later research may allow learning to take place during the lifetime of an individual animat. Context-sensitive gene expression is also planned for future versions.

Clearly, a value of such exploration in the domain of evolving neural networks goes beyond the problem of producing a specific simulation of a life form. Ideally, a general set of parameters for the network could be found which work well regardless of the particular neural connections. A type of network that is suitable at adapting to new situations or environments is important to domains beyond the animat itself.

References

Anderson, J.R., *The Architecture of Cognition*, 1983, Cambridge, Harvard University Press.

Collins, A.M., and E.F. Loftus, "A Spreading Activation Theory of Semantic Processing," *Psychological Review*, 82, 407-428, 1975.

Dawkins, Richard, *The Blind Watchmaker*, 1986, New York, W.W. Norton.

Dress, W.B., "In-Silico Gene Expression," submitted for publication, 1990.

__________, "Synthetic Organisms and Self-Designing Systems," *Telematics and Informatics*, 6, no. 3-4, pp. 351-363, 1989.

Goldberg, David E., *Genetic Algorithms in Search, Optimization & Machine Learning*, 1989, Reading, MA, Addison-Wesley.

Holland, John H., *Adaptation in Natural and Artificial Systems*, 1975, Ann Arbor, University of Michigan Press.

__________, "Escaping Brittleness: The Possibilities of General-Purpose Learning Algorithms Applied to Parallel Rule-Based Systems," in Michalski, R.S., J.G. Carbonell, and T.M. Mitchell, *Machine Learning: Volume II*, 1986, Los Altos, Morgan Kauffman.

Kauffman, Stuart A., "Principles of Adaptation in Common Systems," in Daniel L. Stein, ed., *Lectures in the Sciences of Complexity*, 1989, Redwood City, CA, Addison-Wesley.

Lenat, Douglas B., "The Role of Heuristics in Learning By Discovery: Three Case Studies," in R.S. Michalski, J.G. Carbonell, and T.M. Mitchell, editors, *Machine Learning: An Artificial Intelligence Approach*, 1983[a], Palo Alto, Tioga.

__________, "Eurisko: A program that learns new heuristics and domain concepts," in *Artificial Intelligence*, 21, 1983[b], pp. 61-98.

Maes, Pattie, "How to Do the Right Thing," in *Connection Science Journal: Special Issue on Hybrid Systems*, 1, no. 3, 1990.

Park, Jack, "A Spreading-activation Network Driven Animat," in *Proceedings of the AAAI Spring Symposium*, 1988, Palo Alto, AAAI.

Park, Jack, Frances Abrams, C. William Lee, and Dan Wood, "Using a Neural Network for Process Control," document in preparation.

Shuster, Peter, "Dynamics of Molecular Evolution," *Evolution, Games, and Learning*, 1986, Amsterdam, North-Holland.

Travers, Michael D., *Agar: An Animal Construction Kit*, Master's Thesis, 1988, MIT Media Laboratory.

Wilson, Stewart, "Knowledge Growth in an Artificial Animal," in J.J. Grefenstette, ed., *Proceedings of an International Conference on Genetic Algorithms and their Applications*, 1985, Carnegie-Mellon University.

__________, "Classifier Systems and the Animat Problem," *Machine Learning*, 2, no. 3, 1987.

Winkless, Nels, and Iben Browning, *Climate and the Affairs of Men*, 1975, Burlington VT, Fraser Publishing.

Wood, Dan, "Mary: An Experiment in Piagetian Concept Formation," unpublished manuscript.

Wood, Dan, and Jack Park, "Qualitative Process Discovery: The Next Step Toward Intelligent Systems," Technical report, in press, Dayton, OH, Wright-Patterson Air Force Base.

ARCHITECTURES, ORGANIZATIONAL PRINCIPLES, AND FUNCTIONAL APPROACHES

Challenges for Complete Creature Architectures

Rodney A. Brooks
MIT Artificial Intelligence Laboratory
545 Technology Square
Cambridge, MA 02139, USA
brooks@ai.mit.edu

Abstract

In recent years there has been a move within the artificial intelligence and robotics communities towards building complete autonomous creatures that operate in the physical world. Certain approaches have proven quite successful, and have caused a re-analysis within the field of artificial intelligence of what components are necessary in the intellectual architecture of such creatures. However nothing built thus far yet comes close the dreams that many people hold dearly. Furthermore there has been quite some criticism of the new approaches for lacking adequate theoretical justification. In this paper we outline some of the more obvious challenges that remain for these new approaches, and suggest new ways of thinking about the tasks ahead, in order to decompose the field into a number of manageable sub-areas that can be used to shape further research.

1 Introduction

There is a growing interest in building artificial creatures of some sort. One example is the recent boom in a field known as Artificial Life (see **[Langston 87]** and **[Langston 90]**). While much of the emphasis is on building forms resident in computers, which are agents acting in an information domain, there has also been some interest in physical embodiments of artificial creatures.

This author, at the MIT AI Lab, introduced the subsumption architecture (**[Brooks 86]** and extended in **[Brooks 90]**) with the explicit goal of building mobile robots with long term autonomy. Later the word *creature* crept into the language of the MIT group (e.g., **[Connell 87]**). The goal is to build autonomous mobile robots which operate over long periods of time, completely autonomously, in dynamic worlds. It is envisioned that these worlds are worlds which already exist for some other purpose—not worlds specially built to house the robots. Further, it is envisioned that these robots carry out some task which has some utility for whoever wanted the robots to exist and *live* in this world.

As **[Flynn 87]** points out, there are many components to such creatures, including sensors, actuators, power sources, and intelligence. Over the last five years we have found that all these components are intimately related as we have tried to build prototype creatures (**[Flynn and Brooks 89]**). Choices in any part of the system architecture (e.g., sensor characteristics) have major impacts upon other parts of the system. In general it is very dangerous to think that any one component (such as intelligence) can be isolated and studied by itself.

Our experience with the subtleties of such interactions has led us to our current construction of a very complex robot, named Attila (**[Angle and Brooks 90]**) pictured in figure 1 (in fact we are building multiple copies of Attila). It has six legs, each with three degrees of freedom, an active whisker, a gyro stabilized pan-tilt head carrying a range finder and a CCD camera, 10 onboard processors, and over 150 sensors. We built an earlier six legged robot named Genghis (**[Angle 89]**, **[Brooks 89]**, but its complexity pales in comparison to that of Attila. Many of the issues raised in this paper were brought to our attention as we have tried to work out how to program this complex robot Attila to be an artificial creature.

The bulk of this paper is devoted to the problems and challenges in designing and building the computational architectures for such creatures. However, the reader should not forget that the other aspects of a creature's architecture cannot be considered in isolation from intelligence. In a complete design, all aspects greatly influence each of the others.

We first argue that there are multiple levels of analysis or abstraction with which we must be concerned in designing and building complete creature architectures. There can be no single magic bullet or theory which will tell us all we need to know. Some problems within these levels are well circumscribed and so can be worked on in isolation. However, in order to build complete creatures we need to bridge the gaps between these levels also.

The bulk of the paper then goes on to examine each

Figure 1: A partially constructed Attila robot. It has six legs, 24 motors, 10 processors, and over 150 sensors. Still missing at the time of this photograph are a pan-tilt camera mount, gyroscopes, and an active steerable whisker.

level of analysis from three points of view: *theory*, *relations*, and *practice*. In terms of challenges which must be faced in building complete creature architectures, these can be described as:

Theory: What theoretical tools can be developed, and how will they be useful?

Relations: What existing bodies of work in other disciplines can be drawn upon to assist with analysis and synthesis at this particular level of abstraction?

Practice: How can we go about practically building systems, with or without the help of future developments in relevant theory, or related work?

Finally the paper outlines a tentative agenda for development of complete creature architectures.

2 Levels of Analysis

Traditional science has found it necessary to relate complex biological systems to fundamental properties of matter in terms of a number of levels abstraction. In the coarsest terms these levels correspond to physics, chemistry, and biology. Traditional science tends to work within these boundaries. It is impossible to do good science without having an appreciation for the problems and concepts in the other levels of abstraction (at least in the direction from biology towards physics), but there are whole sets of tools, methods of analysis, theories and explanations within each discipline which do not cross those boundaries.

While there are newer disciplines such as physical chemistry, bio-chemistry, and molecular biology which try to bridge the gaps between the major disciplines there is still no definitive crossing of the major boundaries except in the simplest of cases, and then only very recently. For instance, the best that has been achieved in terms of explaining a chemical reaction in terms of the physics of mass, force and quanta, is (as reported by **[Pool 90]**) a fairly complete understanding of the reaction $D + H_2 \rightarrow HD + H$ (where D is deuterium). This is just about the simplest possible reaction in terms of the number of fundamental particles involved.

As science progresses we can expect a more unified theory to emerge, with more and more bridges between the traditional disciplines being built to the point that the abstraction barriers are completely understood. But that is certainly not the case at this point in history, despite the enormous body of scientific and philosophical work that has been undertaken within the realm of modern science.

As we try to build artificial creatures there is no reason to expect that we should instantly be able to come up with a uniform theory or set of tools of analysis and synthesis that cover all problems which need to be examined. In fact, there seems a quite natural decomposition of the task into three somewhat distinct levels of abstraction:

Micro: The study of how distinct couplings between the creature and its environment are maintained for some particular task. Example tasks are temperature regulation, wall following, climbing, etc. Primary concerns here are what forms of perception are necessary, and what relationships exist between perception, internal state, and action (i.e., how behavior is specified or described).

Macro: The study of how the many micro perceptions and behaviors are integrated into a complete individual creature. Primary concerns here are how independent various perceptions and behaviors can be, how much they must rely on and interfere with each other, how a competent complete creature can be built in such a way as to accommodate all the required individual behaviors, and to what extent appearingly complex behaviors can emerge from simple reflexes.

Multitude: The study of how a multitude of individual creatures interact as they go about their business. Primary concerns here are the relationships between individuals' behaviors, the amount and type of communication between creatures, the way the environment reacts to multiple individuals, and the resulting patterns of behavior and their impacts upon the environment (which would not occur in the case of isolated individuals).

These levels or fields by no means correspond to physics, chemistry, and biology, but there is a certain metaphorical connection in terms of them being somewhat distinct levels.

3 Micro Level

Any artificial creature which we build must be able to interact with its environment. At any particular time, an observer might say that a creature is doing some particular task, or doing many particular tasks in parallel. Examples of such tasks in the robot domain include wandering, avoiding obstacles, wall following, looking for a recharge station, delivering some object, cleaning the floor, following someone, etc. We will refer to what the robot is doing when it is achieving some task as carrying out a behavior.

At the micro level we will be concerned with one particular such task at a time, although the creature may really be carrying out many in parallel. The particular challenges at the micro level are as follows.

Convergence: Demonstrate or prove that a creature is programmed in such a way that its external behavior will indeed achieve a particular task successfully. For instance, we may want to give some set of initial conditions for a creature, and some limitations on possible worlds in which it is placed, and show that under those conditions, the creature is guaranteed to follow a particular wall, rather than diverge and get lost.

Complexity: Deal with the complexity of real world environments, and sift out the relevant aspects of received sensations rather than being overwhelmed with multitudes of data.

Synthesis: Given a particular task, automatically derive a program for the creature so that it carries out that task in a way which has clearly demonstrable convergence.

3.1 Micro Theory

Suppose we are to have some theory about how a creature interacts with the world. The creature both senses the world and acts, to change its own position, and perhaps to change the world itself. The world may also

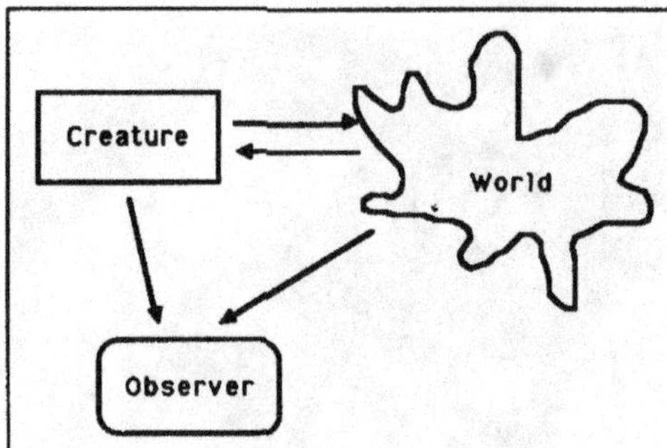

Figure 2: The creature and the world interact. The observer or modeler watches them both.

change for other reasons. As in figure 2, any observer of the system must sense both the creature and the world. A theory may be couched in terms of objective truths, but in order for us to test a theory we must compare it with observations made by an observer. In fact, no theory can be given complete objective truth. It must be given approximations to the true situation supplied by some observer.

The creature itself acts as an observer of the world. It has sensors which provide the only inputs it can have concerning the world. The first question is what should the creature do with the sensor inputs? How should they be organized and manipulated?

We argue below that the traditional idea of building a world model, or a representation of the state of the world is the wrong idea. Instead the creature needs to process only aspects of the world that are relevant to its task. Furthermore, we argue that it may be better to construct theoretical tools which instead of using the state of the world as their central formal notion, instead use the aspects that the creature is sensing as the primary formal notion.

3.1.1 Uncertainty

In the early days of robotics it was noticed that uncertainty is inherently present when sensors are used to glean information to model the real world (e.g., **[Roberts 63]**). The early approaches to this problem were either to ignore it (e.g., **[Lozano-Pérez 76]**) or to engineer the situation so that it could be ignored (e.g., **[Nilsson 84]** and **[Giralt, Chatila and Vaisset 84]**; both are reports on earlier work).

Later attempts were made to explicitly model the uncertainty in the sensors (e.g., **[Moravec 80]**), and hence in the world models built. In the manipulator domain, others (e.g., **[Taylor 76]** and **[Brooks 82]**) tried to reason over these uncertainties, and eventually a whole sub-discipline grew up around planning for manipulator motions in the presence of uncertainty ([**Lozano-Pérez, Mason, and Taylor 84**] provides a foundational paper in this area).

But why is uncertainty a problem, and what is the problem if there is one?

> *The key problem is in trying to build a model of the world.*

By building a model, we mean that the system tries to have some internal representation of an external objective reality. The problem is in correlating the current readings from the sensors with the existing (partial) world model. The original readings used to build the existing model were noisy and introduced uncertainties in the representation of the world. The new readings also include noise. Furthermore, if the robot has moved between sensor readings then there is uncertainty in how the coordinate systems of the two (or more) sets of sensor readings are related. [**Moravec 80**] and [**Chatila and Laumond 85**] describe the two fundamental approaches to these problems. If more than one type of sensor is used, there is also the problem of fusing the different classes of data into a single representation.

If there are no models built, the problem of uncertainty is inherently reduced. This alternative is to operate in a tight coupling with the world through a sensing-acting feedback loop. Instead of relying on inaccurate values returned by noisy sensors, we can rely on the time averaged derivative of these signals as the creature actively changes its state within the world in a way which forces larger changes in the sensor readings than those contributed by noise.

This approach has actually been used by people working in Artificial Intelligence for many years in order to avoid the issues of uncertainty.

The first computer controlled mechanical hand [**Ernst 61**] was programmed in a language (THI for Teleological Hand Interpreter) which lets the programmer specify how sensor readings (Ernst calls sensors "sense organs") should change. The interpreter then executes motion commands to make them happen. Ernst distinguishes this ***outside-in*** mode of control, where the position of a block is indicated by the block itself, from an ***inside-out*** mode where the the position of a block must be stored inside the machine as a set of coordinates.[1]

The robot Shakey [**Nilsson 84**] at Stanford Research Institute in the late sixties is usually viewed as a *tour de force* in the use of explicit internal world models, and indeed was the basis for a whole generation of planning research. The planner planned in terms of atomic actions, without worrying about how they were implemented in the world. Atomic actions were simple commands that just "worked". But in fact, the robot worked as it did because the environment was carefully engineered so only a few possible classes of situations could be encountered, and the atomic actions were actually implemented as tight sensing-action feedback loops which did not use explicit world models at all. They were called *intermediate level actions* (or ILAs) and enabled the robot to handle such things as the differences in location of objects from the world model, and to handle such uncertain and unsensed things such as the exact way a large block would slide when pushed by the robot.

More recent work (e.g., [**Brooks 86, 89**], [**Horswill and Brooks 88**], [**Connell 89**], and [**Ballard 89**]) has dispensed with the higher level planner altogether, but has been able to get robots to operate well in the world without having to explicitly worry about uncertainty.

3.1.2 The State of the World

Can a robot, or creature, be anything but reactive if it does not build world models? We answer this question affirmatively below, but first we need to understand the following.

It turns out that there are more problems with world models than just uncertainty.

Creature researchers who work in simulated worlds (e.g., [**Pollack and Ringuette 90**] is an example; [**Chapman 90**] is an exception) are usually not even aware of the problem of uncertainty. But worse, they are not aware of the current technological impossibility of using sensors and perception systems to build models of the world which reflect an objective reality. Such an objective reality is usually assumed by such researchers (see the collection [**Langston 90**]). In these simulated systems, objective reality is rather simple, and can usually be described by a short vector of bits. The vector encodes the state of the world, and is an implementation of the notion of a logical description of the state of the world (e.g., [**McCarthy and Hayes 69**]). The real world is not so simple, however.

The way we humans, as observers, describe the world is very much biased by our cultural, morphological, and functional knowledge. Extracting such descriptions of the world from sensors, in terms of things like "chairs", "living rooms" and "crowds" is certainly beyond the state of technology. In very controlled situations a small library of CAD-like models can be matched to grey-level or depth images (e.g., [**Grimson 90**]). In less controlled situations, some regions of an image can be labeled as belonging to a certain class (e.g., [**Thorpe, Hebert,**

[1] Admittedly, a rather confusing (to this author) discussion of the nature of analog versus digital control follows, with comparisons to animal control systems. From this discussion it is fair to assume that Ernst, at that time, would not necessarily have agreed with all the arguments being put forth here, in spite of the way he actually made things work.

Kanade and Shafer 88]).

The idea that such complete objective models of the world could be built within a creature is somewhat problematic in any case. What does the creature do with such models? It seems that it simply reduces the problem of building a creature to one of building a homunculus that is to live within the creature and operate on complete world models.

What about simpler objective descriptions, such as surface models of what is in the world? Computational vision research, as outlined by **[Marr 82]**, is predicated on the assumption that such descriptions of the world can be *recovered* from retinal images. These descriptions are known as $2\frac{1}{2}$–D sketches. Recently, there has been a movement away from the belief that such general purpose representations can be computed. **[Ballard 89]** argues that such representations are not available to humans.

Without such representations what are we left with? Recent work suggests that we can still have creatures do many things that have traditionally been thought to require representations such as we are rejecting. For instance, despite having no such representations, **[Mataric 90]** has demonstrated a robot that is much more than reactive, incorporating an ability to modify its behavior based on its experiences within the world. It builds "maps" of the world (topological with some metric annotation, rather than purely metric) as it wanders about indoor environments. But the features in its maps are not objective things in the world. Rather they are the records of certain time averaging *detectors* having fired while the robot moves about. These detectors are based on very noisy sonar readings, but they have the property that they are both positively and negatively repeatable with respect to any particular place in the world over a wide variance in the robot's actual position and trajectory through that place. Such representations of the world are called *deictic* by **[Agre 88]** and **[Chapman 90]**.

3.1.3 Analyzing Behaviors

The discussion above can lead us to a rather simpler way of analyzing the behavior of a creature than if we adopted a stance of having objective world models. In particular this method will let us deal with *uncertainty* entirely at *analysis* time (analogously to dealing with, say, type checking at *compile* time), rather than have the creature deal with it at *run* time.

Given a state of the world and the position of a robot creature within the world, the creature's sensors may deliver different values at different times—i.e., there may be some dynamic noise in the sensors. Worse than that, however, is the system noise, which makes it impossible for the observer of the robot/world system to predict what the sensor readings will be. For instance **[Connell 90]** found that the flux gate compass on his robot delivered different directions for north in different parts of its domain. The actual values were consistent over time for a given place, but without using that sensor itself, or a similar one, at the exact places where the readings were to be taken, there was no way in which the experimenter or observer would know the actual sensor reading. However it was relatively easy to determine empirically the range of possible errors in the sensor reading. This range of errors, then, is the uncertainty we must deal with for this particular sensor. Similar analyses apply to other sensors, such as infrared proximity sensors, or sonar sensors.

Now suppose we characterize a behavior at analysis time in terms of an allowable range of sensor readings for the creature (or more generally allowable ranges on some functions computed on the sensor readings).

For instance, instead of characterizing wall following as staying within some distance of a wall, and within some orientation range, we could characterize it as maintaining some range of values on three sonars pointing north-east, east, and south-east, relative to the robot's orientation being north. We might even say that only readings two had to be within their ranges at any particular time.

Or consider approaching a recharging outlet visually. We could characterize that behavior as forcing the visual image of the outlet to appear at every frame taken by a forward looking camera, and for that image size to be non-decreasing between successive images, and to be bigger than some monotonically increasing sequence.

Now our analysis does not have to be concerned with the creature having uncertain sensor readings. Indeed they will be uncertain with respect to the actual world. But what we have to show is that given those sensor readings, the action computed by the creature is guaranteed to produce a next set of sensor readings that lie within our allowable ranges. To do this reasoning at analysis time, and formally prove that a creature will follow a particular behavior, we may need to consider all possible worlds which could have given rise to the initial set of sensor readings, and take into account uncertainty there, and then consider all possible results of the computed action on those worlds and all possible new sensor readings. But the creature itself does not have to maintain any runtime measure of uncertainty, nor be confused or concerned by it.

The essence of what we are suggesting here is to invert the sensor (or more generally a function of the sensors) readings. Instead of thinking in terms of a particular world state and then analyzing all possible sensor readings that could be generated we are analyzing using an inverse method. Namely, given a sensor reading, consider which possible worlds could have given rise to that reading. Recall that this is done at analysis time.

3.2 Micro Relations

There are two obvious fields which are quite related to the micro level of creature architectures: control theory and animal ethology.

Control theory (e.g., **[Doyle and Stein 81]**), at first glance, seems an attractive tool of analysis for investigating couplings of creatures to the environment. There is a long and rigorous tradition of such analysis of the interaction of mechanical, electrical, and fluid systems with their environments. If a single behavioral mode of the creature can be sufficiently isolated from other aspects of its interaction with the environment, and if the sensors and actuators are well enough understood, then control theory may help to analyze convergence of the behavior. It should be understood, however, that control theory can only be one tool or component of the theories that will be needed for full mastery of engineering artificial creatures.

[McFarland 87] provides a useful introduction to all aspects of animal behavior. Ethology is an approach to studying animal behavior which combines both functional and causal explanations, and in the main concentrates on behavior of animals in their natural environments. To a large extent however, ethology *per se* does not quite get down to the level of analysis we need at the micro level. It does have some interesting observations about *fixed action patterns*, essentially feedback-free (at the gross level in any case) patterns of action that are quite stereotyped. Such patterns may be of use to our artificial creatures in well prescribed environmental niches.

Some literature, closer to neuroscience, does deal with the sensor to action connection we have been discussing at the micro level (e.g., **[Wehner 87]**, **[Cruse 90]**, **[Götz and Wenking 73]**). This literature can inspire us by the robustness of behavior that is generated by relatively simple systems.

3.3 Micro Practice

The practical problems at the micro level in building complete creatures are still immense.

Primary amongst these is perception—how to map from sensor inputs to aspects of the world. There has been a huge amount of research focused on computer vision over the last twenty years. Unfortunately most of it has been aimed, however indirectly, towards object recognition. Recently there has been a move towards vision that operates on sequences of images, but even there, there are strong traditions of regarding the purpose of vision as something to "recover" properties of objects. Only recently (**[Ballard 89]**, **[Chapman 90]**, and **[Horswill and Brooks 88]** has there been a move towards visual processing for a creature that operates continuously in the world, in such a way that appropriate aspects of the world are extracted directly for the particular tasks with which the creature is concerned.

Another approach to useful perception is to make careful choices of sensors, which are well adapted to the task at hand. **[Ernst 61]** chose sensors that were close to the task—touch and proximity sensors to determine when his mechanical hand should grasp. Such sensor choice is also used extensively in the biological world, as described by **[Wehner 87]**—in fact the presentation there has great resonance with the thrust of this complete section.

At this point in time, for actually specifying a single behavior for a creature, our programming tools are rather primitive. We are forced to individually craft programs to generate particular behaviors. It would be good to have almost any form of synthesis tools which could take a goal specification (note that **[Kaelbling 88]** is trying to do this, but at the macro level rather than the micro level), and produce some code or rules to generate the appropriate behavior. In some sense, this is what **[Lozano-Pérez, Mason and Taylor 84]** are trying to do, but only in very well characterized situations.

4 Macro Level

Once we have imbued our creatures with multiple behaviors to handle a variety of circumstances and achieve a variety of tasks, we are faced with the problem of deciding which behavior or behaviors should be active at any particular time. There are two components to this decision—which behaviors are potentially correct in the circumstances, and how to resolve conflicts between behaviors. In more detail, we need to consider the following issues:

Coherence: Even though many behaviors may be active at once, or are being actively switched on or off, the creature should still appear to an observer to have coherence of action and goals. It should not be rapidly switching between inconsistent behaviors, nor should two behaviors be active simultaneously, if they interfere with each other to the point that neither operates successfully.

Salience: The behaviors that are active should be salient to the situation the creature finds itself in—it should recharge itself when the batteries are low, not when they are full.

Adequacy: The behavior selection mechanism must operate in such a way that the long term goals that the creature designer has for the creature are met.

I.e., a floor cleaning robot should successfully clean the floor in normal circumstances, besides doing all the ancillary tasks that are necessary for it to be successful at that.

4.1 Macro Theory

There are at least two fundamental questions of organization for a behavior selection mechanism, at the implementation or algorithmic level.

- Should the behavior selection mechanism be centralized or decentralized?
- Should the conflict resolution scheme be fixed priority or dynamically reconfigurable?

At this point in time we have no theoretical way of linking these questions to the issues of coherence, salience and adequacy raised earlier. Later, we will point to work which has been done in these areas however in terms of practical implementations.

One point of theoretical importance, however, concerns hierarchical planning. Centralized behavior selection mechanisms have a long tradition in the form of AI planning as strongly introduced in the late sixties [**Nilsson 84**]. Recently classical planning has been shown to be intractable even in quite limited circumstances ([**Chapman 87**]), and has fallen from favor amongst creature builders.

4.2 Macro Relations

Ethology (again, see [**McFarland 87**]) has struggled with the problems of behavior selection for many years. Theories of *drive* have come and gone to explain *motivation* for particular behaviors, but there are still no definitive answers, as to exactly what leads to a particular behavior being activated when many are at least not inappropriate in certain circumstances.

Ethologists have also observed and described *displacement* activities—inappropriate behaviors that operate when two conflicting but appropriate other behaviors could reasonably be triggered. It seems that real creatures are by no means bug free.

From the neuroscience side there has been some recent progress in explaining animal behavior selection, at least in simpler animals. [**Kravitz 88**] describes a model of hormonal control of behavior. Following Kravitz' own summary the major points of this model are:

1. Sensor inputs enhance the release of hormones.
2. A hormone may be released from tissues or neurons and may act in isolation or in concert with other substances.
3. The hormone finds receptors where it either stimulates a new behavior pattern or enhances or diminishes an existing pattern.
4. The systems affected by hormones may include sensory elements, groups of neurons in higher processing centers, and motor or hormonal output systems.
5. The method of stimulation is a gain setting mechanism, biasing the output of the organism in particular directions.
6. The change is apparent in that the organism then responds to particular sensory inputs with altered outputs.
7. Individual organisms may respond to particular hormones with different levels of effectiveness—determined by genetic or environmental influences.
8. The affected circuit may itself further enhance or diminish the release of the hormonal substance.

Kravitz uses the lobster as his primary instantiation of this model.

In [**Brooks and Viola 90**] we demonstrated a robotic instantiation of Kravitz' model (programmed in the "Behavior Language" [**Brooks 90**]) which successfully demonstrated seven of the listed eight properties (number seven being the only one not accounted for).

4.3 Macro Practice

If we restrict ourselves to implementations on actual robots there has been very little actual work in behavior selection.

[**Brooks 86**] used a distributed system, called the *subsumption architecture*, with an almost fixed priority scheme to arbitrate between behaviors. There was a small amount of internal communication between the behaviors, but it was not directly related to priorities in conflict resolution or behavior activation. Indeed the behaviors were all active at all times. [**Connell 89**] later purified this scheme so that there was no internal communication between behaviors. A major criticism of the scheme used is that there is no explicit mention of behavior selection or priorities in the programming language for expressing behaviors.

[**Kaelbling 88**] remedies that shortfall by having an explicit programming language for behavior selection. The code generated by her scheme was again decentralized and used fixed priorities determined at compile time.

[**Maes 89**] suggested another decentralized scheme but with more explicit references to behavior activation. She develops the scheme in [**Maes 90**] and [**Brooks 90**] produced a superset language, the *Behavior Language*, which could be used to implement the scheme on mobile robots. This language was used by **[Mataric 90]** in a system where some individual behaviors represented landmarks in the world, and by **[Maes and Brooks 90]** where a learning mechanism was introduced to learn arbitration between behaviors (enabling a six legged robot to learn to walk from negative feedback when it fell down). An interesting thing about the Behavior Language is that it actually compiles into subsumption architecture programs with a fixed priority arbitration scheme. The user of the Behavior Language does not see this restriction however, as it is hidden in an abstraction mechanism.

The Behavior Language also contains features inspired by the work of **[Kravitz 88]** with the hormone system of the lobster. **[Brooks and Viola 90]** show how these features enable almost a full implementation of the Kravitz model.

Two problems stand out with these schemes:

- We do not know how well they will scale. We certainly do not have anything like a Turing equivalence theorem that tells us that at least in principle these schemes can do anything we might ever want them to do.
- We do not have any analytic tools for understanding in advance what sort of conflicts and other unexpected interactions might arise from the ways we paste together behaviors using these methodologies.

These two problems are key research areas, and progress in either would fundamentally help the field.

5 Multitude Level

Once we have physical creatures successfully operating in the world, handling its dynamics and surprises, there naturally arises the question of how such creatures will interact with each other.

This question can be posed in terms of how just two creatures will interact, or inspired by large colonies of social insects such as ants and bees, we can ask whether and how interesting behavior of that class might arise when we have large collections of artificial creatures in contact with each other. It is this latter question that we will concentrate upon for the rest of this paper.

Some of the major issues for social interactions of large numbers of artificial creatures are:

Emergence: Given a set of behaviors programmed into a set of creatures we would like to be able to predict what the global behavior of the system will be, and as a consequence determine the differential effects of small changes to the individual creatures on the global behavior.

Synthesis: As at the micro level, given a particular task, automatically derive a program for the set of creatures so that they carry out the task.

Individuality: Robustness can be achieved if all creatures are interchangeable. A fixed number of classes of creatures, where all creatures within a class are identical, is also robust, but somewhat less so. The issue then is given a task, decide how many classes of creatures are necessary

Cooperation: In some circumstances creatures should be able to achieve more by cooperating—the form and specification of such possible cooperations need to be understood.

Interference: Creatures may interfere with one another. Protocols for avoiding this when it is undesirable must be included in the design of the creatures' instructions.

Density dependence: The global behavior of the system may be dependent on the density of the creatures and the resources they consume within the world. A characterization of this dependence is desirable. At the two ends of the spectrum it may be the case that (a) a single creature given n units of time performs identically to n robots each given 1 unit of time, and (2) the global task might not be achieved at all if there are fewer than, say, m creatures.

Communication: Performance may be increased by increasing the amount of explicit communication between creatures, but the relationship between the amount of communication increase, and performance increase needs to be understood.

This is an area where almost no work has been done with actual physical robots, so the following remarks are rather brief. Some work has been done with a small number of robots (e.g., **[Premvuti and Yuta 90]**) but none with large numbers of robots.

5.1 Multitude Theory

In many ways the aim of controlling multiple creatures to generate a global behavior is similar to the problem of controlling a single creature to generate a single behavior. The difference is in the large amount of communication that goes on through the world, in the case of multiple creatures, where their very presence, besides any changes they may make to the world itself, are signals for changes in behavior of other creatures.

For suitably abstracted creatures it may be possible to describe their group behavior as differential equations and again use classical control theory to understand the conditions under which they will produce a certain global behavior. This sort of analysis will become more difficult as the creatures modify the world—and may turn out only to be useful for understanding grouping and flocking behaviors.

5.2 Multitude Relations

There are a number of disciplines which may shed light on organizations of many creatures.

From a descriptive and mechanistic viewpoint their is much interesting work on the organization of social insects. **[Deneubourg et al 87]** is just one example of how complex patterns of behavior can emerge from simple systems of biological creatures.

The theories on dynamical systems, and self organization and the construction of dissipative structures (**[Nicolis and Prigogine 77]**) may have some relevance to the field also. See **[Steels 89]** for one approach to investigating the behavior of many simple creatures acting in a single environment.

None of this work at this point gives us any handle on how to synthesize creatures which will carry out global tasks that we desire.

5.3 Multitude Practice

There is no real work yet reported on multiple real (physical) robot creatures interacting in non-trivial ways. There is a lot of work on simulations (e.g., see **[Langston 87]** and **[Langston 90]**). It remains to be seen how relevant all this work is to actual physical systems interacting in the world.

6 Conclusion

Building complete creatures is a complex endeavor. No single theory will be adequate to guide us in either synthesis or analysis. We can expect many partial theories to be developed over time, some quite limited in their scope. Likewise, we can expect many programming methods to be developed, again limited in scope. By combining the pieces we will be able to build practical systems. In the meantime, more global theories will emerge, as we have more experimental systems on which we can gain experience.

7 Acknowledgements

Anita Flynn and Maja Mataric provided many valuable comments on earlier drafts of this paper.

Support for this research was provided in part by the University Research Initiative under Office of Naval Research contract N00014-86-K-0685, in part by the Advanced Research Projects Agency under Office of Naval Research contract N00014-85-K-0124.

8 References

[Agre 88] "The Dynamic Structure of Everyday Life", Philip E. Agre, *MIT A.I. Lab Technical Report* 1085, Dec.

[Angle 89] "Genghis, a Six Legged Autonomous Walking Robot", Colin M. Angle, *MIT S.B. Thesis in Electrical Engineering and Computer Science*, March.

[Angle and Brooks 90] "Small Planetary Rovers", Colin M. Angle and Rodney A. Brooks, *IEEE/RSJ International Workshop on Intelligent Robots and Systems IROS '90, July,*, Ikabara, Japan, 383–388.

[Ballard 89] "Reference Frames for Active Vision", Dana H. Ballard, *IJCAI-89*, Detroit, MI, 1635–1641.

[Brooks 82] "Symbolic Error Analysis and Robot Planning", Rodney A. Brooks, *The International Journal of Robotics Research 1:4*, 29–68.

[Brooks 86] "A Robust Layered Control System for a Mobile Robot", Rodney A. Brooks, *IEEE Journal of Robotics and Automation*, RA-2, April, 14–23.

[Brooks 89] "A Robot that Walks: Emergent Behavior from a Carefully Evolved Network", Rodney A. Brooks, *Neural Computation*, 1:2, Summer, 253–262.

[Brooks 90] "The Behavior Language; User's Guide", Rodney A. Brooks, *MIT A.I. Lab Memo* 1227, April.

[Brooks and Viola 90] "Network Based Autonomous Robot Motor Control: from Hormones to Learning", Rodney A. Brooks and Paul A. Viola, *Advanced Neural Computers, Rolf Eckmiller* (ed), Elsevier, 341–348.

[Chapman 87] "Planning for Conjunctive Goals", David Chapman, *Artificial Intelligence 32*, 333–377.

[Chapman 90] "Vision, Instruction, and Action", David Chapman, *MIT A.I. Lab Technical Report* 1204, Aug.

[Chatila and Laumond 85] "Position Referencing and Consistent World Modeling for Mobile Robots", R. Chatila and J. Laumond, *IEEE International Conference on Robotics and Automation*, March, 138–145.

[Connell 87] "Creature Building with the Subsumption Architecture", Jonathan H. Connell, *IJCAI-87*, Milan, August, 1124–1126.

[Connell 90] "A Colony Architecture for an Artificial Creature", Jonathan H. Connell, *MIT A.I. Lab Technical Report* 1151, June.

[Cruse 90] "What Mechanisms Coordinate Leg Movement in Walking Arthropods?", Holk Cruse, *Trends in Neurosciences, 13:1*, 15–21.

[Deneubourg et al 87] "Self-Organization Mechanisms in Ant Societies, II: Learning in Foraging and Division of Labor", J.-L. Deneubourg, S. Goss, J. M.

Pasteels, D. Fresneau, and J.-P. Lachaud, *For Individual to Collective Behavior in Social Insects (Experientia* Supplement, *vol 54)*, 177–196.

[Doyle and Stein 81] "Multivariable Feedback Design: Concepts for a Classical/Modern Synthesis", J. C. Doyle and G. Stein, *IEEE Transactions on Automatic Control, AC-26(1)*, 4–16.

[Ernst 61] "MH-1, A Computer-Operated Mechanical Hand", Heinrich A. Ernst, *MIT Ph.D. Thesis*, Dec.

[Flynn 87] "Gnat Robots (And How They Will Change Robotics)", Anita M. Flynn, *IEEE Micro Robots and Teleoperators Workshop*, Hyannis, MA, Nov.

[Flynn and Brooks 89] "Building Robots: Expectations and Experiences", Anita M. Flynn and Rodney A. Brooks, *IEEE/RSJ International Workshop on Intelligent Robots and Systems IROS '89, Sept*, Tsukuba, Japan, 236–243.

[Giralt, Chatila and Vaisset 84] "An Integrated Navigation and Motion Control System for Multisensory Robots", Georges Giralt, Raja Chatila and Marc Vaisset, *Robotics Research 1, Brady and Paul* (eds), *MIT Press*, 191–214.

[Götz and Wenking 73] "Visual Control of Locomotion in the Walking Fruitfly *Drosophilia*", Karl Georg Götz and Hans Wenking, *J. comp. Physiol.* 85, 235–266.

[Grimson 90] "Object Recognition by Computer: The Role of Geometric Constraints", W. Eric L. Grimson, *MIT Press*, Sept.

[Horswill and Brooks 88] "Situated Vision in a Dynamic World: Chasing Objects", Ian D. Horswill and Rodney A. Brooks, *AAAI-88*, St Paul, MN, August, 796–800.

[Kaelbling 88] "Goals as Parallel Program Specifications", Leslie Pack Kaelbling, *AAAI-88*, St Paul, MN, August, 60–65.

[Kravitz 88] "Hormonal Control of Behavior: Amines and the Biasing of Behavioral Output in Lobsters", Edward A. Kravitz, *Science, Vol 241*, September 30, 1775–1781.

[Langston 87] "Proceedings of Artificial Life", Christopher Langston (ed), *Addison-Wesley*, appeared 1989.

[Langston 90] "Proceedings of Artificial Life, II", Christopher Langston (ed), *Addison-Wesley*, to appear.

[Lozano-Pérez 76] "The Design of a Mechanical Assembly System", Tomás Lozano-Pérez, *MIT A.I. Lab Technical Report* 397, Dec.

[Lozano-Pérez, Mason and Taylor 84] "Automatic Synthesis of Fine-Motion Strategies for Robots", Tomás Lozano-Pérez, Matthew T. Mason and Russell H. Taylor, *Robotics Research 1, Brady and Paul* (eds), *MIT Press*, 65–96.

[Maes 89] "The Dynamics of Action Selection", Pattie Maes, *IJCAI-89*, Detroit, MI, 991–997.

[Maes 90] "Situated Agents Can Have Goals", Pattie Maes, *Robotics and Autonomous Systems*, 6.

[Maes and Brooks 90] "Learning to Coordinate Behaviors", Pattie Maes and Rodney A. Brooks, *AAAI-90*, Boston, MA, August, 796–802.

[Marr 82] "Vision", David Marr, *W. H. Freeman and Co.*, San Francisco.

[Mataric 90] "A Distributed Model for Mobile Robot Environment-Learning and Navigation", Maja J Mataric, *MIT A.I. Lab Technical Report* 1228, May.

[McCarthy and Hayes 69] "Some Philosophical Problems from the Standpoint of Artificial Intelligence", John McCarthy and Patrick Hayes, *Machine Intelligence* 4, Edinburgh University Press.

[McFarland 87] "The Oxford Companion to Animal Behavior", David McFarland (ed), *Oxford University Press*, Oxford.

[Moravec 80] "Obstacle Avoidance and Navigation in the Real World by a Seeing Robot Rover", Hans P. Moravec, *Stanford A.I. Memo* 340, Sept.

[Nicolis and Prigogine 77] "Self-Organization in Nonequilibrium Systems", G. Nicolis and I. Prigogine, *John Wiley and Sons*, New York.

[Nilsson 84] "Shakey the Robot", Nils J. Nilsson *(ed)*, *SRI A.I. Center Technical Note* 323, April.

[Pollack and Ringuette 90] "Introducing the Tileworld: Experimentally Evaluating Agent Architectures", Martha E. Pollack and Marc Ringuette, *AAAI-90*, Boston, MA, August, 183–189.

[Pool 90] "Understanding the Simplest Reaction", Robert Pool, *Science, Vol 247*, January 26, 413.

[Premvuti and Yuta 90] "Consideration on the Cooperation of Multiple Autonomous Mobile Robots", Suparerk Premvuti, and Shin'ichi Yuta, *IEEE/RSJ International Workshop on Intelligent Robots and Systems IROS '90, July,*, Ikabara, Japan, 59–63.

[Roberts 63] "Machine Perception of Three-Dimensional Solids", Larry G. Roberts, *MIT Lincoln Laboratory, Technical Report No.* 315, May.

[Steels 89] "Cooperation Between Distributed Agents Through Self-Organization", Luc Steels, *Proceedings of the workshop on Multi-Agent Cooperation, Cambridge, UK, August*, North Holland.

[Taylor 76] "The Synthesis of Manipulator Control Programs from Task-Level Specifications", Russell H. Taylor, *Stanford A.I. Memo* 282, July.

[Thorpe, Hebert, Kanade and Shafer 88] "Vision and Navigation for the Carnegie-Mellon Navlab", Charles Thorpe, Martial Hebert, Takeo Kanade and Steven A. Shafer, *IEEE Trans. PAMI*, 10(3), May, 362–373.

[Wehner 87] "'Matched Filters' – Neural Models of the External World", Rüdiger Wehner, *J. comp. Physiol. A* 161, 511–531.

Cognitive Action Theory as a Control Architecture

H. L. Roitblat

Department of Psychology
University of Hawaii at Manoa
Honolulu, Hawaii, USA 96822
herbert@uhccux.uhcc.hawaii.edu

The three laws of robotics: (1) *No robot may harm a human being, or through inaction, allow a human being to come to harm.* (2) *A robot must obey the orders given it by human beings except where such orders would conflict with the First Law.* (3) *A robot must protect its own existence as long as such protection does not conflict with the First or Second laws.* Asimov, 1950/1977, p. 40.

Abstract

Standard versions of control theory approach their limits in autonomous robotic control because of their conceptualization in terms of a fixed mapping between environmental variables and behavior. Cognitive action theory, in contrast, views behavior as hierarchically organized by a network of interacting nodes. Nodes at different levels represent action with different degrees of abstraction. Activation of nodes in the hierarchy is controlled by potentiation and inhibition received from other nodes and by environmental stimulus information. Learning consists of the formation of connections between nodes as the result of locally available information about the satisfaction of cybernetic feedback functions. Such networks are capable of planning and executing highly flexible behaviors in complex, uncertain environments.

1. Introduction

Among the difficult problems for designers of autonomous robot systems are control of (a) nonlinear systems, (b) stochastic systems, (c) systems requiring sensor fusion and feedback, and (d) systems with incomplete information about the environment and its structure (Meystel, 1988). Many of these problems are exacerbated by the control-theory-dominated approach that robotic designers have taken. Although control theory has been remarkably successful in many domains, it approaches its limits in autonomous robotic control, because in its standard versions it assumes that behavior can be adequately conceptualized as a system with a fixed mapping between environmental variables and behaviors. It typically characterizes the behavior of the robot in terms of movements or operations that are precisely defined as specific responses to specific environmental conditions (Brooks, 1986). This approach works well in situations in which variability is limited, goals are simple and noncompetitive, uncertainty is minimal (or at least statistically characterizable), and the environment and actions appropriate to it are fairly exhaustively known (e.g., a factory environment). The approach is likely to prove ultimately inadequate, however, in highly variable environments, in the face of complex competing demands, and in situations in which no complete set of situation-response rules is available a priori.

In contrast to these control-theoretic approaches to robotics, Asimov's robot stories (e.g., Asimov, 1950) suggest another approach to robotic control. They describe a theory of "robopsychology" embedded in the three laws of robotics. The laws are vague. They do not describe specific movements or even specific actions, rather they outline a set of high-level goals for the device. The specifics of achieving these goals are left to the robot's "positronic" brain; they are not imagined to be directly programmed. The methods by which the robots in his stories translate the vague laws of robotics into specific behaviors and the ways in which they resolve conflicts among the laws are the primary themes in his stories, and at least part of their continuing interest.

2. Action and movement

The distinction between general descriptions of behavior and specific movements is also a theme in the history of psychology. Wundt (1907), Ach (1910), Lewin (1935), and Tolman (1932) recognized a distinction between action and movement and conceived of action as the basic unit of psychology. A movement description is couched in terms of the motion of the body and the limbs, described as the direct result of patterned muscular contractions, or of external forces or imbalances (e.g., the motion observed when an individual is walking). The language of kinematics is typically sufficient to describe movements. Two movements are equivalent if they involve the same coordinated pattern of muscular contractions and limb trajectories. In contrast, actions are defined functionally in terms of the function, purpose, or goal of the behavior (see Newell, 1978). Although any particular action, such as opening a door, is instantiated by one set of movements or another, a potentially large set of movements may be useful in performing any particular action (Bernstein, 1967) and the same movement may be a component of many actions (Mischel, 1969). Handwriting and typewriting both serve the function of committing a text to paper, but the two forms of writing involve radically different movement patterns. Similarly, opening a door, eating, and typewriting each involve the movement of extending the right hand, but constitute very different actions. Two actions are equivalent when they both perform the same function, regardless of the pattern of movement employed.

Tolman (1932) argued forcibly for an action-based analysis of behavior because he noticed that an organism's behavior appeared to be strongly controlled by its goals. Descriptions couched at the level of movements are inadequate to capture the regularities in the animal's performance. For example, in one experiment, rats were trained to run in a plus maze (Tolman, Ritchie, & Kalish, 1946). The maze consisted of four arms: north, south, east, and west. One arm (e.g., north) was blocked off on each training trial and the rat was started from the opposite arm (e.g., south). The rats were trained always to go to one particular arm (e.g., the east arm) to find a piece of food. They were thus trained to perform *movements* involving right turns (from south to east) and to perform *actions* of going to the east arm. After training, the south arm was blocked off and the rats were started from the north arm. If they had learned to perform a certain specific movement to obtain food, then they should start in the north arm, turn right, and end up in the west arm. On the other hand, if they had learned a more abstract representation of the behavior, such as having the east arm as a goal, then they should start in the north arm, perform a novel movement, and end up in the east arm. This is what Tolman and his colleagues found. The rats were overwhelmingly more likely to go from the north arm into the east arm, rather than to perform the right-turn response and enter the west arm on these tests. Results like these and others indicate that the rats had learned the task relative to actions that were more abstract than the specific movements used to instantiate those actions.

3. Thought and Action

Tolman's abstract descriptions of purposive behavior entailed a certain cost. For example, Guthrie (1935) accused Tolman of being vague, of leaving his animals lost in thought. The same criticism is applicable to Asimov's fictional laws of robotics. Although they specify the robot's goals, neither Asimov's nor Tolman's laws of behavior provide a mechanism to implement a causal chain from its goals to its performance. The usual approach in robotics (Brooks, 1986) is to build a control system that decomposes performance into a chain of operational components beginning with perception and ending with motor control. Similarly, certain behaviorists, such as Hull (1931, 1943) assumed that behavior is controlled by chains of stimulus-response-stimulus-response units. Hull's (1931, 1943) concept of the r_G as a substitute for intentions is a behavioral example of this kind of control mechanism. Hull argued that anticipation and other seemingly goal-directed behaviors were really the animal's response to internal response-produced stimuli. Each response the animal made resulted in certain (e.g., kinesthetic) stimuli and changes in the perceived environment (e.g., by changing location). These sensations were then automatically associated with and were able to elicit subsequent responses, which also produce kinesthetic sensations.

The kinesthetic stimuli were conceived as being just like any other stimulus, which could be associated with subsequent responses until the end of a complex chain of behavior was performed and reward was received. Complex and purposive-appearing behaviors could thus be explained away as the result of long chains of stimulus-response-stimulus-response. Similarly complex robotic behaviors are frequently composed of simple chains of movement in response to specific sensor data. The robot's behavior effects a change in the sensor the data (and perhaps the robot's internal state), which then elicits specific additional behaviors, and so on. The complexity of the behavior is often more accurately characterized as simple responses to complex changes in the environment than as a complexly organized behavior (Bever, Straub, Terrace, & Townsend, 1980; Simon, 1969; cf. Lashley, 1951).

4. Hierarchical cognitive action theory

These chain-like representations of behavior are single-layer representations because the components are joined through a single-linkage linear pattern. Production of one part of the chain elicits the next part in a linear order. The concept of action, in contrast, suggests that behavior can be organized at more than one level at a time in complex multiple-linkage patterns. Hierarchically organized cognitive action systems (Roitblat, 1988) take advantage of this multiple-linkage structure and of the many-to-many mappings between movements and actions to offer a solution to the problems of behavioral organization in the form of a hierarchically organized network system. In this kind of system, nodes at different levels represent behaviors with different degrees of abstraction (see Saridis, 1985; Meystel, 1988; for related views derived from a control-theory perspective; MacKay, 1987 for related views derived from a language-processing perspective, and Gallistel, 1980 for a biological perspective). Behavior is organized at each level and organisms are capable of learning at every level.

4.1 Connectionist models

Network representations of action provide a natural application for connectionist modeling. The last few years have seen an explosive growth in interest in connectionist or neural network models. These models conceive of the mind as a real-time, self-organizing, highly interconnected network of simple, neural-like elements (Grossberg & Stone, 1986; Hinton & Anderson, 1981; McClelland & Rumelhart, 1986; Wasserman, 1989). Information processing occurs by transmitting activation patterns via weighted connections among the nodes of network. These networks have been found to be extremely useful for solving such problems as pattern recognition that have been difficult to solve using more conventional artificial intelligence approaches.

4.2 The cognitive action theory network

Cognitive action theory (Roitblat, 1988) is a kind of connectionist model that views behavior as the product of a layered system of interacting nodes or units. Although the nodes themselves are relatively simple, the complexity of their interconnection allows them to perform complex functions. Layers in the network correspond roughly to levels of abstraction. Nodes at the lowest level directly control patterns of muscular and neuroendocrine activity (e.g., opening the fingers). The specified movement is performed when one of these nodes is activated. Nodes at somewhat higher levels are abstracted representations of the actions that employ those movements (e.g., writing letters). Nodes at the highest levels represent goals, intentions, thoughts, expectancies, and motivations. Superordinate nodes are increasingly abstract representation of actions and goals. Subordinate nodes are generally representations of alternative ways of accomplishing the action specified by their superordinates. Each superordinate node can have many subordinates (including subgoals), and each subordinate node, in turn, can serve many superordinates.

Nodes interact with one another through potentiation and inhibition. Potentiation makes a unit easier to activate and inhibition makes it more difficult to activate. Higher level nodes enable (potentiate) or inhibit lower level nodes; they do not directly activate the lower level nodes that specify the details of the action. For example, a certain node may represent the action of moving food from a plate to the lips. This node is subordinate to one that specifies eating and is superordinate to those that specify how the food will be eaten. There are many ways that food can be moved from plate to lips, including using a fork, using a spoon, lifting the plate, and so forth. These subordinate nodes have their own subordinate nodes that specify more directly which of the various muscles will move or the specific pattern in which they will move.

A particular higher-level node may potentiate a number of lower level nodes, each of which describes an alternative, and potentially incompatible, means of accomplishing the action specified by the higher node. For example, food could be moved to the mouth either by using a fork and a stabbing motion or by using a spoon and a scooping motion. Both behaviors may be simultaneously potentiated, but both cannot be simultaneously activated, because they employ incompatible movements and a merging of the two movements is unlikely to be effective. The unit (e.g., stab or scoop) that is activated depends on the stimulus conditions. If the food is in solid chunks and a fork is available, then a stabbing motion may be more appropriate. If the food is relatively fluid, then a stabbing motion will be entirely ineffective and a scooping motion will be more appropriate. Selection among potentiated nodes is accomplished through mutual inhibition and through control by evidential stimulus inputs.

Nodes at every level are influenced by the stimulus inputs they receive. Each node responds to particular stimulus patterns. Many neural network models are available that describe how raw stimulus inputs can be translated and fused into conceptual categories (see e.g., Kohonen, 1988; Roitblat, Moore, Nachtigall, & Penner, 1989). The degree to which a unit is activated depends on the interaction between the net potentiation or inhibition and the strength or pertinence of stimulus inputs (evidence) it receives. Each node has its own "rules of evidence," or specification of the stimulus features and the degree to which the presence of those features will contribute to its activation. Potentiation lowers the unit's threshold for activation, so that weaker evidence is capable of activating the node and inhibition raises its threshold, so that stronger evidence is necessary to activate the unit.

Hence, activation of a unit is controlled by the net evidence the unit receives from environmental events (including those in the internal milieu) as well as the net potentiation it receives. The interaction between these variables is described by a cusp catastrophe:

$$\Delta A = \frac{2}{1+e^{-(Ev/(1-Po)}} \qquad (1)$$

A is the activation level of the node. *Ev* is the net evidence the unit receives from the perceptual network and is determined by the connection patterns in this network and by the connections between the perceptual system and the unit. *Po* is the net potentiation received by the node. Potentiation is received from superordinate (parent) nodes that represent the action in a more abstract form and inhibition is received from nodes at the same level in the hierarchy (sibling nodes) that represent incompatible actions at the same level of abstraction. For example, activation of a low-level node representing arm flexion inhibits nodes for arm extension. The same kind of "lateral inhibition" occurs at all levels in the hierarchy and is ubiquitous in neural processing.

The kind of evidence to which a unit responds can be quite abstract as a result of processing through self-organizing perceptual networks. Presumably, the abstractness of the evidence relevant to each node is related to the abstractness with which the node represents behavior. Nodes representing abstract properties of behavior, such as "eat dinner" will be responsive across a wide range of possible dinner foods, events, and settings. They are invariant across a wide range of transformations in the situations and in the movements used to accomplish the specified actions.

In addition to specifying the stimulus conditions that activate them, nodes also specify the conditions that "satisfy" them. The evidence that satisfies a node serves to inhibit its activation and also functions in learning, which is described below. The evidence and satisfaction conditions may be mutually exclusive (e.g., the presence or absence of some property) or they may overlap (e.g., the presence of one property may provide sufficient evidence to activate a node, but this property need not be absent to satisfy the node). The evidence conditions that activate a node may be specific to that node, but the satisfaction conditions may be general (e.g., delivering food to the mouth) across a range of mutually inhibiting nodes. The node that represents raising a glass is activated in part by the perception of a glass filled with a drinkable liquid, but it is satisfied when the glass is raised, not necessarily when the glass is emptied.

4.3 Organization of behavioral sequences

Behaviors do not occur individually, but rather in a stream, in the context of the organism's experience, goals, other actions, and a changing environment. Many behaviorists believed that the organization of an animal's actions comes from kinesthetic and other sensory changes produced by the organism's previous actions (see Roitblat, 1987). Many robotic control systems are similarly organized (see above). Behavior on this view is the result of stimulus-response reflexes. A related conceptualization views the organism as a kind of conceptual servomechanism always operating to reduce the discrepancy between its current state and its goal state.

Although both the reflex and the servomechanism approaches have significant utility, they both prove to be inadequate. Animals, for example, have been found capable of performing tasks in which the sensory feedback is only abstractly similar to that present during learning (e.g., MacFarlane, 1930), they respond to certain environmental features, even when these features are incompatible with previous patterns of response (e.g., Olton & Collison, 1979). Servomechanisms, even if conceived as containing an embedded representation of the desired goal state are also inadequate as basic control structures, because they merely respond to perceived differences. These differences, and the system's responses to them can be more complicated than is allowed by simple reflex mechanisms. but, unless the mechanism has the capacity to plan and to derive subgoals as well as global goals, it runs the risk of getting stuck at barriers (e.g., Elfes & Talukdar, 1983). Planning and the ability to carry out a programmed sequence of behaviors would seem to be essential to any organism whether living or artificial.

As Tolman (e.g., 1948) argued, purposive behavior requires cognitive maps. Tolman conceived of these maps as organized families of expectancies about the environment. Expectancies include representations of relations of the form x predicts y (perhaps x is near y in time or space), if x then y (perhaps x and y are contiguous in time or space, or x causes y), and perhaps some others. Programs can be made from a series of expectancies if x predicts y, and x is achieved, then y might predict z, so y is achieved, next, and so on. These expectancies can be organized into a single coherent representation of the environment called a cognitive map, which can control frequently complex behaviors (see Dell, 1986; Gallistel, 1980; MacKay, 1987; Roitblat, Scopatz, & Bever, 1987). The organization of these expectancies can be linear (similar to the linear behavioral chains described above), but the notion of a cognitive map, implies that they can be organized in more complex (e.g., 2- or higher dimensional) patterns. Even linearly organized strings of expectancies entail more complex behavioral structures, because they may include information about the relations among events in a linear sequence, but information about how those events are produced is represented at a subordinate level.

Planning requires the organism to have goals, to be able to form expectancies regarding paths to those goals, and be able to respond to obstacles along those paths. Cognitive action theory provides such mechanisms in its conceptualization in terms of superordinate and subordinate goals. The potential complexity of the stimulus patterns by which competing nodes are selected also provides a mechanism to support planning. Among these stimulus representations can be learned maps of the environment (see e.g., Gallistel, 1980; Tolman, et al., 1946; Woodbury & Oppenheim, 1988).

4.4 Learning

Learning in cognitive action theory occurs at all levels in the network through three mechanisms: the formation of new nodes, the modification of connections among nodes, and the modification of the perceptual networks providing sensory information to nodes (see Roitblat, 1988; Hayes-Roth, 1977). This learning can occur at any level in the hierarchy and some evidence exists that there is a bias for learning in biological systems to be represented at the highest appropriate level (e.g., the experiment by Tolman, et al., described above; see also Bever, 1984).

Each node specifies the conditions under which it will become active as well as the conditions under which once active it will become satisfied. Criteria for these state transitions (e.g., from inactive to active) depend on the degree to which the node is potentiated. Hence each node can be described according to a kind of rule:

Condition: Detect situation in which the node may be effective.

Expectancy/action: Potentiate nodes representing more specific actions or predicted conditions.

Satisfaction: Detect evidence that the node has been effective.

The unit becomes activated when it detects sufficient evidence in the form of sensory information. The amount of evidence required to activate the rule depends on the level of potentiation the node receives from superordinate nodes. If the node is strongly potentiated then little evidence will be necessary. If it is weakly potentiated then more evidence will be necessary before the unit becomes activated. See above.

A node that is activated distributes potentiation to the nodes in the lower layers to which it is connected (its "child" nodes), and distributes inhibition to the "sibling" nodes at the same level to which it is connected. The magnitude of potentiation received by a node depends on the strength of the learned or unlearned connection from all of the parents and all of the siblings to which it is connected.

The connections between two nodes at different levels are strengthened when they are simultaneously satisfied. This is a variation on the Hebb (1949) learning rule:

$$\Delta C_{12} = S_1 S_2 R (1 - C_{12}) \qquad (2)$$

The change in connection strength (ΔC_{12}) introduced by this function depends on the level of satisfaction of the two nodes ($S_1 S_2$) and on the current strength of the connection (C_{12}). Learning allows the network to self-organize its knowledge about the environment and the actions that are appropriate to it. As a result of this ability to self-organize, the organism need not come fully programmed for environmental contingencies, but may develop its own response patterns. The reinforced satisfaction of a node can also function to adjust the connections between the action unit and the perceptual network that feeds it. Kohonen (e.g., 1984) suggests a learning algorithm according to which the connection strength between nodes in the perceptual network and a satisfied node can be adjusted to more closely approximate the conditions that satisfy the node, hence causing that node to more tightly select those stimulus conditions.

5. Summary and Conclusion

Standard single-layer control systems are severely limited in their ability to deal with the kinds of problems faced by autonomous robots. Psychological investigations suggest that organisms employ multi-level representations of their own behavior that include abstract descriptions of the behaviors' goals and alternative means of achieving those goals. They suggest the use of hierarchically organized cognitive action systems that use a multiple-linkage structure. Nodes at different levels in these networks represent behaviors with different degrees of abstraction. Higher level nodes include goals, intentions, thoughts, expectancies, and motivations. Cognitive action theory takes advantage of the many-to-many mappings between movements and actions to offer a solution to the problems of behavioral organization that supports resolution among competing demands, planning, and intelligent, flexible interaction with a changing, incompletely known, environment.

Cognitive action theory is instantiable in real hardware, including distributed networks of biological neurons. Although the system includes a certain degree of ambiguity, it also provides direct means to translate vaguely specified goals and other "ideas" into specific forms of motion. Learning consists of the formation of connections between nodes as the result of locally available information about the satisfaction of cybernetic feedback functions. As a result, cognitive action theory provides a promising architecture both for the development of theories of animal behavior and for the construction of autonomous robotic devices.

References

Ach, N. (1910). *Ueber den Willensakt und das Temperament: Eine experimentelle Untersuchung.* Leipsig: Quelle & Meyer.

Asimov, I. (1950/1977). *I, robot.* New York: Ballantine.

Bernstein, N. A. (1967). *The coordination and regulation of movement.* London: Pergamon.

Bever, T. G. (1984). The road from behaviorism to rationalism. In H. L. Roitblat, T. G. Bever, & H. S. Terrace (Eds.) *Animal Cognition*, Hillsdale, NJ: Erlbaum, pp. 61-75.

Bever, T. G., Straub, R. O., Terrace, H. S., & Townsend, D. J. (1980). Study of serial behavior in humans and non-humans. In P. Jucsyk & D. Klein (Eds.) *The Nature of Thought: Essays in Honor of D. O. Hebb.* Hillsdale, NJ: Erlbaum. pp. 51-93.

Brooks, R. A. (1986). A robust layered control system for a mobile robot. *IEEE Journal of robotics and automation*, RA-2, 14-23.

Dell, G. S. (1986). A spreading activation theory of retrieval in sentence production. *Psychological Review*, 93, 283-321.

Elfes, A. & Talukdar, S. N. (1983). A distributed control system for the CMU rover. *Proceedings of the IJCAI*, 830-833.

Gallistel, C. R. (1980). *The organization of action: A new synthesis.* Hillsdale, NJ: Lawrence Erlbaum Associates

Grossberg, S. & Stone, G. (1986). Neural dynamics of word recognition and recall: Attentional priming, learning, and resonance. *Psychological Review*, 93, 46-74.

Guthrie, E. R. (1935). *The Psychology of Learning.* New York: Harper.

Hayes-Roth, B. (1977). Evolution of cognitive structures and processes. *Psychological Review*, 84, 260-278.

Hebb, D. O. (1949). *The organization of behavior.* New York: Wiley.

Hinton, G. & Anderson, J. A. (Eds.) (1981). *Parallel models of association memory.* Hillsdale, NJ: Erlbaum.

Hull, C. L. (1931). Goal attraction and directing ideas conceived as habit phenomena. *Psychological Review*, 38, 487-506.

Hull,C. L. (1943). *Principles of Behavior.* New York: Appleton-Century-Crofts.

Kamil, A. C. & Roitblat, H. L. (1985) Foraging theory: Implications for animal learning and cognition. *Annual Review of Psychology*, 36, 141-169.

Kohonen, T. (1984). *Self-organization and associative memory.* New York: Springer-Verlag.

Lashley, K. S. (1951). The problem of serial order in behavior. In L. A. Jeffress (Ed.), *Cerebral Mechanisms in Behavior.* New York: Wiley, pp. 112-136.

Lewin, K. (1935). *A dynamic theory of personality.* New York: McGraw Hill.

MacFarlane, D. A. (1930). The role of kinesthesis in maze learning, *University of California Publications in Psychology*, 4, 277-305.

MacKay, D. G. (1987). *The organization of perception and action.* New York: Springer-Verlag.

McClelland, J. L. & Rumelhart, D. E. (Eds) (1986). *Parallel distributed processing: Explorations in the microstructure of cognition*. (2 volumes) Cambridge, MA: The MIT Press.

Meystel, A. (1988). Intelligent control in robotics. *Journal of Robotic Systems*, 5, 269-308.

Mischel, T. (1969). Scientific and philosophical psychology: A historical introduction. In T. Mischel (Ed.), *Human Action*. New York: Academic Press. pp. 1-40.

Newell, K. M. (1978). Some issues on action plans. In G. E. Stelmach (Ed.) *Information processing in motor control learning*. New York: Academic. pp. 41-54.

Olton, D. S. & Collison, C. (1979). Intramaze cues and "odor" trails" fail to direct choice behavior on an elevated maze. *Animal Learning and Behavior*, 7, 221-223.

Roitblat, H. L. (1987). *Introduction to comparative cognition*. New York: W. H. Freeman.

Roitblat, H. L. (1988). A cognitive action theory of learning. In J. Delacour & J. C. S. Levy (eds.), *Systems with learning and memory abilities*. New York: Elsevier. pp. 13-26.

Roitblat, H. L., Moore, P. W. B., Nachtigall, P. E., Penner, R. H., & Au, W. W. (1989). Dolphin echolocation: Identification of returning echoes. R. Hecht-Nielsen (Program Chair). Proceedings of the International Joint Conference on Neural Networks--89. pp. 295-300. Washington, DC: Institute of Electrical and Electronic Engineers.

Roitblat, H. L., Scopatz, R. A., & Bever, T. G. (1987). The hierarchical representation of three-item sequences. *Animal Learning and Behavior*, 15, 179-192.

Saridis, G. N. (1985). Intelligent control for robotics and advanced automation. *Advances in automation and robotics*, Vol 1. JAI Press.

Simon, H. A. (1969). *Sciences of the artificial*. Cambridge, MA: MIT Press.

Tolman, E. C. (1932). *Purposive Behavior in Animals and Men*. New York: Appleton-Century-Crofts.

Tolman, E. C. (1948). Cognitive maps in rats and men. *Psychological Review*, 55, 189-208.

Tolman, E. C., Ritchie, B. F. & Kalish, D. (1946). Studies in spatial learning I. Orientation and the short-cut. *Journal of Experimental Psychology*, 36, 13-24.

Wasserman, P. D. (1989). *Neural computing: Theory and practice*. New York: Van Nostrand Reinhold.

Woodbury, R. F. & Oppenheim, I. J. (1988). Geometric reasoning: concepts and demonstrations. In J. S. Gero (Ed.), Artificial intelligence in engineering: Robotics and processes. (pp. 61-75). New York: Elsevier.

Wundt, W. (1907). *Outlines of Psychology*. London/New York: Engelmann.

TOWARDS A THEORY OF EMERGENT FUNCTIONALITY

Luc Steels
Artificial Intelligence Laboratory
Vrije Universiteit Brussel
Pleinlaan 2, B-1050 Brussels Belgium.
email: steels@arti.vub.ac.be

Abstract

The paper sets the first steps towards a theory of emergent functionality. The theory tries to make explicit what emergent functionality is by contrasting it with hierarchical functionality. It analyzes the principle advantages of this approach and conjectures a formal structure common to systems with emergent functionality.

1. Introduction

Emergent functionality means that a function is not achieved directly by a component or a hierarchical system of components, but indirectly by the interaction of more primitive components among themselves and with the world. Emergent functionality has become one of the main themes in research on Artificial Life (Langton, 1988) and autonomous agents (Brooks, 1989). So far, engineers and scientists have used their intuitions to build systems that exhibit emergent functionality but there is no explicit theory yet on what emergent functionality is, how it can be achieved, when it is appropriate and why. This paper reports on research to understand the principles underlying emergent functionality and how it can be used for designing and building systems. We first discuss hierarchical systems to make the specific properties of systems with emergent functionality stand out more clearly. Then we turn to emergent functionality itself. Examples are discussed and the advantages of emergent functionality are analyzed. The final part of the paper conjectures a formal structure that seems common to systems with emergent functionality.

2. Hierarchical systems

2.1. Characterization of hierarchical systems

In hierarchical systems there is a direct relationship between structure and function. The system consists of a set of components. Each of these components has three aspects: (1) inputs and outputs, (2) a control element to turn the component on or off, (3) a well-determined functionality. A component stands on its own in the sense that its functionality can be tested independently from the other components. Moreover this functionality is a recognizable subfunction of the global functionality of the system. For example, the motor of a car needs fuel. The tank realizes a subfunction of fuel supply, namely to hold the fuel. The tank realizes this function independently from the other components like the pipes that transfer fuel to the motor or the accelerator that regulates the flow.

Some components are specialized in obtaining input from the environment. Others are concerned with output and actions in the environment. There are also components whose major role is the control of the operation of other components. So the different components interact in two ways: (1) There is flow of data through input/output relations between components. (2) There is flow of control when one component turns another component on or off.

Because the components function independently of each other they can be constructed and put together in a modular fashion: The global functionality is decomposed into different subfunctions. A subfunction is either directly realized by a particular component or it is further decomposed into more restricted subfunctions. This decomposition continues until all subfunctions are realized by concrete components.

2.2. Examples of hierarchical systems

Hierarchical systems occur in nature as well as in engineering. Here are just a few examples. Later we will discuss the same examples but show how they are done through emergent functionality.

Example 1: Building a tree

Suppose we have to write a program that builds a tree (without recursive branches). A programmer would instinctively decompose this problem into various components: one for drawing a left-branch, one for drawing a right-branch, and one for drawing the tree-stem. These three components together draw a subtree. Drawing a subtree is then invoked recursively as in the following LISP program. A tree is displayed on the right.

```
(defun draw-tree (angle size delta)
  (if (< size delta) 'done
    (progn
      (draw-left-branch angle size)
      (draw-right-branch angle size)
      (draw-and-go-up size)
      (draw-tree angle (- size delta) delta))))
```

The hierarchical decomposition of this program is given in figure 1. The draw-tree program sequentially invokes each of the subcomponents. Angle, size, and delta are data that are passed through to each of the components. They are the global parameters that determine the shape of the tree.

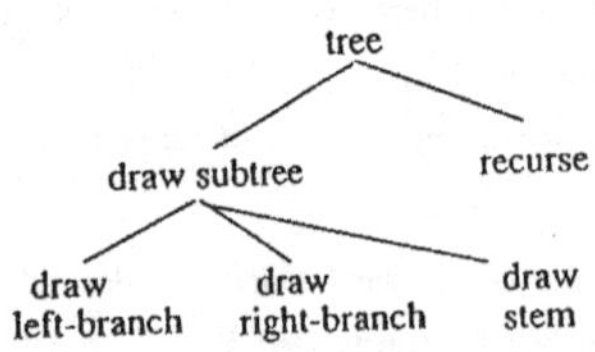

Figure 1. Hierarchical system for drawing a tree.

Example 2. Following a wall.

Suppose we have to build a robot that "follows" a wall. The robot should move towards a wall when there is one in its neighborhood and then move forward at a constant distance from the wall. A hierarchical solution would decompose this global functionality into different subfunctions. One of them is "go to wall". This decomposes further into "detect wall" and "move towards the wall" (figure 2). Detect wall may decompose into "read sensors" and "interpret sensors to see whether it is a wall". Output of "detect wall" is the position of the wall and the distance that needs to be traversed to go towards it. "Move towards wall" decomposes into "turning around in the direction of the wall" and "moving forward" the distance to the wall.

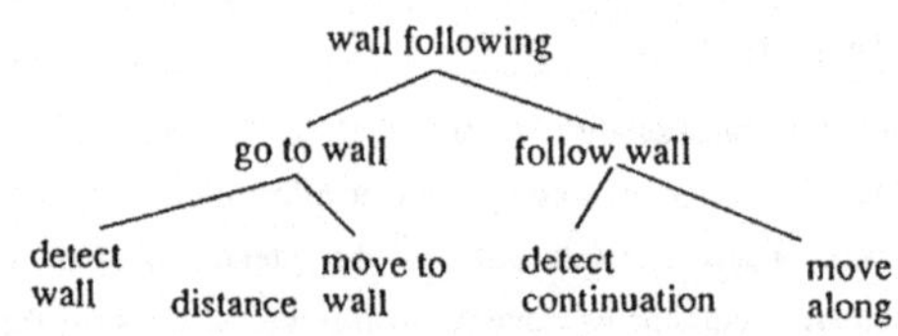

Figure 2. Hierarchical decomposition of wall following

Example 3: Collecting rock samples

Suppose we have to build a system that collects rock samples on a distant planet. The location of the rock samples is unknown in advance but they are typically clustered in certain spots. There is a vehicle that can drive around on the planet and later reenter the spacecraft to go back to earth.

Again it is not difficult to see what a hierachical approach to this problem would look like.

The global functionality (collect rock samples) is decomposed into different subfunctions like detect sample, pick up sample, and move sample into vehicle. Each of these is further decomposed into more primitive functions. For example "detect sample" requires navigation in the environment which may be based on a world map, on components to recognize obstacles and avoid them, and on components to recognize samples. Output of "detect sample" are the coordinates of a sample which are then passed on to the "pick up sample" routine.

2.3. Features of hierarchical systems

There are obvious advantages of hierarchical systems. Because there is a clear relation between structure and function, systems are modular. This helps in design. Very complex systems can be built by patiently putting together components that implement subfunctions. Modularity also helps in diagnosis and repair. When there is a particular subfunction that is no longer realized the component can be relatively easily identified and replaced. But there are also disadvantages:

1. Hierarchical systems are not fault tolerant. As soon as there is a single component that breaks down, the total system fails. For example, if for some reason the component that recognizes obstacles fails, the rock collector breaks down completely. Components can of course be doubled but then the system still fails if both elements fail.

2. There is a critical dependence on prior analysis. The designer has to identify all the assumptions about the environment and carefully delineate the initial conditions. These need to be taken into account when decomposing a function into its subfunctions or when making a choice for primitive components, particularly if the components already exist. For example, if a component is assumed for detecting walls, a prior analysis needs to be made about what will count as a wall. Dependence on prior analysis is responsible for the brittleness of hierarchical systems. Slightly different initial conditions or unforeseen environmental circumstances may lead to complete failure. Programmers are well aware of this phenomenon. Many bugs are due to situations where certain components suddenly get inputs that violate constraints (such as a calculation routine that unexpectedly gets a negative integer).

3. The interaction with the environment is delegated to certain components (such as the "detect wall" component). Other components (such as the "move to wall" component) receive information about the environment as input through these components. This leads to a limited interaction with the environment and hence to a certain slowness with which the system can react to environmental changes.

4. Prior analysis and delegation of interaction with the environment to certain components presupposes categorization of the possible environmental situations that may be encountered. Prior categorization is difficult when a system has to cope with the richness of the world. There will typically always be situations that fall outside the scope of what was foreseen. This problem is well illustrated if we want the growth of the tree to depend also on external circumstances, such as availability of sun light. This would require a quite drastic revision of the draw tree program, particularly if we allow the sun to move as time progresses.

5. There is a lack of robustness which follows from the need for prior analysis and from the many hidden assumptions that are made about how a component realizes a function. Robustness means that the system also keeps functioning in situations that bring it temporarily out of balance. For example, when the wall following robot is suddenly pushed against the wall, it should continue to function.

These observations indicate that a hierarchical approach is more appropriate when it is easy to foresee all the circumstances of usage, when the global functionality is very complex and involves many different subfunctions, or when there is a limited and predictable interaction with the environment.

3. Emergent functionality

3.1. Characterization of emergent functionality

A system with emergent functionality also consists of a set of components but there are some important differences:

• Each component has its own interaction with the environment. This means that the inputs and outputs are not intended for other components but are inputs from the world and outputs or actions into the world. The contact with the environment is continuous. As a result the components are dynamically responding to environmental changes and they get immediate feedback from the environment.

• The activation of a component is directly driven by the environment, as opposed to another component. A component becomes active when its preconditions for activation are met.

• A component has a particular functionality but this is not recognizable as a subfunction of the global functionality. Instead a component implements a behavior whose side effect contributes to the global functionality. The relation between the behavior of a component and the global functionality is similar to the way macroscopic properties (such as pressure) emerge due to the more primitive behaviors of molecules. Consequently a component does not stand on its own. It needs other components to do useful work.

The way a system with emergent functionality is put together is very different from hierarchical systems. Each behavior has a side effect and the sum of the side effects gives the desired functionality. In a hierarchical system, a component stands on its own: it executes a particular subfunction independently of the other components. In a system with emergent functionality all the components need to be together and operate simultaneously to achieve the desired functionality. Not only is the data flow through the environment (and not through internal input/output relations) but also the control flow is driven by the environment.

3.2. Examples of emergent functionality.

Example 1. Building a tree

Nature does not build trees through a hierarchical mechanism. The exact point where a branch will form is not determined in the genetic code of the tree which is then executed by some biochemical mechanism. Instead there are a set of more primitive underlying mechanisms such as cellular growth and reaction-diffusion, and the specific structure of the tree emerges as a consequence of the operation of these mechanisms (see e.g. (Babloyantz, 1987), (Meinhardt,1982)). Steels (1990) describes some computational experiments to build trees this way which are here summarized.

A tree will consist of a series of cells. The cells come in three groups: stem cells, left-branch cells and right-branch cells. Each cell contains a number of substances that allow it to spin off new cells. New cells can either grow on top of a stem cell or to the left or right of it. Once split off, left-branches or right-branches can continue to extend. Recursion is not considered.

The substances that determine whether growth takes place will be called growth substances. The stem cells have a stem growth substance, a left-growth substance and a right-growth substance. A new cell is created when a growth substance exceeds a certain treshold. For example, when the stem growth substance exceeds its treshold a new cell is created on top of the existing stem.

The art of building trees consists in regulating the dynamic evolution of these different substances based on a number of simple components that implement the following behaviors:

1. Food from the root diffuses at a certain rate across the cells.

2. Increase in growth substance depends on food.

3. When the amount of growth substance is beyond a particular treshold a new cell is generated.

4. The presence of a stem cell, a left-branch, or a right-branch, inhibits the increase of growth substance.

All these components gather their own inputs (e.g. food, presence of growth substance) and generate their own outputs (e.g. increase in growth substance). Communication between components is indirectly through the environment (e.g. the presence of growth substance). There are no components responsible for control. A component becomes active when its environmental conditions are met (e.g. growth substance beyond treshold). None of the behaviors talks about the size of the tree, or when and where a new branch needs to be split off, instead each behavior only indirectly has something to do with building trees. The simple behaviors put together nevertheless result in regular trees. Fig 4 contain a few snapshots. Branches regularly split off to the NE and NW of a cell and continue to grow from there.

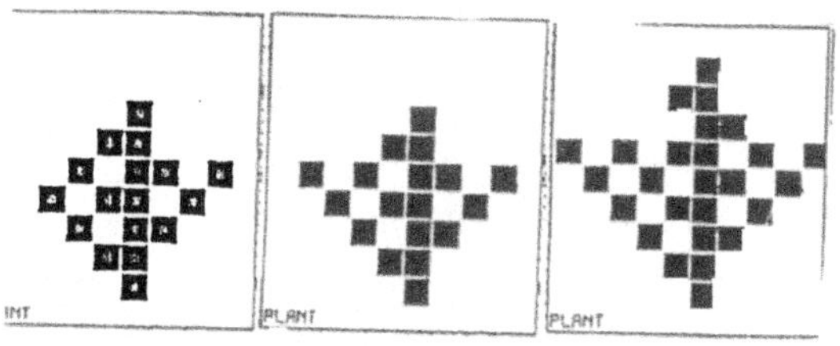

Figure 4. Snapshot of tree growth.

The major characteristics of this tree are emergent:

• The size of the tree or of the branches depends on the amount of food and the rates of diffusion.

• The regular spacing is due to the dynamics regulating the interaction between the left-growth and right-growth substances in a stem. Initially one direction of growth is chosen at random. But when a branch shoots off, growth in the same direction is inhibited so that a branch can grow in the other direction. As soon as this branch has materialized the inhibition switches around and the first direction gets the upperhand. At no point is there an explicit call to a component to create a left or right branch.

Already one major difference with the hierarchical approach is that it is easy to introduce new environmental influences. For example, we can add an additional behavior that the increase of growth substance depends on temperature. Introducing a source of heat then causes a tree to grow in a particular direction. Figure 5 shows first a regular tree (without rule 4), then a tree which is growing more quickly towards the sun. The gradient field created by the sun is visualized on the right.

Example 2. The wall following robot.

An emergent functionality approach to the wall following robot has been proposed by Brooks and realized in various MIT mobile robots (see e.g. Mataric, 1990). It requires a set of quite different components than the ones discussed earlier on for the hierarchical approach. Each component has direct contact with sensors (like sonars) and effectors (stop, go, or turn commands of the robot base). The components implement the following behaviors:

1. *Stroll*: Move forward in a more or less straight line motion when there is no obstacle in front. Stop when there is an obstacle. Back up when the obstacle persists.

2. *Avoid*: Turn right or left when there is an obstacle in front of the robot.

3. *Align*: Turn by a small angle when the distance to the object behind is shorter than the distance in front.

There is no additional control structure to combine these components. They are all simultaneously active which is possible because the conditions triggering them are mutually exclusive. Figure 6. (from Mataric, 1990, p. 53) shows four runs taken by attaching a marker to the base of a mobile robot working along these principles. The experiments show that consistent boundary tracing indeed emerges out of the combination of the 3 behaviors.

This example illustrates particularly well that data flow and control flow go through the environment. The stroll behavior may move the robot forward in such a way that it is no longer aligned which will bring the align behavior into action. There is no data flow nor any control flow between the stroll or align component.

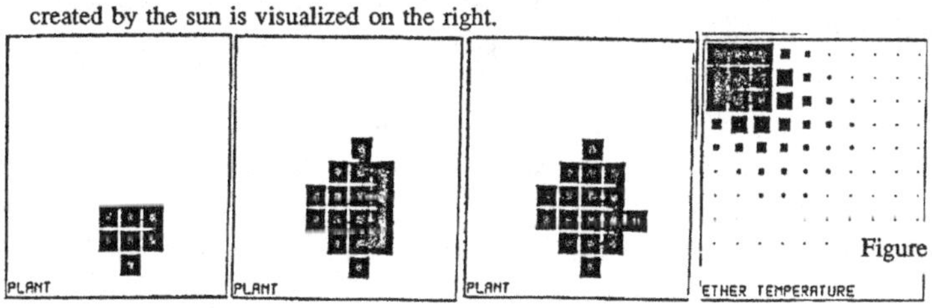

Figure 5. A tree growing towards the sun

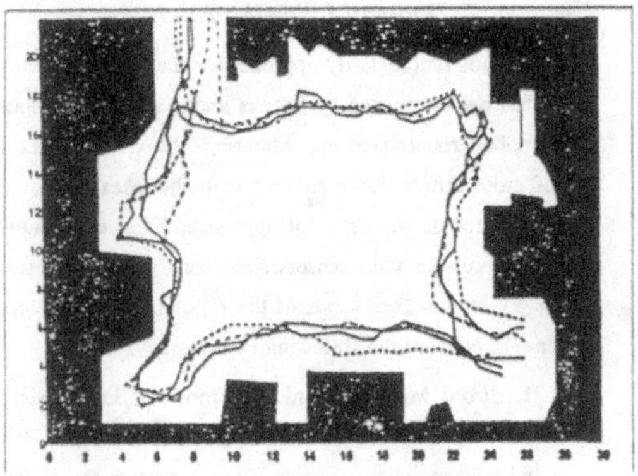

Figure 6. Mobile robot tracing boundaries in real world environment.

Example 3. Collecting rock samples

An emergent functionality solution for collecting rocks was reported in Steels (1990). The solution was inspired by the behavior of ant colonies. Instead of a single robot it is assumed that there is a set of robots which each operate independently. The robots are very simple and know very little. They have no maps of the environment, they do not perform complicated sensing, or they have no capacity to communicate or to plan joint action. There is a region around the vehicle (e.g. created by a physical field) in which the robots are moving. The robots put down crumbs in the area between the source of the food and the nest. This way they establish a zone that attracts other robots which therefore also engage in sample collecting and transport to the main vehicle.

The robots themselves are built out of components that implement simple behaviors. The first set of components determine the direction in which the robot will move.

- Random movement: This component causes random movement of the robot through the environment. It is active when there are no other components active that determine movement.
- Obstacle avoidance: This component becomes active when there is an obstacle. It causes the robot to stop and turn away from the obstacle.
- Path attraction: This component becomes active when crumbs are detected. The robot moves in the direction of the crumbs.

Some additional behaviors regulate sample collection, crumb handling and movement from and to the vehicle:

- Crumb handling: If I carry a sample, I drop 2 crumbs. If I carry no sample and crumbs are detected, I pick up one crumb.
- Sample collection: If I sense a sample, I pick it up.
- Movement: If I am at the vehicle and I carry a sample I drop it and leave the vehicle. If I am at the border of the exploration region I turn around.

Note again the basic features of all these components: direct environmental interaction, data flow through the environment (particularly the crumbs), control through the environment, subfunctions that do not relate directly to the global functionality. Figure 7 contains a few snapshots of a simulation demonstrating that these behaviors together solve the problem. Fig 7 a. is taken when one robot has discovered the samples which are located in the bottom right. Consequently there is a small square in the CARRYING-SAMPLE window at that position. The robots are scattered around randomly in the area to be explored around the vehicle. Figure 7 b. illustrates the formation of a path. A second robot has now discovered the sample and the first one has arrived back at the vehicle (given a first small square in the COLLECTED-SAMPLES window). Note that crumbs have been put down by the first robot and that there is already a grouping of the robots because of attraction to the crumbs path. In figure 7 c. we see several of these robots coming back with samples to the vehicle and the set of collected samples has already increased. Notice however that the crumbs path has been broken down considerably because more robots without samples are walking along the path to the samples (and thus breaking down the samples) than there are robots returning with samples and thus reestablishing the path. Figure 7 d. shows that the crumbs path has been reestablished again by returning robots. We also see that robots now maximally cooperate in the task. There is only one robot that is not on the path. In figure 7 e. the samples are exhausted and the path is being broken down again. Robots are already scattered around in search of new samples. In figure 7 f. the system has returned to the original exploratory behavior.

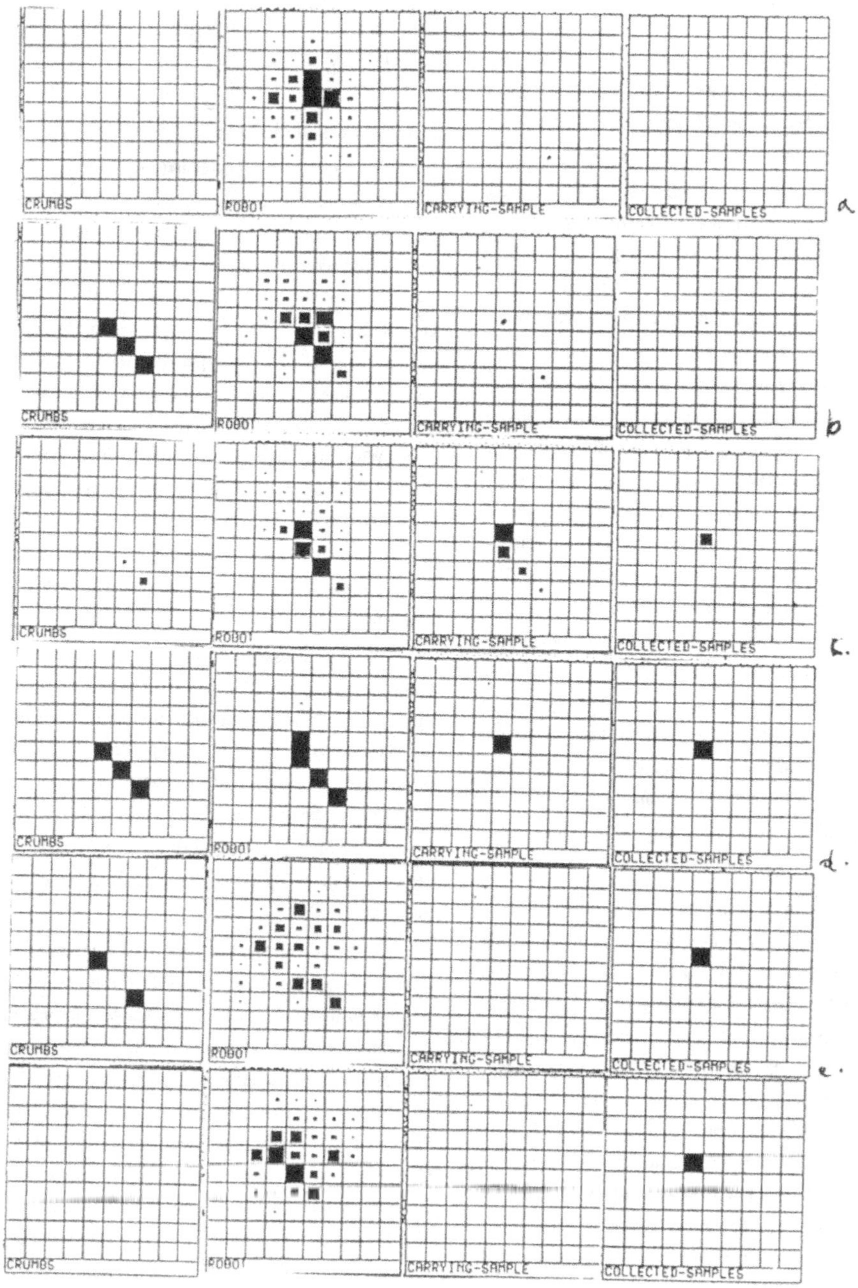

Fig 7

What is emergent in this case is the cooperation between the different robots to establish an efficient path between the source of the samples and the vehicle. This cooperation is achieved by creating a structure (the zone filled with crumbs) that attracts the robots. An important property is that this zone is broken down again when there are no more samples to be collected. Note also that the robots need each other to solve the total problem.

There are obvious differences between these different examples. The components of the wall following robot are all different, whereas the rock collection system consists of identical components. The wall following robot is therefore not fault tolerant, whereas in the other system components can be removed without a total system break down. Another important difference is that the wall following robot directly operates on environmental parameters whereas both the tree growing program and the rock collection system operate indirectly through the creation of an auxiliary structure in the environment (the growth substances, the zone with crumbs). The way this structure is formed is through self-organization in the sense of Prigogine (Prigogine and Stengers, 1986). This topic is highly interesting in itself but cannot be explored further in this paper. Despite these differences all the examples share the properties outlined earlier on: direct interaction of the components with the environment, control through the environment, and a functionality which is only indirectly related to the global functionality.

3.3. Features of systems with emergent functionality.

1. Systems with emergent functionality show a remarkable robustness. For example, the wall following robot can be pushed towards the wall or away from the wall, and it will re-establish its distance. The path formed by robots between the vehicle and the samples can be destroyed but after a while the path will re-establish itself. Parts of the tree can be destroyed and the tree will regenerate itself. The following figures illustrate this. In figure 8 a. we have a partially formed tree. In figure 8 b. some of the left branches of this tree have been removed. Figure 8 c shows that most of the cut off branches have been regenerated while the rest of the tree (to the right and at the top) has continued to grow.

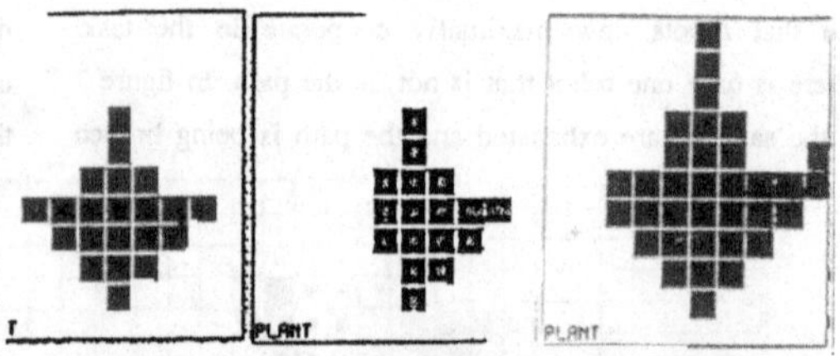

Figure 8. Regeneration after branches have been cut.

2. Systems with emergent functionality are strongly situated, in the sense of being embedded in the environment in which they operate. This is due to the fact that each component has a direct interaction with the environment. Situatedness has a number of important advantages. One of them is a direct and quick reaction to changes in the environment. For example, the wall following robot will immediately respond to changes in the position of the wall, or it will immediately react when pushed out of its path. When the heat source moves from one side to another side, the tree will start growing immediately to the other side. When the collecting robots encounter a new sample, they will take that immediately into account (in fact, with the behaviors described above a dual path will be formed).

3. Situatedness introduces also a new sort of modularity. On the one hand systems with emergent functionality are not modular at all because functionality emerges due to the interaction of the components. This makes the design, construction, and diagnosis of such systems difficult. On the other hand it is a lot easier to add new environmental dependencies as we already illustrated by the ease with which the growing of trees could take heat into account.

4. In hierarchical systems some components are there to supply inputs which are then used by other components to make decisions. The input components classify the situation and decision making components are designed to cover a particular set of situations. This kind of prior analysis is less needed in cases of emergent functionality because each component has its own environmental interaction. Because there is less a need for prior analysis, systems with emergent functionality are less brittle and thus exhibit a more graceful performance degradation.

5. Finally we see fault tolerance in cases where the functionality is due to the collective behavior of many elements, as illustrated in the rock collector. It is possible to remove individual robots and the complete system will continue to function.

These observations indicate that systems with emergent functionality are useful when there is a lot of dependence on the environment and it is difficult to foresee all possible circumstances in advance, when the functionality is not so complex that a hierarchical approach is needed to decompose it hierarchically in many different subfunctions, and when robustness is very important.

4. Formal structure

The discussion so far is useful because it distinguishes systems with emergent functionality from hierarchical systems and because it indicates where each approach is relevant. In practice most natural systems are a combination of hierarchical elements and subsystems that exhibit emergent functionality, so that the advantages of both approaches is combiend. We now go one step further and try to find the formal structure of systems with emergent functionality. The formal structure we propose is at this point more a conjecture than a worked out theory but it may be useful as starting point for further work, both to understand why systems work and to design new systems.

The formal structure of systems with emergent functionality seems to require three properties: (1) systems with emergent functionality are dynamical systems, (2) an emergent property is a particular evolution of a variable which is not controlled and asymptotically stable, and (3) the function describing the evolution of an emergent property can be decomposed into at least two components. Each of these points is now further discussed. Due to space limitations the full mathematical details are left out and only the general nature of the structure is sketched.

1. Systems with emergent functionality are dynamical systems.

Dynamical systems are systems that can be described in terms of a number of variables that continuously change in time. Variables for the wall following robot are the position and the distance to the wall; for the tree the regular spacing of the tree and the size of the branches; for the rock collector the line between the samples and the vehicle, the number of robots carrying a sample, etc. The variables are of three sorts. (1) There are those that are effectively controlled by the system (e.g. the wall following robot is able to control the position by using its motors). (2) There are those that may be used directly or indirectly by the system to make a decision (e.g. output of a sensor that measures the distance from the wall). (3) There are those that are only relevant to an observer (e.g. the number of samples already collected by the vehicle, or the number of branches of the tree). Some of the variables may refer to internal states of a component, others refer to properties of the environment.

Given a set of variables X, the system can be described as a set of equations of the form

$$\frac{dX_i}{dt} = f_i(X_1, \cdots, X_n, t)$$

The function f_i is a complex function which depends on some or all of the variables of the system. In particular, it is expected in this case to depend directly on properties of the environment. Consider the simplified example of a wall following robot in the form of a point which has two behaviors: An attraction behavior that attracts the robot to the wall and a repulsion behavior that causes the robot to move away from the wall. There are three variables: the position of the robot p, the distance to the wall d, and the position of the nearest wall w. Only the first one is controlled. The behavior of this variable can be described by an equation such as the following one:

$$\frac{dp(t)}{dt} = f(a(d(t)), r(d(t)), p(t))$$

a (attraction) and r (repulsion) are functions of distance which map into changes to the position of the robot. f is a function which integrates these changes with the current position. The behavior of the other variables depends on the position of the robot and the structure of the environment. The functions defining the behavior of the controlled parameters can be viewed as the specifications of the system because those are the functions that need to be implemented by the hardware.

2. An emergent property is not controlled and asymptotically stable.

We define an emergent property to be a desired evolution of a variable of the system. This variable has the following properties: (i) it is not controlled by the system, (ii) its evolution is asymptotically stable, (iii) the property is a relevant part of the functionality of the system. For example, keeping an equal distance from the wall is an emergent property in the wall following robot.

Asymptotically stable means that the variable should evolve towards a certain limit and stay there, even if there are perturbations moving it temporarily away from this limit (figure 9). For example, we expect that the distance evolves towards a particular limit and stays equal, and that it moves back to this limit after perturbations (e.g. a very small distance is created when the robot is pushed against the wall).

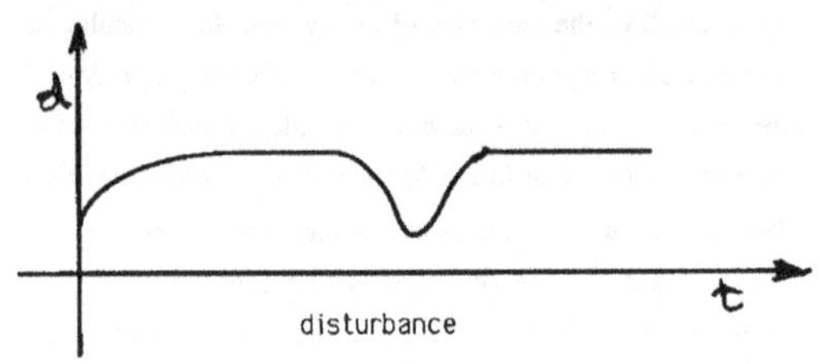

Figure 9. Asymptotic stability of an emergent property.

In the language of complex systems (Nicolis and Prigogine, 1989), we would say that the phase trajectory of the variable converges to a point attractor. Asymptotic stability is directly responsible for the robustness of systems with emergent functionality.

3. The functions for the controlled variables are decomposable.

We now try to restrict further the class of dynamical systems which exhibit emergent functionality by focusing on the functions f_i that define the evolution of the controlled variables. We conjecture that those functions which describe the evolution of variables that indirectly determine the emergent properties (such as the position of the robot which directly determines distance to the wall) are necessarily decomposable into at least 2 components. Let us restrict ourselves to the case of two and call them a positive and a negative component. The positive component pulls the evolution of the variable in one direction (for example away from 0). The negative components pulls it in the other direction (for example towards 0). For the wall following robot attraction acts as the positive component and repulsion as the negative one. The net effect of combining the two functions is that the variable reaches its limit and stays there.

The component functions (and therefore the behaviors that implement these functions) need an additional property which is responsible for positive (or negative) feedback. The closer the robot is to the wall, the greater the influence of the positive component causing repulsion from the wall, whereas the farther away the robot is from the wall, the greater the influence of the negative component.

The positive and negative components can be found in all the examples discussed in this paper. In the case of the tree, the dynamics centers around the evolution of the growth substance. There is a positive component influencing increase (e.g. due to food or heat) and there is a negative component inhibiting growth substance in case there is already a branch or if there is a branch on the other side. In the case of the rock collector, the dynamics centers around the establishment of the zone with crumbs that delineates the area in which the robots should move about. In the crumb handling behavior there is a positive component: If I carry a sample, I drop 2 crumbs, and a negative component: If I carry no samples and crumbs are detected, I pick up one crumb. (The exact number of crumbs does not matter as long as the first amount is larger than the second). The positive component builds up the crumb zone. The negative component breaks it down again. The positive feedback is present due to the interactions of the behaviors: when there are more crumbs, more robots are attracted to the zone, and when they carry samples they will put down more crumbs which will attract even more robots, etc.

From these hypotheses follow a set of concrete guidelines for identifying and designing systems with emergent functionality:

(1) The major variables of the system need to be identified. Some of these will be controlled directly, others will be used in decisions, still others are only there from the viewpoint of the observer.

(2) The dynamic dependencies of the variables need to be defined. The equations defining the behavior of the controlled variables constitute the specifications of the system.

(3) The desired emergent properties show up as non-controlled variables which are asymptotically stable.

(4) The dynamics needs to be such that the controlled variables which directly or indirectly determine the emergent properties are decomposed into at least 2 components which force the evolution of the variable towards its limit value.

5. Conclusions

This paper is a first step towards the development of a theory of emergent functionality. The theory attempts to distinguish emergent functionality from hierarchical functionality and to understand the conditions under which emergent functionality is appropriate. It also conjectures a basic structure that systems with emergent functionality need to have.

6. Acknowledgement.

Several members of the VUB AI laboratory, particularly the "complex dynamics group" have contributed to the ideas of this paper. Jo De Cuypere and Francis Van Aeken have commented on earlier versions.

7. References

Babloyantz, A. (1986) *Molecules, Dynamics, and Life: An Introduction to the Self-Organization of matter.* Wiley Interscience, New York.

Brooks, R. (1989) A robot that walks; emergent behavior from a carefully evolved network. In *Neural Computation,* 1:2.

Lindenmayer, A. (1968) Mathematical Models for Cellular Interaction in Development, parts I and II. In: *Journal of Theoretical Biology* 18, 280-315.

Mataric, M.J. (1990) A distributed model for mobile robot environment-learning and navigation. MIT TR 1228. MIT AI Lab, Cambridge Ma.

Meinhardt, H. (1982) *Models of Biological Pattern Formation.* Academic Press, New York.

Nicolis, G. and I. Prigogine (1989) *Exploring Complexity.* Freeman and Cy, New York.

Prigogine I. and I. Stengers (1987) *Order out of chaos.* Basic Books, New York.

Steels, L. (1990) Cooperation between distributed agents using self-organizat: n. In: Y. Demazeau and J-P. Muller (1990) *Decentralized AI.* North-Holland Pub, Amsterdam.

Steels, L. (1990b) Growing trees without a grammar. Extended abstracts of the 2nd Artificial Life Conference. Santa Fe Institute, New Mexico.

Eight Principles for Building an Intelligent Robot*

David L. Waltz

Thinking Machines Corporation
245 First Street
Cambridge, MA 02142
USA

Brandeis University
Computer Science Department
& Center for Complex Systems
Waltham, MA 02254 USA

Abstract

We cannot expect to know the detailed "wiring diagram" of the nervous system for any intelligent creature for quite a long time. Even then, the true organization is likely to be incredibly complex and tangled. However, in order to build intelligent robots now, we need a plausible interim architecture. A functional model for robot organization is proposed, starting with a basic, first order model, which is gradually refined. In particular, it is proposed that associative memory provides a useful -- and perhaps plausible -- basis for an intelligent system.

0. Introduction

While remarkable progress is being made by neuroscientists in unraveling portions of the nervous system (see, for example, [Kosslyn 89] or [Halgren 87] for insights into the visual system and memory systems, respectively), we are still far from being able to map the wellsprings of action, intention, and decisions. Other researchers have investigated abstract models of adaptation and learning, such as genetic algorithms and classifier systems [Holland 77], or the SOAR system [Newell 87]; abstract models have been used to build explicit models of creatures (e.g. the Animat [Wilson 87]). [Drescher 89] has introduced the "schema mechanism," and his ideas have much in common with the proposals below, especially in his views on chaining, and in his key ideas on identifying and learning reliable schemas, using large amounts of statistical analysis. "Subsumption architecture" researchers in AI (e.g. [Brooks 86], [Maes 90]) hope to arrive at intelligent systems by first building a (layered) system with the abilities of, say, a cockroach, and adding yet more control layers to eventually reach greater and greater intelligence. This work is broadly within a "Society of Mind"-type theory that views intelligence as composed of a very large number of independent agents and "bureaucracies" of agents, each responsive to specific situations or patterns [Minsky 87]. While I subscribe in general to the Society of Mind view, I believe that it is both possible and valuable to divide up the model of mind somewhat differently than is done within subsumption architectures.

I propose here a model of a robot's "mind" whose components are divided up along very different lines, somewhat analogous to principal components analysis: the first component is a general associative memory model that captures general patterns and principles of behavior, while successive components add refinements, culminating in society of mind-like demons that recognize very specific situations or patterns, and then override (by priming or inhibiting) more general behaviors. Intermediate refinements include control structures that allow search and chaining of actions, as well as rote learning and generalization. Such a model fits neatly on any massively parallel computer architecture (e.g. [Hillis 85]), but can also be simulated on serial computers (though perhaps not fast enough to allow real-time performance, except in the simplest of environments).

1. Principle One

Use associative memory as the overall organizing conception.

Basic associative memory operations can capture the essence of what intelligent entities do: select relevant precedents in any situation, and act on them. "Precedents" can be actions, options, remindings, etc. This type of operation, akin to case-based reasoning (CBR) [DARPA 88, 89] and memory-based reasoning (MBR) [Stanfill & Waltz 86] is easily programmed on a massively parallel machine, and has found useful applications [Waltz 90]. A number of techniques can be used to find "relevant" items, including nearest-neighbor algorithms, and majority votes of n nearest neighbors.[1]

If only a single precedent is close to the current situation (as when the robot is operating in a familiar environment on a familiar task), then little more than an associative memory is needed in order to act intelligently. Only when two or

---

*This work was supported in part by the Defense Advanced Research Projects Agency, administered by the U.S. Air Force Office of Scientific Research under contract #F49620-88-C-0058.

[1] What makes a neighbor "near" is a very subtle issue, and the key open problem in CBR and MBR.

more incompatible precedents of roughly equal plausibility are found, or when nothing matches the current situation at all closely, or when all known precedents have negative outcomes, does such a system need to reason (in the ordinary AI sense), as discussed further below.

Thus, the basic action of the robot can be represented as the (behaviorist) schema:

situation --> action

Obviously, what constitutes an "action" or "situation" is also important. I am going to partially duck these issues here. The notion of situation is enlarged upon somewhat below; the notion of action depends upon the particular granularity or chunking of motor sequences built into the system, along with their parameterizations. See also [Agre & Chapman 87] and [Drescher 89].

2. Principle Two

Include innate drive and evaluation systems to provide the robot with moment-to-moment guidance for its actions.

The notion of "current situation" can be enlarged to include not only interpreted sensory inputs, but also "desires" of the robot (potentially time-varying, innate constraints, that could not be learned easily through experience; for example, a particular class of robots should "desire" to go plug themselves into recharging locations with progressively greater intensity as its battery power gets lower, and this desire should overwhelm all others as the power becomes very low). Other examples of this principle include innate wiring of a desire to avoid collisions, or a desire to minimize energy expenditure ("laziness"). Actions that are "desirable" should cause the robot to assign them a positive evaluation when they are carried out. The robot should also store the results of taking the action as part of the schema.

Revised schema:

situation + desire --> action + evaluation
evaluation =: value + results

3. Principle Three

Populate the associative memory system with sequenced rote experiences.

Actions taken, along with their evaluations should become part of the system's repertoire of precedents, to be available for guiding future actions. Storage in memory should be more likely if desires/goals are either met or thwarted, and should be less likely for actions with little or no evaluative content. Storage should also be more likely if expectations from precedents turn out to be violated. Following [Drescher 89], statistical analysis can play an important role in identifying actions that have reliable consequences from the noisy data of actual robot experience. Action sequences should be linked bidirectionally, i.e. with each schema pointing to both the next schema memory, and the previous schema memory. This linking allows control structures (see below) to chain forward from memory items that match current situations, and backward from goals (situations that match desires). Paths that include entities or objects present in the current situation should also be preferred, a kind of "middle-out" chaining.

4. Principle Four

Include mechanisms to automatically generalize across rote memories.

There are several goals here. The main ones are to combine memories of episodes that are frequently repeated, to remove irrelevant preconditions and postconditions (i.e. ones that occur unreliably), and to group actions that lead to the same outcomes (eventually paving the way for "backward chaining"). Memories that have taken part in generalizations can be garbage-collected (possibly probabilistically). Generalized memories, which consist simply of items common to all original memories, can still be matched directly to situations.

5. Principle Five

Include control structures to allow planning.

Two extensions are necessary: 1) associative retrieval of memories must be split apart from the actual performance of their actions; and 2) the retrieved situations and results of memories must be able to be used by the robot as "hypothetical situations" to trigger further retrievals, as though they were actual situations. We can assume that there is an "intention" unit, and that unless this unit is on (activated), retrieved actions will not be actually taken. The intention unit would be on whenever an emergency is detected, when in familiar situations where one precedent dominates all others, etc. Provisions also need to be made to keep track of the search, and to back up and compare values for various possible paths.

Some revised schemas:

situation' + desire(s) --> action' + evaluation'
situation' =: situation | hypothetical-situation
action' + intention --> action

Note that the associative memory can be used as a source of actions to be tried, in order of priority. This in turn can help avoid combinatorial search spaces, since branching factors can be kept much smaller than the total number of schemas.

6. Principle Six

Use specific recognizers ("demons") as censors or primers of memories and actions.

This basically allows generalizations to be very general, while still allowing the robot to cope with exceptions. Specific demons can override general memories, or can "prime" them to make their use more likely. Such demons can be formed whenever expectations are violated, for better or for worse. Here again, the work of Drescher is relevant, in particular in identifying very specific reliable schemas.

7. Principle Seven

Include control structures to encourage only controlled experiments.

If no actions with unknown consequences have been tried recently, make it more likely to try something risky; however, once one or more actions with unknown consequences have been tried, strongly prefer known, reliable schemas until a reward, punishment, or static situation results. This then allows credit to be assigned fairly, since the step/action with the unknown consequences can be assumed to be the likely cause of any surprising outcomes.

8. Principle Eight

In emergencies, or if one schema clearly dominates all others, take the best schema without further evaluation; otherwise, use search to find the best actions.

If there is no danger, and no rush, evaluate schemas by serial exploration of hypothetical actions and situations, and act on the results of this search.

9. Summary

I have argued that associative memory is a useful first-order model of intelligent behavior; in many -- perhaps most -- circumstances, such a system is able to produce appropriate actions. By adding just a few modifications and auxiliary structures one can then account for a wide range of behaviors. Whether or not these ideas have any explanatory power for real intelligence is not clear. However, the proposed model does seem to account concisely for the main phenomena of intelligent action and action selection, a problem generally ignored in AI. (It has been possible to ignore these issues in AI only by sticking with toy problems and microworlds, where the numbers of possible actions or operators that could apply at any given time is small.) The proposed model fits neatly onto the massively parallel machines that seem destined to dominate the high-end supercomputer field for the foreseeable future, and so offers one potential path that can lead to truly intelligent machines.

References

Agre, P. & D. Chapman "Pengi: an implementation of a theory of activity," *Proceedings of the 6th National Conference on Artificial Intelligence,* Los Altos, CA: Morgan Kaufmann, 1987.

Brooks, R. "A robust layered control system for a mobile robot," *IEEE Journal of Robotics and Automation, RA-2,* 1, 1986.

DARPA *Proceedings of the Case-Based Reasoning Workshops,* 1988 (Clearwater Beach, FL) and 1989 (Pensacola Beach, FL).

Drescher, G. "Made-up Minds: a Constructivist Approach to Artificial Intelligence," unpublished Ph.D. dissertation, MIT AI Lab, Cambridge, MA, September 1989.

Halgren, E. "Human hippocampal and amygdala recordings and stimulation: evidence for a neural model of recent memory," In N. Butters & L. Squires (eds.) *The Neurophysiology of Memory.* New York: Guilford, 1984, 165-181.

Hillis, W. D. *The Connection Machine,* Cambridge, MA: MIT Press, 1985.

Holland, J. "A cognitive system with powers of generalization and adaptation," Tech, Report, Dept. of Computer and Communication Sciences, University of Michigan, Ann Arbor, 1977.

Kosslyn, S., R. Flynn, J. Amsterdam, G. Wang, "Components of high-level vision: a cognitive neuroscience analysis and accounts of neurological syndromes," *Cognition 34,* 203-277, 1990.

Maes, P. "How to do the right thing," *Connection Science,* Spring 1990, to appear.

Newell, A. "Unified theories of cognition," William James Lectures, Harvard University, 1987.

Minsky, M. L. *The Society of Mind,* New York: Simon & Schuster, 1987.

Stanfill, C. & D. Waltz, "Toward memory-based reasoning," *Communications of the ACM 29,* 12, 1213-1228, December 1986.

Waltz, D. L. "Massively parallel AI," to appear in *Proceedings of AAAI-90,* Boston, August 1990.

Wilson, S. W. "Classifier systems and the Animat problem," *Machine Learning 2,* 199-228, 1987.

Robot Ethology: A Proposal for the Research into Intelligent Autonomous Systems

Uwe Schnepf
Expert Systems Research Group
German National Research Center for Computer Science (GMD)
P.O.Box 1240
5205 Sankt Augustin 1, FRG
usc@gmdzi.uucp

Abstract

In this paper it is argued that studying the behaviour of (even simple) animals and human beings from an Ethologist's point of view will provide a basis for understanding human cognition. A simple model of the organisation of behavioural sequences in animals as described by Niko Tinbergen [1] and Konrad Lorenz [2] is presented as an starting point to develop intelligent autonomous systems. The relation to current research into behaviour-based Robotics is shown (cf. Brooks [3]) and essential extensions to the behavioural model such as optimisation procedures based on genetic algorithms and evolution technology, a framework to link basic sensory-motor skills to higher-order categorical perception as described by the symbol grounding problem and more advanced models of the organisation of behaviour are presented. In particular, self-organisational processes are advocated as a key feature in achieving intelligent behaviour of autonomous robots.

1 Introduction

Initially, Robotics has been the most attractive field for researchers in Artificial Intelligence (AI) in order to study the whole range of cognitive capabilities of human beings. Defined as the intelligent connection between sensing and acting, Robotics was supposed to naturally pose the questions one has to answer in order to understand human intelligence.

Researchers focussed on various aspects of the sensing-to-acting chain using a knowledge-based approach to specify the information necessary for the robot to perform various tasks. In general, this information consists of a world model and the knowledge about possible transitions from one world state into another due to the actions of such an intelligent agent.

Since a robot is usually located in an unstructured environment, world modelling requires us to impose structure on this environment using an explicit description of the objects surrounding the robot. The problem of which aspect of the environment should be modelled and which aspects of the robot-world-interactions should remain constant or not became one of the most important questions in this domain, well-known as the Frame problem [4].

Another problem which arises when using a world model is the precision of the robot's internal representation of the surrounding world. Since the control algorithms of today's robots are based on the transformation of three-dimensional coordinates in order to perform the required actions, the internal representation of the world which serves to interpret sensor data and to choose the appropriate actions (the representation of which in turn has to be translated into sequences of three-dimensional trajectories of the robot's actuators) must be precise in order to allow an exact mapping from the external world to the internal representation and vice versa.

The data received from the sensors is used to update the internal world model. But as it is difficult to gain precise three-dimensional description from currently available sensors, it was claimed that the current sensor technology was not sufficient to deliver precise updates of the model.

Because of this lack of information, it is argued, robots are not able to perform appropriate and flexible actions. Once more precise sensors have been developed, the problem of incorporating more precise world knowledge into the internal representation could be solved more easily.

We do not agree with this opinion and are convinced that the inflexibility and subtlety of current robot controllers are inherently based on the approach chosen. Making a decision in favour of such a knowledge-based approach to world modelling naturally leads to the problems discovered and cannot be solved by using high-tech equipment.

Another aspect of today's approaches to intelligent Robotics is that because most people are dealing with world modelling, planning, control algorithms, plan revision etc., only little knowledge has been gained during all the years of intensive research into these fields about the principles of the organisation of goal-directed behaviour (by behaviour we refer to "spatially and temporally coordinated patterns of movements", cf. Ewert [5]).

As work done in the development of concepts and cognitive processes in children shows, there is a strong relation between the development of sensory-motor processes and the reasoning capabilities of children (cf. Piaget [6]). Actually, the experience of a body and its interaction with the environment via sensors and actuators seems to provide the basis for the development of categories and higher-order cognitive processes such as decision making, problem solving, language etc.

So, without grounding the concepts human beings use for reasoning purposes via the sensory-motor capabilities acquired earlier, cognitive processes are not possible (apart from the other prerequisites as far as the brain is concerned). Hence, using reasoning processes in a robot in order to coordinate the robot's motions is an endless chain, since these reasoning processes themselves require the robot motions and sensing capabilities to be grounded.

One has to break this chain by introducing representations and control strategies which are not knowledge-based and which do not require the notion of a homunculus which is in charge of controlling the agent in order to achieve rational behaviour.

In this paper we sketch a framework for research into the relation between basic sensory-motor skills and cognition. Although there is only a little knowledge available at the moment about what constitutes rational and goal-directed behaviour in an autonomous agent, the difficulties encountered with the classical knowledge-based top-down approach to intelligent systems suggest the use of a different approach in order to explain basic cognitive skills.

During the search for initial useful steps in this direction, we discovered the rich source of explanation models in the study of animal behaviour developed over a long period of time in a particular subfield of applied Zoology: Ethology. Since we found out that Ethologists (even the very early one's such as Tinbergen) and Robotics researchers show many similarities in dealing with the problem to explain observed behaviour, we would like to give a short introduction to the field and hope to be able to prove the necessity to focus on ethological aspects in the explanation of intelligent behaviour.

2 Ethology - an Overview

So, why should it be useful to focus on the study of animal and human behaviour in order to build intelligent autonomoums systems? Following the long trail of evolution, intelligence is based on the biological development from very simple organisms to the human race. Due to these incremental steps from one species to more intelligent one's, there is still a lot (and much more than many people tend to believe) of animal behaviour present in the everyday behaviour of human beings. But not only the superficially observable behavioural patterns, but also the neural mechanisms responsible for the creation and performance of behavioural sequences are very much the same due to the shared evolutionary development of low-level and high-level beings. Hence, the Darwinism of behaviour is worth looking at since the behaviour of even simple animals contributes to the more complex behaviour of man. Because we strongly believe that the ability to act and react flexibly in the real world determines the ability to higher-cognitive skills, focussing on biological models to explain behaviour is not only worthwhile, it is necessary.

Using Ethology as a source for roboticist inspiration has several advantages:

- A lot of work has been done during the last 80 years in Ethology and related fields in order to explain the organisation of animal and human behaviour as well as to explain learning processes. This knowledge provides a rich source of explanation models of behaviour and should be used to guide future research in the field of robot beings.
- There exists a huge amount of experimental data derived from the observation of behaviour of all kinds of animals. So the description of sequences of animal behaviour can serve as an objective scale when it comes to evaluating the validity of different approaches chosen to actually instantiate animal-level behavioural performance.
- Relating the performance of autonomous agents to the performance of animals which are supposed to have only little or no consciousness about their activities keeps the developer of such an artificial system from taking the "intentional stance" as Dennett [7] suggests.

The last point seems to be especially important and worth discussing separately. In trying to explain animal behaviour without assuming the presence of cognitive capabilities, it is easy to avoid the stages where one starts to talk about goals and goal-directedness in behaviour from a psychological point of view. We believe that to a great extend such concepts exist only in the imagination of the observer of such a system (cf. Braitenberg [8]). By assuming the presence of a rational agent or controller which directs the behaviour of an artificial system (instantiated e.g. by the creation and formation of reflexes or by some cognitive capabilities), one simply neglects the fact that simple animals are incapable of instantiating such a controlling agent. In their case, one cannot distinguish between control and behaviour, as control is performed by the execution of behavioural sequences and behaviour is activated by control (cf. Marken [9]). Together, behaviour and control form an undividable unit based on the reinforcement and inhibition activities of self-organising processes. This is why we believe that one has to explain goal-directedness of behaviour by focussing on these processes, their creation and their coordination, at a level of unconsciousness without assuming reasoning processes.

2.1 The Early Years

Ethology, or the comparing study of behaviour, was initialized as a discipline in its own rights by the work of Charles Otis Whitman and Oskar Heinroth who discovered that related animal species show not only similar body features but also similar behavioural patterns. The main questions Whitman and Heinroth were concerned with were: How was this innate behaviour passed on to the descendants of a species? Is all behaviour inherited? And is there some learning of behaviour involved?

The clear answer to the first question was that only the genes can be responsible for the inheritance of behavioural sequences in one species. The appropriate reactions of a perceived situation was supposed to be embodied by so-called reflexes, and the chaining of those reflexes was considered to constitute goal-directed purposive behaviour. The nature and origin of these reflexes was not of any interest since they were supposed to be granted to each species.

The other two questions were much more difficult to answer and led to a long lasting debate between purposive psychologists, who claimed that all behaviour is innate and based on the chaining of reflexes, and behaviourists, who claimed that there is no innate behaviour at all and everything has been learned.

Konrad Lorenz, one of the grandfathers of Ethology, started to investigate animal behaviour from a different point of view allowing the observing experience from purposive psychologists and the analytical force of behaviourism to emerge into a new explanation of the sources and mechanisms of behaviour. His early studies were also based on the function of reflexes, but he tried to explain their physiological mechanisms.

Lorenz' main interest was to study how such a reflex and a chain of reflexes could be created and activated. At this time, a reflex was considered to be a simple linear relation between the perception of a stimulus and the following reaction of the organism receiving this stimulus. Nevertheless, Lorenz found this explanation as misleading as simple: up to then it was believed that the predetermined functionality of the reflexes constitutes the goal-directedness of the animal behaviour. But how has this determinism of reflexes developed? Could it not be the case that this goal-directedness was much more an artifact introduced by humans who wanted to explain the behaviour observed? Was it not possible to investigate, influenced by the ideas of the Darwinism of species, the origin of reflexes and behaviour from this point of view?

During his observations of animals, Lorenz found out that the behaviour exhibited by an animal is well goal-directed, but not purely reflexive. The goal of the behaviour does not consist in reaching a particular aim or state. It purely consists in releasing the innate behavioural patterns. This goal-directedness becomes observable if an animal is not able to perform a particular behavioural pattern for some time because of the absence of the activating external stimulus. In this case, the animal becomes very nervous and unquiet, walks around, shows weak signs of releasing exactly the pattern which has not been performed for some time. The unrest behaviour of the animal leads into the search for appropriate sensor stimuli.

After some time, the appropriate sensor sensation is not necessary any more, since already vague reminiscence in shape or colour in the sensor information is sufficient to release the behaviour concerned. This steadily increasing unrest behaviour was termed as consummatory action. The final release of the behavioural pattern without the usually necessary sensor stimuli is called vacuum activity.

Since these two important observations, consummatory and vacuum action, could not be explained within the reflex theory, Lorenz concluded that instinct motions are not at all related to the simple chaining of reflexes, but are initialized by processes in the central nervous system which are triggered mainly independently of external stimuli and which repeat themselves continuously. Hence, the physiological basis of these behavioural patterns consists of the internal production and central coordination of stimuli.

It is assumed that the activity of the behavioural patterns is permanently inhibited by some higher instance which is able to release or suppress the various behaviours. But because of the different basic mechanisms involved in the generation of releasing stimuli and motor activities, Lorenz suggested distinguishing these innate behavioural patterns into the processes that are responsible for the selective response to particular stimuli sensations and the behavioural modules which are finally performed by the organism.

To this end, the terms *Innate Releasing Mechanism* and *Instinct Motion* have been coined. Both components constitute a fairly complex model of animal behaviour which serves for the interpretation of observed behavioural sequences in animals, but does not cover an important aspect of animals: the ability to learn.

Only some time after introducing this model it was considered that learning processes are based on a phylogenetically developed program to acquire useful behaviour that serves the survival of the corresponding being.

But how is it achieved that an organism learns the right thing during the modification of its behaviour or the adaptive improvement of its behavioural repertoire? There must be a mechanism which is responsible for measuring the success of the learning procedure and which relates the success or failure of a sequence of behavioural patterns to the initial links of this chain of activity via reinforcement and extinction of behaviour. This information must somehow get into the organism and is done the following two different ways:

- during phylogenetical processes through mutation and evolutionary combination of genes
- during ontogenetical processes through gaining of information in the individual organism

Hence, innate and acquired behaviour are not mutually exclusive as often claimed by the purposive psychology or the behaviourists, but differs in the origin of the information which describes the environment. Nevertheless, both types of behaviour are actually based on evolutionary or optimising processes which govern the consistency between the biological requirements and the behavioural repertoire of the organism concerned. From this point of view an evolutionary based approach to the understanding of the development of behavioural sequences as far as the enforcement of useful behaviour and the extinction of damaging behaviour is concerned is essential.

3 Behaviour-Based Robots

Rodney Brooks [3] was one of the first researchers in AI and Robotics who decided to leave the main roads of knowledge representation and knowledge-based techniques in order to develop intelligent systems. He claimed that, although the representation and the models of the world and the reasoning processes based on them have been a central issue in AI over the last 20 years, there is no evidence that our mental capabilities are actually based on these processes. Although we can talk about the reasoning processes in our heads this does not mean that these processes are really based on verbal representations and inference techniques.

He argued that, looking at the evolution of intelligence during the billions of years of biological development, it seems that the (in evolutionary terms) recent emergence of cognitive abilities such as problem solving, language, expert knowledge etc. are relatively simple once the substrate of being and reacting in the real world is provided.

Furthermore, he claimed that knowledge representation and abstraction are even dangerous methods, since they generally serve to reduce the complexity of the world to a simple toy world, in which all the relevant aspects are supposed to be included, completely and explicitly. Verbal descriptions of the world are considered as a key to success. The user of an AI system is demanded to provide verbal input to the system thus eliminating the hard problem for the system to perform the abstraction from world perception to world description, hence all the intelligent performance of a biological system.

Brooks' approach is to build Creatures which are seen by the human user as intelligent beings in their own right. He is not interested in showing how human beings work, although he considered them, together with animals, as interesting objects as far as their autonomous status is concerned. These Creatures should react robustly to changes in their dynamic environment and should be able to cope with these changes. They should have a purpose for being and maintain various goals as intentionally described by their purpose and the goal-directedness of their behaviour. How did he want to achieve this flexibility and stability?

The classical approach in AI would have been to decompose the system to be developed into various subparts according to their functionality. Hence, there would be a system for sensing, a system for situation interpretation, a system for choosing the appropriate actions according to the specified situations and a system to execute the actions chosen (cf. Fig.1). Each of these subsystems would have been very complex, too. Based on various knowledge-based techniques such as knowledge representation, model-based vision, planning, qualitative reasoning, truth maintenance etc., each subcomponent itself would have been built of various complex components.

We have seen systems emerging like this making intensive use of powerful state-of-the-art computer hardware: one computer to control the vision system, one computer to control the robot arm and one to coordinate the communication between them (cf. Anderson [10]). This computational effort increases steadily with the complexity and number of control tasks involved (e.g. guiding the robot through the environment, identifying various objects, following multiple goals etc.). But is this really what natural cognition and surviving in the world are based on? We doubt it.

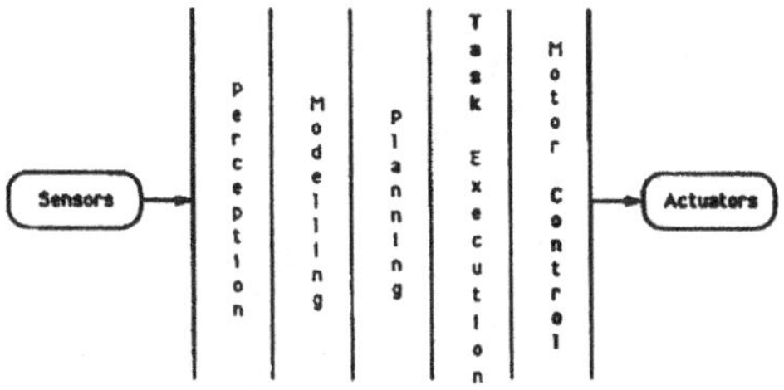

Figure 1: Vertical decomposition into functional modules

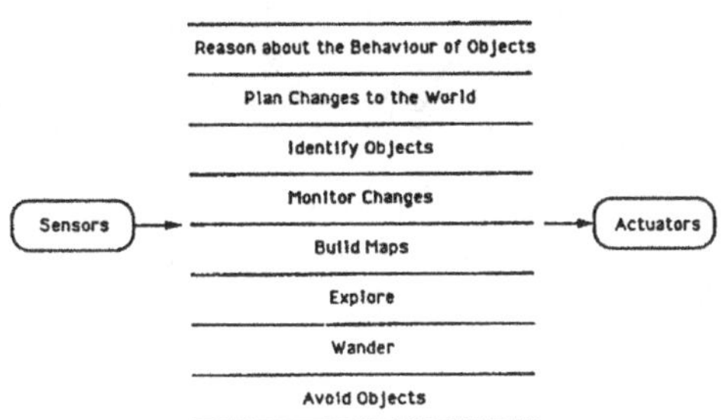

Figure 2: Horizontal decomposition into competence modules

Brooks suggested, instead of building systems which intend to cover the complete range of human reasoning activities, one should follow an incremental path from very simple, but complete systems to complex autonomous intelligent systems. The incrementality of this approach is constituted by a horizontal decomposition of tasks according to various "activities" in an autonomous agent rather than a vertical decomposition based on functional entities (cf. Fig.2). Each activity producing subsystem, or behaviour, is a complete system in itself featuring sensing, reasoning and acting capabilities. Once such a behaviour has been implemented and tested, it remains unchanged and covers a particular aspect of the agent's overall behaviour.

By introducing several layers to enhance the agent's behavioural repertoire, it becomes necessary to define the appropriate organisation and ordering of the layers. Since all layers are running in parallel, they have permanent access to the agent's sensing facilities and try

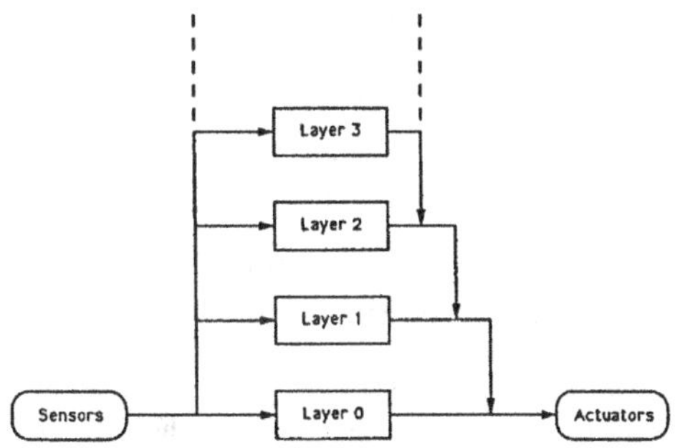

Figure 3: The subsumption architecture

to become active as soon as they receive the appropriate state description. This somewhat chaotic behaviour, where multiple activities are competing in gaining control over the agent, has to be organised by means of suppression of less important behaviours in order to achieve a reasonable global behaviour of the agent.

To this end, Brooks opted in favour of a hierarchically fixed organisation of the behaviours building each new layer on top of already existing ones. Additionally, each layer is equipped with the ability to monitor and to influence the behaviour of the next layer below it (cf. Fig.3). Hence, a community of relatively independent, parallel behaviours is created in which each behaviour constitutes a subtask of the agent's cognitive and non-cognitive skills.

The interaction of the individual competing layers determines the complexity of the overall behaviour of the autonomous system perceived by the observer. The complete control system of the agent consists of various layers and is referred to as the subsumption architecture (the names derives from the fact that the individual layers are combined by suppressing and inhibiting mechanisms). The control flow in the agent emerges from the changing activities of the distributed control flow resident in the various layers, but there is no central control involved.

Although Brooks's approach is mainly based on engineering considerations in the design of his Creatures without focussing on more theoretical aspects of system behaviour, one can say that the robots developed at his lab are much more efficient, reliable, robust, and are able to react more flexibly to the dynamic changes in their environment than the robots developed elsewhere. Nevertheless, the subsumption architecture which represents an implementation of fixed behavioural patterns having a fixed relation to each other lacks some flexibility as it is present in the behavioural organisation of animals.

Putting the notion of behaviour-based robots on more firmer scientific ground using explanation models of behaviour deriving from Ethology will lead into a new class of Creatures which will be more flexible and "intelligent" than their current predecessors.

4 A Simple Model of Behavioural Organisation

In this section we would like to present a model of the organisation of behavioural sequences which is able to explain the characteristics of animal behaviour as described in the section about Ethology. The model itself has been introduced by Niko Tinbergen [1] and Konrad Lorenz [2] to describe the interaction and mutual influence of instinct motions or so-called fixed motor patterns at an abstract level.

As Lorenz describes, there is evidence that in organisms which have multiple systems to perform behavioural sequences, there is a central mechanism which guarantees that not all possible behaviours are released at the same time. This mechanism can be described as a simple backward-directed inhibition activity between the various releasing mechanisms of each fixed motor pattern. Not only in simple vertebrates, but also in human beings is such a mechanism at work in order to guarantee that in accordance with the perceived excitation configuration only the appropriate one of the permanently inhibited fixed motor patterns is released.

The model consists of various layers which are not hierarchically ordered. Each layer is responsible for performing a particular behavioural sequence of the autonomous agent. Conceptually, a layer is built of two main components: one to compute the excitation level of this layer depending on sensor information and internal "motivational" state (innate releasing mechanism, IRM), the other to perform the robot action (fixed motor pattern). The motivation and the sensor information contribute to the IRM's excitation level. The actual activity is blocked as long as the excitation level of each IRM does not reach the IRM's threshold level (hence the IRM serves as a block preventing the motor pattern from becoming active). Once the threshold level has been reached, the motor pattern competes against other behaviours in order to become active. This competition is based on the backward-directed lateral inhibition where each behaviour inhibits other existing behaviours. The strength of the inhibition depends on the threshold and excitation level of the behaviour's IRM. The behaviour having the most excited IRM will actually gain control over the actuators and perform the appropriate motor action. The overall model is shown in Fig. 4.

Technically, the following functional modules must be

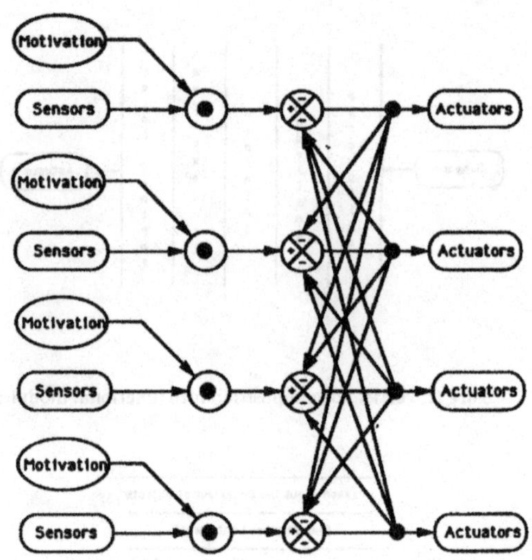

Figure 4: The organisation of behaviours based on lateral backward-directed inhibition

implemented in each layer to achieve this kind of system behaviour:

- The excitation configuration deriving from the sensor input.
- The representation of the agent's internal state such as the role of consummatory actions, stimuli thresholds etc.
- A system summarising both endogen and exogen influences on the excitation level. This system also serves to distribute the local excitation level as inhibitory signals to all other behaviours which themselves remain unknown to the sending behaviour.
- A releasing mechanism which allows the motor pattern having the highest excitational status to pass through and to gain control over the robot.
- The fixed motor pattern responsible for performing the appropriate action.

The computational basis of this architecture resembles to some extent the work described by Maes [11] as far as the activation/inhibition dynamics of the system is concerned, but – as we believe – has several advantages over the approach chosen there:

- In the action selection network, there is a direct link between preceding and succeeding competence modules (i.e. the fixed motor patterns in our terminology). Hence, the model of behavioural organisation used there is more similar to the chaining of reflexes than to the self-organising processes moderating between independent modules. As the robustness of the agent depends on the ingenuity of the system designer and the correct chaining of behavioural sequences, the system described by the action selection network lacks some flexibility.

- The implicit representation of subgoals as activation conditions and expected effects for each competence module requires the representation of the world in order to map it onto the goal descriptions. The more complex the agent's behavioural repertoire becomes, the more detailed the world description has to be in order to allow appropriate mapping. We have already discussed the danger of this kind of world representation and consider this as a major disadvantage.

- The lack of considering the internal state of the agent represented by "motivational" influences as this represents a key feature to be used in the adjusting of activity thresholds.

Nevertheless, we are keen on learning about the application of self-organisation principles in the activation/inhibition framework of action selection as we think that the knowledge about the appropriate use of system parameters (e.g. activation threshold, temporal information about activity/inactivity etc.) is also useful to our approach. This is because the most difficult part of the work on the lateral inhibition network will be to define the correct system parameters to be used by the inhibitional processes. But these system parameters must be based on the agent's "metabolism", "motivation" and "desire" rather than on the definition of goals and their value to the system attributed by the system designer.

5 Higher-Order Cognitive Processes

How is this basic model of sensory-motor interaction described in the previous section related to more complex cognitive skills? How can a model of the organisation of behavioural sequences influenced by motor and sensor excitation contribute to a deeper understanding of cognitive processes? There has been much discussion recently on purely symbolic representations of cognitive processes as used in the so-called knowledge-based approach to modelling cognition in AI, also known as the debate between strong and weak AI. In general, one can refer to this as the "symbol grounding problem" of symbol manipulation. Stevan Harnad [12] sketches this problem the following way:

- How can the semantic interpretation of a symbol system (i.e. the interpretation of object and category names as an instantiation of reasoning processes in the human brain) be made intrinsic to the symbols rather than just parasitic on the meanings in our head?

Harnad suggests a framework to ground these symbol tokens bottom-up in nonsymbolic representations of two kinds:

1. Iconic representations which are analogues of the proximal sensory projections of distal objects and events.

2. Categorical representations which are learned and innate feature-detectors that pick out the invariant features of objects and event categories from their sensory input.

Finally, a third higher-order level of symbolic representations based on the former two nonsymbolic ones is used to describe object relations.

In general, we agree with Harnad as far as the necessity of providing a nonsymbolic substrate for cognitive processes is concerned. Nevertheless, we disagree as far as the scope of the approach chosen is concerned. Because the interpretation of visual sensory input is based on the coincidence and correlation of motor and sensor activity patterns other than vision (as we will see in the next section), Harnad's approach itself is lacking some fundamental grounding, namely of the visual system. We would like to develop a more general framework for grounding symbols without limiting ourselves to sensor information. Instead of using only a visual system for grounding symbols, we would like to use a robot equipped with appropriate actuators and sensors for this purpose.

Nevertheless, we think that Harnad's work highlights a very important problem in AI and sketches the foundation of future work in the field of knowledge representation and machine learning.

6 A General Framework

Using the architecture described in Section 4, we might be able to reproduce the flexibility and ability of an autonomous agent sufficient to survive in a dynamic environment at the level of simple animals' performance.

But how can we bridge the gap between this kind of goal-directed behaviour and more intelligent and rational one? How can we modify the model of behavioural organisation presented so far in order to achieve this goal? Work done in Neuroethology gives us an answer to this and provides us with the ability to enhance our model.

6.1 Analysis

As Manteuffel has described in [13], a child is considered to be born as a blank sheet as far as cognitive abilities is concerned. There is no knowledge about regularities in the surrounding world available to this child. But still, after a while, a baby is able to develop the capabilities to perceive its environment and to use this experience for the interpretation of more complex interactions with the world, such as grasping, walking around etc., later on. The necessary structures and abilities for this learning process are:

- Various sensors.
- Nerve cells.
- Main connections between the sensors and particular regions of the brain.
- A pleasure/pain system (so-called hedonic stimuli) with innate releasing mechanisms.
- An arousal system with innate releasing mechanisms
- Basic motor reflexes and motor skills

The various excitational patterns of the sensors (i.e. eyes, ears etc.) and muscle groups (i.e. legs, arms etc.) are perceived by specific parts of the brain. A baby moving its legs and arms arbitrarily creates excitation patterns in the according parts of the brain. Through the coincidence of the activity and sensing patterns correlation is created in the sensory-motor hemisphere of the baby's brain. Once this is achieved, input from the visual systems delivering various excitation patterns from the retinas is linked to the already existing sensory-motor experience of the child.

Initially, there are just main connections from one particular sensing organ to a particular part of the brain which receives all incoming excitation from this sensing organ. Through the muscle actions and feedback excitations, self-organising processes based on the reinforcement of coincident excitation patterns are initialised. This behaviour leads to a stabilisation of correlated events. Through this process, sensory-motor maps are created in the brain which serve to structure (i.e. to relate coincident excitation during activity and perception) excitational patterns in the brain. The overall goal of this behaviour is to impose structure on the chaotic excitational patterns through self-organisation.

These basic sensory-motor maps serve as a basis for more complex correlations of events and enable the brain to build a more and more complete model of the environment. But this model is by no means a precise reproduction of the environment: it is designed by the brain as the brain tries to impose structure on the chaotic excitational patterns coming from the environment.

A main factor influencing these self-organising processes is the instructor which measures the success and fitness of various excitational patterns in the body and the necessity to restructure the sensory-motor maps. This instructor activity is mainly based on reactions to surviving enhancing or lowering activities of the biological system and is recognised as an essential component in all learning processes.

Due to these structuring processes, a fairly complex world model is constructed and maintained consistent to new incoming information. In this world model all concepts are directly linked to the experience in the sensory-motor system of the human body.

6.2 Synthesis

How can one transform this description of the acquisition of basic cognitive skills into a framework of behavioural organisation? Tinbergen [1] has developed a general architecture introducing a hierarchy of what he called instinct centers (which is equivalent to a layer in Lorenz lateral inhibition architecture [2]). Each instinct center receives input from the world via sensors and from the the agent's internal state representation (such as motivation, desire, metabolism etc.). Once the innate releasing mechanism of an instinct center has been activated, this instinct center is able to influence subordinate instinct centers. In this way, behaviour is decomposed from high-level "goals" into fine-grained activities, down to neurophysiological events in the motor neurons of muscle cells. This architecture is shown in Fig.5.

As shown in Fig.5, each instinct center is actually prevented from becoming active by a block. This block can be lifted by the innate releasing mechanism if the appropriate sensor sensation is present. Once this has been done, the released signal serves as a source of additional excitation for subordinate instinct centers (apart from the excitational signals the subordinate centers receive from their inner and outer sensors, hormones etc.). Once more or less "atomic" or "primitive" instinct centers have been reached, their released signals serve as input for nerve cells (e.g. in muscle groups, muscle strains, and motor neurons). It is possible to map this more detailed model of behavioural organisation onto the description

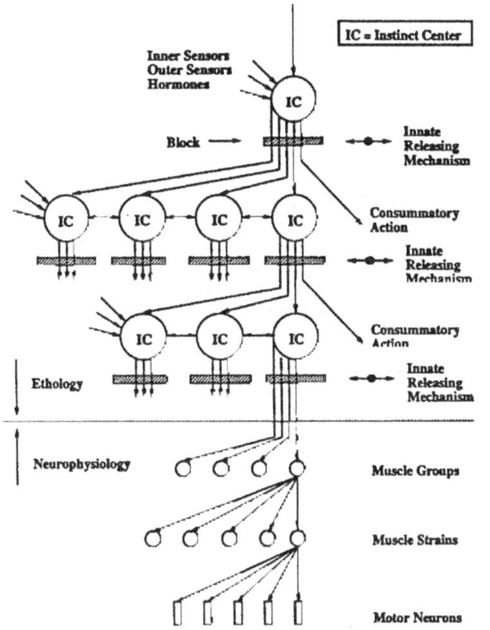

Figure 5: A hierarchy of instinct centers

of map-creating activities described by Manteuffel [13]. The model of the self-organising processes based on the coincidence and correlation of motor and sensor activities can be used to explain the generation of Tinbergen's hierarchy of instinct centers assuming that the terms "instinct center" and "sensory-motor maps" refer to very much the same. Altogether, both components form an initial framework to guide our research into intelligent autonomous systems.

7 Proposition

What has been said so far in this paper about the basic ingredients of intelligent behaviour can be summarized into the following two main problems:

1. How is behaviour organised, how is new behaviour learned by an autonomous agent?
2. How are cognitive processes related to the basic sensory-motor skills which are supposed to provide a grounding semantics for them?

These are the two major point we would like to investigate in order to build intelligent autonomous systems. We are well aware of the fact that this approach will not provide a short-term answer to the question of what intelligence is (probably not even a mid-term answer), but to us it looks more promising than the well-known top-down approaches used in classical AI.

As a framework for the necessary research into intelligent agents, we suggest the decomposition into the following subtasks:

1. To define a model of the primitive motor and "body" sensor (i.e. only sensors which are related to motion sequences) activities in a robot being. This includes the design of appropriate interfaces between the instance which correlates the excitational patterns and the motors and sensors of the agent.
2. To define a model of the permanent and spontaneous creation of motor activities and the feedback via sensors.
3. To define a model of how primitive sensory-motor maps which form the fixed motor patterns can be built.
4. To define a model of the hedonic stimuli (i.e. pleasure and pain) in an artificial creature. These stimuli are considered to be important in the production and modification of behaviour. But how can they be defined for a robot being?
5. To define a model of the inhibition processes moderating between the various fixed motor patterns in order to filter out the appropriate behaviour as a response to particular excitation configurations experienced.

The result of this work will provide us with a deeper understanding of the processes necessary to organise behaviour and will serve to incorporate visual processes. To include some sort of visual sensing facility is considered to be important as it enables the robot to experience the world and to build up a representation of the relations between itself and the surrounding environment. The following steps are considered to be important when focussing on this attempt:

1. The definition of appropriate non-verbal subsymbolic representations of primitives derived from the visual sensors. This representation will be based on the representation of the excitation configurations developed for the sensory-motor skills.
2. The development of categorical representations which serve to identify and structure the invariants of the world perceived by the visual sensor. But not only visual signals should be correlated and grouped into those categories, also the activities and states

of the robot's motors and body sensors should be included to serve as invariant features.

3. The development of a theory of how these invariant features and their coincidence and correlation could serve as a basis for the creation of concepts. We have to admit, that at this time we do not know yet what this theory should look like.

Since we are interested in instantiating the processes present in some behavioural activity and not necessarily the structure which generates and performs it, we are not focussing on any particular computational approach. Instead, we will analyse the requirements imposed by the framework presented and will make a decision in favor of particular hard- and software to be used later on.

8 Conclusion

We have suggested a tentative framework for the research work into autonomous intelligent systems based on the study of behavioural sequences in robot beings. Although we are starting nearly from scratch as far as the cognitive capabilities of the robot are concerned, we strongly believe that this bottom-up approach is necessary to provide a more general theory of what constitutes cognitive processes. The work on these issues will provide a basis for future work in the field of knowledge representation, machine learning, and overall in Robot Ethology.

References

[1] N. Tinbergen. *The Study of Instincts.* Oxford University Press, 1966.

[2] K. Lorenz. *Vergleichende Verhaltensforschung.* Springer Verlag Wien, 1978.

[3] R. A. Brooks. *Achieving Artificial Intelligence Through Building Robots.* Artificial Intelligence Memo 899, MIT, 1986.

[4] The Frame Problem in Aritificial Intelligence. *Proceedings of the 1987 Workshop.* Morgan Kaufmann Publishers, 1987.

[5] J.-P. Ewert. *Neuroethology.* Springer-Verlag Berlin, 1980.

[6] J.Piaget. Piaget's Teory. *Carmichael's Manual of Child Psychology,* ():, 1970.

[7] D. C. Dennett. Intentional Systems in Cognitive Ethology: The Panglosian Paradigm Defended. *The Behavioral and Brain Sciences,* 6(3):343–390, 1983.

[8] V. Braitenberg. *Vehicles. Experiments in Synthetic Psychology.* MIT Press, 1984.

[9] R. S. Marken. The Nature of Behavior: Control as Fact and Theory. *Behavioral Sciences,* 33():196–206, 1988.

[10] R. L. Anderson. *A Robot Ping-Pong Player.* MIT Press, 1987.

[11] P. Maes. *How To Do The Right Thing.* Artificial Intelligence, AI Lab, Vreije Universiteit Brussel, 1989.

[12] S. Harnad. The Symbol Grounding Problem. *Physica D,* ():, 1990.

[13] G. Manteuffel. Neuronale Selbstorganisation als Basis von Wahrnehmung. *Informatik Fachberichte,* ():, 1989.

A Model of the Mechanisms Underlying Exploratory Behaviour

Liane M. Gabora
Center for Research on Concepts and Cognition
Indiana University, Bloomington, IN, U.S.A., 47405
Email: liane@cogsci.indiana.edu

Patrick W. Colgan
Department of Biology
Queens University
Kingston, Ontario, Canada, K7L-3N6
Email: Colganp@qucdn.bitnet

Abstract

A model of the mechanisms underlying exploratory behaviour, based on empirical research and refined using a computer simulation, is presented. The behaviour of killifish from two lakes, one with killifish predators and one without, was compared in the laboratory. Plotting average activity in a novel environment versus time resulted in an inverted-U-shaped curve for both groups; however, the curve for killifish from the lake without predators was (1) steeper, (2) reached a peak value earlier, (3) reached a higher peak value, and (4) subsumed less area than the curve for killifish from the lake with predators. We hypothesize that the shape of the exploration curve reflects a competition between motivational subsystems that excite and inhibit exploratory behaviour in a way that is tuned to match the affordance probabilities of the animal's environment. A computer implementation of this model produced curves which differed along the same four dimensions as differentiate the two killifish curves. All four differences were reproduced in the model by tuning a single parameter: the time-dependent component of the decay-rate of the exploration-inhibiting subsystem.

1. Introduction

Selection tends to favour the evolution of systems whose organization enables more efficient ways of perceiving and interacting with the environment, and greater capacity to cope flexibly with environmental change. This often entails progressive differentiation of the parts of a system into subsystems that are specialized to take care of different aspects of survival. For example, the parts of an animal involved in the detection and avoidance of predators can be thought of as comprising one subsystem, the parts involved in obtaining and metabolizing energy another, and so on.

Thus, an adaptive system can be thought of as a set of specialized subsystems working together to maintain the integrity of the whole. However, when we divide the system into subsystems according to ultimate causal goals such as avoiding predators and obtaining energy, every subsystem that relies on behaviour includes skeletomusculature. Much of the skeletomusculature plays a role in the functioning of *many* subsystems. This makes sense; subsystem-specific limbs are redundant, to the extent that (1) subsystems require only periodic control of skeletomusculature to function effectively, and (2) the skeletomusculature required to meet the needs of one subsystem could also be used to meet the needs of another. The upshot: skeletomusculature is shared amongst subsystems, and though subsystems have to work together cooperatively, they must also *compete* for control over what McFarland and Sibly (1975) refer to as the system's "behavioural final common path" (see for example, Miller, 1971; Baerends and Drent, 1982; Colgan, 1989). Which subsystem wins the competition depends upon the relative need (deviation from homeostasis) of the subsystems, the opportunities and dangers currently afforded by its environment, and the pros and cons of engaging in behaviour that has only indirect or long-term effects. Dawkins (1976) has suggested that subsystem competition is lessened somewhat by the fact that the behaviours associated with various subsystems occupy different positions on an established behavioural hierarchy; given equal need to control behaviour, the behaviour that is higher on this hierarchy wins the competition.

The primary purpose of this research was to explore the extent to which we can explain and predict behaviour in a novel environment by viewing it as the emergent outcome of the continual process of *mutually satisfying the constraints imposed by competing, specialized subsystems*[1]. An early theory of behavioural response to novelty, states that the extent to which an animal explores its environment depends on the outcome of a competition between the tendency to avoid novel stimuli because they evoke fear, and the curiosity-induced tendency to approach (Montgomery, 1955). Subsequent research has shown that when animals encounter a novel environment, exploratory behaviour increases, reaches a maximum, and then decreases; thus, plotting exploration against time yields an inverted-U-shaped curve (Weisler & McCall, 1976). This finding has held across a variety of environmental stimuli and organisms, including humans (May, 1963; Karmel, 1969; McCall, 1971, 1974), chimpanzees (Welker, 1956a, 1956b), monkeys (Mason, 1961; Sackett, 1972), and rats (Dember and Earl, 1957). However, there has been no attempt to employ the Curiosity-Fear theory to account for the inverted-U-shaped curve; in fact little attention has been given to the mechanisms that underlie the phenomenon.

We hypothesize that in a well-adapted animal, a number of subsystems simultaneously excite or contribute to exploration of the environment, including those that can benefit from the finding of food, mates, or shelter, or from opportunities for learning. The subsystem that deals with predator avoidance (and perhaps others) is expected to inhibit exploration. The impact of these subsystems on behaviour should correspond to the relative probabilities of these beneficial and harmful effects; in other words, they should reflect the affordance probabilities of the animal's environment. Note that we are not proposing any particular classification of subsystems; we are simply exploring the utility of this general approach. Note also that the subsystems account is functionally equivalent to the Curiosity-Fear theory, but differs from it in that it invokes an explanation at what biologists refer to as the level of ultimate causation. Animals may use emotions such as curiosity and fear to manifest ultimate goals in their behaviour, but we do not address this issue.[2]

This project consists of two parts: empirical work, and simulation. In the laboratory, we compared the behaviour of two groups of banded killifish (Fundulus diaphanus) from two natural populations in a novel environment. Killifish are cyprinodontids found in the weedy, sandy shallows (less than four meters in depth) of lakes and ponds. The Lake Opinicon population is sympatric (cohabits the lake) with piscine predators; the Fish Lake population is allopatric from (does not cohabit the lake with) predators. In previous studies, Lake Opinicon killifish maintained a significantly greater mean horizontal distance, and were more likely to move to shallow water in the presence of a largemouth bass, than killifish from Fish Lake (Dunlop, 1987). It seems reasonable to expect that in addition to exhibiting differences in behaviour in the presence of a predator, these populations might show differences in behaviour that could affect the likelihood of encountering a predator in the first place. These behavioural differences would be expected to reflect the relative probability of encountering a predator in each of the two lakes. Thus, differences in predation between the two lakes could give rise to intraspecific variation in behaviour in the absence, as well as in the presence, of a predator.

Since the two lakes do not differ, as far as we know, in the probability of yielding something beneficial such as food, but do differ in the probability of yielding a predator, it was hypothesized that while the extent to which certain subsystems *excite* exploration is comparable in the two populations, the extent to which the predator avoidance subsystem *inhibits* exploration is greater in Lake Opinicon killifish than in Fish Lake killifish.

The goal of the simulation experiments was twofold. First, I wished to confirm that the above model of exploration can generate the general pattern of results observed in the laboratory. Second, I wished to isolate the variable(s) that give rise to differences in the behaviour of the model that parallel the behavioural differences observed between the two killifish populations.

2. Laboratory Experiments

2.1 Subjects and Collecting Sites

Nonbreeding killifish were collected from Fish Lake, Prince Edward County, Ontario, Canada, and Lake Opinicon, Leeds County, Ontario, Canada, between May, 1987 and January, 1988. Fish Lake

1. Others (MacArthur and Piankam [1966]; Roitblat [1985]) refer to this process as "optimal decision making". This term is somewhat inappropriate because the behaviour need not be optimal.

2. For discussion of how exploration satisfies curiosity, see (Berlyne, 1950). For discussion of exploration and fear, see (Russel, 1973; Halliday, 1966). See also (Fowler, 1967) on exploration and reduction of boredom, and (Leuba, 1962) on exploration as a means of maintaining a constant level of arousal.

(44° 00' N, 77° 10' W) is a small, irregularly-shaped lake (Terasmae & Mirynech, 1964). Fish Lake killifish may have been isolated from other killifish populations since the lake formed, approximately 10,000 years ago. A less likely possibility is that they were introduced to the lake more recently by human intervention. The fish community of this lake has been sampled since 1965 and has not been found to contain piscivorous fish (A. Keast, pers. comm.) It is likely that Fish Lake is too shallow to support large fish, due to winterkill. Thus, Fish Lake killifish have not been exposed to piscine predators.

Lake Opinicon (44° 30' N, 76° 30' W) is a large shallow lake that is part of the Rideau Lakes canal system (Turnbull, 1975). Lake Opinicon killifish are sympatric with a number of piscine predators (e.g. largemouth bass, _Micropterus salmoides_, northern pike, _Esox lucius_, and yellow perch, _Perca flavescens_).

Killifish are a schooling species (Scott and Crossman, 1973), and isolating individual fish disrupts normal behaviour. Therefore, experiments were conducted on groups rather than individuals. Whether or not there are differences in home range size and social group size between the two populations is unknown. Both populations feed on small crustaceans and insect larvae (Keast and Webb, 1966; Keast, unpublished data).

Fish from Lake Opinicon were caught with a seine net, and fish from Fish Lake were caught either with a seine net or a funnel trap. Only fish between five and six centimeters in total length (which corresponds to approximately one year of age, Scott & Crossman, 1973) were used. Fish were transported to Queens University, Kingston, in large plastic pails containing garbage bags filled with oxygenated lake water.

2.2 Housing Conditions

In the laboratory, fish were housed in groups of 25-50 in flow-through aquaria outfitted with immersion heaters. Water temperature was maintained at 20-23° C. When the temperature of the lake water was lower than this, the temperature of the tanks was increased by approximately one degree per day until the desired temperature was reached, thus minimizing thermal stress. Fish were fed Murex freeze-dried ocean plankton once daily. The light-dark cycle was 12L:12D.

2.3 Experimental Apparatus

Experiments were conducted in a tank (120 x 45 x 45.5 cm deep) divided by sheets of opaque white plexiglass into four chambers: a holding chamber (HC) (30 x 15 cm), an experimental chamber (EC) (120 x 30 cm), and two unused chambers (UC) (45 x 15 cm), (Figure 1). The sheet dividing the experimental chamber from the holding chamber had a door (12 cm wide x 15 cm high). Water depth was 30 cm.

The tank was illuminated by three 95 Watt incandescent lighting fixtures placed 90 cm from the top of the tank. A video camera was positioned directly above the tank, 116 cm from the top of the tank, and was connected to a video camera recorder. The holding chamber contained an airstone, and one of the unused chambers contained a heater.

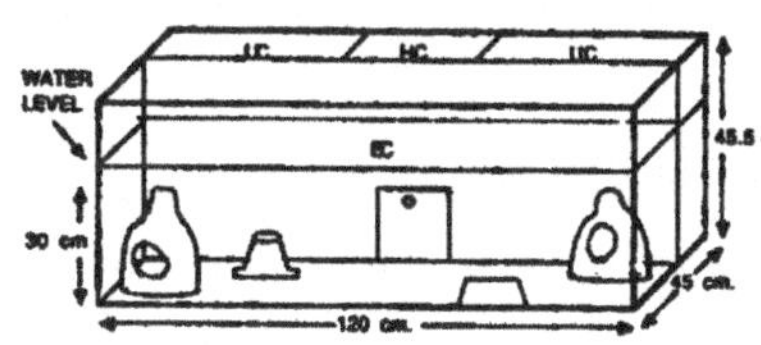

Figure 1. Tank in which killifish were exposed to a novel environment. EC, Experimental Chamber; HC, Holding Chamber; UC, Unused Chamber.

The glass along the sides of the experimental chamber was covered with white plastic, to which four panels of wallpaper of various colors and designs (20 x 15 cm) were attached. The wallpaper designs did not resemble anything the fish would encounter in their natural environment. The experimental chamber contained two empty plastic bottles and a small red clay flower pot. Each of these items had large holes cut out of them. Two slabs of plexiglass (approximately 3 x 10 cm) were glued to the sides of the tank, and another to the floor, with silicon cement. A fourth plexiglass slab (approximately 15 x 15 cm) was leaned against the side of the tank and at an angle of approximately 45 degrees, and fixed into place with silicon cement. The purpose of these items was to ensure that the tank was a novel environment for the fish, and to invite exploration of this environment.

The experimental chamber was covered with a clear plexiglass sheet upon which a numbered grid was drawn. The dimensions of the grid spaces were

10 x 10 cm. Because of refraction and perspective from the camera, equally-sized grid spaces did not correspond to grid columns of equal volume. However, since it was not the use of space per se, but the comparative use of space between the two populations, that was of interest, small differences in the volumes of grid columns had little effect on the conclusions drawn.

2.4 Experimental Protocol

Five fish from one of the populations were placed in the holding chamber. 24 hours later, the door to the experimental chamber was opened. The behaviour of the fish was recorded on the VCR for 20 minutes, starting when the door was opened. *Whole-body activity* — movement of the entire body through space — was measured as the sum total number of grid line crossings by all five fish per minute. Three trials were conducted for each of the two populations; 30 fish were used in total. Experiments were begun at approximately 10:00 am.

2.5 Results

Fish from both populations exhibited inverted-U-shaped activity curves (Figure 2). The fact that activity over time was nonuniform is strong evidence that it reflected the novelty of the environment, and it is therefore not unfitting to refer to it as exploration.

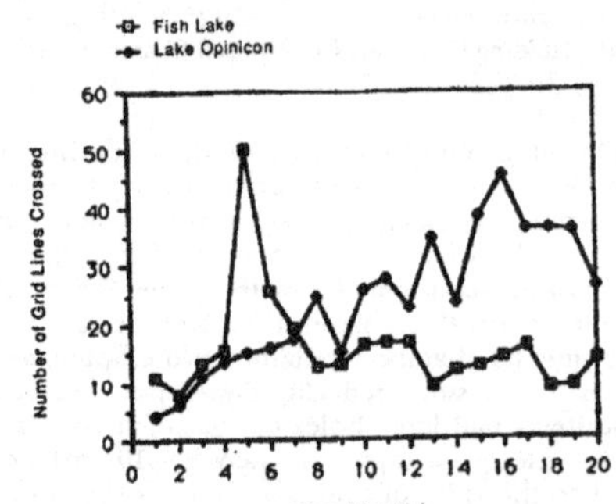

Figure 2. Exploration curves for Fish Lake and Lake Opinicon killifish, plotted as the total number of grid lines crossed by all five fish per minute of exposure to a novel environment.

The shapes of the two curves differ. The activity of Fish Lake killifish rose sharply, reaching a maximum value (51.7 grid spaces/minute) after five minutes of exposure to the novel environment. By eight minutes, it had returned to the initial level (around 12 grid spaces/minute), and it did not deviate from this by more than five grid spaces/minute for the remainder of the experiment. 335 grid lines were crossed in total.

The initial activity level of Lake Opinicon killifish was 6.3 grid spaces/minute. Activity rose much more gradually and erratically, reaching a maximum value (45.3 grid spaces/minute) after 16 minutes of exposure to the novel environment. It subsequently declined, but still had not return to the initial level by the end of the twenty minute trial. The inverted U-shape of this curve was indicated by the near significance of a quadratic term in regression ($p = .057$). The total number of grid lines crossed was 497.

Fish Lake killifish tended to school (swim within two body widths of one another, oriented in the the same direction) more frequently than Lake Opinicon killifish. Fish Lake fish tended to explore the novel environment "region by region", and generally maintained a fairly constant speed. In contrast, Lake Opinicon fish darted quickly from one spot to another, with prolonged periods of immobility interspersed between the periods of rapid movement.

2.6 Discussion

In sum, fish from both Fish Lake and Lake Opinicon exhibited inverted-U-shaped exploration curves, and the major differences between the two curves were: (1) the slope of the Fish Lake curve was greater, (2) it peaked earlier, (3) its peak was higher, and (4) the total amount of exploration was greater for Lake Opinicon fish (i.e. the area under the Lake Opinicon curve is greater). It is not possible to prove conclusively that these differences are related to the difference in threat of predation. Nevertheless, they are consistent with the following theoretical model:

When exploration of a novel environment does not lead to the discovery of affect-laden objects such as food or predators, the animal's (not necessarily conscious) perception of the probability that the novel environment will yield such objects decreases. Thus, exploratory behaviour has a negative feedback effect on both subsystems that excite it and subsystems that inhibit it. There are two additional reasons to expect that exploratory behaviour has a negative feedback effect on excitatory subsystems. First, as exploration proceeds, even as-yet-unexplored space might become less enticing, and be perceived as less likely to yield opportunity for learning, because of its similarity to the space just explored. Second, exploration can lead to fatigue.

For these reasons, were it possible to measure and graph the time-course of the extent to which exploration is (1) excited and (2) inhibited as an animal explores, we would expect to get two negatively-sloped curves.

Inhibition should additionally decrease as a function of the amount of time spent in the novel environment, regardless of the extent to which the environment is explored. This follows from the nature of ecosystems. Harmful things — predators — are relatively likely to come to *you;* beneficial things — prey — are relatively less likely to come your way. Since predators are not stationary, time spent immobile in a novel environment without encountering predators provides evidence that the environment does not contain predators, but does not tell an animal much about whether the environment contains prey. It follows that activation of the inhibitory subsystem decreases both with exploration and with time (independent of exploration), while activation of the excitatory subsystem decreases with exploration only. Therefore, unless (1) exploration has less feedback effect on the inhibitory subsystem than the excitatory subsystem, and the difference between the two is greater than the effect of the inhibition-specific time factor, it follows that (2) the slope of the inhibition curve is greater that of the excitation curve. As it turns out, the second condition is necessary in order for exploratory behaviour to get going when it does not begin as soon as the animal is exposed to the novelty (this will be elaborated upon in Section 3.1).

We propose the following explanation for the inverted U-shaped exploration curve. Upon initial exposure to a novel environment, if subsystems that inhibit exploration are more active than those that excite exploration, the animal is initially immobile. Even if this is not the case, inhibition is still initially high, and exploration is minimal. As time passes and novelty diminishes, excitation decreases. However, assuming the novel stimuli were not harmful, inhibition decreases more rapidly than excitation, and the net effect is an increase in exploration. Exploration peaks when the difference between excitation and inhibition of exploration reaches a maximum. When the animal begins to perceive the environment as a relatively safe place, inhibition reaches a minimum, and since excitation continues to wane, exploration decreases. It eventually bottoms out when the animal becomes habituated to the environment (i.e. the environment has lost its novelty). This model is illustrated in Figure 3.

The fact that the total number of grid lines crossed was greater for Lake Opinicon fish than for Fish Lake fish is somewhat surprising. Since, given

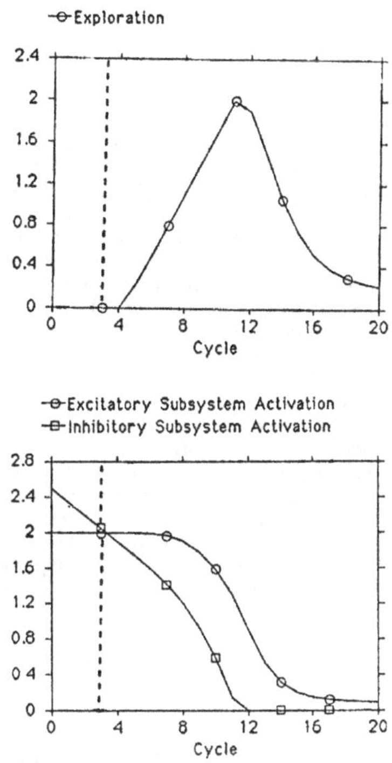

Figure 3. (a) An inverted-U-shaped exploration curve. (b) Activations of the corresponding excitatory and inhibitory motivational subsystems. Dotted line indicates point in time at which excitation surpasses inhibition, enabling exploratory behaviour to begin. Both curves were produced by the computer program described in Section 3. Initial activation of the inhibitory subsystem is 2.5. All other parameters took the default values given in Section 3.1.

sufficient time, Lake Opinicon killifish engaged in more exploration than Fish Lake killifish, the extent to which exploration is excited by subsystems that could benefit from it is at least as great for Lake Opinicon killifish as for Fish Lake killifish. Since Lake Opinicon killifish took longer to *begin* exploring, results are consistent with the hypothesis that the populations differ only in the extent to which exploratory behaviour is initially inhibited. The fact that the slope of the Fish Lake curve is steeper than the Lake Opinicon curve is also consistent with this theory. For Fish Lake killifish, inhibition is virtually nonexistent, and exploration increases more quickly. Since excitation is inversely related to the animal's perception of the amount of novelty left to explore, when exploration proceeds quickly, excitation diminishes faster, and exploration dies out more quickly.

Because of the highly social nature of the species used, laboratory data were based on the behaviour of *groups*; however, what we are interested in is the mechanisms underlying exploratory behaviour in an *individual*. Since the inverted-U-shaped curve has been reported many times in the exploratory behaviour of individuals, we need not worry that the positively-sloped portion of the exploration curve is due simply to individual differences in how quickly group members begin exploration. It is, however, possible that the more gradual slope of the Lake Opinicon curve is due to greater individual variability in how quickly Lake Opinicon killifish increase and decrease their rate of exploration. We hope to find species that are less social, and for which there exist two populations that experience very different predation pressures, so that similar experiments can be conducted with individuals. To keep things simple, our goal in developing the computer program described in the next section was to model the mechanisms that underlie exploratory behaviour in an *individual*.

3. Simulation Experiments

3.1 Architecture

A computer model consisting of two subsystem units and a behavioural output unit was constructed in Common Lisp. One subsystem, the excitatory subsystem is linked by a positive weight to the output unit; it can be thought of as corresponding to a subsystem that could benefit by exploration, such as the subsystem concerned with finding food. The other subsystem inhibits the output unit; it can be thought of as the subsystem that takes care of predator avoidance. Both subsystems are in a negative feedback relationship with the output unit; when activation of the output unit is high enough to cause exploration to surpass a threshold, activation of the subsystems decreases, and consequently their impact on the output unit decreases. Activation of the inhibitory subsystem also decreases as a function of time, independent of exploration, by a constant that will be referred to as the time-factor. The time-factor is a vital component of the model when initial exploration is zero; without it, activation of neither subsystem changes, so the impact of the excitatory subsystem on the output unit never overtakes that of the inhibitory subsystem. The time factor ensures that sooner or later excitation surpasses inhibition and exploration therefore gets underway. (When initial exploration is not zero, the same effect can be achieved by making the feedback weight to the inhibitory subsystem greater than that to the excitatory subsystem.)

Output for each cycle is either a zero, signifying immobility, or a positive number; the larger the number, the greater the amount of exploration prescribed by the system. The relevant variables are:

s_1 = activation of excitatory subsystem
s_2 = activation of inhibitory subsystem
w_1 = weight from excitatory subsystem to output unit
w_2 = weight from inhibitory subsystem to output unit
w_f = feedback weight from output unit to each subsystem
k_1 = decay on output unit
k_2 = time factor
E = amount of exploration

Exploration was calculated as follows:

$$E_{(t)} = \max\{0, k_1 E_{(t-1)} + w_1 s_1 - w_2 s_2\}$$

New subsystem activation values were calculated as follows:

$$s_{1\,(t)} = \max\{0, s_{1\,(t-1)} - w_f E_{(t-1)}\}$$

$$s_{2\,(t)} = \max\{0, s_{2\,(t-1)} - w_f E_{(t-1)} - k_2\}$$

Note that if there was no exploration during the previous cycle, activation of the exciting subsystem does not decrease, but activation of the inhibiting subsystem does.

Three parameters were varied: (1) initial activation of the exploration-inhibiting subsystem, (2) weight from the exploration-inhibiting subsystem to the behaviour unit, and (3) time factor. These parameters affect the impact of the exploration-inhibiting subsystem on the output unit. Earlier we hypothesized that, since exploration could lead to an encounter with a predator for Lake Opinicon killifish but not for Fish Lake killifish, we might expect that the extent to which the predator avoidance subsystem inhibits exploration is greater in Lake Opinicon killifish than in Fish Lake killifish. If this is the case, these parameters are the ones most likely to be functionally similar to the parameters that give rise to the differences in exploration between Fish Lake killifish and Lake

Opinicon killifish. It was hoped that by determining which (if any) of these manipulations give rise to differences in curve shape qualitatively similar to the differences between the Fish Lake and Lake Opinicon curves, we could refine our model of how animals tune the mechanisms underlying exploration in a way that matches the affordance probabilities of their environment.

3.2 Results

Plotting the program's output did indeed yield an inverted-U-shaped curve. This outcome was robust across all sets of values for initial parameters tested. Initial values of parameters used for all data presented in this paper are unless stated otherwise as follows:

$s_1 = 2.0$
$s_2 = 2.0$
$w_1 = 1.0$
$w_2 = 1.0$
$w_f = 0.5$
$k_1 = 0.5$
$k_2 = 0.15$

By decreasing the time factor (k_2) from 0.36 to 0.05, a steep curve with an early, high peak (i.e. a Fish Lake-type curve) was transformed into a more gradually-sloping curve with a lower, later peak, that subsumed a greater area (i.e. a Lake Opinicon-type curve). (These particular time factor values were chosen because they facilitate comparison of program output with laboratory results.) Thus varying a single parameter results in curves that differ by all four of the features that distinguish the exploration curves of the two populations (Figure 4). Total exploration over 20 cycles was for 13.5 Curve 1 and 16.5 for Curve 2.

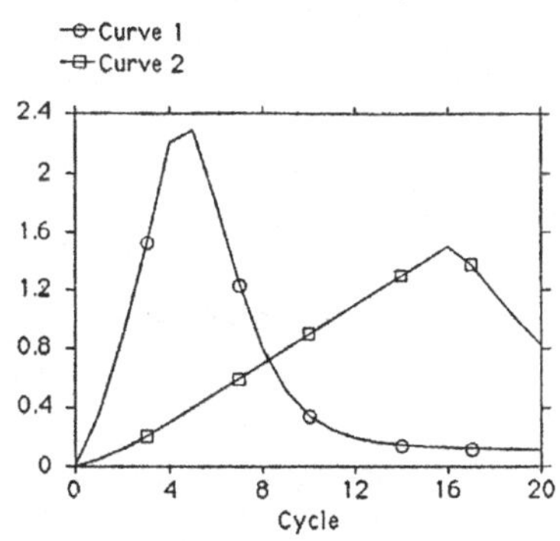

Figure 4. Exploration curves produced by two runs of the computer model. Time factor was 0.36 for Curve 1 (the Fish Lake-type curve), and 0.05 for Curve 2 (the Lake Opinicon-type curve). All other parameters had the same initial values on both runs.

Effects of the varying other parameters were less impressive. Increasing the initial activation of the inhibitory subsystem caused a decrease in peak height, and delayed the timing of the peak, but had no effect on slope. It decreased total exploration. Decreasing initial weight from the inhibitory subsystem produced the same effects. The results of these experiments are summarized in Table 1. A summary of the ways in which the Lake Opinicon curve differs from the Fish Lake curve is presented for comparison.

	Fish Lake	**Lake Opinicon**
Smaller Slope		x
Later Peak		x
Lower Peak		x
More Exploration		x

Table 1a. Summary of the ways in which the Lake Opinicon curve differs from the Fish Lake curve.

	Incr. Act.	Incr. Wt.	Decr. TF
Smaller Slope			x
Later Peak	x	x	x
Lower Peak	x	x	x
More exploration			x

Table 1b. Results of manipulating network variables. Act, activation; Wt, weight; TF, time factor.

3.3 Discussion

By tuning one parameter, the program produced a curve which differed from the original along the same four dimensions as differentiate the two killifish curves. This of course does not prove that the mechanisms invoked are similar, even at a course-grained level. Moreover, the specific features that differentiate the killifish exploration curves presented here might not generalize. However, the findings fit together nicely with the ecological situation. Lake Opinicon killifish are more likely than Fish Lake killifish to have experienced time-slices in which they didn't encounter predators followed by time-slices in which they did. Therefore, it makes sense that spending a given amount of time in a novel environment without encountering a predator decreases the probability that the environment will afford a predator more for Fish Lake killifish than for Lake Opinicon killifish. The time factor, the parameter that accounted for the four differences in the model that paralleled those in the two populations, provides a means of integrating time-dependent probability assessments in to the equation that determines exploratory behaviour. It is possible that this parameter has a functional equivalent in living brains.

In sum, despite limitations, the model implemented in this program has a number of things going for it: (1) It makes sense from an adaptationist standpoint, (2) It is consistent with the available empirical data, (3) It is computationally feasible, and (4) It seems to be the simplest account that meets the criteria of the first three statements.

4. Summary and Conclusions

The premise underlying this research is that an animal's behavioural final common path arises from the dynamics of specialized motivational subsystems. Subsystem parameters are tuned in a way that (1) enables the animal to satisfy constraints imposed by both its body and the environment, and (2) takes into account not only the present situation, but the relative probabilities of possible future situations. The time-course of exploration reflects the dynamics of subsystems that could be benefited or harmed by exploration, and these subsystems are tuned to match the probabilities of these possible beneficial or harmful effects.

We have shown that a computer model based on this theory can generate the pattern of results observed in the laboratory. The amount of exploration prescribed by the program is a function of the activation of subsystems that excite and inhibit exploration. Subsystem activation decreases as a function of exploration; activation of the inhibitory subsystem also decreases as a function of time, independent of exploration. When the program is run, and exploration is plotted against time, the result is an inverted-U-shaped curve typical of the time-course of exploration exhibited by animals in a novel environment.

The exploration curve from data using killifish that are allopatric with predators was steeper, had a higher, earlier peak, and indicated less total exploration than data using a population sympatric with predators. All four of these differences could be reproduced in the model by tuning a single parameter: the time-dependent component of the decay-rate of the exploration-inhibiting subsystem. It is hoped that, as a package, this set of results is illustrative of how empirical work and simulation can complement one another, and thereby deepen our understanding of the mechanisms that underlie adaptive behaviour.

The question arises as to whether the behavioural differences observed between these two populations are learned — result from "close calls" with predators or observation of attacks on conspecifics[3] — or whether they are innate. There is evidence for both. The tendency to explore a novel object can be bred for or against in laboratory rodents (Fuller and Thompson, 1978), and reliable differences in the tendency to explore have been shown in different strains of mice (Fink and Reis, 1981). Csanyi (1988) conducted a behaviour-genetic analysis of the paradise fish (Macropodus opercularis), and found that many behavioural traits, including some related to exploration, such as how long fish took to enter a novel environment, were genetically-based or had a strong genetic component. However, in another study, Csanyi found that paradise fish that had previously been chased by a pike subsequently attempted to escape when placed in the presence of a pike, whereas naive paradise fish did not (1985). This indicates that, in this species at least, antipredator behaviour can be learned.

To the extent that antipredator behaviour is learned, it is likely that differences in exploratory behaviour that arise because of differences in threat of predation are also learned. Learning could be incorporated in the computer model by adjusting the weights using backpropagation, according to whether exploration leads to harmful, neutral, or beneficial stimulus input.

Acknowledgements

The laboratory experiments were financially supported by a grant from the National Research Council of Canada to the second author. Simulation experiments were conducted by the first author, and supported by a grant from Indiana University. Thanks to David Chalmers, Mike Gasser, Stewart Wilson, and Bill Timberlake for helpful comments. Also thanks to Rick Beninger, Rudy Eikelboom, Michael Fox, V. Gotceitas, Dolph Harmson, and Jim Sutcliffe for comments on an earlier draft of section 2.

---

3. Conspecifics are other members of one's own species.

References

Baerends, G. P., and Drent, R. (1982). The herring gull and its eggs. *Behaviour 82*, 1-416.

Berlyne, D. E. (1950). Novelty and curiosity as determinants of exploratory behaviour. *British Journal of Psychology 41*, 68-80.

Colgan, P. (1989). *Animal Motivation*. London: Chapman and Hall Ltd.

Csanyi, V. (1985). Ethological analysis of predator avoidance by the paradise fish (Macropodus opercularis L.): I. Recognition and learning of predators. *Behaviour 92*, 227-240.

Csanyi, V. and Gerlai, R. (1988) Open-field behaviour and the behaviour-genetic analysis of the paradise fish (Macropodus opercularis). *Journal of Comparative Psychology, 102*, 326-336.

Dawkins, R. (1976). Hierarchical organization: A candidate principle for ethology. In P.P.G. Bateson and R.A. Hinde (eds.), *Growing Points in Ethology*. Cambridge: Cambridge University Press, pp. 7-54.

Dember, W. N. and Earl, R. W. (1957). Analysis of exploratory, manipulatory, and curiosity behaviours. *Psychological Review 64*, 91-96.

Dunlop, C. L. (1988). *A comparison of antipredator behaviour in two populations of the banded killifish, Fundulus diaphanus.* M.Sc. Thesis, Kingston, Ontario: Queens University Press .

Fowler, H. (1967). Satiation and curiosity: constructs for a drive and incentive-motivational theory of exploration. In K. W. Spense & J. T. Spense (Eds.) *The psychology of learning and motivation* (Vol. 1). New York: Academic Press.

Fuller, J. L., & Thompson, W. R. (1978). *Foundations of behaviour genetics.* New York: Vintage.

Halliday, M. S. (1966). Exploration and fear in the rat. *Sympos. Zool. Soc. Lond. 18*, 46-59.

Karmel, B. Z. (1969). The effect of age, complexity, and amount of contour on pattern preferences in human infants. *Journal of Experimental Child Psychology* 7, pp. 339-354.

Keast, A. and Webb, D. (1966). Mouth and body form relative to feeding ecology in the fish fauna of a small lake, Lake Opinicon, Ontario. *Journal of Fisheries Research Board of Canada 23*, 1845-1974.

Mason, W. A. (1965). The social development of monkeys and apes. In I. Devore (Ed.) *Primate behaviour: Field studies of monkeys and apes.* New York: Holt, Rinehart, and Winston.

May, R. B. (1963). Stimulus selection in preschool children under conditions of free choice. *Perceptual and Motor Skills 16*, pp. 203- 206.

McCall, R. B. (1971). Attention in the infant; avenues to the study of cognitive development. In D. Walcher and D. L. Peters (Eds.) *Early childhood: The development of self-regulatory mechanisms.* New York: Academic Press.

McCall, R. B. (1974). Exploratory manipulation and play in the human infant. *Monographs of the Society for Research in Child Development 39* (No. 2).

McFarland, D. J., and Sibly, R. M. (1975). The behavioural final common path. *Philosophical Transactions of the London Royal Society 270B*, 265-93.

Miller, N. E. (1971). Neal E. Miller: Selected Papers, 2 vols. Chicago: Aldine.

Montgomery, K. C. (1955). The relation between fear induced by novel stimulation and exploratory drive. *Journal of Comparative and Physiological Psychology 48*, 254-260.

Russell, P. A. (1973). Relationship between exploratory behaviour and fear: a review. *British Journal of Psychology 64*, 417-433.

Scott, W. B. and E. J. Crossman. (1973). *Freshwater fishes of Canada.* Ottawa, Ontario: Fisheries Research Board of Canada, pp. 966.

Terasmae, J. E. and Mirynech, W. A. (1964). Postglacial chronology and the origin of deep lake basins in Prince Edward County, Ontario. *Great Lakes Research Division, The University of Michigan 11*, pp.161-169.

Turnbull, D. A. (1975). *The molluscan fauna of Lake Opinicon.* B.Sc. Thesis. Kingston, Ontario: Queens University Press.

Weisler, A. and McCall, R. (1976). Exploration and play: resume and redirection. *American Psychologist 31*, pp. 492-508.

Welker, W. I. (1956a). Some determinants of play and exploration in chimpanzees. *Journal of Comparative Physiological Psychology 49*, pp. 84-89.

Welker, W. I. (1956b). Variability of play and exploratory behaviour in chimpanzees. *Journal of Comparative Physiological Psychology 49*, pp. 181-185.

THE CAUSAL ANALYSIS OF AN ADAPTIVE SYSTEM: SEX-RATIO DECISIONS AS OBSERVED IN A PARASITIC WASP AND SIMULATED BY A NETWORK MODEL

Marijke Vonk, Felix Putters & Berend-Jan Velthuis
Department of Zoology, Div. of Ethology, University of Leiden,
P.O. Box 9516, 2300 RA Leiden, The Netherlands,
e-mail: SBW1MV@HLERUL57

ABSTRACT
The computer metaphor, based on rule and symbol manipulation, is challenged by our connectionist model. Neural nets, Parallel Distributed Processing systems, *etc.* are capable to achieve goals by way of cooperative activity and can ignore all gadgets that are required in the computational models based on the digital computer metaphor. The consequences of our structure-oriented modelling approach for causal and functional explanations in ethology are discussed.

KEY WORDS
Neural nets, PDP, motivation, adaptation, parasitic wasps, sex allocation

1. INTRODUCTION

Ethologists use the modelling approach in order to reveal the mechanisms underlying overt behaviour. Relevant principles of organization are tested by embodying them in a mathematical model, the behaviour of which is compared to experimental data. The serial computer metaphor and systems theory have set the fashion for some time in modelling information processing, such as involved in problems of decision-making. Models based on the computer metaphor manipulate bits of data in a formal way, according to preset rules and operations which are specified in programs that were designed for a given task (van Rhijn 1977; van Rhijn & Westerterp-Plantinga 1989; Coderre 1989; Travers 1989). This way of modelling involves a distinction between system hardware and software, and postulates a central processor that operates on the data and drives the system. These procedures are attractive from a methodological point of view, in that one knows the complete set of assumptions necessary to make these constructs work as desired. But, what is contributed to causal analysis when we apply this approach? In practice, one postulates homunculi with the self-same capacities that the theory sets out to explain. Functionally defined concepts take up key positions in the computer model, but these concepts lack a proper (albeit potential) backing of causal mechanisms that could perform the assumed tasks. The approach of cybernetics and system theory (Toates 1986) may provide appropriate solutions when dealing with simple phenomena (*e.g.*, orientation), but their theoretical framework is not suited for handling more complex phenomena. Moreover, many models based on this conventional modelling approach happen to be goal-directed.

The condition of an 'explicit goal-representation' makes conventional modelling an inappropriate technique for the assessment of decisions that are made by our experimental animals: parasitic wasps. Many parasitic wasps are known to adjust the offspring sex ratio to characteristics of the environment, to maximize reproductive success. They are capable to optimize the F1 offspring sex ratio: the actual ratio sons - daughters is such that the maximum number of gene copies will be present in the F2 progeny. In order to explain the actual decisions of our wasps (but we believe that this point can be generalized, see McFarland 1989), we require systems which adapt to the environment without an explicit representation of the world, and which can achieve goals without a representation of these goals (*i.e.*, without goal directedness). Conventional explanatory concepts are therefore of no use when applied to the decisions made by our experimental animals.

Recent developments in the field of Artificial Intelligence have expanded the possibilities to develop ethologically relevant models: systems of semi-autonomous information-processing entities, whose local interactions with one another are controlled by a set of simple rules. Such systems do not contain rules for their behaviour at the global level. The observable behavioural output and its complex dynamics are emergent properties, which develop from the local interactions of the low-level entities (Langton 1989). Connectionist models perform without an explicit representation of a goal or an environment. This kind of modelling is the tool for exorcising the homunculi from ethological theory.

2. BIOLOGICAL BACKGROUND

Sex allocation theory presents one of the finest opportunities to study adaptation, because it offers quantitative predictions of those traits that have a considerable and direct effect on fitness. In fact, the evolution of these sex-ratio adjustments is one of the few areas in the study of evolutionary processes where quantitative predictions of adaptations are consistent with observed behaviour. Detailed studies of adaptation are possible in this area. Parasitic wasps in particular provide excellent examples in this respect because of the immense diversity of life histories, which constitute as many opportunities for the evolution of sex allocation strategies. It is their way of sex determination (by haplodiploidy) that provides the physiological mechanism to manipulate the offspring sex-ratio. A female wasp can actively determine the sex of each offspring by fertilizing an egg (daughter) or leaving it unfertilized (son). By doing so, she can propagate a maximum number of genes to the next generations, *i.e.*, she can maximise her own fitness.

A characteristic feature of the wasps' life history is the patchy environment. Each patch may be colonized by one or a number of females. The wasps of a subpopulation that developed in a patch mate locally and randomly upon emergence; inseminated females migrate to new patches. Because of inevitable interactions of wasp progenies that developed in the same patch, the optimal sex-ratio strategy of an individual female depends largely on the strategy of the other females that laid or will lay their eggs in that patch (Hamilton 1967; Werren 1983). As each female sets out to promote her own fitness, and regards the behaviour of her opponents as part of her environment, this situation represents a good example of a non-cooperative game. Current theory of evolution postulates the existense of a non-cooperative equilibrium, the evolutionarily stable strategy (Maynard Smith 1978).

If host quality has a differential effect on the fitness of male and female offspring, then host quality influences sex allocation. Under these circumstances selection will generally favour those females that allocate the sex which is least affected to poor-quality hosts, and the sex most affected to higher quality hosts. Solitary parasitic wasps, like *Lariophagus distinguendus* (Förster) (Hym. Pteromalidae), our experimental animal, produce a single offspring per host. For these species host quality is often related to host size. The size of a host determines the size of the wasp that will develop on it, which, if this wasp is a female, bears heavily on its longevity and life-time egg production. For male offspring the size factor has far less impact on reproductive success (van den Assem *et al.* 1989). Hence, the rule must be: Use the largest hosts for daughters (Charnov 1979, 1982; Werren 1984; Werren & Simbolotti 1989). In general, wasps do prefer the larger hosts for daughters (Clausen 1939; van den Assem 1971; Charnov *et al.* 1981). If the wasp were to put fertilized eggs only in hosts of above a certain fixed size, the offspring sex-ratio would be dictated exclusively by the patch composition (and this probably would not maximize her reproductive success). For any actual distribution of host sizes there is some optimal foraging behaviour for the wasp that will maximize the number of her grandchildren. In fact, what should be treated as 'large' or 'small' relates to the overall quality of the environment. Egg laying females indeed obey this rule of relativity when active in an environment where host quality varies (van den Assem 1971; Charnov *et al.* 1981).

3. THE STRUCTURE-ORIENTED MODELLING APPROACH

Connectionist models, neural nets, or parallel distributed processing systems consist of a large number of simple units, interacting with one another by connections. The strength of a connection between two units is represented by a weight factor. Connections can be inhibitory (negative weight) or excitatory (positive weight). A unit sums the weighed inputs, and transmits the result (*i.e.* its activation) to the units with which it is connected. The processing units are characterized by an internal threshold (the 'bias') and by a kind of transfer function (*e.g.* a sigmoid). The elementary units do not necessarily map directly onto meaningful concepts. An individual unit typically plays a role in the representation of multiple pieces of knowledge. The complex constructs necessary for the performance of a task are encoded in this pattern of interconnections. Input/output transformations are performed without explicit internal representations. The state of a unit results from the synthesis of the states of other units from which it receives input.

Usually the units are used to represent neurons, (in order to model brain functions: learning, perception, pattern recognition (Arbib 1989)). Connectionist models can also be used to represent structures underlying overt behaviour. In fact, all kinds of processors can be represented by the properties of such a model. Furthermore, additional structure can be incorporated, such as recurrence, feedback, fixed wiring, or other physiologically relevant properties. We call this approach structure-oriented (Putters & Vonk 1990), a term borrowed from Hogeweg and Hesper (Hogeweg 1989).

The break-through is the development of algorithms that allow us to train a naive, unspecified network in such a way that it can learn some (any) transformation of input to output. It is learning from examples. So, when modelling animal behaviour, we can train the network with antecedent conditions (stimuli, preceding behaviour, recurrent states, experience, *etc.)* at the input units, and a target pattern (decisions, behaviour) at the output units.

The network we used to model a wasp's decisions, 'ACTIONNET', (Putters & Vonk 1990) consisted of three layers: one layer with input units, a layer with hidden units, and one with output units. Our model is a feedforward network; units are connected to units of the next layer. The weights and biases of the network were adjusted by error-backpropagation. The following steps are involved in this method of training. One starts from a guessed architecture (may be completely unspecified, everything possible) with small and random initial weights. For each input pattern the activation of the output units is calculated. Next, the error between the network's output and the desired output (target) is calculated, and the weights and biases of the network are adjusted in a way that minimizes the error (see Rumelhart *et al.* 1986, in particular for the discussion on the chance of getting stuck). The information relevant for the production of the output pattern is selected by the model, and noise is neglected; the model will not converge if the experimenters have overlooked an essential variable.

Although neural nets learn by incorporating experience into the interconnection patterns, it should be clear that we are not modelling a learning process as such. Our experimental animals must make sex-ratio decisions without a supervisor (they never receive a feedback from their decisions: they do not live to see their offspring, let alone its reproductive success; natural selection is the supervisor in this case). Moreover, the error-backpropagation algorithm is a very unbiological procedure and unlikely to occur anywhere in the central nervous system. However, the method can provide us with the principles of organization which real animals have acquired during evolution. Once the initial training phase is over (when the model lives its 'own life') it must be able to perform properly (optimally) in new environments, (*e.g.*, follow the relativity rule, as required in our wasp example), but the adaptations have to be accomplished without altering the degree of interconnectedness (*i.e.* without changing the weights).

Connectionist modelling provides us with a tool to analyse the principles of information processing. The reasoning is as follows. A wasp's information processing machine has evolved to work properly under different conditions of host-size distributions. It will always produce results, even under the (experimental) condition of only one single host size present. In such an (artificial) situation one may observe traces of the internal machinery. A comparable result would be impossible with conventional ethological models, because (*e.g.* in the case of the conventional computer metaphor with retrievable representations of the host-size distribution) the appropriate subroutine would never be activated. At the same time, the wasp's internal state, that cannot be rigidly controlled, has to be reconstructed. It is this complex problem of how to detect regularity by simultaneously making inference on intervening variables, that can be solved by connectionist modelling.

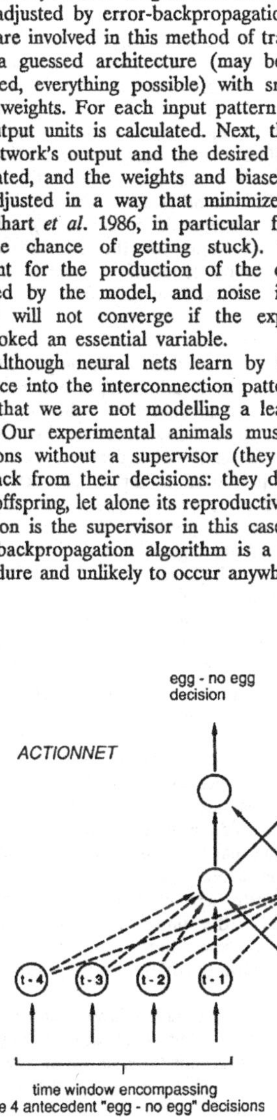

Figure 1. Lay-out of ACTIONNET. Dashed arrows: inhibitory connections; solid arrows: excitatory connections. Host sizes range between 0.9 and 1.8 mm; preceding oviposition decisions may take the value 1 (egg laid) or 0 (host rejected). A logistic transfer function is applied to calculate the activation of a unit.

4. EXPERIMENTAL ANALYSIS

Modelling procedures:

Our model ACTIONNET was trained with samples randomly drawn from a dataset on observed oviposition decisions of *Lariophagus*. Each sample was a training pair of input pattern and output target. Fig. 1 shows the architecture of ACTIONNET.

The input consisted of the current host size, and the preceding behaviour, *i.e.* the temporal pattern of host acceptances and host rejections in a window of 4 preceding time steps. The target of the output was the actual oviposition decision (unfertilized or fertilized egg, or host rejection) that had been observed to occur. Two output units were defined, one represented the 'egg - no egg' decision, the other the 'fertilization - no fertilization' decision. All units had a sigmoid, logistic transfer function, which implies an activation that ranges between 0 and 1, and increases monotonically as a function of its input. Error-backpropagation was applied until consistent responses were produced by the network. We assume that it has learned to respond to 'interesting' patterns in the input by that time.

Wasps:

Newly emerged (inseminated) *Lariophagus* females were collected from culture vials. They were presented with 2 hosts on the day prior to experimentation; on the day of experimentation they were 3 days old.

Host presentation in experiments:
The hosts, larvae of the granary weevil, *Sitophilus granarius* (L.) were exposed in a special device: the oviposition carousel (illustrated in Fig. 1 of Simbolotti *et al.* 1987). Individual females were presented with a sequence of 10 hosts; new hosts appeared at 2 hour intervals. Between sequences, host sizes differed, ranging between 0.9 and 1.8 mm. There were two replications.

5. ACTIONNET

What has been learned?
Under the present experimental conditions, the oviposition decisions at each of the preceding time steps are treated more or less equally: the connections of those units to the hidden units have equal weights. With the state space approach (McFarland & Houston 1981) we can analyse how the behavioural output is affected by its causal factors. The wasp's motivational state can be accommodated in a two-dimensional space, with one axis corresponding to the 'number of eggs laid in the time window', and the other to the 'size of the present host'. How are these states mapped to the oviposition and fertilization tendencies?

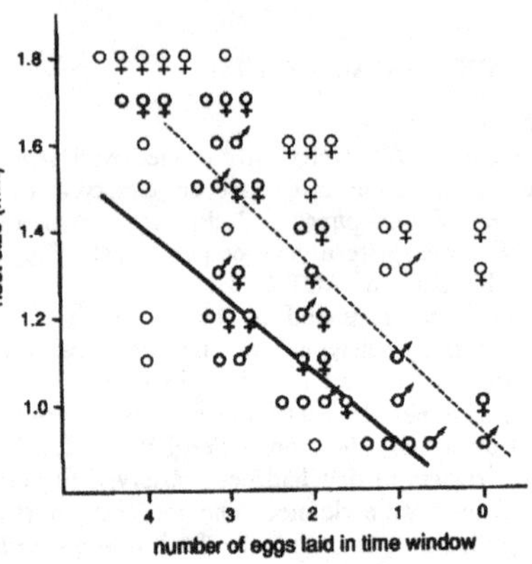

Figure 2. Oviposition decisions of Lariophagus as a function of the causal factors 'number of eggs laid in time window' and 'host size'.
(o = host not accepted; ♂ = unfertilized egg laid; ♀ = fertilized egg laid). The lines represent the limits where the combination of these factors results in rejection, oviposition without a fertilization, and oviposition with a fertilization, respectively, as calculated by the network. (—— = 100% oviposition isocline; - - - = 100% fertilization isocline) (From Putters & Vonk 1990).

Fig. 2 depicts the decisions made by the wasps in the experiments. Identical decisions appeared under the conditions 'wasp has laid few eggs, and encounters a small host' and 'wasp has laid many eggs, and encounters a large host'. We presented the network with these states, to assess the degree of the activation at the output units. The activation on the output units is at the maximum value (1.0) in the areas that are delineated by the 100% host acceptance and 100% fertilization isoclines in Fig. 2. For the other areas, this network (which is essentially a deterministic construct), calculates probabilities, which represent the strengths of corresponding behavioural tendencies. Indeed, the model mimics the wasp's decisions successfully.

Principles of organization:
Fig. 3 illustrates the activation of the output units when ACTIONNET was given various input patterns.

The network extracts successfully the significance of the factor 'host size'. Activations are low when the current host size is small and activations are high when a large host is presented as input. The activations of the units also depend on the preceding ovipositions: activations are low if ovipositions have been performed, whereas activations are high if hosts have been rejected (*i.e.* more eggs are kept in store) (Fig. 3). In other words, there appears to exist some kind of bookkeeping that keeps track of the number of eggs that were laid during the previous period, or of its complement: 'egg load'. The existence of this 'egg load' factor was predicted on theoretical grounds by Mangel (1987); it had already been recognized as a causal factor in egg laying decisions before (van den Assem *et al.* 1984), and its role had been modelled accordingly (Simbolotti *et al.* 1987; Putters & van den Assem 1988).

The factor host size is combined with antecedent behaviour by the hidden units, which might be interpreted as meaning that the sensitivity to external stimulation depends on antecedent behaviour. ACTIONNET discriminates poorly between hosts of different size if few eggs are laid in the time window. Also, the number of eggs laid in the time window doesn't have much influence on the activation of the output units, if a large host is presented.

The two units of the output layer both receive input from the hidden units, but to a different degree. The 'egg - no egg' decision unit 'fires' at lower activations of the underlying units. This suggests a thresholdtype of mechanism (Fig. 3).

Identification of the internal state variable:
In order to verify the role of the 'egg load' factor, a number of experiments have been done (Putters & Vonk 1991). For the present purposes, the effects of different egg loads, under the condition of identical experience are the most relevant ones.

Wasps were exposed to different temperature regimens; this results in different numbers of available,

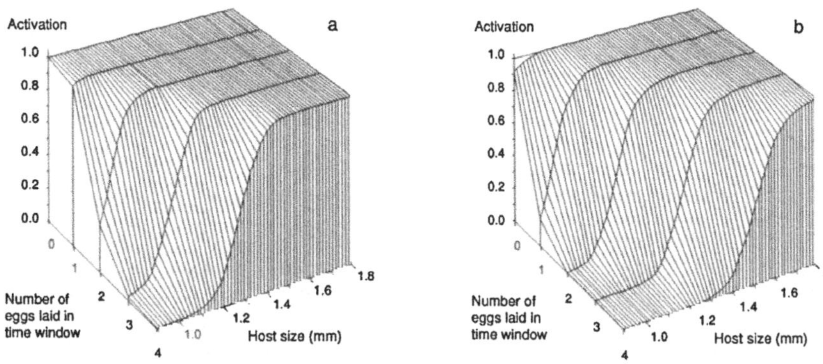

Figure 3. Activations of the output units of ACTIONNET as a function of the causal factors 'number of eggs laid in time window' and 'host size'. a: activation of the 'egg - no egg' decision unit. b: activation of the 'fertilization - no fertilization' decision unit.

mature eggs. Next, they were presented with a host. 42% of the medium sized hosts was rejected if the hosts followed a cold pause, and only 19% if they followed a warm pause. Large hosts were accepted equally well by both groups, but the proportion fertilization was much higher for the group with a warm pause (82 *vs* 37%).

Egg load may vary without experimental manipulation, even under conditions of similar experience. This variation has been used to analyse the effect of egg load on oviposition decisions in the following way: Wasps were presented with one host. After two hours, the wasps' remaining egg load was counted. Eventually egg load and the sex of the offspring that had been produced, were correlated. It appeared that the females that produced daughters had a larger egg load on average, than those that produced sons and, in turn, they had a larger egg load than those that did not oviposit. We can conclude that egg load has a predominant effect on the 'egg - no egg' and 'fertilization - no fertilization' decisions. It can probably be identified as the factor that was detected by the network as the effect of preceding ovipositions. This internal factor, in combination with host quality, can account for the observed decisions (Putters & van den Assem 1988). The model (Fig. 2) does not account for the 'no egg' decisions of the wasps under the condition: 'many eggs laid in an environment with large hosts'. The network would require additional units to reproduce this phenomenon (which does not result from rejections due to the state of the causal factors, but because of depletion of the egg store; the rate of oviposition overtakes the rate of egg maturation).

6. FACULTATIVE ADAPTATION: THE EMERGENCE OF FUNCTIONAL PROPERTIES

The target of any organism is to maximize its darwinian fitness. In this respect, it can achieve success, because it possesses adaptive, optimal traits that have been selected for in the course of evolution. As far as the optimal behavioural traits are concerned, the relevant problems could be solved by the computer metaphor or so it seemed: information was stored, retrieved, and combined by a rule-driven and controlled system, and the appropriate decisions were made. At first sight, the structure-oriented modelling approach seems unable to handle such rules. In connectionist models, the complex, high-level dynamics and the patterns of behavioural output emerge from the interconnection network as an inherent property. The fixed strengths of the connections, moulded by natural selection, set the pattern of activation and determine the eventual output of the system. The problem of how to maximize fitness has been solved by natural selection. The knowledge and beliefs necessary to perform this task, and obedience to rules should therefore be represented in the pattern of interconnections. Did the model manage to obtain these properties?

The proof of the model is to set it free in an environment, *i.e.*, we present the model with dynamic input: changes in the state variables occur as a consequence of a) changes in the environment (in practice we arranged encounters with hosts of different sizes: hosts of a certain size (here called the target host size) alternated with hosts that were either 0.4 mm smaller or 0.4 mm larger), b) the animal's own behaviour (the output 'egg laying' is fed back to the input in the next

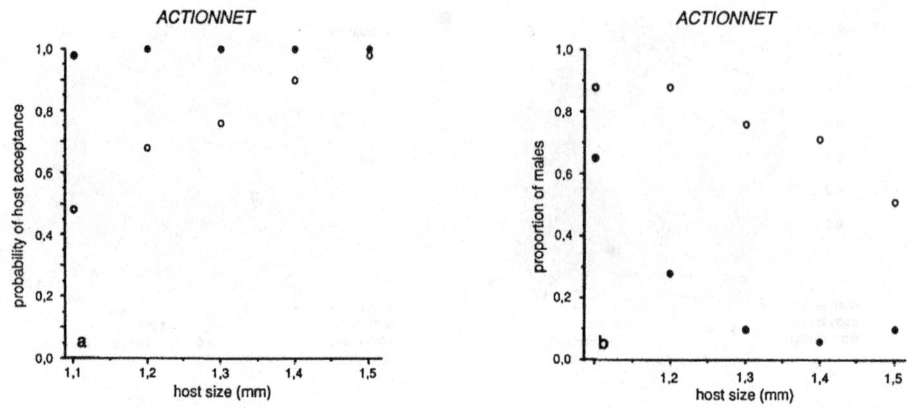

Figure 4. The dynamic properties of the network when presented with a sequence of two alternating host sizes (difference 0.4 mm). a: Influence of the size of the alternating host on the probability of host acceptance (o = target host size was the smaller of the two; • = target host size was the larger of the two) b: Influence of the size of the alternating host on the proportion of males (symbols as in a). (From Putters and Vonk 1990).

time step), and c) the moving time window (*i.e.*, a limited memory). Experimental evidence showed that the wasps' decisions on how to parasitize the target host size depends on the 'richness' of the environment (Charnov *et al.* 1981): when it was the smaller of the two, the proportion of males was high; when it was the larger one, the sex ratio decreased markedly.

Fig. 4 illustrates the output that was obtained by the simulation. Obedience to this rule of relativity turns out to be an inherent property of the model's structure. It is explained in Fig. 5 as the consequence of a simple mechanism: it appears that the target host size (1.3 mm in this case) was frequently encountered with different egg loads (compare, in Fig. 5, egg loads at the states R and P, in the rich and poor environment, respectively); consequently, different decisions appear.

7. UNIFIED EXPLANATION IN ETHOLOGY

Animals never act on a collection of explicit rules to solve problems that are posed by the environment, they only seem to do so. In reality, organisms work because they have been designed to work as they do, just as our model. Goals are not represented, and they do not direct behaviour, nor do the so-called 'rules-of-thumb'. Rules are a redescription of the system's output which is the result of the activation patterns of the underlying structure. The real rules that cause an animal to act upon, are the very simple ones, that were described by ACTIONNET. The information required to perform as if 'knowing the rules' is contained in the connection strengths between the processors. Likewise, the ability to store and retrieve information is provided by the fix-

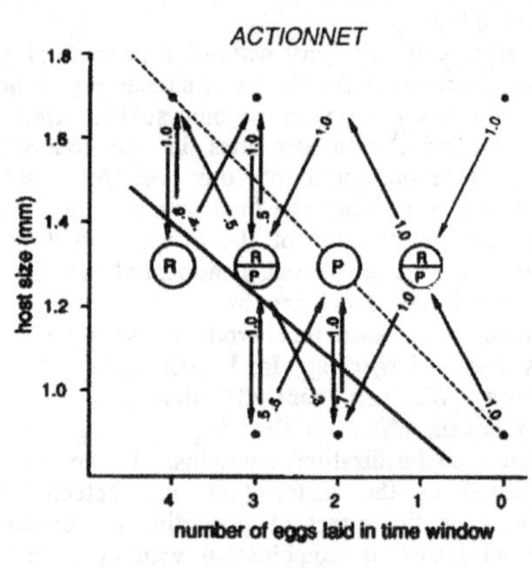

Figure 5. Transition probabilities between states of the network when it operates under conditions of alternating host sizes. R: target host encountered in a 'rich' environment (alternating host sizes 1.3 and 1.7 mm) P: target host encountered in a 'poor' environment (alternating host sizes 0.9 and 1.3 mm).(— = 100% oviposition isocline; - - - = 100% fertilization isocline) (From Putters & Vonk 1990).

ed connections: Experience is not represented as a stored image of the past; events are specified by states of activity that result from the system's internal organization and dynamics. Natural selection has moulded the connections and thereby adapted the system to the properties of the environment. Therefore, a holistic approach is required to solve the problems of simultaneous and dynamic decision making (McFarland 1977). This is exactly what our structure-oriented modelling approach can achieve!

ACKNOWLEDGEMENTS

Data were collected by R.L. Los-den Hartogh, who was subsidized by BION-grant 14.10.18. We wish to thank F.X. van Berge Henegouwen for his experimental work, M.J. Brittijn for preparing the drawings and G.G. Dijkstra for formatting the manuscript. We especially thank J. van den Assem for his helpful comments and the numerous improvements on previous drafts.

REFERENCES

Arbib M.A. (1989). *The metaphorical brain 2: Neural networks and beyond.* John Wiley & Sons, pp 1-458

Assem J. van den (1971). Some experiments on sex ratio and sex regulation in the pteromalid *Lariophagus distinguendus. Neth. J. Zool.* 21: 373-402

Assem J. van den, Putters F.A., Prins Th.C. (1984). Host quality effects on sex ratio of the parasitic wasp *Anisopteromalus calandrae* (Chalcidoidea, Pteromalidae). *Neth. J. Zool.* 34: 33-62

Assem J. van den, Iersel J.J.A. van, Los-den Hartogh R.L. (1989). Is being large more important for female than for male parasitic wasps? *Behaviour* 108: 160-195

Charnov E.L. (1979). The genetical evolution of patterns of sexuality: Darwinian fitness. *Amer. Nat.* 113: 465-480

Charnov E.L. (1982). *The theory of sex allocation.* Princeton University Press, pp 1-335

Charnov E.L., Los-den Hartogh R.L., Jones T.W., Assem J. van den (1981). Sex ratio evolution in a variable environment. *Nature* (London) 289: 27-33

Clausen C.D. (1939). The effect of host size upon the sex ratio of Hymenopterous parasites and its relation to methods of rearing and colonization. *J. N. Y. Ent. Soc.* 47: 1-9

Coderre B. (1989). Modeling behaviour in petworld. In: *Artificial life* (Langton C.G., ed). Addison-Wesley, p 407-420

Hamilton W.D. (1967). Extraordinary sex ratios. *Science* 156: 477-488

Hogeweg P. (1989). Mirror beyond mirror, puddles of life. In: *Artificial Life* (Langton C.G., ed). Addison-Wesley, p 297-316

Langton C.G. (1989). Artificial life. In: *Artificial Life* (Langton C.G., ed). Addison-Wesley, p 1-47

McFarland D.J. (1977). Decision-making in animals. *Nature* (London) 269: 15-21

McFarland D.J. (1989). *Problems of animal behaviour.* Longman Scientific and Technical, pp 1-158

McFarland D.J., Houston A. (1981). *Quantitative ethology: The state space approach.* Pitman, London, pp 1-204

Mangel M. (1987). Oviposition-site selection and clutch size in insects *J. Math. Biol.* 25: 1-22

Maynard Smith J. (1978). *The evolution of sex.* Cambridge University Press, pp 1-222

Putters F.A., Assem J. van den (1988). The analysis of partial preference in a parasitic wasp. *Anim. Behav.* 36: 933-935

Putters F.A., Vonk M. (1990). The structure-oriented approach in ethology: Sex-ratio adjustments in parasitic wasps. *Behaviour* 114: 148-160

Rhijn J.G. van (1977). The pattern of preening and other comfort behaviour in a herring gull. *Behaviour* 63: 71-109

Rhijn J.G. van, Westerterp-plantinga M.S. (1989). *Ethologie: Veroorzaking, ontwikkeling, functie en evolutie van gedrag.* Wolters-Noordhoff, Groningen, pp 1-425

Rumelhart D., Hinton G., Williams R. (1986). Learning internal representations by error propagation. In: *Parallel distributed processing, Explorations in the microstructure of cognition Vol 1: Foundations* (Rumelhart D., McClelland J., and the PDP Research Group, eds). MIT Press, Cambridge, p 318-362

Simbolotti G., Putters F.A., Assem J. van den (1987). Rates of attack and control of the offspring sex ratio in the parasitic wasp *Lariophagus distinguendus* in an environment where host quality varies. *Behaviour* 100: p 1-32

Toates F.M. (1986). *Motivational systems.* Cambridge University Press, Cambridge, pp 1-188

Travers M. (1989). Animal construction kits. In: *Artificial life* (Langton C.G., ed). Addison-Wesley, p 421-442

Werren J.H. (1983). Sex ratio evolution under local mate competition in a parasitic wasp. *Evolution* 37: 116-124

Werren J.H. (1984). A model for sex ratio selection in parasitic wasps: local mate competition and host quality effects. *Neth. J. Zool.* 34: 81-96

Werren J.H., Simbolotti G. (1989). Combined effects of host quality and local mate competition on sex allocation in *Lariophagus distinguendus*. *Evol. Ecol.* 3: 203-213

Evolution as pattern processing: TODO as substrate for evolution

P. Hogeweg & B. Hesper
Bioinformatica Padualaan 8,
3584 CH Utrecht the Netherlands
EMAIL: HESPER@RULCRI.Leiden.Univ.nl

Abstract

Evolution is most often viewed (and formalised) as an optimisation process. In this paper we suggest that a useful alternative heuristic may be to view (and formalise) evolution as a self-enhancing pattern transformation, pattern detection and pattern generation process. This suggestion is based on the growing awareness of the emergence of complex behaviour from simple environmentally dependent action rules when they operate in a structured environment, and the structuring of the environment by such behavioural patterns. We describe paradigm worlds, which suggest that such emergent behaviour may underlie behaviour patterns observed in various species. We show how the emergence of macro-behaviour patterns can be interpreted as a form of pattern processing by the action rules on the environment. We suggest that these emergent patterns function as a prepattern for evolutionary processes: evolution fixates and enhances these patterns.

1.Introduction

Ever since Darwin's profound insight in equating 'survival' and 'fitness' for self-replicating entities, evolutionary theory has had a strong footing in terms of optimisation processes. This optimisation viewpoint pervades most biological thinking about evolution and adaptation. Population genetics is entirely formulated in such terms, and traits of organisms are customarily 'explained' in terms of their 'fitness', recently in particular in sociobiology and behavioural ecology. Genetic Algorithms (Holland 1976, Goldberg 1989) have used basic 'genetic mechanisms' for solving general optimisation problems. Only relatively recently have quantitative studies begun to expose the constraints on a 'mutation selection' process leading to appreciable optimisation (after all: 'optimisation by 'survival of the fittest' is not a tautology). Eigen and Shuster (1979) exposed the 'error threshold', i.e. they showed that only a limited amount of mutation is compatible with evolutionary optimisation. Kauffman (Kauffman & Smith 1986 Kauffman 1989a,b) stressed that optimisation is only possible in not too rugged fitness landscapes, i.e only if similar genotypes have,in general, similar fitnesses. Rugged fitness landscapes result from extensive coupling between genes, by which the system exhibits strong selforganising properties. Such selforganisation thus seems to be a constraint on evolutionary optimisation.

All these approaches use an external, apriori, 'user' imposed fitness criterion. Only a few models are studied, in which only survival determines the evolutionary process in a coevolutionary context. (Conrad & Rizki 1989, Packard 1988, Holland 1990, Kauffman 1989b). In such models 'fitness' is not clearly defined, and fitness landscapes are wildly dynamic entities, if they can be visualised at all.

In this paper we propose that it may be worthwhile to view evolutionary processes not as primarily an optimisation process, but instead as pattern processing (i.e. as pattern detection, pattern transformation or pattern generation). Such a view is of course entirely compatible with the optimisation viewpoint, but provides different heuristics for studying such processes. In particular we should like to study what may be called the generation of 'fitness' dimensions, rather than walks through fixed fitness landscapes, although we prefer a terminology like 'pattern of survival' rather than fitness. The insight that pattern recognition and pattern detection can be studied in terms of energy minimisation ('optimisation' as used in evolutionary theory) (Hopfield 1984, Ackley et al. 1985 and the extensive literature which followed this conceptualisation) has led to important new models and machines for pattern recognition/ pattern detection. We hope that a similar, but reversed, change of viewpoint with respect to evolution (i.e. from optimisation to pattern detection) will likewise lead to new models elucidating (and possibly machines, exhibiting) innovative evolutionary adaptation.

2. TODO and emergent behaviour

2.1 Introduction to the TODO principle.

The potential of local rules to generate complex behaviour in interaction with a structured environment was first hinted at by Simon (1969) in his phrase: "an Ant viewed as a behaving system is quite simple, the apparent complexity of its behaviour in time is largely a reflexion of the complexity of the environment in which it finds itself..."; "a Man viewed as a behaving system is quite simple, the apparent complexity of his behaviour in time is largely a reflexion of the complexity of the environment in which he finds himself...". Simon, and with him most Artificial Intelligence research, have concentrated entirely on humans, and have in practice dismissed this phrase as an irony. Nor have those studying animal behaviour taken the hint seriously, they have continued to study behaviour virtually independent of the environment or they have paid attention to the environment as a constraint on (optimising) behaviour only. By contrast our own research has been in the direction of Simon's pointer, but has gone beyond it by studying emergent self-regulating behaviour generated by entities governed by local rules in apriori lowly structured environments. We call the basic mechanism underlying this self-structuring the TODO principle: the entities 'do what there is to do'; what there is to do changes by their doings. In the following subsections we briefly summarise some insights which emerged from studying worlds governed by the TODO principle.

2.2. Flexibility vs adaptation.

Food seeking behaviour is a paradigm which has been frequently used in studying various forms of (adaptive) behaviour specification (e.g. Booker (1982), Wilson 1985), Coderre (1988), Travis (1988)). It is instructive to compare the behaviour of an extremely simple TODO specification with that obtained by a sophisticated learning scheme. We did this for ANIMAT (Wilson 1985) (whose behaviour is governed by a classifier system (Holland 1986)) and compared it with SMARTY (with TODO based behaviour, which has the form "if you see food eat it; otherwise go straight") (Hogeweg 1989). We used the same environment, the same 'sensors', and the same experimental procedure in both cases(i.e. those reported by Wilson 1985). This experimental procedure admits of only half of the basic possibilities of TODO based entities: the environment is kept constant and therefore 'what there is to do' is not influenced by 'what is done'. This exercise makes clear that:

1. It is much easier to formulate SMARTY than it is to formulate ANIMAT and much less adhoc choices and tuning are necessary; 'hidden assumptions' play less crucial and unexpected roles (e.g. global or 'egocentric' coding of direction).
2: SMARTY is a more efficient foodseeker than the best ANIMATs obtained.
3. When SMARTY is brought in a moderately different environment its behaviour degenerates less than that of ANIMAT. Also, in the long run, ANIMAT has a hard time to 'adapt' to the new environment, whereas SMARTY's behaviour is flexible enough to cope.

This example illustrates a conflict between adaptability and flexibility. Information-rich representations of the environment (such as ANIMAT builds), which are reflected in local behavioural rules, impairs flexibility to cope with new environments. Coping with new environments is of course of paramount importance. It is an open question how to evolve flexibility: a variable environment is not enough (Rizki & Conrad 1986).

2.3. Emergent behaviour patterns.

The previous example already suggests that information-poor (flexible) local behaviour specification leads to larger scale behaviour in which patterns in the environment are reflected, e.g. in a poor environment SMARTY will move straight over long distances, whereas it will zigzag and remain in a small area in richer environments. In this case this is not very striking because the largescale behaviour is expressed in the same type of terms as the local behaviour. However the macrobehaviour generated is not restricted to behaviour dimensions represented in the local rules.

Using again the foodgathering paradigm we studied such emergent patterns in MIRRORworlds inhabited by CROWNs (Hogeweg & Hesper 1990), CHIMPs (Hogeweg & te Boekhorst 1990), and ORANGs (te Boekhorst & Hogeweg in prep). These worlds were shaped such that they resemble in some aspects the worlds of Crown of Thorns (a starfish from the coral reefs), chimpanzees and orang-utans respectively. In all cases individuals move to the nearest foodsource and eat it; moreover they may move towards each other. The resulting macrobehavioural patterns resemble closely those found in the corresponding REALworlds:

1.The CROWNs are outbreak prone: i.e. they are inconspicuous and barely influence their environments upto a certain density: slightly above that density they aggregate into large groups and entirely deplete the environment within a short time. Likewise Crown of Thorns depletes coral reefs, changing the entire ecosystem drastically for long periods (e.g. Endean & Stablum, 1975; Moran, 1988). Sudden large increases of the population size is the most often assumed explanation. In contrast, in the CROWNworld the potentially large travelling speeds and the size of the CROWNs the usually large abundance of food (coral) and the local coral regrowth are the features responsible for this dramatic phenomenon.

2. In the CHIMP world we find travel groups of MALEs (who in search for mates approach every CHIMPlike entity in view), and lone FEMALEs. This group structure resembles closely the group structure (= social structure as defined by Rowell 1972) found in chimpanzees (Halparin 1979). The explanation of the occurrence of all male groups is usually sought in terms of defense against neighbouring groups. In the CHIMP world it is due to the food availability and the mutual CHIMPseeking behaviour.

3. In the ORANG world we find travelling bands which form in small fruit trees, rather than in large ones. Travelling bands were found in orang utans in the season in which many small trees have fruits (Sugardjito et al 1987), which was explained in terms of a compromise between a 'social urge' and avoidance of food competition. In the ORANG world it is a side effect of going to the nearest food source and eating until satisfied. This behaviour leads to travelling bands when food sources are depleted before the eaters are satisfied, and when enough food sources are available. Reexamination of the data (te Boekhorst & Hogeweg in prep.) shows that this is indeed the way in which travelling bands are formed in the orang utans.

These examples show:

1. the large impact of the interplay between simple TODO rules and a specific environment; it creates the environment in which the entities regularly find themselves: the depleted coral, the all male CHIMP groups and the seasonal travelling

ORANG bands.
2. How the patterns found in the TODO based artificial worlds reflect patterns found in REALworlds.
3. The apparently superfluous mechanisms and concepts used in the common explanation of these REALworld patterns.

The emergence of new environmental features by the interaction of TODO rules with the environment can be viewed as the emergence of new 'fitness' dimensions.

2.4. Selfregulation of TODO based emergent patterns.

The patterns discussed above, emerging from food seeking behaviour, are fairly robust to changes in the environmental conditions. This invariance is caused by a subtle interplay between changes in the environment and the resulting changes in the actions of the entities (see Hogeweg & te Boekhorst 1990, Hogeweg & Hesper 1990). However, the selfregulation is more clearly demonstrated in our older work on socioinformatic processes, i.e. TODO based interactions restricted to initially identical entities, which may, however, differentiate due to the interactions in which they participate. We studied socioinformatic processes in BUMBLE's (Hogeweg & Hesper 1983,1985). BUMBLEs are entities who reflect the lifehistory patterns of bumble-bees: EGG, LARVA, PUPA, WORKER, QUEEN and DRONE stages are recognised. The TODO is triggered by the type (stage) of BUMBLE met: a LARVA is fed when a WORKER or QUEEN meets it etc. When adult BUMBLEs meet they engage in a so called DODOM interaction: with a probability proportional to the value of their respective DOM parameter one of them 'wins'. The outcome of the interaction results in an updating of the DOM parameters: The DOM parameter of the winning BUMBLE is increased with an amount inversely proportional to the probability of winning (and vice versa for the losing BUMBLE). Thus, a damped positive feedback of DOM values ensues. The DODOM interaction together with the maintenance interactions generate many features observed in life bumblebee nests: a stable differentiation between elite and common workers, a 'rebellion' of the worker at the end of the season, chaotic interactions afterwards, egglaying by only the elite workers.

Moreover, the emergent features are strongly selfregulating. For example, 'compensatory feeding' (Pendrel & Plowright 1981) is observed after the removal of part of the worker force. In the BUMBLE world this is a simple consequence of the TODO structure: WORKERS meet more often a LARVA when there are fewer WORKERs, and therefore more often feed them; in bumblebees an inspection behaviour was thought to be responsible for this (adaptive) behaviour. Even more interesting, the timing of the worker rebellion is invariant for a large range of growth rates of the nest. This is clearly important because, according to ergonomic optimality theory (Oster & Wilson 1978), the rebellion, and therewith the switch to the production of generative offspring, should occur at the end of the season, when there is just enough time left to rear the generative offspring. The invariance of this 'socially generated clock' works via the differentiation in common workers and elite workers. In fast growing nests this differentiation is less pronounced than in slow growing nests. Therefore a larger worker force is needed to remove the QUEEN from the nest. Thus, in very slow growing nests where ergonomic consideration do no longer apply, rebellion occurs earlier, as is indeed observed in REAL nests.

This example clearly demonstrates the tendency of TODO, 'do what there is to do', to generate automatically 'what should be done', i.e. to generate automatically 'adaptive behaviour'.

2.5.TODO as basis of building, behaving BEINGS.

The TODO based worlds, discussed so far, are all entirely symbolic, and are simulated on digital computers. Such a symbolic representation can both obscure and generate 'hard' problems (Pattee 1988, Brooks 1989) Profoundly new insights can therefore be obtained by actually building TODO based entities which live in the REALworld. Brooks (Brooks 1986,1989 Brooks & Flynn 1989, Flynn & Brooks 1990) is doing just this; his goal is to build autonomous BEINGS which can subsist in some real environment, pursueing their own 'goals'. Unlike previous robotics research, but similar to the pre-history of robotics, i.e. the Grey Walter turtles (Grey Walter 1953), and to the above discussed line of research, Brooks' BEINGs do not try to build a global model of the world in which they live, but instead use the actual (local) world as the best possible model of the world(Brooks 1989). They react on the input provided by their sensors in a TODO-like manner. However, a real environment is much more complex than the simulated environments discussed so far; behaviour of a BEING is therefore governed by a number of simultaneously operating TODO like processes, each using their own observations of the local environment. They interact at the action level, rather than cooperating to build a more complete representation of the situation in which the BEING finds itself. Apart from state changes in the environment (which trigger other TODOs) Brooks uses for the interaction of the TODO like ACTORs his 'subsumption network' in which lower level processes (e.g. 'do not bump into objects') can inhibit higher level processes (e.g. 'explore').

An important design feature is the autonomy of the lower layers of the subsumption network: they are functionally independent of the presence of the higher layers. Moreover the higher levels are also in a sense 'autonomous' but in such a way that the world with which they interact is 'shaped' by the lower level processes (e.g. is shaped such that bumping into things does not occur). Modularity facilitates the design enourmously. Similar modularity apparently exists in biotic systems. However models of evolution usually do not exploit such modularity (see section 4).

2.6 Multiple TODO based worlds.

The TODO principle, as discussed sofar, applies the TODO rules on the basis of information that is locally available in the environment in which the entities dwell. Environments in which the entities can extract this local information and can survive on the basis of it, may be called 'benign' for the entities under consideration. In less benign environments, information obtained from the environment has to be supplemented, in order to survive, by information obtained from an internal model of the environment. Thus an internal model may serve as a 'crutch' for the inability to extract the information directly from the environment, and/or can serve as a means to obtain more global information. This can be seen by comparing the mechanisms for chemotaxis in bacteria and protozoa. The latter observe a chemotactic gradient directly by comparison of concentrations 'in front' and 'behind'. The size of the former is too small to elevate this difference above noise level, and bacteria therefore use an internal model, i.e. a memory of the concentration, to decide whether to go straight or to turn. (This suggests that good observers can afford to be bad theoreticians and vice-versa! compare drowsing rod practicioners, whose successes are probably due to keen local observation). It should be noted that in the case of the CHIMPs and ORANGs, discussed above, the observations used (i.e. the distance in which fruittrees were spotted) is probably beyond the observational capacities of chimpanzees and orangutans; instead these animals are assumed by fieldworkers to have an internal model of the environment such that they 'know' whether a fruittree is present at such distances (Goodall 1986, Rijksen 1978).

The internal model needed in, in this sense, less 'benign environments might take a similar form as the TODO based worlds discused so far (e.g. Hogeweg & Hesper 1986,1988). For example, Maes (1989,1990) proposed as an alternative to the (hierarchical) subsumption architecture of (Brooks (1986), a more 'heterarchical', but also low information content interaction between TODO-like ACTORS These ACTORs change each others environment in the sense of adding and deleting preconditions; moreover they pass activity levels toward each other. Maes (1990) claims that by such an architecture the ACTORs collectively 'do the right thing', and continue to do so when new ACTORs are added to the network.

The similarity between such an ACTOR network and the TODO based models discussed above suggests a similarity with respect to emergence of macro-patterns (compare Hogeweg & Hesper 1988). Thus, multiple levels of TODO based world can be envisioned as the basis for 'complex' behaviour.

3. TODO based pattern processing.

All examples reviewed above show that collections of simple TODO based entities can be viewed as pattern processing devices. This pattern processing can be viewed as pattern detection, pattern transformation or pattern generation whichever image one prefers. Although these are formally equivalent, they indicate different impacts of the pattern processing: we examine them in turn.

1. TODO as pattern detection. The robustness of the macropatterns studied above for some differences in the environment can be interpreted in terms of pattern detection (classification): environments are classified into those generating the macropatterns discussed, and those generating some other pattern (compare e.g. CHIMPs vs ORANGs, although the TODOs are not exactly identical). The generated classes of the environment are 'labelled' with the macropattern generated, i.e. a qualitative difference is added to the initially present gradual differences.

2. TODO as pattern transformation (pattern association). Patterns in the environment are transformed via TODO rules into what is actually done. Equivalently one might say that patterns in the environment are associated with behaviour patterns. As we have seen the robustness of some macrobehavioural features is obtained by subtle and multiple changes in what is actually done. Thus, in viewing TODO based behaviour as a pattern transformation process, we do not focus on invariant macropatterns as is the case with pattern detection, but focus on the variation in the macrobehaviour, which is caused by changes in the environmental conditions. Moreover, we focus on the change in representation of the patterns: from environmental patterns to patterns of behaviour. Such transformation potentially renders the patterns observable for entities equipped with other sensors (and other TODOs).

3. TODO as pattern generation. TODO based pattern processing tends to be selfenhancing. Environmental patterns which are transformed into behavioural patterns are doubly represented and therefore will be preferentially detected. This is in particular true for invariant features generated by a TODO in a set of environments: invariances are easily detectable. This selfenhancement, together with the transformation of scale of the patterns, and the introduction of new representations of the pattern makes that the pattern processing can often best be regarded as pattern generation: similar pattern processing may underlie processes primarily observed as pattern generation processes.

Examples discussed in section 2 show how TODO based pattern processing viewed as pattern detection, pattern transformation, or pattern generation, provides alternative explanations for patterns of behaviour usually explained in terms of evolutionary optimisation. TODO based pattern processing may, moreover, provide prepatterns on which evolutionary processes operate (section 4).

4. TODO as substrate for evolution.

The emergent adaptive processes we have considered operate at a short (behaviour) time scale, whereas evolutionary adaptation is of course a long term process. Long term processes are usually considered to determine the short term behaviour. Coupling between 'genes' results in emergent patterns and rugged fitness landscapes. This constraints optimisation of an apriori chosen fitness criterion (expressed in terms of 'gene'states) (Kauffman 1986, 1988a,b). Thus, emergent patterns and evolution seem to oppose each other. In particular for autonomous evolutionary processes (i.e. evolutionary processes without a-priori specified fitness criterion) it is important to consider the influence of short term emergent patterns on long term evolutionary processes as well. This influence is multifacetted:

1. (Spatial) Patterning changes basic properties of evolutionary processes: e.g. spiral patterns emerging from hypercyclic interaction (Eigen & Shuster 1979) in poorly mixed media cause positive selection for 'altruistic' traits (hypercycles become resistent to parasites) (Boerlijst & Hogeweg 1990)

2. The 'pattern detection' or 'pattern generation' properties of TODO rules (as described above) create (new) fitness dimensions, and (new) fitness criteria. As mentioned e.g. 'social' behaviour becomes a fitness dimension when feeding interactions create (large) parties. At least as important: stable emergent patterns will lead to evolutionary loss of TODO's for situations never (seldom) encountered. This leads to the formation of the generated (detected) pattern becoming itself a fitness dimension.

3. The 'pattern transformation' properties of TODO rules (as described above) create prepatterns of regulatory 'adaptations' to produce the same emergent (generated/detected) patterns in different environments. Long term evolutionary processes may tend to strengthen such prepatterns (viz 2. above). Interestingly, Flynn & Brooks (1989) note that layers added in the subsumption hierarchy to improve the walking (and climbing) abilities of an 'insect' with 6 separately controlled legs, sometimes only strengthens emergent behaviour already present in the system due to e.g. gravitation (e.g. "bend the leg on the ground when an other leg cannot reach a foothold").

Evolutionary (and learning) processes are most often studied in isolation: the device is supposed to be unstructured relative to the problem to be learned, i.e. it is initialised with as little prepattern as possible, i.e. the device is initialised 'randomly'. (viz. classical evolution theory (one trait, one de novo gene), genetic algorithm (random initial population, but see Huynen and Hogeweg (1989)), neural networks (random initial (internal) state in pattern recognition; random or uniform initial weights in learning)).

Absence of 'prepatterns' is certainly not the case for any 'real world' evolutionary (or learning) process. It always starts in a 'viable' state in which its interactions with the environment can be seen as governed by TODO rules. At the very least gross physical TODO's are present (e.g. "if no support then fall down" (Flynn & Brooks 1989)). As discussed, TODO rules lead to prepatterns which form a substrate for (autonomous) evolutionary processes.

Prepatterns are ignored when learning systems are studied in isolation. Such an approach can be called 'holistic' or 'reductionistic', opposites which are equivalent in the sense that interesting structures are left unanalysed. The pattern processing image of simple TODO rules sketched above, and the image of emerging patterns as substrate for evolution may lead to a modular description of evolving and evolved systems.

References

Ackley, D.H. Hinton, G.E. & Sejnowski, T.J. (1985) A learning algorithm for Boltzmann machines. Cognitive Science 9: 147-169

teBoekhorst I.J.A. & Hogeweg P. (1990) Tree size and travel bands in orang-utans: data-analysis suggested by a model (in prep).

Boerlijst, M.C. & Hogeweg, P. (1990) Spiral wave structure in prebiotic evolution: Hypercycle stable against parasites. Physica D (submitted)

Booker, L.(1982) Intelligent Behaviour as an adaptation to a task environment PhD. Dissertation. The Univ. of Michigan

Brooks, R.A. (1986) A robust layered control system for a mobile robot. IEEE Journal of Robotics and Automation. RA-2: 14-23.

Brooks R.A. (1989) The whole Iguana. in: Brady M. (ed.) Robotics science MIT press pp: 432-458

Brooks R.A. & Flynn A.M. (1989b) A robot being. in NATO workshop on robotics and biological systems, Tuscany, Italy.

Coderre, B. (1988) Modelling Behaviour in Petworld. in: C. Lanton (ed.) Artificial Life. Addison Wesley Publ Co pp:407-420

Conrad M. & Rizki M. (1989) The Artificial world approach to emergent evolution. Biosystems 23: 247-260

Eigen M. & Shuster P. (1979) The Hypercycle, a principle of natural selforganisation Springer Verlag.

Endean R. & Stablum, W. (1975) Population explosions of Acanthaster planci and associated destruction of the hard coral cover of reefs in the Great Barrier Reef, Australia. Environm Conserv 2: 247-259.

Flynn A.M. & Brooks R.A. (1989) Building Robots:

expectations and experiences

Goldberg, D.E. (1989) Genetic Algorithms in search, optimisation and machine learning. Addison Wesley Publ. Co.

Goodall J. (1986) The Chimpanzees of Gombe. Belknap, Cambridge (MA)

Halperin S.D. (1979): Temporary association patterns in free -ranging chimpanzees: an assessment of individual grouping preferences. In: Hamburg DA, McCown E.R. (eds) The great apes. Benjamin/Cummings, Menlo Park (Calif.), pp: 491-499.

Hogeweg P. & Hesper B. (1983) The ontogeny of the interaction structure in Bumblebee colonies: a MIRROR model. Behav. Ecol Sociobiol. 12: 271-283.

Hogeweg P. & Hesper B. (1985) Socioinformatic processes, a MIRROR modelling methodology. J Theor Biol. 113: 311-330.

Hogeweg P. & Hesper B. (1986) Knowledge seeking in variable structure models. in Modelling & Simulation in the Artificial Intelligence Era (ed. Elzas, Oren & Klir) North Holland pp: 227-243

Hogeweg P. & Hesper B. (1987) Simulation Modelling Formalism: Heterarchical Systems. in Systems & Control Encyclopedia (ed. M.G. Singh) Pergamon Press pp: 4350-4353

Hogeweg P. (1988) MIRROR beyond MIRROR, puddles of Life. Artificial Life, SFI Studies in the Sciences of Complexity (ed C. Langton) Addison Wesley Publ. Co pp: 297-315

Hogeweg P. (1989) Simplicity and Complexity in MIRROR universes. Biosystems 23 231-246

Hogeweg P. & Hesper B. (1990) Crowns crowding, an individual oriented model of the Acanthaster phenomenon. The acanthaster phenomenon. (ed. R. Bradbury) Springer Lecture Notes in Mathematical Biology (in press)

Hogeweg, P. & te Boekhorst I.J.A. (1990) On the natural history of an artificial CHIMP world: group structure as a side-effect of feeding behavior. (submitted)

Holland J.H.(1976) Adaptation in natural & Artificial systems Univ. of Michigan Press, Ann Arbor

Holland, J.H. (1986) Escaping Brittleness: the possibilities of general purpose learning algorithms applied to paralell rule-based systems. in: R.S. Michalski et al. (eds.) Machine Learning Vol. 2 Morgan Kauffman, Los Alamos Ca.

Holland, J.H. (1990) Oral presentation at Artifial Life II Santa Fe.

Hopfield J.J. (1984) Neurons with graded response have collective computational properties like those of a two-state neurons PNAS 81: 3088-3092

Kauffman, S.A. & Smith, R.G. (1986) Adaptive automata based on darwinian selection. Physica D 22: 68-82

Kauffman, S.A. (1989a) Adaptation in rugged fitness landscapes. in: D.L. Stein (ed.) Lectures in the sciences of complexity Addison Wesley Publ. Co pp: 527-618

Kauffman, S.A. (1989b) Principles of adaptation in complex systems. in: D.L. Stein (ed.) Lectures in the sciences of complexity Addison Wesley Publ. Co pp: 619-712

Maes, P. (1989) The dynamics of action selection. IJCAI 1989.

Maes, P. (1990) How to do the right thing.

Moran, P.J. (1988) The Acanthaster Phenomenon, Australian Institute of Marine Science Monograph series no 7. Townsville, Australia.

Oster G.F.& Wilson E.O.(1978) Caste & Ecology in the social insects. Princeton Univ. Press Princeton.

Packard, N. (1989) Evolving bugs in a simulated ecosystem. in:Langton C. (ed.) Artificial Life. Addison Wesley . pp:141-156.

Pattee H.H. (1988) Simulations. Realisations and theories of life. in: C. Langton (ed.) Artificial Life, Addison Wesley, pp:49-63

Pendrel, A.A. & Plowright, R.C. (1981) Larval feeding by adult bumble bee workers (hymenoptera: Apidae) Behav. Ecol Sociobiol 8: 71-76

Rizki M.M. & Conrad M. (1986) Computing the theory of evolution. Physica D 22: 83-99.

Rowell T.E. (1972) The social behaviour of monkeys. Pinguin books Baltimore

Rijksen, H.D. (1978) A Field Study on Sumatran Orang-utans (Pongo pygmaeus abelii, Lesson 1982): Ecology, behaviour & conservation. H. Veenman & Zonen, Wageningen (Netherlands).

Simon, H. (1969) The Sciences of the Artificial, MIT press.

Sugardjito J, teBoekhorst I.J.A.& van Hooff J.A.R.A.M. (1987) Ecological constraints on the grouping of wild orang-utans (Pongo pygmaeus) in the Gunung Leuser National Park Sumatera, Indonesia. IntJ. Primatol 8:17-41.

Travers, M. (1988) Animal construction kits. in: C. Langton (ed.) Artificial Life. Addison Wesley pp. 421-442

Walter, W.G. (1953) The Living Brain. Norton, New York.

Wilson, S.W. (1985) Knowledge growth in an artificial animal in: J.J. Grefenstette (ed.) Genetic Algorithms. (Carnegie Mellon, Pittsburg PA).

Matching, maximizing and melioration as alternative descriptions of behaviour

Alasdair I. Houston
NERC Unit of Behavioral Ecology
Department of Zoology
University of Oxford

Abstract

This paper reviews some issues concerning the use of optimality principles in the context of animal behaviour. After a brief discussion of constraints, I look at matching and maximizing as accounts of behaviour. As long as maximization is used descriptively there is no conflict between these approaches. They are alternative forms of description; in general it is possible to convert one form of description into the other. Melioration has been suggested as a priciple that underlies matching. When faced with two alternatives melioration requires that the animal increases its allocation of time to the alternative with the higher local rate of reinforcement. When the local rates become equal, an equilibrium is attained at which matching occurs. I present a simple model of melioration in which this equilibrium is not necessarily stable. It is shown that time allocations can be periodic or chaotic, and that matching does not necessarily occur.

1. Introduction

The use of optimality principles in biology remains controversial (see for example Maynard Smith 1978, Gould and Lewontin 1979, Williams 1985, Ollason 1987, Dupré 1987). In this paper I make no attempt to review all the issues. After some remarks on the problem of constraints, I concentrate on alternatives to optimality that have emerged in operant psychology. It is argued that some of the confusion in this area arises because of ambiguities concerning the status of optimality as an account of behaviour. I then consider a dynamic principle called melioration. This is a local optimization principle that may not always result in optimal behaviour. I show that under some circumstances a form of melioration can result in chaotic behaviour.

2. Constraints

Gould and Lewontin (1979) use the central dome of St Mark's, Venice as a starting point for their attack on optimality. The dome rests on four pendentives, which Gould and Lewontin refer to as spandrels. (The term "spandrel" is usually restricted to 2-dimensional surfaces of the sort shown in Figure 1.) Each pendentive is decorated with a mosaic of one of the four evangelists, together with his associated river. Gould and Lewontin suggest that the design could result in the view that the architecture is secondary to it. They contrast this with the proper path of analysis's that 'begins with an architectural constraint: the necessary four spandrels and their tapering triangular form'. (p582)

It is important to realise that this is not an argument against an optimality analysis, but merely an illustration of the importance of asking the right question. As Gould and Lewontin (1979) state, the mosaics clearly serve the function of expressing the Christian faith. We can look at the decoration as a whole in this light, and we can compare the decoration of restricted triangular spaces in various domed churches. The architecture specifies the constraints within which an "optimality" analysis of the decoration operates.

Florence cathedral had to be constructed to cover a crossing designed by other architects (e.g. Murray 1986, White 1966), and William Wynford's nave at Winchester encases its Norman predecessor (Harvey, 1978), yet both buildings are much admired. (To take the point even further, perhaps all art can be seen as constrained either by previous tradition or by self-imposed limits, such as Jane Austen's "little bit ... of ivory on which I work" - Letter to J Edward Austen, Mon Dec 16th 1816.)

Gould and Lewontin (1979) give the impression that a dome above a square requires pendentives. There are, however, two basic solutions to this problem - pendentives and squinches (see Figure 1). Reviewing the evidence, Talbot Rice (1968) suggests that the squinch arose in Persia in the third century, whereas the pendentive is probably of Syrian origin. From its invention there in the second or third century, it spread to Byzantium, Italy and France. (See Maidstone (1988) for the background to the dome of Hagia Sophia, Istanbul.) This example illustrates the danger of inferring a constraint from observed uniformities. If our knowledge is limited to churches like Hagia Sophia and St Mark's, then it may be tempting to see pendentives as a necessary constraint. Such a notion would be dispelled by the squinches of Daphni, which are decorated to dramatic effect (Talbot Rice 1967). It is all too easy to invoke constraints to justify existing structures or behaviours - indeed the quote from Candide that Gould and Lewontin (1979) use to ridicule the adaptationist program is much more appropriate as a description of constraints. By its very nature, an optimality approach must consider alternatives. The flipside of this is that with enough constraints, anything can be made (trivially) optimal. It is probably not a coincidence that the examples attacked by Gould and Lewontin (1979) involve verbal suggestions about function rather than mathematical analysis. Merely suggesting that a certain structure or behaviour serves a certain function does not require alternatives to be considered.

It can be argued that far from ignoring constraints, a formal optimality analysis forces one to be explicit about them (Maynard Smith 1982). The architectural example showed that an assumed constraint (spandrels) was not the only possibility . Similarly, what is assumed to be a constraint under one set of conditions may turn out to show systematic variation across a wider range of conditions (Newman et al 1988). In such cases one can perform a "local" optimality analysis against the background of the rest of the animal's behaviour in a given condition (Houston and McNamara 1985).

3. Matching and Maximizing

The matching law (Herrnstein 1961, 1970, Baum 1974) gives a simple characterisation of how behaviour is allocated between alternatives once equilibrium has been reached. At the heart of matching is the idea that some sort of local rate is equalised across alternatives. Another approach, based on economics, assumes that behaviour is allocated in such a way that something ("utility") is maximized (see for example, Rachlin 1980, Rachlin et al 1981). There has been considerable debate concerning the relative advantages of these approaches (eg Staddon and Motheral 1978, Heyman and Luce 1979, Baum 1981, Rachlin et al 1981, Prelec 1982, Boelens 1984, Rachlin et al 1988).

McFarland (1977) points out that the application of optimality principles to behaviour can be justified in two ways:

(i) an consistent decision-maker can be described by an optimality principle

Fig. 1a

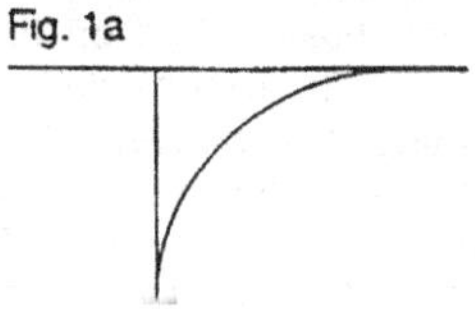

A spandrel

Fig. 1b

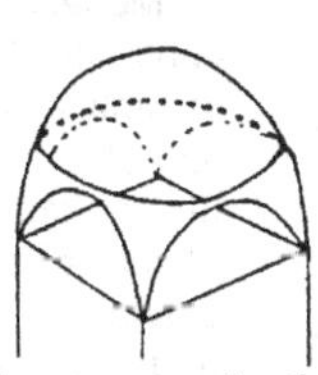

A dome on pendentives

Fig. 1c

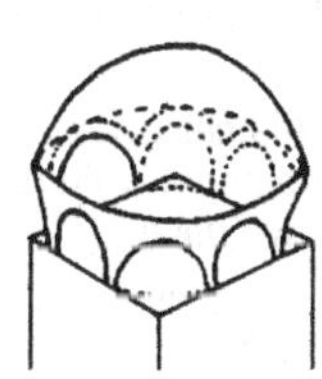

A dome on squinches.

(ii) the action of natural selection tends to produce behaviour that maximizes fitness.
(i) is a descriptive use of optimality: I will use "utility" as as the name for what is maximized in this case. There is no conflict between accounts of behaviour based on matching and those based on maximizing utility. In fact it is shown below that one can convert one sort of account into the other. Any tension that does exist in this area can probably be attributed to assumptions about the form that utility functions should have.

3.1 Notation

During an experimental session, an animal can allocate its time between n alternatives. The animal's behaviour can be measured in terms of the total number of responses made to an alternative or the total time allocated to an alternative. The time allocated to alternative i will be denoted by T_i. When I do not wish to specify the measure of behaviour, I will denote the behavioural allocation to alternative i by B_i and denote the total allocation by $B=(B_1, B_2, \ldots \ldots B_n)$. The total number of reinforcements obtained from alternative i will be denoted by R_i. I assume for simplicity that all reinforcements are essentially identical.

3.2 Matching

Matching arose as an empirical finding about choice (Herrnstein 1961) and was then developed into a general principle that defines reinforcement value. Following Killeen (1972), I will use the term "empirical matching relation" to refer to the matching of behaviour to reinforcement. When there are only two alternatives, the empirical matching law can be written as

$$B_1/B_2 = R_1/R_2 \qquad (1)$$

The theoretical matching law says that behaviour matches reinforcement value; when there are two alternatives we have

$$B_1/B_2 = V_1/V_2 \qquad (2)$$

This equation can be thought of as defining reinforcement value (Rachlin 1971, Killeen 1972) . Deviations from equation (1) then indicate that the value of reinforcement is not given by just the amount of reinforcement obtained. When behaviour is measured in terms of time allocation, equation (1) can be written as

$$R_1/T_1 = R_2/T_2 \qquad (3)$$

which states that the local rates of reinforcement are equal. The equivalent version of equation (2) is

$$V_1/T_1 = V_2/T_2. \qquad (4)$$

We can thus define a local rate $v_i = V_i/T_i$ that is equalised across alternatives.

3.3 Maximizing

Models based on maximization have been used in both behavioural ecology and operant psychology. In behavioural ecology the use of maximization is based on the action of natural selection , and the function that is maximized is fitness or an indication of fitness (see Maynard Smith 1978 for further discussion). In the context of foraging behaviour, it has often been assumed that the appropriate function is the net rate of energetic gain, γ. Although this assumption clearly cannot be correct in all circumstances, it has been supported in some cases (see Stephens and Krebs 1986 for a review). I will refer to this approach as rate maximizing. If the energetic costs of behaviour are ignored, rate maximizing says that the animal should maximize the rate at which it obtains food. This currency has been used in models of choice in concurrent schedules (eg Heyman and Luce 1979, Houston and McNamara 1981).

Models based on the maximization of utility are conceptually different from those based on the maximization of fitness or a currency related to fitness. Utility is something that is revealed by the animal's behaviour. In other words. an animal cannot fail to maximize utility; utility is defined as what is maximized by behaviour. Luce and Raiffa (1957) make the point in the following way. "In this theory it is extremely important to accept the fact that the subject's preferences among alternatives and lotteries came prior to our numerical characterization of them. We do not want to slip into saying that he preferred A to B because A has the higher utility; rather, because A is preferred to B, we assign A the higher utility." In contrast, an animal may not maximize γ, and furthermore it may not maximize fitness. As a trivial example, consider a moth that flies into the flame of a candle and dies. The moth's behaviour can be described in terms of maximizing its closeness to the flame, but the behaviour is not maximising the moth's fitness. It may, however, be possible to understand the moth's behaviour in terms of rules for flying that are adaptive in other contexts.

Another example is provided by the fact that many birds will prefer to sit on large model eggs rather than on their own eggs. In a detailed series of experiments, Baerends and Kruijt (1973) investigated the effect of various stimulus dimensions on the decisions of incubating herring gulls. It is possible to produce a consistent ranking of eggs in terms of their ability to induce incubation, but this ranking is not directly related to fitness.

3.4 Matching and Maximizing

In defining matching and maximizing, I have attempted to make a clear distinction between objective rates at which an animal obtains food and theoretical construction such as reinforcement value and utility. Reinforcement value is a function of reinforcement that is, by definition, matched by the animal's behaviour (equation (2)). Alternatively, equation (4) gives us a local rate that is equalised over alternatives. Utility is a function of reinforcement that is, by definition, maximized over alternatives. The main point of this paper is that reinforcement value and utility are alternative rather than competing descriptions of behaviour. Indeed the existence of one sort of description more or less guarantees the existence of the other sort of description. The justification of this claim is based on the following results.

3.5 Maximizing implies Matching.

Let the utility be denoted by $U(B_1,B_2,.... B_n)$, where the behaviours are subject to some constraint of the form

$$\sum_{i=1}^{n} B_i = \beta \qquad (5)$$

Then the behavioural allocation that maximizes utility is given by the condition

$$\partial L / \partial B_i = 0 \qquad (6)$$

where the Lagrangian L is given by
$L = U(B_1,B_2,....B_n) + \lambda(\beta - B_1 - B_2 ... - B_n)$.
As long as all optimal allocations are non-zero (see Houston and McNamara 1981 for an example of zero stay time under the optimal policy)

$$\partial u / \partial B_i = \lambda \qquad (7)$$

and so marginal utility is equalised across alternatives. This is a standard result in optimality theory. Prelec (1982) pointed out that it means that maximizing is a form of matching.

3.6 Matching implies Maximizing

Denote the behavioural allocation vector $(B_1,B_2,...B_n)$ by B. Then the matching condition states that

$$v_1(B) = v_2(B) = ...v_n(B). \qquad (8)$$

Now consider a utility function of the form

$$U(B) = -\Sigma (v_i - v_j)^2 \qquad (9)$$

It is clear that this function is maximized when equation (8) holds. This sort of utility function suggests a function of the form

$$U(B) = -(R_1/T_1 - R_2/T_2)^2 \qquad (10)$$

as a suitable "empirical" utility function. Indeed, this function must perform as well as the empirical matching relation (equation (1)) in describing behaviour. It hardly needs pointing out that there is no obvious connection between this utility function and fitness.

Equation (10) is not the only utility function that will describe the data: any even power of the differences in the local rates will also be a suitable function. There are also possible functions of a different form.

Given the descriptive use of optimality advocated by Rachlin et al. 1981 (see especially pages 385 and 406) it is hard to see why the utility function given by equation (10) is not adequate. Perhaps it is too obvious a restatement of the matching law. A less obvious function can be obtained by the following procedure. If it is observed that equation (3) holds, then equation (7) tells us that

$$\frac{\partial U}{\partial T_i} = \frac{R_i}{T_i} \qquad (11)$$

We then integrate the right-hand side of this equation with respect to T_i in order to obtain a utility function. We clearly require that doing this for each T_i (i = 1, n) will give the same function. We can obtain a necessary condition for this by noting that we require

$$\frac{\partial^2 R_i}{\partial T_j \, \partial T_i} = \frac{\partial^2 R_j}{\partial T_i \, \partial T_j}$$

which results in the following condition

$$\frac{1}{T_i}\,\frac{\partial R_i}{\partial T_j} = \frac{1}{T_j}\,\frac{\partial R_j}{\partial T_i} \,. \qquad (12)$$

It follows by an application of Stokes' theorem that this is also a sufficient condition.

I illustrate this equation in the context of two schedules (i.e. i=1,2). If schedule 1 is a VR and schedule 2 is a VI then

$$\partial R_1/\partial T_2 = 0 \text{ and } \partial R_2/\partial T_1 > 0,$$

so that equation (12) cannot be satisfied for positive T_1 and T_2. Now let both schedules be constant probability VIs with programmed rates λ_1 and λ_2. If the animal always spends a time t_1 on schedule 1 and t_2 on schedule 2 and the changeover delay (COD) is τ then the expected reward obtained on a visit to side 1 is

$$r_1 = \lambda_1 T_1 + 1 - \exp(-\lambda_1(T_2 + 2\tau))$$

and the corresponding equation for side 2 is

$$r_2 = \lambda_2 T_2 + 1 - \exp(-\lambda_2(T_1 + 2\tau)) \,.$$

(See Houston and McNamara 1981 for further discussion.) Because t_1 and t_2 are used on all visits, $T_i = nt_i$ and $R_i = nr_i$ for some positive constant n. Thus it follows that

$$\frac{1}{T_1}\,\frac{\partial R_1}{\partial T_2} = \frac{\lambda_1}{\tau + T_1}\exp(-\lambda_1(T_2 + 2\tau)) \qquad (13a)$$

and

$$\frac{1}{T_2}\,\frac{\partial R_2}{\partial T_1} = \frac{\lambda_2}{\tau + T_2}\exp(-\lambda_2(T_1 + 2\tau)) \qquad (13b)$$

If $\tau = 0$ and matching to programmed rates holds (i.e. $\lambda_1/T_1 = \lambda_2/T_2$) then the right-hand side of equation (13a) becomes equal to the right-hand side of equation (13b), so that the consistency condition given by equation (12) holds. A corresponding result holds for the Markov switching model of Heyman and Luce (1979).

4. Melioration and Stability

Herrnstein and Vaughan (1980) discuss a general principle for describing how an animal allocates its time to two alternative sources of reinforcement. The principle, which they call melioration, requires that more time should be allocated to the alternative that provides the higher local rate of reinforcement. There are various ways in which melioration can be achieved. In this section I present one simple formulation and show that it can produce a wide range of behavioural sequences.

4.1 The model

There are two schedules of reinforcement, denoted by the subscripts 1 and 2. Consider a cycle that consists of a visit of duration x_1 to schedule 1 followed by a visit of duration x_2 to schedule 2. Let the reinforcement obtained from schedule i be denoted by r_i $(i = 1,2)$. Herrnstein and Vaughan (1980) and Vaughan and Herrnstein (in press) do not specify how the local rates on each side are calculated. At one extreme, all previous visits to a side could be used, with the local rate being the cumulative total of rewards obtained from a side over the cumulative total time spent on a side. At the other extreme, only the most recent visit could be used. This is the procedure that I adopt. I define the local reinforcement rate on side i, LR_i, as r_i/x_i and take melioration to require that x_1 on the next cycle should increase if and only if $LR_1 > LR_2$.

To obtain an explicit implimentation of melioration, it is necessary to specify how x_1 and x_2 change as a function of LR_1 and LR_2. As a simple example, I remove one degree of freedom by constraining the total length of a cycle to be T. We can now concentrate on how x_1 changes because x_2 is given by $T-x_1$. Let $x_1(n)$ be the time spent on schedule 1 on the nth cycle. I assume that the time $x_1(n+1)$ spent on schedule 1 on the next cycle is give by the equation

$$x_1(n+1) = x_1(n)\ LR_1 / LR_2, \qquad (14)$$

subject to the constraint on cycle length i.e.

$$x_1(n+1) = T \text{ if } x_1(n)\ LR_1 / LR_2 > T$$

Previous discussion of melioration has concentrated on the equilibrium state at which $LR_1 = LR_2$, i.e. the matching law holds (Herrnstein & Vaughan, 1980; Vaughan & Herrnstein, 1987). I now explore the stability of this equilibrium. I describe a simple choice procedure in which the matching equilibrium is not necessarily stable. When it is unstable x_1 may follow a simple periodic pattern or may be"chaotic" in the sense that is described below. I assume that on side 1 $r_1 = px_1$, whereas on side 2 $r_2 = 1 - \exp(-\lambda x_1) + \lambda x_2$, where p and λ are positive constants that characterize the schedules. The equations for r_1 and r_2 are actually the mean reward obtained when a VR schedule is on side 1 and a VI schedule is on side 2. Thus the procedure can be thought of as a VI - VR in which the animal always obtains exactly the mean reward for a given stay time. This formulation can be made more plausible by considering a cycle to consist of a series of visits x_1 and x_2. Putting the equations for r_1 and r_2 into equation 14, and making use of the constraint equation $x_2 = T-x_1$, we obtain

$$x_1(n+1) = \frac{\text{Min } T,\ x_1(n).p.(T-x_1(n))}{(1 - \exp(-\lambda x_1(n)) + \lambda(T-x_1(n)))} \qquad (15)$$

Equation (15) is a non-linear first order difference equation. The properties of such equations have received considerable attention in the context of population dynamics (e.g. May, 1976; May & Oster, 1976). Information about the equilibrium and its stability can be obtained by plotting $x_1(n+1)$ against $x_1(n)$. This function is denoted by F. The equilibrium occurs where F intersects the line $x_1(n+1) = x_1(n)$. The equilibrium is locally stable if the tangent to F at the equilibrium has a slope that is between

45° and -45° (May, 1976; May & Oster, 1976). An example, corresponding to p = 0.048/sec and λ = .0167/sec (i.e. VI 1 min) is shown in Figure 2a. As p is increased, F becomes more steeply humped and eventually the equilibrium becomes unstable. This is illustrated in Figure 2b, where p = 0.053. Although the melioration process no longer settles down to a fixed point, there is still a simple characterisation of its behaviour. Once any transients have died away, there is a regular alternation between two stay times, as is shown in Figure 3. The reason for this pattern can be seen from Figure 4, which illustrates the function F^2 that gives $x_1(n+2)$ as a function of $x_1(n)$. The line $x_1(n+2) = x_1(n)$ intersects this function in three places. The middle intersection is the unstable fixed point $x_1(n+1) = x_1(n)$. The other two intersections are the period 2 solutions that have arisen by bifurcation. The processs is described by May (1976) - note however that his Figures 2 and 3 have been transposed. As p increases, the period 2 solution bifurcates, giving rise to a period 4 solution (see Figure 5). This bifurcation process continues, giving rises to stable cycles of period 8,16,32,64 ... (May, 1976; May & Oster, 1976). In general, a smaller and smaller increase in p is required to bring about each successive bifurcation. Beyond a certain value of p, there are an infinite number of periodic solutions, although only a finite number of them are stable. There are also an infinite number of initial stay times from which the subsequent stay times are aperiodic. Such a sequence of stay times appears to be stochastic, although it is generated by a completely deterministic process. Li and Yorke (1975) call such behaviour "chaotic". Two examples are given in Figure 6.

In the chaotic region, a slight change in initial conditions can lead to a radical change in subsequent behaviour. This sensitivity to initial conditions is associated with a stretching and folding of the state space that is produced by the repeated action of the function F (see Schaffer, 1984, 1985; Schuster, 1984 for further discussion).

It is of interest to investigate whether the meliorating rule produces matching even when the fixed point is not stable. Let R_i be the total reinforcements from side i in a session and T_i be the total time allocated to side i in a session. Table 1 gives R_1/R_2 and T_1/T_2 for four values of p. (The session length is about 1 hour.) When p = 0.048, the fixed point is stable and so matching occurs. When p = 0.053, x_1 alternates between 7.2 sec and 11.4 sec, and there is a slight deviation from matching. As p increases through the period four cycle region to the chaotic region, the deviation from matching increases.

The simple model of melioration that I have presented has several interesting features. It does not necessarily settle down to either fixed or regular behaviour and it does not necessarily result in matching. The model is formulated in terms of a one-dimensional state-space that is transformed by the function F, but chaotic behaviour is not confined to such a system (e.g. Guckenheimer, Oster & Ipaktchi, 1977; Schuster, 1984). The one-dimensional model has the advantage of simplicity, and it also can be used to characterise systems of higher dimensionality (Schaffer, 1985; Shaffer & Kot, 1985).

I have obtained chaos for a particular model of melioration. One important feature of the model is the deterministic equations for the rewards obtained. If a stochastic formulation is used, behaviour can still be irregular but will not show the precise chaotic pattern that I have described - cf. May 1985.

The constraint $x_1(n)+x_2(n) = T$ requires discussion. At first sight it is in conflict with the findings of Heyman (1979). From his work on the behaviour of pigeons on conc VI VI, Heyman suggested that $t_1 t_2/(t_1+ t_2)$ is constant, where t_i is the average stay time on schedule i. It must be pointed out, however, that Heyman's constraint is concerned with average stay times as schedule parameters are varied. It does not necessarily dscribe the possible relationship between stay times within a session for given schedule parameters. To illustrate the transition from a stable fixed point to cycles of period 2 and 4 and

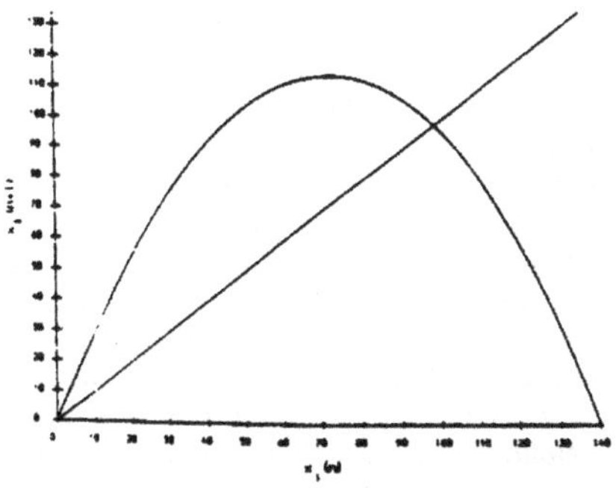

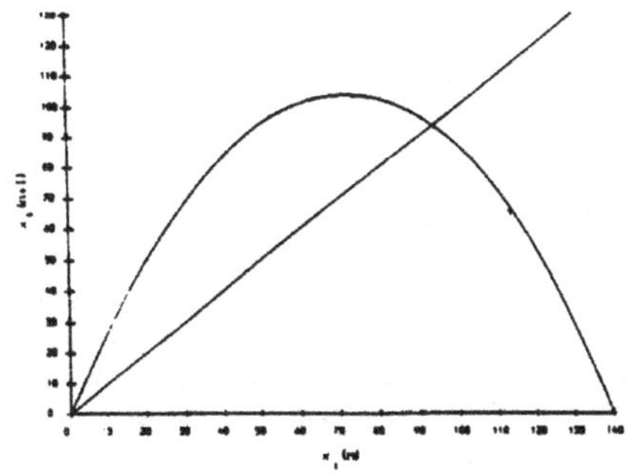

Figure 2 (a) The function F that relates $x_1(n)$ and $x_1(n+1)$ when $p = 0.048$ (curved line) together with the straight line $x_1(n+1) = x_1(n)$. ($T = 14$ sec, $\lambda = 0.0167\ sec^{-1}$). The intersection of these lines is a fixed point, and in this case the fixed point is stable.

(b) The function F when p is increased to 0.053. The fixed point is now unstable. Other parameters as in a.

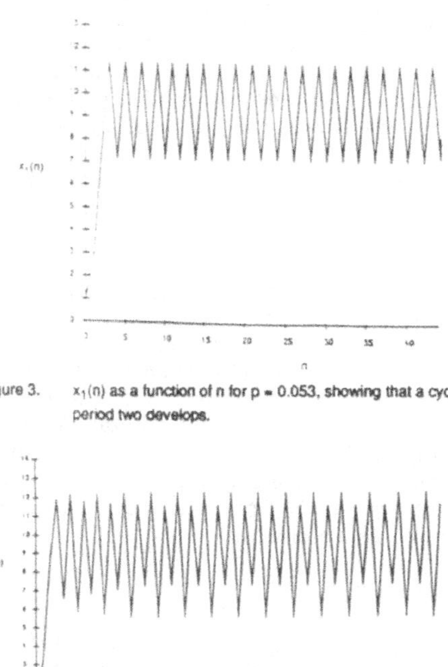

Figure 3. $x_1(n)$ as a function of n for $p = 0.053$, showing that a cycle of period two develops.

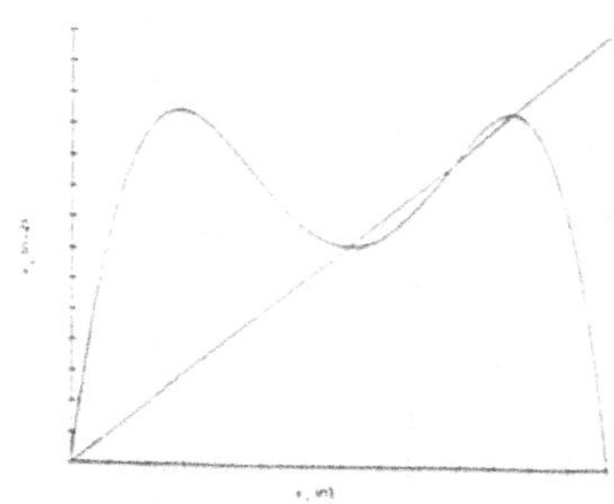

Figure 4. The function F^2 that relates $x_1(n)$ and $x_1(n+2)$ when $p = 0.053$, together with the straight line $x_1(n) = x_1(n+2)$. This line intersects the curve F^2 in three places. The middle intersection corresponds to the fixed point $x_1(n) = x_1(n+1)$ and is unstable. The other two intersections correspond to the stable cycle of period two. (Other parameters as in Figure 2).

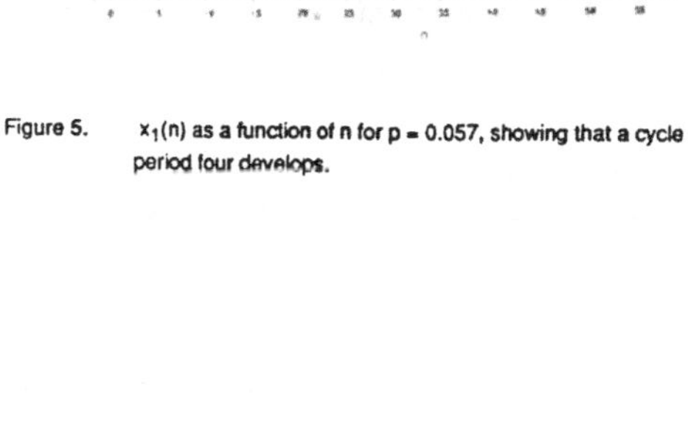

Figure 5. $x_1(n)$ as a function of n for $p = 0.057$, showing that a cycle of period four develops.

Fig. 6a

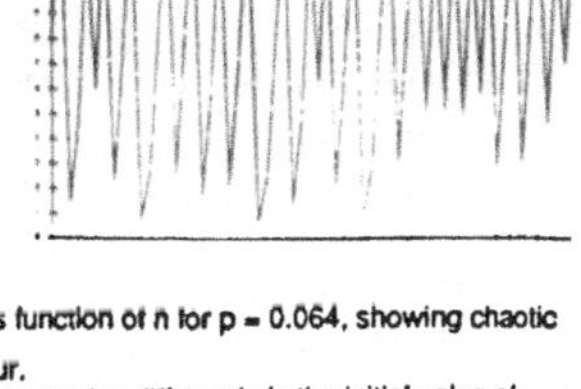

Fig. 6b

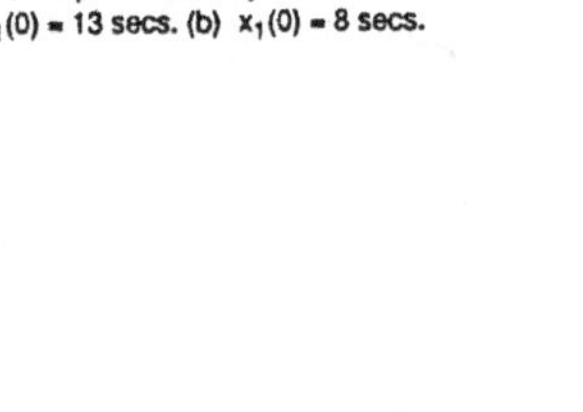

Figure 6. $x_1(n)$ as function of n for $p = 0.064$, showing chaotic behaviour.
The two examples differ only in the initial value of x_1(a) $x_1(0) = 13$ secs. (b) $x_1(0) = 8$ secs.

ultimately to chaos, I have developed a model in which this sequence is generated by changing one parameter, the reward probability on the VR schedule. It is possible, however, that factors outside my model can change, say, the cycle duration T as the schedule parameters are changed. This might complicate the sequence that I have presented, but chaos would still be possible. Heyman's constraint might hold across sessions, while a constant cycle time might hold within sessions. Despite this defence of the constraint, I do not wish to argue that the fixed cycle time is completely realistic. It is used to provide a simple illustration of chaotic behaviour, and thus to show that the behaviour that results from melioration is richer than has been thought.

5. Discussion

Given the disputes between matching and maximizing as accounts of behaviour, it might be thought absurd to argue that the two are equivalent. It is important to be clear about what has been established in this paper. I have argued that if reinforcement value and utility are not known *a priori* but are inferred from data, then there can be no conflict between them. They are alternative descriptions. What I am not claiming is that any particular functions are alternative descriptions. To give a simple example, it is well established that if local rate of reinforcement is equalised over alternatives, then overall reinforcement rate is not necessarily maximized (eg Prelec 1982, Houston 1983) It follows that if reinforcement value is equal to reinforcement rate, then a utility function based on overall rate will not necessarily describe the data.

It is perhaps worth pointing out that although the argument that I have described is simple, it gives us a stronger result than the obvious point that the result of any single experiment can be described either in terms of matching or in terms of maximizing. Any series of results that can be described in one way can be redescribed in the other way. It follows that a machine designed to equalize something in a range of environments can be described as optimizing something. The example of chaos arising from melioration shows that a simple deterministic machine using a local optimization rule can either reach a stable equilibrium, show regular oscillations or apparently random behaviour, depending on its environment.

A function such as that given by equation (10) describes behaviour, and thus should meet the requirements of Rachlin et al. (1981). From the point of view of behavioural ecology it is inadequate, in that it does not relate the animal's behaviour to its reproductive success. McNamara and Houston (1986) describe a general approach for relating behaviour over a short period of an animal's life to the animal's life-time reproductive success (see also Mangel and Clark 1986, 1988). Applying this approach to laboratory experiments makes it clear that the value of food (in terms of its contribution to reproductive success) depends on both the the animal's state and its expectations about the future (Houston and McNamara 1989). According to the viewpoint of descriptive optimality, an animal is by definition maximizing its utility, even if it kills itself by flying into a candle-flame or fails to reproduce because it prefers wooden model eggs to its own eggs. It is possible to use optimality principles to study behaviour without adopting such a view. Instead of assuming that animals always optimize with respect to their interests, it is assumed that animals follow relatively simple rules. These rules perform well in some environments, but not in others (see McNamara and Houston 1980, Houston 1987 for further discussion). Some form of melioration is a good example of such a rule. Rather than asking what is optimal about any particular example of matching, we should ask why natural selection should favour a rule such as melioration.

TABLE 1

Deviations from matching develop as the parameter p that determines reward on side 1 is increased (λ = 0.0167). R_i= total reinforcements from side i, T_i = total time allocated to side i.

p	.048	.053	.057	.064
R_1/R_2	2.03	2.20	2.36	2.30
T_1/T_2	2.03	1.94	1.89	1.33

References

Baerends, G.P. & Kruijt, J.P. (1973). Stimulus selection. In R.A. Hinde & J. Stevenson - Hinde (Editors) *Constraints on Learing.* Academic Press, London

Baum, W.M. (1974). On two types of deviation from the matching law: Bias and undermatching. *Journal of the Experimental Analysis of Behavior,* 22, 231 - 242.

Baum, W.M. (1981). Optimization and the matching law as accounts of instrumental behaviour *Journal of the Experimental Analysis of Behavior,* 36, 387 - 403.

Baum, W.H. and Rachlin, H.C. (1969). Choice as time allocation. *Journal of the Experimental Analysis of Behavior* ,12, 861 - 874.

Boelens, H. (1984). Melioration and maximization of reinforcement minus costs of behaviour. *Journal of the Experinental Analysis of Behavior,* 42, 113 - 126.

J. Dupré (Editor) (1987). *The Latest on the Best,* MIT Press,

Gould, S. J. and Lewontin, R. C. (1979). The spandrels of San Marco and the Panglossian paradigm: a critique of the adaptationist programme. *Proceedings of the Royal Society of London, Series B* 205, 581-598.

Guckenheimer, J., Oster, G. & Ipaktchi, A. (1977). The dynamics of density dependent population models. *Journal of Mathematical Biology* , 4, 101-147.

Harvey, J. (1978). *The Perpendicular Style.* Batsford, London.

Herrnstein, R.J. (1961). Relative and absolute strength of response as a function of frequency of reinforcement. *Journal of the Experimental Analysis of Behavior,* 4, 267 - 272.

Herrnstein, R.J. (1970). On the law of effect. *Journal of the Experimental Analysis of Behavior* ,13 , 243 - 266.

Herrnstein, R.J. & Vaughan, W.Jr. (1980). Melioration and behavioural allocation. In Staddon, J.E.R. (ed.) *Limits to Action.* Academic Press, New York. pp. 143-176.

Heyman, G.M. (1979). A Markov model description of changeover probabilities on concurrent variable-interval schedules. *Journal of the Experimental Analysis of Behavior,* 31, 41-51.

Heyman, G.M., and Luce, R.D. (1979). Operant matching is not a logical consequence of maximizing reinforcement rate. *Animal Learning and Behavior,* 7, 133 - 140.

Houston, A. I. (1987) The control of foraging decisions. In: M.L. Commons, A. Kacelink, S.J. Shettleworth (Editors). *Quatinitaive Analyses of Behavior.* Harvard Symposium.

Houston, A. I. (1983) Optimality theory and matching. *Behaviour analysis Letters* , 3, 1-15.

Houston, A.I. and McNamara, J.M. (1981) How to maximize reward rate on two variable interval paradigms. *Journal of the Experimental Analysis of Behavior,* 35, 367 - 396.

Houston, A. I. and McNamara, J. M. (1985). The variability of behaviour and constrained optimization. *Journal of Theoretical Biology* 112, 265-273.

Houston, A. I. and McNamara, J.M. (1989). The value of food: effects of open and closed economies.*Animal Behaviour,* 37, 546-562

Killeen, P. (1972). The matching law. *Journal of the Experimental Analysis of Begavior,* 17, 489 -495.

Li, T.Y. & Yorke, J.A. (1975). Period three implies chaos. *American Mathematical Monthly* , 82, 985-992.

Luce, R.D. and Raiffa, H. (1957). *Games and Decisions*. John Wiley, New York

Maidstone, R. J. (1988). *Haiga Sophia.* Thames and Hudson, London.

Mangel, M. and Clark, C. W. (1988). *Dynamic Modelling Behavior in Ecology.* Princeton University Press.

Mangel, M. and Clark, C. W. (1986). Towards a unified foraging theory. *Ecology* , 67 , 1128-1138.

May, R.M. (1976). Simple mathematical models with very complicated dynamics. *Nature* , 261, 459-467.

May, R.M. (1985). Regulation of populations with non-overlapping generations by micro-parasites: a purely chaotic system. *American Naturalist,* 125, 573-584.

May, R.M. & Oster, G.F. (1976). Bifurcations and dynamic complexity in simple ecological models. *American Naturalist* 110, 573-599.

Maynard Smith, J. (1978) Optimization theory in evolution. *Annual Review of Ecology and Systematics.* 9, 31 - 56.

Maynard Smith, J. (1982). *Evolution and the Theory of Games.* Cambridge, Cambridge University Press.

McFarland, D. J. (1977). Decision making in animals. *Nature* . 269, 15-21.

McNamara, J. M. and Houston, A. I. (1980). The application of statistical decision theory to animal behaviour. *Journal of Theoretical Biology* 85, 673-690.

McNamara, J. M. and Houston, A. I. (1986). The common currency for behavioral decisions. *American Naturalist* , 127, 358-378.

Murray, P. (1986). *The Architecture of the Italian Renaissance (3rd edition).* Thames and Hudson, London.

Newman, J. A., Recer, G. M., Zwicker S M and Caraco, T. (1988). Effects of predation hazard on foraging "constraints": patch-use strategies in grey squirrels. *Oikos* 53, 93-97.

Ollason, F. G. (1987). Learning to forage in a regenerating patchy enviornment: can it fail to be optimal? *Theoretical Population Biology* 31, 13-32.

Prelec, D. (1982). Matching, maximizing and the hyperbolic reinforcement feedback function. *Psychological Review.,* 89, 189 - 230.

Rachlin, H. (1971). On the tautology of the matching law. *Journal of the Experimental Analysis of Behavior,* 15, 249 - 251.

Rachlin, H. (1980) Economics and Behavioral Psychology. In :Staddon, J.E.R. (Editors) Limits to Action , *Academic Press, New York,* pp 205 236.

Rachlin, H., Battalio, R.C., Kagel, J.H. & Green, L. (1981). Maximization theory in behavioural psychology. *Behavioral and Brain Sciences,* 4, 371 - 388.

Rachlin, H., Green, L., & Tormey, B. (1988). Is there a decisive test between matching and maximizing? *Journal of the Experimental Analysis of Behavior,* 50, 113 - 123.

Rice, D. T. (1967). *A concise History of Painting from Prehistory to the 13th Century.* Thames and Hudson, London.

Rice, D. T. (1968). *Byzantine Art.* Harmondsworth, Middlesex, England.

Schaffer, W.M. (1984). Stretching and folding in lynx fur returns: evidence for a strange attractor in nature? *American Naturalist,* 124, 798-820.

Schaffer, W.M. (1985). Order and chaos in ecological systems. *Ecology,* 66, 93-106.

Schaffer, W.M. & Kot, M. (1985). Do strange attractors govern ecological systems? *Bioscience*, 35, 342-350.

Schuster, H.G. (1984). *Deterministic chaos*. Weinheim, Physik-Verlag.

Staddon, J.E.R. & Motheral, S. (1978). On matching and maximizing in operant choice experiments. *Psychological Review*, 85, 436 444.

Stephens, D.W., & Krebs, J.R. (1986). Foraging theory. *Princeton University Press, Princeton.*

Vaughan, W. Jr., & Herrnstein, R.J. (1987). Stability, melioration and natural selection. *Advances in Behavioral Economics Vol. 1.* L. Green and J. Kagel (eds.). Norwood, N.J., Ablex.

White, J. (1966). *Art and Architecture in Italy 1250 to 1400.* Penguin Books, Harmondsworth.

Williams, G. C. (1985). A defense of reductionism in evolutionary biology. *Oxford Surveys in Evolutionary Biology*, in: R. Dawkins, and M. Ridley (Editors), OUP, Oxford.

AN ADAPTATION ANOMALY OF A GENETIC ALGORITHM

Yuval Davidor
Department of Applied Mathematics and Computer Science
The Weizmann Institute of Science
Rehovot 76100, Israel
yuval@wisdom.weizmann.ac.il

Abstract

The purpose of this paper is threefold. To analyze an adaptation anomaly observed in a specific genetic algorithm designed to optimize robot trajectories, to propose an explanation for this unusual adaptive behaviour by drawing an analogy with some elementary mechanisms in nature, and to suggest that a much more robust adaptive strategy is to allow concurrent adaptation of both the information content and the representation structure by the genetic plan. This latter strategy results in the optimization of the ability to adapt.

This paper suggests that when artificial life or machine learning applications attempt to capture the essence of natural adaptation, it is important they allow selection to operate on all levels of the system. Furthermore, it is essential to expose both the structure of the system and its information content to selection.

1 A Genetic Algorithm Designed to Optimize Robot Trajectories

Most robot applications are based on a motion trajectory composed of a movement sequence of a robot arm. Mechanically, a robot arm is an open kinematic chain comprising relatively stiff links with a joint between adjacent links. Each link represents one degree of freedom and can be commanded to move independently of all other links. Standard systems have six degrees of freedom to obtain full spatial flexibility. Since a robot arm performs a task through the motion of its end-effector attached to the last link, the last link is the primary component of the whole structure.

An *arm-configuration* is a unique arm structure defined by a set of link positions (Fig. 1). Given the positions, the end-effector's position is uniquely determined. A robot trajectory is defined by specifying a sequence of spatial positions the end-effector is required to visit. Thus, any given path is approximated with a finite number of intervals specified by the motion vectors joining the discrete end-effector positions. In other words, any sequence of end-effector positions (>2) defines a valid trajectory which will approximate the desired path with a measurable error.

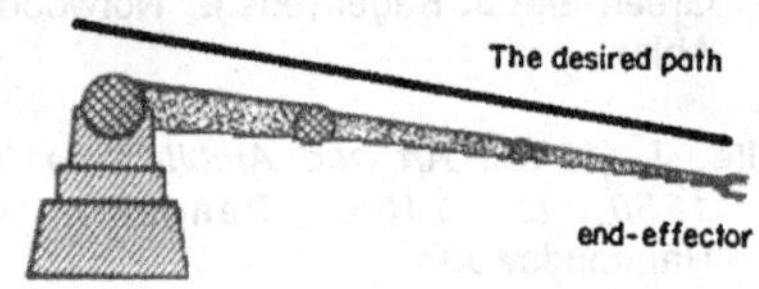

Fig. 1 - 3-link planar robot arm commanded to a fully stretched position, but the end-effector exhibits a steady-state positioning deviation from the horizontal due to limited positioning control accuracy.

Therefore, the optimization of a robot trajectory means the identification of both the optimum combination and number of end-effector positions, and that means a great many alternative trajectories that should be considered. The size of the trajectory space grows substantially even further when the robot used is of a redundant structure and most end-effector positions instantiate a multitude of different arm-configurations.

The complexity of programming a trajectory can be appreciated by examining the vertical plane in which the end-effector is required to follow the straight line connecting points A and B (Fig. 2a). One robot trajectory may be specified by sites 1 and 2 the end-effector should visit (Fig. 2b), while another trajectory might consider sites 3, 4, 5 and 6 as an alternative specification (Fig. 2c). The performance resulting from the different trajectories might be quite resulting from the different trajectories might be quite different. Too often robot systems are incorrectly assumed to have a predetermined performance (see Fig. 3 for illustrative example of one of the ambiguities in robot motion).

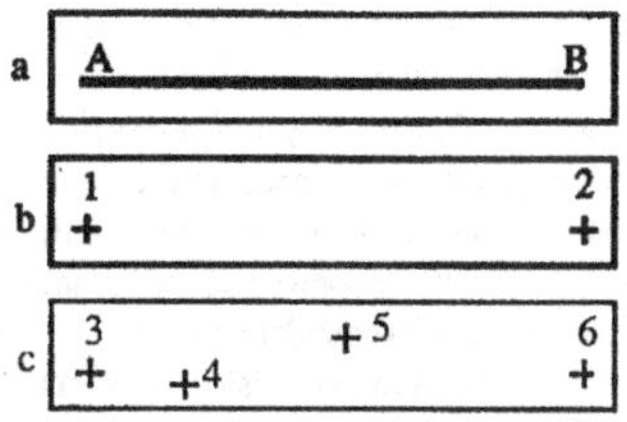

Fig. 2 - Illustrating the complexity in generating a trajectory. (a) The desired straight line path between the starting end A and the terminating end B. (b) One possible trajectory specified by two arm-configurations that correspond to the end-effector being at the two ends of the desired path. (c) Another alternative trajectory with two intermediate end-effector positions along the desired path.

In order to apply GA to the domain of robot trajectories, trajectories have to be represented by a string of parameters. The set of link positions which defines an arm-configuration is the basic string building block in this discussion. The decision regarding this representation structure was made primarily because it coincides with the natural structure of trajectories used by the robot controller; a sequence of end-effector positions defined by the corresponding arm-configurations.

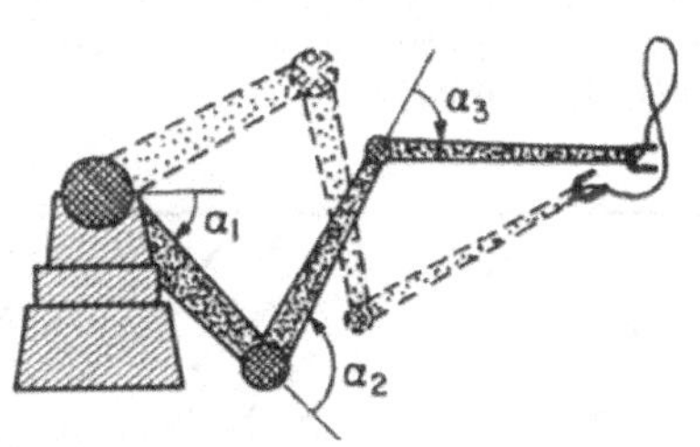

Fig. 3 - The redundant 3-link robot structure used in this study. Three Euler angles define the first arm-configuration. The motion between the two arm-configurations results in a path drawn by the end-effector.

It is now possible to construct the full string (Fig. 4). A string is composed of link positions (angles) grouped in n-tuples. Each n-tuple represents a single arm-configuration, where the first angle in the tuple corresponds to the position of the first link, and so forth for all n links in the arm. In a way, the basic unit of the string is the n-tuple which repeats l tines to define the l arm-configurations used to define a given trajectory.

$$\underbrace{\overbrace{\alpha_{1,1}; \alpha_{1,2}; \ldots; \alpha_{1,n}}^{n\text{-tuple}}; \alpha_{2,1}; \alpha_{2,2}; \alpha_{2,3} \ldots ; \alpha_{l,n}}_{l \text{ times } n\text{-tuple}}$$

Fig. 4 The trajectory representation as a string of link angles. A complete trajectory is composed from l n-tuples each defining an arm-configuration.

The discussion here of the GA model for the optimization of robot trajectories is minimal, and a detailed discussion can be found else where.[1] At each generation two strings are selected probabilistically proportionally to their relative fitness. The selected strings are replicated and two cross sites are selected along one string, and then are matched to their analogous cross sites in the second string. The two replicated strings crossover the counterpart sub-strings marked by the two cross sites.[2] The resulted strings are the offspring string and define a new trajectory which is a combination of sub-trajectories defined by the two parent strings.

2 An Adaptation Anomaly of the Genetic Algorithm

This paper focus on the optimal number of arm-configurations used to define a trajectory, and on the study of the dynamics of this number as a function of time during the optimization process. In general, what one would expect if the number of arm-configurations defining a trajectory has no substantial effect on the performance of trajectories is that when starting from an arbitrary trajectories, the average number of arm-configurations in the whole population would not change during the optimization process (not change drastically beyond the sampling noise). Individual trajectories may loss few arm-configurations, but others would gain, and on average this will balance out. This expectation should still hold even when the deletion and addition operators are introduced, due to their equal and low frequency.[1]

However, if the number of arm-configurations does affect performance, then one should expect a monotonic change (either a growth or a decline) because neither the algorithm nor the fitness function change during the optimization cycle. If having more arm-configurations introduce some novelty to the trajectory, and thus improving its performance, then the average number of arm-configurations in the population should increase monotonicly until some saturation level is reached.

The same arguments hold for the opposite case. If having less arm-configurations improves performance, then the average number of arm-configurations in the population should decline monotonicly to some threshold level. Whatever effect the number of arm-configurations has on performance, it seems not logical to expect that a characteristic behaviour of the optimization process will include an initial increase and later a decrease.

In contrast to the above intuitively founded arguments the GA behaves differently. The average number of arm-configurations characteristically declines during initial stages of the optimization, but at some stage stabilizes and starts to grow (see Fig. 5 for qualitative description). This phenomenon where the average number of arm-configuration is being reduced due to selection favouring short trajectories, but at some stage and due to the same selection pressure starts to reward long trajectories which consist more arm-configurations, is termed here *adaptation anomaly.*

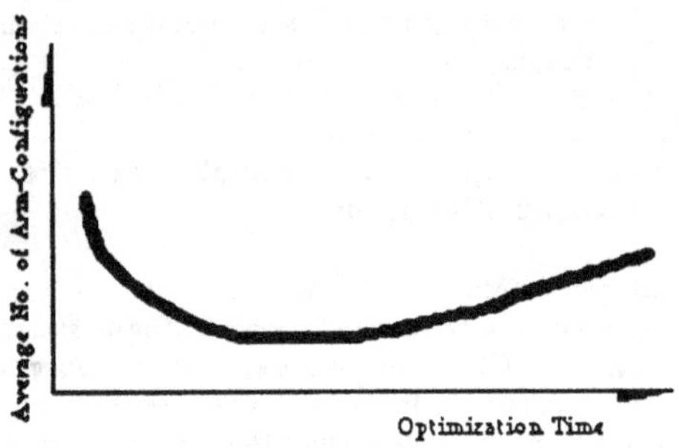

Fig. 5 - A qualitative Description of the adaptation anomaly in the trajectory-GA.

3 Adaptation Anomaly in Nature

Before suggesting an explanation for the anomaly presented by the GA, and attempting to draw some general behavioural conclusions about this model, three natural adaptation phenomenon are presented. The three examples cover both micro- and macro-systems (many other natural systems exhibit a similar behaviour). In a sense, because the adaptation anomaly is so fundamental in nature, this behaviour should probably not be regarded as anomalous. The purpose in describing these processes as anomalous is solely to emphasize their counter-intuitive aspect.

3.1 Pre-Attentive Human Vision

The example of pre-attentive human vision is a good example of what the anomaly is about because it is clear and very illustrative. Though it is not a development process *per se*, it does show one aspect of the adaptive anomaly very clearly, the fact that the amount of information as a parameter should really be viewed as the balance between the quality of the data and its abundance.

Recent research in human vision suggests that vision is composed of two distinctly different processes, pre-attentive which operates in parallel in the sense that it can process visual stimuli in several ways independently, and attentive which is mainly a serial process which process limited visual stimuli. This example focuses only on one characteristical aspect of the pre-attentive vision - the ability to detect an object. Namely, to detect a line segment against a background of line segments of a different orientation (Fig. 6).

It was shown that the ability to detect an object depends on the number of background objects in the view (or retina) in such a way that when only few background lines are presented, it is becoming more difficult to distinguish the object from the background when the number of the background lines is increased (up to 10 background lines in Fig. 7).[3] The correct response declines inversely to the increase in the number of background stimuli up to a certain threshold, and becomes proportional to this number thereafter. Fig. 7 portrays the adaptation anomaly as presented qualitatively in Fig. 5.

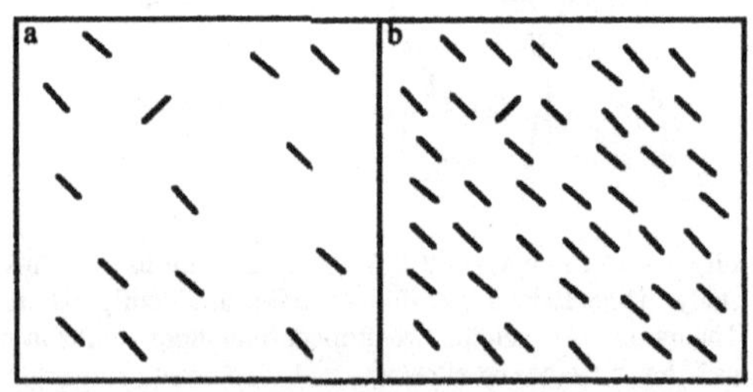

Fig. 6 - Two illustrated displays used as stimuli in measuring the response of an observer which had to detect the presence of the line having a different orientation from the others. (adopted by permission from [3].)

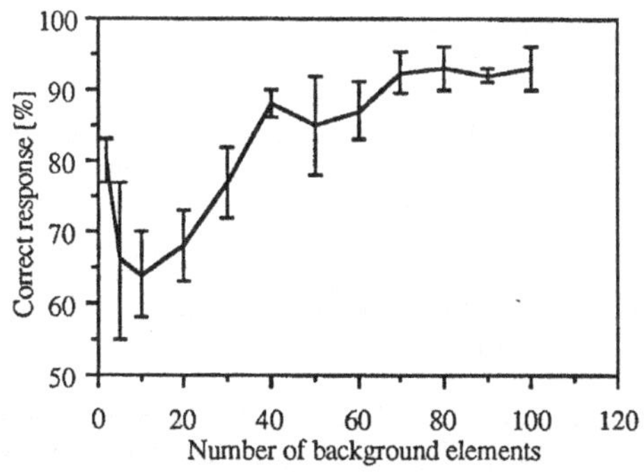

Fig. 7 - Average ability to detect an object for a different element presented on the retina (horizontal among vertical lines, diagonal lines, and horizontal and diagonal lines) of three observers. (adopted by permission from [3].)

3.2 Population Genetics

The adaptation anomaly can be viewed as a 'bottleneck'. This term originates from population genetic, where it describes a phenomenon in which a population's diversity during the population's adaptation or specialization is initially reduced but later grows again. This phenomenon is frequently observed when there is a major change in the selective pressure acting on the population, a drastic change such as experienced when the population settles on a new habitat. Let us follow a flock of birds arriving at an island which they have not been before and observe the diversity of phenotypic traits interacting with a new habitat. What is observed is that the genotypic diversity drops rapidly (within few generations), and then is reversed until a new equilibrium is reached.

There are several explanations for this phenomenon, and in general it is a result of several evolutionary forces acting simultaneously. It is not essential for the present discussion to analyze this phenomenon in greater detail beyond the following illustrative description. It is intuitively clear that a population of sexually reproducing species which is in equilibrium, has a phenotypic (as well as a genotypic) diversity. The fitness distribution is normal, and its mean corresponds to the environmental conditions. When the environmental conditions change, the two distributions no longer overlap. As a result of this shift, phenotypes which previously were in equilibrium find themselves below the survival level and therefore become extinct, and genotypic diversity is consequently reduced. With time, reproduction creates new genotypes and the population diversity grows until a new equilibrium is reached.[4]

3.3 The Development of a Nervous System

The development of a nervous system is a perfect example of the adaptation anomaly. There are two widely reported facts concerning the development of the nervous system. One, that neurons are committed to become a specific type of neurons prior to their growing and subsequently, are committed to develop certain connections with specific loci in their target. Two, that the developing nervous system initially incorporates redundancy of staggering proportions. These two facts seem to be in conflict, especially since this redundancy is eliminated, either at the embryonic or postnatal phase.

During normal development of the nervous system of a variety of vertebrates, a significant number of neurons migrate and undergo cellular degeneration at an early developmental stage. The final size of the neuronal population is adjusted to match the size of the field they innervate.[5,6,7,8] It is commonly assumed that the limiting factor determining the final number of neurons is the number of functional contacts available to the axons of the developing neurons. In order to eliminate the superfluous number of neurons, the development of the nervous system is followed by a process of selective cell death (Fig. 8a). In those parts of the brain where cell death has been documented, it involves the elimination of 15-85% of the initial neuronal population.[9]

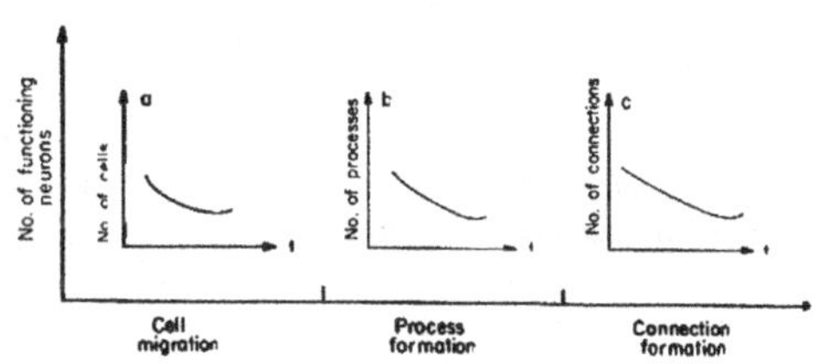

Fig. 8 - The development of the brain. The long-term development (outer time frame) is composed of three consecutive processes: cell migration (a), process formation (b), and connection formation (c). Each of the above processes involves an initial redundancy which is subsequently reduced due to competition and selection.

The neuronal population is not the only level in which an initial redundancy and subsequent elimination can be observed. Once the migrating neurons reach their destination, elaboration of processes commences. Most neurons are multipolar with several dendrites, which generally function as receptive processes, and a single axon which serves as the main effector process. Though the information required for neurons to generate their dendritic branching is partly genetically-determined, the number of processes is greater than is eventually needed to innervate the target (Fig. 8b). This stage of tremendous proliferation of dendrites and their subsequent elimination was already reported by Ramon Y Cajal.[10]

The wasteful development of the nervous system can also be observed at the connection formation level. More connections are initially formed than are actually needed. Therefore, some connections are withdrawn (Fig. 8c). There appears to be a transient phase in which the system is loosing stability by loosing connections in order to eventually exploit more efficiently the ones that remain. Selective elimination of axons is an additional phase responsible for cell death and is part of the mechanism which is involved in developing the specificity of the nervous system (such as in the segregation of binocular visual connections[11]).

Due to the high commitment of the nervous system prior to development, and to the limiting growth constraints presented by the target field, there is no need for substantial subsequent growth at the end of the three elimination phases. There is some evidence, however, that subsequent to loss of dendrites the mature remaining dendrites are in a metastable configuration for a while and undergo extension and branching.[6]

The fact that biological systems develop and grow is not the difficult part of the fundamental development cycle we have just described. Rather, it is the massive simplification of the system which precedes growth and increase in complexity which is puzzling. Why would it be advantageous for an organism to develop its nervous system in this seemingly wasteful fashion? In that respect, Oppenheim comments:

> "The death of some cells during normal development is perhaps not an entirely unexpected phenomenon in that a certain amount of genetic and epigenetic imprecision is bound to exist in any biological system, giving rise in some instances to circumstances (either intrinsic and/or extrinsic in origin) that may be lethal to developing cells. When the extent of such cell loss reaches massive proportions, however, as is the case for many parts of the developing nervous system [Oppenheim, 1981], the occasional imprecision that gives rise to small-scale cell death is no longer a sufficient rationale for such a large-scale phenomenon and one must, instead, search for alternative explanations involving biological adaptiveness and survival value... The occurrence of massive cell death during ontogeny is not unique to neurons, but rather it appears to be a phenomenon that has evolved so as to comprise a normal part of the development of many different types of cells, tissues and organs..." (Oppenheim, 1985, p. 487).

Oppenheim pertains to the adaptation anomaly and points out a possible explanation to the anomalous phenomenon which we shall investigate shortly.

4 The Evolution of Path-Planning Programs

In this section, the adaptive behaviour of the GA for the optimization of robot trajectories is analyzed in details. An explanation for the adaptive anomaly is proposed both in terms of the specific trajectory environment and in terms of the adaptation phenomenon observed with natural systems.

As presented earlier, the average number of arm-configurations during an optimization of a population of trajectories, initially decreases and later increases (Fig. 9). What causes this behaviour? The quality of a trajectory and consequently its fitness is affected by various factors which have a varying importance. The performance of a trajectory is the sum effect of these factors each one having its own weight (proportional contribution), and an absolute quality. The importance of the various fitness factors is not equal and is not even fixed. At different stages of the optimization the GA directs its processing resources to the different factors according to their weight. This dynamic shift in search effort is not explicitly controlled by the GA, but rather implicitly (the central aspect of the Darwinian theory). The average number of arm-configurations demonstrates this dynamics.

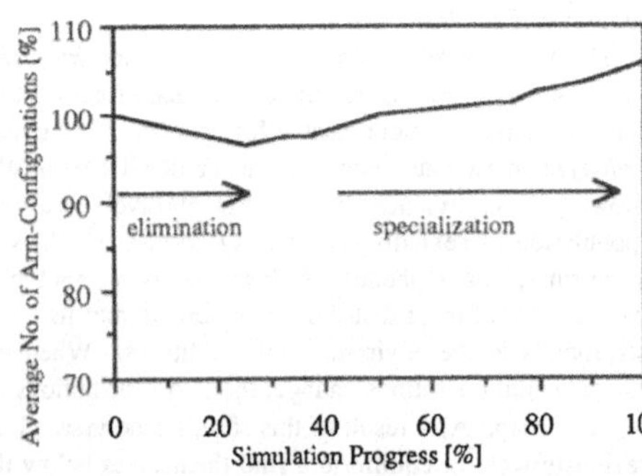

Fig. 9 - The average number of arm-configurations and the number of arm-configurations in the best trajectory as a function of the search progress (both the abscissa and the ordinate are normalized).

At the initial stages of the search, where trajectories are formed from random collection of arm-configurations paying no attention neither to the streamline aspect of the sequence of arm-configurations nor to the end-effector positions, the main source for bad performance is the movement between unharmonious arm-configurations (Fig. 3). A displacement vector which is composed from two considerably different arm-configurations irrespective to the position of the end-effector, will result in a vigorous end-effector movement which inflicts a large deviation from the specified desired path. At this stage, short trajectories with few arm-configurations have an advantage over trajectories with a large number of arm-configurations because the probability of a trajectory with few arm-configurations becoming harmonious is greater than that of trajectories composed from many arm-configurations. Thus, at the initial stages of the search, it is expected that the average number of arm-configurations will tend to drop.

After a period of relaxation, during which the trajectories become harmonious (no vigorous arm displacements occur between adjacent end-effector positions) due to crossover and the emergence of co-adapted arm-configuration sequences, performance can be improved further in two ways:

1) Optimizing the current arm-configurations within a trajectory so that they will be more harmonious and accurate.
2) Increasing the number of arm-configurations composing a trajectory.

As the search progresses, trajectories become more harmonious and the 'unharmoniousness' factor becomes less significant. Substantial improvements in performance can then be gained by increasing the accuracy of the trajectory. An increase in the number of arm-configurations is directly linked with the frequency with which the robot (and the simulation) motion controller can sample the path and the amount of feedback it receives on its progress (this sampling occurs at each discrete end-effector position in a trajectory). The accuracy of path following is proportional to the amount of feedback and thus, to the number of arm-configurations. At the stage the unharmoniousness and accuracy have an equal effect over performance, the average number of arm-configurations stabilizes, and subsequently the average number of arm-configurations increases to allow greater accuracy.

In the next section, the underlying assumptions of the model and the resulting simulations results are formulated into a development–theoretic model.

5 A Development–Theoretic Model

In section 3 it was shown that the adaptation anomaly is a 'popular' strategy in nature. In this section a formulation of a behaviour of an adaptive system subject to selection is attempted. A development–theoretic model is presented which offer a possible explanation to the puzzling question "why so many natural systems exhibit the adaptation anomaly in spit of its seemingly wasteful character?".

The GA exhibited a consistent behaviour in respect to adaptation in an environment incorporating uncertainty. In all simulations, which covered a diverse range of test conditions, the population always experienced a proportional simplification followed by development and specialization.

The results of the simulations can be explained by assuming the following development-theoretic model:[10]

1) Natural selection operates concurrently on all levels.
2) If an environment exhibits a finite degree of uncertainty, a system which incorporates redundancy would be more adapted.

It follows from the above model that:

1) Any system capable of development which incorporates a finite degree of redundancy will encounter an initial simplification before it specializes.
2) The degree of redundancy is proportional to the uncertainty and the extent of the initial simplification is proportional to the degree of redundancy.
3) During the specialization phase, a system will experience further growth if such growth is possible and constitutes an improvement.

6 Redundancy, Entropy and Adaptation

A basic concept of thermodynamics, *entropy*, is introduced in its almost literal meaning. The 1st law of thermodynamics – the law of energy conservation – states the that during any cycle or process a system undergoes, the cyclic integral of the heat is equal to the cyclic integral of the work. Though the 1st law quantifies the state changes during any process, it does not specify the direction in which a process will occur (any cyclic process can occur in two directions, producing either positive or negative work). In the eyes of the 1st law, a cup of hot coffee can heat up by further absorption of heat from a cool surroundings equally as it cools by virtue of heat transfer to it.

It is clear from daily experience that some processes can

occur while others cannot. The 1st law does not indicate which processes are plausible and which are not. It is therefore, experimental evidence that led to the formulation of the 2nd law of thermodynamics, and to the consideration of an abstract thermodynamic property entropy which symbolizes, when loosely interpreted, the potential of a system to change or to undergo a thermal process. Another popular interpretation of entropy is that it indicates the amount of disorder in the system. It is universally accepted that no process can occur if this process produces net entropy change (net change refers to a close system). Putting it in a different way, the 2nd law leads to the principle of the increase of entropy which states that no process can occur if it would result in a total decrease of entropy.

The picturesque meaning of entropy as a measure of disorder is important for our discussion, and an illustrative paradigm of entropy in terms of disorder is presented – the freshman-graduate paradigm. Freshman's table is a good example of disorder (no offence, of course, to the sparse exceptions). Due to few years of experience, the graduate's table looks very different. The freshman exercise excess effort for any studying task he is required to perform because of the little space resources available. Many books are opened and more are piled, loose paper is scattered all over. Little additional disorder can be introduced to the table of our prototypical freshman since it has so much of it as it is. Often, the freshman realizes that to many notes and books are in use and there is absolutely no more room on his/hers desk for additional studying operations.

At this stage, any new studying task (such as an unexpected exam) will require some housekeeping operations and tiding up before the new task can be approached. In contrast to our freshman, the graduate's table can endure much disorder (many unexpected tasks if one wants to be consistent in our paradigm) because there is so little of it. If one considers the disorder aspect associated of entropy, than one can say that the more a system is in order the more flexible it is to adapt to new requirements. Like the cooling cup of hot water, the hotter it is the less heat absorption from the surrounding it is capable to accommodate. At the state of maximum disorder (and the reader is urged to view this statement at the intuitive level it is made), a system cannot undergo any further processes and lost its ability to change (adapt) by itself with the resources available to it.

Entropy is an abstract concept that has been devised to aid the description of physical processes. It is concluded that entropy of a system is a measure of the system's ability to change so that when the system's entropy decreases, it is increasing its potential to take part in more processes and therefore increases its adaptability. But what is entropy in adaptive natural information systems which has the same meaning as just presented for the thermodynamic meaning of entropy?

Any information system has two basic operations it can perform: it can acquire or discard information, and it can manipulate the information it already has. For artificial systems the two operations are independent and practically have zero cost (the cost of acquiring more information is often an artificial aspect of the system). In natural system however, the cost of any operation is a real and meaningful factor. In the case of a nervous system for example, an organism has to invest valuable time and energy to generate neural cells and connections. Furthermore, the development of any organ has to be balanced and integrated with the functioning and available resource of the whole organism. It is clear that ultimately any growth or change in one organ is affecting the whole organism in terms of resources in a reciprocating fashion, the more complex an organ is the more demanding its needs are on resources.

Entropy highlight the advantage in optimizing adaptability. What follows from the analogy between entropy and natural systems is that an increase in the entropy means that the system is using more resources and that it is closer to the limit of its available resources. Reducing the entropy which is a result of reducing the resources the system requires enables the system to devote greater resources for development. Yet again, the paradox about the adaptation anomaly is manifested. What gain would a natural system acquire by simplifying itself and thus increase its potential for development, if it is going to grow again and loose that potential? Well the answer to this question was already answered in the previous sections of the paper. It is selection which 'realizes' that optimizing the adaptability is a rewarding strategy in the presence of uncertainty, and it is redundant which allows this optimization to occur.

7 Conclusions

This paper postulated that general adaptive systems have a characteristic behaviour in which adaptation involves the optimization of the system's ability to adapt. In order to optimize the ability to adapt, the system simplifies its information structure and organization. Natural systems achieve structural adaptation by manipulating the degree of redundancy in an indirect manor by way of natural selection operating on redundant parts of the system. A similar adaptive behaviour can be observed with a GA. This postulate was based on biological evidence both on the micro- and macro-level.

The conclusiveness of the biological evidence leave little room for creative interpretation other than that adaptation of natural information systems in environments exhibiting finite degree of uncertainty characteristically starts from a relatively chaotic and redundant state so that they may adapt to diverse condition through the simplification and re-structuring of the system before final specialization. This phenomenon was termed adaptation anomaly.

This paper also proposed an explanation to this phenomenon in terms of increased fitness due to robust adaptability. The explanation is based on strong analogies between thermodynamics and information, and on experimental results obtained in an artificial environment of a GA which reproduce the adaptation anomaly. Though the adaptation anomaly is characteristic to a specific model of GA and is repeatable under wide test conditions, it is recognized that the robot model does not constitute a general model of natural adaptation. For this reason, one should be cautious in projecting the results of this model on other artificial adaptive systems.

What are the lessons one might draw from the evidence and analysis provided herewith about the adaptation anomaly and the development-theoretic model? The most important lesson is that natural selection operates concurrently on all levels of a system — from the micro to the macro. One may sometimes not be able to observe the effect of selection on certain levels due to too strong selection effects on other levels, but sure enough selection is there. Another important lesson is the advantage redundant systems have in that they can afford to discard less adapted parts, and by that to optimize their ability to adapt. Redundant representations in GAs were not popular so far due to theoretical and practical difficulties. However, the advantage encompassed by the redundant models is potentially rewarding to the extent that such representations should be further investigated.

The author wishes to conclude this paper with two personal notes. One, that it is recognized that the ideas presented in this paper come through with a speculative tone to them. It is not intended. The main objective was to highlight the importance of allowing free rein for selection and adaptation when the objective is general adaptive systems and the optimization of an adaptive process. Two, though customized GAs are justified for particular applications, it is vital for general purpose GAs to leave as much plasticity as possible to the algorithm itself to define its own optimum representation and operators through the exposure of all the GA components to selection.

Acknowledgements

Preparation of this article was supported in part by the Centre for Absorption in Science, The Ministry of Immigrant Absorption, The State of Israel. Y. Davidor is a recipient of a Sir Charles Clores Fellowship.

Bibliography

[1] Davidor, Y. (1990) *Genetic algorithms and robotics: A heuristic strategy for optimizion*, World Scientific

[2] Davidor, Y. (1989) *3rd Int. Conf. on Genetic Algorithms*, 98-103, Morgan Kaufmann

[3] Sagi, D. and Julesz, B. (1987). Short-range limitation on detected of feature differences. *Spatial Vision*, 2(1), 39-49.

[4] Maynard-Smith, J. (1989). *Evolutionary genetics*, Oxford University Press.

[5] Oppenheim, R. W. (1985) *Trends NeuroSci.* 8, 487-493

[6] Smith, Z. D. Y. (1981) *J. Comp. Neurol.* 203, pp. 309-333

[7] Oppenheim, R. W. (1981) in *Studies in Developmental Neurobiology* (Cowan, W. H. Ed.), pp. 74-133, Oxford University Press

[8] Cowan, W. M., Fawcett, J. W., O'Leary, D. D. M. and Stanfield, B. B. (1984) *Science* 225, 1258-1265

[9] Cowan, W. M. (1979) *Scientific American*, 241(3), 106-117.

[10] Ramon Y Cajal, S. (1909) *Histologie du Systeime Nervoux*, Vol. I (1952 reprint), Instituto Ramon Y Cajal

[11] Rakic, P. and Riley, K. P. (1983) *Science* 219, 1441-1444

Autonomous Agents, AI and Chaos Theory

George Kiss

Human Cognition Research Laboratory
The Open University
Milton Keynes
England, MK7 6AA
email: gr_kiss@vax.acs.open.ac.uk

Abstract

Agent theory in AI and related disciplines deals with the structure and behaviour of autonomous, intelligent systems, capable of adaptive action to pursue their interests. In this paper it is proposed that a natural reinterpretation of agent-theoretic intentional concepts like knowing, wanting, liking, etc., can be found in process dynamics. This reinterpretation of agent theory serves two purposes. On the one hand we gain a well established mathematical theory which can be used as the formal mathematical interpretation (semantics) of the abstract agent theory. On the other hand, since process dynamics is a theory that can also be applied to physical systems of various kinds, we gain an implementation route for the construction of artificial agents as bundles of processes in machines. The paper is intended as a basis for dialogue with workers in dynamics, AI, ethology and cognitive science.

1 Introduction

Agent theory is a branch of artificial intelligence (Kiss, 1988). Its domain is the theory, design and implementation of artificial systems, similar to animals or people, that are capable of autonomous, rational actions through which to pursue their interests and goals. Aspects of this theory cover, among other things, how actions are related to knowledge, how plans for actions to reach goals can be formed, how goals are formed, what the role of intentions for action is, how the state of the world is perceived, and many others.

The abstract formulation of agent theory can be stated in many different languages, both informal and formal. Much current work in this field has made use of formal logical languages (Georgeff and Lansky, 1986). Although these specialised logics are convenient and expressive, often it is dificult to formalise their semantics, or the semantics that have been offered have undesirable properties. An example of this is the possible-world semantics of epistemic logics which unfortunately makes agents omniscient.

The implementation of theories expressed in such formal languages has additional problems. When agent implementation is done by direct mechanisation of the logic, for example as a theorem-prover, the resulting systems turn out to be inefficient. This is a natural consequence of the expressiveness of the language. On the other hand, the languages are sometimes not expressive enough to deal with some concepts that seem needed to describe agents. An example is the expression of quantitative magnitudes for describing strength of belief in an agent.

Refinement of these logics and their formal semantics, and their efficient implementation, is of course an ongoing enterprise. This paper is intended as an informal preliminary to such work, offering some intuitions about the interpretation of agent theory through the general theory of process dynamics.

Such an interpretation can also provide a strategy for implementation. The situation is analogous to the relationship between the abstract Boolean algebra of classes, the propositional calculus, and hardware logic circuits. The abstract algebra is defined in terms of classes and operations on them; intersection, union, complementation, etc. One interpretation of the Boolean algebra is propositional logic, where the variables range over propositions and the operations are truth-functional manipulations, etc. The possibility of implementation arises from the fact that another interpretation of Boolean algebra can be found in the operation of physical electrical circuits. Because of this, the operation of the circuits can thus be described by propositional logic, or stated conversely, the circuits are an implementation of the logic.

Let us represent this by the following schema:

Propositional logic -> Abstract Boolean algebra -> Electrical circuits

This suggests that we should look for an abstract mathematical theory such that both agent theory and some suitable physical systems can be interpretations of it. The abstract theory can then be used as an intermediary between agent logic and the physical systems that can be used as efficient implementations. This paper explores the possibility of using abstract dynamics as such an abstract theory and physical dynamic systems as implementations of agents, as shown by the schema:

Agent theory -> Abstract Dynamics -> Physical dynamic systems

A theoretical foundation has already been laid by Rosenschein (1985, 1986, 1989) for the epistemic (informational) analysis of agents regarded as processes. We propose that Rosenschein's framework is to be extended in two senses. First, the process-based interpretation of agents is to be extended from the epistemic to praxiologic (action-related) and axiologic (value-related) concepts. Second, processes are to be analysed in terms of their state space dynamics in addition to the correlative relationships between states of a process and an environment.

2 Concepts in Agent theory

The main idea an agent theory attempts to capture is purposiveness: that agents execute actions in order to reach goals. Refinements of the theory are concerned with optimality issues. Rational agents are often described as executing actions which maximally satisfy their goals.

To provide more structure, agents are also often described in mentalistic, intentional terms by attributing to them (propositional) attitudes. Some of the main examples of such attitudes, perhaps a minimal set of them, are wanting, knowing, liking (related to preferring), and intending.

The common-sense interpretation of these concepts is briefly as follows. Knowledge characterizes how the world is, from the agent's point of view. Likes characterise how the agent likes the world to be. Wants characterise the agent's commitment to reach a goal. Intentions characterise the commitment of an agent to an action.

3 Concepts of Abstract Dynamics

The main concepts of dynamics deal with the structure of state spaces. Abstractly, the theory can be formulated in terms of functional iteration. The functions which define dynamical systems are also called mappings or maps. The main concern of the abstract theory is with the asymptotic behaviour of iterative mappings. The iteration of a function is a discrete process. If the process is continuous, the description is often given in the form of differential equations to describe the behaviour of the solution over time.

In a geometric interpretation, the iterative process maps points into points. The points correspond to the states of the process. The process is then said to go through a trajectory or orbit of points. The main concern of dynamics is to understand the nature of all trajectories of a system and to classify them as moving to a fixed point, being periodic, asymptotically periodic, etc. We shall now turn to an informal summary of some of these concepts. For more detail, see, for example, Abraham and Shaw (1981), Devaney (1986), Thompson and Stewart (1986), or Cvitanovic (1984). Cvitanovic also contains an extensive bibliography. The field is developing very rapidly under the designation of chaos theory, which is a specialised branch of dynamics.

The *state space* of a system is generally a topological surface (manifold) on which the possible states of the system are located. This can be just three-dimensional space, or some curved surface, for example, like a doughnut (torus).

It is normally assumed that there is a *vector field* acting at all points of the state space. This vector field determines the *dynamics* of the system by constraining the *trajectories* to certain directions at each point of the state space. When typical or many trajectories of the system have been drawn, we get a *phase portrait* of the system.

Closed trajectories produce *cyclic behaviour*. Trajectories can otherwise take many shapes, like spirals, straight lines or any kind of curve.

The focus of interest is in the *asymptotic behaviour* of trajectories. *Limit sets* of state spaces are sets of points towards which the trajectories move asymptotically. Limit sets may be solitary points, or cycles, or more complicated distributions of points. Limit sets which are solitary points, are called *fixed points*.

Fixed points of functions are points x for which $f(x)=x$. That is, the fixed points are mapped into themselves by the function. Fixed points are important in dynamics, because they correspond to equilibrium (steady) states of systems. Once a system has somehow got to a state which is a fixed point, it will not move from that state under the iteration of the function f.

It is of interest to ask how a system may get to a fixed point. The simplest case is that the system may start from an initial state that is a fixed point, and there will be no further change. More interestingly, trajectories starting at other states may lead to a fixed point after a number of transitions. In such cases we say that the fixed point *attracts* the trajectory. The set of states from which trajectories lead to an attractive fixed point are called the *basin of attraction* of the fixed point. It turns out that a fixed point is attractive if the slope (derivative) of the function f is less than 1 at the fixed point. The magnitude of the slope characterizes the strength of the attractor: the greater the strength, the faster the trajectory approaches the fixed point. Two different kinds of behaviour in the neighbourhood of a fixed point are illustrated in Figure 1.

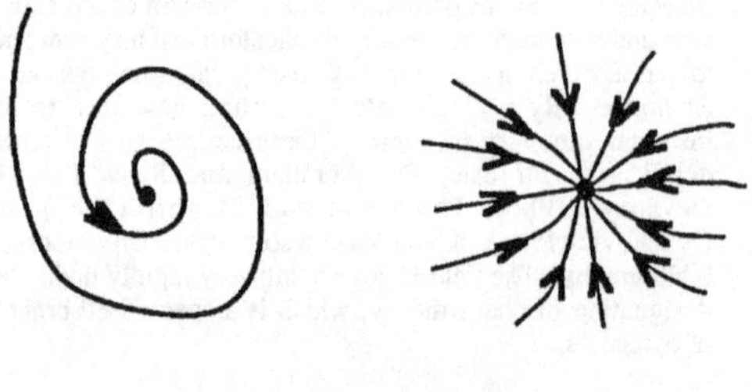

Figure 1. Attractive fixed points

A periodic point is a generalisation of the concept of the fixed point to the case when a trajectory cyclically visits a point after every *n* iterations of the function *f*.

If the iteration is run backwards, trajectories would appear to diverge from an attractive fixed point. In this situation the fixed point is called a *repellor*. Such fixed points correspond to unstable equilibria in physical systems. Slight disturbance from the equilibrium starts the system on a trajectory leading away from the equilibrium state. Conversely, attractive fixed points correspond to stable equilibria.

An interesting situation is shown in Fig. 2. There is a limit point which is attractive for points on the left and repelling for points on the right. The two heavy lines show trajectories which are not in the basins of attraction and repulsion, and are deflected by the presence of the limit point without going through it. This illustrates how the global attractors and repellors influence the local shape of the trajectories of the system even when the trajectories do not touch them.

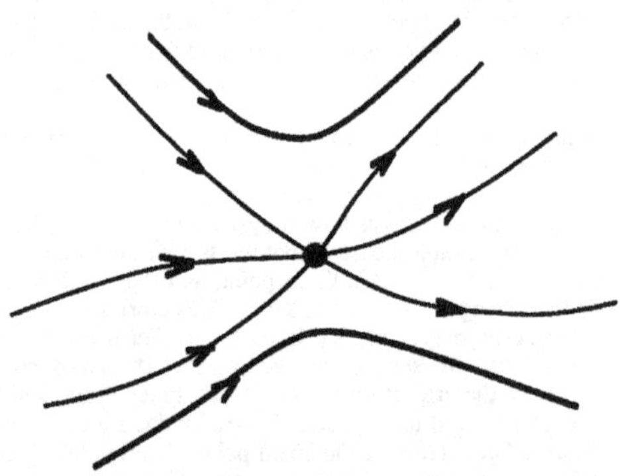

Figure 2. Deflected trajectories

4 Dynamics and Information

Dynamics and information theory are connected. The connection with the concept of information can be found when we consider the information needed to *specify* or *measure* the state of a dynamic system at a point in time. For example, when we want to start up a dynamic system from some initial state, we need to specify that starting state to some degree of precision. Conversely, if we try to find out what the state of the system is at some later instant of time, we can only measure its state to some limited degree accuracy, determined by the resolution of our measuring instruments or sense organs. We can then speak of the amount of information needed to make the specification, or the amount of information gained from the measurement.

The convergence and divergence of trajectories near limit sets are associated with information gain and loss. Divergence of trajectories is associated with greater uncertainty about the actual state of the system and therefore with information loss, as can be seen from Figure 3. Consider the trajectories that go through the smaller circular area on the left. In order to specify or locate a particular trajectory, we have to specify the position of a point within this area. The accuracy of specification or measurement defines a coordinate grid over the area. The amount of information involved is that needed to select a cell in this grid. The larger circular area on the right shows the effect of divergence in the phase space. If the accuracy of specification or measurement remains constant, a larger amount of information is needed to specify the location of a trajectory, and a larger amount of information is yielded when the state is measured, because there are a larger number of cells in the grid.

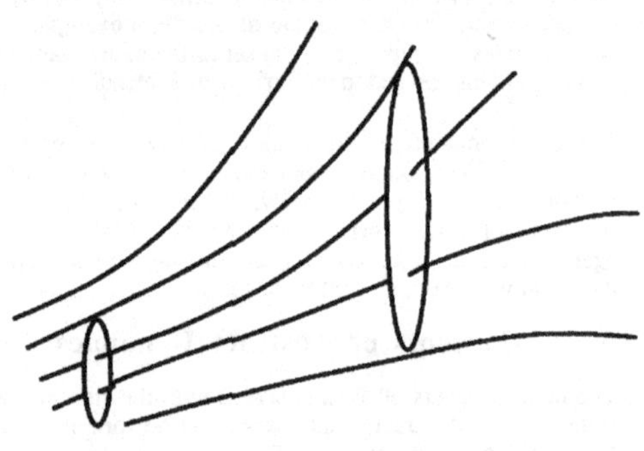

Figure 3. Information loss in divergent flow

Since diverging flows involve a progressive loss of information about the state of the system they lead to sensitive dependence on the initial conditions (the so-called "butterfly effect": the flapping of a butterfly's wings can cause a cyclone later). Although the system is *deterministic*, it becomes *unpredictable* in the long run due to this information loss.

As far as practical applications are concerned, this phenomenon can be used for *amplification and control* purposes in information processing: a small change in initial conditions can produce a large change in later behaviour. The converse effect, gain of information in convergent flows, produces insensitive dependence on the initial conditions and can therefore be used for *classification or recognition* processes in information processing. The convergence of trajectories to a fixed point serves as the recognition of the starting points in the basin of attraction as belonging to the same class.

5 Agent Attributes and Dynamics

We now turn to the central proposal of this paper, which is to interpret agent-theoretic concepts in terms of abstract dynamics.

Rosenschein (1985) has already interpreted agents as processes and offered an operationalisation of epistemic attitudes in terms of states of a process. The epistemic interpretation of states hinges on there being a correlation between states of the environment (the world) and states of the agent. When such a correlation exists, the agent is said to know a fact that can be stated as a description of the corresponding world state.

This interpretation of knowledge as states of a process does not yet offer any insight into why the correlations exist and what their nature is, nor is there any principle that would describe the dynamics of the process that is said to be the agent.

Complex agents are architecturally compositional both structurally and behaviourally. The complex agent structure is produced by assembling simpler component elements. Complex agent behaviour is produced through the interactions between the simpler component behaviours. A vital point is that agents are *nonlinear* systems. The importance of this is in the fact that *only in nonlinear systems* will *qualitatively* new phenomena arise through the interactions between components. In linear systems these interactions will be merely additive accumulations of size (scaling), without new features. Nonlinearity leads to the appearance of new features in the *global* behaviour of the system which were not present in the behaviour of the *local* components and are not just the additive accumulation of the component behaviours.

Concurrency, parallelism and distributed systems become important because structural and behavioural complexity only emerge as the result of iterative accumulation of local components. The local components can be identical and simple in the computations they carry out. Complex structures appear as the result of the many interactions. For real-time operation the execution of the numerous *simple identical steps* can be, and needs to be, parallel. Long behavioural trajectories can then be computed in essentially a single step in the limit of parallelism, or increasingly serially between a single-processor and the maximally parallel system. Proportional speed-up is thus obtainable by adding more processors. This contrasts with the rather more limited gains that can be obtained in the case of carrying out complex structured computations on multiprocessor systems.

The dynamics of the agent process are of importance in relation to all three aspects of agent theory: epistemological, praxiological and axiological.

The epistemic issues are concerned with the way states of the environment process determine states of the agent process and hence produce the correlations referred to above. The trajectory of a part of the environment process (abstractly in a phase-space, concretely in physical space) enters into the region within the boundary of an agent system and then continues as part of the agent process. Since this part of the agent process is causally determined by the environment process by means of the state-to-state transitions, the required epistemic correlations will be produced. In automata-theoretic terms these are inputs. In agent-theoretic terms these are sensations (at the boundary) or perceptions (further inside the agent) or cognitions (deep inside the agent). Looking at this another way, the agent process is simply a part of a global world process and its states are parts of the world states. An agent is just a *local* phenomenon within the *global* world process.

Praxiological issues are concerned with the way agent processes eventually (at the boundary of the agent) exert causal influence on the environment process. In automata-theoretic terms, these are outputs. In agent-theoretic terms, these are actions (at the boundary), volitions, willings or intentions (progressively deeper within the agent).

Axiological issues are concerned with the dynamics of the agent-process trajectory in its phase-space and, in particular, with the directional nature of the process. The teleological nature of agent behaviour is one of the central examples of such issues. In terms of nonlinear system theory, the dynamics can be described in terms of the movement of the system state *towards* stable equilibrium states and *away from* unstable equilibrium states. Teleological agent behaviour is to be identified with movement towards stable equilibria which are in this sense *preferred* states of the system: the agent *likes* to be in these states. Aversive agent behaviour is to be identified with movement away from unstable equilibria which are in this sense *disliked* by the agent. In the terminology of dynamic system theory, these states are *attractors and repellors.* Unstable equilibria arise mainly through competition between attractors and represent boundaries between the basins of attraction of those attractors.

The correspondence between the various main characteristics of *purposive* agents and dynamic process-related concepts can be sketched as follows:

- Agents correspond to atomic or structured processes.

- What the agent knows corresponds to the information contents of a state.
- An agent's actions correspond to a change of state in the state space.
- The likes and dislikes of an agent correspond to the global (high-dimensional) attractors and repellors of the state-space. In complex agents these are elements of a value system.
- A goal of an agent corresponds to a local (low-dimensional) attractor in a basin of attraction.
- A want of an agent corresponds to a *trajectory* which converges to a local attractor in its basin of attraction.
- Hedonic states of pleasure and pain correspond to satisfaction and dissatisfaction of the constraint system constituted by the attractors and repellors of the state-space, i.e. to *distance* from them.

Following Rosenschein's conceptual framework, processes will be regarded as trajectories over time and spatial locations. A trajectory is specified by a function $w : L \times T \rightarrow D$, where T is a set of times, L is a set of locations of some physical system and each location a can take on values from some set D_a. D is the union taken over all sets D_a. An agent will be taken to be a process or a structured set of processes. The structuring is defined in terms of constraints between the process states.

Interpreting goals as local attractors can be seen as a more special case of trajectories being shaped by the existence of attractors and repellors in the state space. Attractors and repellors determine the direction of movement, i.e. the direction of agent action. It is natural to interpret the pro- and anti-attitudes of agents with this kind of directionality. Movement according to attractors and repellors leads to the notion of the movement *satisfying* the constraints applied by them. We propose to tie this notion of constraint satisfaction to the notion of hedonic satisfaction in complex agents. Pleasure and pain constrain agent action in the same structural sense as attractors and repellors constrain dynamic systems. We assume that due to the physiological structuring of living organisms attractors and repellors are created in their behavioural space. By analogy, it should be possible to create attractors and repellors in non-living computational systems through appropriate construction or programming.

Wants in agent theory express the notion that an agent is committed to carrying out some action or series of actions to reach a goal. This commitment can be naturally interpreted as embarking on a trajectory towards a local attractor in a basin. Weaker forms of desire, like wishing, can be interpreted as a belief that some goal state is attractive. without commitment to actions.

For simplicity we do not distinguish here between belief and knowledge as epistemic states of an agent. Adopting the theory proposed by Rosenschein (1985, 1986), we interpret knowledge as the information content of an agent's state. The information content expresses the relationship between an agent's internal state and the corresponding state of the environment. An agent is said to know a proposition p in a situation in which its internal state is s, if in all possible situations in which the agent is in state s, p is satisfied by the environment. Recall that agents are interpreted as processes, so that agent states correspond to process states. An agent can only tell what state the environment is in by examining parts of its own internal state, i.e. the state of its sensory apparatus. Since there are causal constraints between the state of the environment and the state of the agent, the agent's internal state contains information about the environment. What the agent knows is the discriminatory power of this information in distinguishing between possible states of the environment. The agent's knowledge defines what states of the environment are indistinguishable from each other. The more the agent knows, the smaller these indistinguishable equivalence classes.

In this framework reasoning can be shown to be the concentration of information from a set of locations to a smaller set of locations, and also to be the making explicit of the information contained implicitly in the larger set (Rosenschein and Kaelbling, 1986). Agent states can thus be thought of as being ordered in terms of the amount of explicit infomation they contain. The greater the explicit information content, the more precisely the state of the world is known.

The usefulness of knowledge for an agent is, of course, in guiding action towards a goal. In process dynamics terms the agent needs knowledge in order to tell what trajectory to follow. In areas of the state space where trajectories diverge, there is sensitive dependence on the current state, so even small errors in determining a trajectory may mean missing the goal later. In areas where the flow converges, little knowledge is needed: the agent can't help taking the right action.

6 Brief Notes on Other Issues

This section presents brief notes on some conjectures relating agent theory to dynamics. These notes are intended as a basis for discussion about these issues with others both in the areas of dynamics, AI and cognitive science. The notes will be revised and expanded as a result of discussions.

6.1 Learning

Learning consists in shaping the phase-space of the system. For example, goal-directed and aversive behaviour are produced by introducing attractors and repellers in the phase-space. Shaping of the phase-space is produced by changing the computational mapping between environmental inputs, internal state, and agent outputs. The behavioural trajectories of an agent are produced by the iterative computation of this mapping using the current environmental inputs and current internal state. The attractors and repellers can be both implicitly and explicitly represented in the structures that define this computational mapping. It is reasonable to assume that low level learning of the kind produced by conditioning is implicitly represented, while learning procedures from linguistic descriptions is explicitly represented.

6.2 *Linking Cognition to Action at Fixed Points*

Concepts of dynamics are also applicable to some of the internal processes taking place inside agents. In this section I briefly discuss two proposals: (a) that the process of constructing increasingly abstract representations has fixed points, and (b) that for maximal generality the two processes of cognition and action are best linked to each other at fixed points of the cognition process.

One role of the cognitive and perceptual processes is to obtain representations of the environment. I regard representations as approximations, because they contain less information than the objects they represent. Such approximations are more economical than the original objects, because they abstract away from irrelevant detail. It is often pointed out in the literature of psychology and AI that perceptual processing proceeds through a number of levels in terms of the abstractness or generality of the representations used. I assume that the iterative progression through such levels of mapping into more general representations has limits which are fixed points. These fixed points are irreducibly abstract concepts. I want to emphasize, that I do not mean irreducibility to properties. If reduction to properties is possible, the properties are themselves represented at the most abstract level. My most abstract conception of a dog has still properties like legs, but the legs are the most abstract kind that I have available.

Action patterns are produced by using a mapping to produce patterns on the "surface" of the agent starting from representations inside. This surface is the set of actuators or transducers of the agent. In terms of directionality, actions are an "inside-out process" in terms of causation from agent to environment.

Cognitive and action patterns are best linked together at increasing levels of condensed representation in order to achieve economy and generality. The most economic way of doing this is to place the linkage at fixed points. The progression from cognition to action (and back to cognition again via the environmental feedback loop) is via the focus of fixed points where the representation from cognition and the representation for action are at the same level: when looked upon from the cognitive point of view, it is an appropriately abstract representation of a situation *as it is*; when looked upon from the action point of view, the representation is of a situation *as it is to be* as the result of the action, when appropriately expanded. The result of condensation through cognition is a fixed point. This fixed point is then used as the starting point (or parameter value) for expansion into an action pattern. The usefulness of the fixed-point representation is in its economy, both in size, and also in its abstractness: many similar situations are conceptually classified together for the purposes of identical or similar action. Generalisation over similar situations is thus obtained, producing economy.

6.3 *Problem Solving*

Problem solving can be interpreted as a trajectory towards stable equilibria, as has been discussed in relation to goals earlier in this paper. Alternative paths to an attractor are disjunctive solutions, while points on a single path are conjunctive subgoals.

The problem-solving metaphor can possibly be extended to logical inference, as is done in intuitionistic logic. It may be natural to interpret the existence of a proof as convergence to a maximal informational state.

Axioms could be interpreted as fixed points since they require no further proof process.

Reasoning is seen as finding a path from the point corresponding to a formula to a fixed point corresponding to an axiom. The path is the proof process, consisting of a sequence of transformations of the formula according to derivation rules. The derivation rules correspond to reasoning actions of an agent.

Alternatively, formulas to be proved can be regarded as problems to be solved, with constituent subformulas as subgoals. The informational state of the agent increases as more and more constituents are proved. Such an interpretation has been used by Kripke (1965) to provide a semantics for intuitionistic logic. It will be of interest to analyse the dynamics of changes in informational states along the lines suggested in this paper.

6.4 *Process Dynamics as an Implementation Strategy*

The theoretical framework proposed represents agents as processes (or bundles of processes) and constraints acting on these processes, producing the appropriate dynamics. This suggests that the implementation should be in the form of machines recursively composed of parts. Machines can generally be described as bundles of processes, with the machine structure acting as constraints on the processes.

Rosenschein and Kaelbling (1986) has offered a detailed formalisation of this concept by modelling machines as a pair of processes (input and output), subject to behavioural constraints. The constraints impose a structuring on the state-space of the output process that can occur in the machine, producing what we call a dynamics for that process. Having a dynamics *means* being constrained, so that not all possible state trajectories are possible, and those that are possible must take a shape that satisfies the constrains.

Convenient high-level languages need to be developed for expressing the constraint system in order to facilitate the design of artificial agents. Rosenschein and Kaelbling (1986, 1989) have developed one example of such a language, REX, and its extension to the declarative specification of goals, GAPPS. This language has been used in robotics by Rosenschein and by the author in developing agents as intelligent interfaces to computer software.

7 Conclusions

This paper has taken the first steps in providing an interpretation of agent theoretic concepts in terms of process dynamics concepts. If successful, this approach could have the double advantage that it provides both a mathematical theory which is already known to be successful in other application domains like physics and biology, and it also gives a hint at implementation strategy.

8 References

Abraham, R.H. and Shaw, C.D. (1981) Dynamics, the Geometry of Behavior, Parts 1-4. Santa Cruz, CA: Aerial Press.

Cvitanovic, P. (ed) (1984) Universality in Chaos. Bristol: Adam Hilger.

Devaney, R.L. (1986) An Introduction to Chaotic Dynamical Systems. Menlo Park, CA: Benjamin/Cummings.

Georgeff, M. and Lansky, A. (eds) (1986) Proceedings of the Conference on Reasoning about Actions and Plans. Los Altos: Kaufmannn.

Kiss, G.R. (1988) Some aspects of agent theory. Project Report HLD/WP/OU/GRK/24.

Kiss, G.R. and Brayshaw, M. (1989) Agent architecture and implementation strategy. Project Report HLD/WP/OU/GRK,MB/25.

Kripke, S. (1965) Semantical analysis of intuitionistic logic I. In: J.N. Crossley and M.A.E. Dummett (eds) *Formal Systems and Recursive Functions.*

Rosenschein, S. (1985) Formal theories of knowledge in AI and robotics. New Generation Computing, 3, 345-357. Also SRI International Technical Note 362.

Rosenschein, S. and Kaelbling, L. P. (1986) The synthesis of digital machines with provable epistemic properties. In: Halpern, J. (ed) Theoretical aspects of reasoning about knowledge. Proc. of the 1986 Conference. Los Altos, CA: Morgan Kaufman.

Rosenschein, S. and Kaelbling, L. (1989) Integrating planning and reactive control. In: Proc. NASA/JPL Space Telerobotics Conference, Pasadena, CA. January, 1989.

Rosenschein, S. (1989) Synthesizing information-tracking automata from environment descriptions. In: Proc. Toronto Conference on Knowledge Representation, 1989.

Thompson, J.M.T. and Stewart, H.B. (1986) Nonlinear Dynamics and Chaos. Chichester: Wiley.

INCREMENTING INTELLIGENT SYSTEMS BY DESIGN

Brendan McGonigle
Laboratory for Cognitive Neuroscience
Edinburgh University
Scotland

SUMMARY

Our operating hypothesis can be stated simply enough. It is that the class of non-trivial solutions to the problem of information management is small (see McGonigle, 1987). If true, it follows that intelligent systems (however instantiated) are solution constrained *a priori* and must converge on a similar design logic if they are to succeed. In the case of biologically instantiated intelligence, this convergence is seen more as a primary consequence of such solution constraint (for example, the power-generality trade-off), and less one of genetic affinity (although it may be secondarily a factor).

Our general goal is the characterisation of intelligent systems in the broadest sense, informed in particular by converging research in comparative, developmental and cognitive psychology situated within a neuroscience framework (see McGonigle and Chalmers, 1990a). More recently, we have opened a dialogue with roboticists at Edinburgh (Smithers, Malcolm and Donnett, in particular) and are delighted to see the growing basis for productive dialogue. The position outlined here also intersects with one espoused by Brooks (1986) but has an independent origin and rationale.

The goal of this paper is to summarise some of the characterisations which emerge from work on biological systems in an attempt both to cross-check with designers of artificial systems and to exchange concepts of possible mutual benefit. To limit the vast area under review, I shall concentrate on the incremental aspect of intelligent systems as this is, as I see it, the key issue.

CHARACTERISATION

Although there are (and have been) many documented approaches to the study of evolutionary intelligence, most have failed on the key issue of how systems 'invest in complexity' or grow (ontogenetically or phylogenetically) from 'weak to strong'. Classical ethology, for example, has worked best with simpler 'reactive' systems which 'wear their adaptation' on their sleeve[1]. Certainly, their domain of inquiry has precluded the study of human problem solving and intelligence either from a comparative, a cognitive or a developmental standpoint. As a consequence, ethologists may offer some solace to designers interested in making simple reactive agents as a first step. However, and crucially for our current agenda, they offer few general characterisations of intelligent systems that could afford strong clues as to the ways and means

1 A psychologist looking at these reactive systems as a group, however, might characterise their group adaptation as an example of the power-generality trade-off, both in terms of niche/habitat selection and in the specialisation of subsets within the group as in (say) bees. However, a study of learning mechanisms *per se* as an alternative to reactive agents, fares little better. Motivated by black box behaviourism, the search for universal laws of learning and memory has failed to provide any stable correlates of species differentiation according to phylogeny or brain architectures. Although learning appears early in the evolutionary process, it is not the fact of learning, but WHAT is learned that differentiates animals. And if simple habit formation is mainly what has been achieved in most of these experiments, it is not surprising that they have not picked up direct implications of advanced nervous systems, for, as Mishkin (1985) has claimed, such habit mechanisms are mediated by the motor cortex (an 'old' area) of the brain.

of moving beyond reactive knee-jerk intelligence to the more interesting, commercially attractive 'scaling-up' stage that would be necessary to deliver dynamic agents with broad ranging competences.

We believe that progress can and will be made when two related scientific exercises are concatenated.

The first of these we might dub a failure of 'commonsense' characterisations of intelligent systems - vivid because of their ubiquity, and as lethal to any subsequent science/engineering enterprise as was the common misperception of bird flight to early plane makers (the birds' integration of lift and thrust led to the now ridiculous-looking early attempts to construct wing flapping airplanes).

The second requires a new biological synthesis incorporating clinical and experimental neuroscience, together with experimental, cognitive, comparative and developmental psychology.

COMMON LETHAL ASSUMPTIONS

1. 'Intelligence' is reflective not reflexive. Unified, monolithic, and symbol-using, it is controlled by a central executive capable of accessing all of the main procedures for decision making, including a general, trans-situational capacity for general problem solving. Out of this characterisation falls the *piece de resistance* - a central representational device conceived of as the stronghold of the 'rational', 'abstract', 'symbol-using' agent. Given this end-state characterisation of 'intelligence', it is perhaps not too surprising that both developmental and comparative psychologists have invested such effort (and to little or no avail) in tracking the 'development' of abstract thought. In fact our own comparative and developmental research on inferential mechanisms (McGonigle and Chalmers, 1977, 1986; Chalmers and McGonigle, 1984) indicates that what had previously been taken as indexical of 'high level' abstract thought is basically a robust biological mechanism dedicated to 'rational' choice control.

2. Systems are studied by processes like perception, learning, memory, thinking and reasoning, action, etc. (see Brooks, 1986). These processes are envisaged as collaborating serially to make up a total system, each participating at its own level of functioning in the analysis of the flow of information through the system from sense to action (McGonigle, 1989). In such a scheme, it is not asked (explicitly) which competences particular systems support; in short, what perception/memory acts are FOR. As a consequence, laboratory tasks have frequently been of the 'toy-world' variety, invoking behaviours of little or no ecological validity. As a consequence, the assumed competence of a system is often too informally characterised to be useful.

HUMAN COGNITIVE DEVELOPMENT AND BRAIN GROWTH

One of the most enduring characterisations of human cognitive development is that it is stage-like and discontinuous. For long periods of development, a progressive unfolding of competences can be observed in both linguistic and non-linguistic domains. Hemispheric specialization, for example, is a case in point where children develop analytic abilities with progressive lateralisation of hemispheric function. These reveal themselves in such apparently straightforward tasks as looking at a visual pattern with a view to committing it to memory. As the Russian investigator Zinchenko has shown, for example, subjects as young as four years are essentially holistic in their encoding, betraying long (visual) fixations and little dynamic interrogation of the target stimuli, with poor retention as a consequence. Later, analytic, serial control develops (usually) in the left hemisphere, thus adding a new competence (logically) presupposing a spatial, holistic one for its success.

This phrased or staged view of human cognitive growth has received strong support from research in the neurosciences. Thatcher *et al* (1987), using EEG phase and coherence measures, have picked up a pattern of growth spurts and consolidations in the development of cortico-cortico connections between hemispheres in human brain development seen by them as evidence of genetically preprogrammed additions of new neural systems. If so, the results imply that many of the important cognitive leaps forward in development are the main consequence of such changes in the nervous system and are thus by 'design' and not a consequence of earlier antecedent behaviours. The particular competences supported by these spurts in brain growth is the subject of a research programme in my laboratory (McGonigle and Chalmers, 1990b).

MODULARITY

Whatever the sequential and staged character of development, recent and converging work in the neurosciences strongly suggests that systems decompose into a number of psychobiologically distinct subsystems working in parallel, each dedicated to a limited set of competences (of the total set which is supported by the sytem as a whole). Thus in clinical studies of memory for example, studies of amnesia have revealed syndromatic expressions of trauma and disease which strongly implicate different memory systems, such as those to support procedural knowledge (as in riding a bicycle) as distinct from those of episodic encoding or the retention of 'meaning'. Thus patients suffering from amnesia may learn a motor-skills task without showing any retention whatsoever of the situation (episode) or the clinician involved in the testing and training.

Such syndromatic presentations showing highly selective impairment of function contrast markedly with black box memory characterisations which see memory as merely an enabling, general retentive device to be hooked to any input at will. The very partiality of degradation, however, indicates that memory is a collection of systems that have been engineered by evolution to solve specific sets of requirement. As Sherry and Schacter (1988) argue, furthermore, such requirements may be functionally incompatible one with the other. So instead of seeing the multiplicity of mnemonic devices as merely an example of a simple division of labour, the division may also reflect competing requirements which no one unencapsulated system would be able to sustain.

LOGICAL HIERARCHY

Modularity apart, however, the hierarchical relationships between these competences may also have important engineering implications. It is clear from the biology, for example, that the use of vision 'to avoid hitting things' is an on-line system based (for example) on optic flow and whole visual field displacements (parallax). Primarily egocentric, such a system requires no memory of the visual properties of objects sensed. The very movement of an agent, however, inevitably induces the problem of locative 'tagging' - if only to find a former position in space. This problem, an immediate consequence of free movement by the agent, requires some sort of memory to help extract some of the invariants in a given lay-out or niche. Yet this competence, made necessary by locomotor competence, does not in itself underwrite the ability to code and identifiy objects within the map space. Instead, objects which are free to migrate within that space require a different sort of treatment, often involving detail vision. Despite this independence of function, however, there is nevertheless, a logical dependence amongst these competences. For it would seem utterly pointless to solve the (detail) vision object recognition problem before solving the spatial reference one: equally it would be foolish to solve the spatial reference one without benefit of the on-line control mechanisms which allow the system to avoid obstacles and pitfalls wherever these are encountered.

What is true for the biological case should also be true for robots! The designer should have a principled hierarchy of this sort in mind, where each layer is independent in so far as the mechanism or module can be designed as a separate competence. However, when installed in the agent, specification of a hierarchy in these terms specifies the prioritisation in the control of the agent. On this view, what is functionally presupposed (albeit not directly causally implicated) with respect to later competences must have priority (e.g., not walking over a cliff edge whilst reading a book!).

INCREMENTING BY DESIGN: A TWO-VECTOR TRAJECTORY

Combining the concepts of modularity and hierarchy has led us to a bi-vector specification of trajectories over which systems invest in complexity or become more competent. Figure 1 (McGonigle, 1990; McGonigle and Chalmers, 1990a) features a vertical vector depicting increments in competences by design where, for example, some aspects of a visually-based adaptation such as egocentric, on-line obstacle avoidance are arranged hierarchically with respect to other dependent but not causally determined competences, such as 'map'

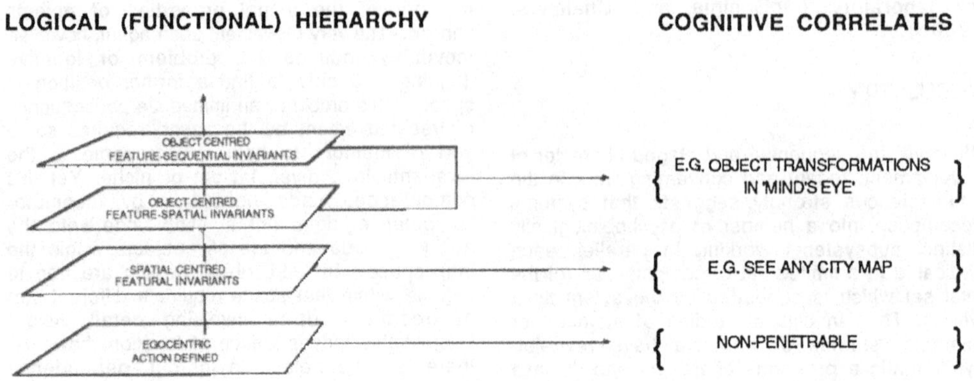

Figure 1. Two-vector incrementation.

construction involving memory. However, the basis for the solution is not constrained by this hierarchy, and is free to vary from task to task and from species to species along a horizontal axis. Thus, for example, a rat in a maze may recall previous routes as an important competence, and one very different from its ongoing on-line control problems with visual space which are route-independent. However, the manner by which the route is recalled could vary from urine marking to the extraction of some of the invariant visual features of the terrain explored during route learning. Along the same dimension of complexity incrementation, we might place human map representation and map reading, as in a city map designed for a large scale environment where the costs of exploration and route learning by other means are too high. Crucial to this representation of the incremental trajectories is the notion that increments along one vector have no direct implication for increments along another. It is further assumed that the basic unfolding of competences along the vertical vector is not driven by learning mechanisms or other factors in 'self-incrementation': these, however, have a much fuller expression along the trajectory represented by the horizontal coordinate.

EPISTEMOLOGICAL ISSUE

A central epistemological question concerns both the staging and sequencing of cognitive growth. Are these merely accidental features of a biological system developed in perhaps an *ad hoc* unprincipled manner by blind selection, or do they reveal (when known at all) an important epistemological constraint which applies to any system designed to mimic human competences ?

For the designer this issue is surely a crucial one, as to follow what Brooks dubs "the oxentrail of evoloution in design" might prove unnecessarily tedious if the biological solution, although adequate, is not by any means a unique or necessary one. Yet it is through the design and engineering of artificial systems that offers the best prospect of adjudication on this crucial question. On these grounds alone, therefore, it is important that, as biologists, we do our best to characterise systems in ways which will be transparent to the designer.

SUMMARY DESIGN POINTS

It would appear from our biological analyses that three major requirements have to be met in the design and (competence) incrementation of intelligent systems.

1. The systems are modular not monolithic.

2. Incrementing is done by design, not by a self-boootstrapping agent which 'takes off' from a set of primitives and learns to be more intelligent! However, elaborations of such solutions can be represented along a trajectory orthogonal to the former.

3. Major steps or stages of incrementation are managed on a logical principle or logical hierarchy (McGonigle and Chalmers, 1990a) wherein primitive or 'first-strike' behaviours, although important preconditions for the development and elaboration of the system as a whole, are not in themselves causal determiners of later competences.

PRESCRIPTION

BUILDING AN ARTIFICIAL SYSTEM

Several prescriptions emerge from this analysis for the design of robots.

a. Limiting the domain over which the system is to be competent. This takes two forms of constraint: the first is determined by considerations of what the behaviours are FOR, the second by the particular niche within which the system is to be competent. Thus a moving agent capable of some form of navigation may be best considered at the design stage in terms of whether the niche is an office environment, for example or a larger scale open terrain.

b. Dividing up the spectrum of competence(s) into 'rational' sub-systems each capable of achieving a coherent set of goals.

c. Establishing a (logical) hierarchy based on what needs to be established as a competence before a further one is added. This is a fundamental issue for the fraught question of 'scaling-up' the system without requiring the elaboration of earlier competences.

IMPLEMENTATION

A HEARING ROBOT

Recently, collaborative work with Jim Donnett of the Department of Artifical Intelligence has led to a project which I shall describe briefly here. Intended for an office/corridor niche with a possible role in mail service, a robot has been constructed which uses infra-red sensors to avoid objects in its path. However, in addition to object avoidance, a preliminary and crucial competence, the robot is designed to locate objects in its niche. One of these is the enery supply source: the machine has a state evaluator which allows it to estimate its residual run time. However, the primary targets are possible mail locations. Given the sensors for object avoidance, all the robot would do without the addition of further competences would be a series of random walks triggered by the sensing of impediments. However, we have built in 'purpose' and directionality by investing the niche with sound beacons. These are distributed around the robot's environment, transmitting at different frequencies.The robot has a moving head, fitted with two ears selectively tuned to each bandwidth of frequencies in turn. Direction and location are determined by keeping the time and intensity discrepancy between the ears to a minimum. As rotation of the head is low cost, relative to movement of the whole vehicle, hunting for signals can be engaged in whilst the vehicle is stationary. Once a signal is located located, however, and directional symmetry achieved, the body can be rotated to dead-lock into a pre-determined head-body position. The robot can then move towards the beacon. However, this activity is interruptible by any 'object-approach' signal from the object avoidance module which has priority. Advance towards the beacon is therefore the default condition.

In this way the vehicle incorporates both a principled hierarchy (vertical) and a set of pragmatic solutions for 'navigation' within a particular niche (horizontal) which are orthogonal, one to the other.

TECHNICAL SPECIFICATION

Our experimental vehicle is an autonomous robot consisting of a plexiglass platform on which are mounted two independent DC motors for locomotion, a third motor which drives the vehicle's head, control and sensor electronics, and a battery power supply. The head is an aluminium disk sporting two 'ears' (condenser microphones), and a pair of infrared emitter/detectors used to alert the head to possible collision with bridge-like obstacles not detected by body-mounted sensors. Electronic control comprises two networked microprocessors; amplifiers, filters, and analog-to-digital conversion circuits for the ears; speed and direction regulators for the head and body drive motors; and a battery charge state evaluator. One of the microprocessors is primarily responsible for beacon tracking, the other for collision avoidance. An elaborated description of hardware and software appears in Donnett and McGonigle (1990a).

Beacon tracking presupposes competent collision avoidance, but the two systems share nothing except that both express themselves in the motion of the robot; there is no inter-system competition, however, as obstacle avoidance preempts control temporarily if the vehicle's path is obstructed, although the distribution of control between navigation and avoidance is such that avoidance can be suppressed when this is sensible in context (in the immediate presence of a beacon marking an energy source, for example).

Furthermore, the interaction between beacon navigation and collision avoidance leads to an optimal fitting of vehicle to operating niche. Vehicle speed is continuously adjustable between 'stopped' and a maximum of about three metres per second. From a theoretical model of optimum speed, which is a function of obstacle avoidance response time and relative obstacle density, and behavioural simulations, a speed-to-travel look-up table is derived (Donnett and McGonigle, 1990b), tuned where necessary on the actual machine. Current 'best speed' may then be selected by the control processors to fit environmental circumstances: the vehicle decelerates or accelerates as specified in the look-up table, subject to past experience and power supply energic state, where the relative obstacle density index into the table is simply the frequency of interrupts from the collision avoidance system.

FUTURE

The first robot is still under trial, but the systems approach we have taken allows us to pursue a number of related avenues of work simultaneously. First, the niche engineering aspect of the enterprise allows us to predict reasonably accurately the basic environmental features the robot is likely to encounter without succumbing to a 'toy-world' charge! Second, 'investing' that niche with cues and signals to fit the robot seems a lot easier and cheaper than designing a robot to 'learn' about its environment *de novo*. The particular signals we are using, furthermore, give the robot both locative and serial information and allow us to program the environment or niche in place of programming the robot. Third, although sound has its limits and the object sensors demand an upper bound on the velocity at which the robot can traverse its environment, there are many plausible developments within these operational limits which will teach us a great deal about the control of multi-tasking systems. Surely it is here that animal and animat converge!

REFERENCES

Brooks, R. (1986). Achieving artificial intelligence through the building of robots, MIT AI Laboratory, AI Memo 899.

Chalmers, M. and B. McGonigle (1984). Are children any more logical that monkeys on the 5-term series problem? *J. Exp. Child Psychol.*, **37**, pp. 335-377.

Donnett, J. and B. McGonigle (1990a). A 'local area' autonomous robot, Part 1: Acoustic navigation system. In preparation.

Donnett, J. and B. McGonigle (1990b). A 'local area' autonomous robot, Part 2: Automatic adaptation to operating environments. In preparation.

McGonigle, B. (1987) Non-verbal thinking by animals. *Nature*, **325**, pp. 110-112.

McGonigle, B. (1988). Is it the thought that counts? *Behavioural and Brain Sciences*, **11** (4).

McGonigle, B. (1989) Dall'engramma al percetto. *Nuova Civilta delle Macchine*, **1** (25), pp. 25-30.

McGonigle, B. and M. Chalmers (1977). Are monkeys logical? *Nature*, **267**, pp. 694-696.

McGonigle, B. and M. Chalmers (1986). Representations and strategies during Inference. In *Reasoning and Discourse Processes*, eds. T. Myers, K. Brown, and B. McGonigle, London: Academic Press.

McGonigle, B. and M. Chalmers (1990a). *Intelligent Systems: A Cognitive Analysis.* In preparation for publication (Columbia University Press).

McGonigle, B. and M. Chalmers (1990b). An experimental analysis of ordering skills in children, project funded by the ESRC of GB.

Sherry, D.F. and D.L. Shacter (1987). The evolution of multiple memory systems. *Psychol. Rev.*, **94**, pp. 439-454.

Thatcher, R.W., R.A. Walker, and S. Giudice (1987). Human cerebral hemispheres develop at different rates and ages. *Science*, **256**, pp. 1110-1113.

ANIMATS IN EDUCATION

Animal Simulations with *Logo: Massive Parallelism for the Masses

Mitchel Resnick
Massachusetts Institute of Technology
Cambridge, MA 02139 USA
mres@altdorf.ai.mit.edu

Abstract

Experimentation with animal simulations is limited, in large part, by the difficulty of converting ethological ideas into computer programs. *Logo is a new programming language that aims to make it easier for non-expert programmers (researchers as well as students) to develop and modify their own simulations. *Logo is designed specially for simulating colony-level behaviors—that is, group behaviors that emerge from interactions among hundreds or thousands of individual creatures. Unlike most simulation languages, *Logo gives the creatures' environment an equal computational status to the creatures themselves. Users write rules for creatures and for "patches" of the environment, then observe the higher-level behaviors that result. A sample *Logo program shows how local, parallel actions among ants can lead to spatial-extended and temporally-sequential patterns in the colony-level behavior.

1. Introduction

During the past several years, a growing number of researchers have begun creating computer-based simulations of animal behavior. Some are motivated by ethological goals: they hope to gain a better understanding of the mechanisms underlying the behaviors of real animals. Others are motivated by engineering goals: they hope that simulations of animals will provide ideas (or at least inspiration) for building computers and robots that function more effectively in the world.

Unfortunately, designing and programming animal simulations typically requires significant programming expertise. Most animal simulations are developed as customized programs, by experienced programmers. Although there are several new languages and tools designed to facilitate the development of animal simulations, even these tools are meant primarily for experienced programmers.

As a result, it is difficult for non-experienced programmers to convert ethological ideas into computer simulations. Certainly, novice programmers can change parameters or initial conditions on existing simulations. But they are not able to make more serious modifications or create entirely new simulations. In short, animal simulations are still not for the masses.

What is needed is a new type of programming language that allows people to more easily create and experiment with animal simulations. This paper describes a language that aims to do just that. The language, called *Logo (pronounced *star-logo*), is designed especially for simulating "colony-level behaviors"—that is, group behaviors that emerge as large numbers of individual animals interact with one another, as in bird flocking or ant foraging. These types of simulations are particularly difficult to construct using traditional, sequential programming languages. Indeed, simulations of colony-level behaviors highlight the need for a new "massively parallel" approach to programming, in which many "computational creatures" act in parallel (at least conceptually, if not in reality).

Section 2 describes the audience for *Logo. Although *Logo is designed primarily to help students explore self-organizing phenomena, it could serve equally well as a tool for ethologists. Section 3 discusses the computational requirements for programming colony-level animal simulations. Section 4 discusses the central ideas underlying the design of *Logo, including the decision to treat the creatures' "world" as an active computational actor, equal in status to the creatures themselves. Section 5 presents examples of *Logo simulations. A simulation of ant foraging, for example, shows how local, parallel actions by hundreds of individual ants can result in spatially-extended and temporally-sequential behaviors by the colony as a whole. Section 6 describes future directions for *Logo.

2. Who Is It For?

*Logo is designed primarily as a language for students. Why bring animal simulations to the classroom? Of course, animal simulations could supplement existing curricula in biology or computer science. But I believe that ideas from animal behavior can play a much more fundamental and radical role in education, cutting across disciplinary boundaries and providing students with a new framework for thinking about all types of systems in the world.

Consider, for example, the idea of *self-organization*—the idea that large-scale patterns and structures can form without any centralized design or coordination (Prigogine & Stengers 1984). This idea can act as a unifying theme, helping students understand not only biological systems, but physical and social systems as well. I recently showed a simulation of a simple self-organizing system to four researchers. One researcher was reminded of galaxies forming, another of cities growing, the third of honeybees swarming, the fourth of excess arsenic atoms clustering within gallium arsenide. In fact, it was a simulation of slime-mold cells aggregating into clusters. But it is no accident that the simulation evoked such varied images. Self-organizing systems share certain common features, whether they come from physics, economics, ethology, or ecology.

Traditional school curricula erect rigid walls between various science (and social-science) disciplines. By examining animal behaviors in terms of self-organization, students can start to break down those walls. They can start to make links between the seemingly disparate phenomena they learn about in the classroom (and see in the world). A student who creates a simulation of an ant colony foraging for food, for example, can learn about much more than ants. By viewing the ant colony as a self-organizing system, the student can develop a better framework for thinking about not only social-insect systems, but also immune systems, evolutionary systems, and even economic systems.

Unfortunately, learning to think in terms of self-organization is not an easy task. Despite all of the examples of self-organization in the world, most people (students and scientists alike) do not think about phenomena in terms of self-organization. One reason: people are more familiar with systems organized by a central designer. When people see neat rows of corn in a field, they assume (correctly) that the corn was planted by a farmer. When people watch a ballet, they assume (correctly) that the movements of the dancers were planned by a choreographer.

On the other hand, when people see self-organizing phenomena in the world, they tend to assume centralized control where none exists. When people see a flock of birds or a school of fish, many of them assume (incorrectly) that there must be a leader of the group. And when people see the diversity of complex creatures in the world, many of them assume (incorrectly) that there must be a single Designer of Life. This "centralized mindset" is quite resistant to change. People need better models and metaphors for thinking about self-organization. And they need a way to "play around" with self-organizing phenomena. That is what *Logo aims to provide.

While *Logo is designed with students in mind, it is certainly not limited to the classroom. Many ethologists would like to create animal simulations, but do not have the programming skills to do so. With *Logo, ethologists could more easily test hypotheses, altering the rules for individual creatures and observing the group behaviors that result. Equally important, *Logo (and other languages like it) could provide a new and better means for ethologists to express their ideas about animal behaviors. In many cases, programming languages are a better medium of description than either mathematics or natural language. As Abelson and Sussman (1985) explain: "... a computer language is not just a way of getting a computer to perform operations but rather it is a novel formal medium for expressing ideas... programs must be written for people to read, and only incidentally for machines to execute."

3. Computational Requirements

It is possible, of course, to create computer-based simulations of self-organizating phenomena in almost any programming language (and, indeed, programmers have written simulations in many different languages). But to explore self-organizing systems in a creative way, people (experts and non-experts alike) need a language that specifically supports such explorations. For example, the language should make it easy for the programmer to:

- Program the *parallel* actions of (at least) *hundreds* of "creatures"

- Program *local interactions* among the creatures

- Program interactions between a creature and its *local environment* (e.g., an ant depositing a pheromone, or an ant sniffing a pheromone trail)

- Program changes in the environment itself (e.g., the growth of food, or the diffusion of a chemical)

In addition, the language must run sufficiently fast so that users can observe interesting phenomena in real time (and make multiple runs with varying parameters). Moreover, the environment should be based on metaphors and constructs that make it easy for non-expert programmers to create (and understand) their own simulations.

Cellular automata environments (Toffoli & Margolus 1987) fit many of these requirements. But CA's are too low-level an abstraction for many types of explorations (particularly for novices). Common properties like particle conservation must be modelled as emergent phenomena. To model a collective phenomenon like flocking, for example, the user has to deal with two levels of emergence: creatures emerging from cellular rules, and flocks emerging from creatures.

A higher level of abstraction is needed. Several new programming systems—such as Agar (Travers 1988), RAM (Taylor et al. 1988), and RDL (Stells 1989)—move in that direction. Each provides tools for programming the local actions of creatures in a simulated world. But these languages seem better suited to the needs of expert programmers, not novices. *Logo aims to provide many of the same facilities, but special consideration for non-expert programmers.

4. Core Constructs of *LOGO

There is an old saying that goes something like this: If a person has only a hammer, the whole world looks like a nail. Indeed, a person's perceptions and models of the world are strongly shaped by the tools and objects that exist in the world. The same is true for computational systems. The way people interact with (and think about) a computational system depends strongly on the *objects* that compose the system. If the objects are well-chosen for the intended task, even novices will use the system productively. If the objects are not well-chosen, even experts will struggle.

*Logo is designed around two primary types of objects, chosen specifically to facilitate the modelling of colony-level behaviors and other self-organizing phenomena. One type of object, not surprisingly, is a *creature*. Traditional versions of the Logo programming language include a graphic "turtle" on the screen. Students can command the turtle to move and draw lines on the screen (Papert 1980). *Logo creatures go beyond the standard Logo turtle in several ways. Most important, *Logo programmers can control *hundreds* or *thousands* of creatures, and all of the creatures can act in parallel. In addition, *Logo provides programmers with built-in primitives to control local interactions between the creatures, and to create new creatures dynamically.

Programmers can use *Logo creatures to represent birds or ants or molecules or automobiles—or any other type of object that moves in the world. By creating different "breeds" of creatures, programmers can simulate situations with multiple object types—for example, foxes and rabbits in an ecology model, or antibodies and antigens in an immunology model.

The second type of *Logo object, called a *patch*, is a more striking departure from traditional Logo (and other programming languages). A patch is a small section of the "environment" in which the creatures live. (The term *patch* is borrowed from Hogeweg (1988).) If, for example, the creatures are programmed to release a "chemical" as they move, each patch can keep track of the amount of chemical that has been released within its borders. Like creatures, patches are "active" objects, capable of computation. So, as the creatures move, each patch could diffuse some of its chemical into neighboring patches, or it could grow "food" based on the level of chemical within its borders. *Logo includes several special-purpose primitives (such as follow-gradient) to facilitate interactions between creatures and patches, and others (such as diffuse) to facilitate interactions between neighboring patches.

Of course, other "creature-oriented" programming languages allow programmers to control both creatures and the environment. But few treat the environment as an equal-status object. They tend to treat the environment as a passive entity, manipulated by the creatures that move within it. This view, not surprisingly, matches the way many people view the Earth itself.

In *Logo, the environment is given equal status to the animals that inhabit it. Animals (in the form of creatures) and the environment (in the form of patches) are both computationally active objects. They can mutually influence one another, somewhat in the spirit of James Lovelock's Gaia (1979). By reifying the environment, *Logo aims to change the way people think about creature-environment interactions—perhaps leading to new and richer ways of thinking about self-organizing phenomena.

The current version of *Logo is implemented on the Connection Machine, a massively-parallel computer with at least 16,000 processors (Hillis 1985). In this implementation, each Connection Machine processor controls one creature and several patches. But massively-parallel hardware is not theoretically important to the *Logo approach. *Logo aims to provide the programmer with a clear *conceptual model* for massive parallelism. The programmer writes programs *as if* the creatures and patches are all acting in parallel, regardless of the underlying implementation.

5. Examples

This section describes two sample *Logo programs. These are the types of programs that high-school students might create, play with, and modify. Each program is inspired by some real-world example of self-organization. But the goal is *not* to create perfect simulations of real-world behavior. Rather, the programs serve as "conceptually clean" examples: simple programs that highlight key ideas about self-organization.

5.1 Aggregation

Under certain conditions, the larvae of slime mold aggregate in clusters. The mechanism for this self-organizing behavior involves a chemical pheromone that the larvae emit. As described by Prigogine and Stengers (1984):

> The pheromone diffuses in space, and the larvae move in the direction of its concentration gradient. Such a reaction provides an autocatalytic mechanism since, as they gather in a cluster, the larvae contribute to enhance the attractiveness of the corresponding region. The higher the local density of larvae in this region, the stronger the gradient and the more intense the tendency to move toward the crowded point.

This self-organizing behavior can be simulated in a simple *Logo program. Each *Logo creature follows this rule: "Sniff" for pheromone in your current patch. If you detect some pheromone, follow the gradient. If not, keep going in the same direction (with a little randomness). Drop a little pheromone into the patch, then take a step forward.

```
to creature-rule
ifelse ask-my-patch [pheromone] > smell-threshold
  [set-heading (follow-gradient pheromone) + noise]
  [turn noise]
demand-my-patch [make "pheromone pheromone + 1]
forward 1
end
```

Meanwhile, each patch follows this rule: Let a little of your pheromone evaporate, then let the pheromone diffuse into neighboring patches.

```
to patch-rule
make "pheromone pheromone * (1 - evaporation-rate)
diffuse pheromone
end
```

Based on these simple (and totally local) rules, the creatures (generally) aggregate into clusters. As long as there is a bit of "noise" in the system, the clusters appear as dynamic entities, with creatures moving in and out of the clusters, and nearby clusters sometimes uniting into super-clusters.

The number and sizes of the clusters depend on the values of certain parameters: the number of creatures, the smell-threshold for each creature, the evaporation-rate, the amount of "noise." For certain parameter values, clusters will not form at all. One interesting activity for students is to explore the parameter space, looking for regions without clusters. Such explorations can lead to the concept of "critical density" (since a critical density of creatures is needed for aggregation). Of course, this critical density will vary depending on the values of other parameters.

This simulation suggests many "what-if" questions, some of which can be explored through qualitative reasoning. For example: What would happen if the creatures had a better sense of smell (that is, a lower smell-threshold)? It is rather straightforward to deduce that there will be more and smaller clusters, and that the clusters will form more quickly. (And indeed, this happens.)

Interestingly, this result seems to run counter to many people's intuitions. In a small initial study, most people predicted (correctly) that clusters would form more quickly, but most predicted (incorrectly) that the clusters would be larger than before. Why do people have this false intuition? Part of the problem, I believe, is a confusion of levels. People imagine that the creatures are "trying" to form clusters—so creatures with a better sense of smell will do a "better job" and form larger clusters.

5.2 Social insects

With a few changes in the programs, we can turn the slime mold into a colony of ants foraging for food, much like the simulations by Deneubourg et al. (1989, 1990) and Travers (1988). We begin by creating a simulated world with an ant nest and a few randomly-placed food sources. Ants begin by searching (randomly) for food. Once an ant finds some food, it brings the food back to the nest, laying a pheromone trail as it returns. Other ants can use this pheromone trail to find their way to the food source. When they return to the nest, they reinforce the pheromone trail.

The rule for each ant can be written as follows:

```
to creature-rule
ifelse my-food = 0
  [look-for-food]
  [look-for-nest]
end

to look-for-food
ifelse ask-my-patch [patch-food > 0]
  [demand-my-patch [make "patch-food patch-food - 1]
   make "my-food 1
   right 180]
  [set-heading (follow-gradient pheromone) + noise]
forward 1
end

to look-for-nest
ifelse ask-my-patch [nest?]
  [demand-my-patch [make "patch-food patch-food + 1]
   make "my-food 0
   right 180]
  [set-heading (follow-gradient nest-scent)
   demand-my-patch [make "pheromone pheromone + 1]]
forward 1
end
```

At first thought, an ant's journey back to the nest might seem to require some non-local knowledge. But, in fact, the behavior can be implemented with purely local rules. The trick is to make the nest emit a characteristic "nest scent" that diffuses through the world. If an ant is carrying food, it follows the gradient of the nest scent to find its way back to the nest, just as an ant without food follows the gradient of the pheromone trail to find its way to the food.

Patches are responsible for the diffusion of both the pheromone and the nest-scent:

```
to patch-rule
make "pheromone pheromone * (1 - evaporation-rate)
diffuse pheromone
if nest? [make "nest-scent constant]
diffuse nest-scent
end
```

The ants' pheromone trail is an example of a large-scale, orderly structure, created entirely through local interactions. The trail lasts only as long as the food source. Once the food source is fully exploited, ants no longer reinforce the trail to the source. So the trail gradually disappears through evaporation and diffusion.

With properly chosen parameters for evaporation, diffusion, and noise, this simple simulation exhibits a striking colony-level behavior. The colony exploits the food sources systematically and sequentially. The ants start with the source closest to the nest, then (after that source is fully depleted) they begin to exploit to the next closest source, and so on. As in the aggregation simulation, this behavior can be analyzed qualitatively. The idea of "critical density" is again useful. For each food source, there is a critical density of ants needed to form a stable trail. The critical density depends on the distance of the food source from the nest (along with other factors, like the evaporation and diffusion rates). A small number of ants can not create a stable trail to a distant food source, since the pheromone trail evaporates and diffuses before the ants find their way back to the food.

One can imagine a "competition" among food sources, each trying to attract a stable trail of ants. In this competition, the closest food source to the nest has two advantages. It is the most likely to be "discovered" in a random walk from the nest. And it requires the lowest density of ants in order to form a stable pheromone trail. As a result, as the ants march out of the nest, they will typically start by forming a trail to the closest food source. But as long as the ants are busy exploiting that food source, it is difficult to reach the critical density for any of the other food sources. Only after the ants have fully exploited the closest food source will there be enough "free" ants to form stable trails to the other sources. In this way, the colony exploits the food sources one by one, moving outward from the closest source to the most distant.

This simulation also provides a good context for exploring the role of randomness in self-organizing behavior. Because of the "noise" in the ant rules, an ant will sometimes lose its way as it follows a pheromone trail. As others have noted (Deneubourg et al. 1989), this "error" can be helpful to the colony as a whole. These wayward ants sometimes discover as-yet unexploited sources of food, forming faint secondary trails. When the colony finishes exploiting one food source, these secondary trails help the colony shift to a new food source with little delay.

6. Looking Ahead

Clearly, *Logo is not yet a language for the masses. The current implementation runs only on an expensive supercomputer. It is used by only a small handful of researchers and students.

Nevertheless, *Logo (and other languages like it) can play an important role in making massive parallelism accessible to the masses. Computer hardware technology is advancing rapidly, and the cost of computational power continues to drop. There is no doubt that, in the next decade, many people (scientists as well as students) will have access to computers with significantly improved processing power. But without new programming languages, and new metaphors for thinking about programming and simulations, these new computers will be useful primarily to a small cadre of expert programmers. To make it possible for a broader range of researchers and students to develop their own animal simulations, we must start developing programming languages for tomorrow's computers today.

This strategy has a strong precedent in the development of the original Logo programming language. Researchers at Massachusetts Institute of Technology began developing Logo in the late 1960's. By introducing a graphic "turtle" for drawing geometric shapes, the developers of Logo aimed not only to make programming more accessible to non-expert programmers, but also to provide people with a new metaphor for thinking about geometry. At first, Logo ran only on minicomputers costing hundreds of thousands of dollars. Many people mocked the efforts by the Logo group, arguing that children (and other non-expert programmers) would never have access to such computers. But a decade later, the situation had changed dramatically. When personal computers were introduced, Logo researchers had already developed a rich conceptual foundation. Logo was quickly adopted by schools: it is now used in roughly one-third of the elementary schools in the United States. Although Logo has never won a large following among professional mathematicians, its idea of "turtle geometry" has led to some strikingly innovative mathematical research (Abelson & diSessa 1980).

Just as some people scoffed at early versions of Logo running on minicomputers, some now question the use of the Connection Machine for developing an animal-simulation language for students and other non-expert programmers. But such criticism misses the point. The Connection Machine, though inaccessible to most people today, is an excellent prototyping environment for the ideas of tomorrow. By observing people as they write *Logo simulations, I hope to gain insights into how people think about and learn about complex phenomena like self-organization. These insights, in turn, can contribute to improvements in the conceptual foundations and design of *Logo.

Some of the ideas in *Logo could be transferred to less expensive computers in the short term. Other *Logo ideas will become widely accessible only as powerful computers become more affordable. But that will inevitably happen. Ultimately, the development of innovative animal simulations will be limited not by hardware costs (which will surely fall), but by the ease or difficulty with which people can develop, experiment with, and think about their own simulations.

Acknowledgements

Hal Abelson, Seymour Papert, Brian Silverman, and Ryan Evans have provided encouragement, inspiration, and ideas for the *Logo project. The LEGO Group and the National Science Foundation (Grant #851031-0195) have provided financial support.

References

Abelson, H., and A. diSessa (1980). *Turtle Geometry: The Computer as a Medium for Exploring Mathematics.* Cambridge, MA: MIT Press.

Abelson, H., and G. Sussman (1985). *Structure and Interpretation of Computer Programs.* Cambridge, MA: MIT Press.

Deneubourg, J.L., et al. (1989). "The Blind Leading the Blind: Modeling Chemically Mediated Army Ant Raid Patterns." *Journal of Insect Behavior,* **2**: 719-725.

Deneubourg, J.L., et al. (1990). "The Self-Organizing Exploratory Pattern of the Argentine Ant." *Journal of Insect Behavior,* **3**: 159-168.

Hillis, W.D. (1985). *The Connection Machine.* Cambridge, MA: MIT Press.

Hogeweg, P. (1988). "MIRROR beyond MIRROR, Puddles of LIFE." In *Artificial Life,* edited by C. Langton. Redwood City, CA: Addison-Wesley.

Lovelock, J. (1979). *Gaia: A New Look at Life on Earth.* New York: Oxford University Press.

Papert, S. (1980). *Mindstorms.* New York: Basic Books.

Prigogine, I., and I. Stengers (1984). *Order Out of Chaos.* New York: Bantam Books.

Steels, L. (1989) *The RDL Manual.* AI Memo 89-11. University of Brussels.

Taylor, C., D. Jefferson, S. Turner, and S. Goldman (1988). "RAM: Artificial Life for the Exploration of Complex Biological Systems." In *Artificial Life,* edited by C. Langton. Redwood City, CA: Addison-Wesley.

Toffoli, T., and N. Margolus (1987). *Cellular Automata Machines.* Cambridge, MA: MIT Press.

Travers, M. (1988) "Animal Construction Kits." In *Artificial Life,* edited by C. Langton. Redwood City, CA: Addison-Wesley.

Lego Vehicles: A Technology for Studying Intelligent Systems

Jim Donnett and Tim Smithers[†]

Department of Artificial Intelligence
University of Edinburgh
5 Forrest Hill
Edinburgh EH1 2QL
SCOTLAND
Telephone: 031 667 1011 ext 2517
Fax: 031 225 9370
E-mail: donnett@uk.ac.ed.edai tim@uk.ac.ed.edai

Abstract

During the summer of 1989 we established the *Intelligent Sensing and Control Laboratory* in the Department of Artificial Intelligence at Edinburgh. This laboratory is designed to support both post-graduate teaching and basic research into intelligent sensing and control. In this paper we present the motives for setting up the Intelligent Sensing and Control Laboratory and the design and implementation of a Lego Technic^{TM} based technology used to build simple autonomous vehicles intended to support the teaching and research activities of the laboratory. We report on some of the experiences gained from its first year of operation and relate these to the requirements of our biologically oriented research programme into intelligent behaviour and its development in autonomous artificial mobile systems.

1 Introduction

During the summer of 1989 we established the *Intelligent Sensing and Control laboratory* (ISC lab) in the Department of Artificial Intelligence at Edinburgh. This laboratory is designed to support both post-graduate teaching and basic research into intelligent sensing and control. In the autumn term (October to December, 1989) it was used for the practical sessions of a new ten week module called "Intelligent Sensing and Control"[1] during which students worked in pairs to build a series of Braitenberg-like ([Braitenberg 1986]) mobile vehicles of increasing motor-sensory sophistication. It is now being used for a number of other projects in a long term research programme to investigate the acquisition, maintenance, and adaptation of task achieving competences in autonomous mobile robots.

In this paper we present the motivation for setting up the ISC lab and the design and implementation of the technology used to build the autonomous vehicles. We also report on some of the experiences gained from its first year of operation in both teaching and research.

2 Background and Motivation

There are two different motives for setting up the Intelligent Senisng and Control Laboratory within the Artificial Intelligence Department at Edinburgh. The first concerns the nature of intelligent behaviour and the way we want to investigate it. The second concerns the kind of sensing and control we want to teach to our intelligent robotics students.

[†]Names appear in alphabetical order, with both being principal authors on this occasion.

[1]This forms part of the Intelligent Robotics theme of the Department's MSc in Knowledge-based Systems.

2.1 The Nature of Intelligent Behaviour

Since its earliest days the main stream of Artificial Intelligence (AI) research has been concerned with what might be called 'mental skills' which can be distinguished from 'motor-sensory skills': playing chess, proving theorems in first order logics, abstract planning, diagnosis, and design, for example, have been and still are major themes of research in AI. This emphasis on problems which can be constructed in purely formal domains has led to a very narrow view of what intelligent behaviour is, and consequently of what it takes to be an intelligent system. The ability to play good chess is a competence displayed by very few people. The ability to move about safely, navigate, and otherwise manipulate the world is, by contrast, an ability shared by almost all people and many animals. We believe that understanding motor-sensory skills is fundamental to understanding mental skills (a view we share with other roboticists like Hans Moravec, [Moravec 1984], Rod Brooks, [Brooks 1986], and early cyberneticians like Grey Walter, [Walter 1961]). We are therefore interested in solutions to the general *problem of acting appropriately in a world in which* the array of alternate actions and their consequences is typically large, complex, and variable. We call this the *appropriate action problem* for short.

Since biological systems offer a very wide range of examples of the kind of intelligent system we are interested in, our research programme is firmly grounded in a biological stance. Though not the result of a design process (as some have mistakenly tried to define them), biological systems can be viewed as example designs. This, taken together with our belief that developmental constraints dominate operational constraints in intelligent systems, means that it is to the biological sciences that we look for the competence theories that we need, and for hints of possible engineering solutions to the problem of incrementing intelligence in artificial systems, (see [McGonigle 1990]). In return, we expect our robotics activities both to test and inform the biological activities in ways that are not possible within the subjects themselves—we can try to build what we want to study, not just study what nature presents us with.

2.2 Sensing and Control in Robotics

The usual treatment of sensing and control in robotics teaching focuses on negative feedback techniques, often to the complete exclusion of anything else. Yet, classical control techniques have very little to offer the mobile robot designer looking for ways of solving the appropriate action problem referred to above. Negative feedback control is fine when used in tight control loops, such as in servomechanisms, but is quite inappropriate when it comes to preventing a mobile robot from colliding with obstacles in its path, or attempting to navigate its way from place to place, for example. Classical control techniques work well if the mapping from required task achieving behaviour to action is simple and remains the same (or within certain parametric variations). In situations in which action sequences are required to achieve tasks (not just single actions), and worse, when the necessary actions and their ordering change due to environmental variations, the control problem goes well beyond the scope of classical techniques. Ballistic control, in which actions are selected and initiated on the basis of initial conditions (and not subsequently modified), offers a simple and cheap way of selecting appropriate action sequences, and of parametrically configuring them to particular environmental conditions. The trick, of course, is in determining the necessary and sufficient initial conditions and an effective action sequence. Feedforward techniques can be used to embody a 'predictive' element in an agent's control system, whereby information about the environment, *task status, and agent can be used to predict future* conditions which are in turn used to select actions or action sequences. Using feedforward techniques thus enables a system to react to situations before they cause detrimental effects on its task achieving behaviour—negative feedback systems only react to adverse influences after they have affected the system. The behaviour of relatively simple animals has been shown to be governed by mixtures of feedback, feedforward, and ballistic strategies, [Staddon 1983]. Understanding how by attempting to replicate similar kinds of behaviour in artificial systems, in *animats*, forms the main purpose of the ISC laboratory.

In designing the *Intelligent Sensing and Control* module for our MSc course we were keen not just to teach a broader set of control concepts and techniques, but also to expose students to the reality of the problems involved in putting the techniques to work. The aim is to develop in the students an understanding, based upon direct experience, of what the different control ideas and techniques can realistically be used for.

2.3 A Common Technology

Having been inspired by Valentino Braitenberg's delightful book, *Vehicles*, [Braitenberg 1986], and being convinced of the relevance of his "law of uphill analysis and downhill invention"[2] we realised that our two

[2]This law states that it is much easier to invent machines which behave in interesting ways than it is to analyse from observations of the machines' performance how such behaviour

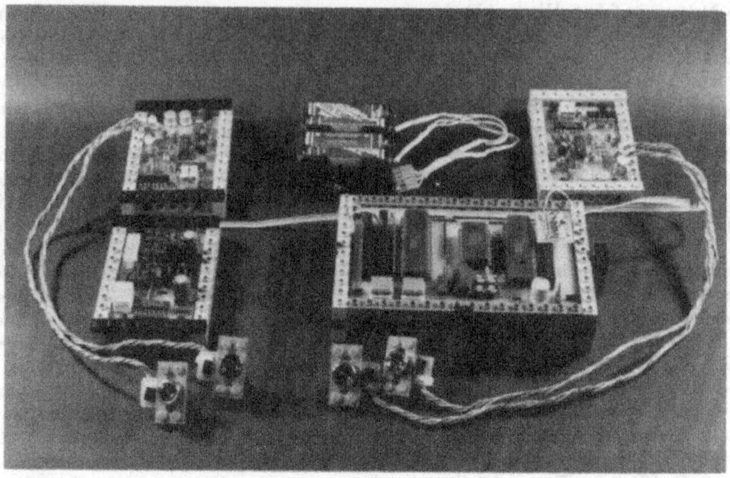

Figure 1: The supplementary sensing and control bricks used to construct Lego vehicles. The brain brick is centre right with the light brick mounted on it in the top right corner. The two IR-processing bricks are top left and top right, with the two pairs of IR sensors in the foreground. The two battery sets and connector are top center, and the motor control board centre left. The hall-effect switch (less magnet) can be seen just to the left and between the motor control brick and IR-processor brick. The four microswitches are not shown.

motives for establishing the ISC lab could be supported by a common technology.

We felt that to set students the problem of building small mobile robots, equipped with simple motor-sensory components, would present them with the right level of real sensing and control problems, and thus expose them to many of the important and basic issues of real intelligent sensing and control, as opposed to just the theoretical ones. We also realised that the technology required to support these kinds of practical exercises would be what we need to investigate the basic competences required for mobile robots and how they might be made to acquire such competences themselves—an essential part of our research interests introduced above.

3 Lego Vehicles

Following the Braitenberg Vehicles idea, by which a variety of behaviours can be obtained by varying the connections between simple sensors and motors, we wanted a technology which would support the construction of small vehicles whose mechanical and structural aspects could be easily reconfigured, together with the connections between the motors and sensors. For the former we chose Lego Technic^{TM} and for the latter we chose a microprocessor-based scheme[3]. The Lego Technic, with its variety of structural elements, motors, gears, etc., offers a convenient mechanical and structural modularity which allows a wide range of vehicle types and configurations to be constructed quickly and easily[4]. By adopting suitable construction styles it is also possible to build vehicles of more than adequate strength and stiffness.

To provide a complete kit with which to construct fully autonomous vehicles (i.e., which do not use an umbilical cable to supply power and/or instructions), the Lego Technic is supplemented with a number of other "bricks" and a specially designed programming language to facilitate the construction of modular sensing and control capabilities.

is engendered.

[3]Similarities with the work of Resnick et al at MIT were subsequently brought to our attention, [Resnick 1989], [Martin et al 1990], and [Resnick 1990], where the educational value of Lego-based vehicles, this time for children, is also being investigated.

[4]Relative to the use of other technologies, such as MECCANO^{TM}, or other non-modular approaches, for example.

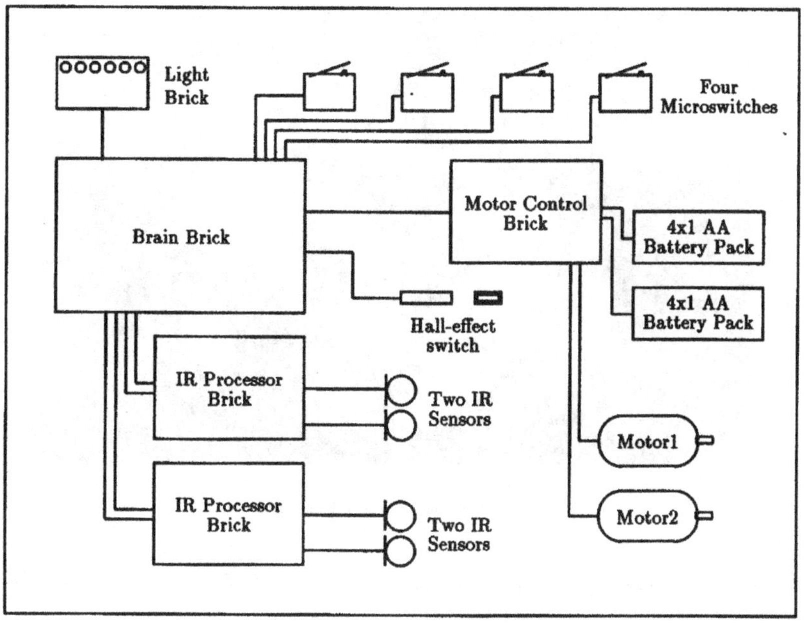

Figure 2: Block diagram of the supplimentary sensing and control bricks for Lego vehicles.

3.1 Sensing and Control Bricks

The main supplementary brick is the *brain brick*. This consists of a Rockwell 6502 8-bit microprocessor and input/output (I/O) devices. The processor addresses 32KBytes of writable memory (RAM), and 8KBytes of read-only memory (ROM). The 40 I/O lines provided can be used as input paths from sensors, or as output paths to control motors and indicator lights (or other kinds of simple output devices). A serial port interfaces the brain brick to an external computer on which control programs are developed and subsequently downloaded (see more on this below).

For sensing, four active *infra-red (IR) sensing bricks* are provided, together with four microswitches and a Hall-effect switch. The microswitches can be mounted on Lego structural elements and thus incorporated into bumper mechanisms. The Hall-effect switch is intended to be used for simple motion sensing and odometry by fixing the magnet in a non-driven wheel and mounting the active element of the switch so that it can detect the magnet passing by as the wheel rotates.

Each IR-brick consists of an emitter producing a modulated beam of IR-light, and a detector (an IR-sensitive diode) intended to catch reflections of the emitted light from objects in the environment, or IR emissions from other sources (such as the emitters of other IR-sensor bricks on the vehicle, or other similarly-endowed vehicles). Each pair of IR-bricks is powered and served by an *IR-processor brick*. Each IR-processor brick supplies the emitters of the two IR-bricks with a drive signal chopped at about 1KHz, and filters and amplifies the signals from the two IR detectors. By chopping and filtering in this way the IR-bricks are made effectively blind to other background sources of IR-light, such as the overhead fluorescent lighting typically found in institutional buildings—the ISC lab being no exception. The sensitivity of each IR-brick can be adjusted (by hand) via a trimming resistor on its associated IR-processor brick.

On our first generation of IR-bricks, the detected light level was thresholded manually using the adjustment available on the IR-processor bricks. This meant that each IR-sensor brick operated only as a binary sensor. On the newer generation of IR-bricks, which we are introducing, this thresholding has been replaced by an analogue to digital conversion which enables a range of detection levels to be read as an eight-bit word by the brain brick. With this new IR-processor brick it is now also possible to turn the emitters on and off under software control. This facility both enhances the information each sensor can generate

Figure 3: An Example Lego vehicle—front view.

and extends the way in which the IR-bricks can be used in combination on one vehicle. It also provides the possibility for communication between different vehicles via program controlled IR signalling.

The next supplementary brick is the *motor control brick*. This provides a clean interface to the Lego motors, driving them at two different speeds (full speed and half speed) in either direction. This brick is powered by eight AA rechargeable nickel cadmium cells (arranged in two packs of four). It also provides a regulated voltage to the brain brick and two IR-processor bricks. This 'two sets of four' arrangement of the batteries provides a convenient way of supplying the various voltage levels required. Each set of four serially connected batteries is used to drive one motor, with four in series supplying the 'full speed' voltage and three in series supplying the 'half speed' voltage, with the switching being done by low current consumption relays on the motor-control board. Both sets of four batteries are connected in series to supply the voltage regulator which in turn supplies all the electronics.

Though the 500mAh provided by the rechargeable AA cells currently used has proved adequate, we are considering alternative energy schemes in an attempt to extend the life of the vehicles between charges.

The last supplementary brick is the only one not to use Lego components to make it integrateable with any of the other Lego structural elements. It does, however, preserve the plug-together style of the Lego by being plugged into the output ports of the brain brick. This is the *light brick*. It provides six LEDs (light emitting diodes) in three different colours which can be used to provide visible signals of important internal conditions of the brain brick. Though not directly involved in the behaviour of the vehicle, this brick has proved invaluable for the interface between vehicles and their often bemused builders.

Most of the supplementary bricks were designed by one of us (JD), and constructed in-house using readily available electronic components. They use CMOS technology wherever possible to mimimise power consumption and have been kept as compact and as light as possible. As can be seen from figure 1, they are all made to be Lego-like by using appropriate Lego structural elements to provide supporting frames for the PCBs (Printed Circuit Boards).

Figure 1 illustrates the brain brick, IR-processor bricks and IR-sensor bricks, and motor control brick. Figure 2 presents a block diagram of the connectivity of the supplementary bricks. Figures 3 and 4 illustrate a complete Lego vehicle. This example is fitted with three IR-sensors on the front and uses two microswitches, one in each bumper. The brain brick is mounted on the top, the motor control brick is at the back, mounted vertically, and the two IR-processing boards are hidden away under the brain brick at the front. The two battery sets are mounted at the back, one on each side.

Figure 4: An Example Lego vehicle—rear view.

3.2 The JTL Programming Language

Having decided to use a microprocessor to provide the means of achieving easily reconfigurable connections between the motors and sensors (and light brick), we were faced with the problem of how this processor should be programmed. During the development of the prototypes of the supplementary bricks we programmed in 6502 assembler code. While this was bearable for prototyping and early development work it is not desirable for teaching and research purposes. What was required was some kind of high-level programming language, but what? We were aware of Brooks' behaviour compiler (in LISP), [Brooks 1987], Resnick's Lego/Logo combination, [Resnick 1989], and the previous Turtle-Logo developed in the Department, [Chung 1987], but it was clear that none of these approaches could meet all of our requirements. Our major concerns stem from a combination of hardware and programming convenience considerations. The simplicity of the processor brick meant that programs would have to be written on an external computer, translated there into machine code, and downloaded to the vehicle through a serial data link. To minimize downloading time and because the amount of RAM addressed by the processor is limited, we sought to mimimize the translated length of vehicle programs. Since, in addition, any notion of adaptive behaviour requires some kind of modifiable or 'plastic' component, we also wanted some RAM space to be available for this purpose.

We also desired to specify the behaviour of the vehicles in a modular way. Where the interaction between motors and sensors involves separate, concurrent, and possibly independent components, we wished a means to specify these in separate modules. We therefore designed a new language which we call JTL. JTL has a Pascal-like structure, but its syntax has been heavily tailored to our requirements, our aim being to make JTL easy to learn and use. For example, it uses `<variable> becomes <value>|<expression>` instead of the more usual `<variable> = <value>|<expression>`, and, `keep doing {statements} until <test>` instead of the more usual `repeat {statements} until <test>`. The definition construction usually called `function` is renamed `Action`, and is used to package frequently used pieces of code in a module. The aim of such syntactic modifications is to encourage the development of an appropriate operational semantics by programmers—i.e., an understanding of what kinds of vehicle behaviour can be obtained using which kinds of program constructions. JTL programs are written and translated on an IBM (or IBM equivalent) PC. The translation process is a combination of compilation and interpretation to keep the length of the downloaded code to a minimum.

Much of the functionality of the language resides in the brain brick's ROM—this includes the obvious things like functions for controlling motors and reading sensors, but also includes code for implementing what might be considered innocuous syntactic tasks like assignment of values to variables; this is so frequent an operation, any code compression is bound to pay dividends. Translated code is therefore a much-abbreviated stream of tokens (essentially a string of function calls), resulting in efficient use of writable memory, and fast downloading. This scheme is essentially a ROM interpreter-based system.

A JTL program consists of a collection (called a 'Gang') of modules (called Behavioural Modules). There are no implicit orderings or relationships between the Behavioural Modules in a gang, although there are mechanisms available in the langauge for explicitly relating the operations in different modules if desired. These include global variables, semaphores used to protect global variables from undesired read/write interactions across modules, and commands for inhibiting and restoring module execution from other modules[5].

Pseudo concurrent execution of the modules specified by a program is effected by a simple multi-tasking kernel in the brain brick's firmware. This time slices the processor over the gang of modules giving each module 1.67 milliseconds of processing time before suspending the module and loading in the next one. By cycling round the modules in this way each one is given sufficient processing time in each second to give the illusion of true parallel execution. Such a simple model of concurrency does mean, however, that the sequence of operations actually executed is very difficult to predict—it *is* deterministic in principle, just typically difficult to do in practice since a module can be suspended at any point in its execution and replaced by another. If more than one module is used to send signals to a motor, say, it is possible to get some strange behaviour as a result of contradictory signals arriving in quick succession—one module might be signalling full forward, while another one is signalling full reverse, for example. By careful program design such problems can be avoided, either by only having one Behavioural module controlling each motor, and not several, or by arranging for such interactions to give rise to effective (and appropriate) behaviour.

3.3 An Example JTL Program

To illustrate the use of JTL to specify the connections between motors and sensors, and thus the behaviour of a vehicle, we present a program intended to control a vehicle with two motors (called motor1 and motor2), and two IR sensors of the binary type (called sensor1 and sensor2); see Figure 5. This is a Braitenberg vehicle 3b, [Braitenberg 1986, figure 4, page 11]. The JTL

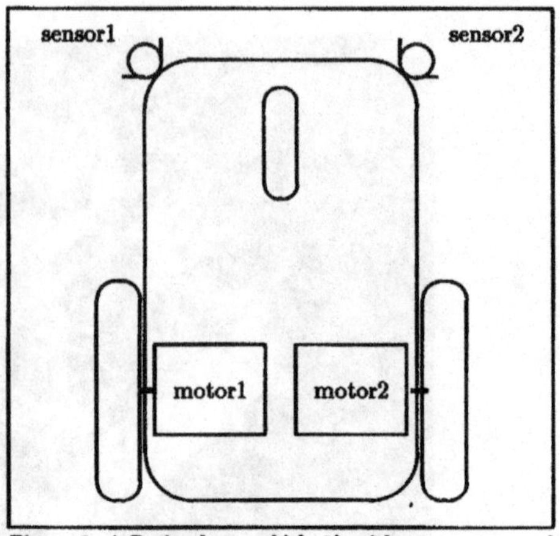

Figure 5: A Braitenberg vehicle 3b with two motors and two binary IR sensors.

program which specifies the required motor-sensory connectivity is given in Figure 6.

The behaviour of this vehicle is as follows: If the vehicle detects something to its right (sensor2), the left motor (motor1) is reversed, and if it detects something to its left (sensor1), the right motor (motor2) is reversed. This vehicle thus tends to swerve away from obstacles it detects on either side of its direction of motion. Of course, since the modules are concurrent, detection in both sensors means that both motors are reversed, and so the vehicle will stop and move backwards. It will continue to move backwards until either one or both sensors stop detecting any reflected IR. Usually this 'emergent' behaviour resulting from the combination of the two specified modules working in conjunction is good enough to prevent a vehicle from becoming trapped in an oscillatory type of behaviour—first reversing away from an obstacle in front of it, then moving straight towards it again, and then backing away again. This is because it only requires small differences in the switching of the motors by the two different modules for the vehicle not to move forward in quite the same direction as it moved back in, thus making one sensor 'see' the obstacle before the other and causing the vehicle to swerve away, rather than reverse away. If this kind of emergent behaviour is not sufficient, then a third module, testing for the condition in which both sensors detect something, could be added to force a turn to the left or right.

By careful selection and design of the gang of Behavioural modules specified by a JTL program, it is possible to engender quite sophisticated, yet reliable and robust, behaviour in a Lego vehicle intended to move about the normal Laboratory environment which

[5]The inhibit action takes a parameter which specifies the number of time steps the named module is to remain suppressed.

```
Gang vehicle3b                              { name of JTL program }

  Behavioural Module left
    Keep Doing
      If sensor(1)
        Do motor2rev                        { reverse motor2 }
        Otherwise Do motor2for              { forward motor2 }
      End If
    Until false                             { i.e., forever }
  End Behavioural Module

  Behavioural Module right
    Keep Doing
      If sensor(2)
        Do motor1rev                        { reverse motor1 }
        Otherwise Do motor1for              { forward motor1 }
      End If
    Until false                             { i.e., forever }
  End Behavioural Module

End Gang
```

Figure 6: A JTL program using two Behavioural Modules to specify the motor-sensory connectivity for a Braitenberg vehicle 3b.

consists of table legs, chairs (which get moved around), people, and other vehicles. The JTL programs for this kind of behaviour are, of course, more complicated than the simple example presented here.

4 Some Current ISC Lab Projects

Our MSc students undertake full time practical projects lasting five months (May to September) in order to put into practice some of the techniques they have learned during the first (taught) part of the course, and to give them an opportunity to develop some of the skills required to make effective use of these techniques in applications and research. Recently two students who took the Intelligent Sensing and Control module have started MSc projects using the Lego Vehicle kits and the facilities offered by the Artificial Life Laboratory.

One student is undertaking a careful and detailed study of the nature and form of the relationship between the behaviour of a Lego vehicle, its motor-sensory configuration (hardware and software), and the aspects of the Laboratory which forms the vehicle's environment—walls, cupboards, table legs, swivel chair bases, other Lego vehicles, people, etc. This *ecological* study is intended to try to understand better what the limitations (and possibilities) are of the motor-sensory behaviour that can be achieved using the existing Lego vehicle technology.

The motivation for this project was the realisation, while observing the students' efforts in the practical sessions of the ISC course, that many factors can significantly affect the success, reliability, and robustness of a vehicle's behaviour. These factors typically involve details such as the shape of parts of the vehicle, the heights above the ground of sensors, the way Behavioural modules are specified and combined in a JTL program, and probably others as yet undiscovered. Consequently constructing vehicles designed to achieve even quite simple tasks in the Laboratory environment can prove to be quite difficult unless these detailed interactions are appreciated and understood. Often the problems involved require ingenuity and inventiveness on the part of the builder before success is achieved. It is not simply a matter of getting the program right; the morphology and dynamics of the vehicle and the environment also significantly influence the actual behaviour of a vehicle—a fact often conveniently forgotten by people who build simulations of agents and environments in software.

The second project student is investigating how simple cooperative task achieving behaviour can be set up between about five Lego vehicles. One of the advantages of having such a cheap and easy to use robot building technology is that it is possible to build more than just one robot. The aim of this second project is to make a start on the barely touched subject of cooperative behaviour in *real* robots[6]. It is a subject that we feel is likely to offer us many interesting problems, and many challenges for our Lego vehicle technology, apart from also being great fun.

A third project, which forms part of our research programme into the acquisition, maintenance, and adaptation of competences in a mobile robot, [Nehmzow and Smithers 1990], involves using the Lego vehicle to replicate the self acquired obstacle avoidance and map building competences already demonstrated in our Fischertechnik^{TM} based robot. The aim of this project is to test further the control techniques developed using one robot by building them into an interestingly different robot.

5 Discussion

It is now a year since we started designing the Intelligent Sensing and Control module for the MSc course and its associated laboratory, and developing and testing the Lego vehicle technology described above. During that time we have been continually impressed by the effects of trying to build real robots designed to work in a real environment, albeit quite simple ones. Towards the end of the ISC course students were able to build and program Lego vehicles which were capable of visiting all four corners of the laboratory in one run despite considerable variation in the positions of chairs, the presence of people, and the occasional other vehicle going by. Despite the relative simplicity of this kind of motor-sensory behaviour the experience has brought home to us, and we believe to the students as well, the detailed and intricate nature of the ballistic and feedforward control techniques that are used. A measure of this is how hard it is to say why a particular vehicle works in terms other than simply describing in detail how it is built and programmed—a theory of Lego vehicles seems a long way off still.

This explanation problem supports Braitenberg's law of uphill analysis and downhill invention (see footnote 3, on page 2). Our Lego vehicle technology has proved to be sufficient to expose the difficulty of solving the appropriate action problem while remaining flexible and easy to use—exactly what we require for teaching and research purposes.

Although the sensing and processing capabilities we have provided are limiting, as is the modularity and design of the Lego Technic, it has been shown by ourselves and the students that a lot of worthwhile knowledge and experience can be gained by using this technology. This is important not just from a pedagogic point of view, but also from a basic research point of view. We understand so little about how biological systems solve the general appropriate action problem that we are forced to start by studying relatively simple behaving systems. An ability to build artificial systems of this kind, quickly and easily, is therefore essential. We believe that the Lego vehicle technology described here provides us with the means to do this, and that it will therefore be able to play an important role in our biologically inspired Artificial Intelligence research.

6 Conclusions

In this paper we have presented the two motives for establishing the Intelligent Sensing and Control Laboratory and the design and development of the Lego vehicle technology described: a desire to research real world intelligent behaviour, and to teach a broader range of control concepts and techniques in a practical way. We have described the hardware and software extensions to the Lego Technic kits that we have used as a starting point for the technology. We have briefly described some of the current projects using the technology, and finally, we have suggested that the building and studying of Lego vehicles is a fruitful way to do research into biologically oriented Artificial Intelligence.

Acknowledgements

We gratefully acknowledge the help and support we have received from other members of the Intelligent Robotics Group within the AI Department and from the technical staff who have been involved in the manufacture of much of the electronic components used, and in setting up the Laboratory. In particular we would like to thank Professor Jim Howe, Gill Hayes, Paul Ardin, Jim Dallas, Tom Alexander, David Wyse, Hugh Cameron, Mike Moran, Angus Kenyon, and Douglas Howie, for their commitment and effort. We would also like to acknolwedge the students who took part in the first ISC course; their enthusiam

[6] We strongly believe that building symbolic simulations cannot result in all the understanding required to build reliable and robust robots for the real world.

and ingenuity provided much encouragement and was the source of much insight. We also acknowledge Dr Brendan McGonigle, Director of the Laboratory for Cognitive Neuroscience in Edinburgh, who helps to supervise and direct some of the work of the ALL. Lastly, but not least, we acknowledge the financial support of Edinburgh University in the form of a studentship to Jim Donnett.

References

[Braitenberg 1986] Valentino Braitenberg, 1986, *Vehicles: Experiments in Synthetic Psychology*, MIT Press.

[Brooks 1986] Rodney A. Brooks, 1986, *Achieving Artificial Intelligence Through Building Robots*, MIT AI Laboratory Memo 899, May 1986.

[Brooks 1987] Rodney A. Brooks, 1987, *A Hardware Retargetable Distributed Layered Architecture for Mobile Robot Control*, Proceedings of the International Conference on Robotics and Automation, Raleigh, NC, March 1987, pp 106–110.

[Chung 1987] Paul Chung, 1987, *Learning Control Applications Through Programming*, PhD Thesis, Department of Artificial Intelligence, University of Edinburgh.

[McGonigle 1990] Brendan McGonigle, 1990, *Incrementing Intelligent Systems by Design*, Proceedings of SAB90—"Simulation of Adaptive Behaviour: From Animals to Animats", Paris, September, 1990.

[Moravec 1984] Hans Moravec, 1984, *Locomotion, Vision and Intelligence*, in Brady and Paul (Eds.), *Robotics Research 1*, MIT Press, pp 215–224.

[Nehmzow and Smithers 1990] Ulrich Nehmzow and Tim Smithers, 1990, *Mapbuilding using Self-Organising Networks in "Really Useful Robots"*, Proceedings of SAB90—"Simulation of Adaptive Behaviour: From Animals to Animats", Paris, September, 1990.

[Staddon 1983] J.E.R. Staddon, 1983, *Adaptive Behaviour and Learning*, Cambridge University Press.

[Resnick 1989] Mitchel Resnick, 1989, *Lego, Logo, and Life*, in Christopher G. Langton, (Ed.), *Artificial Life*, Addison-Wesley Publishing Company, Inc., pp 397–406.

[Martin et al 1990] Fred Martin, Mitchel Resnick, and Brian Silverman, 1990, *Braitenberg Bricks: A Lego-based Creature-Construction Kit*, paper presented at the second *Artificial Life* conference, Frebuary 5–9, 1990, Centre for Nonlinear Studies, Santa Fe Institute, Santa Fe, New Mexico.

[Resnick 1990] Mitchel Resnick, 1990, **Logo: Animal Simulations and the Restructuring of Science Education*, Proceedings of SAB90 —"Simulation of Adaptive Behaviour: From Animals to Animats", Paris, September, 1990.

[Walter 1961] W. Grey Walter, 1961, *The Living Brain*, Middlesex: Penguin Books.

AUTHOR INDEX

Made in the USA
Monee, IL
13 September 2026

62977350R10308